# Stream Hydrology

# Stream Hydrology
## An Introduction for Ecologists
### Second Edition

**Nancy D. Gordon**

**Thomas A. McMahon**

**Brian L. Finlayson**
*Centre for Environmental Applied Hydrology,*
*The University of Melbourne*

**Christopher J. Gippel**
*Fluvial Systems Pty Ltd*

**Rory J. Nathan**
*Sinclair Knight Merz Pty Ltd*

John Wiley & Sons, Ltd

*Other Wiley Editorial Offices*

John Wiley & Sons Inc., 111 River Street, Hoboken, NJ 07030, USA

Jossey-Bass, 989 Market Street, San Francisco, CA 94103-1741, USA

Wiley-VCH Verlag GmbH, Boschstr. 12, D-69469 Weinheim, Germany

John Wiley & Sons Australia Ltd, 33 Park Road, Milton, Queensland 4064, Australia

John Wiley & Sons (Asia) Pte Ltd, 2 Clementi Loop #02-01, Jin Xing Distripark, Singapore 129809

John Wiley & Sons Canada Ltd, 22 Worcester Road, Etobicoke, Ontario, Canada M9W 1L1

Wiley also publishes its books in a variety of electronic formats. Some content that appears in print may not be available in electronic
books.

*Library of Congress Cataloging-in-Publication Data:*

Stream hydrology : an introduction for ecologists / by Nancy D. Gordon ... [et al.].– 2nd ed.
    p. cm.
Rev. ed. of: Stream hydrology / Nancy D. Gordon, Thomas A. McMahon, Brian L. Finlayson. c 1992.
    Includes bibliographical references and index.
    ISBN 0-470-84357-8 (cloth : alk. paper) – ISBN 0-470-84358-6 (pbk. : alk. paper)
    1. Rivers. 2. Hydrology. 3. Stream ecology. I. Gordon, Nancy D. Stream hydrology.
GB1205.G65 2004
551.48′3–dc22                                                                2004005074

*British Library Cataloguing in Publication Data*

A catalogue record for this book is available from the British Library

ISBN-13 978-0-470-84357-4 (H/B)
ISBN-13 978-0-470-84358-1 (P/B)

Typeset in 9/11pt Times by Thomson Press (India) Limited, New Delhi
Printed and bound in Great Britain by CPI Antony Rowe, Chippenham, Wiltshire

# Contents

# Preface to the Second Edition

One of the main purposes in writing the first edition of this book in 1992 was to help improve communication between the disciplines of stream ecology and river engineering and to foster a sense of co-operation in these interdisciplinary efforts. We would like to think that we played a part, however small, in assisting the tremendous growth in interdisciplinary and multidisciplinary research and application that followed over the next decade. But this phenomenon was inevitable anyway; academics, policy makers and managers alike had recognised that river management could not take the next step forward unless the various experts got together and problems were assessed and solved from a broad perspective. An engineer, geomorphologist and ecologist will still have a different emphasis when conceptualising a stream, but these days each viewpoint is cognisant of, and is informed by, the others.

This second edition was a long time in its gestation. The field of river research and management has been evolving so rapidly that it was difficult for us to decide when was an appropriate time to update the book. The background information did not present a real problem, as most of it is grounded in long established principles of hydrology and fluid mechanics. However, the real growth area was in the application of science to stream management; the trial and error approach is no longer acceptable. We feel that now is a good time to take stock of these developments. Many countries have implemented new river laws that require managers to at least maintain the current levels of stream health and be highly accountable for their actions. The ecosystem concept, which originated in ecology as a research paradigm, has now been transferred to the realm of public policy; physico-chemical characteristics are still important, but we now speak of "stream health" and measure it in terms of water quality, habitat availability and suitability, energy sources, hydrology, and the biota themselves. Introduction of the European Union Water Framework Directive in December 2000 has already led to widespread changes in assessment of stream health in Europe. Stream classification is now a routine first step that simplifies the inherent complexity of stream systems, helping to facilitate many aspects of the management process. Research has clearly established the impacts of flow regulation, and the last decade has seen considerable growth in research and assessment of environmental flow needs. River rehabilitation is now one of the central themes of the river management industry. One of our objectives in writing this second edition is to bring some methodological order to these developments. Another objective is to critically evaluate the level of success and failure in efforts to rehabilitate streams. This could not have been done in the first edition, because so few examples existed at that time.

In this second edition we maintain an emphasis on the physical environment. Information has been drawn from the fields of geomorphology, hydrology and fluid mechanics, with examples given to highlight the information of biological relevance. Chapters 1–8, which include tools for studying and describing streams, have been updated by the original authors. Chapter 9, which reviews river management applications, has been totally re-written by Dr. Chris Gippel of Fluvial Systems Pty Ltd. In this final chapter, we could not avoid venturing a little further into the biological realm, and we also drew on a much wider range of source material. Readers expecting mathematical derivations will still be disappointed; we concentrate on presenting principles and demonstrating their practical use.

The software package, AQUAPAK (readers of the first edition will be familiar with the original version) has been completely updated by the original author Dr Rory Nathan of Sinclair Knight Merz Pty Ltd. AQUAPAK can be downloaded at http://www.skmconsulting.com/aquapak and runs in Windows. AQUAPAK has been tailor made for the readers of this book and assumes no prior knowledge on the part of the user other than basic computer keyboard skills. More advanced users may wish to investigate the Catchment Modelling Toolkit available on-line from the Cooperative Research Centre for Catchment Hydrology at http://www.toolkit.net.au/cgi-bin/Web Objects/toolkit.

It is clear now that many mistakes have been made in stream management in the past, leading to what we now call stream degradation. This is a retrospective view, because at the time, river managers were acting under the impression that their work would improve the value of the river from the perspective of the prevailing dominant social view. Other failed works were simply ill-informed from the technical perspective. Streams are now managed for a wider range of values, and advances made in stream management technology certainly hold the promise of ecologically healthier and economically more valuable streams for the future. But we have to remember that stream management is far from simple, and an ill-informed approach, regardless of the best intentions, can fail to produce the expected outcomes. So, as well as learning more about river processes, developing methods for rehabilitation and playing a leading role in implementing science-based management, river professionals have a responsibility to provide honest evaluations of the relative success of works. New knowledge so generated can then be used to improve the next generation of river management. At one extreme some may still hold the view that fundamental science does not have much of a role to play in the practical domain of the on-ground river manager, while at the other extreme, some researchers might still be content to explore rivers with little thought about the implications of how the new knowledge might assist practical management or policy development. We hope that this book provides a resource and inspiration to fellow river management professionals, academics and students whose outlook and passion lies some where between, or who are working to bridge these perspectives.

Web sites referenced in the book are current as of the date of publication but may be subject to change in the future. Any mention of commercial web sites does not constitute endorsement of a product.

## ACKNOWLEDGEMENTS

The authors would like to express appreciation to a number of individuals for their contributions during the evolution of this text. Mr. Andrew Douch created many of the original drawings and diagrams for Chapters 4–6 in the first edition. The diagrams in this edition were re-drawn or newly prepared by Chandra Jayasuriya and Fatima Basic of the School of Anthropology, Geography and Environmental Studies, the University of Melbourne. We are grateful to Dr Michael Keough of the Department of Zoology, University of Melbourne, for producing the realistic examples of Section 2.5. Many individuals, listed in the first edition, provided general guidance and reviews of draft materials that helped focus the scope of the book and greatly added to its accuracy and applicability. Their assistance is again acknowledged. Several reviewers provided suggestions for improvement which added to the quality of this second edition.

We are very appreciative of the professionalism and helpfulness of the people at John Wiley & Sons, especially Sally Wilkinson, Keily Larkin and Susan Barclay and to the team of Thomson Press (India) Ltd.—Jyoti Narula, Sanjay Jaiswal, M.S. Junaidi, C.P. Kushwaha and Udayan Ghosh—who greatly contributed to the quality of the final product.

The work which led to the production of the first edition of this book was supported by funds provided by the then Australian Water Research Advisory Council, now Land and Water Australia.

## COPYRIGHTED MATERIALS

We are grateful to the following organizations, individuals and publishers for permission to reproduce copyrighted material(s):

Addison Wesley, Alberta Environment, American Association of Petroleum Geologists, American Fisheries Society, American Geophysical Union, American Society of Civil Engineers (ASCE), American Water Resources Association, ANZECC, Birkhäuser Verlag AG, Blackwell Science, British Columbia Forest Service, Butterworth-Heinemann, Canadian Water Resources Association, Carfax Publishing Ltd., Colorado State University, Crown (State of Victoria), CSIRO Publishing, Department of Primary Industries, Geological Survey of Victoria, Department of Sustainability and Environment, Department of the Environment and Heritage, Diputacio de Barcelona, East Gippsland Catchment Management Authority, Elsevier, EPA, Federal Environmental Agency, Freshwater Research Unit, Geological Society of America, Harper-Collins Publishers, Hodder and Stoughton, Houghton Mifflin Company, Institution of Engineers, John Wiley & Sons Ltd., John Wiley & Sons Inc., Land and Water Australia, Macmillan, McGraw-Hill, Michigan Department of Natural Resources, National Park Service, NRC Research Press, NSW Department of Infrastructure, Planning and Natural Resources, Ontario Ministry of Natural Resources, Oxford University Press, P. Cardiyaletta, Pearson, Pollution Control Department, Princeton University Press, Rellim Technical Publications, R. L. Folk, Royal Society for the Protection of Birds, Scottish Environment Protection Agency, SEPM Society for Sedimentary Geology, Springer-Verlag GmbH & Co. KG, Stanley Schumm, Swedish Environmental Protection Agency, Texas Water Development Board, The Living Murray, The Royal Society of New Zealand, The University of Melbourne, UNEP, UNESCO, USDA, U.S. Army Corps of Engineers, U.S. Federal Interagency Subcommittee on Sedimentation, U.S. Fish and Wildlife, USGS, Van Nostrand Reinhold International, Washington Department of Ecology, Water Resources Publications, LLC, Water Studies Centre, Wildlands Hydrology, WHO, World Meteorological Organisation, Worldbank.

# Preface to the First Edition

In interdisciplinary applications of stream hydrology, biologists and engineers interact in the solution of a number of problems such as the rehabilitation of streams, the design of operating procedures and fishways for dams, the classification of streams for environmental values and the simulation of field hydraulic characteristics in laboratory flumes to study flow patterns around obstacles and organisms. One of the main purposes in writing this book was to help improve communication between the two disciplines and foster a sense of co-operation in these interdisciplinary efforts.

On the surface, the definitions of ecology and hydrology sound very similar: *ecology* is the study of the interrelationships between organisms and their environment and with each other, and *hydrology* is the study of the interrelationships and interactions between water and its environment in the hydrological cycle. In general, ecology is a more descriptive and experimental science and hydrology is more predictive and analytical. This fundamental difference influences the way streams are studied and perceived in the two disciplines.

For example, a diagram of an ecologist's view of a stream might appear as follows:

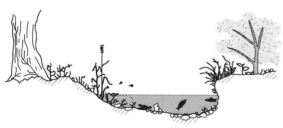

# benthic macro-
invertebrates = 10 000/m$^2$
pH = 7.2
TDS = 220 mg/l
DO = 8.3 mg/l

Here, the focus is on the aquatic biota, their interrelations, and the physical and chemical factors which affect them. An engineering hydrologist, on the other hand, might 'view' the same stream much differently, perhaps more like this:

V = 0.4 m/s
A = 2.5 m$^2$
n = 0.08
$\tau$ = 125 N/m$^2$

In this image, the physical dimensions of the stream have been simplified into a few numbers from which estimates can be made of how the stream will respond under different flow conditions.

Neither view is superior to the other; each represents only a fraction of 'all there is to know' about the stream. Interdisciplinary interaction offers a way of merging the information contained in the different views into a more complete picture. It is often at the interface between disciplines, in fact, that new ideas are generated and progress is made. Perhaps like a stereo pair, new 'dimensions' will be revealed when the images are successfully superimposed.

The emphasis of this text is on the physical environment. Information has been drawn from the fields of geomorphology, hydrology and fluid mechanics, with examples given to highlight the information of biological relevance. Mathematical derivations have been omitted; instead, the intent was to provide an intuitive understanding of the principles, demonstrate their practical use and leave the mathematics to a computer. A software package, AQUAPAK, has been provided for this purpose. Omissions and simplifications were necessary in conveying the wide range of subject matter. We can only resort to the blanket statement that everything is more complicated than our description of it, and that ours is merely another 'view' of streams.

A practical approach has been taken, with the chapter on field techniques forming a central part of the text. In other chapters, examples have been given so that the principles can be applied more readily. Field studies in the Acheron River Basin, located approximately 100 km to the east of Melbourne, provided information for examples throughout the book. We did this to maintain continuity, as well as to illustrate how we went about 'getting to know' this river system.

The process of getting to know a stream is not unlike that of a doctor learning about a patient and his or her health. A conscientious doctor will look beyond the charts, images and the results of various tests to obtain a sense of what causes the patient's health to be what it is. In the same manner, hydrological data, aerial photographs, channel surveys and water quality analyses only measure 'symptoms' of a stream's condition, and, as with human health, the underlying causes are complex and nebulous.

Just as patients are more than the sum of their connective tissues and blood vessels, streams, too, should be viewed 'holistically' as a continuum from source to sea and as systems which interact with the surrounding environment. This book presents methods for 'diagnosing' the physical condition of streams. Criteria for establishing what constitutes physical 'health' are yet to be developed. As Leopold (1960) advocated over 30 years ago, benchmark stations free from grazing and other human influences are needed in order to evaluate the effects of humans on ecologic and geologic change. Interdisciplinary studies are essential for establishing these baseline conditions, for determining the sensitivity of a given stream to 'stress' and for developing appropriate rehabilitation procedures to 'cure' those streams which are found to be in poor condition.

# 1

# Introducing the Medium

## 1.1 Water as a Fluid

Water is a widespread, life-sustaining substance, comprising some 50–90% of living materials and covering nearly three-fourths of the Earth's surface. Of the Earth's total moisture, however, about 97% is contained in the oceans and less than 0.0002% flows through its rivers and streams. Water is recycled globally, with the relative proportions of ice, water vapour, fresh water and salt water changing as the earth warms and cools. Scientists formerly believed that the total amount of water on Earth was essentially constant, but new evidence points to a small influx of water from 'snowball' comets (Pielou, 1998).

Water is a substance with many unique chemical and physical properties. Unlike most substances that contract when frozen, water expands, allowing ice to float on the surfaces of lakes and streams. It is found as a liquid at temperatures common to most places on Earth. With its great heat capacity it can absorb or lose a large amount of energy before showing a change in temperature. As a universal solvent, it dissolves gases, nutrients and minerals. Its internal cohesion gives rise to surface tension, which allows water striders to traverse a pool's surface or even run upstream. Because of its physical properties, a quite different set of environmental conditions is presented to amoebae and fish that both live in the same waters.

Depending on the temperature, water can exist as either a liquid, a gas (water vapour) or a solid (ice). Combinations such as steam-air mixtures or water with entrained air fall into a specialized category called *two-phase flows*.

The general term *fluid* describes both gases and liquids, examples being oxygen, motor oil, liquid glass and mercury. The differences between fluids and solids are not always obvious. Fluids flow readily under the slightest of forces; they do not have a definite shape and vessels are required to contain them. *Solids* are substances that are considered to have both a definite volume and a definite shape. Thus, the line is drawn between molasses as a fluid and gelatin as a solid.

*Liquids* are distinguished from gases by their cohesiveness. Whether sitting in a laboratory beaker or in a frog pond, a liquid will have a definite volume. It will also have a free surface, which is horizontal when the fluid is at rest. A *gas*, in contrast, does not have a definite volume, and will expand to fill a container enclosing it.

The next section will introduce some basic principles of physics and the system of units used in the text. These concepts are applied to the description of physical properties of water in Section 1.3.

## 1.2 The Physics of Fluids

The properties and motion of a fluid, such as water, are measured in terms of four basic quantities: mass, length, time and temperature. The magnitudes of these quantities (e.g. how hot or how large) are expressed in *units*. In the International System of Units (SI), the *fundamental* or *base units* are given as

- kilogram (kg)—mass,
- metre (m)—length,
- second (s)—time,
- Kelvin (K)—temperature.

In studies of aquatic systems, absolute temperatures are not normally of interest, and for the purposes of this text, temperature will be expressed in °C (Celsius), where $273.15 \text{ K} = 0\,°C$, and a change of $1\,°C$ is the same as a change of 1 K.

The metre was originally proposed as $10^{-7}$ of the length of the meridian through Paris (Blackman, 1969). It is now defined in terms of the wavelength of a specific type of orange light. The unit of time, the second, is defined by

Stream Hydrology: An Introduction for Ecologists, Second Edition.
Nancy D. Gordon, Thomas A. McMahon, Brian L. Finlayson, Christopher J. Gippel, Rory J. Nathan
© 2004 John Wiley & Sons, Ltd ISBNs: 0-470-84357-8 (HB); 0-470-84358-6 (PB)

an atomic standard based on caesium. The unit of mass was originally based on the mass of a certain volume of water at prescribed conditions. Thus, conveniently, a litre (0.001 m$^3$) of water at 4 °C has a mass of about 1 kg.

Whereas these base units are all defined in reference to some standard, there are other quantities, such as velocity, for which standards are impractical. These quantities have units that are defined in terms of the base units and are thus called *derived units*. Some of the quantities associated with the area of physics known as 'mechanics', which are relevant to the study of water, will be discussed. A summary of both fundamental and derived units, their dimensions and associated symbols, is given in Table 1.1. For tables of conversion factors and other information relevant to water resource studies, Van Haveren's (1986) handbook is a highly useful reference.

### Velocity

*Motion* is defined as a change of position. *Speed* refers to the rate at which the position changes with time, i.e. if a raft floats 500 m downstream in 5.5 min, then its average speed is about 1.5 m/s. Technically, *velocity* refers to the speed in a given direction; however, in ordinary speech, no distinction is usually made between velocity and speed.

### Discharge or Streamflow

*Discharge*, or *streamflow*, is the rate at which a volume of water flows past a point over some unit of time. In the SI system it is expressed in metres cubed per second (m$^3$/s). For example, if a small spring filled a 0.01 m$^3$ bucket in 2 s, its discharge would be 0.005 m$^3$/s. Discharge is normally symbolized by $Q$.

### Acceleration

*Acceleration* is the rate at which velocity changes with time. An object dropped off a cliff on Earth will accelerate at 9.807 m/s$^2$ (this *gravitational acceleration* ($g$) varies slightly with position on the Earth's surface). The distance $h$ covered by a dropped object (starting at zero velocity) is

$$h = \tfrac{1}{2}gt^2 \tag{1.1}$$

where $t$ is the time in seconds from when it was dropped and $h$ is in metres.

### Force

*Force* is described in terms of its effects. It may cause an object to change its direction of motion, to stop or start, to rise or fall. By Newton's second law of motion, force is proportional to mass multiplied by acceleration. In the SI

**Table 1.1.** *Common quantities used in the description of fluids. Adapted from Vogel, Steven; Life in Moving Fluids. © 1981 by Willard Grant Press, 1994 revised Princeton University Press. Reprinted by permission of Princeton University Press*

| Quantity | Symbol | Dimensions[a] | SI units |
|---|---|---|---|
| Length, distance | $x, y, d, r, h, k, w, \delta$ $H, D, W, L, R, P$ | $L$ | Metre (m) |
| Area | $A$ | $L^2$ | Square metre (m$^2$) |
| Volume | $\forall$ | $L^3$ | Cubic metre (m$^3$) or litre (L) |
| Time | $t$ | $T$ | Second (s) |
| Velocity | $V, v$ | $L/T$ | Metre per second (m/s) |
| Discharge | $Q$ | $L^3/T$ | Cubic metres per second (m$^3$/s) |
| Acceleration | $g$ | $L/T^2$ | Metre per second squared (m/s$^2$) |
| Mass | $M$ | $M$ | Kilogram (kg), tonne (1000 kg) |
| Force | $F$ | $ML/T^2$ | Newton (N or kg m/s$^2$) |
| Density | $\rho$ | $M/L^3$ | Kilogram per cubic metre (kg/m$^3$) |
| Heat, work, energy | $\Omega$, KE, PE | $ML^2/T^2$ | Joule (J or N m) |
| Power | $\omega$ | $ML^2/T^3$ | Watt (W or J/s) |
| Pressure | $p$ | $M/LT^2$ | Pascal (Pa or N/m$^2$) |
| Shear stress | $\tau$ | $M/LT^2$ | Pascal (Pa or N/m$^2$) |
| Dynamic viscosity | $\mu$ | $M/LT$ | Newton second per square metre (N s/m$^2$ or kg/m s) |
| Kinematic viscosity | $\nu$ | $L^2/T$ | Square metre per second (m$^2$/s) |
| Surface tension | $\sigma$ | $M/T^2$ | Newtons per metre (N/m) |
| Temperature | $T$ | — | Degrees Celsius (°C) |

[a] Dimensions are the following: $L$, distance; $M$, mass; $T$, time.

system, the unit of force is the *Newton* (N), defined as the force necessary to accelerate 1 kg at 1 m/s$^2$:

$$\text{Force (N)} = \text{Mass (kg)} \times \text{acceleration (m/s}^2) \quad (1.2)$$

A very small 'Newton's' apple with a mass of 0.102 kg experiences a gravitational force on Earth of about 1 N.

The term 'weight' does not appear in the SI system, and can create confusion particularly when converting from the Imperial to the SI system. *Mass* is an expression of the amount of matter in something, whether a brick, a balloon or a bucket of water. *Weight* is a gravitational force. If Newton's apple were taken to the moon, it would still have a mass of 0.102 kg, but its weight (the force due to gravity) would be considerably reduced. On Earth, if an American buys 2.2 pounds (lb) of apples at the supermarket to make a pie and an Australian buys 1 kg of apples at the greengrocer to make apple slices, they will both get the same amount of produce. In this case, the distinction between mass and weight does not matter. However, to a researcher studying the behaviour of fluids, the distinction is essential!

### Pressure

The *pressure* at any point is the force per unit area acting upon the point. For example, a human of 70 kg standing on the top of an empty aluminium can with a surface area of 0.002 m$^2$ would exert a pressure of

$$\left(\frac{70 \times 9.807}{0.002}\right) \approx 343\,000 \text{ N/m}^2 \text{ or } 343 \text{ kilopascals (kPa)}$$

–probably sufficient to crush it.

### Shear Stress and Shear Force

*Shear stress*, like pressure, is force per unit area. The difference is in the direction in which the force is applied. In pressure, the force acts *perpendicular* to a surface, ⬐, whereas a *shear force* acts *parallel* to it, ▢. For example, a glob of liquid soap rubbed between the hands experiences shearing forces. Shear stress is the shearing force divided by the area over which it acts. For the soap, the shearing force acts over the surface area where the soap contacts the hand. Shear stress, symbolized by $\tau$ (tau), has the same unit as pressure, N/m$^2$.

### Energy and Work

Energy and work have the same units. *Work* is a quantity described by the application of a force over some distance,

measured in the direction of the force:

$$\text{Work (N m or J)} = \text{Force (N)} \times \text{distance (m)} \quad (1.3)$$

For example, if a force of 500 N is required to push a waterlogged log 10 m across a pond, then the amount of work done is 5000 N m or 5 kilojoules (kJ).

*Energy* is the capacity for doing work. Thus, the quantity of work that something (or someone) can do is a measure of its energy; e.g. it would take about 700 kJ for a person of average ability to swim 1 km. Energy is usually symbolized by $\Omega$ (omega).

### Power

*Power* is the amount of work done per unit time:

$$\text{Power (J/s or Watts)} = \text{Work (J)/time (s)} \quad (1.4)$$

Power is usually symbolized by $\omega$ (lower case omega). For a flow of water, $Q$, falling over a height, $h$, the relevant formula for calculating power is

$$\omega = \rho g Q h \quad (1.5)$$

where $\omega$ has units of Watts, $Q$ has units of m$^3$/s, $h$ is in metres, $\rho$ (rho) is the density of water (kg/m$^3$) and $g$ is the acceleration due to gravity (m/s$^2$). As an approximation, this can be simplified to

$$\omega = 10Qh \quad (1.6)$$

with $\omega$ in kilowatts. Thus, if a waterfall of 10 m height is flowing at 1.0 m$^3$/s, the power of the falling water is 100 kW. If the flow were diverted into a small hydroelectric plant rather than over the waterfall, much of this water power could be converted to electrical power. Because of losses associated with the turbine, electrical generator and diversion works, efficiencies of 70% are common. In this example, then, approximately 70 kW of electricity could be produced.

## 1.3   Physical Properties of Water

### 1.3.1   Density and Related Measures

### Density

Because the formlessness of water makes mass an awkward quantity, *density*, or mass per unit volume, is typically used instead. Density is normally symbolized by $\rho$ and in the SI system it is expressed in kilograms per cubic metre (kg/m$^3$).

**Table 1.2.** *Values of some fluid properties at atmospheric pressure. Adapted from Douglas et al. (1983) and Vogel (1981), by permission of Longman Group, UK, and Princeton University Press, respectively*

| | (°C) | Density, $\rho$ (kg/m$^3$) | Dynamic viscosity, $\mu$ (N s/m$^2$) | Kinematic viscosity, $\nu$ (m$^2$/s) |
|---|---|---|---|---|
| Fresh water | 0[a] | 999.9 | $1.792 \times 10^{-3}$ | $1.792 \times 10^{-6}$ |
| | 4 | 1000.0 | $1.568 \times 10^{-3}$ | $1.568 \times 10^{-6}$ |
| | 10 | 999.7 | $1.308 \times 10^{-3}$ | $1.308 \times 10^{-6}$ |
| | 15 | 999.1 | $1.140 \times 10^{-3}$ | $1.141 \times 10^{-6}$ |
| | 20 | 998.2 | $1.005 \times 10^{-3}$ | $1.007 \times 10^{-6}$ |
| | 25 | 997.1 | $0.894 \times 10^{-3}$ | $0.897 \times 10^{-6}$ |
| | 30 | 995.7 | $0.801 \times 10^{-3}$ | $0.804 \times 10^{-6}$ |
| | 40 | 992.2 | $0.656 \times 10^{-3}$ | $0.661 \times 10^{-6}$ |
| Sea water[b] | 0 | 1028 | $1.89 \times 10^{-3}$ | $1.84 \times 10^{-6}$ |
| | 20 | 1024 | $1.072 \times 10^{-3}$ | $1.047 \times 10^{-6}$ |
| Air | 0 | 1.293 | $17.09 \times 10^{-6}$ | $13.22 \times 10^{-6}$ |
| | 20 | 1.205 | $18.08 \times 10^{-6}$ | $15.00 \times 10^{-6}$ |
| | 40 | 1.128 | $19.04 \times 10^{-6}$ | $16.88 \times 10^{-6}$ |
| SAE 30 oil | 20 | 933 | 0.26 | $0.279 \times 10^{-3}$ |
| Glycerine | 20 | 1263 | 1.5 | $1.190 \times 10^{-3}$ |
| Mercury | 20 | 13 546 | $1.554 \times 10^{-3}$ | $0.115 \times 10^{-6}$ |

[a] Ice at 0 °C has a density of 917.
[b] Sea water of salinity 35 ‰. The salinity of sea water varies from place to place.

Pressure can be assumed to have an insignificant effect on the density of water for most hydrological applications. However, water density does change with temperature, decreasing as the temperature increases above 4 °C (i.e. tepid water floats on top of colder water). Water density reaches a maximum at 4 °C under normal atmospheric pressure. As the temperature decreases below 4 °C, water becomes less dense, and upon freezing, it expands (ice floats). The densities of selected fluids at different temperatures are listed in Table 1.2.

Materials dissolved or suspended in water, such as salt or sediment or air, will also affect its density. Thus, fresh water will float above salt water in estuarine environments or where saline groundwater enters a stream. Density is reduced in the frothy whitewater of rapids, under waterfalls or in other areas where large quantities of air are entrained in the water. Swimmers have more trouble staying afloat or propelling themselves in these regions; hence, fish tend to 'jump' towards their upstream destinations from less-aerated areas (Hynes, 1970).

### Specific Weight

*Specific weight* is a non-SI measure, but is commonly used in practice in the Imperial system in place of density. Usually symbolized by $\gamma$ (gamma), specific weight is equal to the product of density and gravitational acceleration,

$\rho g$. Thus, in the Imperial system, where the specific weight of water (at 4 °C) is 62.4 lb/ft$^3$, one can calculate the weight of water in a 10 ft$^3$ aquarium as $62.4 \times 10 = 640$ lb. This measure will not be used in this text, and is included here only because it appears so often in the literature.

### Relative Density

*Relative density* is usually defined as the ratio of the density of a given substance to that of water at 4 °C. It is thus a dimensionless quantity (it has no units). For example, the relative density of quartz is about 2.68. Relative density is equivalent to *specific gravity*, used in the Imperial system, where specific gravity is defined as the ratio of the specific weight of a substance to that of water.

*Example 1.1*

Calculate (a) the mass of a 5 L volume of 15 °C fresh water and (b) the gravitational force (weight) it experiences on Earth:

(a) $(5\,\mathrm{L})\left(\dfrac{.001\,\mathrm{m}^3}{\mathrm{L}}\right)\left(999.1\,\dfrac{\mathrm{kg}}{\mathrm{m}^3}\right) = 5.0\,\mathrm{kg}$

(b) $5.0\,\mathrm{kg}\left(9.807\,\dfrac{\mathrm{m}}{\mathrm{s}^2}\right) = 49.0\,\dfrac{\mathrm{kg\,m}}{\mathrm{s}^2} = 49.0\,\mathrm{N}$

### 1.3.2   Viscosity and the 'No-slip Condition'

Viscosity is a property that is intuitively associated with motor oil and the relative rates with which honey and water pour out of a jar. It is related to how rapidly a fluid can be 'deformed'. When a hand-cranked ice cream maker is empty the handle can be turned relatively easily. If it is then filled with water, the amount of effort increases, and if the water is replaced with molasses, the handle becomes extremely difficult to turn. Viscosity, or more precisely, *dynamic* or *absolute viscosity*, is a measure of this increasing resistance to turning. It has units of Newton seconds per square metre ($N s/m^2$) and is symbolized by $\mu$ (mu). Of interest to aquatic organisms and aquatic researchers is the fact that there is almost no liquid with viscosity lower than that of water (Purcell, 1977).

The dynamic viscosity of water is strongly temperature dependent. Colder water is more 'syrupy' than warmer water. For this reason, it takes less effort for a water boatman to row across a tepid backyard pond in summer than the equivalent distance in a frigid high-country lake. It also takes more work for wind to produce waves on a water surface when the water is colder. Dynamic viscosity of fresh water can be calculated directly from temperature using the Poiseulle relationship, given as follows (Stelczer, 1987):

$$\mu = \frac{0.0018}{(1 + 0.0337T + 0.00022T^2)} \qquad (1.7)$$

Eq. (1.7) will give slightly different values than those listed in Table 1.2 for fresh water. It should be noted that salt water has a higher dynamic viscosity than fresh water at the same temperature. Vogel (1981) describes instruments for measuring the viscosity of fluids for which published values are not available.

The influence of viscosity is perhaps most significant in the region where fluids come into contact with solids. It is here that fluids experience the equivalent of friction, which develops entirely within the fluid. When a solid slides across another solid, like shoes across a carpet, friction occurs at the interface between the two solids. When a fluid encounters a solid, however, the fluid sticks to it. There is no movement at the interface. According to this *no-slip condition*, at the point where a viscous fluid contacts a solid surface like a cobble on a streambed or a scale on a fish, its velocity is the same as that of the solid.

Thus, when water flows by a stationary solid object, the velocity of the water is zero where it contacts the solid surface, increasing to some maximum value in the 'free stream'—the region 'free' of the influence of the solid boundary.

*Kinematic viscosity*, symbolized by $\nu$ (nu), is the ratio of dynamic viscosity to density:

$$\text{Kinematic viscosity } (\nu) = \frac{\text{Dynamic viscosity } (\mu)}{\text{Density } (\rho)} \qquad (1.8)$$

where $\nu$ has units of $m^2/s$. This ratio shows up frequently in important measures such as the Reynolds number, and is another way of describing how easily fluids flow. The quantity was introduced by engineers to simplify the expression of viscosity (kinematic viscosity has dimensions only of length and time).

From Table 1.2 it can be seen that the kinematic viscosities of air and water are much more similar than their relative dynamic viscosities. The similarities in the behaviour of air and water make it convenient to model air currents, chimney plumes or aircraft in water tanks (after applying appropriate scaling factors).

### 1.3.3   Surface Tension

A whirligig beetle darting across the surface of a pool, beads of dew on a waxy leaf, the curve of water spilling over a weir and the creep of water upwards from the groundwater table into fine-grained soils—are all illustrations of the phenomenon, *surface tension*. Surface tension can be regarded as the stretching force per unit length (or energy per unit area) required to form a 'film' or 'membrane' at the air–water interface (Streeter and Wylie, 1979). It is symbolized by $\sigma$ (sigma), and has units of Newton per metre (N/m).

Surface tension of water in contact with air results from the attraction of water molecules to each other. Within a body of water, a water molecule is attracted by the molecules surrounding it on all sides, but molecules at the surface are only attracted by those beneath them. Therefore, there is a net pull downwards which puts tension on the water surface. The surface region under tension is commonly known as the *surface film*. Because this film is under tension, any change in shape which would add more surface area (and further increase the tension) is resisted. Water drops and submerged air bubbles, as examples of air–water interfaces, are almost perfectly spherical because a sphere has less surface area per unit of volume than other shapes.

The surface tension of water is temperature-dependent. It decreases as temperature rises by the following relationship (Stelczer, 1987):

$$\sigma = 0.0755 - 0.000\,156\,9T \qquad (1.9)$$

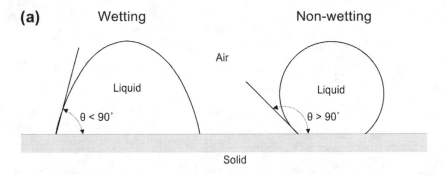

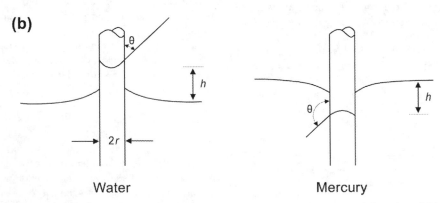

**Figure 1.1.** *Effects of surface tension (a) on the angle of contact, θ, in wetting and non-wetting liquids and (b) on capillarity in circular glass tubes of radius r, where h is capillary rise or depression*

Surface tension also affects whether a droplet will bead up or spread out on a solid surface. The angle of contact between a liquid and a solid is related not only to the *cohesion* of the water molecules (attraction to each other), but also to the *adhesion* of the liquid to the solid. If this contact angle ($\theta$ in Figure 1.1(a)) is less than 90°, the liquid is said to 'wet' the solid. If the angle is greater than 90°, the liquid is 'non-wetting'.

Water is wetting to a clean glass surface or a bar of soap but does not wet wax (White, 1986). Non-wettable objects with a higher density than water can be supported by the surface film up to a certain point. For example in water at 18 °C, a dry sewing needle of 0.2 g will 'float', whereas at 50 °C, it will sink. Near sandy streambanks, patches of fine dry sand may likewise be supported by the water surface. Insects that dart around on the water surface tend to have a waxy coating which functions as a water repellant (Vogel, 1988).

Adding a wetting agent such as detergent to the water will reduce the surface tension, making it more difficult for mosquitoes to 'attach' to the surface film from the underside or for water-striding insects to walk across it. If a baby duck is placed in a tub of soapy water, the water-repelling oil in its feathers dissolves, releasing air trapped within the feathers, and it sinks (Bolemon, 1989). Cormorants do not have water-repellent feathers, and must spread their wings out to dry after diving for fish.

Wetting agents are added to liquid pesticides to make them spread out and cover more surface area on plant leaves. Similarly, laundry detergents reduce surface tension, allowing water to penetrate more readily through dry clothes (Vogel, 1988).

Another important implication of surface tension is that pressure within a droplet of water in air—or within an air bubble under water—is higher than the pressure outside. The increase in pressure is given by (White, 1986)

$$\Delta p = \frac{2\sigma}{r} \qquad (1.10)$$

where $\Delta p$ is the increase in pressure (in N/m$^2$) due to surface tension and $r$ is the radius (in metres) of the droplet or bubble. It can be seen that the pressure becomes larger as the radius gets smaller. Because of the increased pressure, the air held in small bubbles will tend to go into solution and the bubble will shrink. Thus, very small air bubbles will quickly collapse and disappear. Vogel (1988) offers a fascinating discussion on how bubbles form at scratched surfaces in beer glasses and other biologically related implications of surface tension.

*Capillarity* is another phenomenon caused by surface tension. Capillarity, which causes water to rise in plant stems, soil pores and thin glass tubes, results from both adhesion and cohesion. Its height is positive (capillary rise) if liquids are wetting and negative (capillary depression) if liquids are non-wetting, as shown in Figure 1.1(b). Also, the meniscus (curve of the liquid's surface) is concave for wetting liquids and convex for non-wetting.

The formula for capillary rise (or depression), *h* (m), is (White, 1986)

$$h = \frac{2\sigma \cos \theta}{\rho g r} \quad (1.11)$$

where *r* is the radius (m) of the tube or the mean radius of soil pores, $\rho$ is the density of the water (kg/m$^3$) and the other symbols have been explained earlier in this section.

From Eq. (1.11) it can be seen that capillarity decreases as the tube or pore radius gets bigger. For water in glass tubes with diameters over about 12 mm, capillary action becomes negligible (Daugherty *et al.*, 1985). It can also be seen that *h* is positive for $\theta < 90°$ (wetting liquids) and negative for $\theta > 90°$ (non-wetting). For open-water surfaces and soil pores the simplification $\theta = 0$ is usually made so that the $(\cos \theta)$ term drops off (Stelczer, 1987). In soils, organic matter and certain mineral types can increase the contact angle above 90°, in which case the soil will not wet. For example, soils can become 'hydrophobic' after intense fires, preventing water from infiltrating (Branson *et al.*, 1981).

### 1.3.4  Thermal Properties

Temperature has an effect on other properties of water, such as density, viscosity and dissolved oxygen concentration. Temperatures vary seasonally in streams and with the water source (e.g. snowmelt or industrial outfall). Because of turbulence, the thermal stratification characteristic of lakes is uncommon in streams and they respond more quickly to changes in air temperature. Biologically, temperature has an important influence on decomposition and metabolic rates, and thermal cues may exist for reproduction or migration; therefore, aquatic organisms will survive and thrive within specific temperature ranges.

Streams, as a rule, exist between the temperature extremes of ice floes and boiling hot springs. Pure water freezes at 0 °C and boils at 100 °C. The presence of dissolved solids raises the boiling point and depresses the freezing point as compared with pure water. Since aquatic organisms normally concentrate salts in different proportions to those in the surrounding solution, their 'boiling' and 'freezing' temperatures will be different from those of the surrounding medium, and some primitive organisms such as blue-green algae and bacteria can tolerate great extremes of temperature.

In studies of aquatic systems, temperature data are sometimes converted to *degree-days* to correlate temperature with snowmelt, plant germination times or developmental times for aquatic insects, where

$$\text{degree-days} = nT_{\text{avg}} \quad (1.12)$$

Here, *n* is the number of days from a given starting date and $T_{\text{avg}}$ is the mean daily temperature above some base, usually 0 °C (Linsley *et al.*, 1975b). Four days with individual mean temperatures of 20, 25, 25 and 30 °C would therefore represent 100 degree-days above 0 °C.

The amount of heat a body of water absorbs depends on the amount of heat transferred to it from the air and streambanks, as well as the *thermal capacity* of the water. The thermal capacity of water is very high in comparison with other substances, meaning that it can absorb a large amount of heat before its temperature increases substantially. Thermal capacity, $T_c$, has units of J/°C, and is defined as (Stelczer, 1987)

$$T_c = cM \quad (1.13)$$

where *M* is the mass of water (kg) and *c* is the specific heat of the water (J/kg °C). *Specific heat* is the amount of heat required to raise the temperature of a unit mass of water by 1 °C. As shown in Table 1.3, it is

**Table 1.3.** *Values of specific heat for water at various temperatures. From Stelczer (1987), Reproduced by permission of Water Resources Publications, LLC*

| Water temperature (°C) | Specific heat (J/kg °C) |
|---|---|
| Ice | 2.039 |
| 0 | 4.206 |
| 10 | 4.191 |
| 20 | 4.181 |
| 30 | 4.176 |
| 40 | 4.177 |
| 50 | 4.183 |

temperature-dependent, reaching a minimum at 30 °C. Thus, a kilogram of water at 10 °C would require 4.19 J of heat energy to raise its temperature to 11 °C.

Energy is released when water freezes, a fact known by citrus fruit growers who spray their trees with water to protect them from frost damage. The *latent heat of fusion*, the energy needed to melt ice or the energy which must be taken away for it to freeze, is 335 kJ/kg for water at 0 °C.

At the other extreme, additional energy is required when water reaches the boiling point to get it to vaporize. Vaporization reduces the temperature of the remaining water. The *latent heat of vaporization* for water at 100 °C is 2256 kJ/kg. These latent heat values are relatively high in the natural world and are caused by hydrogen bonding. Hydrogen bonding is also responsible for the unusual behaviour of water density near the freezing point.

### 1.3.5  Entrained Air and Dissolved Oxygen

Dissolved oxygen (DO) is actually a chemical property of water, but is included because it is affected by physical properties such as temperature and turbulence, and because of its biological relevance. Oxygen enters water by diffusion at the interface between air and water at the surface of a stream or at the surface of air bubbles. It can also be produced from the photosynthesis of aquatic plants.

**Table 1.4.** *Dissolved oxygen saturation concentrations at atmospheric pressure 760 mm Hg and zero salinity. Generated from USGS DOTABLES program (USGS, 2001), Reproduced by permission of U.S. Geological Survey*

| Water temperature (°C) | Oxygen saturation (mg/L) |
| --- | --- |
| 0 | 14.6 |
| 5 | 12.7 |
| 10 | 11.3 |
| 15 | 10.1 |
| 20 | 9.1 |
| 25 | 8.2 |
| 30 | 7.5 |
| 40 | 6.4 |

*Entrainment* of air under waterfalls and in the frothy whitewater of rapids increases the amount of interface area where diffusion can occur. Most of this entrained air soon escapes, however, and it is the escape of these air bubbles which produces the roar of rapids and the murmur of meandering brooks (Newbury, 1984). The amount of air remaining in the water is determined by the gas-absorbing capacity of water, which is dependent upon temperature and ambient pressure.

Under normal atmospheric pressure and a temperature of 20 °C, water will contain about 2% (by volume) dissolved air (Stelczer, 1987). As temperatures rise, the gas-absorbing capacity of water decreases rapidly, reaching zero at 100 °C. Although the concentration of oxygen ($O_2$) in the atmosphere is about 21%, oxygen is more

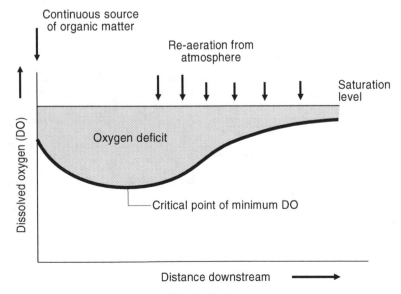

**Figure 1.2.** *Dissolved oxygen sag curve*

soluble in water than nitrogen, and dissolved air contains from 33% to 35% $O_2$, depending on the temperature—a fact which has no doubt played an evolutionary role in the dimensions of gills and other respiratory mechanisms.

The maximum amount of dissolved oxygen that water can hold at a given temperature, atmospheric pressure and salinity is termed *oxygen saturation*. Table 1.4 gives oxygen saturation values for a range of water temperatures.

The amount of DO actually present in the water can be expressed as a percentage of the saturation value (% sat)

$$\% \, sat = \frac{Actual \ DO \ concentration}{Saturation \ concentration} \times 100 \qquad (1.14)$$

Concentrations are usually given in milligrams per litre (mg/L). For example, if a 20 °C water sample has a DO content of 8.2 mg/L, then % sat = (8.2/9.05) × 100 = 90.6%.

Organic matter in streams is assimilated by bacteria that use dissolved oxygen for the aerobic processing of organic materials. An increase in the amount of organic matter (e.g. sewage, detritus stirred up by dredging or an overload of autumn leaves in temperate climate zones) stimulates bacterial growth. If the organic load is extremely excessive, nearly all the dissolved oxygen can be used up by the bacteria, leading to anaerobic conditions. In these streams, conditions can become unfavourable to forms of aquatic life sensitive to oxygen levels (Best and Ross, 1977).

The process of de-oxygenation and re-aeration of streams produces a pattern in the DO concentration known as the *dissolved oxygen sag*, first described by Streeter and Phelps in 1925 (Clark *et al.*, 1977). A sag curve is illustrated in Figure 1.2, representing the dissolved oxygen deficit (amount below saturation level) as it varies with distance downstream. A light organic load and adequate aeration will only cause a slight dip in the curve with a quick recovery, whereas a heavy load and low re-aeration rate may cause DO to decrease to 0%, from which it recovers only slowly. Equations for estimating the sag curve are given by Clark *et al.* (1977, pp. 296–298), and require field studies to determine the degree of organic pollution and the re-aeration characteristics of the stream.

# 2

# How to Study a Stream

## 2.1 Focusing on Physical Habitat

Before beginning, the definitions of the terms *river, stream* and *catchment* should be clarified. In general, rivers are larger paths of moving water (i.e. too large to wade or jump across) and streams or creeks are smaller. This relative definition will be retained, but the word 'stream' will be used as a generic term for flowing waters throughout the text. A *catchment* is the area above a specific point on a stream from which water drains towards the stream. Catchments and their characteristics will be described further in Chapter 3.

At the interface between aquatic ecology and hydrology, studies of streams fall roughly into the following categories:

1. Description or classification of aquatic habitats based on their biota and environmental characteristics. Descriptions of the flowing environment are also needed for simulating the same conditions in laboratory flumes.
2. Monitoring programs to determine variability in the natural environment over time or to detect some trend due to environmental deterioration or recovery (Green, 1979).
3. Comparison of conditions at one place/time with conditions at another place/time; e.g. comparing effects of management or of some experimental treatment, either between sites or at the same site at different times (Platts *et al.*, 1987).
4. Development of relationships between variables, e.g. local water velocity and blackfly larvae populations, or catchment area and stream width, in order to estimate or predict one from the other(s).

A variety of factors control the abundance, distribution and productivity of stream-dwelling organisms such as competition for space, predation, chemical water quality, temperature, nutrient supplies, the presence of waterfalls or dams and flow variability. An individual species will have a range of tolerance to any given factor, with some factors more critical than others. Thus, studies of streams may involve the measurement and analysis of biological, chemical and physical parameters. The emphasis throughout this text, however, will be on physical habitat: those factors which form the 'structure' within which an organism makes its home. Physical factors are generally more predictable, less variable and more easily measured than biological or chemical ones, and are thus preferable for general, consistent descriptions of streams.

Following is a discussion of the physical factors which are of the most ecological significance: streamflow, current (velocity), channel shape, substrate and temperature. Dissolved oxygen and salinity, although chemical factors, have been included since they are influenced by physical factors, they have high ecological relevance and they can be measured relatively easily. Vegetation has also been included as a related factor because of its influence on the physical nature of streams. These factors should be considered when planning a stream study (Section 2.2) and developing a sampling design (Sections 2.3–2.5).

More information on the effects of physical factors on the distribution of biota can be obtained from texts on aquatic biology or stream ecology such as Allan (2001), Barnes and Mann (1980), Bayly and Williams (1973), Brown (1971), DeDeckker and Williams (1986), Fontaine and Bartell (1983), Goldman and Horne (1983), Hynes (1970), Maitland (1978), Moss (1988), Resh and Rosenberg (1984), Townsend (1980), Uhlmann (1979) and Welch (1935).

Stream Hydrology: An Introduction for Ecologists, Second Edition.
Nancy D. Gordon, Thomas A. McMahon, Brian L. Finlayson, Christopher J. Gippel, Rory J. Nathan
© 2004 John Wiley & Sons, Ltd ISBNs: 0-470-84357-8 (HB); 0-470-84358-6 (PB)

*Streamflow*

As a general trend, streamflow increases and channels get larger in the downstream direction. Patterns of physical habitats are created along a stream and within the pools, riffles and boulder clusters within particular stream reaches. These habitats and their inhabitants vary with patterns of streamflow. Ephemeral streams (Section 4.1.3), for example, usually (but not always) support different species than perennial streams. Snowmelt-fed and spring-fed streams generally have more predictable and less variable flow patterns than rainfall-fed ones, and their flora and fauna will be different. In highly variable streams, organisms may require more flexibility in their feeding, growth and/or reproductive behaviours. Fish community patterns also tend to be influenced more by the streamflow in variable streams, and more by biological factors such as competition, predation or food resources in stable streams (Pusey and Arthington, 1990).

Floods and droughts can have significant impacts on riverine species. Periodic scouring of banksides and inundation of floodplains regulate plant growth and nutrient input to the stream. Patterns of flooding affect the distribution of plant species within the stream and along a gradient from the river's edge to upland areas. Prolonged flooding of wetlands is needed for waterbirds to feed, rest and reproduce. The survival of juvenile fish may also depend on the inundation of floodplains, billabongs and backwaters. Moreover, floods serve as a signal for some fish species that it is time to spawn. Floods turn over rocks, altering the configuration of the streambed and 'resetting' the ecosystem by allowing a succession of organisms to re-colonize the substrate.

During low flows, temperature and salinity levels may rise, and plant growth within the channel can increase. Some species may rely on low flow periods for a part of their life history; for others, it is a time of stress. The stream may dry to a series of scattered pools connected only by sub-surface flows, limiting movement and increasing predation and competition for nutrients and space within the remaining waters. Intermittent streams experience a greater range of physical and chemical variation (e.g. in temperature and dissolved oxygen levels), and thus support unique biologic communities. These streams generally have a lower species richness as compared to perennial ones (Lake *et al.*, 1985). To cope with temporary waters, some organisms have developed special adaptations, such as dormant phases, which allow them to survive. Some larvae of aquatic insects, for example, burrow deep into the streambed to find sufficient moisture. Other species have drought-resistant eggs or spores and quick regenerative powers for the time when

favourable conditions return. If only part of a stream runs dry, the affected reaches can be re-populated from remaining pools or upstream tributaries. Williams (1987) provides additional information on the ecology of intermittent streams.

Streamflow is a particularly important factor in the study of regulated waters, where modifications to natural flow patterns can have marked effects on the stream's flora and fauna.

*Current*

As stated by Hynes (1970, p. 121), 'current is the most significant characteristic of running water, and it is in their adaptations to constantly flowing water that many stream animals differ from their still-water relatives'. Some species have an innate demand for high water velocities, relying on them to provide a continual replenishment of nutrients and oxygen, to carry away waste products and to assist in the dispersal of the species. At a given temperature, the metabolic rate of plants and animals is generally higher in running water than in still water (Hynes, 1970). However, it takes a great deal of energy to maintain position in swift waters, and most inhabitants of these zones have special mechanisms for avoiding or withstanding the current.

On average, water velocity tends to increase in the downstream direction, even though mountain torrents give the impression of high speeds in comparison to the more sluggish-looking lowland streams. Within a particular region, however, local variations create a mosaic of patterns which support species with different preferences. The velocities actually encountered, then, are of more relevance than average velocities (Armour *et al.*, 1983). As flow levels increase, velocity patterns will shift, forcing organisms to find refuge in calmer backwaters, behind rocks or snags, within vegetation stands or beneath the streambed.

At a finer scale, the leaves and stems of plants or the arrangements of rocks can vary the local flow environment, creating 'micro-habitats' for other organisms. Moving even closer to the surfaces of these features, very small animals such as protozoans can live in a thin fluid layer of near-zero velocity. Complex communities of bacteria, fungi, protozoans and other microscopic organisms form 'biofilms' on surfaces within the stream, which constantly grow and slough off under the influence of the current.

A factor related to velocity patterns is turbulence, which is important in the aeration of waters and the ability of a stream to carry sediments. Turbulence has a 'buffeting' effect on organisms exposed to the current. Near the streambed, organisms may feed at the edges of small turbulent vortices that stir up the substrate and circulate foodstuffs.

Current affects the distribution of sediments on the streambed through its influence on lift and drag forces. Organisms subject to these same forces often show morphological adaptations such as streamlined or flattened bodies or the presence of hooks or suckers for clinging to the substrate. Blackfly larvae (Diptera: Simuliidae), characteristic of very fast waters, attach to the substrate with hooks and have a silk 'lifeline' with which to reel themselves in when they are dislodged. Stream-dwelling mollusc species have heavier, thicker shells than still-water forms, perhaps for ballast as well as protection from moving stones (Hynes, 1970). Species of fish which must negotiate strong currents tend to be streamlined, whereas those which spend almost all of their time near the streambed are more flattened from top to bottom (Townsend, 1980).

Species unable to tolerate high currents may use behavioural mechanisms to escape by burrowing into the streambed, hiding under rocks or building shelters. Fish and eels utilize the dead water regions behind rocks for shelter, moving in short bursts from one to the next.

The distribution of current within a stream can be considerably affected by channel modifications such as de-snagging or straightening. The comparison of velocity distributions in modified and unmodified streams is valuable in stream rehabilitation work.

### *Water Depth and Width*

A stream's depth and width is related to the amount of water flowing through it. However, variations in channel form such as pools and riffles, wide meander loops and sand bars will create variation in water width and depth even where the streamflow is the same.

Water depth has an influence on water temperature, since shallow water tends to heat up and cool down more rapidly. It affects light penetration (more so in turbid waters), influencing the depth at which aquatic plants have enough energy for photosynthesis. Hydrostatic pressures also increase with depth, affecting the internal gas spaces in both plants and animals.

Depth affects the distribution of benthic invertebrates, with most preferring relatively shallow depths (Wesche, 1985). Both depth and width affect the physical spacing between predator and prey species. In general, larger fish prefer to live in pools and smaller fish in shallower water. Depth may become a limiting factor for fish migration when the water is too shallow for passage. Changes in water level can also affect the survival of species (e.g. by stranding fish or eggs, or inundating seedlings at the wrong time).

According to Pennak (1971), the width of a stream determines much of its biology. Migrating birds may require open riverine corridors for navigation and a certain width of water not obstructed by trees for landing and take-off. For resting and nesting, some waterbirds need a water 'barrier' of a certain width to protect them from predators. Mammals such as beavers also have specific requirements for width, depth and slope (Statzner *et al.*, 1988). For terrestrial animals, width and depth will affect their ability to migrate from one side of a stream to the other.

Geomorphologically, width increases in the downstream direction and the shading of streams by overhanging streamside vegetation decreases. Thus, as streams increase in width downstream, organic input from riparian vegetation becomes less significant and instream photosynthesis increases.

### *Substrate*

In a stream, 'substrate' usually refers to the particles on the streambed, both organic and inorganic. Inorganic particle sizes generally decrease in the downstream direction. On a more local scale, larger particles (gravel, cobble) are associated with faster currents and smaller particles (sand, silt, clay) with slower ones. Generally, streambeds composed of smaller particles are less stable, but this will also depend on the mix of particle sizes and shapes. Studies of substrate composition should consider the average and range of particle sizes, the degree of packing or imbeddedness and the irregularity or roundness of individual particles.

Substrate is a major factor controlling the occurrence of benthic (bottom-dwelling) animals. A fairly sharp distinction exists between the types of fauna found on hard streambeds such as bedrock or large stones and soft ones composed of shifting sands. Slow-growing algae, for example, require stable substrates such as large boulders (Hynes, 1970). Additionally, a whole complex of microfauna can occur quite deep within streambeds (in the *hyporheic* zone). These may carry out most or all of their life histories underground.

In general, lowland streams with unstable beds tend to have a lower diversity of aquatic animals. Aquatic plants, however, may prefer finer substrates, the plants then becoming substrates for other organisms. Silt substrates may support high populations of burrowing animals, particularly if the silt is rich in organic matter. Freshwater mussels, for example, mostly occur in silty or sandy beds. Clay substrates typically become compacted into 'hardpan' (Pennak, 1971), supporting little except encrusting algae and snails.

The greatest numbers of species are usually associated with complex substrates of stone, gravels and sand. The mix of coarser particles in riffles provides the richest aquatic insect habitat, and is considered the 'fish food' production zone in upland streams. Larger substrate materials provide firmer anchoring surfaces for aquatic insects and more shelter is available in the form of crevices and irregular-shaped stones. Crustaceans such as freshwater crayfish also require rock crevices for shelter.

Fish require substrates that provide shelter from the current, places for hiding from predators and sites for depositing and incubating eggs. Salmonids, for example, dig nest-like redds in gravel substrates by lifting particles with a vacuum-generating sweep of their tails. Successful incubation of the deposited eggs depends on circulation of water through the gravels to supply oxygen and carry away waste products.

Streambed particles are subject to dislodgement during floods, from dredging or other human disturbances and to a much lesser extent when the activities of bottom-feeding fish or burrowing animals stir up sediments. Suspended sediments reduce light penetration and thus plant growth; they can also damage the gills of insects and fish. Larger grains in suspension can have a 'sandblasting' effect on organisms, and rolling stones can crush or scour away benthic plants and animals.

The composition of stream substrates can be altered by sediment influxes from upland erosion and by channel modification. Excessive siltation of gravel and cobble beds can lead to suffocation of fish eggs and aquatic insect larvae and can affect aquatic plant densities. This, in turn, can result in changes in mollusc, crustacean and fish populations. Generally, these changes tend to cause a shift towards downstream conditions (i.e. unstable beds of fine materials), effectively extending lowland river ecosystems further upstream.

### Temperature

In general, water temperature increases in the downstream direction, to a point where the water reaches an equilibrium with air temperatures. Water temperature changes both seasonally and daily, but to a lesser degree than air temperature. Seasonal fluctuations tend to be more extreme in lowland streams whereas daily fluctuations may be more extreme in the smaller, upland ones, especially where they are unprotected by vegetation or other cover. In temperate or cold climates, upstream reaches may actually remain warmer in the winter than those downstream, particularly when these upper reaches are spring-fed.

Local variations in shade, wind, stream depth, water sources (e.g. hot springs) and the presence of impoundments will alter the general trends caused by geographical position. Many organisms take advantage of these local variations. For example, Wesche (1985) cites studies which have shown that some trout species select spawning sites in areas with groundwater seepage, where the warmer waters protect eggs from freezing and reduce hatching times.

When water cools (above 4 °C), it becomes denser and sinks. In most stream reaches, turbulence keeps the water well mixed, but temperature stratification can occur where waters are more stagnant, such as in deep pools. One of the unique properties of water is that it is less dense as a solid. In winter, ice and snow can form an insulating blanket over streams, under which aquatic life can continue. Ice usually starts forming when the entire water mass nears 0 °C, beginning with low-velocity areas near the edges of streams, along the streambed and on the underwater surfaces of plants.

The temperature of a stream is critical to aquatic organisms through its effects on their metabolic rates and thus growth and development times. With the exception of a few aquatic birds and mammals, most aquatic animals are cold blooded, i.e. their internal temperatures closely follow that of the surrounding water. As a general rule, a rise of 1 °C increases the rate of metabolism in cold-blooded aquatic animals by about 10%. Thus, these aquatic organisms will respire more and eat more in warmer waters than in colder ones. Each organism will have maximum and minimum temperatures between which they can survive, and these limits may change with each life stage. For fish, unusually high temperatures can lead to disease outbreaks, cause the inhibition of growth and cause fish to stop migrating (Platts, 1983).

Water temperature is thus an important factor in regulating the occurrence and distribution of riparian vegetation, fish, invertebrates and other organisms. Because temperature also affects other properties of water such as viscosity and concentrations of nutrients and dissolved oxygen, it can be difficult to separate the direct effects of temperature from the indirect effects of these other properties. Stream classification systems (Chapter 9) often include water temperature as a factor.

### Dissolved Oxygen

Dissolved oxygen (DO) is essential for respiration in aquatic animals as well as being an important component in the cycling of organic matter within a stream. Since gas solubility generally decreases as the water temperature rises, this can lead to lower DO levels during the summer.

As mentioned in Section 1.3.5, the oxygen concentration of air dissolved in water is higher than it is in the atmosphere. Some organisms use this property indirectly, such as beetles that carry bubbles of air underwater to function as gills—oxygen diffuses into the bubble from the surrounding water as it is used (Hynes, 1970).

In the turbulent, well-mixed waters of upland streams, dissolved oxygen concentrations are usually near saturation levels. As these turbulent reaches gradually give way to more poorly mixed waters downstream, biological sources of oxygen become more important. Photosynthesis can actually super-saturate the water with oxygen during the day. Then, night-time respiration and decomposition demands for oxygen can lead to diurnal changes in DO level. Diurnal variations in DO have been used as a basis for computing primary production in streams (Bayly and Williams, 1973).

The distribution of DO influences the patterns of species found along streams. Carp, tench and bream, for example can survive in low-oxygen waters, whereas trout require higher levels. The amount of oxygen will also affect the activity of organisms. As mentioned by Hynes (1970), the swimming speed of young salmon varies with DO concentration.

Oxygen is supplied continuously to stream organisms by the current. In fact, some swift-water invertebrates rely on the current to carry water across their gills, being unable to produce their own currents for respiration. Many of these organisms can tolerate lower DO concentrations if the velocity is sufficiently high (Hawkes, 1975).

Low-oxygen conditions can result from both natural and artificial causes. When organic matter such as sewage or detritus undergoes aerobic decomposition by bacteria, oxygen is removed from the water. Oxygen deficiencies can be found in stagnant waters at the edges of streams or when the stream is totally covered by ice or mats of water weeds. Reductions can also be caused by influxes of de-oxygenated groundwater or releases of anoxic water from lower depths of reservoirs. These effects can create conditions unfavourable to those aquatic species sensitive to oxygen levels.

### Dissolved Salts

Salinity refers to the concentration of ions dissolved in water. The ions most commonly contributing to salinity are the cations sodium, magnesium and calcium, and the anions chloride, sulphate, carbonate and bicarbonate. Generally, the concentration of salts increases in the downstream direction, especially if streams originate in areas with resistant rock and flow into regions with sedimentary rocks that erode more rapidly (Townsend, 1980). Salts can also enter a stream from saline ground-water, from sea salts dissolved in the rainwater of coastal areas and from agricultural runoff. In general, water originating as groundwater tends to have a higher dissolved salt concentration than surface runoff.

Since saline water is denser than purer water, stratification can occur in slow-moving pools fed by saline groundwater or in estuarine regions. Here, a layer of fresher water floats over the denser saline water. Salinity levels are generally (although not universally) inversely related to discharge levels: the highest salinities occur during low flows, with higher flows having a diluting effect.

As with other factors, tolerance of saline conditions can influence the distribution and abundance of stream biota. Most freshwater plants and animals are unable to maintain their internal ionic balances in saline waters. Water then diffuses out of cells, leading to dehydration, or excessive amounts of ions can diffuse into cells, producing toxic conditions. The concentration of individual ions is also important; for example, increases in water hardness favour some groups such as molluscs and crustaceans (Hawkes, 1975). Some organisms, particularly those that have evolved from marine species, have special mechanisms for restricting salt movement and/or excreting salt against an osmotic gradient. Hart *et al.* (1990, 1991) provide a comprehensive review of the effects of salinity on Australian plants and animals.

### Vegetation

Vegetation influences the physical habitat of streams by providing shade and altering the structure of channels. The flexibility of plants can absorb erosive forces directed against the streambanks by water currents and ice and debris flows. Overhanging vegetation can help keep streams cooler in summer and warmer in winter. Shade affects the growth of algae in streams and the distribution of other organisms that have shade or sun preferences. Dappled shade also has a camouflaging effect on fish, which helps prevent predation.

Trees and shrubs are both important for bank stability, for shading the banks and streams, for providing nutrient input to the streams in the form of litter, and as wildlife habitat. Standing dead trees offer hollows for breeding and shelter for birds and animals. When trees fall into a stream, they provide cover for fish and insects, and dams of logs and other debris form pools and restrict the downstream movement of sediments. Shrubs, grasses and other plants growing along the sides of streams slow the current and encourage silt deposition.

Aquatic macrophytes and algae grow within the stream, especially in less-turbulent and less-shaded waters. The presence of large quantities of aquatic plants is often

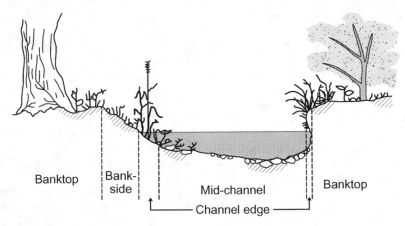

Banktop    Bank-    Mid-channel    Banktop
          side
          Channel edge

**Figure 2.1.** *Channel cross section showing vegetation zones as defined by Lewis and Williams (1984). Comparing this diagram with Figure 8.1 can give an indication of the frequency of inundation of the various zones*

considered undesirable since they retard the water flow, causing waterlogging of adjacent lands and increasing the risk of flooding. Mechanical cutting of aquatic vegetation occurs in many countries. The retarding effect of plants depends on their density and flexibility, and will thus vary with species and life stage.

As an aid to separating 'zones' of a riparian ecosystem, Lewis and Williams (1984) suggest dividing the river cross section into mid-channel, channel edges, banksides and banktops (Figure 2.1). Aquatic macrophytes, algae, and mosses may inhabit the *mid-channel zone* if scour and shade are not excessive. At the *channel edges*, pond-like conditions can develop in backwater areas and during low-flow periods, allowing rich plant growth including floating, un-rooted plants such as duckweed. Rooted plants with floating leaves, and reeds, rushes, liverworts, mosses and other species tolerant of inundation may also inhabit this zone. *Banksides* form a transition between the channel and the floodplain, with a merging of river and terrestrial plant species. Tree species may grow in this zone, and, with grasses, rushes, sedges and other herbs, they form a dense mat of roots that protects banks from erosion. Annual plants often spring up in eroded or unshaded areas, their flowers and seeds being important food sources for insects and birds. *Banktops* are primarily dry, and vegetation is usually influenced by the adjacent land use. Some plants characteristic of this zone, such as ferns and berry bushes, may also be found in upland areas.

The removal of streamside vegetation can lead to bank instability and changes in aquatic biota. Revegetation is therefore an important component of stream rehabilitation work.

## 2.2    The Planning Process

### 2.2.1    General

When charting the course of a new study, the object is to choose an approach that allows one to 'miss the snags', so to speak. It may be tempting to head out to a stream with a Surber sampler, a current meter and a vague notion that a statistician can work magic with the results, but a carefully thought-out plan greatly increases the chances of obtaining meaningful results. As Hamilton and Bergersen (1984, p. I-7) put it, 'no amount of data juggling will make the results of a poorly planned study useful'.

A well-organized study plan will maximize the amount of useful information collected while minimizing wasted efforts and worthless data. Expending up to 20% of the total study effort on the planning process should not be considered unreasonable. The time is well spent and can prevent wasted efforts in field, even though, as Green (1979, p. 31) says, 'we all find time saved in the hand to be more attractive than time saved in the bush.'

In the planning process, all phases of the study should be considered, from purpose through to the presentation of results. The process follows the scientific method, and can be summarized by the following steps:

1. Define a question (or series of questions);
2. Choose a method for answering the question(s);
3. Collect the appropriate information;
4. Analyse the information;
5. Answer the question(s).

The steps are interconnected and the process requires a delicate balancing of statistical ideals, choice of measurement techniques, the unpredictability of nature and available resources of time, money, equipment and personnel. The stages of the planning process are presented sequentially, but an 'iterative' approach should be taken in developing a final plan.

## 2.2.2 What Is the Question?

The first question to be asked (after 'is this study necessary?') might be: 'What is the object?' Are high, low or average flows of interest? Will the study be localized or representative of a larger region? Will the results be used to classify habitat, to test a hypothesis about the effects of some treatment or to make predictions? Fundamental research on the habitat of freshwater mussels, for example, will have a different objective than studies on the reclamation of a dredge-mined stream.

*Objectives* should be defined in the initial stages of a project in order to set goalposts for the study. The objectives should be clearly stated, communicating the nature and depth of the problem, and they should be achievable within the limitations of time and budget. Excessive ambition and vagueness are common faults at this stage (Platts *et al.*, 1987). It is better to successfully carry out a small project than to leave a larger, more complicated one unfinished.

The *study question* is an outgrowth of the objectives. A study question may start out in a general form, such as 'Does the length of crocodiles vary with river width?' However, in the process of defining boundaries for the study, it may be necessary to make the question more specific, e.g. 'Is the average length of freshwater crocodiles (*Crocodylus johnstoni*) captured by netting during summer months in the main stem of the Wildman River different between lower and upper reaches?' It is important to clearly formulate the study question based on the extent to which generalizations (inferences) can be made from the answers. The physics of flowing waters is essentially the same anywhere on earth; crocodile lengths have a less universal relevance.

Four types of stream studies are listed at the beginning of this chapter. In the first two categories, the question is 'What's out there?', whereas the third asks 'How is this different from that?' and the fourth 'How is this related to that?' On the surface, there are similarities between the different types of studies, but they each represent different objectives, sampling designs, analysis procedures and methods of presenting results.

A little preliminary or 'exploratory' work can help to focus both the design and scope of the project. In this preliminary stage, maps and photographs are collected, the study area is defined and tentative sites for data collection are identified. If possible, a trip to the site or a fly-over should be made. An appropriate period of study is chosen, based on whether the study is for existing conditions or long-term trends. Hydrological and meteorological data should be obtained for the study area, and/or for nearby or similar regions. Additionally, a literature search can provide life-history data on species to be evaluated, as well as information on pertinent ecological and hydrological measurement techniques. In the course of this preliminary process, initially fuzzy boundaries should sharpen into a clear definition of the study question(s).

## 2.2.3 Choosing Your Method: The Study Design

'Methods' include data collection techniques, statistical analysis methods and computer models: what to measure, when and how to measure it and how to analyse the results. The choice of methods may be patterned after previous studies to allow the comparison of results, or modified based on recommendations of those who have been there before. If the results will be used in a model such as PHABSIM (Chapter 9), specific data or methods may be required.

When first considering the possible number of environmental variables to measure, all may seem critical to the study. However, the variables should be carefully chosen so that the study gives relevant information rather than just being an exercise in data collection. Data requirements will depend on the stream(s) to be studied, the study objectives, the level of detail required and budget restrictions.

The length of study and sampling times will also be based on study objectives. Monitoring studies may extend over a period of years, whereas other studies may coincide with a particular life stage of a species such as the spawning season. Alternatively, a period of high or low flow may be of interest. Automated sampling, either continuous or triggered by a rise in water level, is helpful in the sampling of flow-dependent variables.

A tentative approach should be developed for collecting field information. This should consider the practicality of techniques and their potential for achieving appropriate levels of accuracy and precision (Platts *et al.*, 1987). Units, scale of measurement and taxonomic level of identification should be selected. Advantages and disadvantages of possible methods in terms of equipment prices, time requirements, and the expertise and availability of personnel should be balanced. The 'glamour factor' associated with complex, more precise, expensive methods makes them attractive, but often a simpler method will suffice. It would be overkill, for instance, to try to measure a stream's discharge with a laser Doppler anemometer.

Statistics has affected the basic philosophy of biological sciences to such a large extent that study design, analysis and the conclusions drawn are now virtually inseparable from statistical methods. Each study will utilize statistics to a different degree. Different procedures are needed if the purpose is to measure something specific (e.g. length of a certain crocodile named 'Jaws') or to make inferences from the results about a larger population (e.g. lengths of all freshwater crocodiles in Australia).

In a well-designed study the method of statistical analysis will be built in at the start. The optimal design for environmental studies will differ from that of agricultural or laboratory studies because the same degree of control over variables is not possible (Armour *et al.*, 1983). Care should be taken in the application of statistical methods. Techniques such as the analysis of variance are only powerful if sampling is structured properly and certain assumptions are met. If possible, a statistician's advice should be sought early on in the planning phase, especially for long-term, large-scale studies.

A useful way to check the study design for flaws is to write out a 'mock' analysis detailing the statistical analyses that will be used, and their limitations and assumptions. Some assumptions such as random sampling are crucial, and the study design should ensure that these assumptions are satisfied. Others, such as equality of variance or the presence of an underlying normal distribution, cannot always be checked ahead of time. In this case an alternate plan should be formulated in case the assumptions are not met. For instance, can the data be transformed? What are the consequences of not meeting the assumptions?

Also, there will be limits to the number of samples or measurements that can be taken and analysed. A well-designed study will ensure that enough samples are taken to meet the study objectives. At the same time, over-sampling should be avoided since it wastes time and energy that could be spent on something else, like another problem or playing volleyball.

A completed study design should, therefore, include a plan map showing the study sites, a listing of the variables to be measured, a choice of approximate dates for field surveys and the number of samples or measurements to be collected. More details on statistical sampling methods and methods of site selection are given in the following sections.

### 2.2.4    Collecting Information: The Value of a Pilot Study

As a general rule, it is a good idea to test the water before plunging into a new study. Reviewing pertinent literature will provide background information, but a pilot study in the field is advisable for checking out the proposed methodology. This 'sneak preview' can provide advance estimates of biological and physical variability, which are needed for designing an efficient sampling program. Pilot studies may point out the need for adjustments in the size and number of sampling sites, the number of samples required and the sampling design. Excessive variability, created by either the methods or by the object of study, may indicate the need to select an alternate variable or develop new sampling methods (Armour *et al.*, 1983).

Practical difficulties will often show up during pilot studies. For example, if a site is not easily accessible the sampling equipment must be portable. New species may be collected for which the taxonomy is not known. A pilot study is also useful for familiarizing field personnel with equipment and methods. The feasibility of the study itself may be questioned during this stage, and a trip 'back to the drawing board' indicated.

A pilot study is well worth the investment. Problems can be detected and corrected at lesser expense in this stage than after the full-scale study is launched.

### 2.2.5    Analysing and Presenting the Results

Methods for drawing conclusions from results should be outlined during the planning process. This is a good time to consider the purchase of statistical computer software (see Appendix), and to become familiarized with it before data are collected. If these preliminary steps are taken and statistical methods are integrated into the study design, the analysis should be just a matter of entering numbers into a computer and interpreting the output.

Graphic output is particularly valuable for revealing patterns in the data and summarizing results. The format for presenting results should also be considered during the planning phase. Publishing results in a refereed journal, for example, may require a different approach than presenting a report to a river management agency or presenting evidence in court. Huff (1954) provides an entertaining look at how to distort and exaggerate graphs to promote one's cause—and how to avoid this practice, while Hay (2002) provides a comprehensive guide to the presentation of your material in written and graphical form.

Conclusions can be only as good as the study design, the accuracy of measurements and the appropriateness of statistical tests. The odds of obtaining legitimate conclusions can be improved with careful planning. For large-scale, longer-term studies, the cooperative judgement, experience and knowledge of a larger group of professionals is needed. At any scale, plans must be customized to the individual study; there are no 'cookbook methods'. For further information, Green (1979) is an excellent reference for guiding the development of environmental study designs, and Downes *et al.* (2002) provide a comprehensive guide for assessing and monitoring ecological impacts.

## 2.3   Strategic Sampling

"There are three kinds of lies: lies, damned lies, and statistics."

(Disraeli, quoted in Huff, 1954).

### 2.3.1   Population, Sample and Other Vocabulary of the Trade

Wardlaw (1985) points out that science rarely gives a complete and final description. Although we may never arrive at the Absolute and Complete Truth about anything in nature, statistics can provide some useful approximations and descriptions.

Statistics can be either descriptive or inferential. *Descriptive statistics* summarize the information contained in data in terms of indices that describe central tendency, spread, 'clumpedness' and so forth. The objective of *inferential statistics* is to draw conclusions about the characteristics of a large (even infinite) group from a small sampling of that group. In environmental studies, both types are commonly of interest. Data are collected, described and related, and generalizations made from the results.

*Statistical analysis* is a tool for extracting useful information from observations and measurements. In discussing statistical analysis methods it is helpful to develop a common language. A few of the more common terms are described as follows.

A *population* refers to the entire collection (group) of observed or measured *elements* (items) about which one wishes to draw conclusions. Populations can be finite (e.g. the number of logs in a short stream reach), or infinite (e.g. the width of a stream, measured everywhere along its length), or so large as to be considered infinite (e.g. the number of crocodiles in Australia). The qualities or characteristics of interest, such as length, width or counts, are termed *variables*. Variables that cannot be expressed

numerically (for example the quality of pools, colour or a property like 'dead or alive') are sometimes referred to as *attributes*. Attributes can be treated statistically when combined with frequency of occurrence (Sokal and Rohlf, 1969).

If the population is small enough, it may be possible to measure every element in it (a *census*). For example, one could count (perhaps on one hand) the number of ecology journal citations found in a mechanical engineering journal during a certain time period. Each citation would be an element of the total population of citations. In most cases, limitations on time, money, site access or the destruction of an element during measurement (e.g. when measuring the life of a light bulb) make a 'total census' impractical. Therefore a *sample* is collected in such a way that it represents the whole population. A sampling *unit* is a collection of elements (e.g. a selected stream reach, within which measurements are taken). From information contained in the samples, *inferences* (leaps of faith bounded by levels of confidence) are made about the population from which the samples were drawn.

*Parameters* describe some fixed characteristic of the population, such as the total, mean or variance. An estimation of a population parameter is called a *statistic*, and is calculated from sample measurements. As the sample size approaches the population size, a statistic will become a better *estimate* of the parameter it is estimating. For example, a sample of crocodiles might be chosen from all the crocodiles in a region to obtain a mean crocodile length. The mean crocodile length calculated from the sample is the statistic and estimates the mean length for the region. If *all* the crocodiles in the region are measured, then the computed mean value *is* the population mean. *Efficiency* describes how well the statistic approximates the parameter being estimated (Zar, 1974).

A brief mention of *hypothesis testing* should be made, as it is a frequent application of statistics in biological research. A *null hypothesis* is a formal answer to a study question, stated in a way that is 'testable and falsifiable' (Green, 1979). The study question about crocodile lengths in the Wildman River given in Section 2.2.2 can be restated as a null hypothesis ($H_O$): crocodile lengths are the same in upper and lower river reaches, with the *alternative hypothesis* ($H_A$): lengths are different. Statistical tests of hypothesis verify or reject the null hypothesis at a certain *level of significance*.

A test designed to have significance level 0.05 will reject $H_O$ with a probability of 0.05 when $H_O$ is actually true. For the given example, this means there is only a 5 in 100 chance that we will incorrectly conclude that crocodile lengths are different when they are actually the same. Such an error is called a *type I error*. Depending on the seriousness of making such an error, we may wish to use a test with a significance level smaller or larger than 0.05. A certain amount of loyalty to one's initial choice of significance level and null hypothesis is necessary. One should not rationalize unexpected or unfavourable results to come up with a 'better' hypothesis and significance level (Green, 1979).

Some trade-off exists between a type I error and a *type II error*, which is the acceptance of a false null hypothesis (i.e. there really is a difference in crocodile lengths, but the study did not detect it). Associated with the type II error is the *power* of a test. This is the probability of rejecting the null hypothesis when it is false. Statistical power depends on (1) the variation in the data (the more variable the data, the lower the power), (2) the actual difference between experimental treatments (the larger the difference, the easier it is to detect) and (3) the number of replicates taken within each category. Commonly, the power of tests applied to field data is low because of large variability and small actual differences. In general, the more the replicates, the more powerful the test. However, there is a 'point of diminishing returns' once a certain level is reached, where further sampling does not substantially increase the power.

Hypothesis testing and the vocabulary of statistical sampling is covered in most statistical texts. Although focusing on aquatic insect studies, Allan (1984) is a particularly good reference on hypothesis testing in environmental work. An example of hypothesis testing is given in Section 2.5.

### 2.3.2  The Errors of Our Ways

Error can creep into a study through faulty sampling design, sloppy measurements, improperly calibrated instruments and poor sample site location. It will affect how closely the statistics from the sample data resemble the population parameters. *Error of estimation* refers to the distance by which an estimate misses the 'true' population value. The error of estimation can be defined by a level of confidence. For example, we might state that the mean freshwater crocodile length in a given region is 1.5 m, with 95% confidence that the 'true' mean length is between 0.9 and 2.1 m. This confidence is based on sample variability, which reflects the natural variability in crocodile length and the investigator's ability to measure it.

*Measurement error* refers to the difference between a recorded measurement and its true value. *Accuracy* is the closeness of a measured or computed value to its true value. *Bias* is a term used when measurements are 'off' due to an introduced source of error (e.g. inaccurate

instrument calibration or the stretch of a measuring tape). A constant bias is insidious because it will not be revealed by any manipulation of the sample data (Cochran, 1977). *Precision* is the closeness of repeated measurements to each other. The precision of field measurements will depend on the methods used. As Hamilton and Bergersen (1984, p. I-4) say, 'don't measure it with a micrometer, mark it with a grease pencil, and cut it with an axe'. Care and consistency in collecting, recording and processing data can greatly improve the quality of information collected.

*Statistical error* includes both measurement error and the inherent natural variability of the observed phenomenon. It refers to the difference between the parameter of interest and its estimate derived from sampling. Assumptions about the structure of errors are sometimes necessary in order to calculate confidence intervals, perform tests of significance and draw valid inferences from information contained in the measurements. *Parametric* statistical techniques, which assume that the population follows some statistical distribution (see Appendix), require errors to be (1) additive (homogeneous in terms of variance), (2) normally distributed and (3) independent (Green, 1979).

Although the natural world almost never satisfies these requirements, all is not lost. Pilot sampling or the literature may reveal a simple transformation that can reduce violations of the first two assumptions, which commonly occur together. Alternatively, one can proceed without a transformation but with an awareness of the effects on statistical analysis. Distribution-free, *non-parametric* techniques do not have the same restrictions. These techniques are discussed in a number of statistical texts, and if selected, should be built into the initial study design rather than used as a last-ditch effort to salvage data (Green, 1979).

Independence of errors is the only assumption for which violation is impossible to cure by transformation of the data. Random sampling, built into some level of the sampling design, is the only way to prevent this violation. In agricultural and laboratory settings, random sampling is more easily employed than in riverine environments, where, in practice, there is almost always a certain degree of subjectivity involved in the collection of data.

It might be said that true randomness exists only in theory, the real world being a mixture of order and randomness. Even the notion of 'random' coin-tossing fails in the hands of statistician-magician Persi Diaconis, who can control which side lands up (Kolata, 1985). Randomization will be discussed further in the following sections.

Large errors can lead to *outliers*, values that depart significantly from the bulk of the data. Outliers should be eliminated from an analysis only if there is knowledge about why they are abnormal (e.g. someone had been giving growth hormones to Jaws). Graphical exploration of the data can quickly reveal the presence of outliers as well as the homogeneity of error and other patterns that indicate a need for transformation. Statistical tests are also available for testing the validity of assumptions and for determining whether an outlier should be rejected. The reader is referred to statistical texts for these techniques.

### 2.3.3 Considerations in Choosing a Sampling Design

Hydrological and ecological variables follow their own patterns of ebb and flow in space and time, and sampling designs should reflect these differences. It might be of interest to measure populations of benthic invertebrates in a riffle, for example, yet a riffle may not be the best location for measuring average stream discharge. Similarly, measurements of discharge or bedload movement from a bridge or cable car may be feasible at high flows but drift diving for fish observation during floods would be unwise. Often, some variables can be measured to give information about how to sample others more efficiently. Statzner (1981), for example, found that precision was increased when abundance of benthos was first correlated with hydrodynamic factors.

The object of sampling is to estimate population parameters from information contained in a sample. A good sample survey design will maximize the amount of information for a given cost. Replication and randomization are key considerations for obtaining useful results.

#### *Replication*

Replicates are taken in order to quantify variability within a stream, between different streams and/or over time. To monitor change, it is best to use the same study site across time. However, if the study variable is sensitive to sampling methods (e.g. electro-fishing), other sites may be necessary. The channel itself will change shape over time, in which case pool-riffle locations may change and new reaches should be chosen.

The number of replicates required will vary with sampling design. It is extremely important to replicate at the appropriate level. For example, if we want to make inferences about a region, taking ten samples on different streams is preferable to taking ten at one site on one stream. In the latter case, the results would only be applicable to the one site! Hurlbert (1984) provides a discussion on 'pseudo-replication'. This point is also illustrated in the example of Section 2.5.

Pooling of samples (combining several samples into one) can be used in some cases to reduce costs. Pooling several small samples is preferable to taking one large one. For example, several water samples can be collected over an hour or across a section, and then combined before analysing for suspended sediment. The pooled sample would be considered a single replicate.

#### *Randomization*

Randomization means that the selection of any member (e.g. sample or sample site) must not influence the selection of any other member (Zar, 1974). Efforts should be made to *randomize* at some level of a study, whether in randomly selecting a set of rivers, randomly locating a series of transects, placing samplers randomly within a transect or randomly placing a wire grid within a reach. Randomization need not take place at all levels; in fact if a recognized variation exists (e.g. pools and riffles), a stratified design (Section 2.3.5) with randomization within strata will be more efficient.

Accessibility sampling, haphazard sampling and representative sampling are examples of non-random sampling. *Accessibility sampling* means that the samples or observations are those that are the most easily obtained. Easily accessible sites close to roads are often chosen because thick overgrowth, cliffs, non-portable equipment and/or private land ownership make sampling of other sites impractical. *Haphazard sampling* refers to the use of one's own judgement to 'randomly' select a sample. An example would be the common practice of tossing a wire quadrat, where sampling is not random but subject to the personal bias (and arm strength) of the tosser. *Representative sampling* involves choosing a sample that is considered to be 'typical' or 'representative' of the population. These are all forms of *judgmental sampling* and are subject to investigator bias. Their use may invalidate statistical tests and 'lead to estimators whose properties cannot be evaluated' (Scheaffer *et al.*, 1979, p. 32).

However, if a researcher's intuition is sound, judgmental sampling may actually yield better estimators than random sampling, especially if a random sample includes an extremely odd condition. People constantly form associations and determine averages, and a certain degree of 'natural history intuition' develops from years of exposure to the literature, the field environment and the opinions of others. According to Konijn (1973), a judgmental sample may be better than a random sample if the correlation is high between what is being measured (e.g. invertebrate populations) and the characteristics used to select a sample (e.g. eye estimates of substrate size).

The experienced eye will integrate a number of characteristics; unfortunately, the 'view' is inconsistent—a biologist will see things differently from an engineer. Thus, judgement may eliminate anticipated sources of distortion but introduce others because of personal prejudices or lack of knowledge about the population (Barnett, 1982). The appearance of bias in the selection process may also mean the data will not hold water if challenged in court.

Some compromise is needed to balance the demands of statistical perfection with the practicality of data collection. In the selection of sampling sites it might be feasible to have personnel from different disciplines (e.g. engineering, hydrology, biology) each select 'representative' reaches. A random sample could then be taken from their choices. Another alternative would be to use judgement to eliminate unacceptable sites rather than to preselect sites. An excess number of candidate sites (which can be streams, reaches or grid plots) can be selected at random, and judgement used to guide the ultimate selection of the study site(s), or criteria can be imposed to weed out atypical sites. If all the candidate sites are essentially equal with respect to the variables being studied, then perhaps it can be reasoned that an easily accessible site is as good as any other.

Judgmental sampling can be used to obtain 'suggestive information' during pilot studies (Konijn, 1973). It is recommended, however, that randomness be introduced at some level of a full-scale study, tempered by judgement, so that statistical analysis techniques can be applied and valid conclusions drawn.

### 2.3.4  Partitioning the Stream

The selection of measurement sites begins with decisions about the extent to which results are to be generalized, e.g.

catchment or continent. Is the objective to study a critical site, to sample only tributaries of a certain size or to study changes in a stream from headwaters to mouth? Investigations of streams can take place over a range of scales from micro-habitat to stream reach to stream system, as discussed by Frissell *et al.* (1986).

Selected streams are normally divided into relatively homogeneous sections, based on topography, geology, slope, streamflow and/or biological characteristics (Bovee and Milhous, 1978). This type of division is called *stratification.* Selection of the number and lengths of segments in a stream reaches in a segment and/or transects or plots within a reach is ultimately a problem in sampling design. This must take into account the variability of the stream, the precision required, and budget and time limitations. Some basic guidelines for partitioning a stream are as follows (see Figure 2.2):

1. *Identify and eliminate anomalous areas.* Bridge or dam sites, road crossings, waterfalls, channelized stream segments or other atypical features should normally be eliminated from the study area unless they represent a large portion of the stream or are the object of study. These areas would be set aside for separate study rather than included in a statistical analysis for making inferences about the stream.
2. *Divide a stream into segments.* A segment of a stream is a section where the flow and morphology are fairly uniform. Bovee (1982) recommends locating segment boundaries at locations where the average streamflow changes by more than 10%, such as at major tributaries or diversions. He also recommends placing boundaries where abrupt changes occur in slope, sediment input, bank materials and channel morphology. Where changes are gradual, boundary locations become more subjective. Stream ordering (Section 4.2.5) is

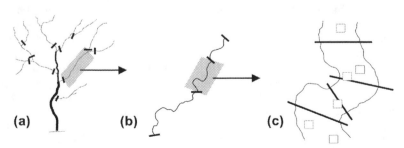

**Figure 2.2.** *Location of sampling sites: (a) division of stream system into homogeneous segments, (b) division of a segment into reaches and (c) location of transects (lines) and plots (boxes) within each reach*

another method of segmenting a stream. Based on the study design, a certain number of segments are chosen for sampling, from which inferences are made about the river system(s). Maddock (1999) recommends a habitat mapping or reconnaissance survey of stream segments, in order to extrapolate results from detailed studies back to this larger scale (e.g. for management purposes).

3. *Divide stream segments into reaches.* A segment can be further subdivided into reaches, some of which are sampled to make inferences about the whole segment. Reach boundaries can be located on a uniform basis (all reaches are of equal length) if the segment is homogeneous or changes smoothly from one end to the other. Alternatively, reaches can be divided into riffles and pools or at other easily discernible changes in channel shape. For example, two cycles of riffles and pools or two meander loops constitute a workable reach length for hydraulic geometry surveys (Section 7.2.6).

A common practice in stream surveys is to select 'representative' reaches which are considered to be representative of conditions in the segment. An aerial reconnaissance survey of the stream segment using a video camera may be helpful in the selection of representative reaches. Selection is usually based on the ability of the investigator to integrate knowledge about the stream segment and choose a reach which contains the same features in the same relative proportions. To introduce randomness at this level, several candidate reaches can be located and one or a few chosen at random for further study.

4. *Select transect, plot or point measurement sites.* A reach may be measured in its entirety (for example, in counting fish populations or the number of debris dams) or sites might be located within a reach for measuring bed material sizes, cross sectional profiles, aquatic insect numbers, etc. A *transect* is a line along which measurements or samples are taken. The term *plot*, from agricultural terminology, refers to an area within which samples are taken. *Point* measurements are taken over an area determined by the size of the sampling instrument (e.g. rainfall gauge or substrate sampler).

Transects, plots or point measurement sites can be located at random, uniformly, at representative sites or within noticeable micro-habitats. These may constitute sub-samples at either the segment or reach level. For instance, stream cross sections might be surveyed at uniformly spaced transects along an entire segment but crocodile populations sampled only within a few reaches.

Most of the steps in partitioning a stream can be done in the office if adequate streamflow records, topographical maps and recent aerial photographs are available. Anomalous or critical areas, segment and reach boundaries and measurement sites should be marked on maps and/or aerial photographs. The precise location of reaches or measurement sites should be done in the field, possibly in conjunction with pilot sampling.

Prior establishment of study sites is an essential part of planning. This eliminates confusion and time lost in the field, improves the validity of results and reduces the impact of the study itself on the stream.

### 2.3.5   Basic Sampling Designs

Statistical texts normally present three basic sampling designs: (1) simple random sampling, (2) stratified random sampling and (3) cluster sampling. Variations on these designs include two-stage sampling, systematic sampling and sequential sampling. Table 2.1 compares the various methods and their potential applications. Platts *et al.* (1987) give pertinent examples for several of the designs. Tests of significance and other statistics associated with each sampling design will not be covered in this text, but information can be found in statistical texts and in manuals for statistical computer packages (see Appendix).

Selection of a sampling design is based on prior knowledge of the stream, the variable(s) being measured, relative costs, time available for sampling and distance between sites, statistical efficiency and ease of computation, and compatibility of sampling designs with measurement techniques. The statistical efficiency of a design is influenced by the type of variable being sampled. Bovee (1986) gives the example that aquatic macro-invertebrates tend to be more prevalent in streams than fish, so random sampling may be efficient for the former group, but may not pick up a rare, highly mobile fish species. Spacing of sites is also an important consideration when sampling at one site can affect conditions at adjacent sampling locations. For example, control plots should almost always be located upstream from treatment plots. By being aware of the various designs and their limitations, researchers can structure their studies more efficiently.

### *Simple Random Sampling*

Simple random sampling requires that every possible sample of the same size has an equal probability of

**Table 2.1.** *Comparison of statistical sampling techniques. Based on Green (1979) and Platts et al. (1987)*

| Sampling design | Description | Considerations | Potential field use |
|---|---|---|---|
| Simple random | Each element of the population has an equal chance of being sampled. Random sampling is necessary to meet assumptions of independence | If used without judgement, anomalous samples can lead to poor estimates. Also, if elements are widely scattered throughout a large area, costs can be excessive | Relatively small, homogeneous areas with a randomly distributed population |
| Stratified random | A population is divided into uniform 'strata' and random samples are taken from within each stratum. More samples should be taken in more variable strata. The method yields estimates for strata as well as for the population | Variation between strata should be high and within strata low. Finding the optimal size of strata and sample size within strata requires trade-offs | Areas with recognizable patterns of heterogeneity in channel form and habitat (e.g. pools and riffles) |
| Cluster sampling | A population is divided into clusters, each of which has heterogeneity similar to that of the population. A random sample of clusters is taken and all or a sub-sample of elements within a cluster are measured | Clusters should be somewhat alike, but heterogeneous within each cluster. Again, trade-offs exist between number and size of clusters. Intensive sampling within a few clusters can be an economical approach | Areas where populations are dense so that the distance between elements is small |
| Two-stage sampling | This is a variation on the other methods, where a sample or cluster or stratified sample is further sub-sampled | Sub-samples should be well mixed (if applicable) and of sufficient quantity to characterize the sample | When analysing or counting a large sample is cumbersome; when elements are so similar that counting all of them is inefficient or when it is desirable to use different precisions at different levels |
| Systematic sampling | Samples or sampling units are spaced uniformly | Method is simple to use, but standard error of mean cannot be evaluated. Also, periodicities in environment may coincide with spacing interval | Where gradual, regular changes in the channel occur or if uniform spacing is required for modelling |
| Sequential sampling | Samples are collected until a desired level of precision is reached | Flexible approach, but population distribution must be constant throughout study. Impractical if costs are high for travel to site or if there is a large time lag between sample collection and analysis | Not recommended for ecological sampling unless population distributions are stable over time |

being chosen from the population, and that no population member is included more than once.

Each element of the population should be identifiable as an individual (e.g. a fish or a stream reach). To select elements at random from the larger population, a random number table from a statistical text, numbered slips of paper or even the last two digits of the numbers in a phone book can be used. However, in practice, selecting a random sample is more difficult than riffling through a phone book. A common approach is to superimpose a numbered grid system over an area, then randomly choose grid numbers. The 'grid' may be one dimensional, as in dividing a stream segment into uniform-length reaches; two dimensional, as in gridding a gravel bar for estimating particle size or perhaps multi-dimensional. Grid size will depend on the desired intensity of sampling and/or the area sampled by a measurement device.

In the types of studies performed in aquatic environments, simple random sampling will be most successful in relatively small, homogeneous areas. Otherwise it may result in samples that (1) are widely dispersed, causing considerable travel expense, and (2) leave some areas totally unsampled (Platts *et al.*, 1987). The advantage of simple random sampling lies in its relative simplicity of design and ease of statistical computation.

### Stratified Random Sampling

Stratified random sampling is more efficient than simple random sampling if there are obvious patterns (e.g. in stream form, vegetation or substrate) which can be separated into 'strata'. Each stratum should be more homogeneous than the whole population. Pools and riffles, for example, can be separated into different strata not only for statistical purposes but because sampling techniques may be different. In addition to being more efficient for estimating population parameters, stratified random sampling provides separate estimates for each stratum. A summary of this sampling procedure is given by Ching (1967) as follows.

1. Group the population into several sub-populations called strata. Each item in the population is put into one and only one stratum and no item of the population is left out. The strata are said to be 'exclusive and exhaustive'.
2. A random sample of items is then drawn from each stratum; these samples form the stratified sample.
3. The estimates (e.g. mean, variance or total) are calculated separately from each of the strata and combined to get estimates of the population parameters.

The stratified sample mean and variance are weighted by strata size. Weighting factors might be based on area, volume or counts. Stratified sampling is most efficient if the variation between stratum means (e.g. substrate size in riffles versus pools) is large compared to the within-strata variation (substrate size within riffles) (Barnett, 1982). The sample and strata sizes should be based on the comparative costs of sampling within and between strata and the acceptable precision.

*Proportional allocation* is a type of stratified sampling in which the sample size is proportional to the relative *size* or *variance* of the stratum. With this method, stream reaches with many different habitat types, for example, would be sampled more intensively than more homogeneous reaches (Bovee, 1986). *Optimum allocation*, another alternative, minimizes the amount of error for a given *cost* of a sample (Hamilton and Bergersen, 1984).

### Cluster Sampling

Cluster sampling is similar to stratified sampling in that the population (or area) of interest is divided into units (sub-areas), from which samples are taken. In cluster sampling, each sampling unit is a cluster of items or elements. All or a further sub-sample of the elements are measured. Platts *et al.* (1987) give the example of measuring all the trees within several 2.5 hectare plots. The plots are considered the 'clusters' and a sample of clusters would be selected rather than a sample of individual trees.

The difference between cluster sampling and stratified sampling is that in the latter, strata should be as homogeneous as possible within each stratum but one stratum should differ as much as possible from another with respect to the characteristic being measured. Clusters, on the other hand, should be as heterogeneous as possible within the cluster and one cluster should look very much like others (Scheaffer *et al.*, 1979). Cluster sampling is most effective when the physical distance between elements (like trees) is small and only a few clusters are needed, so that travel expenses and time required for sampling are minimized.

The steps involved in cluster sampling are given by Ching (1967) as follows.

1. Separate population into mutually exclusive, heterogeneous subpopulations called clusters.
2. Choose a random sample of clusters.
3. A cluster sample may include all the items in clusters, or a sample of items can be drawn from each of the chosen clusters.

4. Estimates are calculated separately for each cluster and then combined to give an estimate of the population parameter.

The cluster sample mean and variance are influenced by both the sampling of clusters and the sampling of items from within the selected clusters. The size and number of clusters sampled will affect the efficiency of this sampling design. As a general rule, each cluster should be representative of the diversity of the entire population. When measurements of elements within a cluster are highly correlated, a large cluster size is inefficient, and it is better to take more samples of smaller clusters (Scheaffer *et al.*, 1979).

In general, sampling a larger number of small clusters is more efficient than sampling more items within each cluster. However, the larger number of clusters increases both travel expenses and sampling time, so a compromise must be made. Compared to either simple random or stratified random sampling, cluster sampling offers one major advantage: the cost per element sampled is lower. The disadvantages are that the variance among elements tends to be higher, and the required computations are more extensive.

### Two-stage Sampling

Two-stage sampling is a variation on the above methods, in which samples are again 'sub-sampled'. For example, it may be desirable to subsample a one litre sample of water to analyse the zooplankton count or sediment concentration. In cluster sampling, sub-sampling is used when it is obvious that many near-identical elements are included within a cluster sample, and to sample them all would be a waste of effort (Platts *et al.*, 1987).

Two-stage sampling is also useful when a variable is difficult or expensive to measure precisely. In this case, a 'quick and dirty' method can be applied to a large number of sites (e.g. remote sensing) and a more precise method (e.g. vegetation surveys) applied to only a small number of the same sites. The imprecise method can then be calibrated against the results of the precise method by a ratio or regression method (see Armour *et al.*, 1983).

### Systematic Sampling

Systematic or uniform sampling is another method of sub-sampling. The first unit in a sample is randomly selected, followed by the selection of the other units at fixed (uniform) intervals (e.g. transect locations in a stream reach). This design is particularly relevant if the data are to be input into a model requiring a fixed spacing interval. Bovee (1986) outlines a variation on this method, 'the

systematic random walk', in which the distance between sampling locations is uniform but the bearing from one location to the next is randomly selected.

Uniform spacing is most effective when gradual, regular changes occur. However, this design is inefficient if the sampling interval coincides with some periodicity in the stream geometry or aquatic populations, much like sampling only the valleys between evenly spaced mountains, or only the mountaintops. Also, there is no valid way to estimate the standard error of the sample mean (Elliot, 1977). The method is probably not applicable to most habitat studies.

### Sequential Sampling

In sequential sampling, described by Green (1979), statistics are initially calculated from preliminary samples (such as those taken during a pilot survey), and sampling is then continued until a desired level of precision is reached. For the design to be effective, the population distribution must remain the same during the study (perhaps unrealistic in ecological work). It is impractical if most of the cost is in travel to the site rather than the cost per sample at the site.

### Stream Surveying in Practice: The Acheron River

Despite cautions about the need for randomization, it may be impractical for physical descriptions of streams because of the large number of samples required. In a study of the channel geometry of the Acheron River near Melbourne, Australia (Gordon, 1996), used as the basis for examples throughout this text, measurement sites were selected using representative sampling (sampling sites are shown in Figure 4.7). The stream network was first divided into segments using stream ordering, and the stream segment lengths were measured on maps. One fifth-order, two fourth-order, four third-order, eight second-order and sixteen first-order streams of average length were then selected, with several alternates, based on the assumption that segments of average length were 'representative'.

Within each segment, a 'typical' reach was selected judgmentally after walking upstream and downstream. The reach included at least two pools and two riffles, and was located well upstream of roads and stream junctions if possible. Two cross sections were measured in riffles and two in pools, again at 'typical' locations that reflected average local conditions.

In this study, the intent was to describe the average channel geometry for each stream order and separately for pools and riffles. Relationships were developed for the stream network, but results were not extrapolated to other basins. The study was primarily descriptive in nature.

### 2.3.6  Sample Size

After getting an idea of the expected variability (e.g. from pilot studies or from the literature), an estimate of the number of required samples can be calculated, based on the chosen level of precision. If there is much variation— either natural or due to measurement error—the number of samples may seem unrealistic. Statzner (1981) cites several examples to illustrate that hundreds of samples are necessary for estimating abundance of benthos, even when considering only a small, uniform section of a stream.

For random sampling, assuming the data follow a normal distribution, the number of samples can be estimated as

$$n = c\left(\frac{s}{\bar{x}}\right)^2 \qquad (2.1)$$

where

$n$ = sample size
$s$ = sample standard deviation
$\bar{x}$ = sample mean

and $c$ is a factor. The term $s/\bar{x}$ is also called the coefficient of variation ($C_v$; see Appendix 1). Based on assumptions of an infinite population, a reasonably large sample, and 95% confidence limits on the error of estimation of the mean, $c$ can be approximated by

$$c = \frac{4}{\varepsilon^2} \qquad (2.2)$$

where $\varepsilon$ is the percentage error of the mean, expressed as a decimal (the probability is 95% that the true mean lies within the range $\bar{x} \pm \varepsilon\bar{x}$). The factor $c$ would therefore vary from 4 for an estimate of the mean within $\pm 100\%$ to 1600 at $\pm 5\%$. If the standard deviation and mean are approximately equal (i.e. $C_v = 1$), then the sample size is the same as $c$; thus 1600 samples would be needed to obtain an error of $\pm 5\%$. Obviously, if the variation is higher, even more samples will be needed for the same level of error.

Equation (2.2) is based on 95% confidence limits; however, $c$ can be calculated for other levels of confidence. For confidence limits of approximately 68%, $c$ would equal $1/\varepsilon^2$. Thus, the required sample size would be smaller but we would be less certain about our estimate of the mean.

Realistic confidence levels must be chosen in order to avoid collecting an overwhelming number of samples. 'Acceptable' may mean a boundary on the true value of the mean of $\pm 40\%$ or more. Since the power of a test is affected by sample size (Section 2.3.1), one should also be realistic about the magnitude of effects that can be detected. In sampling benthic invertebrates, Allan (1984) recommends 10–20 replicates for modest precision and at least 50 for high precision. In the analysis of streamflow data, we are limited by the number of years of record.

There will be a trade-off between precision, budget and the number of samples taken at different levels (e.g. the number of reaches versus the intensity within each reach). A common approach is to first determine the maximum sample size based on time, personnel and budget and then decide whether this will yield an acceptable level of precision (Armour *et al.*, 1983). In general, it is preferable to

- Sample the study area by taking many small samples rather than a few large ones;
- Have more main plots at the expense of less within-plot sampling;
- Draw a larger sample from a stratum which is larger in size and/or more variable;
- Balance the expense of taking samples at each site with travel expenses.

## 2.4  Know Your Limitations

Platts *et al.* (1987) and Armour *et al.* (1983) discuss several 'confounding factors' which can affect the results of a study: institutional, political, biological, statistical and those relating to equipment and personnel. To these can be added effects imposed by weather, flow conditions and site accessibility.

Institutional and political factors may lead to a study being terminated before its time because of budget cuts or the unpopularity of potential results or of the study itself. If a study extends across a long period of time it is important that management of the study area remains under constant administration. Land-use changes during the period of study (such as logging) can seriously affect results if unforeseen.

Budget also affects the number and expertise of personnel. Although sometimes unavoidable, especially if seasonal workers are hired, changes in personnel can have potentially drastic effects on measurements. The quality of sampling by the same personnel may also vary over time (e.g. they may become more efficient with experience). Administrators and investigators both need to ensure a continued commitment of time, personnel and money until the study is finished. If it seems that a study will go over budget, further funding should be sought or a scaled-down plan developed early in the planning process.

Weather and flow conditions and the accessibility of a site affect both stream biology and field workers. More remote sites may be more representative, but will perhaps receive less intensive or careful attention by workers who have had to 'bush bash' to get to them. Although directed towards stream gauging, AWRC (1984, p. 11) nicely sums up the difficulties in working in a stream environment:

> Gauging streams under high flow conditions is expensive, dangerous, uncomfortable and requires skilled and dedicated staff. Cuts in funding usually result in less resources being available for high flow gaugings, loss of morale of staff trying to do a professional job and an eventual unsatisfactory set of records, which rely heavily on subjective and tenuous extrapolations of rating curves.

The natural variability of biological and physical factors can be substantial. If the objective is to detect the effect of some treatment, such as a management practice or a point pollution source, excessive natural variation may mask treatment effects. A treatment may also only cause a response if a certain threshold is reached, rather than causing a linear response. Clumped distributions of biota, fishing pressures or migration of a species out of an area may cause problems with sampling. Additionally, variability which causes 'noise' in one study may be the information of interest in another (Green, 1979).

Statistical confounding factors are common. Some variables may be assumed independent when they are not. The sampling size may not be sufficient to give adequate power to a test. Rounding errors or errors in recording information can also affect statistics. Thus, methods and sampling procedures should be efficient and consistent throughout the study. To reduce bias, instruments should be calibrated as needed and personnel made aware of the effects of different conditions on the precision and accuracy of readings.

Additionally, the inferences from statistics must be tempered by sanity. A significant relationship is not proof of causality; similarly, results from a small area should not be extended beyond reasonable boundaries. If a significant relationship is found between the frequency of frog croaking and the number of eddies along a stream reach, it does not mean that one caused the other or that the relationship is a universal one.

## 2.5   Examples of How to and How Not to Conduct a Study (by M. Keough)

Three different ways of answering the same question will be presented. The study question is:

Do streams with trout have lower densities of blackflies (Diptera: Simuliidae) than streams lacking trout?

Blackflies live in fast-flowing riffle areas, where they feed on micro-algae growing on rock surfaces. They are absent from slow-moving sections of streams. All the sampling will therefore be done within riffles. Rocks will be removed from the stream and all of the blackflies counted. These stream insects are large enough to see with the naked eye.

*The null hypothesis ($H_O$).* There is no significant difference in blackfly abundance between the two types of stream.
*The statistical method.* Analysis of variance. A significance level of 0.05 is selected for testing $H_O$.
*Data to be collected.* The number of blackflies on the surfaces of rocks, each having an area of approximately 0.009 m$^2$.

In each case, it will be assumed that time and money allow a minimum of 100 rocks to be sampled, and that travelling time between streams or riffles is insignificant compared to the time needed to collect and examine the rocks. The data will be analysed using the computer program SYSTAT (Wilkinson, 1986).

### Sampling Program 1

Two streams are chosen, one with brown trout and one which has lacked these fish for as long as records have been kept. In each stream, a riffle area is selected and 50 rocks sampled from it.

The data are analysed using a one-way analysis of variance. We find that the mean number of blackflies on rocks from the trout stream is 20.94, with a standard deviation ($s$) of 7.49, while the estimates from the trout-free stream are 24.36 and 6.00, respectively.

The following table is generated by SYSTAT:

*Analysis of variance*

| Source | Sum-of-squares | DF | Mean-square | F-ratio | p |
|--------|----------------|-----|-------------|---------|-------|
| Fish   | 292.949        | 1   | 292.949     | **6.703** | **0.011** |
| Error  | 4282.793       | 98  | 43.702      |         |       |

From the analysis (*F*-ratio and its associated probability, in boldtype) we conclude that our two samples differ (i.e. that the fish have some effect).

The problem with this sampling method is that only two streams were sampled, one with fish and one without. Further, only one riffle from each stream was sampled. What would happen if we had sampled a different riffle from each of these streams? Would we have gotten the same result? Would we have gotten the same result if we had used a different pair of streams? We know that there is variation between the streams, but with only one stream of each fish type, we cannot know whether the results reflect a true effect of fish, or whether the two streams would have differed regardless of the presence of fish. *There is no way to identify the effect of fish from this design!*

### Sampling Program 2

A better approach would be to sample at least two streams of each type. Suppose, then, that four streams had been chosen from a map: two with trout and two without. The 100 rocks are again sampled, 25 from each stream. The sampling now has two levels: rocks within streams, and streams within each fish category.

The mean values for rocks in each stream are 22.49 and 19.39 for the two trout streams and 23.17 and 25.55 for the streams without trout. (Note: the same data have been used in all examples, so if the data were pooled from the streams in each category, the same mean and standard deviation would be obtained as in the first example.) The mean for the two trout streams is 20.94 ($s = 2.19$), and 24.36 ($s = 1.68$) for the trout-free streams.

The analysis in this case must take account of these sources of variation and so a nested analysis of variance is used. The SYSTAT output is as follows.

*Analysis of variance*

| Source | Sum-of-Squares | DF | Mean-Square | F-ratio | p |
|---|---|---|---|---|---|
| Fish | 292.949 | 1 | 292.949 | **3.060** | **0.222** |
| Streams within fish categories | 191.488 | 2 | 95.744 | 2.247 | 0.111 |
| Error | 4091.305 | 96 | 42.618 | | |

In this case, we can see from the analysis that there is no evidence (the F-ratio and probability in boldtype) for an effect of fish.

The problem in this case is not in the analysis but in the sampling design. Practically, much effort has been spent counting blackflies on 25 rocks in each stream, only to use an average value for that stream in our test about trout. Unfortunately, we then have a statistical test with very low power: the F-ratio has one degree of freedom for its numerator and only two for the denominator.

Did we need to look at 25 rocks in each stream? Are there better ways to do the sampling? As might be guessed, one method is to use more streams and fewer rocks per stream.

### Sampling Program 3

Instead, suppose ten streams with trout and ten without had been chosen (preferably randomly), and only five rocks from each stream sampled. The mean for rocks in the ten streams with trout are 21.14, 21.37, 24.34, 22.42, 23.18, 15.45, 19.30, 22.05, 17.86, 22.23, while for the streams lacking trout, the values are 25.70, 26.02, 24.72, 16.84, 22.57, 23.22, 26.46, 27.93, 25.29, 24.87. For the two stream categories, the means are 20.94 ($s = 2.67$) and 24.36 ($s = 3.06$), respectively.

The analysis this time shows a significant effect of fish:

*Analysis of variance*

| Source | Sum-of-Squares | DF | Mean-Square | F-ratio | p |
|---|---|---|---|---|---|
| Fish | 292.949 | 1 | 292.949 | **7.118** | **0.016** |
| Streams within fish categories | 740.858 | 18 | 41.159 | 0.930 | 0.547 |
| Error | 3541.935 | 80 | 44.274 | | |

What is different about this example? The key change is that more effort has been devoted to examining the variation between streams, and less to documenting variation among individual rocks. The statistical test is more powerful because the F-ratio used to test whether fish have an effect now has 18 degrees of freedom for its denominator.

In this example, with fewer rocks per stream, higher standard deviations are obtained than in the second example. Statistical power can be calculated for each case. Assuming that trout really do decrease the number of blackflies per rock by 5 (the means are 20.9 and 24.4), we can then ask if we are likely to detect such a difference if we sample, say, two, five, 10 or 500 streams. If we use the observed standard deviations we find that the chance of detecting an effect of this magnitude with only two streams of each type is only 30%, but when the number of streams is 10, we are 94% certain of detecting this fish effect.

This example shows the benefits of prior planning of sampling effort. In the first two cases, efforts were largely wasted. In the first sampling program no conclusions could be drawn from the sampling because we did not account for the fact that trout vary at the level of whole streams—some streams have fish and others do not. If we had done a mock analysis we would have seen that there are actually two levels of variation—rocks within streams and streams within fish categories. In the first design there was replication at the lower level but no replication at the level of whole streams.

In the second design streams were properly designated as the sampling units, but the test of hypothesis about trout was not very powerful because most of the effort was spent on looking at rocks. Again, if we had done a mock analysis, we would have seen that the power of the statistical test depended on the number of streams, not the number of rocks.

The third design demonstrates the benefit of advance planning of field surveys. More information was obtained by concentrating sampling efforts at the proper level. This was done by minimizing the number of rocks from each stream and maximizing the number of streams.

This example is a very simple one, but the task for which these sampling programs are designed is fairly common. All three designs are, at face value, quite plausible, but one is clearly better than the other two. The design flaws illustrated are, unfortunately, also common. The term 'pseudoreplication' has been applied to this error by Hurlbert (1984).

Field sampling programs like this cannot determine whether trout *cause* changes in blackfly density, but can only *test* whether the blackfly density is higher or lower in streams with trout than in those without them. If we need to show causality, we would need to experimentally manipulate the abundance of fish. In such an experiment, we would still need to guard against errors in the sampling program.

One cannot emphasize too strongly the importance of careful thought and planning of the statistics *before* going out and collecting the data. The three sampling programs described above all require approximately the same field and laboratory effort, but the first two represent poor allocation of time and money.

# 3

# Potential Sources of Data
# (How to Avoid Reinventing the Weir)

## 3.1 Data Types

A researcher typically starts out on the path to discovery by first retracing someone else's footsteps. An investigative phase can provide valuable background information on the study stream and its species, it can reveal existing sources of data, and it prevents the unwitting replication of someone else's work.

Finding the information, however, can be a time-consuming task. Computer-assisted literature searches are of immeasurable help in sifting through the profusion of written "fallout" dispersed by the information explosion. However, the search for data may require more detective work, as much of it is not catalogued or readily obtainable.

Maps and data summaries are usually published in a form compatible with the requirements of the majority of users. The quality, quantity and availability of data will vary from one place to another and from time to time, depending mainly on who has collected them and their budgets and priorities. It is best to approach any form of data with a healthy sense of skepticism and a realistic attitude about how much information can be drawn from it.

Although the concepts of study design introduced in Chapter 2 relate to scientific investigations in general, the remainder of the text will emphasize the study of physical properties of flowing water. Relationships between hydrological and biological variables will be discussed, but methods of collecting and sorting through biological samples will not be covered.

Physical data usually take the form of (1) a time series or (2) spatial data. *Time-series data* are collected at regular or sporadic intervals, and may either be instantaneous values or an accumulated or average quantity measured across some time period. Hydrological and hydraulic data, sediment data, water quality data and climatic data can be included in this category. *Spatial data* are observations made across a line, an area or a space, and include maps, photographs and other remotely sensed images. When temporal and spatial data represent different views of the same variable they can form a powerful combination for site analysis. For example, maps from different dates can be compared to detect river channel changes over time, and annual rainfall averages from different gauges can be combined to create regional maps.

## 3.2 Physical Data Sources, Format and Quality

### 3.2.1 Where to Look for Data and What You Are Likely to Find

When looking for data, one thing to bear in mind is that the agencies responsible for data collection are not necessarily the same as those disseminating the data. National data centres and principal data-collection agencies will be primary sources of data, and should be the starting point of a search for information. Other potential sources include water authorities at national, state and local levels, environmental protection agencies, departments of agriculture or lands, power companies involved in hydroelectric generation, research centres and universities. Highway or railway departments or consulting firms may also collect data for specific projects such as design of bridges, culverts, or drainage systems. A shrewd investigator will turn over as many rocks as possible to uncover data hidden in such unlikely places as newspapers, diaries of pioneers, files of state and local historical societies and interviews with local residents.

Stream Hydrology: An Introduction for Ecologists, Second Edition.
Nancy D. Gordon, Thomas A. McMahon, Brian L. Finlayson, Christopher J. Gippel, Rory J. Nathan
© 2004 John Wiley & Sons, Ltd ISBNs: 0-470-84357-8 (HB); 0-470-84358-6 (PB)

Increasingly, data are being supplied on a commercial basis from private firms. An example is Hydrosphere Data Products (www.hydrosphere.com) which supplies climatic, stream-flow, and water quality data from the USA and Canada on compact disk, along with software for analysing the data.

The availability of data will depend on restrictions imposed by a nationality or by an individual agency. Free access may not be allowed, especially while the data are still in a provisional state. The density of data collection sites will also vary. Streamflow gauging networks, for example, may average one gauge per 1–2000 km$^2$ in Europe and North America, but only one per 10 000 km$^2$ or more in tropical countries (Budyko, 1980). Rain gauge networks tend to be more dense and water quality stations less so.

Each establishment will have its own way of collecting, processing, storing and presenting data. A researcher may thus be faced with the problem of assimilating information acquired by a variety of methods and recording tech-niques, especially if the study area extends across state or national boundaries. For example, the units of measure-ment may vary. Methods of recording data range from a pencilled number written down by an observer to the automatic digital recording and telemetry of sensor data obtained at a remote site. Records will vary in quality from the original 'raw data' to the finalized form in which records have been refined, polished, and missing values estimated and annotated. When collecting data one should be sure to obtain notes on station operation, its location, instrument type(s), historical changes and estimates of data quality. Data formats will also vary; for example, 'flags' on data values (e.g. to indicate that precipitation fell as snow or a runoff value was obtained from an extrapolated rating curve) may be kept or suppressed.

Corrected and summarized data will typically be avail-able in published report form such as the Water Supply Papers of the US Geological Survey (USGS) or volumes of *British Rainfall* (variously titled), on computer storage media, or in a format that can be downloaded from the Internet. Relevant Internet sites include those of the USGS (water.usgs.gov), US National Weather Service (www.nws.noaa.gov), the Met office of the UK (www.met-office.gov.uk) and the Bureau of Meteorology in Australia (www.bom.gov.au).

A drawback of processed data is the long time lag (two years is common) between collection and release of the data. Raw data may be available but will usually contain blemishes that must be repaired by the user. Older, unprocessed records on streamflow, climate and morpho-logic and hydraulic characteristics of rivers may be mothballed in dark closets of various state, federal or private and university institutions. The reduction of unprocessed data is a specialized and time-consuming task, but worthwhile if they are the only data available for a site. Practical references for streamflow data proces-sing include Beven and Callen (1979), Brakensiek *et al.* (1979) and Woolhiser and Saxton (1965).

As standards rise for both research and design work and as competition for water increases, more pressure will be placed on data collection agencies to provide higher-quality information in a shorter period of time, collected more frequently and at more locations. Advances in instrument and computer technology will continue to increase the efficiency, accuracy and consistency of data collection and processing. The following is a description of some of the types of physical data that may be available.

### Hydrological and Hydraulic Data

Streamflow data are normally obtained at either natural sections or gauging structures (e.g. weirs). Typically, the water level (stage) is recorded and related to streamflow (discharge) by a stage-discharge relationship called a rating curve. Readings may be taken on an intermittent basis (e.g. weekly readings of a staff gauge) or as a continuous record.

Data are usually available on a daily basis, with stream-flow values expressed as a total daily volume or an average daily discharge. Instantaneous peak discharges and daily maximum and minimum flows are often pro-vided as well. Monthly and annual summaries are com-monly published for established stations. In the USA, the cubic foot per second (ft$^3$/s or 'cusec') is the basic discharge unit; in Australia, megalitres per day (Ml/d) is more typical; however, in the SI system, the cubic metre per second (m$^3$/s or 'cumec') is the basic unit for dis-charge. Monthly or yearly totals may also be expressed in volume units such as million m$^3$.

Information on the regulation of dams and irrigation diversion works will be available from agencies in charge of the structures. Current operation procedures should be verified by direct correspondence.

Paleohydrological data may be of interest for obtaining long-term trends in climate or improving estimates of flood frequencies. Data from tree rings, fossil pollen, sediment deposits, fluctuations in levels of closed lakes or glacial movement may be available commercially or through research agencies. References on the interpreta-tion and analysis of these data include Chow (1964b), Fritts (1976), Gregory (1983) and Jarrett (1990).

### Channel Characteristics

Data on channel characteristics will most likely be obtained from agencies involved in detailed studies on water surface profiles, local scour, or channel changes, particularly in the vicinity of bridges, dikes, dams, or in channelized or dredged reaches. Size distributions of bed and bank material may also be available.

Water data agencies are often able to supply cross-sectional profiles at stream-gauging sites. During and after high flows, river-regulation agencies may collect data on streambed roughness, magnitude of scour, and changes in channel form. Observations of flooding made by either professionals or local residents, including high-water marks, lateral distribution of flow, stages achieved by ice dams, or the amount of sediment or debris in the river, are valuable in evaluating the response of a channel to high flows.

### Climatic Data

For the purposes of this text the only climatic data that will be emphasized are those of precipitation and temperature. A researcher carrying out environmental studies may also wish to collect data on evaporation, wind, solar radiation, dewpoint, humidity and other climatic variables.

Precipitation data are normally available as daily readings of water depth in a rain gauge, published as monthly or yearly summaries. Information may also be available about the amount of snow that fell over a day, or snow depths on the ground. A history of each rain gauge station should be obtained in order to evaluate inconsistencies in the data. If the gauge is read daily by an observer, the actual time of observation should be noted, as should missing data or rainfall amounts which represent an accumulation of several days' precipitation. Near-continuous records are sometimes available, either in digital form (for example, from a tipping bucket gauge), or as data processed from a pen trace on a chart. Agencies will sometimes process the data to derive maximum depths of rainfall per unit of time, the minimum duration of certain rainfall depths, or estimates of the 'probable maximum precipitation' (the maximum amount likely to occur over a given time during extreme meteorological events). Regional maps are commonly published from the results, such as the one shown in Figure 3.7.

Temperature data are commonly recorded as daily maxima and minima. Summaries of monthly and yearly averages are often available. Both precipitation and temperature data can be obtained from meteorology agencies.

### Water Quality

Water quality data may be available from national or state agencies (typically environmental branches), and from municipal water and sewage treatment facilities or indus-

trial plants sited along rivers. Some water-quality sampling stations on main rivers in the USA and the UK now have many years of record, but it is more typical to find short, discontinuous records. Because of the cost and number of possible parameters to measure, the length of record and type of data available are highly inconsistent. The quality of laboratory analyses probably does not vary as much as the type of analyses done and the frequency of sampling.

Water-quality data may include electrical conductivity, pH and concentrations of heavy metals, ions (e.g. ammonium, chloride or sulphate), organics such as pesticides, and dissolved oxygen. Daily records of turbidity, salinity and temperature may be available if the site is automatically monitored, but it will be more common for samples to be collected manually at fixed intervals of time (for example, once or twice per month) without regard to the magnitude of stream discharge. If the aim is to determine seasonal variability, data should be reviewed to see if the range of discharges is adequately sampled (AWRC, 1984).

### Sediment

Sediment data provide useful information for analysing sediment influxes and throughput, identifying potential sources and evaluating problems of sediment deposition in reservoirs, channels and irrigation diversions. Decisions about the management of land resources can be assisted if the effects of land-use options on sediment production are known. Sediment data are collected at least as widely and as often as those on water quality, and may be obtained from the same sources. Data can also be found in engineering reports and in flood control and other water resources investigation reports.

Sediment data may be given as an instantaneous concentration or as a yield (daily, monthly or yearly). Yield is usually calculated by combining concentrations with flow data to obtain the total amount of sediment transported past a station over a given time period (Section 7.5). Occasionally, sediment data will be analysed by particle size. The usefulness of published data will vary. If 'daily' sediment samples are only taken at a designated time once per day, the peak events will often be missed; thus the 'representativeness' of one instantaneous value is questionable. This problem is not as great in larger streams, which are more consistent over a day. In all cases, however, one should investigate the method of data collection before using sediment data.

### 3.2.2  Data Quality

The data record is only a sample of the total population of values that have occurred and may be expected to occur. It

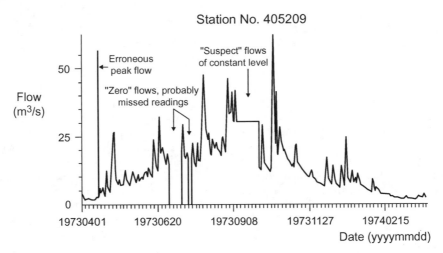

**Figure 3.1.** *A time series plot of daily streamflow data over a year, with artificial errors to show problems which might occur in a data record*

is thus subject to statistical errors resulting from the fact that it is only a sample of a larger population, as well as measurement errors. Wide variations may be found in the incidence of missing or inaccurate records and the frequency with which records are taken at each station.

For statistical analyses to provide projections into the future based on past properties of hydrological data, data must have tolerable measurement errors, be of sufficient length to give a good representation of the total population and be 'homogeneous', meaning that the data are from the same population. The following is a discussion of measurement error, representativeness and homogeneity, including methods of error detection and correction.

### Measurement Errors

Measurement quality depends on the precision and accuracy of instrumentation, its installation, site characteristics, and the conscientiousness of the observer (Shaw, 1988). In a high-quality monitoring program, personnel will calibrate and maintain measuring devices and apply corrections to the data to adjust for problems such as slow clocks, leaky rain gauges or the growth of mosses or algae at stream-gauging sites. The checking of records and processing of data is best carried out by the actual observer who is most likely to know about the causes of errors, how to detect them and how to fix them.

Unfortunately, it is often difficult to assess the quality of data. Once hydrological data have been stored in a computer or published in a report, it is a common tendency to accept these figures as true and accurate. The greatest danger in using the data is not in the actual

quality of the data collected, but in ignorance about what are good or bad data. Digital data especially give a false sense of validity; a pressure transducer may give precision readings of stream levels but these readings will be of marginal value if the transducer is bouncing on the streambed. If possible, an investigator should take a visit to a measurement site to learn about the quirks of the particular site and instrument. Enough inquiring should be done to develop a sense about how much attention is given to detection and correction of error in order to establish the degree of 'faith' one can have in the data.

A quick visual scan of the data time series to detect gross errors should be an initial step in data analysis. First, a quick scan of yearly or half-yearly data series using automatic scaling should be made to detect order-of-magnitude jumps in the data (e.g. from a misplaced decimal point), long periods of missing records and other erroneous readings. Figure 3.1 illustrates some of these errors. The data should then be examined at a smaller time interval (e.g. 30 days), and the maximum vertical axis (*y* axis) limit set to a lower value so that truncations of low flow values, short-term missing values and erratic data can be detected. When erroneous data are discovered, a researcher can then reject that part of record altogether or attempt to reconstruct it. Methods for infilling gaps in data are covered in the next section.

### Representativeness

The degree to which a record is representative of the total population is difficult to determine, since, geologically, records have only been collected for a relatively short period of time. The longest record of river stage in the world is over 1200 years for the Nile River at Roda. For most rainfall and runoff records, 100 years would be an exceptionally long record and 20 years more typical.

Representativeness may be affected by the inherent properties of a stream such as its 'flashiness' or by sampling deficiencies. Palaeohydrology records can be used to get an idea of long-term trends, but should be used with caution since they depend on subjective interpretation.

Data should cover a representative range of values rather than being exclusively from unusually wet or dry periods—unless, of course, these are the periods of interest. Records of only a few years in length are not likely to be representative of the long-term variability at a site (Yevjevich, 1972). One has little control, however, over the 'sample' provided by nature.

The number of years of record required for statistical analysis is related to sampling design. If a normal distribution can be assumed, Eq. (2.1) can be used to approximate the number of years of data needed to calculate a mean value of annual runoff or rainfall within a certain margin of error. Section 8.2 gives information on the length of record needed for flood frequency analysis. For habitat studies, Bovee (1982) recommends a minimum of 10 years of record, encompassing at least one high and one low sequence.

### Homogeneity

A 'homogeneous' record is one which is drawn from the same population or statistical distribution (Bovee, 1982), an analogy being homogenized milk versus its non-homogeneous form where cream and skim milk have separated. In hydrology, non-homogeneity can result from changes in the hydrologic environment, which can occur slowly or abruptly as a result of various human or natural activities (Yevjevich, 1972).

Although for most purposes a long record is preferred to a short one, changes in the physical conditions of the stream and its catchment, or in methods of data collection, are more likely to occur during a longer time period (Searcy and Hardison, 1960). Causes of non-homogeneity in streamflow records include (Bovee, 1982; Linsley *et al.*, 1975b; Yevjevich, 1972):

- Movement of a gauge;
- Change in observer;
- Change in data recording method (e.g. a staff gauge manually read on a daily basis may be replaced by a recording gauge);
- Change in channel configuration at the gauging site;

- Installation of a dam, irrigation works, trans-basin diversion works or levees; or the pumping of large quantities of groundwater upstream (and in some cases, downstream) of the gauging site;
- Sudden changes in hydrologic parameters from catastrophic natural events (e.g. wildfires, landslides, earthquakes or large floods) or land use changes (e.g. urbanization or deforestation).

For rainfall records the first three causes of non-homogeneity are applicable. Additionally, the erection of a building or fence, or the growth of vegetation around a rain gauge, can affect the 'catch'.

The *double-mass curve* is a useful method for detecting non-homogeneities in a record. Searcy and Hardison (1960) provide a definitive reference on the technique. The double-mass curve is based on the concept that a graph of the cumulative data of one variable versus the cumulative data of another is a straight line as long as the relation between the variables is a fixed ratio. Data from streamflow gauges, for example, can be compared with data from other streamflow gauges from the same general area to detect non-climatic changes in flow regime. A sample graph is shown in Figure 3.2. Streamflow is converted to millimetres by dividing by catchment area to make records from different sites comparable.

When only two stations are plotted against each other, as illustrated in Figure 3.2, it is not possible to determine which station is inconsistent. It is preferable to average values for a group of surrounding stations since the average is less affected by an inconsistency at any one station. This average becomes the 'base station' against which individual stations can be compared.

Streamflow should not be plotted directly against rainfall, as the relationship between them is rarely linear. A more effective approach is to first develop a relationship between streamflow (mm) and precipitation (mm). Estimates of streamflow derived from this relationship are used for comparison with the suspect station (Linsley *et al.*, 1975b).

When an inconsistency is noted, further investigation into its cause is warranted. The inherent variability in data causes spurious breaks in the curve, and these should be ignored. Major breaks in slope, such as the one illustrated in Figure 3.2, indicate that the data are non-homogeneous. If breaks are discovered, historical records for the gauge and/or the catchment area should be carefully reviewed to find causes corresponding with the time and direction of the changes. The researcher can then either eliminate all but the part of the record which is of interest (e.g. that portion representing 'present' or 'natural' conditions) or adjust the record to make it homogeneous.

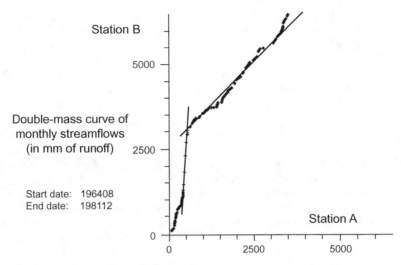

**Figure 3.2.** *Double-mass curve, showing cumulative monthly streamflow data from Station B (suspect station) plotted against data from Station A (reference station). In this illustration, lines have been fitted through the data by eye to highlight the break in slope*

Methods of adjustment differ for rainfall, runoff and sediment, and Searcy and Hardison (1960) should be consulted for details. Rainfall records can be brought back into alignment by using a ratio of the slopes of the two segments, since the slope is the constant of proportionality between the stations.

With streamflow records the double-mass curve should be used only for detection of inconsistencies. Adjustment of the record should be made using techniques which take into account its underlying cause. This may be a simple matter if, for example, a gauge is moved downstream and an adequate adjustment can be made by using a ratio of the new and former catchment areas, or it may be a complex process if reservoirs, diversions and levees have modified the flows. If only coarse monthly estimates are needed, attempts can be made to add diversion amounts back into a flow record and adjust for evaporation differences (e.g. from reservoir surfaces rather than vegetation). If daily 'natural' flow estimates are required for a stream which is now highly regulated by reservoirs, the only solution may be to reconstruct a flow series using rainfall-runoff simulation methods (Linsley *et al.*, 1975b).

### 3.2.3   How to Fill in a Streamflow Record

The problem of missing data is widespread. Gauges can be damaged during floods, power supplies may run low, pens may run out of ink, inlet pipes to gauging stations can clog with sediment, or an observer might go walk-

about. Various techniques are available for filling in estimates of missing data, and some of these can be used to extend a record back in time for statistical analyses. Ideally, the gaps in data will be filled by the data-collection agency and identified as estimated values.

Short gaps can sometimes be filled from straight-line interpolation (graphical or numerical) between correctly recorded discharges. A formula for linear interpolation is given in Figure 3.3.

Longer periods of missing records or those which include rainfall or snowmelt events should be reconstructed by establishing a regression relationship with a nearby gauge or group of gauges. Gauges should be located in a 'hydrologically homogeneous region' (see Chapter 8), and preferably on the same stream. It is advisable to develop regression equations using data from time periods during which conditions were similar to those for the period of missing data (e.g. storm type, water level, condition of the channel). For short periods of daily record, linear regression may be sufficient. Searcy (1960) advises the use of exponential (log-log) regression to normalize the data and linearize the relationship. For higher flows he demonstrates that this relationship between two basins is generally parallel to the *equal yield line*, a line of 1:1 slope drawn through the origin. Estimates of monthly or yearly values will typically have less error than daily ones, and estimates for more stable regimes will be better than those for flashy streams. Regression analysis is covered in the Appendix.

Errors will always remain in data no matter how carefully they are scrutinized and corrected. Measurement

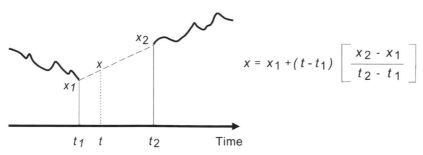

$$x = x_1 + (t - t_1) \left[ \frac{x_2 - x_1}{t_2 - t_1} \right]$$

**Figure 3.3.** *Linear interpolation by graphical and numerical methods. The points $x_1$ and $x_2$ represent values of a variable (e.g. water level) at times $t_1$ and $t_2$, respectively. The unknown value x is interpolated for time t*

errors are usually assumed to be random and normally distributed for the purpose of computing statistics. However, systematic errors that show up as false trends in data can also occur—e.g. from the accumulation of bed materials at a gauging site or the growth of vegetation around a rain gauge (Yevjevich, 1972). A decrease in the magnitude of errors from improvements in measurement techniques over time may be desirable; however, it also means that a constant variance cannot be assumed.

These cautions are not meant to discourage potential investigators from using data or performing conventional statistical analyses. A little wariness is advisable to ensure that conclusions are based on changes in the parameter of interest and not the quirks of an instrument or the biorhythms of the observer.

### 3.3 Maps: Finding Those Spatial Places

#### 3.3.1 What Types of Maps Are Useful?

When working with a riverine system a bird's eye view of the situation can provide information about the stream in relationship to its surroundings. A map is a graphic depiction of an area in two-dimensional form: its roads, soils, land use, vegetation, topography, geology, etc. Maps relate information such as distance, direction, shape, position, and relative size, according to a given scale and method of projection. They give a 'filtered' version of the real world, in which information is presented symbolically as aerial coverage or as 'isolines' of equal value (e.g. contour lines on a topographic map). Each country will have its own type of mapping system, criteria and standards for mapping. As with time-series data, maps have their limitations, and field checks may be necessary for detailed studies.

In hydrology, maps are useful for assessing land characteristics that influence runoff patterns and erosion rates, and as tools for predicting these variables when hydro-

logical data are unavailable. In conjunction with field surveys, maps are helpful for locating the extent of a floodplain, the position of a river channel and points of diversion for irrigation, hydropower or municipal and industrial water use. They are also useful for calculating channel slopes and other measures associated with channel morphology. Enterprising individuals have taken old maps developed before construction of a dam to develop 'fishing maps' which describe the contours of the inundated land. Finally, but certainly not least important, maps provide a means of orienting oneself in the field to obtain the geographic position of data-collection sites and to avoid getting lost.

Maps are basically divided into two types: *topographical* and *thematic*. A topographical ('topo') map describes a site in terms of its horizontal and vertical position. The leg muscles of anyone who has walked the tracks and trails of the world no doubt have a 'feel' for the meaning of closely spaced contour lines on a 'topo' map. 'Thematic' literally means theme; thus, thematic maps follow a special theme such as population, natural resources or economics. They are an excellent source of standardized, cheap, readily obtainable and easily captured land data. The classification of 'themes' is simple in concept; for example, if one soil or vegetation type is more common in an area than any other, then that area is classified by the prevailing type. However, there are rarely sharp demarcations between types. As with any map type, the larger the ground surface represented, the greater the generalization.

Maps have specific purposes and applications. Bell and Vorst (1981) point out a need for an international scheme of mapping for hydrology. UNESCO (1970) has developed an international legend for hydrogeological maps to provide this consistency. The stream hydrologist may use a number of different map types, some of which are described as follows.

*Topographical maps* portray the relief of the land, its water features, developments, roads and tracks. An example is given in Figure 3.4 and an aerial photograph of the

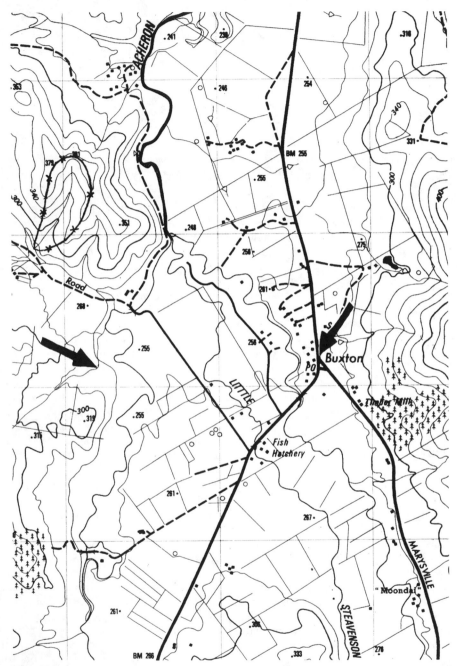

**Figure 3.4.** *A section of the Buxton (Victoria, Australia) map, scale 1:25 000. Contour intervals are 20 m and grid lines are 1000 m apart. The map was compiled from 1972 aerial photography. The ground distance between the arrows is the same as in Figure 3.5 (see discussion of Eq. 3.2). A catchment boundary is indicated by the line x—x in the upper left section of the map. Black and white reproduction of colour original, © 1978, Crown (State of Victoria) copyright, reproduced by permission; not to be sold or copied without the written permission of the Surveyor-General*

*Figure 3.5.* A section of a black and white aerial photograph, taken over Buxton, Victoria, on 5 February 1984. Height of the aircraft was 16 900 ft (5150 m) above sea level and focal length of the camera was 152.57 mm. The ground distance between the arrows is the same as in Figure 3.4. © Crown (State of Victoria) copyright, reproduced by permission; not to be sold or copied without the written permission of the Surveyor-General

same region is shown in Figure 3.5. Relief is sometimes supplemented with hill shading, which makes recognition of landform features much easier. Topo maps are essential for outlining catchment areas and evaluating drainage patterns. 'Crenulations' in contour lines can give clues to the presence of watercourses even if they are not marked as blue lines on the map.

Topographical data are now available in digital form, as horizontal and vertical coordinates on a grid system called a digital elevation model (DEM) or digital terrain model (DTM). Computer software packages have been developed for computing aspect and slope. National cartographic or survey agencies should be contacted for information on these products.

*Orthophoto maps* are a relatively new product in which topographic maps are plotted over a black and white photomosaic base, which has been corrected (ortho = 'correct') to true geometric precision. The advantage of this product is that it is as accurate as a map for obtaining measurements but identification of ground objects is enhanced because of the rich detail of the photographic image. Orthophoto maps have limited availability, but have great potential for use in hydrology.

*Geological maps* indicate geophysical features such as the age and distribution of rock formations and glacial and river deposits. Cross-sectional diagrams showing subsurface composition are often included, from which patterns of groundwater movement can be inferred (Wilson, 1969). Since geology is a major factor in stream formation, geological maps are important to hydrologists for interpreting the evolution of channel patterns and stream characteristics. Channel geometry, roughness, and the source and size of streambed materials may be associated with geological features shown on maps. A black and white reproduction of a segment of a colour-coded geologic map is shown in Figure 3.6. The light regions near the centre with drainage-like patterns depict alluvial deposits.

*Morphological maps* symbolically indicate breaks of slope in the landscape—the convex/concave boundaries.

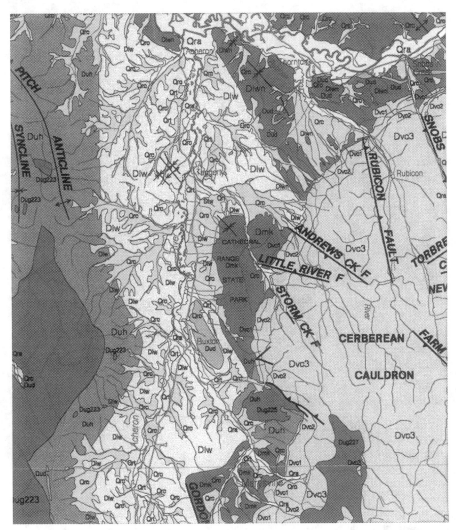

**Figure 3.6.** *A section of the Warburton (Victoria, Australia) geological map, scale 1:250 000; 1997 edition. Buxton is just below centre in the area shown. Reproduced by permission of Department of Primary Industries, Geological Survey of Victoria*

They thus give an indication of the form and steepness of landscape, which can be useful in locating areas that are too steep for transport or agriculture. *Geomorphological mapping* relates information on surface form, giving the location of features formed by fluvial, glacial, fluvioglacial and aeolian (wind) processes (Gerrard, 1981).

*Soils maps* may be helpful in determining erosion potential of lands. The soil type and its ability to absorb water will also affect the amount entering streams (CSU, 1977). Soil classifications differ, however, and some may relate more to crop-growing potential than to hydrology.

*Vegetation maps* often depict 'theoretical' or 'potential' natural vegetation, although in extensive areas of the world the natural vegetation has been removed or replaced. Global maps in particular should be used with care, as concepts of major plant associations and 'potential' vegetation types may not be relevant on a more local scale.

*Climate maps* are useful for assessing patterns of temperature, rainfall, evaporation and other related variables. Climate maps are of interest in making generalizations about the behaviour of streams and erosion rates and about an area's potential for supporting vegetation or

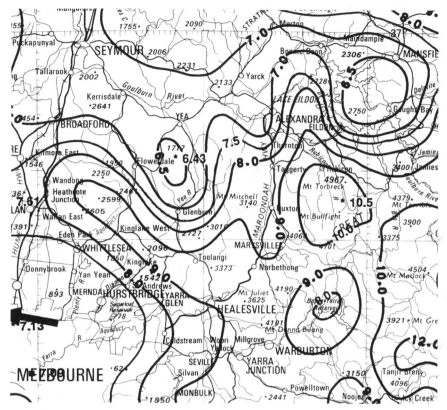

**Figure 3.7.** *A section of a 1:1 000 000 map showing design rainfall isopleths of 12 h duration and 50-year recurrence interval. Isopleths are lines of equal rainfall intensity in mm/h. Buxton is located to the right of centre. Black and white reproduction of original, © 1987, Institution of Engineers, Australia; reproduced by permission*

people. As well as general climate zones, maps may also depict individual climatic elements such as rainfall and evaporation. Rainfall maps may show average depths, frequencies and/or intensities. Figure 3.7 gives an example of a rainfall intensity map developed specifically for flood analysis.

*Runoff maps* usually display the mean annual runoff for a region and are useful for obtaining a general picture of geographical variations in runoff (Linsley *et al.*, 1975a). Depending on the amount of detail included in the maps, they may prove useful for extrapolating data from one region to regions of similar runoff, for general classification studies and for study site selection.

*Custom mapping* may be appropriate if available maps do not contain sufficient detail for a particular analysis. Finding the headwaters of streams or obtaining accurate contours of a streambed or eroded gully may require the addition of detail to commercial maps or the drafting of topo maps for very small areas. Special aerial photo-

graphy and photogrammetric mapping can be commissioned for this purpose at varying degrees of cost and accuracy.

### 3.3.2 Map Interpretation

#### *Marginal Information*

Some of the most valuable information on a map is contained in its margins. This normally includes the name, scale and date of the map; the units of measurement; grid and contour intervals; an explanation of symbols; and arrows giving true, grid and magnetic north. Dates of the original map and revisions are particularly important, especially if the map is assumed to represent current conditions. Symbols will differ with the map type, country of origin, and scale, and should be examined before making interpretations. For example, different symbols may be used to indicate road or river type, or the locations of mining claims or boat docks. International symbols for many hydrological features have been standardized, such as permafrost, groundwater contours, karst formations, the disappearance point of a stream and springs. Attributes such as vegetation or soil type will often be noted on a map by a specific colour or label, with a key provided as part of the legend.

### Location and Orientation

Latitude and longitude provide a coordinate system for pinpointing any location on the earth's surface. Both latitude and longitude are measured in degrees: 360 degrees in a circle, 60 minutes to the degree, 60 seconds to the minute (e.g. $37°19'30''$). Of historical interest, the system of using a base of 60 originated with the Babylonians, who discovered that the number 12 had more divisors (2, 3, 4, 6 and 12) than the number 10 (2 and 5). The number 60 combines the advantages of both base 10 and base 12 systems, and is used throughout most of the world as a basis for measuring time and degrees.

'Orienting' oneself by locating a position with respect to the points of the compass literally means finding the East or 'Orient'. Coordinates are usually indicated at corners of maps and by tick marks along margins. Besides latitude and longitude, maps may also be gridded in kilometres or quadrants. These lines will either intersect at right angles or converge towards the top or bottom of a map, depending on the type of projection used in making the map and the distance away from the equator; i.e. this effect will be much more apparent in Alaska than in Tonga.

*Latitude* is the angular distance measured north and south of the earth's midriff. The Equator is 0°, the North Pole 90° N, and the South Pole 90° S. A line connecting all points of the same latitude is referred to as a *parallel*. True North differs from the direction a compass points as the needle orients itself within the earth's magnetic field. Magnetic north drifts slightly east each year. The amount of correction (declination) and its annual rate of variation are normally indicated in the marginal information on topographic maps. If the map is more than about 10 years old, the declination should be adjusted by the factor: (annual variation) × (number of years since map was made).

*Longitude* is the angular distance measured west and east of the prime meridian, where *meridians* are lines connecting equal longitudes. The prime meridian passes through the Royal Observatory at Greenwich, just east of London. This location, chosen in 1884, had more of a political than a scientific basis, influenced mainly by the prominence of British map makers and the power of Great Britain at that time (McKnight, 1990). Longitude is measured east or west of a plane through the prime meridian, to 180° in either direction. The *international date line*—where today changes to tomorrow or yesterday—generally follows the 180th meridian, with a few jogs around island groups so that places like Fiji and Tonga remain on the same side of the line.

### Scale

The scale of a map is the relationship between the distance measured on a map and the corresponding horizontal distance on the ground. It is usually given in the margin area of a map as a graphic scale and/or as a fractional scale. The *graphic scale* is a line marked off in graduated distances, and remains correct even if the map is reproduced in a larger or smaller size. A *fractional scale* compares the map distance with the ground distance as a ratio (e.g. 1:10 000) or fraction (e.g. 1/10 000). A ratio of 1:10 000, read 'one to ten thousand', means that a measure of one unit (whether millimetres, inches or spans) on the map represents 10 000 of the same units in the real world. The use of metric units considerably simplifies conversions.

Scale may not be constant over a map because of the projection and method of mapping, but can normally be assumed constant if the map covers a relatively small area. It is helpful when reading maps to know the ground distance represented by 10 mm or some other convenient unit, so quick estimates of distance can be made.

*Large scale* and *small scale* are comparative terms. Just as a 'large' cup of coffee at one fast-food restaurant may be classified as 'small' at another, the difference between large and small scales is also dependent on the viewer. The distinction can be confusing because 'large' and 'small' refer to the ratio or fraction rather than the amount of earth surface covered by a map. For example, 1/100 is a larger fraction and thus a larger scale than 1/100 000. The International Cartographic Association (ICA, 1984) uses this classification:

| | |
|---|---|
| Large scale | greater than 1:25 000 |
| Medium scale | 1:25 000 to 1:250 000 |
| Small scale | 1:250 000 to 1:2 500 000 |
| Very small scale | less than 1:2 500 000 |
| (e.g. maps in atlases) | |

Each scale of map is suitable for different purposes; i.e. there is no 'best' scale. A comparison of Figures 3.4, 3.6 and 3.7 illustrates the effects of scale, with the larger-scale maps showing more detail than the smaller-scale ones. For hydrological work, a medium- or small-scale map is useful for identifying the location of a project within a riverine system, obtaining information on the stream network and regulating structures, and calculating parameters such as sinuosity. Larger-scale maps are needed to obtain detailed characteristics of the stream reaches under consideration.

### *Elevation*

Elevation is given on most maps in terms of distance above a certain level, usually a standardized mean sea level such as Mean Sea Level (MSL), or a national height datum such as the Australian Height Datum (AHD). Heights of significant mountains, benchmark points and other features are often shown as spot elevations.

On topographic maps the elevation is represented by *contour lines*: lines joining points of equal elevation. The *contour interval*, or vertical distance between contours, may be altered to suit the terrain or may be fixed for a given map series. This interval is normally specified in the marginal information on maps, and elevations are marked every few contour lines. Contour lines may contain error when aerial photography is used to develop the maps, especially where vegetation obscures the ground surface. Jennings (1967) cites an example of a map made from surveys taken between the two world wars. During later surveys, the nature of the errors in the map made it clear that much of the mapping had been done from ridge crests and the valleys were 'eyeballed' in.

### 3.3.3   Revision, Accuracy and Standardization of Maps

When using maps it is important to realize that they have limitations in terms of accuracy and representation of the landscape. Map users cannot expect mapping agencies to produce map series for specialized needs, such as large-scale maps with close contours for drainage analysis. On the other hand, one should not be timid about questioning map accuracy. 'As all who have made maps know, as finished products they have a definiteness about them which somehow conveys an accuracy even greater than that of the printed word and as with the latter it is not always fully justified' (Jennings, 1967, p. 80).

Representation of natural features on maps has become more accurate over time as the amount of quantitative information extracted from them has increased. The depiction of streams on maps, however, is fraught with problems of interpretation. In limestone terrain, segments of streams may be completely underground at low stage and above ground at high stage. Even when stream patterns are more typical, the classification of a stream as perennial or intermittent is a difficult one and map makers may choose to avoid the distinction altogether (Drummond, 1974).

Erosion and deposition normally cause slow change over historical time, but other landforms such as watercourses, glacier snouts, sand dunes and coastal spits change rapidly (Jennings, 1967). New settlements, roads and areas of timber harvest also cause the landscape to change quickly, and repeated surveys and map revisions are necessary to keep up with the changes.

The use of aerial photography and computer-supported photogrammetric systems has greatly increased the accuracy of topographical maps, as well as enabling more efficient revision. A danger in using maps of different editions, however, is that the revised map may represent only the changes in man-made features and not in the surrounding environment. Before maps are used for quantitative analysis, information on mapping criteria and standards should be obtained from the mapping agency. Field surveys and aerial photography can also help to verify map accuracy. Mostly, one should be aware of how far a map can be 'stretched' to provide information— whether getting an idea about how hard a hill will be to climb or calculating bifurcation ratios of river systems.

### 3.4   Photographs and Other Remotely Sensed Data

#### 3.4.1   What Is Remote Sensing?

Remote sensing, a term coined in 1960, is the 'measurement or acquisition of information by a recording device that is not in physical contact with the object under study' (McKnight, 1990). This definition includes more than just imagery, such as the remote acquisition of data on snow water content or wildlife movements. Another form of remote sensing is the use of laser scanner systems to obtain precise measurement of elevations on the earth's surface (Drury, 1990). For example, Bollweg and Van-Heerd (2001) found that laser altimetry was accurate within a few centimetres when used to measure the height of a 1998 flood wave on the Rhine River.

In the environmental sciences, remote sensing is defined more precisely as the recording of images of the environment using electromagnetic radiation sensors, and their interpretation (Carter, 1986; Curran, 1985). This definition includes conventional aerial photography and aerial photograph interpretation.

The amount of information which can be gained about a region is greatly enhanced by the acquisition of imagery. Imagery is attractive, too, where field survey costs are high or where areas are physically inaccessible. River patterns are dynamic, and aerial survey and satellite image analysis methods have contributed to the ability of hydrologists to monitor changes. Environmental and resource surveys, crop conditions, snow cover, extent of urban area development or timber harvest and the assessment of the extent of bushfires, flooding or drought are some of the practical uses of remotely sensed images.

In contrast to maps, where interpretations have been made by a map maker, images can be interpreted directly. The amount of information on an image may at first seem overwhelming because of this lack of prior interpretation, but with experience and repeated association with field conditions, qualities can be quickly inferred.

Maps, photographs and remotely sensed images each provide their own brand of information. High-altitude space imagery will cover the entire face of the planet, whereas large-scale colour or black-and-white photographs are more appropriate for detailed terrain studies. Knowledge of each product and its advantages and limitations can help in the selection of the best type of imagery.

### 3.4.2   Photographs

Compared to a map, a photograph is much closer to reality. Photographs can be used *before* a field trip to familiarize oneself with an area, to select suitable study sites, or to identify major landforms or drainages. They can be used *while in the field* to locate and orient oneself, to check the identification of features on photographs and to discover what lies beyond the immediate field of view. Additionally, they can be used *after* a field trip to refresh one's memory or to extend results from one site to similar areas nearby (Chapman *et al.*, 1985).

Photographs provide a unique record of the past, and are especially valuable when it is difficult to revisit a site. They can be re-examined—sometimes many years later— when techniques (and time) for extracting more information become available. Baseline photographs can be compared with monitoring photographs taken five to ten years later to evaluate environmental changes, either natural or due to management efforts (Platts *et al.*, 1987).

The earliest aerial photographs were taken from hot-air balloons in the mid-1800s (McKnight, 1990). Conventional aerial photography remains a primary technique when it is necessary to resolve the detail of ground conditions. Features that are not marked on maps, such as sinkholes and the fine patterns of upper stream drainages, may be observable on aerial photographs. Aerial photography is versatile because of the wide range of scales available and the large combinations of films and spectral filters which can be used. Limitations of aerial photography are those of cost and quality, which are dependent on the type of film, camera and aircraft, the distance above ground and the suitability of flying and sensing conditions.

For small riparian area studies, Platts *et al.* (1987) suggest less expensive methods of acquiring aerial photographs with a 35 mm camera and a small helicopter. However, for obtaining precise measurements of distance or aerial coverage, higher-quality images are needed. The timing of aerial photography is critical, not only for optimizing lighting or avoiding cloud cover but also for obtaining the best resolution of the features of interest. Soil-tone changes, for example, are more distinct in winter and spring, crop patterns in summer (Barrett and Curtis, 1992). Aerial photographs are useful in the analysis of river reaches: the sizes and locations of sandbars, river-control structures, changes in channel form over time, sediment concentrations and seasonal variations in vegetation along banks or changes in floodplain vegetation following inundation. Hooke and Kain (1982) provide a reference for using historical information to study changes in the physical environment.

Although 'remote' implies a great distance, photographs taken from hand-held cameras with one's feet firmly planted on terra firma (or slightly higher, from a ladder or 'cherry picker' boom truck) are of great value in environmental studies. For stream studies, 35 mm colour print snapshots should be taken laterally, upstream and downstream at a site (Platts *et al.*, 1987) to provide a historical record. Photographs of substrate are useful for direct analysis of particle size, roughness and percentage cover.

Photographs are either taken vertically (meaning the optical axis is perpendicular to the ground) or obliquely (at an angle). Oblique photographs give a more familiar point of view, but because of perspective effects, measurements are more difficult. *Photogrammetry* is the science of obtaining measurements from photographs. Measurement errors can be caused by distortions in the image caused by the lens system (e.g. at the edges of photographs), the relief of the terrain, or the tilt of the aircraft. For point-to-point comparison with maps, an image must be adjusted ('rectified') to ground distances with the use of previously placed and surveyed markers ('control points') or established features such as highway intersections or small water bodies. Agencies developing maps from aerial photographs of hilly terrain must allow for the fact that closer objects appear larger on a photograph. Computerized rectification is commonly used to create corrected orthophotos of high precision.

Vertical as well as horizontal measurements can be calculated from photographs if stereo pairs—two photographs of overlapping areas taken at slightly different angles—are available. Viewing these with a stereoscope yields a three-dimensional mental image. Using the same process, stereo-plotters are used commercially for tracing contours. Methods of using close-range photographs for describing and measuring surface microtopography have been compared by Grayson *et al.* (1988), and additional information on measuring heights from stereo pairs is given by Curran (1985), Lillesand and Kiefer (2000) and other texts on photogrammetry.

Although researchers may require custom, up-to-date aerial photography for special studies, photos can also be obtained from various land-use, military, weather service and cartographic agencies for more general use (e.g. examining stream channel changes over time). For example, the aerial photography from which published maps are made can usually be purchased from the mapping agency. Sources of aerial photography in the US include the USGS EROS Data Center (edcwww.cr.usgs.gov), the National Archives and Records Service (www.nara.gov) and local offices of federal and state land-management agencies.

### Scale

The scale of a photo has the same definition as for maps, and as with maps, it is helpful to work out the distance represented by 10 mm or some other unit length. Approximate scale can be calculated for aerial photographs by

$$\text{Scale factor} = \frac{\text{Focal length of camera}}{\text{Height of camera (or aircraft) above surface}}$$

$$(3.1)$$

The camera height and the focal length must both be expressed in the same units. This scale applies to the negative and contact print, and does not apply if the photograph has been enlarged or reduced.

Information on the intended aircraft elevation above sea level (not the ground surface) and camera focal length, the date and time when the photograph was taken, the general location, roll and exposure number, and name of the agency responsible is normally printed in the margin of commercial aerial photographs. The aerial photograph shown in Figure 3.5 was taken at an elevation of 16 900 ft (5150 m) above sea level, and the focal length of the camera was 152.57 mm. In Eq. (3.1) the elevation of the ground surface is needed to compute the height of the camera, yet the ground surface is not a consistent elevation. Using an average elevation of 275 m for the photograph the scale comes out 1:31 950, which is somewhat smaller than that of the topographic map shown in Figure 3.4.

A more practical method of obtaining scale is to measure a distance between two landmarks on a photo-graph and compare it to the equivalent distance obtained from a map. The scale of the photograph is then obtained by

$$\text{Scale factor} = \frac{\text{Distance measured on photo}}{\text{Distance measured on map}}$$

$$\times \text{ scale of the map (as a fraction)}   (3.2)$$

Distances must have the same units. Using measurements taken between the arrows shown in Figures 3.4 and 3.5, a scale of (46 mm/58 mm) × (1/25 000) = 1/31 520 is obtained. This method of calculating scale is preferable for obtaining more accurate measurements, and is essential if the photograph has been enlarged or reduced.

Platts *et al.* (1987) recommend a scale of 1:2000 as 'acceptable and achievable' for photographing riparian areas on small streams. Channel patterns are best obtained from smaller-scale photographs such as the one in Figure 3.5.

### Film Types

The four main types of film are: panchromatic black and white, colour, and both black and white and false-colour infrared (IR). Panchromatic film has typically been used in the past, and remains the least expensive of the four types. Black and white IR is also relatively inexpensive. When a red filter is used, IR film cuts through haze more effectively than panchromatic. It is also sensitive to soil moisture and provides high contrast for the tracing of watercourses. Colour film allows the discrimination of a larger number of hues and is more easily interpreted since the human eye is not used to viewing the world in black and white.

False-colour IR film was introduced as the 'camouflage detection film' because it could discriminate between living vegetation and the withering vegetation used to hide objects. For the same reason, both black and white and false-colour IR are useful for determining the health of plants. Astroturf, as an example of unhealthy vegetation, is blue on a false-colour IR image, but healthy grass appears red—even though the two types of 'grass' would look the same with panchromatic film (McKnight, 1990). False-colour IR film has also proved to be suitable for monitoring sediment transport and identifying sediment sources in streams, being very sensitive to the concentration of suspended material. Tones change from a very dark blue for clear water to a very light blue for waters with high sediment concentrations. These tones should be calibrated against field samples.

### 3.4.3 Other Remote Sensing Imagery

Remotely sensed images can be divided into two types: data collected by (1) passive systems which sense natural radiation (e.g. light reflected off the Earth's surface) and (2) active sensing systems in which electromagnetic radiation is emitted and the reflected signal detected by a sensor. Passive systems detect radiation emitted within a specific wavelength range such as visual or infrared, and are most often associated with satellites such as Landsat. Active systems include sonar (sound navigation ranging) and radar (radio detection and ranging). Microwave sensors have also been investigated and found to be sensitive to subsurface soil moisture (McKnight, 1990), although cost has limited their use.

### Passive Systems: Landsat

Remotely sensed data are largely used for mapping land uses and for monitoring changes in characteristics of the Earth's surface (Curran, 1985). As with aerial photographs, satellite data are used for environmental and resource surveys at up to a global scale. For example, Kotwicki (1986) used Landsat satellite imagery to document the drying of Lake Eyre following flooding in January 1984. Lake Eyre is a large inland lake in Australia which is dry for large periods of time and highly saline when it fills.

The science of remote sensing was rapidly propelled into the future with the launching of the first Landsat satellites in the 1970s. Since then, millions of images have been obtained for the evaluation of earth resources. The Landsat satellites have all carried a multispectral scanning system (MSS). LANDSAT 7, launched in 1999, carries an eight-band MSS (Enhanced Thematic Mapper Plus). Sensors record reflected light of various wavelengths as 'grey levels', transmitted from the satellite to ground stations in digital form. Computers process the data to re-create a false colour composite image. LANDSAT 7 repeats coverage of the same area every 16 days. The potential amount of data is mind boggling, and not all of the possible images are collected or processed. The development of new products and software has grown rapidly in response to the need for efficient methods of filtering, error-checking, classifying and interpreting the digital data.

MSS data from LANDSAT 7 consists of eight images of a scene taken in a panchromatic band; blue/green, green/yellow, red, near- and short-wave IR wavebands; and a thermal infrared band. The 'pixel' or *pi*cture *el*ement size is the unit of resolution of an image—15 m for the panchromatic band, 60 m for the thermal IR band and 30 m for the other bands. Pixels are combined into scenes which each cover a ground area of 183 km by 170 km.

Different properties of a surface can be ascertained from each waveband or from combinations of wavebands. For example, blue/green, green/yellow and near-IR are good for detecting vegetation reflectance, and near-middle-IR and thermal IR are moisture sensitive. Thermal IR

scanners, developed in the 1940s for 'nocturnal snooping' by the military, provide a measurement of emitted heat. They have special applications related to temperature differences such as location of frost hollows, estimation of plant moisture stress, location of thermal water pollution or hot springs, and the location and extent of vulcanism or above- and below-ground wildfires (Curran, 1985; Lyon, 1995; McKnight, 1990).

### Active Sensing Systems: SLAR

SLAR ('Sideways-Looking Airborne Radar') senses terrain to the side of an aircraft's path by sending out long wavelengths (up to radio wavelengths) and recording the returned pulses (Curran, 1985). Radar imagery is particularly useful for terrain analysis. It yields images with long 'shadows' which enhance microtopography and relief and is also sensitive to surface soil moisture. The major advantage of radar is that it has better than 99% cloud-penetration ability, so it is helpful where cloud cover restricts the use of conventional photography or satellite data (Barrett and Curtis, 1992).

The relationships between image tone and characteristics of the land surface are different than on conventional photography, and SLAR imagery has not yet gained wide acceptance. Although the cost is relatively high, SLAR imagery has potential in hydrological studies for detecting drainage patterns, wetlands and subsurface features.

### 3.4.4   Sources of Imagery

Remote sensing guidebooks such as Carter (1986), Cracknell and Hayes (1988) and Rees (2001) are excellent references for sources of aerial photographs and satellite sensor data, software for processing the data, and addresses of consulting, mapping and research establishments. Other good references on remote sensing are the companion volumes written by Curran (1985) and Lo (1986), texts by Avery and Berlin (1992), Barrett and Curtis (1992), Campbell (1996), Drury (1990), Lillesand and Kiefer (2000) and Thomas *et al.* (1987), and the 'bible' of remote sensing published by the American Society of Photogrammetry and Remote Sensing (Colwell, 1983). The best way to obtain current information in this rapidly changing field is to check Internet sites, make direct inquiries to the establishments involved with the collection and processing of imagery, and refer to recent issues of major remote sensing journals and symposia.

There is a vast amount of imagery available, although some governments may place restrictions on the use of coverage within their territories. Landsat is operated in international public domain, and data are available from the USGS (edc.usgs.gov). Satellite images of snow cover and climatic conditions are available from the US National Oceanic and Atmospheric Administration (NOAA) at: www.noaa.gov as well as www.goes.noaa.gov and www.ncdc.noaa.gov/oa/ncdc.html. Data are also available from earth-sensing satellites of other countries such as the SPOT (Système Probatoire D l'Observation de la Terre) satellite, developed by the French space agency CNES (e.g. at www.spot.com). SPOT collects panchromatic data (with 10 m spatial resolution) and MSS data (20 m resolution), its orbit repeating every 26 days. The meteorology offices listed in Section 3.2.1 are other sources of satellite imagery.

Satellite data may be acquired as a false-colour 'photograph' or as digital data. Digital data are normally available on computer storage media; however, complete coverage of an area over several periods of time can be quite costly. A number of computer-assisted techniques for mainframe and microcomputers are now available for rectifying an image, contrast stretching, colour and edge enhancement, filtering and other types of image processing and analysis.

In using remotely sensed data, the same cautions are advised as for maps and hydrologic data. One should know how the data were collected, whether from the same or different sensors or satellites; be aware of possible errors from noise in the signal, variations in attitude, altitude and velocity of the satellite, or prevailing environmental conditions; and only push the data as far as its quality will allow.

### 3.4.5 Interpretation, Classification and 'Ground Truthing' of Imagery

Remote sensing cannot be practiced without knowledge of field conditions ('ground truth'). Otherwise the patterns on a photograph or numbers on a computer will have little meaning. Although satellite data are increasingly treated by semi-automatic computer analysis, most operational uses of remote sensing still involve manual procedures.

Interpretation of products in photographic format (called *photointerpretation*) is more art than science, based on the ability of an interpreter to integrate the patterns on an image either consciously or subconsciously. It is a deductive process to go from qualitative information such as tone and/or color hue, location, shape, size, pattern, shadow, texture and depth (from stereoscopic views) to the identification of relief features, vegetation types or Tasmanian devils. Brightness and texture may be more accurately quantified with the assistance of a computer, but the eye is generally better at interpretation.

The conditions of resources as well as their identification can be assessed by inference. For example, soil moisture conditions may be inferred from the tone of vegetation. Experienced photo-interpreters work from familiar features to that which is unknown, utilizing other evidence such as topographical and thematic maps, photographs or data taken at ground level, or a visit to the site itself. The darker patches of vegetation in Figure 3.5, for example, are pine tree plantations that contrast with the native eucalyptus forest. The plantations are also designated on the map of Figure 3.4. Photographic keys are sometimes available to aid in interpretation of vegetation, landforms, forest sites or soil conditions.

Since remote sensing is generally less expensive than field surveys, the trend to replace 'contact' methods of assessment with remote sensing will no doubt continue, especially for reconnaissance-level studies (Lyon, 1995). By correlating remote sensing with field data, measurements such as vegetation densities or shallow water-table levels can be extrapolated to similar areas using remote sensing alone. In the same manner, remote sensing can be used to monitor changes in sites over time once a connection has been made between image patterns and what they represent on the ground. Although changes can be detected from imagery, the cause of the change must be identified or inferred by the user (Platts *et al.*, 1987).

Geographic information systems (GIS) provide a relatively new tool for classification. The concept is the same as flipping backwards and forwards between several maps and photographs in the hope that one's brain will combine the information. Unfortunately, when more than two or three maps are visually compared, the mind tends to overload (Curran, 1985). A computer-based GIS can digitally superimpose spatial data from thematic maps, digital elevation models and imagery to yield new maps. For example, climate, topography, vegetation and soils information might be combined to delineate areas with high erosion potential. Other data such as wildlife populations or water-quality data can also be added as GIS 'layers'.

Collecting ground truth data for correlation with imagery becomes a 'question of how much ground checking will be required to produce a result comparable in objective accuracy to a ground survey' (Barrett and Curtis, 1992, p. 138). A stratified sampling design is preferred, (see Chapter 2), with samples taken within each identifiable class of surface features (e.g. substrate or vegetation type). In practice, sampling techniques may range from randomly selecting sites on an image for detailed study in the field, to casual verification from the window of a car. For extrapolation of ground truth data to other areas of the same class, more refined sampling techniques are required so that results are statistically valid.

# 4

# Getting to Know Your Stream

## 4.1 General Character

### 4.1.1 Preliminary Introductions

The process of getting to know a stream is not unlike that of a doctor learning about a patient and his or her health. Preliminary observations, standard questions and routine measurements of temperature and dimensions are followed up by remote imagery and tests on various samples. A conscientious doctor will then look beyond the charts, images and test results to obtain a sense of the underlying causes of a patient's health. In the same manner, ecologists often use indices of catchment characteristics and streamflow patterns to obtain an indication of a stream's physical condition and suitability for various organisms.

This chapter covers the preliminary stage in which first impressions of a stream system are obtained from initial visits and from existing sources of information such as maps, aerial photographs and hydrological and climatic records. Methods of general, qualitative description are presented within this section. Section 4.2 gives techniques for describing catchments and stream networks and Sections 4.3 and 4.4 cover methods of describing streamflow patterns.

### 4.1.2 Putting the Stream Channel and Its Catchment into Context

Most texts on hydrology begin with a picture of the hydrologic cycle such as the one shown in Figure 4.1, and then go into an explanation of each component. In fact, much of the effort in hydrology relates to the modelling of water movement both above and below ground. In studying a stream and its biota, it is important to maintain a perspective on the water's origins. Snowmelt-derived streams, for example, will have a different hydrological and biological character than the temporary streams of arid regions.

Only a small part of the water flowing in streams is a result of precipitation landing directly on the stream. Most of the water is derived from surface and sub-surface runoff that results from rain or snowmelt on upland areas. The region from which water drains into a stream is termed the *drainage basin* or *catchment area*. The boundary or 'rim' of the drainage basin is called the *drainage divide*, and follows the highest points between two drainage basins. The somewhat ambiguous term *watershed* is variously used to describe either the drainage basin or the drainage divide. Langbein and Iseri (1960) provide a compact and useful reference for definitions of these and other hydrological terms.

It is important to be aware of the hydrological, geological, morphological and vegetational setting of a stream. Climate is a major factor controlling streamflow patterns and the shaping of landforms and vegetation communities. It provides the energy and water necessary to drive catchment ecosystems. Geology influences the shape of drainage patterns, bed materials and water chemistry. Catchment soils are the weathering products of rock materials, which influence upland erosion potentials, water-infiltration rates and vegetation types. Vegetation is a source of biological production, and affects channel bank stability, upslope resistance to erosion, water loss through evapo-transpiration and runoff rates.

Adjustment of a stream to its climate and geology takes place continuously, leading to changes in slope, rate of sediment transport and channel configuration. Associations of stream organisms are established in harmony with this dynamic nature of the channel's physical conditions (Cummins, 1986; Vannote *et al.*, 1980).

From either a biological or hydrological viewpoint, the characteristics of a stream are dependent on the downstream transfer of water, sediment, nutrients and organic

Stream Hydrology: An Introduction for Ecologists, Second Edition.
Nancy D. Gordon, Thomas A. McMahon, Brian L. Finlayson, Christopher J. Gippel, Rory J. Nathan
© 2004 John Wiley & Sons, Ltd ISBNs: 0-470-84357-8 (HB); 0-470-84358-6 (PB)

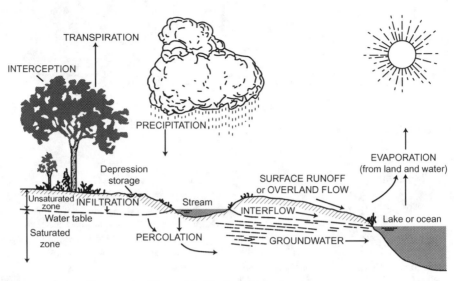

INTERCEPTION

TRANSPIRATION

PRECIPITATION

EVAPORATION
(from land and water)

Depression
storage

SURFACE RUNOFF
or OVERLAND FLOW

Unsaturated
zone  INFILTRATION    Stream

INTERFLOW

Water table

Lake or ocean

Saturated
zone    PERCOLATION    GROUNDWATER

**Figure 4.1.** *The hydrological cycle*

debris (Petts and Foster, 1985). Progressive changes in temperature, stream width, depth, channel pattern, velocity, sediment load and instream biota occur from headwaters to mouth. Vannote *et al.* (1980) hypothesize that the continuous gradient of physical conditions within a stream system results in a predictable structuring (a 'continuum') of biological communities (see Chapter 9).

Life in streams is not necessarily constrained by a stream's bed and banks. The sub-surface character of streams has been investigated by several researchers including Bencala (1984), Bencala *et al.* (1984), Fortner and White (1988), Triska *et al.* (1989a,b) and White *et al.* (1987). Their work demonstrates that the interstitial zones in streambeds are important in the storage of dissolved gases and nutrients, and that for ecological purposes, the stream 'boundary' may lie deep within the streambed. Triska *et al.* (1989a, b) define this boundary as the interface between groundwater and channel water, which may be located using piezometers or tracer injections. Water may enter the streambed, travel for some distance underground and then re-enter the stream. These recharge, underflow and discharge processes are dependent on the proximity of the water table to the channel bed surface, streamflow level, bed permeability and topography. Patterns of flow movement can thus affect the distribution of hyporheic organisms, rooted aquatic plants and spawning areas for fish.

Amoros *et al.* (1987) developed a methodology for considering fluvial hydrosystems as interactive ecosystems over four dimensions: (1) the upstream-downstream progression; (2) the interconnections between the main stream, side arms, flood plain and marshes; (3) the vertical interchange between regions above (epigean) and below (hypogean) the channel bed surface and (4) the changes in a river's dynamics and ecosystems over time. The method was applied to the Rhone River, France, to develop predictive scenarios for the impact of engineering works on channel morphology and ecology. Examples of some of the geomorphic patterns and associated biological functions used as spatial units in the study are given in Figure 4.2. Thus, there is no reason to draw the boundaries of a stream at the water's edge (Cummins, 1986). It is preferable to use the catchment area as the basic ecosystem unit (Lotspeich, 1980; Moss, 1988), especially if it is extended in the vertical direction to include sub-surface processes and considered over time.

### 4.1.3  Initial Assessments of the State of a Stream

Natural and human modification of the stream or upland areas can have profound effects on the state of a stream. Before attempting to develop 'cures' for adversely impacted streams, the current state of a stream and its ecology must first be ascertained and compared with what is considered 'healthy'.

In 'getting to know' another, first impressions are noted and then traits are evaluated through a dialogue consisting of questions which are basically 'are you this or that?' By 'asking questions' of a stream, one can diagnose its condition, put its measurements into perspective against those of other streams and classify it according to its

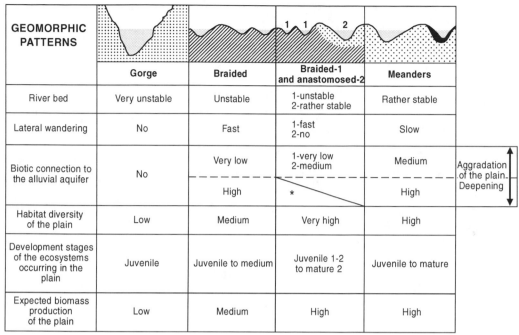

| GEOMORPHIC PATTERNS | Gorge | Braided | Braided-1 and anastomosed-2 | Meanders | |
|---|---|---|---|---|---|
| River bed | Very unstable | Unstable | 1-unstable 2-rather stable | Rather stable | |
| Lateral wandering | No | Fast | 1-fast 2-no | Slow | |
| Biotic connection to the alluvial aquifer | No | Very low | 1-very low 2-medium | Medium | Aggradation of the plain. Deepening |
| | | High | * | High | |
| Habitat diversity of the plain | Low | Medium | Very high | High | |
| Development stages of the ecosystems occurring in the plain | Juvenile | Juvenile to medium | Juvenile 1-2 to mature 2 | Juvenile to mature | |
| Expected biomass production of the plain | Low | Medium | High | High | |

*Anastomosed pattern occurs only with aggradation

**Figure 4.2.** *Associations of geomorphic patterns and their ecological implications. Redrawn from Amoros et al. (1987), Reproduced by permission of John Wiley and Sons Ltd.*

individual characteristics. In the evaluation process, some questions that might be asked include the following.

### Perennial, Intermittent or Ephemeral?

The terms perennial, intermittent and ephemeral are related to the terms influent and effluent. *Influent* ('losing') streams are those in which the stream feeds the groundwater as compared to *effluent* ('gaining') streams in which the stream receives water from it (see Figure 4.3). A stream may change from effluent to influent across its length depending on the geological formations crossed. Even large streams may disappear completely, reappearing as springs many miles away.

Boundaries between the definitions of perennial, intermittent and ephemeral streams are vague. They apply to the general nature of a stream's water flow under average conditions. *Perennial* streams are those that essentially flow year round. Perennial streams are primarily effluent, and consist of baseflow during dry periods. Most large streams and streams in humid regions will be perennial, although a continuous low flow may be maintained in well-shaded channels with a source of sub-surface water. *Intermittent* streams are those which only flow during the times when they receive water from springs or surface

runoff. They are thus either influent or effluent, depending on the season. During dry years they may cease to flow entirely or they may be reduced to a series of separate pools. *Ephemeral* streams are influent, with channels that are above the water table at all times. They carry water only during and immediately after rain. Most of the streams in desert regions (called arroyos in the western US) are ephemeral. Some of these channels are dry for years at a time, but are subject to flash flooding during high-intensity storms.

Determining the permanence of flow is more than a matter of distinguishing a solid line from a dashed line on a map, as definitions used by map makers will vary. If the stream is gauged, records will provide insight into conditions at the gauge site. Local residents familiar with the stream's behaviour may be able to give a fair estimate of how often a stream reach dries up, and in what manner. Vegetational clues such as the presence of cattails or cottonwoods may also help to determine the boundaries between temporary and permanent waters.

### Bedrock-controlled or Alluvial?

*Alluvium* is a general term for stream-deposited debris. Streams can be separated into the two major groupings,

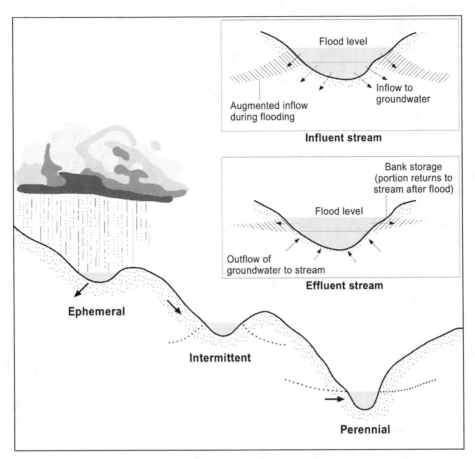

**Figure 4.3.** *Descriptions of flow permanence. Water table level is shown as a dotted line in the main figure. Inserted diagrams illustrate how water enters the banks as the stream swells during a flood; in effluent streams, much of this bank storage returns to the stream. Inserted diagrams are redrawn from Wilson (1969), by permission of Macmillan*

bedrock and alluvial, based on whether the channel form is predominantly controlled by geology or streamflow, respectively. In bedrock-controlled channels the flow is confined within rock outcrops and the channel morphology is determined by the relative strength and weakness of the bed material. Alluvial channels, by comparison, are free to adjust their dimensions, shape and gradient, and bed and bank materials are composed of sediment transported by the river under the present flow conditions (Schumm, 1977).

### Headwater, Middle-order or Lowland Stream?

The general downstream trends of energy input, water quality and physical conditions lead to a longitudinal succession of fish, benthic invertebrates, plants and other organisms. These general changes can be divided into three zones: headwater, middle-order and lowland. This

classification is similar to one of the oldest and most famous three-zone classifications of Davis (1899), who used the categories 'youth', 'maturity' and 'old age'. Although distinctive differences exist between upland, middle-order and lowland streams, wide variations in physical and biological characteristics can occur within each division and over time. Characteristics of the three major types are summarized as follows:

1. *Headwater zone.* Typically, the upper, 'young' reaches of a stream are incised into V-shaped valleys, with steep slopes and few, short tributaries. The channel bed material consists of coarse gravels, boulders and rock outcrops. Water temperatures are relatively cool and stable. With their narrow widths, upland streams are more shaded by riparian vegetation and there is proportionately more material entering the stream as

leaves and logs. Shading and the scouring action of coarse sediments restrict the growth of algae and other plants. Thus, organic matter from outside the stream supplies most of the food for fungi, bacteria, macroinvertebrates and others, which in turn become food for higher organisms such as small fish. Habitat diversity may be low because of the restricted temperature range and low input or production of nutrients.

2. *Middle-order zone.* Stream slopes lessen in the middle, 'mature' reaches of a river. These reaches transport sediment from bank erosion and from upstream supplies, and have highly variable physical characteristics. Floodplain development begins, and bank cutting replaces downward cutting (Schumm, 1977; see Figure 9.6). Channels are wider and aquatic plants contribute organic nutrients and oxygen to the stream, augmenting supplies transported from upstream. The coarse substratum, diversity in channel form, diversity of nutrient sources, variable discharge and wider range of temperatures favour a diverse fauna since the range of conditions encompass the optimum conditions for a large number of species (Petts and Foster, 1985). Vannote *et al.* (1980) relate this to a convergence of organisms that evolved downstream from terrestrial origins (e.g. insects) and upstream from marine origins (e.g. crustaceans).

3. *Lowland zone.* In the lowland, 'old age' zone, bed materials are composed of fine sediments, discharges are relatively stable, and temperature fluctuations are buffered by the large volume of water. Valleys are very broad, deeply filled with alluvium, and marked with the evidence of frequent channel changes: meander scars, oxbow lakes (billabongs) and swamps. Deposition of sediment occurs through this zone and to the terminus of the river, where sediment may deposit out on an alluvial plain, delta or in an estuary. Natural levees may border the stream, and if sediment deposits between them, a stream can actually flow at a level higher than the floodplain. Increased turbidity and depth in lowland zone streams may restrict the growth of aquatic plants, and the macroinvertebrate populations tend to be dominated by those which collect fine particles of organic matter received from upstream. Overall biotic diversity may be low, although fish species diversity may increase with the presence of larger fish, which feed on smaller ones.

### 'Stable, Aggrading or Degrading?'

A *stable* channel is one that does not exhibit progressive changes in slope, shape or dimensions, although short-term variations may occur during floods (Schumm, 1977). *Degradation* refers to the downcutting of a stream into its

bed materials and *aggradation* is the accumulation of bed materials. Bedrock streams do not degrade very quickly, but alluvial channels can change rapidly in response to alterations of flow, sediment supply or base level.

A river is a delicately balanced system, and any alterations will cause the stream to attempt to re-establish an equilibrium condition (Richards, 1982; Sear, 1996). Disequilibrium results from changes in sediment or runoff conditions, as from diversion of a major amount of a stream's discharge or from modifications to upland conditions. Stream stability is also important biologically, as mentioned in Section 2.1. When a stream aggrades, fine sediments can impede the exchange of water, organic matter and organisms between bed surface and hyporheic habitats, whereas during degradation, fine sediments are washed out, enhancing exchanges (Amoros *et al.*, 1987). The concept of equilibrium is discussed further in Chapter 7.

Evaluation of channel stability from either aerial photographs or in the field is relatively difficult, although partly buried fences or other permanent landmarks may indicate changes. Changes in bed cross-section and elevation is best accomplished by comparing old channel surveys to recent ones.

An associated question is: *narrowing or widening?* Trees falling into a river from both banks may indicate widening; however, narrowing occurs more slowly and may only be detected by increases in vegetation on islands or a trend from braided to meandering condition. The stability and direction and rate of change of a channel are important when evaluating habitat potential, designing stream rehabilitation works and determining minimum instream flow recommendations.

### 'Regulated or Natural?'

Dams and diversion canals can be easily located from maps, aerial photographs or ground surveys, but their effects are much more difficult to evaluate. Thus, 'regulated or natural' is not an either/or question, but a matter of degree. Effects depend on whether regulated releases are returned to the river further downstream, and on the operating policy of reservoir storages (McMahon, 1986).

Cadwallader (1986) has indicated that regulation may change the seasonal distribution of flow (the regime), reduce the incidence and severity of flooding and decrease long-term average flows downstream. Regulation can have a marked influence on low-flow behaviour by increasing the duration and frequency of low-flow extremes (including periods of zero flow). In addition, reservoirs can influence sediment movement, stream temperatures and water quality. Changes in flow regime and sediment supply can lead to changes in downstream channel dimensions.

Once a river is regulated, it is likely to remain so, although dam decommissioning on salmonid streams in the Northwestern US has become a serious topic of debate. A goal of river management should be to determine the habitat value of modified streams and the degree to which regulation structures and their operation can be modified to optimize it. The effects of dams and reservoirs on stream habitat and the provision of reservoir releases for habitat improvement are discussed in Chapter 9.

### 'Channelized or Non-channelized?'

The clearing and straightening of a stream to improve water conveyance is termed channelization. This usually increases channel slope and thus water velocity and sediment transport capacity, which causes scour within the modified section and sediment deposition in flatter sections downstream. Slow-water refuges for fish and other aquatic life, e.g. in backwaters and behind rocks, are often eliminated. Vegetation and shading is usually reduced in channelized sections, leading to an increase in stream temperature and a lowered supply of organic nutrients for biota. Water tables in the region of channelization may also be affected.

### 'What is the Condition of the Upland Catchment?'

Streamflow and ecology are both affected by catchment conditions. Because the catchment and stream system are integrated, a change in one part of the system will be felt elsewhere (Morisawa, 1985, Brizga and Finlayson, 1994). Responses to a change may be immediate, delayed, or dependent upon a critical factor reaching some threshold level.

Vegetation removal, as from fire, logging or conversion from forest to pasture, can change the natural drainage system and the rate at which water and sediment runs off the land surface. Road construction can be a major point source of sediment. Urbanization has a more drastic effect on hydrology—roads, parking lots and houses prevent infiltration, increasing total runoff and the magnitude of peak flows. These changes may result in alterations to the channel's slope, stream pattern and bed materials.

Regulation, channelization and catchment conditions were all considered by Macmillan (1986, 1987) in the development of five categories for evaluating the 'naturalness' of a catchment, as presented in Table 4.1. Leopold and O'Brien Marchand (1968) also describe a system for ranking the quality of river landscapes.

### 4.1.4 An Example of a River Condition Survey: The Index of Stream Condition, Victoria, Australia

The Index of Stream Condition (ISC) was developed in Victoria, Australia, to provide a guide for the managers of the State's rivers in prioritising resource allocation. It is widely recognised that the State's rivers have been degraded to various degrees since the arrival of European settlers in the early part of the nineteenth century and the degradation processes persist with the increasing impact of land use and water extraction on the condition of rivers. This problem is not confined to Victoria and there have been over 100 river 'health' indices developed world wide (Greenwood-Smith, 2002). The concept of stream health is thoroughly explained and many of these methods are reviewed in Section. 9.3.

The purpose of the ISC is to assist waterway managers to

- Provide baseline and comparative information on the condition of streams;

*Table 4.1. Criteria for evaluating 'naturalness' of a catchment. Modified from Macmillan (1986), Reproduced by permission of Water Studies Centre*

*Pristine.*    The entire catchment system represents an unmodified ecosystem, which can act as a baseline reference area.

*Slightly modified.*    Catchment processes are largely intact. The flow regime has been modified to only a minor extent, and the only input of pollution is sediment. There are no barriers to the movement of instream biota.

*Moderately modified.*    Catchment processes, hydrology and instream biota have been noticeably altered. There may be direct manipulation of the flow regime by impoundment, and sediment input may have altered the stream substrate. Levels of biostimulants may be elevated but not other toxic inputs.

*Heavily modified.*    Catchment processes, riparian and instream biota have been substantially modified. The flow regime may be highly manipulated, sediment input may be substantial and levels of biostimulants are substantially elevated. Toxic substances may be present at significant levels.

*Severely degraded.*    Major modification of the stream has taken place, leading to severe degradation of riparian and instream biota. Examples would be streams grossly affected by elevated levels of salinity, heavy metal pollution and/or enclosure within a concrete channel.

- Enable the long term effectiveness of management intervention to be assessed;
- Assist waterway managers to set objectives (Ladson *et al.*, 1999).

While the ISC was developed by a group of scientists from a range of relevant disciplines, it is designed to be applied by people with limited scientific training in order to make it widely accessible and to keep the cost of its application to a manageable level. The ISC is intended to draw attention to problem areas which would then be followed up with a more detailed analysis.

The ISC is a single numerical index based on an holistic assessment of the stream and its surroundings. The index is composed of five sub-indices (Figure 4.4, Table 4.2) and is always quoted together with the values of the sub-indices in the form of a bar graph scaled internally for the value of each sub-index (Figure 4.4). In this way, users can make their own assessments of the relative importance of each individual component of this assessment of the stream's overall condition.

The essential components of each sub-index are shown in Table 4.2. The maximum for each sub-index is 10 and this constructed on the basis of the indicators listed in Table 4.2. The hydrology sub-index is scored initially in the range 0–4 on the basis of the percentage of hydrologic deviation, at the monthly level, from natural. This score is then adjusted by subtracting 1 if the catchment is more than 20% urbanized and also by subtracting 1 if the reach being assessed is downstream of a hydropower station. The score is then multiplied by 2.5 to make it a score out of a maximum possible of 10.

Each of the four indicators for the physical form sub-index, bank stability, bed aggradation and degradation, density and origin of coarse woody debris and influence of artificial barriers are rated on a scale on 0–4 using visual assessment with the help of photographs which show typical examples of each rating level.

There are six individual indicators for the streamside zone sub-index and the overall rating is based on the sum of the six ratings, scaled to lie between 1 and 10. These six indicators cover the quantity (width and continuity) and quality (structural intactness, percentage of cover provided by indigenous species, and condition of billabongs) of the riparian vegetation.

The water quality sub-index is based on only four water quality parameters—phosphorus, turbidity, electrical conductivity and pH—using the median value of monthly measurements over one year scaled to index form. For many sites assessed in Victoria, the data to calculate this sub-index could be obtained from routine water quality measurements carried out by government agencies such as the Environment Protection Authority.

The aquatic life sub-index uses the SIGNAL (stream invertebrate grade number average level) developed by Chessman (1995) for eastern Australia (see Section 9.6 for detailed explanation). The SIGNAL value is assessed by sampling macroinvertebrates at the family level and converted to a sub-index score in the range 0 to 10.

The whole of the State of Victoria was assessed using this index in 1999 and the results of that baseline survey can be found at http://www.vicwaterdata.net. Repeat surveys are planned at intervals of between 5 and 10 years and the results will also be available on the website.

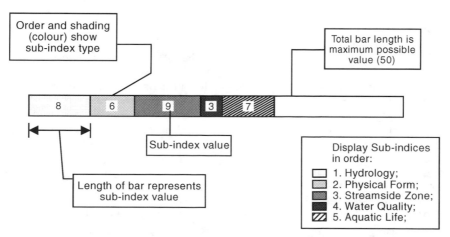

**Figure 4.4.** *The construction of the Index of Stream Condition and the diagrammatic reporting of its sub-indices (Ladson and White 1999). © State of Victoria, Department of Sustainability and Environment, 1997*

*Table 4.2.* The components of the Index of Stream Condition (Ladson et al., 1999). Reproduced by permission of Blackwell Science

| Sub-index | Basis for sub-index value | Indicators |
|---|---|---|
| Hydrology | Comparison of the current flow regime with the flow regime existing under natural conditions | Hydrologic deviation (comparison of monthly flows with those that would have existed under natural conditions)<br>Percentage of catchment urbanized<br>Presence of any hydropower stations that cause water surges |
| Physical form | Assessment of channel stability and amount of physical habitat | Bank stability<br>Bed aggradation and degradation<br>Presence and influence of artificial barriers<br>Density and origin (i.e. exotic or native species) of coarse woody debris (only assessed in plains streams) |
| Streamside zone | Assessment of quality and quantity of streamside vegetation | Width of vegetation<br>Longitudinal continuity of vegetation (a measure of the number and significance of gaps in streamside vegetation)<br>Structural intactness (comparison of overstorey, understorey and groundcover density with that existing under natural conditions)<br>Proportion of cover that is indigenous<br>Presence of regeneration of indigenous species<br>Condition of wetlands and billabongs (only assessed in plains streams) |
| Water quality | Assessment of key water quality parameters | Total phosphorus concentration<br>Turbidity<br>Electrical conductivity<br>pH |
| Aquatic life | Presence of macroinvertebrate families | Presence of macroinvertebrate families using the SIGNAL index (Chessman, 1995) |

## 4.2  Catchment Characteristics

### 4.2.1  General

A number of factors affect the way water and sediment move from upland areas to the stream and from there to its terminus. Many geomorphic descriptors have been developed which are related to catchment hydrology. The general term *morphometry* is applied to the measurement of shape and pattern. Measures given in this section can be relatively easily obtained from maps.

Because of the inter-relationships between factors, one (usually the one most easily measured) can often serve as a surrogate for others. The selected factors can be used in the prediction of a catchment's hydrologic response to rainfall and for distinguishing one catchment from another for comparative or classification purposes.

### 4.2.2  Delimiting and Measuring the Catchment Area

Catchment area is one of the more important descriptors of a basin since it influences the water yield and the number and size of streams. It includes all of the upstream land and water surface area that drains to a specific location on the stream. We can speak of catchment areas for a whole stream system, or for a particular point on a stream (e.g. at a gauging station or study site). The area is delimited by the topographic divide, a theoretical line which passes through the highest points between the stream system and those neighbouring it.

Catchment boundaries are located by using the contour lines on a topographic map. These can be supplemented with stereo pairs of aerial photographs. Boundaries are drawn by following the ridge tops, which appear on topographical maps as downhill-pointing V-shaped crenulations.

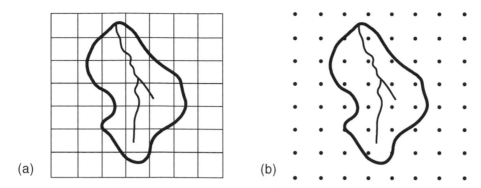

**Figure 4.5.** *Methods of measuring catchment area using (a) square grid and (b) dot grid*

The boundary should be perpendicular to the contour lines it intersects. The tops of mountains, often marked as dots on a map, and the location of roads which follow ridges are other clues. In areas of little relief, it will be difficult to locate boundaries precisely. A line has been drawn around the catchment of a small stream in the upper left-hand section of Figure 3.4.

Catchment area can be measured directly from the marked maps using a planimeter or digitizer with appropriate software. Other methods include superimposing a grid of squares or a dot grid over the map and then counting the number and fractions of squares or the number of dots which fall within the catchment area (Figure 4.5). Any of the methods must be calibrated to the map scale. It is a good idea to check the calibration by first measuring a section of known area (e.g. bounded by grid lines).

Another technique suggested by Gregory and Walling (1973) involves tracing the catchment onto a high-grade paper and then carefully cutting it out using a sharp cutting tool. A square of known map area is also cut from the same paper. Both pieces are weighed on an accurate balance. The catchment area is then obtained from Eq. 4.1.

$$\text{Catchment area} = \frac{\text{Catchment 'cutout' weight}}{\text{Weight of square 'cutout'}}$$
$$\times \text{Area represented by square} \qquad (4.1)$$

where the catchment area and square map area have the same units. For example, if a catchment 'cutout' weighed 3.5 g and a square cutout representing 16 km$^2$ weighed 10.5 g, the catchment area would be: $(3.5/10.5) \times 16 = 5.3$ km$^2$.

The topographic divide may not represent the true area from which water in a stream is derived. Sub-surface water may move from one drainage basin to another, as illustrated in Figure 4.6. The distinction between the topographic divide, which determines the direction of surface drainage, and the phreatic divide, which determines the direction of sub-surface drainage, is an important consideration in hydrologic studies.

For the purposes of this text, the topographical definition will be used. The stream network and catchment area

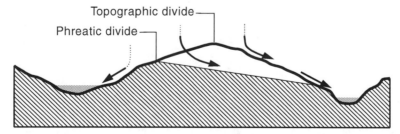

**Figure 4.6.** *Description of topographic and phreatic divides. Re-drawn from Gregory and Walling (1973), by permission of Hodder and Stoughton Ltd*

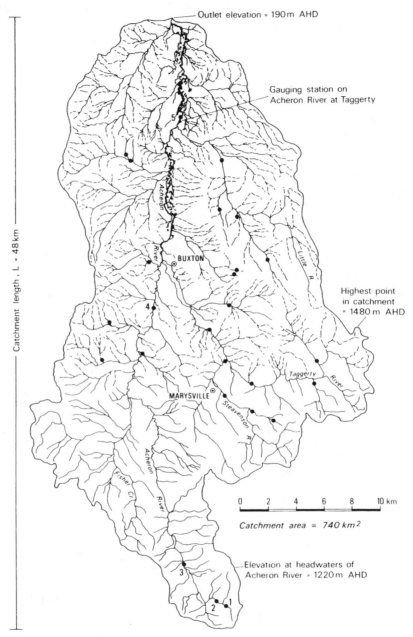

**Figure 4.7.** *Catchment for the Acheron River, Victoria, Australia, traced from 1:100 000 maps. Both map and scale have been reduced. Dots designate stream sampling sites, discussed in other chapters. The catchment area for the gauging station at Taggerty is 619 km²*

for the Acheron River, Victoria, which is used as an example throughout the text, is shown in Figure 4.7.

In addition to measuring the whole catchment area, the area covered by a vegetation or soil type or other thematic feature may also be of interest. Measurements are made using the same methods.

### 4.2.3  Stream Length

Stream length will influence the amount of stream habitat area in a catchment, the travel time of water in a drainage system and the availability of sediment for transport. Stream length is most often obtained from measurements of the blue lines on topographical maps. The actual length of channel containing surface water can change; thus the map measure should only be considered a standardized index.

Measurements of stream length may be made by following the line with a rule or scale graduated in map units, a digitizer or a pair of dividers. Map wheels (opisometers) which measure distance as they are rolled across a map are also helpful for measuring distances, although they are difficult to use when a stream line is highly tortuous.

Differences in map scales and accuracy will lead to variations in measured stream lengths. In fact, one can pose the philosophical question: What is the true length of a stream? Is it the distance measured down the middle of the stream channel or down its deepest path (the *thalweg* distance)? Does it extend to the drainage divide? Should it represent the path of a mass of water or a particle of organic matter as it passes around boulders and travels across and through the streambed?

Gan *et al.* (1992) suggest the use of a *fractal stream length* as a standardized measure. Using maps of a single scale, the stream is stepped off repeatedly with dividers set at different spacings. In this manner, a stream length ($L$) is obtained for each spacing ($X$). Using simple regression, the data are fitted with a log-log equation of the form:

$$L = aX^b \qquad (4.2)$$

where $a$ and $b$ are regression constants. A log-log plot of stream length against step size for the Acheron River is shown in Figure 4.8 with the fitted regression line. The fractal stream length ($L'$) is simply equal to the constant $a$

$$L' = a \qquad (4.3)$$

which is 55.0 km for the example illustrated.

The fractal dimension ($f$) is given by

$$f = 1 - b \qquad (4.4)$$

If an average value of $f$ is computed for a basin, the fractal length ($L'$) of any stream in the basin can be calculated as

$$L' = X^f N \qquad (4.5)$$

where $X$ is again the step length and $N$ is the number of steps needed to 'walk' the length of a stream.

For example, if the fractal dimension of 1.0534 obtained for the Acheron River is applied to a segment measured as 9.3 steps of size 2 km, then $L' = (2)^{1.0534} \times 9.3 = 19.3$, as compared to a conventional measure of 18.6 (9.3 × 2).

The advantage in using the fractal stream length is that it is a robust measure, independent of both map scale and finiteness of measurement. In contrast, the conventional measure of stream length tends to increase as the step size (in map units) decreases (Gan *et al.*, 1992). It should be noted that this technique 'standardizes' measurements to the chosen step size unit (km in Figure 4.8), which should be consistent for comparison of stream lengths.

### 4.2.4 Stream Patterns

Catchments can be described according to their stream channel patterns, as viewed from maps or from the air. All drainage patterns 'are tree-like, but different patterns resemble the branchings of different kinds of trees' (Horton 1945, p. 300). Each stream has its own individual characteristics, based on the particular topographical and geological obstacles encountered as it seeks the path of least resistance in its journey towards the sea. Stream patterns may develop randomly on uniform soils, or in response to weaknesses in the underlying geology. Some

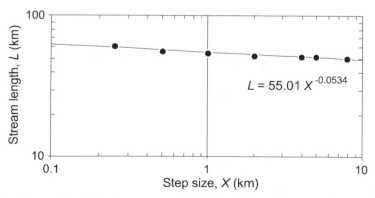

**Figure 4.8.** *Log-log plot of stream length (km) versus step size (km) for the Acheron River, as measured from headwaters to mouth on 1:100 000 scale maps. A power equation and fitted line obtained by simple regression are shown*

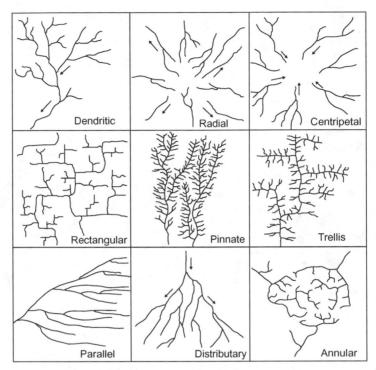

**Figure 4.9.** *Basic drainage patterns*

of the basic drainage patterns are (McKnight, 1990; Morisawa, 1985):

*Dendritic:* found in areas of relatively uniform geologic structure

*Trellis:* usually develops on alternating bands of hard and soft strata

*Pinnate:* forms in very fine-grained surfaces

*Rectangular:* common in areas with right-angled faults and/or joints, such as some types of granitic bedrock

*Radial:* forms where streams flow outward from a dome or volcanic cone

*Centripetal:* results from a basin structure where streams converge centrally

*Annular:* develops around a dome or basin where concentric bands of hard and soft rock have been exposed

*Parallel:* occurs in areas of pronounced localized slope

*Distributary:* refers to divergence of channels (for example, in deltas or alluvial fans)

Examples of the drainage types are shown in Figure 4.9. These are only some of the more common and easily recognizable regional patterns of drainage. The patterns in the Acheron River drainage are mainly dendritic.

The development of branching networks is a fascinating topic in itself. A variety of phenomena which distribute or collect matter or energy exhibit branching patterns: root systems, tree branches, veins in leaves or in animals; lightning strikes, fern-like precipitations of manganese in rocks, and highway and telephone systems. Knight (1984) provides a discussion on the evolution of drainage networks on land surfaces, and Jarvis and Woldenberg (1984) have compiled a number of papers on river networks.

So, given a certain area, what is the most efficient method of draining water from its surface? As an illustration, we can consider a number of strategies for 'draining' water from all points in Figure 4.10(a). To study the relative efficiency we will look at the total 'channel' length and the average length from each point to the delivery point. Figure 4.10(b) shows a spiral 'snail-shell' configuration, Figure 4.10(c) a 'starburst' pattern, and Figure 4.10(d) a branching system. It can be seen that the pattern of Figure 4.10(c) provides the maximum possible drainage efficiency because each point has a direct path to the outlet. This is reflected in the low average channel length. However, this is the 'big budget' alternative, with its large total channel length. In the

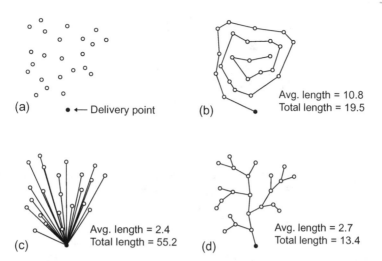

**Figure 4.10.** *A theoretical comparison of drainage pattern efficiencies: (a) points to be drained through the delivery point; (b) a spiral drainage pattern; (c) a 'starburst' configuration; and (d) a branching pattern. Average length = average distance from each point to the delivery point and Total length = sum of all channel segment lengths (both in arbitrary units). See text for discussion*

branching system (Figure 4.10(d)) the average length is only slightly higher yet the total channel length is greatly reduced. Thus, the branching system requires less channel length to maintain a near-maximum drainage efficiency (Newbury, 1989).

In natural streams it is typical for major tributary branches to drain about half the total catchment area, with the remaining area draining directly into the main stem of the stream. For the Acheron River system (Figure 4.7), 52.6% of the area is drained by the main tributaries, the Little River, Steavenson-Taggerty and Fisher, leaving 47.4% for the main stem. The implication of this for flood-control or water supply is that dams built on these major side tributaries would catch water from only about one half of the basin area, in comparison to one dam built at the mouth which would control runoff from the whole basin (Leopold and Langbein, 1960).

### 4.2.5 Stream Orders

Stream ordering is a widely applied method for classifying streams. Its use in classification is based on the premise that the order number has some relationship to the size of the contributing area, to channel dimensions and to stream discharge (Strahler, 1964). This premise is often criticized; however, because of its simplicity, stream ordering is a rapid 'first approach' method of stream classification and a convenient means of stratification for sampling designs.

### Stream Ordering Methods

Stream orders provide a means of ranking the relative sizes of streams within a drainage basin. In most stream-ordering methods, the smallest tributaries are 'first-order', and major streams and tributaries have higher orders. Catchments can also be classified by order number (for example, the area draining a second-order stream would be labelled a second-order basin). Figure 4.11 shows four different ordering systems.

Horton (1932) first introduced the concept of stream order in the United States, after reversing the European practice of giving main streams an order of one and the fingertip tributaries the highest order. Strahler's (1952) method is a slight modification of Horton's, which has been widely accepted and is commonly used by stream biologists. In this system, all of the small, exterior streams—those which 'carry wet weather streams and are normally dry' (Strahler, 1952, p. 1120)—are designated as first-order. A second-order stream is formed by the junction of any two first-order streams; third-order by the junction of any two second-order streams. A limitation of Strahler's method is that a large number of minor tributaries may intersect a larger-order stream, adding substantially to its discharge but not to its stream order. To overcome this drawback, Shreve (1967) and Scheidegger (1965) give alternatives which are based on a summation of upstream orders. These latter two methods are computationally more difficult and an ordering must be completely reworked if any streams are missed the first time through.

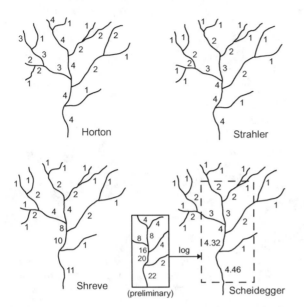

**Figure 4.11.** *Stream ordering systems. Strahler's method is most widely applied*

marked on maps (Drummond, 1974; Mark, 1983). In a fisheries study, Platts (1979) defined first-order streams as being the first recognizable drainage on 1:31 680 scale maps, whereas Lotspeich (1980, p. 582) gives a biological definition of a first-order stream as one 'with sufficient continuous flow to support an aquatic biota at all seasons'. It is essential to state the map scale and method used when citing a stream order.

### Relationships between Stream Orders and Other Measures

There are a number of relationships that can be demonstrated using stream orders. Several laws of drainage basin geometry have been developed which state that as stream orders increase: (a) the number of streams decreases, (b) the average stream length increases, (c) catchment area decreases and (d) average slope decreases, and these relationships are geometric (Selby, 1985).

A stream ordering using Strahler's method was applied to the entire stream system shown in Figure 4.7, including both intermittent (dotted lines) and perennial (solid lines) streams. Stream lengths were measured from 1:100 000 maps with a ruler and the relationships derived therefrom are shown in Figure 4.12.

Figure 4.12(a) demonstrates that there are more small-order streams than large-order ones, and that the relationship approximates a geometric progression. It can also be seen in Figure 4.12(b) that stream length increases with stream order. However, for the stream and mapping system used, the data do not follow the expected semi-logarithmic trend, perhaps due to a lack of resolution at the level of order-one streams or to the nature of the drainage system.

Stream ordering, although a relatively quick and simple means of classification, has some major drawbacks. These include the variability in mapping standards, the problem of deciding which map scale is appropriate, and inconsistent definitions of first-order streams (Hughes and Omernik, 1983). A first-order stream interpreted from a map may turn out to be as much as third- or fourth-order when interpreted in the field. Some researchers have used contour crenulations (V-shaped contours) and the location of the headwater divide to extend the stream patterns

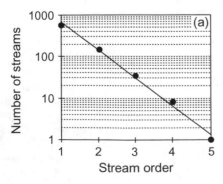

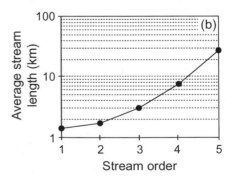

**Figure 4.12.** *Semi-log plots of (a) the number of streams and (b) average stream length versus stream order (by Strahler's method) for the Acheron River system (Figure 4.7). A regression line has been fitted to the data in graph (a); in graph (b) the lines simply connect the data points*

These simple, regular relationships provide a quantitative description of drainage development in a basin. Sampling designs should take these relationships into account; i.e. if the object is to describe all streams in a basin, then more small streams should be sampled than large ones. The relationships for a basin can also be extrapolated from large streams to their smaller tributaries. For example, the number or average lengths of first- and second-order streams might be estimated for large basins where the sheer number of streams make measurements impractical (Leopold *et al.*, 1964).

### *Bifurcation Ratio*

Horton (1945) introduced the term bifurcation ratio ($R_b$), where bifurcation means dividing in two. The data plotted in Figure 4.12(a) can be described by this index, given as

$$R_b = \frac{\text{Number of stream segments of given order}}{\text{Number of stream segments of next highest order}}$$

$$(4.6)$$

The average $R_b$ for the Acheron River system is approximately 4.0. This compares to a US average of about 3.5 (Leopold *et al.*, 1964). Bifurcation ratios normally range between 2 and 5 and tend to be larger for more elongated basins (Beaumont, 1975). As a matter of interest, the average $R_b$ for trees is about 3.2, for lightning strikes 3.5 and for blood vessels 3.4, perhaps implying that ratios in this range approach a natural optimum (Newbury, 1989; Stephens, 1974).

### 4.2.6 Miscellaneous Morphometric Measures

### *Drainage Density*

Drainage basins with high drainage densities are characterized by finely divided networks of streams with short lengths and steep slopes. In contrast, basins with low drainage densities are less strongly textured. Stream lengths are longer, the valley sides flatter, and the streams further apart. A comparison of high and low drainage density patterns is shown in Figure 4.13.

Drainage density ($R_D$) is calculated by dividing the total stream length for the basin ($\sum L$) by the catchment area ($A$):

$$R_D = \frac{\sum L}{A}$$

$$(4.7)$$

Drainage density has units of 1/length; thus its value will vary with the chosen units. It is preferable to express it in units of length/unit area. For example, the Acheron

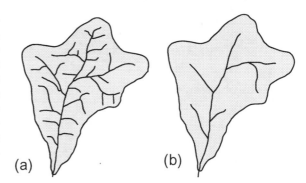

**Figure 4.13.** Drainage basins with (a) high and (b) low drainage density

basin (Figure 4.7) has a drainage density of 1209 km/740 km$^2$ = 1.63 km of channel per square kilometre.

Drainage density thus represents the amount of channel required to drain one unit of catchment area. The inverse, as the amount of drainage area needed to maintain one unit of channel length, was termed the *constant of channel maintenance* by Schumm (1956).

The density of a stream network reflects the climate patterns, geology, soils and vegetation cover of a catchment. Drainage density is highest in semi-arid areas where surface runoff from intense thunderstorms erodes sparsely vegetated slopes. More highly developed channel systems typically have high sediment yields (Knighton, 1984); thus the relationship between drainage density and precipitation is similar to that shown in Figure 7.20 for sediment yield.

### *Relief Ratio*

Schumm (1956) gives a simple expression for describing relief, the relief ratio ($R_r$)

$$R_r = \frac{h}{L}$$

$$(4.8)$$

where $L$ is the maximum length of the basin (see Figure 4.7), and $h$ is the difference in elevation between the mouth of the basin and the highest point on the drainage divide. Units of $h$ and $L$ should be equal so as to make $R_r$ dimensionless. For the Acheron River basin, $R_r = (1.480 - 0.190)$ km/48 km = 0.0269.

Drainage density and slope of the upland areas are both related to basin relief. In a study on small basins in the western United States, Hadley and Schumm (1961) demonstrated that annual sediment yields increase exponentially with the relief ratio.

### Mean Stream Slope

Channel slope is one of the factors controlling water velocity. Mean channel slope ($S_c$) is given by

$$S_c = \frac{(\text{Elevation at source} - \text{Elevation at mouth})}{\text{Length of stream}} \quad (4.9)$$

Using the fractal stream length for the Acheron River, this value is: $(1.220 - 0.190)/55 = 0.0187$ or $1.87\%$ or $1.07°$ (see Section 5.2.4 for slope conversions).

### Mean Catchment Slope

The slope of the catchment will influence surface runoff rates, and is related to drainage density and basin relief. Van Haveren (1986) gives several definitions of mean basin slope ($S_b$), the simplest method of computation being

$$S_b = \frac{(\text{Elevation at } 0.85L) - (\text{Elevation at } 0.10L)}{0.75L} \quad (4.10)$$

where $L$ is again the maximum length of the basin, as defined for Eq. 4.8, and measurements are taken along this line ($0.10L$ near the lower part of the catchment, $0.85L$ towards the upper end). For the Acheron River basin, $S_b = (1.080 \text{ km} - 0.200 \text{ km})/(0.75 \times 48 \text{ km}) = 0.0244$ or $2.44\%$ or $1.40°$.

Another method of computing average catchment slope is to superimpose a grid with approximately 100 points over the catchment. The slope at each point is tabulated and an average computed. This is similar to the methods described for aspect and hypsometric curves.

### The Longitudinal Profile

The longitudinal profile of a stream describes the way in which the stream's elevation changes over distance, as shown in Figure 4.14. The x-axis represents the distance along a stream, as measured from some outfall point such as a stream junction, a lake, or an ocean.

For many streams, the longitudinal profile shows a characteristic concave shape, with slope decreasing from the upper 'eroding' reaches to the lower 'depositional' ones. This shape is associated with both an increase in discharge and a decrease in sediment size in the downstream direction (Schumm, 1977). In catchments of high rainfall the concave shape is more pronounced, whereas in rivers such as the Nile, which derive their flow from headwater regions and do not increase in discharge downstream, a concave profile does not develop (Petts and Foster, 1985). There have been a large number of attempts to quantify the shape of the longitudinal profile, including the use of exponential-decay equations (Morisawa, 1985).

The general profile will be modified by local topography, bedrock features, changes in bed material, etc. In general, the profile will tend to be steeper on harder rock types and flatter when the streambed is less resistant to erosion. Abrupt changes in slope called *knickpoints* may occur at stream junctions or where the geology changes. For example, waterfalls in the Acheron River system show up as knickpoints in Figure 4.14.

### Hypsometric (Area-elevation) Curves

A curve can also be developed for describing the distribution of catchment area with elevation, called a hypsometric or area-elevation curve. These are useful for describing hydrologic variables that vary with altitude, for example, rainfall or snow cover (Beaumont, 1975).

A hypsometric curve is derived by measuring the area of contour 'belts' from a topographic map (the amount of basin area between two contours). The cumulative area above (or below) a given elevation is then plotted against the elevation. It may be convenient to express areas as percentages of the total catchment area in order to

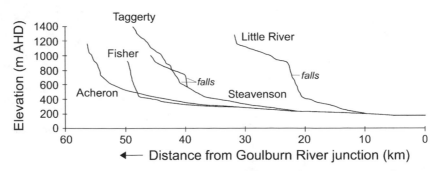

**Figure 4.14.** *Longitudinal profiles for major streams in the Acheron River basin, Victoria, Australia (see map, Figure 4.7). AHD = Australian Height Datum*

compare curves between different catchments. Elevation can also be expressed as a relative height from 0 at the stream outlet to 1 at the highest point on the drainage divide.

Linsley *et al.* (1975a) suggest an alternate method of deriving hypsometric curves, which reduces the amount of effort required. A grid is superimposed on a topographic map to obtain at least 100 points within the catchment. The curve is then plotted using the number of points falling within each elevation range, expressed as cumulative percentages. The technique is similar to the development of cumulative frequency curves (see Appendix).

### Drainage Basin Shape

Drainage basin shape is difficult to express unambiguously (Gregory and Walling, 1973), and various authors have suggested a number of quantitative indices. Selby (1985), for example, lists seven different measures.

Horton (1932) gives a simple *form ratio* ($R_f$) for describing basin shape:

$$R_f = \frac{A}{L^2} \qquad (4.11)$$

where $A$ is catchment area and $L$ is the length of the basin, as defined for Eq. (4.10). Units should be chosen to make the ratio dimensionless. For the Acheron River basin, $R_f = 740 \text{ km}^2/(48 \text{ km})^2 = 0.32$.

Morisawa (1958) determined that the *elongation ratio* ($R_e$) given by Schumm (1956) had the best correlation with hydrology:

$$R_e = \frac{D_c}{L} \qquad (4.12)$$

where $L$ is the same as for Eq. (4.8) and $D_c$ is the diameter of a circle with the same area as that of the basin. Working backwards from an area of 740 km$^2$, the diameter of a circle with the same area is $\sqrt{(4 \times \text{area})/\pi} = \sqrt{(4 \times 740)/\pi} = 30.7$ km. Inserting this value into Eq. (4.12) gives a value of $R_e = 0.64$ for the Acheron River basin.

### Aspect

The aspect of a hillslope is the direction it faces (e.g. southwest). Aspect influences vegetation type, precipitation patterns, snowmelt and wind exposure. To determine catchment aspect, a uniform or random grid is superimposed over a map of the catchment and the aspect at each grid point is noted. Aspect is measured as the bearing (in the downhill direction) of a line drawn perpendicular to the contour lines at the grid point. The number of points within each aspect category (e.g. 330° to 30°) is tabulated

to estimate the percentage of catchment area within each aspect interval. The distribution of aspect in a basin is usually plotted as a polar or 'rose' diagram, with the aspect shown as an angle (0–360°, with zero representing North), and percentage area as distance from the origin (Linsley *et al.*, 1975a).

## 4.3   Streamflow Hydrographs

### 4.3.1   Definitions

A hydrograph is a graph of water discharge or depth against time. The term hydrograph can refer to the pattern of streamflow that occurs over a season or over a year, e.g. the pattern of daily flows shown in Figure 3.1. In this section, however, we will consider hydrographs produced by a single runoff event resulting from snowmelt, rainfall or both. These are sometimes referred to as 'flood' hydrographs or 'storm' hydrographs. For flood analysis, it is preferable to plot these using near-instantaneous readings; however, average daily values will be used here.

Engineers analyze the response of a catchment to rainfall when designing structures such as the overflow spillways on dams, flood-protection works, highway culverts and bridges. *Rainfall-runoff* and *hydrograph analysis* are important and highly-researched aspects of engineering hydrology, and readers are referred to hydrology texts such as Chow (1964a), Chow *et al.* (1988), Linsley *et al.* (1975a,1982), Maidment (1993) and Shaw (1988) for a more detailed coverage of the subject. Stream biota are also affected by how quickly a stream rises and falls. Thus, hydrograph characteristics are useful in the classification of streams for biological purposes (Hawkes, 1975).

Figure 4.15 shows a flood hydrograph for the Acheron River, developed from daily streamflow data. The section of the hydrograph where the flow is increasing is called the *rising limb* and the section where the graph falls off the *falling limb* or *recession curve*. Some catchments will produce narrow, peaked hydrographs which rise and fall quickly. Streams which characteristically have this type of hydrograph are termed *flashy*. In contrast, *sluggish* streams are those which have wide, rounded hydrographs, and the runoff is spread over a longer time period. In general, streams from smaller catchments will be more flashy than those from larger ones, and hydrographs from sudden, intense thunderstorms will be more flashy than those from snowmelt events or low-intensity rainfall.

When precipitation falls on a catchment or when snow melts a certain amount of time elapses before the stream level begins to rise. The water may be intercepted by vegetation, trapped by depressions in the land surface,

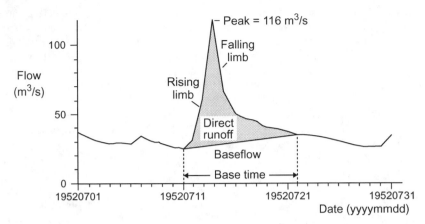

**Figure 4.15.** *Flood hydrograph for the Acheron River at Taggerty, July, 1952, developed using daily data. Hydrograph components are labelled*

absorbed by the soil, or it may evaporate. Any excess water will make its way to the stream as overland flow (surface runoff) or subsurface interflow (see Figure 4.1). Overland flow will reach the stream more quickly than interflow, and it may take many days for all of the upslope water to reach the stream. A measure of the catchment response time, the *time of concentration* or *lag time*, has a number of definitions, usually based on the elapsed time between the rainfall event and the runoff peak.

When the channel fills, groundwater levels near the stream are temporarily raised. The portion of the water entering the bank is termed *bank storage*. When the hydrograph drops off, some of this stored water again re-enters the stream. Bank storage is important hydrologically because of its effects on the shape of the hydrograph. It affects bank stability when saturated soils slump after streamflow levels drop. It also has ecological relevance, since dissolved salts and nutrients can be leached from the soil and introduced to the stream in the temporarily stored waters.

### 4.3.2  Hydrograph Separation

A hydrograph can be separated into two main components:

1. *Direct runoff or quickflow*—the volume of water produced from the rainfall or snowmelt event,
2. *Baseflow*—the volume of water representing the groundwater contribution.

The relative contributions of direct runoff and baseflow will vary between events. For rainfall events, the amount of direct runoff corresponds to the amount and rate of precipitation falling on the catchment. Direct runoff is sometimes divided into surface runoff (the water flowing over the land surface) and interflow (the water which moves through the soil) (see Figure 4.1).

Methods of separating hydrographs into components are commonly based more on characteristics of the hydrograph shape than the actual origin of the streamflow. Thus, the division is somewhat arbitrary. Choice of a method is less important, however, than the consistent use of one method. Linsley *et al.* (1975b) suggest the use of a straight line which tapers up slightly from the point of rise on the hydrograph to a point on the other side (as in Figure 4.15). The angle of the line is a matter of choice, based on what 'looks right'. More objective methods exist, as discussed by Nathan and McMahon (1990b). Methods that use water quality data as a basis for separating the components are presented by Gregory and Walling (1973). A program for hydrograph separation, Hysep, is available on-line from the USGS at http://water.usgs.gov/cgi-bin/man_wrdapp?hysep. It is based on the procedures described by Pettyjohn and Henning (1979).

The *base time* is the base width of the direct runoff portion of the hydrograph. A simple measure of hydrograph 'peakedness' is the peak to base time ratio (Gregory and Walling, 1973). For the hydrograph of Figure 4.15 this value is $116/11 = 10.5$ m$^3$/s/day. Again, the ratio will vary with the units used, so the units should be consistent for comparative purposes.

Another ratio of interest is the ratio of direct runoff to baseflow (with both expressed as volumes). This ratio will be higher for flash floods in arid lands and lower for events from gradual snowmelt.

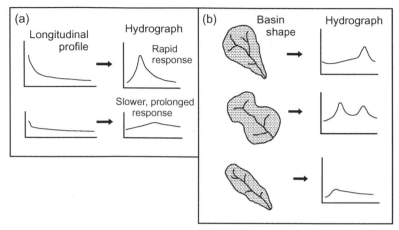

**Figure 4.16.** *Effect of various factors on hydrograph shape: (a) stream slope and (b) catchment shape. Modified from Gregory and Walling (1973), based on (a) Schumm (1954) and (b) DeWiest (1965) and Strahler (1964), by permission of Hodder and Stoughton Ltd*

### 4.3.3 Factors Influencing the Hydrograph Shape

The rate at which runoff moves towards the stream is dependent on the drainage efficiency of the hillslopes. Drainage efficiency is influenced by the slope and length of the upland surface, its microtopography (i.e. existence of a drainage pattern and depressions), the permeability and moisture content of the soil, sub-surface geology and vegetation cover. The hydrograph shape will be affected by these factors as well as catchment shape, drainage density, channel characteristics and storm patterns. A catchment with a well-developed drainage system will have a shorter time of concentration than one with many marshy areas, lakes, reservoirs and other surface depressions (Petts and Foster, 1985). Stream systems with low bifurcation ratios tend to produce flood hydrographs with marked peaks, while those with high ratios produce lower peaks, which are spread over longer time periods (Beaumont, 1975). In addition, the channel's configuration and pattern, and the presence of vegetation in and along the channel will also have an effect on hydrograph shape.

As shown in Figure 4.16(a), a stream with a steeper longitudinal profile will show a more rapid response and will produce higher peak discharges than one that is not as steep. Figure 4.16(b) illustrates the effect of basin shape on the shape of the hydrograph. Shorter, wider catchments will produce a faster stream rise and fall than longer, narrower ones because of the shorter travel times. Basin shape factors, however, may not correlate well with hydrograph shape because they do not account for the direction of storm movement. If a rainstorm moves across the catchment from top to bottom, the peak flows from all streams tend to 'compress', arriving at the outlet at nearly the same time. If the storm moves upstream, the water in lower sections of the catchment runs off long before the water from the upper catchment arrives, 'stretching out' the hydrograph. This effect is more pronounced in elongated catchments.

In general, a hydrograph will move downstream as a 'wave' of increasing then decreasing discharge. In small headwater streams the hydrograph responds quickly, and then may drop back to baseflow levels before the flow peaks at downstream sites. The hydrograph lengthens and becomes rounder as it progresses downstream, collecting the volumes of water contributed at varying rates by tributaries. This 'damping' of the hydrograph shape is called *attenuation* (Shaw, 1988). The behaviour of a hydrograph can be numerically followed downstream using *flood-routing* techniques given in most applied hydrology texts.

Vegetation cover and land use within the catchment will also have an effect on hydrograph shape. Since vegetation affects infiltration rates, its removal can cause direct runoff to increase and hydrographs to become more peaked. It is well known that runoff rates increase after wildfires until vegetation becomes re-established. Grassland and agricultural land may exhibit a larger range of flows and an earlier, more rapid hydrograph rise than woodlands (Gregory and Walling, 1973). Branson *et al.* (1981) give a number of examples of studies which illustrate the effect on runoff of converting from one vegetation type to another. Urbanization leads to increased total runoff, higher peak discharges, more frequent flooding and shorter times of concentration (Morisawa, 1985).

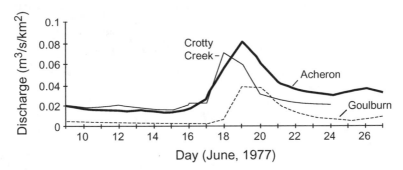

**Figure 4.17.** *Downstream progression of hydrograph shapes. Crotty Creek is a small tributary of the Acheron River, and the Acheron flows into the Goulburn. Discharge values are mean daily flows per unit area, where catchment areas for the three basins (at the gauging sites) are 1.2, 619 and 8601 km², respectively*

It should also be borne in mind that only part of the basin may produce runoff at any given time. The maximum flood discharge per unit area typically decreases with basin size because of the fact that storms have a limited extent. In Figure 4.17 a hydrograph from the Acheron River is presented with hydrographs from upstream and downstream sites, with all discharge values divided by catchment area. In this case, the hydrograph from the furthest upstream site (Crotty Creek) does not show a larger maximum per unit area, possibly because precipitation did not occur as intensely in that catchment as in other parts of the Acheron basin. Complex storm patterns and variations in catchment characteristics may also lead to multi-peaked hydrographs which bear little similarity to textbook examples.

### 4.3.4   Recession Curve Analysis

The rate of rise of the hydrograph, and thus the shape of the rising limb, is influenced mainly by the character of the event (e.g. snowmelt or rainfall). In contrast, the shape of the recession curve is based on groundwater flow patterns and tends to be fairly consistent for a particular location on the stream. Thus, if a flood peak has passed through, the time required for the flood waters to recede can be predicted, given that no additional runoff occurs. Linsley *et al.* (1975b) give a simple exponential decay equation for describing the shape of the recession curve:

$$q_t = q_0 K_r^t \qquad (4.13)$$

where $q_t$ is the discharge at $t$ time units (e.g. 1, 2, 30 days) after $q_0$ and $K_r$ is a recession constant which is less than 1.0. This equation will plot as a straight line on semi-log

paper, with $q_t$ on the logarithmic axis and $t$ on the linear one. Since $q_0$ is the starting value, it is also the $y$-intercept value. Usually, $K_r$ changes gradually, approaching 1.0 as the flow levels off. Thus, Linsley *et al.* (1975b) recommend the use of one curve to represent baseflow and another to represent the recession from direct runoff.

For perennial and intermittent streams, this technique can also be used to describe the seasonal fall of groundwater levels (Todd, 1959; WMO, 1983). For example, it may be of interest to analyse the general downward trend following the snowmelt period to predict the time when a stream will reach a certain level (e.g. zero flow).

If a number of hydrographs are analysed, an average or *master recession curve* can be developed. It is preferable to develop separate curves for snowmelt and rainfall events. Recession analysis is treated in more detail in engineering hydrology texts. Nathan and McMahon (1990b) also present and discuss several techniques.

## 4.4   How Does This Stream Measure Up?

### 4.4.1   General

When two personalities come into contact for the first time a certain amount of 'sizing up' takes place. Similarly, a great deal of information about the characteristics of a stream can be obtained from some very basic analyses of streamflow data. Annual, seasonal and daily patterns of streamflow determine many of the physical and biological properties of streams. A description of this 'hydrological habitat' is necessary for interpreting changes in communities of stream organisms (Hughes and James, 1989). Statistical measures can reveal differences between catchments or between locations on the same stream, and changes due to natural trends or land-use modifications.

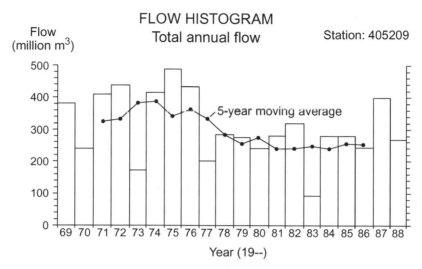

**Figure 4.18.** *Histogram of annual streamflow totals from the Acheron River at Taggerty. A 5-year moving average has been added which smooths year-to-year fluctuations*

Preliminary analyses of streamflow data might include the calculation of total volumes, averages, maxima and minima, the degree of variability, seasonal distributions and/or trends over time. Statistical definitions and formulas are given in the Appendix. Commercial statistical and spreadsheet software packages provide various methods for analysing and presenting data. For example, Bren *et al.* (1988) used box-and-whisker plots to show the monthly statistical distribution of the fraction of red gum forest flooded for pre- and post-dam conditions on the Murray River, Australia.

This section will only present methods for characterizing the runoff patterns of a stream. For descriptive purposes, no assumption about the underlying population distribution is necessary for computing statistics (e.g. mean, coefficient of variation). However, for making inferences from samples to a population, and for putting confidence limits on estimates of population parameters, the distribution of error must be known (see Section 2.3).

In these preliminary computations it will also be assumed that the streamflow data are independent. In reality, rainy days tend to occur together and dry days together. In the same manner, days of high flow tend to follow other high-flow days, and low flows follow other low flows. This tendency is called *persistence*, and the implication is that more data are actually required for estimating the mean than if the data were truly random (Leopold, 1959). One measure of persistence is the autocorrelation coefficient, which is described in the Appendix.

Because of variability, the length of a data series will affect the computed statistics. For comparative purposes,

therefore, it is best to use the same time period for analysis.

### 4.4.2 Annual Statistics

Before computing annual statistics, one should first decide on what is meant by 'annual'. Results can depend on which starting month is chosen (McMahon and Mein, 1986). The calendar year is a fairly standardized annual measure, and most precipitation data will be published as calendar-year summaries. However, in most hydrologic studies it is preferable to use another interval, called a *water year* or *hydrological year*. This is defined such that the flood season is not split between consecutive years. Especially in regions where runoff originates from winter snowmelt, the cycle of snow accumulation and melt should be contained in one interval.

In the United States and Britain, the usual water year runs from 1st October to 30th September. Water year 1990, for example, would end on 30 September 1990. McMahon and Mein (1986) reviewed several methods for determining the start of a water year, and concluded that the most appropriate starting month for a hydrological year is the one with the lowest mean monthly flow.

Plots of annual totals (Figure 4.18) show their range and distribution over time. The pattern is determined by climatic conditions, soil moisture and changes in land use. A line illustrating the 5-year moving average has been superimposed on the graph of annual totals. This average is calculated by averaging the value for the given year with values from the two previous years and the two following years. It has the effect of smoothing year-to-year fluctuations to reveal longer-term trends. For the Acheron River it can be seen that the 1970s were generally wetter than the 1980s.

*Mean annual flow* (either as a total volume or an average discharge) gives an indication of the size of a catchment, its climate, and the 'typical' amount of water delivered from it. Mean annual flow, as a discharge in cubic metres per second ($m^3$/s) can be computed by averaging daily data from complete years. It can also be estimated (in $m^3$/s) by dividing the average annual volume (in million $m^3$) by 31.5576. Table 4.3 lists discharges of selected rivers of the world.

*Mean annual runoff* (as a depth) represents the difference between annual precipitation and evaporation. It is useful for obtaining gross estimates of the water resources of a catchment. Mean annual runoff (in mm) is obtained by dividing the mean annual flow volume (million $m^3$) by the catchment area ($km^2$) and multiplying by 1000. For the Acheron River, this figure is: $(337/619) \times 1000 = 544$ mm. Figure 4.19(a) shows the distribution of mean annual runoff by continent. As a general rule, mean annual runoff decreases as basin area increases; however, there is a great deal of variability in this relationship worldwide (McMahon, 1982). Hughes and Omernik (1983) recommend the use of mean annual flow per unit area and catchment area as an alternative to stream ordering for classifying stream and catchment size.

Since streams of high variability experience a rapid turnover of organisms, one statistic of particular interest as an index of hydrologic variability is the annual coefficient of variation ($C_v$). A high $C_v$ may be indicative of high

**Table 4.3.** *Selected rivers of the world, as ranked by average discharge measured at the river mouth. Data sources: Holeman (1968), and Knighton (1984); converted to SI units*

| River | Catchment area ($10^3$ $km^2$) | Discharge ($10^3$ $m^3$/s) |
|---|---|---|
| Amazon, Brazil | 6,130 | 181.0 |
| Congo, Congo | 4,010 | 39.6 |
| Orinoco, Venezuela | 950 | 22.7 |
| Yangtze, China | 1,940 | 21.8 |
| Brahmaputra, East Pakistan | 666 | 20.0 |
| Mississippi, USA | 3,220 | 17.8 |
| Yenisei, USSR | 2,470 | 17.4 |
| Mekong, Thailand | 795 | 15.0 |
| Parana, Argentina | 2,310 | 14.9 |
| St. Lawrence, Canada | 1,290 | 14.2 |
| Ganges, East Pakistan | 956 | 14.1 |
| Danube, USSR | 816 | 6.2 |
| Nile, Egypt | 2,980 | 2.8 |
| Murray-Darling, Australia | 1,060 | 0.7 |
| Acheron River, Australia | 0.619 | 0.01 |

disturbance and low predictability (Lake *et al.*, 1985). In studies on streams in Tasmania, Australia, Davies (1988) concluded that inter-annual flow variability was likely to be a principal factor limiting trout abundance in streams. In general, streams in arid or semi-arid regions and in areas affected by tropical cyclones are characterized by high variability (AWRC, 1984). In these areas, the runoff tends to be 'all or nothing'.

Based on the work of McMahon (1982), Figure 4.19(b) shows a plot of the $C_v$ of annual flows against the mean annual runoff (MAR). It can be seen that variability

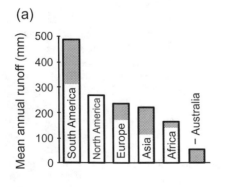

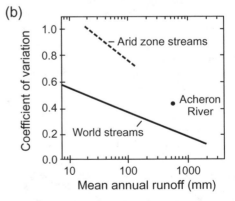

**Figure 4.19.** *(a) Mean annual runoff for world streams, by continent (redrawn from McMahon and Mein, 1986, Reproduced by permission of Water Resources Publications, LLC), and (b) relationship between coefficient of variation of annual flows and mean annual runoff for world rivers and arid zone streams (From Hydrological Characteristic of Selected Rivers of the World, By Thomas A. McMahon, © UNESCO 1982. Reproduced by permission of UNESCO)*

generally decreases with an increase in the yearly runoff. However, the streams of arid zones show a much higher variability at all levels of yearly runoff. McMahon *et al.* (1987) demonstrate that the annual coefficient of variation of runoff for Australia and Southern Africa is nearly twice that of the other continents. Papers by McMahon (1982) and Finlayson and McMahon (1988) provide useful information on the hydrological characteristics of many of the world's rivers.

In intermittent and ephemeral streams it may also be of interest to calculate the amount of time each year that the stream is 'dry', where 'dry' refers to the lack of surface flows. Water and subterranean life may persist below the channel bed. The length of the aquatic phase will vary with geographic location and local hydrology. Statistical or spreadsheet programs can be utilized to calculate the number of days of zero flow (or below some minimum), which can be expressed as a percentage of time during the year.

### 4.4.3  Monthly Statistics

Monthly averages, coefficients of variation and other statistical measures are of interest in studying seasonal variations in discharge, which are controlled by climatic patterns and channel and catchment characteristics. Some regions, for example, will show a strong snowmelt contribution in spring. In studies of rivers in Victoria, Australia, Hughes and James (1989) found that streams in high-rainfall areas had less variable monthly flows than rivers in dry regions.

A simple index can be derived from monthly statistics for describing average low-flow (baseflow) conditions in perennial streams:

$$\text{Baseflow index} = \frac{\text{Lowest mean monthly flow}}{\text{Mean annual flow}} \times 100$$

$$(4.14)$$

where the flows should be expressed as average discharges, in cubic metres per second. An index value near one would indicate that the flow remains fairly constant over the year, whereas a value of zero would be indicative of an intermittent or ephemeral stream (Hamilton and Bergersen, 1984).

The *regime* of a river refers to its seasonal pattern of flow over the year. River regimes have been identified by Hynes (1970) as having an important influence on instream biota, along with average and extreme water temperatures. The classification of river regimes has a wide range of potential applications, including the description of natural regime type in regions of extensive river regulation, the identification of potential water avail-

ability or shortage in particular seasons, the extrapolation of hydrological predictions within like regions and the classification of stream organisms or riparian vegetation by regime type.

The effect of regulation on regime patterns can be demonstrated by comparing regimes from pre- and post-regulation periods or by comparing reservoir inflows with outflows. For example the regulation of streams for irrigation may cause the regime to turn 'upside down' due to winter storage of water and the augmentation of low flows in summer.

Using mean monthly flows from a global data set of 969 stream-gauging stations from 66 countries, Haines *et al.* (1988) developed 15 regime classes for perennial streams, as illustrated in Figure 4.20. Month 1 is the first summer month (December for the Southern Hemisphere, June for the Northern). Average monthly flows are expressed as a percentage of the average annual flow.

The limitations of this technique as a method of stream classification are that the absolute volume of flow is not considered in the classification—only its pattern. Also, the method does not address the regularity of the regime type from year to year. However, it does provide a method of classifying streams based solely on streamflow data. Other hydrological classification methods are discussed in Chapter 9.

### 4.4.4  Daily Statistics

As mentioned in the discussion of annual statistics, daily data can be analysed to obtain the mean daily discharge for a stream. The daily $C_v$, like the annual $C_v$, also characterizes hydrological variability. Horwitz (1978), for example, demonstrated a relationship between fish community structure and the $C_v$ of daily discharges.

An average annual hydrograph can also be developed using daily data. Points on the hydrograph represent the average value for each day, computed as an average over all complete years of interest. It is simply another view of the runoff regime at a finer scale of resolution. To give some indication of the amount of variability, maxima and minima for each day and/or the standard deviation can also be plotted. The average hydrograph is probably most useful for describing runoff patterns, which are fairly consistent from year to year.

### 4.4.5  A Method for Describing Hydrological Predictability: Colwell's Indices

The predictability of flow patterns from year to year, from one season to another, and even from day to night is

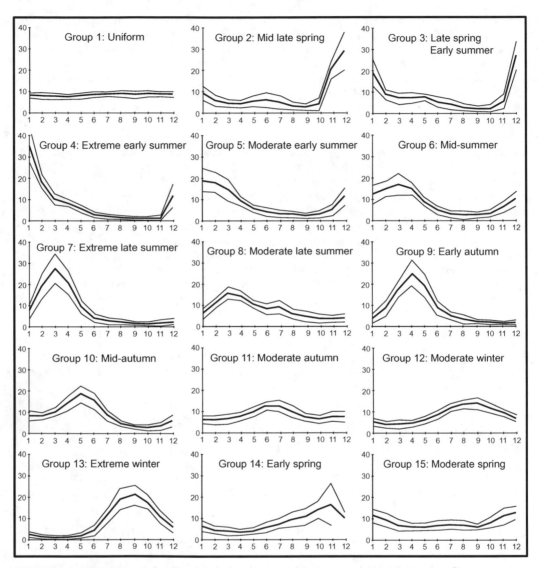

**Figure 4.20.** *River regime patterns for the global classification of Haines et al. (1988). Average flows are expressed as percentages of the mean annual flow, and are shown with bands of plus and minus one standard deviation. Month 1 in the classification is the first month of summer. Reprinted from Applied Geography, Vol 8, Haines, A.T., Finlayson, B.L. and McMahon, T.A., 'A global classification river regimes', pp 255–272, Copyright 1988, with permission from Elsevier*

ecologically important for a number of reasons. For example, flow predictability is thought to influence the evolution of behavioural mechanisms and the timing of life-history stages in stream biota (Resh *et al.*, 1988). If a seasonal pattern were to be repeated in exactly the same way each year it would be totally predictable. Then, knowing the time of year, we could exactly predict the state of the phenomenon (e.g. a certain fish species will spawn for exactly 10 days from 12 May to 21 May).

Resh *et al.* (1988) state that a quantitative technique for comparing the predictability of flow patterns among streams must address both the frequency and intensity of flows as well as the contribution of seasonal phenomena to the annual runoff pattern. Colwell (1974) presents a method based on information theory which satisfies these requirements.

Colwell defines three simple measures for describing fluctuations in physical and biological phenomena over

**Table 4.4.** *Colwell's indices based on monthly streamflow data from the Acheron River at Taggerty, Victoria. Matrix entries are the number of months in which the flow falls within the given interval*

| Class | Jan | Feb | Mar | Apr | May | Jun | Jul | Aug | Sep | Oct | Nov | Dec |
|---|---|---|---|---|---|---|---|---|---|---|---|---|
| < 13.629 | 26 | 37 | 38 | 37 | 24 | 12 | 5 | 1 | 1 | 2 | 6 | 12 |
| to 27.257 | 13 | 4 | 4 | 4 | 10 | 19 | 10 | 5 | 4 | 7 | 16 | 21 |
| to 40.886 | 1 | 1 | 0 | 1 | 4 | 5 | 11 | 8 | 10 | 15 | 11 | 6 |
| to 54.514 | 2 | 0 | 0 | 0 | 2 | 1 | 4 | 7 | 12 | 1 | 1 | 1 |
| to 68.143 | 0 | 0 | 0 | 0 | 1 | 3 | 7 | 8 | 7 | 12 | 4 | 0 |
| to 81.771 | 0 | 0 | 0 | 0 | 1 | 0 | 1 | 7 | 3 | 3 | 1 | 2 |
| > 81.771 | 0 | 0 | 0 | 0 | 0 | 2 | 4 | 6 | 5 | 2 | 3 | 0 |

Number of years of data: 42
PRED = 0.38   CONST = 0.17   CONTING = 0.21

time: predictability, constancy and contingency. *Predictability* is a measure of the relative certainty of knowing a state at a particular time. It is the sum of two components; constancy and contingency, where all measures have a range of 0 to 1. *Constancy* is a measure of the degree to which the state stays the same (e.g. the year-round leaf drop of eucalyptus trees would have higher constancy then the seasonal leaf drop of maples). *Contingency* is a measure describing how closely the different states correspond to different time periods (e.g. for many trees, fruiting occurs in late summer, leaf drop in autumn and bud formation in spring).

For streamflow data, Colwell's method can yield indices that are biologically meaningful, since periodicities in flow have an important influence on riverine ecosystems. For instance, the indices can be used to characterize flow seasonality, patterns of drought or occurrence of peak flows. 'Predictable, seasonal' streams can be separated from, say, 'unpredictable, unseasonal streams' for classification purposes. Streamflow indices calculated from 'before' and 'after' time periods can indicate the effects of river regulation or catchment changes.

To apply the technique, a matrix is constructed with the states of the phenomena represented by rows and the time periods within some cycle represented by columns. An example is shown in Table 4.4, where months form the time periods and the 'state' is designated by different streamflow intervals (classes). Entries in the matrix are the number of months in which the streamflow falls within the given interval. Constancy reaches a maximum (value of 1) when only one row has values (i.e. there is only one state). Contingency is zero when all columns are identical, and it reaches a maximum when there is only one non-zero entry

in each column and each row. As stated previously, predictability is simply the sum of the other two components. These indices are shown in Table 4.4 for the Acheron River data.

Both the length of record and the way in which continuous records (such as streamflow data) are partitioned into 'state' classes can affect the computed values. Gan *et al.* (1991) investigated the use of Colwell's indices for analysing periodicity in monthly rainfall and streamflow data. They found a tendency for predictability and contingency to be biased towards high values for short periods of record. From data-generation methods they concluded that 40 years of data are needed to stabilize these measures. The authors also commented that the lack of a consistent classification of states for continuous hydrological records was a major shortcoming of the method. They adopted seven classes for their study: $<0.5\bar{Q}, 0.5\bar{Q} - 1.0\bar{Q}, 1.0\bar{Q} - 1.5\bar{Q}, \ldots, >3\bar{Q}$, where $\bar{Q}$ is the mean monthly flow. These classes were used in the analysis shown in Table 4.4.

Bunn and Boughton (1990) suggest that a log-2 scale (i.e. 0.5,1,2,4,8,...) be used rather than a linear one, on the premise that a larger change in streamflow constitutes a disturbance in larger rivers whereas only a small change would be needed to produce an equivalent effect in smaller streams. They argue that trade-offs are necessary in selecting both the number of states and the length of record. Too few categories can produce high constancy, the extreme example being one category which would lead to a constancy of 1.0. In contrast, too many categories can lead to low predictability but high seasonality. Further comparative studies are needed to develop consistent methods which provide indices of the greatest ecological relevance.

# 5

# How to Have a Field Day and Still Collect Some Useful Information

## 5.1 Venturing into the Field

Velocity, water depth, substrate, discharge, sediment concentrations and channel configurations all interact to form the hydrologic and hydraulic environment of stream-dwelling organisms. The subject of this chapter is the measurement of these variables in the field. There are numerous methodologies for collection of physical habitat data and some of the more common techniques are presented here. For further information, Dackombe and Gardiner (1983) is a compact guide to standard geomorphological field techniques. Publications of the U.S. Geological Survey (USGS) provide time-tested methods of stream data collection, many of which are now available online at http://water.usgs.gov/pubs/twri. Other manuals on field data collection include those by Hamilton and Bergersen (1984), Harrelson *et al.* (1994), Newbury and Gaboury (1993a) and Platts *et al.* (1983).

Before the first visit to the field, decisions should be made about which variables are to be measured and in what order. This helps in the choice of equipment and methods, estimation of the number of people required and their duties and development of field data forms. It is a good idea to run through the entire field procedure as a 'thought exercise' to smooth out any rough spots.

Field data forms (preferably printed on waterproof paper) should allow efficient and systematic recording of measurements and observations. Each form should have spaces for the date, time, site name and personnel names. If the information is to be computer processed, the format and coding requirements should be considered. Dataloggers and other electronic versions of the field notebook can accept data which can be directly downloaded onto a computer. A pencil and notebook should always be carried for jotting down random notes and sketches. A field notebook developed by Luna Leopold and David Rosgen is available at Wildland Hydrology, 1481 Stevens Lake Road, Pagosa Springs, CO 81147, USA.

Equipment should be checked and calibrated thoroughly and batteries charged as needed before venturing into the field. It is a good idea to put together a small tool kit with some spare parts, pliers, a few screwdrivers, small adjustable wrench, a roll of duct tape ('100 mile-an-hour' tape) and some bailing wire for minor repairs and adjustments. Cameras, calculators, waterproof gear, camping equipment and emergency and first aid supplies are other considerations. Checklists are invaluable for remembering all the bits and pieces required, such as the one provided by Thorne (1998) for reconnaissance-level surveys.

Field personnel should be told which observations are critical and why, trained in data entry and field measurement procedures, and instructed in the use, care and repair of instruments. Safety instruction should form an integral part of training sessions, since there are many hazards in working on or in streams, drowning being only an obvious example. It is a well-known rule of thumb that the depth (in metres) times the velocity (in metres per second) should not exceed 1.0 for safe wading (see Abt *et al.*, 1989). Life preservers should be included on the equipment list if conditions approaching this level are anticipated. A safety rope or cable can also be strung across swifter streams for workers to hold on to. As a minimum, field crews should have two people and preferably three to four, with at least one trained in first aid. The location and phone numbers of emergency services should be noted, and if study sites are highly remote, communication equipment should be considered.

Stream Hydrology: An Introduction for Ecologists, Second Edition.
Nancy D. Gordon, Thomas A. McMahon, Brian L. Finlayson, Christopher J. Gippel, Rory J. Nathan
© 2004 John Wiley & Sons, Ltd ISBNs: 0-470-84357-8 (HB); 0-470-84358-6 (PB)

Field personnel should be taught to be keen observers—to look beyond the depth, breadth and speed of a stream to notice other factors appropriate to the study. Each person will filter out different information from their surroundings, biased by the pre-conditioning of past learning. Particularly if measurements have a qualitative nature, field workers and researchers should be able to make the same interpretation; e.g. what does 'degraded' or 'undisturbed' mean? For example, logs in a stream may be viewed as good habitat by biologists, whereas to another, they make the stream look 'messy'. In a book on nature study, Pepi (1985, p. 54) gives the following illustration of preferential perception:

> A New York banker was giving a tour of the city to an upstate entomologist. As they walked along a crowded street, the entomologist tapped the banker on the shoulder, pointed to a potted shrub, and said, "Listen, there's a field cricket singing in there". The banker was amazed that the entomologist had heard the cricket's song above the city's roar and complimented the scientist on his "good ears". A short time later, the banker stopped at a crack in the sidewalk and picked up a dime.

Perception can be improved through the practice of asking questions. Why is it this way? Is it affected by the season, the weather, the time of day? Is it a nice day, and what do you mean by that? What if? What else? Why not? Researchers should constantly be on the lookout for unidentified hypotheses hurtling through the spaces between known facts.

Training and re-training of field personnel is thus essential to make sure that all share the same 'vision' in terms of what is observed and the goals of the project. Avenues of communication should be constructed prior to the study to guide the smooth flow of information between everyone involved. As mentioned in Section 2.2, planning should be done with care to ensure that the study runs smoothly and yields information of quality.

## 5.2 Surveying: A Brief Introduction

### 5.2.1 General

The objective of any surveying technique is to establish the horizontal and/or vertical location of a given point. This location can be referenced to map co-ordinates and a national height datum, or used directly, as for constructing channel cross sections or measuring the heights of waterfalls. Whereas measurements taken from maps and aerial photographs will suffice for studies of larger scope and lower detail, surveying becomes more essential

as the study shrinks from continental to micro-habitat scale.

Some surveying techniques will be more appropriate than others. For example, a measuring tape and a meter rule may be accurate enough for cross-sectional profiles of smaller streams, whereas more precise methods are needed for measuring water surface slopes. Larger rivers, too, will require more sophisticated equipment simply because of the distances covered. The techniques employed will depend on the purpose of the work, the accuracy required, the equipment available and how easily it is transported, time constraints, and ultimately the sensitivity of the hip pocket nerve.

Some of the more common methods of measuring horizontal and vertical distances and slope are presented in Sections 5.2.2, 5.2.3 and 5.2.4, respectively. In Section 5.2.5, techniques which yield more than one of these measures are given. Specialized literature on surveying techniques includes references by Brinker and Minnick (1995) and Uren and Price (1994).

### 5.2.2 Horizontal Distance

#### *Pacing*

With practice and care in taking consistent steps, pacing can be surprisingly accurate. Two natural steps constitutes one pace; i.e. the number of paces equals the number of times the right foot touches the ground. One's pace length will naturally shorten on steeper slopes, whether traversing uphill or downhill. Thus, the pace must be calibrated by stepping off a known distance such as 100 m and re-calibrated on different terrains. String-operated pedometers or 'hip chains' are commercially available, which record the length of a biodegradable string as it is pulled out of the device (Thorne, 1998).

#### *Rangefinder*

A rangefinder is a relatively inexpensive device for indirectly measuring distances. Some units have a focusing mechanism, whereby two images of a 'target' object are brought together, and the distance between the rangefinder and the target is then read from the instrument. Other units project a laser beam. Accuracies and prices will vary. Rangefinders require calibration with objects at known distances prior to use. It is advisable that readings be taken only by the person who has done the calibration. It is also best to take several measurements each time and average them.

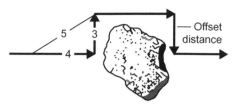

**Figure 5.1.** *Offset technique for avoiding obstacles when chaining*

### Measurements using a Tape (Chaining)

The direct measurement of distance using a tape is sometimes called 'chaining' because a chain made of 100 links was historically used for distance measurement. It is now more commonly accomplished by stretching a tape horizontally between two points. Tapes may be made of steel, fibreglass, cloth or plastic and various lengths are available. Metal tapes will normally give the most accurate measurements; however, other materials will be lighter for the backpacking researcher.

For most environmental field work, high precision surveying is not necessary. Some sag of the tape or other deviation from horizontal is acceptable if the error is less than about 5%. It may be preferable to use tachometry under conditions where the error is greater, such as when measuring across a wide gorge or gully, or when measuring horizontal distances on steep slopes.

Two or three people are normally required for chaining—one on each end of the tape and one to record measurements. If the distance to be measured is longer than the tape, some technique must be used to keep the measurements on line. This can be done by lining up ranging poles or flagging left at each station or by using a compass bearing and a landmark.

If an obstacle is encountered it may be necessary to use an offset technique of some kind to make a precisely engineered detour, as illustrated in Figure 5.1. A few steps are taken at a right angle to the original path (by using some type of optical square or by using the tape itself and properties of the 3-4-5 triangle) and the same distance is measured back after passing the obstacle.

### 5.2.3 Vertical Distance

#### Measurement with a Tape or Rule

A graduated tape or rule can also be used for vertical measurements. The simplest example is the use of a meter rule or surveying staff to measure from the streambed to the water surface to obtain water depth. If needed, a simple spirit level can be used to ensure that the measurement is taken vertically. A tape weighted on one end can also be used to take vertical measurements over longer distances—for example, to measure water depth in a well.

#### Levelling

Levelling is a procedure for determining the relative heights of a number of points. The procedure involves the use of a device which allows one to sight along a horizontal (0°) line. In this section the levelling device is assumed to be a hand-held telescopic level, an Abney level or a clinometer. Levelling with a dumpy level is described in Section 5.2.5. Magnification is often needed for taking readings over distances longer than about 30 m, particularly in shaded locations.

The eye height can be kept constant by resting the level against a staff or rule (e.g. a 2 m carpenter's rule) or by using a pole with a mechanism to lock the level into place. A sighting is taken through the level to a survey staff to obtain a reading. At longer distances where the numbers on the staff become illegible, the person with the staff can hold a finger horizontally in front of the staff and raise or lower the finger until the level height is reached. In this case, the staff bearer would record the reading. Measurements taken at one location are compared to those taken at other locations, or to the 'eye height' of the level, to determine the difference in elevation. This is shown in Figure 5.2(a), and the relevant calculations are given in the caption.

A simple arrangement for levelling consisting of a spirit level attached to a graduated staff by a sliding bracket is shown in Figure 5.2(b). This apparatus can be used to quickly obtain water and channel bed elevations by

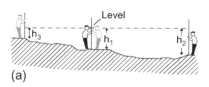

(a)

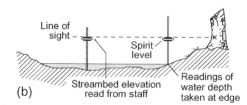

(b)

**Figure 5.2.** *Levelling: (a) vertical distance can be measured as the distance between the level height and the height at another point (e.g. $h_2 - h_1$) or the difference between elevations at two points (e.g. $h_2 - h_3$); (b) a simple method for measuring channel and water elevations*

sighting to some fixed point such as a mark on a boulder or a nail in a tree. This technique was used by Bren (personal communication, 1988) to rapidly survey 100 cross sections in one day on a fifth-order stream in Oregon, USA. Laser levels are now available which can be used in a similar manner. These instruments project a laser beam in a level circular plane, and the surveyor uses a level rod with a detector that is moved up or down until the beam intersects it.

### Indirect Methods for Measuring Height

Vertical heights of features such as trees or waterfalls may require indirect methods. The properties of right triangles can be used to obtain vertical distances by measuring one side and one angle of the triangle (see Figure 5.3). Vertical angles can be measured upward or downward from horizontal with a clinometer or Abney level. These instruments are relatively inexpensive and easy to use and carry. To obtain a vertical angle, the reading is normally taken by looking through a viewing window with one eye while 'aiming' the instrument with the other. An Abney level is operated in a similar manner, although the angle must be read from a protractor-like device.

To measure the height of a feature, a specific distance, $L$ (e.g. 50 m), is first measured out from its base. A sighting is made through the clinometer or Abney level to the base of the feature and then to the top, and each vertical angle is recorded. The vertical height ($H_v$) is then calculated as

$$H_v = L(\tan\theta_1 + \tan\theta_2) \qquad (5.1)$$

where $\theta_1$ and $\theta_2$ are the upper and lower angles as shown in Figure 5.3 and $L$ and $H_v$ are in metres.

For example, if the top reading was 12.2° above horizontal, the lower reading 2.3° below horizontal and the measured distance, $L$, was 50 m, then the feature would be $50(0.256) = 12.8$ m tall. If the person is totally above or below the object, a little more trigonometric juggling is needed to come up with an appropriate formula, and this is left to the reader.

### 5.2.4  Slope

The *slope* or *gradient* of a stream, road or hillside is the amount of vertical drop per unit of horizontal distance. Measurement of slope, then, involves either a direct measurement of the slope angle or measurements of vertical distance over some length.

Slope can be expressed as a fall (or rise) per unit distance (e.g. metres per kilometre). If the distances are

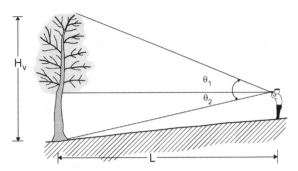

**Figure 5.3.** *Measuring the height of a tree using vertical angles (see Eq. (5.1))*

measured with the same units, the slope can be expressed as a ratio (e.g. 1 in 10 or 1:10), a decimal (e.g. 0.10), a percent (e.g. 10%) or an angle in degrees. To convert from one form to the other is easily done with a scientific calculator:

$$\text{Percentage slope} = \text{slope (in decimal form)}$$
$$\times 100 \qquad (5.2)$$
$$\text{Slope (in decimal form)} = \tan(\text{slope in degrees}) \qquad (5.3)$$
$$\text{Slope (in degrees)} = \tan^{-1}(\text{slope in decimal form}) \qquad (5.4)$$

For example, a 45° slope would be a '1 in 1' slope, and a 0.009 ('9 in 1000') slope would be 0.9% or $\tan^{-1}(0.009) = 0.52°$.

### Measuring Slope with a Clinometer or Abney Level

Slope can be read to about one degree with a clinometer or Abney level (described in Section 5.2.3). To obtain a direct reading of slope, the instrument can be roughly aligned with a hillslope by eye. For local slopes, such as streambanks, the clinometer can be placed on a board which is set on the ground. This method is not accurate enough for measuring water surface slopes in streams since these are normally very close to zero.

For more accurate measurements of slope, two people are required: one to use the clinometer and one to hold a staff or a pole marked at the clinometer 'eye height'. Measuring the slope is then a simple matter of looking through the clinometer with one eye and viewing the staff with the other, and raising or lowering the clinometer until the crosshair lines up with the mark on the staff (see Figure 5.4(a)).

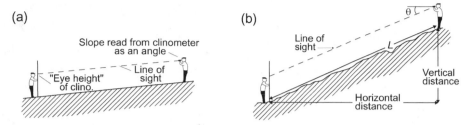

**Figure 5.4.** *Using a clinometer to (a) measure slope and (b) measure slope and horizontal and vertical distance. L is the distance measured along the ground and θ is the angle read from the clinometer*

By including a measurement along the ground between the two people ($L$), horizontal and vertical distances can also be determined, as shown in Figure 5.4(b). The vertical angle ($\theta$) measured with the clinometer is used to calculate distance as follows:

$$\text{Horizontal distance} = (L)(\cos\theta) \qquad (5.5)$$
$$\text{Vertical distance} = (L)(\sin\theta) \qquad (5.6)$$

where $\theta$ is the slope angle in degrees and distances are in metres.

### Hydrostatic Levelling

Hydrostatic levelling, a principle familiar to most carpenters, is an inexpensive, light and yet accurate method of slope measurement. A manometer for levelling can be simply made from two metre rules and a long length of hose marked at convenient intervals with indelible ink (Figure 5.5). 20 metres of 10 mm i.d. (inside diameter) clear tubing is a practical size. The hose is filled with water, with care taken to exclude all air bubbles.

At the site to be measured the hose is extended over the length of the slope. The hose may be curled up as

necessary, as this will not affect the readings. After the water level stabilizes in the manometer a measurement is taken at the bottom of the meniscus at each end and the difference in readings determines the vertical drop. An adequate measure of the horizontal distance between the two rules can be made by pulling the calibrated hose taut and using it as a tape measure. Slope is then calculated by the formula given in the caption for Figure 5.5.

In comparing slopes measured with a manometer to measurements with a surveyor's level and stadia rod, LaPerriere and Martin (1986) reported agreement within 2%. An added advantage of this equipment is that it can be used to measure slope around bends, meaning that neither a straight stretch nor a clear line of view are required (in this case, the horizontal distance is measured along the bend, not as a straight line). In streams, the slope of both the water surface and the streambed can be measured at the same time.

### Estimation of Slope

Slopes can also be estimated within about plus or minus 10° or better by eye or by 'feel'. It is a good idea to 'calibrate' oneself each time an estimate is to be made and to recognize that overestimation is common. There are few natural slopes of 90°. Main roads seldom exceed a slope of 1 in 7 (8°), the maximum gradient for a four-wheel drive vehicle is about 1 in 2 (27°), and a 35° slope is about where it becomes difficult to stand upright on a hill.

### 5.2.5  The Full Contingent of Co-ordinates, Including Methods of Mapping

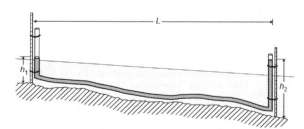

**Figure 5.5.** *Hydrostatic levelling. Slope of channel bed is $(h_2 - h_1)/L$, with all measurements in metres*

In the previous sections, methods were presented for measuring vertical and horizontal distances and slope.

This section presents methods for fixing the measurement points in space. For many applications an estimate of position using landmarks and a map or aerial photograph is sufficient. In other cases, such as sketching a scaled map of a stream reach, a compass or plane table may be used to locate one's position more accurately. When even higher accuracy is needed (for example, when re-surveying a stream cross section at a historically surveyed site) more sophisticated surveying techniques will be appropriate.

### Tachometry using Level or Theodolite

Tachometry is a traditional surveying method of optical distance measurement. In tachometry, a sighting instrument is used to obtain readings from a staff. From the staff readings and the horizontal distance to the staff, the surface elevation and location of points in reference to others can be derived.

The instruments most commonly employed in tachometry are dumpy levels or theodolites. A dumpy level is basically a telescope fixed in the horizontal plane. The advantage of a theodolite over a level is that it can be tilted vertically. It is more versatile in irregular terrain or deep cross sections where a dumpy level must be moved frequently to cover the range of elevations.

The level or theodolite should be positioned to give the clearest possible view of the area of interest. The instrument is preferably located near the middle of the measured distances (e.g. stream reach or hillslope) to reduce any bias introduced by the instrument or the way in which it was set up. The instrument is set on a tripod and levelled by adjusting the tripod leg heights and tilting screws on the instrument platform (tribrach). On hillslopes, the tripod should be set up with one leg uphill and two downhill to improve stability. In other situations, aligning one leg in the main direction of surveying will keep the operator from having to straddle the tripod.

The levelling instrument should be checked periodically by using a 'peg test'. This involves setting up the instrument halfway between two pegs set at a measured distance apart (e.g. 100 m). Readings are taken to both pegs. The instrument is then set up outside the pegs, in line with them, and measurements of the two pegs are repeated. The differences in readings between peg 1 and peg 2 should be the same from both positions, otherwise the instrument should be re-calibrated.

Before taking measurements, the elevation of the instrument should be determined with reference to a real or assumed benchmark elevation. The zero azimuth point for the horizontal angle is then set with reference to some origin (e.g. a tree or benchmark) or to a compass direction

**Figure 5.6.** *View through the theodolite or level, showing metric levelling staff plus centre and stadia crosshairs*

(e.g. North). The horizontal angle is usually read on a scale viewed through an eyepiece.

With either a level or theodolite, the staff bearer holds the staff vertically at the point to be surveyed and the instrument is focused on the staff. As shown in Figure 5.6, the view is of a centre crosshair bracketed by two stadia crosshairs. All three readings and the vertical and horizontal angles to the staff should be recorded. The stadia reading is the distance between the stadia crosshairs, as read from the staff (or twice the distance between one stadia hair reading and the centre line). Thus in Figure 5.6 the vertical reading is 2.70 m, and the stadia reading is $2.76 - 2.64 = 0.12$ m. All readings should be recorded in a systematic manner. Special booking forms are available for this purpose from equipment suppliers.

With a *level*, all sights are horizontal and the relevant measures are shown in Figure 5.7(a). The vertical distance is simply the staff reading at the middle crosshair. For some surveys a vertical reading is taken at a benchmark in order to relate all other measurements to that elevation. In cases where only a difference in elevation is needed (as for calculating slope), this is not necessary.

The horizontal distance, $L$, is obtained from

$$L = 100s \qquad (5.7)$$

where $s$ is the stadia reading and both $s$ and $L$ are in metres. The multiplying constant, 100, should be verified from the instrument reference manual or by calibration over a known distance.

With a *theodolite*, the calculations become more complex. The zenith angle ($\theta_z$), or angle of inclination ($\theta_v$), shown in Figure 5.7(b), is obtained from a scale on the

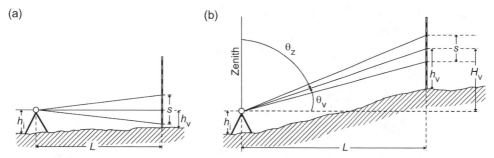

**Figure 5.7.** *Measures used in tachometry: (a) with a level and (b) with a theodolite. Shown are the horizontal distance (L), vertical distance ($H_v$), zenith angle ($\theta_z$), angle of inclination ($\theta_v$), stadia reading (s), vertical reading ($h_v$) and instrument height ($h_i$). All distances are in metres and angles in degrees*

instrument. A quick tilt of the instrument will reveal whether it is measuring $\theta_z$ or $\theta_v$. To simplify computations, the instrument can be set to horizontal (in which case it becomes a level) and the vertical adjustment used only when needed. The horizontal distance, $L$, is calculated as

$$L = 100s \left( \sin^2 \theta_z \right) \qquad (5.8)$$

or

$$L = 100s \left( \cos^2 \theta_v \right) \qquad (5.9)$$

where 100 is the multiplying constant of the instrument and should be verified. All other terms are defined in Figure 5.7. When the vertical angle is horizontal, the equations reduce to Eq. (5.7). The vertical distance, $H_v$, is calculated as

$$H_v = 100s \left( \sin \theta \cos \theta \right) \qquad (5.10)$$

or, equivalently,

$$H_v = 100s \left( \tfrac{1}{2} \right) (\sin 2\theta) = 50s \left( \sin 2\theta \right) \qquad (5.11)$$

Eqs. (5.10) and (5.11) are applicable whether $\theta$ is $\theta_z$ or $\theta_v$. $H_v$ will be negative if the theodolite is pointed downwards (below horizontal) and positive if pointed upwards. The calculations may be programmed using a small scientific calculator or read approximately from a 'stadia slide rule', available from surveying equipment suppliers.

The horizontal and vertical distances and the horizontal angle establish the $x$, $y$ and $z$ co-ordinates of the surveyed point with respect to the instrument position. These can then be referenced to a reference origin or to map co-ordinates. These calculations are based on simple arithmetic and trigonometry and are left to the reader.

*Example 5.1*

If the stadia reading is as shown in Figure 5.6 and the measured zenith angle ($\theta_z$) is 80 degrees (the angle of inclination, $\theta_v$, is 10°), find the horizontal and vertical distances with respect to the instrument.

*Horizontal*

$$\begin{aligned}
L &= 100(0.12 \text{ m}) \sin^2(80°) \\
&= 100(0.12 \text{ m})(0.9848)^2 \\
&= 11.64 \text{ m}
\end{aligned}$$

*Vertical*

$$\begin{aligned}
H_v &= 100(0.12\text{m})(1/2) \sin 2(80°) \\
&= (6 \text{ m})(\sin(160°)) \\
&= 2.05 \quad \text{m; so elevation of the ground is} \\
&\quad 2.05 - 2.7 = -0.65 \text{ (0.65 m below} \\
&\quad \text{'eye level' of instrument)}
\end{aligned}$$

**Plane Table Mapping**

In plane table mapping a scaled map is constructed on-site as the readings are taken. The plane table consists of a drawing board that may be levelled on top of a tripod, and a sighting instrument or alidade. A piece of sturdy paper is taped to the drawing board for constructing a map. There are two main types of alidade used: a simple 'peep sight' type with two folding vertical sights attached to either end of a straight line, and a 'telescopic' alidade, which is similar to a theodolite but is attached to a flat base that acts as a rule on the plane table (Figure 5.8). With the simple alidade the position of objects on the map must be determined either by using a tape or by a method of intersection which will be described. No vertical elevations may be determined with the simple alidade, but can

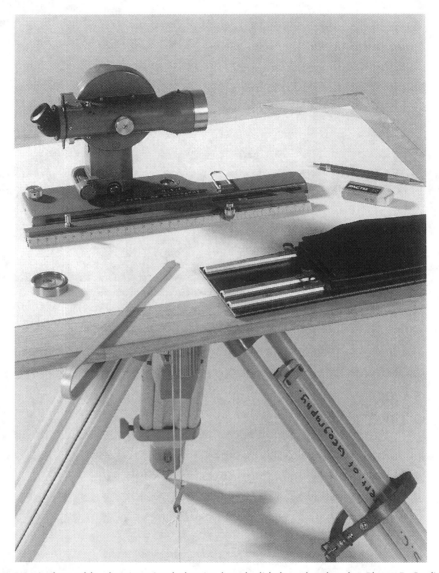

**Figure 5.8.** *A plane table, showing tripod, drawing board, alidade and scale ruler. Photo: P. Cardiyaletti*

be obtained with a dumpy level set up near the plane table. Both horizontal and vertical measurements may be obtained with the telescopic alidade using the same methods described for the theodolite.

To map on the plane table, a convenient location is selected to one side of the site and the table is levelled. If a *peep sight alidade* is used, a measured baseline (e.g. 100 m) is established along the longest axis of the sight to a second position for the plane table. The baseline is plotted on the map, its length determining the map scale. The alidade is then set on the baseline and the plane table is rotated horizontally until it can be locked with the sight

aimed towards the other end of the baseline. With the table now levelled and locked, points to be mapped are sighted in the alidade and a line is drawn along that bearing from the initial position (station) at one end of the baseline (see Figure 5.14). The radiating lines from this position are labelled. When all points have been located, the plane table is moved to the other end of the baseline and levelled. With the alidade along the baseline, the table is then rotated and locked with the initial station sighted. The same mapping points are now sighted again. The location of a point may now be found from the intersection of the alidade bearing and the bearing line drawn from the first station. More baselines may be added to the map to cover larger areas. Depending on the care taken, the simple alidade can produce a scaled map with unsophisticated equipment. An even simpler version of this technique can also be used to develop a scaled map using two compasses, where the maps would be drawn on graph paper with a rule and a protractor.

If a *telescopic alidade* is used, points may be located along the bearing line of the alidade from a single position by using the stadia method for distances and elevations described for theodolite surveys. Because a scale ruler is attached to the alidade, the point can simply be plotted on the drawing paper at the appropriate distance. If the elevation of the plane table is established from a real or assumed benchmark, the elevation of the points may be noted as well. Contours of equal elevation may then be sketched in the field with additional points added as necessary. If the plane table must be moved to cover the entire site, at least two, and preferably three, points of reference should be established which can be sighted from the next station. A new piece of drawing paper is used at the next station, and the two maps are later joined together by overlaying the reference points.

In general, if good weather prevails, a plane table survey requires two or three times the field time required for other surveys. However, a higher level of detail is possible with more on-site observations of features such as soil types, pools, boulders, plant species and possum nests, and the mapping does not require further plotting and drafting. An added advantage is that a map close to the finished product can be checked while still at the site. More information on plane tabling is given by Higgins (1965) and Low (1952).

### Electronic Distance Measurement (EDM)

Electronic distance measurement (EDM) is more precise but also more costly than a theodolite. The instruments, called EDMs or total stations, measure the distance from a station to a target by the time or phase difference between a transmitted and reflected beam of radiation. Microwaves, visible light, lasers or infra-red light can be emitted. The target is normally a retroprism or reflective surface of some kind. As with theodolites and levels, calibration and careful levelling of the instrument is required prior to use.

Most EDM instruments include a microprocessor for calculating distance and height automatically. In some models the data can be stored on a datalogger which is downloaded onto a computer for processing. If data can be plotted in the field using a notebook computer, errors can be corrected before leaving the site. One of the major advantages of using an EDM or total station is that it can

be used over long distances (up to several kilometres). This considerably simplifies surveys of large rivers since the equipment does not need to be moved as frequently. It should be noted, however, that obstructions of any type can result in a weak signal. Heat waves can also cause interference; thus early morning or evening surveys are preferable on hot days.

The major disadvantage of these instruments is the price. They also require some prior training and experience in both care and use. Charged batteries and thus some logistical forethought are needed. They are also more difficult to pack into the field when compared to a clinometer, compass and tape.

### Traversing: Moving from One Station to Another

With a level, theodolite, plane table or total station, the instrument will often need to be moved because of obstructions such as hills, stream bends or vegetation or due to the limitations of the instrument. When the instrument is moved, each new location must be referenced to the previous one. This procedure is called *traversing*. To make sure that crucial measurements are not left out, a little redundancy and overlap is a good idea. For an effective survey, the following precautions should be taken:

- At each position a marker should be placed in the ground directly below the instrument. The instrument height should be measured; with an EDM, the target height is measured as well.
- A sighting should be made to a reference point—if possible, one of known elevation such as a national survey marker. This can be used as an origin to which all points are referenced.
- Staff positions used for relocating the station are called *turning points*. Before moving from one position to the next, a *foresight* to the turning point should be taken. After the instrument is set up at the new location, a *backsight* is taken to the same turning point (see also Figure 5.14(b)).
- The last position should always be tied back to the starting point by surveying in a loop to complete *closure*. This allows an estimate to be made of the error. The error should be calculated in the field to determine whether or not the survey should be repeated.

### Global Positioning System (GPS)

The Global Positioning System (GPS) can provide accurate position locating and precise time data to worldwide users. Originally developed in 1973 by the US Department of Defense for military purposes, GPS is now available at no cost to civilians for air, land and water navigation, emergency response, surveying and environmental research. The GPS is based on a 'constellation' of satellites (24 originally) that orbit the Earth and continuously transmit radio signals to earth-based receivers. Signals contain data on the satellite's position in space and exact time from a precise internal atomic clock. The position of a GPS receiver unit on Earth is determined by how much time it takes for the satellite signal to reach the receiver.

Three satellites are sufficient to fix a receiver's location by triangulation; however a fourth is needed to correct for variations in clock time between satellites and the receiver. To improve the accuracy of readings some GPS units combine a base station receiver and a roving unit to correct for atmospheric conditions. Many permanent base stations continuously collect data (e.g. at universities) which can also be used for this 'differential correction'. Averaging data over time also provides more accuracy.

Unlike surveying techniques that require a clear line of sight, GPS units can be used nearly anywhere there is a clear view of the sky, and under variable weather conditions. GPS also requires only one operator for surveying. Hand-held units may be accurate to several metres, and surveyor-grade instruments are accurate to less than 1 cm horizontally. Vertical accuracy is typically less than horizontal.

Until May 2000, the US Department of Defence introduced 'selective availability' (SA) into GPS signals to degrade its accuracy for security reasons. After that date, the accuracy of hand-held units improved up to tenfold. Systems such as WAAS (wide area augmentation system), which utilize signals from base stations and satellites, are becoming available to provide even better accuracies.

The smaller hand-held units do not have sufficient resolution for detailed stream surveys, but can be very useful for finding study sites again and for locating one's position on maps. Many GPS units can be connected to a computer and some have built-in mapping features to plot where one has travelled. This is a rapidly advancing technology, with new applications constantly evolving and units becoming smaller and more economical. One source of additional information is a tutorial by Trimble Navigation at www.trimble.com/gps/.

## 5.3  Methods of Measuring Areal Extent

### 5.3.1  General

The areal extent of various attributes such as different soil types, vegetation densities, fish cover or substrate composition can be measured by several sampling techniques. For example, density of paddock grasses might be measured in the field using a point intercept method and the size of a paddock measured from aerial photographs. With the advent of computer aided drafting (CAD) software, measurement of areal coverage from photographs is as simple as drawing a line around the area of interest. The general methods that follow are primarily for application in the field.

If ground measurements are to be extrapolated to a larger area, a stratified sampling design (Section 2.3.5) should be used, with samples taken within a relatively homogeneous area. As mentioned in Section 2.3, randomness should be introduced at the proper level of sampling. For example, if an investigator wishes to sample sites meeting the spawning requirements of a particular fish species, it is much more effective to have a fisheries biologist identify spawning sites from which a random sample is taken than to sample randomly across a whole reach (Hamilton and Bergersen, 1984).

### 5.3.2  Visual Estimation of Percentage Cover

In stream hydrology, visual estimates of substrate sizes, vegetative canopy, snow or ice cover or amount of woody debris are often made. For example, one might look up through a tree canopy and estimate the percentage of the sky which is blocked from view. 'Calibration' of one's eyes is crucial. This can be accomplished using visual charts such as the one shown in Figure 5.9.

It is important to first define the area within which an estimate is to be made; e.g. the area inside a grid cell, a hoop of known diameter thrown on the ground, or an area 'as far as the eye can see' from a helicopter or car window. An investigator can become fairly proficient at visually estimating areal cover, reducing the subjectivity of this technique. It should also be realized that proficiency will change with experience, and periodic 're-calibration' is a good idea.

### 5.3.3  Point Intercept Method

The point intercept method is used for surveying along a transect. At set points along a horizontal line, the object or feature on which the point falls—or 'intercepts'—is noted. One method of doing this is to lay out a tape and use a constant interval between points. The 'point' can be established simply by looking vertically downward over the tape, or, more precisely, by vertically lowering a thin, sharpened metal rod from the edge of the tape, or by using

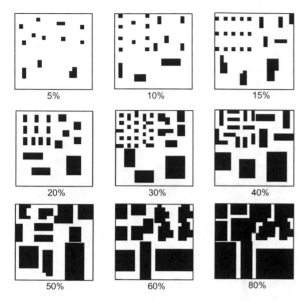

**Figure 5.9.** *Chart for visual estimation of areal coverage. Modified from Northcote (1979), Reproduce by permission of Rellim Technical Publications*

a crosshair telescope aimed downwards. A less rigorous technique consists of walking across a gravel bar, streambank or reach and stopping every few paces to note the feature intercepted by the toe of a shoe (a mark or notch

can be added for more precision). In taking measurements, one must guard carefully against 'selective drift' (e.g. to intercept more aquatic plants and fewer beer cans).

With a point frequency frame, illustrated in Figure 5.10(a), pins are dropped and the 'hits' recorded. The height of each pin can also be measured to give an indication of surface topography. If vegetation is being measured, the pin may 'hit' several layers as it is dropped. In estimates of percent cover only the first hit is needed, whereas all hits might be recorded if the density of the plant itself is to be estimated.

The number of hits in a certain category is calculated as a percentage and used as an estimate of the relative amount of area covered:

Percentage cover for category
$$= \frac{\text{Number of hits in category}}{\text{Total number of hits}} \times 100 \qquad (5.12)$$

### 5.3.4   Line Intercept Method

The line intercept method, illustrated in Figure 5.10(b), is another type of transect survey. This is normally conducted by extending a tape between two points and measuring the distance along the tape intercepted by each category type. Examples of category types are different plant species, overhanging or eroded banks, different soil

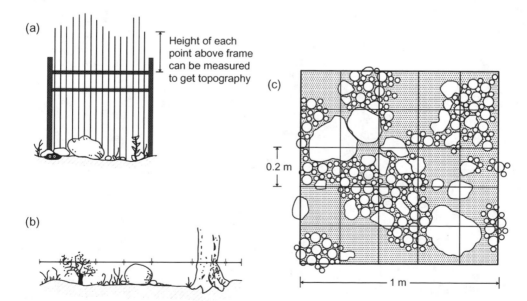

**Figure 5.10.** *Methods of estimating areal cover: (a) a point frequency frame, (b) line intercept method and (c) grid for measurement of substrate types*

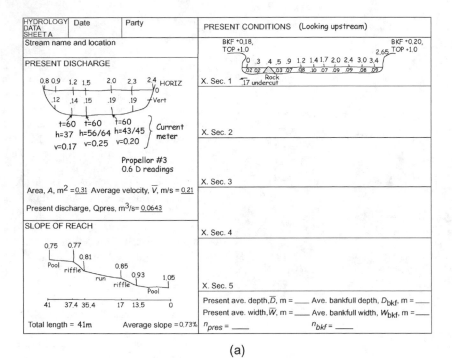

(a)

(b)

**Figure 5.11.** *Partially filled forms for collection of physical data at stream survey sites. These forms may be freely copied. Modified from Newbury and Gaboury (1988), Reproduced by permission of Canadian Water Resources Association*

types, and pools or riffles. The distance intercepted is expressed as a percentage of the total distance to obtain an estimate of cover:

Percentage cover for category

$$= \frac{\text{Distance intercepted}}{\text{Total distance}} \times 100 \qquad (5.13)$$

### 5.3.5  Grids

Grids can be used to define regions of known area for estimating areal cover. Super-imposing a grid on an area before taking a photograph, for example, allows one to later analyse the photograph for percent cover.

As shown in Figure 5.10(c), a 1 m square grid, sectioned into 200 mm squares, might be used to establish the percentage of different bed material sizes. Within each 'cell' the average or dominant size would be measured or estimated. Alternatively, the areal coverage of each substrate category (e.g. boulder, gravel or sand) can be estimated within each cell. The areal coverage of algal mats or other aquatic plant species can be estimated in the same manner. Bunte and Abt (2001) describe a 600 × 600 mm sampling frame for estimating streambed surface particle sizes.

The average area of cover for a category is calculated as the average of the cover in all cells:

Percentage cover for category

$$= \frac{\Sigma(\text{Area of cover in each cell})}{\text{Total area of grid}} \times 100 \qquad (5.14)$$

If cell sizes are the same, then percentage cover can be calculated as an average across all cells.

## 5.4  Surveying Streams

### 5.4.1  General

To describe the physical characteristics of a stream reach, a basic survey should include a measurement of channel slope, several cross-section profiles representative of the channel form, a description of bed materials and vegetation, and a sketch of the reach. The reach selected should be long enough to include a full meander amplitude with two sets of pools and riffles, generally 12–15 times the bankfull width. Field data forms for recording stream survey data are given in Figure 5.11. A space is also provided for discharge measurements, which will be described in Section 5.6.

The site location should be marked on a topographical map of the area. Photographs of the study area are also valuable for confirming observations and monitoring changes over time. For example, in an inventory of rivers in Victoria, Australia (Mitchell, 1990), photographs were taken of upstream, downstream and oblique views, special features and bank types. A code printed on a card was photographed as the first frame in each film, and photographs were checked against recorded data. Digital cameras can provide instantaneous processing and images require less storage space.

### 5.4.2  Cross-sectional Profiles

Once a representative reach has been chosen, 3–5 sites for measuring cross sections within the reach are selected. Fewer are needed in uniform reaches, more in complex channels. Sites can be located randomly, spaced uniformly or selected as representative of a smaller area of the reach. For example, two sites in pools and two sites in riffles might be selected, with each site reflecting average conditions in terms of width, depth and bed topography. Figure 5.12 gives some of the terminology for describing channel dimensions at a cross-section.

A cross-sectional profile can be obtained with a dumpy level or theodolite, or in smaller streams, by using a measuring tape and metre rule or survey staff. If the stream has water in it, the water surface provides a horizontal surface from which to take vertical measurements. If it is dry, vertical measurements can be taken in reference to a pole or tape held horizontally over the streambed or by using a level.

For those who prefer 'streamside' surveys, Molloy and Struble (1988) describe a method of tossing a marked nylon string to the opposite bank to define horizontal distance, and lowering another marked, weighted string from a pole to measure water depth. This method would be appropriate for small, deep stream reaches of low velocity.

With any type of equipment, the procedure of surveying cross-sections is basically the same: vertical measurements are taken at several points along the horizontal line. The horizontal distance to the measurement point, and vertical distances to the streambed and water level are recorded. Measurements should be taken at each break in slope along the bed and at the water's edge, as shown in Figure 5.13. If the study involves monitoring of channel changes the survey should be continued past the edges of the active channel and out to permanent markers.

Bank slope and bank overhang can also be recorded. Both are indicators of bank stability, and bank overhangs

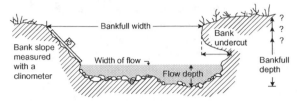

***Figure 5.12.*** *Field measurement of a stream cross section. The definition of bankfull depth is based on field interpretation or geomorphological definitions (see Section 7.2.5)*

are particularly important as shelter areas for fish. *Bank slope* is best measured using a staff and clinometer (Platts *et al.*, 1987). The clinometer is held against the staff which is set against the bank, and the angle is read directly from the clinometer. *Bank overhang* is measured with a staff or metre rule from the farthest point of undercut to the most distant point of overhang. At measured sites on the Acheron River (Figure 4.7), overhangs of 0.30 m were quite common.

*Bankfull width* and *depth* (see Figure 5.12) may actually be more important than the area presently occupied by the stream and these measures provide a more standardized description of channel dimensions. At the edges of each cross section, vertical measurements should be made from the water surface to the bankfull level (or, if the channel is dry, these levels are surveyed). The bankfull width should also be noted at each cross section. In the field, the bankfull elevation is identified by scour lines, vegetation limits, changes between bed and bank materials, the presence of flood-deposited silt or abrupt changes in slope. This involves subjective judgement; however, training and experience will lead to remarkably consistent interpretations. Section 7.2.5 contains a discussion of bankfull definitions, and a video for identification of bankfull level has been developed by the US Forest Service (USFS, 1995).

If a flood has recently passed through a reach, it may be of interest to estimate the peak discharge from *high water marks* or *trashlines*. These are indicated by telltale signs such as flattened vegetation, truncation of lichen growth, debris (driftwood, small seeds, strands of dried grass, plastic bags, etc.), stain lines on walls and/or coarser bedload sediments deposited at a well-defined elevation along the bank. Flood marks should be interpreted with caution; for example, water will 'pile up' on the upstream face of walls or trees and rain can wash marks away. The slope-area method for estimating discharge from high water marks is given in Section 5.6.6.

### 5.4.3   Channel Slope and Thalweg Profile

When measuring channel slope it is important to differentiate between the slope of the streambed and that of the water surface, as the two are not necessarily the same. Also, depending on the detail of the study, the localized slope of a water surface over a rock or a pool may be measured rather than the average slope over a reach.

The average slope of the water surface or channel bed can be surveyed by taking measurements at the upstream and downstream ends of a reach. If the channel has pool-riffle or pool-step sequences, slopes can also be measured separately in each type of channel form. A sample of how slope is measured in a stream reach with pool-riffle topography is shown on the data sheet in Figure 5.11(a).

The *thalweg* profile, or the path of the deepest thread of water, can be surveyed by having the staff bearer follow this path downstream. Similar to the procedure for cross-sectional profiles, measurements of channel bed and water surface elevation should be taken at distinct vertical breaks, including those created by logs, boulders or dams, and at changes in direction of the flow path. The data can be plotted as a longitudinal profile, showing both water and streambed elevations. Newbury and Gaboury (1993a) recommend surveying the thalweg, water surface, top of streambank and terrace elevations on both sides of the channel and plotting them together with observations of bankfull levels.

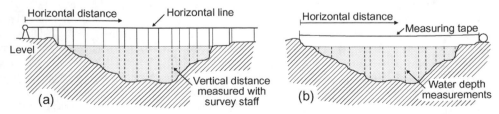

***Figure 5.13.*** *Surveying stream cross-sections: (a) with a level and staff and (b) with a measuring tape and rule, using the water surface as a horizontal line. Measurements are taken at 'breaks' in the slope of the channel bed and at the edges of the stream*

### 5.4.4  Bed Surface Materials

On the second field data sheet shown in Figure 5.11 spaces are provided for recording measurements of rock sizes to estimate channel bed resistance. About 25–30 rocks may be sampled by walking upstream and measuring the larger rocks at each pace or two (the distance is adjusted appropriate to the reach length) (Newbury, 1984). Measurements are normally made along one intermediate axis or all three axes, as described in Sections 5.7 and 5.8. Again, the number of rocks sampled is a problem in sampling design, with more rocks needed where sizes are more variable. For bed roughness sampling, boulders and logs that offer resistance to the flow should also be included (Newbury and Gaboury, 1993a).

The measurements can be used to describe the surface streambed materials in terms of particle shape to estimate streambed roughness coefficients and to estimate whether these particles will move at different flow levels.

### 5.4.5  Mapping the Stream Reach

Maps of a stream reach can provide a useful store of information from which changes in stream morphology can be documented and the extent of various habitat types determined. Mapping can depict habitat features that are not adequately described by gross stream measurements such as depth and average velocity. These might include specific habitats such as larger boulders, areas of spawning gravels, or "hot spots" such as where streams impact cliff faces. Maps can be constructed in the field by freehand sketching or plane table mapping or in the office by plotting survey data by hand or computer. Copies of aerial photographs and/or topographic maps can also be taken into the field and details such as the location of sampling sites or unmapped features can be added in.

A freehand sketch can provide a wealth of information about a stream reach and its surroundings. Just the process of drawing will force people to become better observers, and notes made in the field are often valuable when sifting through data back in the office. The drawing scale should be adjusted to fit the stream reach onto the chosen size of paper. For example, for a 300 mm standard sheet, a 1 m wide stream could be mapped at a scale of 1:60 and a 10 m wide stream at a scale of 1:600, with 50 mm for margins. A sketch might include the location of benchmarks and cross sections (x.s.), riffles and pools, eddies and backwaters, large woody debris, bedrock outcrops, gravel bars, abandoned channels, undercut or trampled banks and visual assessments of vegetation type (aquatic and riparian) and cover, substrate size and the character of surrounding slopes. Arrows can be added to depict flow directions. The scale, orientation, date and names of field personnel should be included on the map. An example sketch is shown in the second data sheet of Figure 5.11.

Another method of mapping is to use a longitudinal transect. A tape is extended down the axis of the stream, as shown in Figure 5.14(a), and the distance from the transect line to the bank, thalweg or other feature is measured from it at right angles. The right angle can be fixed by carrying a plastic drafting triangle, or simply estimated, depending on the precision required. A compass reading should be taken along the direction of the transect tape each time it is moved so the plotted map can be oriented properly.

With a level, theodolite or EDM, the distances and angles are measured to various points in the stream reach,

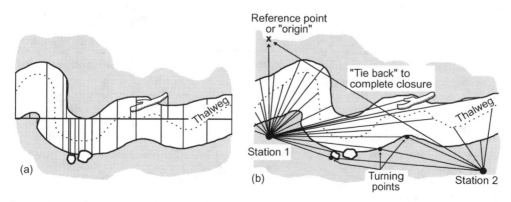

**Figure 5.14.** Mapping the stream reach using (a) longitudinal transect or (b) a level, theodolite, total station or plane table

as illustrated in Figure 5.14(b). Each set of measurements is recorded with a label (e.g. 'beginning of snag', 'end of pool'). These data, together with a freehand sketch of the reach, are later used to construct the scaled map. The same procedure is used with a plane table, although the map is constructed on-site.

If a sufficient number of horizontal and vertical measurements are taken in a stream reach a topographical map of channel topography can be developed by drawing contour lines between points of equal elevation. This information can also be depicted by plotting 'isobaths', lines of equal water depth which correspond to a particular stage. In this manner the distribution of shallow and deep areas can be illustrated.

## 5.5    Measurement of Water Level or Stage

### 5.5.1    General

Existing water levels and estimates of bankfull conditions in streams can be obtained from cross-section measurements; however, water level measurements are needed over a period of time in order to more fully describe the hydrologic environment. This monitoring can be extended to include measurement of groundwater levels.

The water level in reference to some arbitrary datum is called the *stage*. Periodic measurements of stage in streams are converted to streamflow values with the use of a stage-discharge relationship (Section 5.6.5).

### 5.5.2    The Staff Gauge

A *staff gauge* is simply a graduated staff which is installed vertically at a relatively stable site on the streambank (see control section, Section 5.6.5), e.g. upstream of a culvert or bedrock outcrop. Several staff gauges may be installed, progressing up the bank at fixed intervals of elevation, as shown in Figure 5.15. The gauge should be located where wave action and turbulence is minimized, and where it will be protected from damage by boats, vandals and flood-borne debris. It may be fixed to a wall, bridge pier, post, boulder, etc., as long as part of the scale is immersed at the lowest expected water level. The gauge should be located where future erosion and deposition will be minimal. Locations near stream bends, in pools or steep reaches are particularly vulnerable. Maintenance of staff gauges involves cleaning, replacement and checking the datum of the gauge scale.

Staff gauges must be read manually. At flood stages, binoculars may be helpful. Daily readings of stage form the basis for many stream-gauging records, older records in particular. They can also provide useful information before permanent stream-gauging locations are established.

### 5.5.3    Maximum Stage Recorders or Crest Gauges

Unless one lives next to the gauging station, it will not always be possible to manually record peak stages during floods. An estimate of the maximum water level can be obtained with a *maximum stage recorder* or *crest gauge*. Somewhat like a max-min thermometer, some type of substance floats upwards when the stage rises, and then remains in place when the water level drops again. Cork grains, coffee grounds and plastic or styrofoam chips are inexpensive substances for this purpose.

A length of transparent plastic pipe can be used to contain the floating material. Depths are marked on the pipe, corresponding to the same datum as the staff gauge. The pipe is set vertically in the streambank and perforated at the base to allow water to enter. To keep debris out of the pipe, the holes can be covered with a screen. The top of the pipe is also capped, but not sealed airtight. When a reading is taken, the height of the floating material in the pipe is recorded and the gauge 'reset' by washing the material back down inside the pipe. Shaw (1988) describes a standard crest gauge made of a 55 mm steel tube, with a removable graduated rod inside. The US Geological Survey standardized crest gauge is a capped, perforated two-inch galvanized pipe, with a redwood or aluminium shaft that is pulled out, the cork adhering to it (USGS, 1977).

### 5.5.4    Automatic Recorders

Permanent gauging stations are typically equipped with *water-level recorders* allowing the continuous recording of stage. A *stilling well*, connected to the stream by one or more pipes, is used to 'still' wave action and thus provide a more stable reading of stage. In most installations a float sits on the water surface and its vertical movements are recorded digitally or by an ink trace on a chart. At some stations digital readings are telemetered to a base station for near-instantaneous access.

Bubbler gauges are another method of measuring flow depth. In these gauges, nitrogen gas is bubbled slowly out of a tube, and the pressure required to force out the gas is directly related to water depth. Pressure transducers can also be used to measure hydrostatic pressure.

**Figure 5.15.** *A series of staff gauges for measuring water level over a range of elevations. Photo: N. Gordon*

Continuous readings from a stage recorder can provide useful insight into daily and seasonal fluctuations. To obtain good-quality data the gauge should be carefully located, checked frequently (once every week or two) and records processed in a timely manner and compared with other gauging records during events.

### 5.5.5  Depth to the Water Table: Piezometry

Piezometers are useful for monitoring the profile of the water table near a stream to establish whether it is influent (feeding the groundwater) or effluent (being fed by groundwater). A piezometer is simply a piece of pipe, driven vertically into the ground to form a 'well'. Water levels in the piezometer(s) can be measured and related to the water level in the stream to determine the direction of water movement.

The length of pipe should cover the range of water table levels expected. 1–4 m of 13–25 mm i.d. PVC pipe is a workable size. A hole just slightly larger than the pipe diameter is first drilled into the soil with an auger or by driving a metal pipe with a fence-post-type driver. The piezometer is placed such that the lower end is below the lowest expected groundwater level. A cavity of 70–100 mm in length should be left below the pipe. To keep the pipe from plugging with soil, a wooden rod can be inserted inside the pipe when it is installed, and then removed. The cavity should be flushed out by lowering a plastic tube to the bottom of the pipe and gently siphoning water into the piezometer. Loosened soil will be carried out of the top of the pipe when it overflows.

The response rate of the piezometer is tested by observing the rate at which the water level drops after water is added to the pipe. In sands and gravels the rate will be so great that no overflow will occur during flushing; in clays, the drop will just be perceptible. To improve response times in clay soils, the piezometer can be installed in sand lenses (layers) if any can be located. Piezometers should be capped (not air-tight) to keep out dust, leaves and insects. They should be flushed and re-tested periodically.

The piezometer water level is allowed to come to equilibrium with the groundwater level before measurements are taken. Depth from the top of the pipe to the water surface can be measured by lowering a chalked tape into the piezometer (the chalk washes off to indicate the water level). Reeve (1986) also gives instructions for making a 'bell sounder' (also called a 'fox whistle'), a weight with a hollowed base which makes a sound upon contact with the water. Another alternative given by Reeve (1986) is to lower a length of plastic tubing, marked at length increments, until it encounters the water. Contact can be discerned by blowing into the tube and listening for the sound of bubbles.

The heights of the stream water surface and the tops of all piezometers should be tied back to a reference datum by surveying (see Section 5.2). If two piezometers are installed, one close to the stream and another some distance away, the relative heights of the water table will indicate whether the stream is influent or effluent. In an influent stream, shown in Figure 5.16, the water level will be highest near the stream. The converse would be true if the stream were effluent.

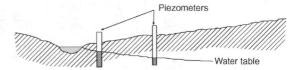

*Figure 5.16.* Piezometers installed next to an influent stream

## 5.6    Measurement of Discharge (Streamflow)

### 5.6.1    General

Stream discharge is the rate at which a volume of water passes through a cross section per unit of time. In the SI system it is usually expressed in units of cubic metres per second ($m^3$/s), although very small flows might be recorded in litres per second (L/s). It should be remembered that the measured discharge includes not only water but any solid dissolved or suspended in it. Especially during flood flows, this may affect discharge totals by a significant amount.

Methods of measuring streamflow generally fall into four categories: (1) volumetric measurement; (2) methods involving some measure of average stream velocity and cross-sectional area; (3) dilution gauging methods using a salt or dye and (4) stream-gauging methods where water-level measurements are made at natural sites or artificial control sections such as weirs, and a stage-discharge relationship is established. The selection of a good measurement site requires careful planning and evaluation. Volumetric measurements are most appropriate for small flows, dilution gauging for turbulent flows and artificial structures or natural control sections for permanent gauging sites. The velocity-area method using current meters is the recognized standard for direct discharge measurement. Streamflow can also be estimated indirectly from channel characteristics. Indirect estimates are commonly made after a flood has passed, or for estimating flows at a specific level such as bankfull. Stream-gauging procedures are given extensive treatment in manuals developed by BS (1973), ISO (1983), USGS (Corbett, 1962) and WMO (1980). Herschy (1978, 1985) also provides a summary of streamflow measurement methods, and the USGS has recently developed a CD-ROM with video clips showing discharge measurement procedures (Nolan and Shields, 2000).

### 5.6.2    Volumetric Measurement

Volumetric measurement is the most accurate method of measuring very small flows such as those from a spring. It involves measuring the time taken to fill a container of known volume. Volumetric measurements are usually made where the flow is concentrated (for example at a weir, overfall or pipe). A small temporary dam may be built with a trough or pipe for collecting the streamflow. In this case the water level behind the dam should be allowed to stabilize before a sample volume of water is collected.

Discharge ($Q$) is simply calculated as

$$Q = \frac{\forall}{t} \qquad (5.15)$$

where $Q$ = discharge in m³/s (or L/s), $\forall$ = volume in m³ (or litres) and $t$ = time (s).

The container should be calibrated by putting graduated marks on the side so that readings can be readily taken in the field. Alternatively, the container can be weighed in the field and the weight of the water (after correcting for container weight) is related to volume $\forall$ by dividing it by density (see Table 1.2):

$$\forall (m^3) = \frac{\text{Weight of water (kg)}}{\text{Density of water (kg/m}^3)} \qquad (5.16)$$

Several measurements should be taken to check for consistency of both the streamflow and the measurer.

### 5.6.3  Velocity-area Method

This method requires measurement of the area of a stream cross section and the average stream velocity. Discharge is then calculated as

$$Q = VA \qquad (5.17)$$

where $Q$ = discharge (m³/s), $V$ = average velocity (m/s) and $A$ = cross-sectional area of the water (m²). Area is calculated from cross-section measurements. Velocity measurements may be made using a current meter or by observing the rate of travel of a float or dye.

*Floats*

Water velocity can be measured by observing the time required for a floating object to traverse a known distance downstream. This method is appropriate for coarse estimates of discharge, particularly during floods, and requires little time or equipment. The 'float' may be a specially designed surface or sub-surface float, a chunk of ice, drifting branches or logs, coffee stir sticks, half-filled bottles or oranges. Even leaves can be used in emergencies, although these are apt to be affected by wind. Ideally, the float should move with the same velocity as the water

just below the stream surface. Objects such as oranges, orange peels or ice cubes which are only slightly buoyant are preferred as they are less sensitive to air disturbances. A fishing line can be attached for the retrieval of floats or they can be caught in a net.

A suitable, straight reach with a minimum amount of turbulence should be chosen, and an interval selected, measured and marked on the bank at each end by pegs, trees or rocks. These marks should be far enough apart to allow a travel time of at least 20 s. The interval should also overlap one or more surveyed cross sections to determine the cross-sectional area.

A float is introduced a short distance upstream of the reach so it can reach the speed of the water before passing the first mark. A stopwatch is used to measure the time of travel between the marked sections. In larger streams (greater than 10 m width) the cross section should be divided into three or more sub-sections. Floats are introduced at the midpoint of each section. Several runs should be made to obtain an average.

The surface velocity ($V_{surf}$) is calculated for each section as

$$V_{surf} = \frac{L}{t} \qquad (5.18)$$

where $L$ = measured reach length (m), $t$ = travel time (s) and $V_{surf}$ is in metres per second. Since the surface velocity is higher than the mean velocity, $\bar{V}$ (see Section 6.6.3), a correction coefficient ($k$) must be applied:

$$\bar{V} = kV_{surf} \qquad (5.19)$$

The correction coefficient generally ranges from 0.80 for rough beds to 0.90 for smooth artificial channels. However, in mountain streams, Jarrett (1988) calculated a value as low as 0.67. The commonly used coefficient is 0.85.

Discharge is calculated from Eq. (5.17) or, if more than one sub-section is used, from Eq. (5.20). Under favourable conditions and with repeated observations, float measurement may be accurate to within an error margin of 10%. In non-uniform sections or where wind is excessive, measurements may be in error by 25% or more.

*Dye*

Coloured dye can be poured into the water and its movement timed, as for floats. Fluorescein dye is useful for this purpose. The major disadvantage with this technique is the inaccuracy in identifying the middle of the mass of dyed water. However, as compared to the float method, a measure of the average stream velocity is obtained rather than the surface velocity. Hence, a correction factor is not necessary.

**Figure 5.17.** *Electromagnetic current meter, showing readout display unit and base of top-setting wading rod. Photo: N. Gordon*

### Current Meters

Current meters may be used in streams of almost all sizes. In larger streams with swift waters and debris flows these meters are larger with heavy attachments. There are three types of current meters commonly used in the measurement of streamflow: (1) propeller, (2) cup and (3) electromagnetic. A listing of current meters is provided by Hamilton and Bergersen (1984).

The first two types of meter work somewhat like a wind gauge, where the fluid moving past a vane causes it to rotate and the velocity is calculated from the rate of rotation. The *propeller-type meter* has a horizontal axis rotor. Propellers come in various sizes for operation over different ranges of velocity. These meters generally disturb the flow less and are less likely to become entangled with debris than the *cup-type meters*. These have a vertical axis rotor and are typically more sensitive at lower velocities. However, they will also rotate when buffeted by non-horizontal currents, leading to error in turbulent sections. A recent development mentioned by Jarrett (1988) is a meter with solid cups which reduces this error.

With the rotary-type meters the number of revolutions is counted either by some type of recording apparatus or by listening to clicks heard through headphones. For low speeds (and as a backup system) a mark can be made so that revolutions can be visually observed. Counts are made over some time interval, which is measured with a stopwatch, and the rotor speed is calculated as the number of revolutions per second. Current meters are normally supplied with a rating curve or table giving the relationship between rotor speed and flow velocity. Modern instruments may have direct readouts which display velocity. Maintenance includes keeping the meters cleaned and lubricated, and, when necessary, re-calibrated.

*Electromagnetic meters* (Figure 5.17) work on the principle that a conducting fluid moving through a magnetic field will induce a voltage (White, 1986). Since salty water is a high-conductivity fluid, the meters are commonly used in oceanography. More sensitive instruments are needed for measurements in fresh water, and this type of meter may be impractical for extremely pure waters. These meters provide a direct reading of velocity. They are also very durable and can be used in situations where rotary meters cannot be operated such as within clumps of vegetation. The bulb portion should be kept smooth and polished as nicks can cause flow separation.

Current meters should be re-calibrated on a routine basis or whenever poor performance is suspected. Commercially, this is accomplished by using a towing tank, in which the meter is attached to a carriage and towed at a set speed. This procedure can be roughly imitated by towing the meter through a pool of still water. A standard length would be measured off alongside the pool and the person holding the meter would walk or run over the length while being timed. This procedure would be repeated at varying travel speeds to check the calibration.

Current meters can be operated from a bridge, boat, cableway or by wading. The type of current meter chosen should be appropriate for the size and velocity of the

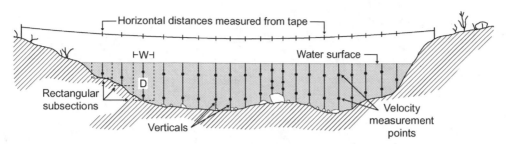

**Figure 5.18.** *Definition of terms used in computing discharge from current meter measurements (see text). Note variable spacing of verticals*

stream. Most standard meters require a minimum flow depth of 0.15 m and velocities over 0.10 m/s. A fairly straight reach should be chosen in which flow depth, width, velocity and slope are relatively uniform. Sites should be avoided which have extreme turbulence, upstream obstructions, eddies, dead-water zones, divided channels or regions where the flow path is noticeably curved. A reach which is slightly contracting in the downstream direction is preferable to one which expands outwards. Reaches can be 'improved' by moving rocks to make conditions more uniform and to eliminate backwater areas at channel edges.

After the cross section has been selected, a measuring tape or *tagline* is strung across the section perpendicular to the flow. The stream is divided into sub-sections within which velocity is measured along *verticals*, shown in Figure 5.18. Each sub-section should have roughly the same amount of flow; thus verticals will be spaced more closely if the water is faster and deeper. Additional verticals are added where sudden changes in depth or velocity occur. Hydrographers calibrating stream gauging stations typically use a minimum of 20 verticals (Braken-siek *et al.*, 1979; Goudie, 1981; WMO, 1980). This is an excessive number for most surveys for biological purposes, especially on smaller streams. A practical guideline is to use about one vertical per metre of channel width, with more if the section is irregular and less if it is uniform.

In the measurement of streamflow it is the *mean velocity* within each vertical which is of interest. Under the assumption that water velocity varies logarithmically from zero at the streambed to a maximum near the stream surface, the mean occurs at about four-tenths of the depth, as measured upwards from the streambed. Most texts and hydrographers refer to this as 'six-tenths depth', in which case it is measured downwards from the water surface (0.6*D*), and this will be the convention adopted here.

For measurements in small, uniform streams or where ice cover is present, or when time is limited, the single measurement at six-tenths depth (0.6*D*) is used. Because velocity profiles are often not exactly logarithmic, it is preferable to take two measurements at 0.2*D* and 0.8*D*, again measured from the water surface downwards. As a rule of thumb, the uppermost (0.2*D*) velocity should be greater than the 0.8*D* velocity, but less than twice as great. Three measurements should be taken if the velocity profile is distorted by overhanging vegetation contacting the water or by large submerged objects. Even more measure-ments can be taken for more precision or for developing a velocity profile. If for some reason it is difficult to lower the meter into the water a surface velocity measurement can be made and adjusted as for the float method. These depth settings and equations for calculating mean velocity at a vertical are summarized in Table 5.1.

For most stream survey work the meter is attached to a graduated wading rod, with a sliding mechanism that allows for depth adjustment. A *top-setting wading rod* can be purchased which greatly facilitates setting the current meter at six-tenths depth.

To take readings the meter is set to the appropriate depth and then placed in the stream, with the rod held vertically and the meter facing upstream. The observer stands as far to the side of the meter as possible and slightly downstream. The meter should be allowed to stabilize before readings are taken. If clicks are counted by ear, a stopwatch is started with the first click, counted as 'zero', not 'one'. For rotary meters, measurement time is typically between 30 and 60 s. Enough time should be allowed to count 40 'clicks'. For high velocities, as little as 10 s may be adequate.

At each vertical, the horizontal distance, the water depth and the current meter reading(s) should be recorded. To define the cross-section additional measurements should be recorded at the water's edge, the deepest point and any other points where the topography changes (see

*Table 5.1.* *Applicability and averaging procedures for determining mean velocity at a vertical. D is the vertical distance between the water surface and the streambed, and is measured downwards from the water surface. Adapted from Goudie (1981) and WMO (1980)*

| Number of points in vertical | Distance of measurement (from water surface) | Application | Equation for mean velocity in vertical $\bar{v}$ |
|---|---|---|---|
| 1 | 0.6$D$ | When $D$ is small ($<0.5$ m) or when a measurement must be made quickly. | $\bar{v} = v_{0.6}$ |
| 2 | 0.2$D$ and 0.8$D$ | This method is preferable to the single 0.6$D$ method if the size of the meter allows both measurements (normally where $D > 0.5$ m) | $\bar{v} = 0.5(v_{0.2} + v_{0.8})$ |
| 3 | 0.2$D$, 0.6$D$ and 0.8$D$ | Where irregularities distort the velocity profile and the stream depth is sufficient | $\bar{v} = 0.25$ $\times (v_{0.2} + v_{0.8} + 2v_{0.6})$ |
| 1 | Just below water surface ($\sim$0.6 m, or lower to avoid turbulence) | In swift streams or in high flows when it is difficult to lower meter into the water | $\bar{v} = kv_{\text{surface}}$, where k depends on shape of velocity profile, normally taken as 0.85. $k$ is best determined by correlating surface and 0.6$D$ volocities |
| Many | A range of depths, including 0.2$D$, 0.6$D$ and 0.8$D$. | When high precision or the shape of the volocity profile is of interest. | $\bar{v}$ is determined by integrating the area bounded by the profile and dividing by $D$. |

entry under 'present discharge' in the first data sheet of Figure 5.11). If the angle of the flow across the tape differs appreciably from 90° the angle should be recorded and the reading corrected by multiplying by the cosine of the angle.

To calculate discharge, the current meter readings are first converted to velocities, if necessary, and averages for each vertical are calculated using the equations in Table 5.1. The mean velocities and areas for each sub-section are multiplied together and then summed to obtain the total discharge. One way to do this is to assume that the sub-sections are rectangular, defined by the depth at each vertical and divisions halfway between each vertical, as shown in Figure 5.18. The 'lost' discharge in the triangular areas at the edges is usually assumed negligible. Discharge ($Q$) is then calculated as

$$Q = w_1 D_1 \bar{v}_1 + w_2 D_2 \bar{v}_2 + \cdots + w_n D_n \bar{v}_n \quad (5.20)$$

where $w$ is the sub-section width in metres, $D$ is the depth of the vertical in metres, $\bar{v}$ is the average velocity at each

vertical (m/s) and $Q$ is in m³/s. This method simplifies hand calculations.

Because of the variation in velocity within a stream, the accuracy of discharge measurement depends largely on the number of points at which velocity and depth readings are taken. Turbulence will also create 'noise' in the readings, although it is assumed that this error is random and will not cause bias with the large number of measurements taken. Most gauging authorities cite accuracies of plus or minus 5–10%. In turbulent mountain streams, however, Bren (personal communication, 1988) estimated an error of about 25% in current meter readings. During rapidly rising flood stages the error will be even higher. WMO (1980) gives recommendations for measurement of quickly changing flows such as flash floods.

### 5.6.4  Dilution Gauging Methods

Dilution gauging methods involve introducing a chemical 'tracer' substance such as a salt or dye into the stream and then monitoring changes in its concentration at some point downstream. These methods are especially useful in highly turbulent rock-strewn streams which provide rapid, complete mixing of the tracer and make conventional current metering difficult. Several such methods have been developed and are reviewed by Gregory and Walling (1973). Church (1974) and Finlayson (1979) also provide details on several of the techniques.

In addition to providing a valuable means of measuring total discharge in turbulent mountain streams and braided rivers, chemical gauging techniques also lend themselves to specialized discharge measurements. These include locating 'dead water zones' where there is little transfer of flow with the main stream, monitoring water flow through the hyporheos and tracing the downstream movement of nutrients or heavy metals. Zellweger *et al.* (1989), for example, compared discharge measurements made by tracer-dilution and current meter methods in a small gravel-bed stream. They concluded that a significant portion of the total stream discharge moved as underflow through the streambed. Triska *et al.* (1989a, b) also give examples of ecological applications. Additionally, it may be possible to use natural physical properties such as temperature, salinity, ion concentrations and sediment to study the origin of stream water (e.g. as baseflow) or the mixing of two merging streams.

The two most frequently used methods of dilution gauging are *slug-injection* and *constant rate-injection*. The types of tracers will be discussed first, followed by procedures for these two techniques. With either method the selected reach is preferably straight, with well-developed turbulence, and of sufficient length to allow complete mixing and dispersion of the tracer fluid throughout the stream water. This length can be surprisingly long. It is recommended that a dye such as fluorescein, a red powder which produces a green colour when dissolved in slightly alkaline water, be used to visually assess mixing.

### Tracers

A tracer is a substance that can be followed in its downstream course by monitoring some characteristic such as its concentration. There are three main types of tracers used: chemical, fluorescent and radioactive. Radioactive tracers require specialized training and will not be covered here. For hydrological purposes, the tracer should preferably be (Church, 1974):

- Highly soluble in water at stream temperatures;
- Stable in the presence of light, sediment or other substances found in natural waters;
- Easily detected at low concentrations;

- Absent from the stream itself or present in concentrations which do not interfere with measurements;
- Non-toxic to stream biota and without permanent effect on water quality;
- Relatively inexpensive.

Chemical tracers are commonly electrolytes or solutions of specific ions (e.g. chloride). For small-scale field studies in streams of low salinity, common table salt (NaCl) is an inexpensive tracer. The salt is dissolved in water and the electrical conductivity of the solution is measured before introducing it into the stream. About one litre of 20% solution is needed for each cubic metre of discharge. If salt is added to a container of water until no more will dissolve, the concentration is approximately 20% (Finlayson, 1979). A well-cleaned large plastic rubbish bin is a suitable container for mixing up a solution of known volume.

Dilution gauging has been successfully used on larger rivers using Rhodamine WT, a dye which can be detected by fluorometric methods at very low concentrations. Another compound, Rhodamine B, is a known carcinogen and should not be used for water tracing (Smart, 1984). Wilson *et al.* (1984) is a source of more information on fluorometric procedures.

Detection of the tracer can be done in the field, as with an EC meter, a specific ion electrode or a fluorometer. Alternatively, samples can be collected and later analysed in the laboratory.

### Slug-injection Method

In this method a solution of known volume and concentration is added to the stream in one 'slug' or 'gulp'. A cloud of marked fluid is produced which continuously disperses as it moves downstream. Concentration of the tracer is monitored downstream as the 'wave' of marked fluid passes by. Background concentrations of the stream should always be measured before the tracer is introduced.

Discharge is calculated from integration (calculating the area under the curve) of the 'concentration hydrograph', as shown in Figure 5.19. The equation for computing discharge is (Shaw, 1988)

$$Q = 1000 \frac{\forall c_t}{\int_{t_1}^{t_2} (c - c_0) dt} \qquad (5.21)$$

where $\forall$ = known volume of tracer (L), $c_t$ = concentration of tracer in introduced solution, $c_0$ = background concentration of stream (may be negligible), $c$ = changing concentration of tracer measured downstream, $Q$ = discharge ($m^3/s$) and $t_1$ and $t_2$ are the initial and final

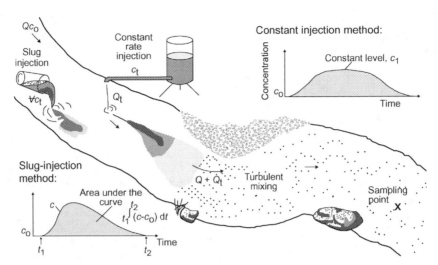

**Figure 5.19.** *Two basic methods of dilution gauging. Modified from Shaw (1988), by permission of Van Nostrand Reinhold International*

times of measurement (in seconds). The units of concentration must be consistent (e.g. mg/L), but the actual units are not important since they cancel each other in the equation. In practice, the area under the curve (the integration term) is measured by numerical methods (e.g. adding up sub-areas for each time step) or by drawing out the curve on graph paper and planimetering or digitizing the total area.

### *Constant-injection Method*

In the constant-injection method a solution of known concentration is fed into the stream at a constant rate. Downstream, the concentration of the tracer will rise and then stabilize at a constant value which can be related to discharge. The discharge is calculated from the equation (after Shaw, 1988)

$$Q = 1000\frac{(c_t - c_1)}{(c_1 - c_0)}Q_t \qquad (5.22)$$

where $c_1$ is the final, constant concentration of the tracer in the stream (see Figure 5.19), $c_0$ and $c_t$ are as defined for Eq. (5.21), $Q$ is in m$^3$/s and $Q_t$ is the injection rate of the tracer in litres per second (L/s).

The tracer is normally injected at a constant rate using commercially available pumps of high flowrate accuracy. Smith and Stopp (1978) also give instructions for a simple apparatus which siphons a tracer solution into the stream. To maintain a constant injection rate the operator must continuously adjust the height of the outflow end of the

siphon tube to maintain a constant elevation drop between it and the fluid level in the container. Researchers may also be able to purchase constant rate-injection tanks which make this continuous adjustment unnecessary.

### 5.6.5  Stream Gauging

#### *Choosing a Control Section*

To obtain the best possible accuracy when a staff or recording gauge is installed for monitoring streamflows over time, careful selection of a gauging site is crucial. Ideally, the beds and banks of the cross section should be stable to ensure a constant relationship between water depth and streamflow. The site should also provide good sensitivity in terms of the response of stage readings to changes in discharge.

The selection of a site requires the identification of a channel section with particular physical characteristics called the *control section*. The control section regulates and stabilizes the flow so that for a given stage the discharge will always be the same.

Controls can either be artificial (e.g. weirs) or natural. A *section control* exists at a cross-section which constricts the channel, or where a downward break in slope occurs (for example, at the brink of a rock ledge, a boulder-covered riffle or a weir). A *channel control* exists when the friction along a section of the stream controls the stage-discharge relationship. Controls may govern this relationship throughout the entire range of stage, or a compound

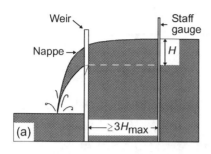 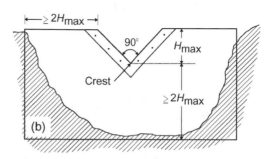

**Figure 5.20.** *Dimensions associated with V-notch weirs: (a) side view and (b) front view. $H_{max}$ is the maximum expected depth through the notch, and H is the depth of water above the bottom of the 'V', as measured in the upstream pond*

control may exist where the control changes. For example, a section control at low stages might change to a channel control at higher ones when the constriction, weir or break in slope is 'drowned out'. Thus, the flow profile is affected by the control section, yet the flow determines whether a section is controlling or not (Henderson, 1966).

The gauging site is situated upstream of the control section, in a relatively uniform reach. The reach should be straight and of fairly uniform slope, bed material and cross section for a distance upstream of about four to five times the channel width, and downstream for about twice the channel width. Improvements can be made to the site by re-arranging rocks, plants and debris to make the reach more uniform. The site should not be located just above a confluence, since varying discharges in the tributary may influence gauge readings through backwater effects. At permanent installations, the stream channel should also be deep enough to contain the larger flood flows.

Since accuracy will vary, the control section and gauging site should be chosen so that the rating curve is sensitive and stable over the region of maximum interest: high, medium or low flows. More information on control sections for gauging sites can be found in Corbett (1962).

### Artificial Structures

Pre-calibrated devices such as flumes and weirs are available for measuring discharge under a wide range of conditions. Their use is recommended when a permanent station is to be established for monitoring streamflows on smaller streams. The cost of construction may prohibit their use on large streams. Existing structures such as highway culverts, bridges, or concrete sills at road crossings or bridges may also be used for routine discharge measurements if a rating curve can be developed. Portable devices may also be useful for the measurement of small flows when current metering is impractical.

When selecting the type and size of the device it is important to consider the magnitude of peak flows, the amount of debris and sediment carried, icing conditions, straightness of the stream reach and soil and bed sediment conditions. Cutoff walls extending downwards and laterally into the earth are required at most sites to prevent flow from bypassing the measurement device. A full description is beyond the scope of this book, and the interested reader is referred to Ackers *et al.* (1978) and agency manuals on stream gauging.

Triangular or V-notch thin plate weirs are probably the most appropriate for work in small streams, as the V-shape allows very small flows to be measured accurately. The notch is machined with a sharp upstream edge so that water springs clear of the notch edge as it passes through (see Figure 5.20(a)). The angle of the 'V' is normally 90°, although weir designs of other angles have been developed (see Brater and King, 1976).

With V-notch weirs the 'V' should be situated some distance above the streambed. Water will pond behind the weir, which reduces and smooths the approach velocity. A control section exists at the notch, meaning that it controls the upstream depth. Measurements of this depth, as a vertical distance from the bottom of the 'V', are taken upstream of the nappe (Figure 5.20(a)). Alternatively, a staff gauge can also be installed on the upstream face of the weir, away from any drawdown effect from the notch.

Brakensiek *et al.* (1979) cite an equation derived from experimental work on 90° V-notch weirs, which relates upstream depth or head ($H$) to discharge ($Q$) as follows:

$$Q = 1.342H^{2.48} \qquad (5.23)$$

where $Q$ is in cumecs and $H$ is in metres. The equation is accurate to within 1% when $H$ is between 0.06 and 0.6 m. Volumetric measurement using the weir as an overfall can be used for even smaller flows or for calibration. The

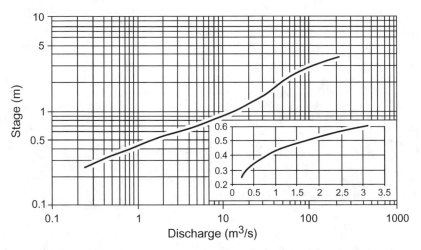

**Figure 5.21.** *Rating curve for the Acheron River at Taggerty. Inset shows lower part of the rating, plotted using linear axes. Data supplied by the Rural Water Commission, Victoria, Australia*

notch should be kept clear of debris so that depths are not affected. V-notch weirs are not appropriate for measuring flows with high sediment or debris loads.

A weir can be constructed out of metal, plywood, perspex or concrete, but the notch should be carefully machined using stainless steel. A guide for sizing the notch is that the depth of the 'V' should be smaller than 20% of the width of the approach channel (Dackombe and Gardiner, 1983). Other recommended proportions are shown in Figure 5.20. For best results, the approaching flow should have a Froude number (see Section 6.6) below about 0.5. As a permanent installation, a V-notch weir may have undesirable ecological side effects such as alterations to the site from sedimentation or scour and blockage of the channel to the passage of fish.

### Stage-discharge Relationships

If a site is visited regularly and simultaneous discharge and stream depth observations are made over a large range of discharges, a *stage-discharge relationship* can be developed. Discharges can then be determined by monitoring the water level. This is the technique by which most of the published data from gauging stations are derived (see Carter and Davidian, 1968).

The stage-discharge relationship is expressed as either a rating curve, a rating table or a rating equation. If the control section is a flume or weir, a standardized equation will be available. In natural streams a relationship must be developed by plotting measured stage against discharge

(Figure 5.21). The accepted procedure of plotting the dependent variable (discharge) on the $X$ axis and the independent variable (stage) on the $Y$ axis is somewhat unconventional, but makes the plot follow the direction of the rising and falling stages.

At gauging stations, curves are developed from a large number of data points, usually collected on the falling limb of a flood hydrograph. For environmental studies this precision may not be necessary, although it is still important to obtain data for ratings at high flows. Measurement errors are likely to be greatest in this range, due to high water velocities, rough, turbulent water surfaces and shifting bed and bank conditions. From data on several streams in the Western USA, Bovee and Milhous (1978) demonstrated that developing a stage-discharge relationship using only three points produced more reliable results than Manning's equation (Eq. (5.25)) when extrapolated within the range 40–250% of the calibration flow. They also showed that little improvement was gained by adding more than three points.

If the control section performs adequately and the cross-sectional profile is fairly regular, the resulting relation on linear axes will be smooth and parabolic, as follows:

and an equation can be easily fitted to the curve. However, if abrupt changes in the profile exist or if the control section becomes submerged at some stage, the resulting curve may look more like this:

In the latter case, it is better to express the stage-discharge relationship as a curve or a rating table from which values can be interpolated. A plot on arithmetic coordinate paper is preferable for this application. It is also better for plotting the lower part of the rating curve (inset of Figure 5.21) because it improves accuracy, and because the zero flow stage cannot be plotted on log-log paper.

The use of logarithmic graph paper is preferred in most cases since parabolic curves will plot as straight lines on log-log paper. This makes it easier to objectively extrapolate the curve to points beyond the observed range of stage heights, if needed, and to identify changes in slope, which may be due to changes in control. It also allows the fitting of a rating equation of the form

$$Q = a(h - z)^b \qquad (5.24)$$

where $Q$ = discharge (m³/s), $h$ = gauge height of the water surface (m), $z$ = gauge height of 'zero flow' (m) and $a$ and $b$ are coefficients. This equation can be fitted to the data by simple regression methods (see Appendix). It should be noted that the regression equation is fitted with $(h - z)$ as the independent variable and $Q$ as the dependent variable, even though the rating curve is plotted with the axes reversed.

The value of $z$ (stage height at zero flow) must be derived by trial and error. The 'true' value of $z$ is assumed to be the value which makes the stage-discharge relationship plot as a straight line on logarithmic paper. It is thus an 'effective' zero flow rather than the stage at which the channel goes dry. Successive values of $z$ can be assumed and the relation plotted. If the assumed value of $z$ is too small the rating curve will be concave upward; if it is too large the rating curve will be concave downward. The assumed value of $z$ is increased or decreased until a value is found which results in a straight-line plot (WMO, 1980).

If changes to the gauging section or the downstream control section occur, the stage-discharge relationship will also change. For example, if the width of the section increases, the coefficient $a$ will increase, causing the rating to shift to the right. If the section scours, $z$ decreases and the depth for a given gauge height increases, causing the relation to curve downwards on log-log paper. Temporary adjustments to the rating curve may be needed to account for vegetal or ice growth or other seasonal effects. WMO (1980) also points out that changes can occur from

summer holidaymakers who pile rocks in the control section to create a deeper pool for swimming. Conditions of *shifting control* are particularly common in sand bed streams.

Ideally, new stage and discharge measurements over a range of flow levels are taken to re-rate the section after a change occurs. More commonly, however, only one or two measurements are obtained. These measurements can be plotted as points alongside the original rating on arithmetic graph paper, and the whole curve shifted left or right so it passes through these points. Corbett (1962) and WMO (1980) both offer a more complete description of shifting control problems and adjustment of rating curves.

Leopold *et al.* (1964) proposed the use of a *dimensionless rating curve* after studies of river data suggested that rating curves from various stations were similar in form. Depth and discharge are both scaled by bankfull values to make the numbers dimensionless. This approach may be useful for deriving regional stage-discharge relationships for use at ungauged sites.

### 5.6.6  Slope-area Method of Estimating Discharge

At ungauged sites, peak discharges during flood events are often indirectly estimated after the fact from the height of water marks left behind. The same techniques can be used to estimate discharges for other water levels, e.g. bankfull depth or the depth at which the flow just covers the streambed in the main channel. Indirect methods can also be used to extend the upper part of a stage-discharge relationship. Herschy (1985) is a source of general information on these techniques.

The *slope-area method* is a commonly used technique for indirect estimation of discharge. Dalrymple and Benson (1967) provide a comprehensive description. Typically, Manning's equation is used for the calculation of discharge:

$$Q = \frac{1}{n} A R^{2/3} S^{1/2} \qquad (5.25)$$

where $Q$ = discharge (m³/s), $n$ = "Manning's $n$", $A$ = cross-sectional area of the flow (m²), $R$ = hydraulic radius (m), and $S$ = slope. These terms are explained further in Section 6.6.5, along with a discussion of the equation and its limitations.

For the method to be valid a reach must be carefully selected such that uniform flow conditions are approximated. This means that the width and depth of flow, the water velocity, the streambed materials and the channel slope remain constant over a straight reach, and further, that the channel slope and water slope are parallel. A straight, fairly homogeneous reach should be selected with a length at least five times the mean width (Dackombe and Gardiner, 1983).

The first step is to identify the water level of interest, for example, bankfull depth or recent high water marks. These levels can be flagged with fluorescent survey ribbon to define the water surface slope. Surveys should be made of at least three cross-sectional profiles (taken at right angles to the flow direction) and of the average bed and water surface slope. The inclusion of more cross sections and greater spacing generally minimizes some of the errors associated with this method (Jarrett, 1987).

Surveyed information is used to calculate the values of $A$ and $R$ in Eq. (5.25). The slope, $S$, is actually the slope of the energy line (Section 6.6.4); however, in practice, the energy slope is assumed to be parallel to the water surface slope and the bed slope. The more closely the reach approximates uniform conditions, the better the results. In highly turbulent sections and in steep streams, particularly those with a pool-step structure, this assumption may not be valid. Jarrett (1985), Jarrett and Petsch (1985) and WMO (1980) provide information on adjusting the value of $S$ for non-uniform reaches. The last variable is Manning's $n$, which is basically a composite factor which accounts for the effects of many forms of flow resistance, as will be discussed. Software is available from the US Geological Survey at http://water.usgs.gov/software for channel geometry analysis and calculation of Manning's $n$.

One of the greatest difficulties in applying this method is the accurate estimation of Manning's $n$. In general, $n$ increases with turbulence and flow retardance effects. In a reach where the slope is uniform and the roughness of the bed and banks is similar (e.g. an artificial channel), Manning's $n$ can usually be assumed to be a constant. However, in natural streams, it will often vary with flow depth, generally decreasing as the heights of protruberances such as snags or rocks become submerged by the flow. For example, Bovee and Milhous (1978) give data from a gravel-bed stream, Oak Creek, Oregon, demonstrating a variation in Manning's $n$ from 0.35 for flows less than 0.03 m³/s to 0.05 for flows greater than 0.84 m³/s. This trend may reverse, however, in streams where the channel bed is smooth and the banks are densely vegetated. Dawson and Charlton (1988) provide a source of information on vegetation-related resistance. For overbank (flood) flows, $n$ values should be evaluated separately for the main channel and floodplain. Discharge in these compound cross sections is computed by summing the discharge in sub-areas, as discussed in Section 6.6.5.

Estimates of Manning's $n$ can be made by choosing a value from a table such as Table 5.2 or by making a visual comparison with photographic keys, such as those provided by Barnes (1967) and Chow (1959). Barnes' book, produced by the US Geological Survey, contains colour pictures and $n$ values that have been verified at near-bankfull discharges from gauged readings. In fact, most of the published values refer to bankfull flows.

Another method given by Cowan (1956) provides a means of estimating Manning's $n$ by considering the individual effects of various roughness components. Cowan's method is suitable for small- to mid-size

**Table 5.2.** *Manning's n values for small, natural streams (top width at flood stage < 30 m). From Chow (1959), by permission of McGraw-Hill*

| Description of channel | Minimum | Normal | Maximum |
|---|---|---|---|
| Lowland streams | | | |
|   a.  Clean, straight, no deep pools | 0.025 | 0.030 | 0.033 |
|   b.  Same as (a), but more stones and weeds | 0.030 | 0.035 | 0.040 |
|   c.  Clean, winding, some pools and shoals | 0.033 | 0.040 | 0.045 |
|   d.  Same as (c), but some weeds and stones | 0.035 | 0.045 | 0.050 |
|   e.  Same as (c), at lower stages, with less effective slopes and sections | 0.040 | 0.048 | 0.055 |
|   f.  Same as (d), but more stones | 0.045 | 0.050 | 0.060 |
|   g.  Sluggish reaches, weedy, deep pools | 0.050 | 0.070 | 0.080 |
|   h.  Very weedy reaches, deep pools or floodways with heavy stand of timber and underbrush | 0.075 | 0.100 | 0.150 |
| Mountain streams (no vegetation in channel, banks steep, trees and brush on banks submerged at high stages) | | | |
|   a.  Streambed consists of gravel, cobbles and few boulders | 0.030 | 0.040 | 0.050 |
|   b.  Bed is cobbles with large boulders | 0.040 | 0.050 | 0.070 |

channels of hydraulic radius less than 5 m. Manning's $n$ is calculated from

$$n = (n_0 + n_1 + n_2 + n_3 + n_4)m_5 \qquad (5.26)$$

where the sub-factors $n_0$ through $n_4$ and $m_5$ separately account for the effects of various influences on $n$, as given in Table 5.3. Selected factors should apply to the entire study reach, not only to the cross-sections. The basic value, $n_0$, should represent the smoothest reach of that bed material type, and then adjustments are made using the other sub-factors (Jarrett, 1985). Under extremely rough conditions (e.g. steep mountain streams or during flood conditions or very low stages) even larger adjustments may be needed.

If the roughness of the banks is very different from the streambed (e.g. dense shrub growth on banks) a composite $n$ for the channel can be derived using a weighting method (Jarrett, 1985):

$$n = \frac{P_1 n_1 + P_2 n_2 + \cdots + P_m n_m}{P} \qquad (5.27)$$

where $P_1$, $P_2$, etc. are the amount of wetted perimeter with $n$ values $n_1$, $n_2$, etc. and $P$ is the total wetted perimeter for the section.

When using Cowan's method, care should be taken to avoid double-counting the contribution of one type of roughness under more than one component. This approach is still very subjective, although it at least forces one to

**Table 5.3.** *Component values for Cowan's method of estimating Manning's n. Sinuosity is defined in Section 7.3. Adapted from Cowan (1956) and Jarrett (1985)*

| Basic n value, $n_0$ | | Surface irregularity, $n_1$ | |
|---|---|---|---|
| Earth | 0.020 | Smooth | 0.000 |
| Rock | 0.025 | Minor (slightly eroded or scoured) | 0.005 |
| Fine gravel | 0.024 | Moderate (moderate slumping) | 0.010 |
| Coarse gravel | 0.028 | Severe (badly slumped, eroded | 0.020 |
| Cobble | 0.030–0.050 | banks or jagged rock surfaces) | |
| Boulder | 0.040–0.070 | | |
| | | | |
| *Variation in cross-section shape causing turbulence*, $n_2$ | | *Effect of obstructions (debris deposits, roots, boulders)*, $n_3$ | |
| Change occurs gradually | 0.000 | Negligible (few scattered | 0.000 |
| Occasional changes from | 0.005 | obstructions) | |
| large to small or side-to-side shifting of flow | | Minor (obstructions isolated, 15% of area) | 0.010–0.015 |
| Frequent changes | 0.010–0.015 | Appreciable (interaction between obstacles which cover 15–50% of area) | 0.020–0.030 |
| *Vegetation*, $n_4$ | | Severe (obstructions cover | 0.040–0.060 |
| None or no effect | 0.000 | >50%, or cause turbulence over most of area) | |
| Supple seedlings or dense grass/weeds | 0.005–0.010 | | |
| Brushy growths, no growth in streambed; grass height of flow | 0.010–0.025 | *Meandering (multiplier)*, $m_5$ | |
| | | Minor (sinuosity 1.0–1.2) | 1.00 |
| Young trees intergrown with weeds; grass twice depth of flow | 0.025–0.050 | Appreciable (sinuosity 1.2–1.5) | 1.15 |
| | | Severe (sinuosity > 1.5) | 1.30 |
| Brushy growth on banks, dense growth in stream; trees intergrown with weeds; full foliage | 0.050–0.100 | | |

observe and consider the various factors affecting Manning's $n$ at different water levels.

*Example 5.2*

(a) Calculate Manning's $n$ using Cowan's method and the following estimates of component values: $n_0 = 0.028$, $n_1 = 0.005$, $n_2 = 0.010$, $n_3 = 0.020$, $n_4 = 0.045$, $m_5 = 1.05$ and (b) calculate the discharge if $A = 5.2$ m$^2$, $R = 0.8$ m and $S = 0.03$.

(a) $n = (0.028 + 0.005 + 0.010 + 0.020 + 0.045)$
$\times (1.05) = 0.113$

(b) $Q = \dfrac{1}{0.113}(5.2)(0.8)^{2/3}(0.03)^{1/2} = 6.87 \, \text{m}^3/\text{s}$.

A 'calibration' value of Manning's $n$ for present conditions can be computed by working backwards from measured values of discharge and channel dimension. This value is then adjusted up or down as appropriate to estimate discharge at other levels, based on field observations of how roughness changes with water depth. One can rely on judgement and experience from evaluating Manning's $n$ at different discharges or by applying Cowan's method to make the adjustments. Based on a review of verified channel roughness data, Jarrett (1985) found that in uniform channels, the base $n$ value, $n_0$, does not vary with depth of flow over the range:

$$5 < \frac{R}{d_{50}} < 276$$

where $R$ is the hydraulic radius, as before, and $d_{50}$ is the median particle diameter of the streambed materials (Section 5.8). A value less than 5 for the ratio $R/d_{50}$ represents mountain streams with large bed materials and a value greater than 276 represents sand bed streams which may have significant variations in $n$ due to bedforms.

The slope-area method using Manning's equation has been verified on low-gradient streams of relatively tranquil flow with good results (Jarrett, 1987). For good hydraulic conditions, peak discharges can be estimated with an error of 10–25%, but under poor conditions errors of 50% or more may occur. A few measurements of channel geometry and a guess at Manning's $n$ from Table 5.2 or a picture book will be adequate for rough estimates. For more accurate predictions, one might want to consider recruiting someone experienced in selecting channel sections and estimating Manning's $n$—if not for the length of a research study, at least for a day or two of training. 'Getting acquainted' with streams on which Manning's $n$ has been verified is also helpful.

In an attempt to sidestep the issue of estimating Manning's $n$ altogether, Riggs (1976) gives a simplified slope-area method which does not require estimation of a roughness coefficient. His method is based on observations that slope and roughness tend to be related in natural channels and that hydraulic radius is closely related to cross-sectional area. From an analysis of the data from Barnes' (1967) book of verified Manning's $n$ values at bankfull conditions, Riggs developed the following equation, given in SI units as

$$\log Q = 0.191 + 1.33 \log A + 0.05 \log S - 0.056(\log S)^2$$
$$(5.28)$$

where $Q$ is in m$^3$/s, $A$ is in m$^2$ and $S$ is dimensionless. The standard error of the equation is about 20%. The equation only applies to conditions similar to those for which it was derived.

Jarrett (1987) reviews many of the problems of estimating peak discharge in mountain rivers using the slope-area method. He states that mis-application of the method in higher-gradient mountain streams (slopes over about 0.01) has tended to overestimate discharge, leading to erroneous 'record breaking' flood values. Jarrett recommends checking the average Froude number (Section 6.6), which will normally be below 1.0 in streams even during peak flows. A higher value may indicate that sources of energy loss have not been properly identified. Wohl (2000) mentions other indirect discharge estimation methods including flow-triggered video cameras to monitor water surface profiles during floods.

## 5.7 Substrates and Sediments: Sampling and Monitoring Methods

### 5.7.1 General

Information on the size and distribution of inorganic particles, whether in motion or forming part of the channel bed or banks, is often needed in studies of the ecology and hydrology of streams. Particles can be described in terms of shape, size, mineralogy, colour, concentration in the water column and orientation and degree of compaction in the bed. Samples of bank materials might be collected for determining soil texture, moisture content, the percentage of roots and other organic matter, and other factors related to bank stability. Samples of bed materials can provide information on surface roughness and benthic habitat type. Cores of deposited sediments in lakes and stream backwaters have been used to evaluate historical changes in the inputs of heavy metals and other substances.

Additionally, a researcher may wish to sift through samples of either bed and/or bank materials to count the number of invertebrates either in the whole sample volume or at different depths within the substrate. Sources of sediment, distribution of sizes transported and estimates of sediment yield can also be evaluated by sampling sediments in motion.

Methods of sediment sampling range from simply scooping up a handful or shovelful of material to more sophisticated coring and collection techniques. Sample sizes may range from 100 g of finer clays, up to hundreds of kilograms for coarser-grained sediments (Brakensiek *et al.*, 1979; Gee and Brauder, 1986). A larger sample size is needed to describe variable bed materials than uniform ones. The methods of sampling and analysis presented are those commonly accepted for soil and sedimentological work. Bunte and Abt (2001) provide a more detailed summary of methods used for bed material assessments in gravel and cobble-bed streams. The US Federal Interagency Sedimentation Project (FISP) offers information on sediment sampling techniques and equipment at http://fisp.wes.army.mil/.

### 5.7.2   Bank Material Sampling (Soil Sampling)

Streambank materials can be sampled to obtain information about bank stability and erosion potential. Soil samples are taken as cores or grab samples using conventional tools such as shovels or augers (spiral, tube or bucket)—either hand or power driven.

To obtain measurements such as texture, moisture content or frost depth, samples should be taken at several locations on the bank and/or where the soil type noticeably changes. The depth of sample taken is typically from 10 to 30 cm, but this will depend on the type of analysis. The representativeness of an augered sample will diminish as the size or number of stones in the sample increases.

If soil samples are to be analysed for water content they should be immediately placed in a leakproof plastic bag or soil can for transport to the laboratory. Sample containers should be labelled with the date, site and depth of sample.

### 5.7.3   Bed Material Sampling

#### *Sampling Strategies*

Bed material samples can be collected at different locations within a stream reach (e.g. riffles, pools, deposits behind logs or boulders, etc.), and at varying depths, depending on the purpose of the study. Often, the surface layer will contain larger particles than those in the sub-

surface, and the two zones have different implications in terms of sediment transport (see Section 7.4). Accumulations of fine sediments may be of interest in studies of spawning habitat. A detailed examination of sediment deposits downstream of boulders will use a different sampling design than the general characterization of a stream reach.

Samples can be collected from the surface by hand, or from different depths by collecting a volumetric sample of known size, e.g. as a core. If the stream is dry or partially dry, the researcher should decide how much of the stream width to sample; for example, in streams where bars are exposed, should the low flow or bankfull channel be sampled?

Surface bed materials can be described using the grid, transect and other methods described in Section 5.3. However, 'pebble counts' are a more popular approach, especially in wadable streams with cobble or gravel beds. Wolman's (1954) method, variously modified (Leopold, 1970, and Bevenger and King, 1995), involves walking along a sampling reach in a straight or zigzag pattern and stopping at a consistent interval such as one or two paces, to measure the rock encountered by the toe of one's boot. The standard method for streambed characterization involves measurement of at least 100 particles. Particles must be selected randomly; therefore the investigator must be careful not to selectively pick (or not pick) larger boulders or small grains.

A procedure for pebble counts in sub-surface materials has been developed by Buffington (1996), which involves first removing surface particles from one or more areas, and mixing the subsurface materials to the depth of a shovel blade before pebble counting at random or with a grid (Bunte and Abt, 2001).

Collected particles are measured by direct measurement methods given in Section 5.8.4. For embedded particles, Harrelson *et al.* (1994) recommend measuring the smaller of the two exposed axes.

In the study of the Acheron River (Figure 4.7), surface materials were sampled at all study sites using pebble counts with a sample size of 25–30 particles. At five sites, samples of surface materials were also collected by hand for particle size analysis. Roughly similar volumes were collected from four locations, two in pools and two in riffles, then combined. Rocks larger than 8 mm were graded in the field and weighed wet; the remainder of the sample was analysed by dry sieving and a laser particle sizer (Gordon, 1996).

#### *Manual Methods and Bed Material Samplers*

For most stream studies, samples of bed materials will most likely be collected *by hand*, e.g. for pebble counts.

For volumetric samples, a *shovel* and pry bar can be used where the water current is sufficiently low so that finer particles are not swept away. A wire grid or hoop can be placed on the bed surface prior to digging to define the area sampled.

*Cores* of bed materials can be isolated by several means. *Piston-type bed material hand samplers* can be purchased which are much like tube-type augers but have a piston which is retracted as the sampler is forced into the streambed. This creates a suction which holds the sampler in place. A *hollow cylinder* (or metal can or barrel) can also be driven into the bed materials to isolate a core. After digging down to expose one side, a thin metal plate is slid under the cylinder, assisted by a hammer. The cylinder is then pulled out with the plate held firmly against its base. Bunte and Abt (2001), Hamilton and Bergersen (1984) and Platts *et al.* (1983) describe a more sophisticated version of this method using a stainless steel McNeil-Ahnell hollow-core sampler, and the US Federal Interagency Sedimentation Project (http://fisp.wes.army.mil) has recently developed the US RMBH-80 hand-held rotary scoop sampler.

*Bed material samplers* have been designed for the collection of samples from the channel bed in large rivers. They are usually quite heavy and are operated from a crane on a boat or bridge. Typically, they have some type of scoop or sampling bucket which may be dragged along the channel bed or triggered on contact with the streambed to trap a sample of the surface bed material.

### Freeze Coring

One of the better methods for sampling bed sediments is *freeze coring*. As illustrated in Figure 5.22, a hollow probe is driven into the streambed and is subsequently filled with a cryogenic medium such as liquid nitrogen or liquid carbon dioxide ($CO_2$). Liquid $CO_2$ generally yields smaller samples, although it is less expensive. After allowing a prescribed period of time for freezing, the probe and the adhered core of frozen sediment is dug out or extracted with a hoist or jack.

The sample is removed from the probe by chipping with a rock hammer or by thawing. The thawing process can be expedited by running a blowtorch back and forth over the sample. Vertical stratification can be analysed by first laying the frozen core over a segmented box. Since benthic organisms will be collected in the frozen sample, freeze coring thus provides a way of observing their stratification within the streambed.

When using this technique, Marchant (personal communication, 1990) found that it was necessary to enclose the sampled area with a section of a metal drum to

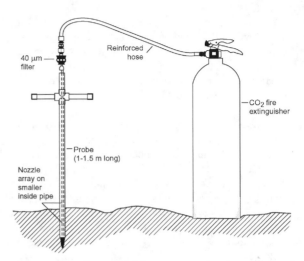

**Figure 5.22.** *Schematic of freeze corer, showing probe and liquid $CO_2$ source. Modified from Hamilton and Bergersen (1984)*

minimize the amount of bed material swept away by the current. He found the technique very effective for sampling particles up to the size of large rocks and for studying the numbers of invertebrates within the streambed.

Platts and Penton (1980) provide information on the construction and operation of freeze corers. Lotspeich and Reid (1980) and Everest *et al.* (1980) describe an improved version with three evenly spaced prongs built into a template. The size of the sample taken using this method, and thus representativeness, is increased substantially. However, the three-pronged sampler is more difficult to drive into the substrate since the chance of hitting a large rock is tripled, and it is more difficult to extract the larger sample.

### 5.7.4   Sampling Suspended Sediments

When sampling streams for suspended sediment it is important to obtain a sample which accurately reflects the stream's sediment load. Since the majority of sediment transport occurs during high flows it is essential that samples be taken during these periods in order to develop long-term averages. Samples must also be taken in such a way that the concentration represents an average for the cross section, as sediment concentration will vary with depth and across the section depending on the particle size (see Section 7.1.4). Sediment concentrations will also vary from one instant to the next as 'pulses' of sediment pass through a cross section, followed by gaps of lower concentration. In data collected by Horowitz *et al.* (1989) on the Cowlitz River, Washington (affected by the Mount St Helen eruption), samples collected at 20 min intervals showed an average deviation from the mean of 4%, with individual samples varying as much as 20% from the mean. Variability was found to be higher towards the centre of the stream than towards the edges.

It should thus be recognized that sampled concentrations will be extremely variable within a section and from one moment to the next, particularly during high flows. Therefore, numerous samples should be taken near the peak discharges to establish the error margin. Fewer are needed during periods of low or stable flow. Sampling schemes represent a compromise between precision and the time and cost of sampling.

Sampling sites will generally be the same as those used for stream-discharge measurements (Section 5.6). At stream-gauging sites samples should be taken upstream of stilling ponds which act as sediment traps. It is also important to obtain samples which correctly integrate the distribution of moving sediment within the channel. This can be accomplished by taking samples at several locations across a transect, as will be described. Alternatively, the mixing effects in turbulent sections created by artificial or natural structures can be exploited so that only one relatively homogeneous sample is needed.

If only the fraction smaller than 0.0625 mm (silts and clays) is important, a grab sample taken near the water surface in the centre of the stream is considered to be sufficient, since this fraction is assumed to be evenly distributed at a section. During high stages it may be necessary to take a surface sample because of difficulty in lowering the sampler. For streams in India, Singhall *et al.* (1977) related mean sediment concentration ($c_{avg}$) to surface concentration ($c_{surf}$) by the equation

$$c_{avg} = 2.353 c_{surf} \qquad (5.29)$$

The sample volume is adjusted according to the estimated concentration of sediment. Recommended minimum volumes for analysis of concentration by drying and weighing are given in Table 5.4. Filtration techniques are more sensitive, only requiring about one-tenth of the listed volumes. If samples are to be analysed for particle size much larger volumes are needed.

The site, date and time of collection (or numbering sequence in an automatic sampler) should be recorded on the sample container with either a waterproof marker or a soft lead pencil and waterproof label tape. For sediment sampling the type of container is not particularly critical as long as it is leakproof. A mark should be made to

**Table 5.4.** *Recommended sample volume for analysis of suspended sediment concentration by drying and weighing. After WMO (1981)*

| Expected concentration of suspended sediment (ppm) | Required sample volume (L) |
|---|---|
| >100 | 1 |
| 50–100 | 2 |
| 20–30 | 5 |
| <20 | 10 |

indicate the water level inside the container after sampling, to monitor leakage or evaporation during transport of the sample.

### Suspended Sediment Samplers

If the stream is sufficiently small and/or well mixed, a *grab-sample* of part or all of the flow can be collected with a cup or bucket. Suspended sediment samplers are, however, preferable. There are two basic types: depth integrating and point integrating.

*Depth-integrating samplers* are commonly used in smaller streams. These samplers, as described by Brakensiek *et al.* (1979) and Skinner (2000) are available in a range of sizes. They are normally made of cast aluminium or bronze and are streamlined, with tail fins to orient the sampler so that its intake nozzle points into the oncoming flow. Nozzles are available in a range of sizes to accommodate different particle diameters. Originally developed in the 1940s, the samplers were designed to hold glass pint milk bottles, which were cheap and readily available at the time. Plastic bottles can be used in some of the samplers, and Teflon nozzles can be installed to minimize contact of the sample with metal when water quality samples are collected.

The samplers are designed such that the velocity at the entrance of the nozzle is the same as the local stream velocity. They generally will only sample to within about 90 mm of the streambed; the unsampled zone is sometimes considered to be the region of bedload transport (Section 7.1.4). Depth-integrating samplers are designed to extract a sample continuously as the sampler is lowered at a constant speed (the 'transit rate') from the water surface to the streambed and back. This gives a velocity-weighted sample over the vertical since more water will enter the nozzle at depths where the velocity is greater. These samplers should only be used for sampling depths shallower than about 5 m, because of the compression of air trapped in the sample bottles (Skinner, 2000).

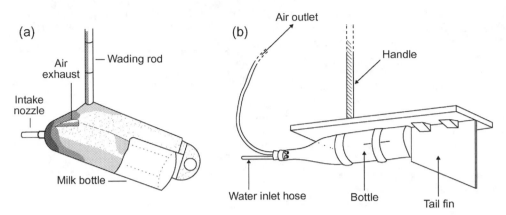

**Figure 5.23.** *(a) Hand-held depth-integrating suspended sediment sampler, US DH-48 model and (b) home-made depth integrating sampler*

In wadeable streams, the hand-held US DH-48 sampler, illustrated in Figure 5.23(a), is the standard tool of the trade. Finlayson (1981) gives a design for a home-made depth-integrating sampler, shown in Figure 5.23(b), which uses a carbonated beverage bottle. The bottle opening is fitted with a two-holed stopper, the lower opening has a straight metal or rigid plastic tube through which the sample enters the bottle, and a flexible hose for the escape of air is attached to the other. A handle and tail fin are also attached. Samples are transferred to storage bottles after collection.

When taking a sample with the US DH-48 or equivalent a person should stand off to one side and slightly downstream of the transect to avoid disturbing the oncoming flow or the bed sediments. Any sediments disturbed by wading should be allowed to settle before the sample is taken. The sampler is washed with stream water and the nozzle checked for clogging before starting.

The rod is held vertically and the sampler lowered at a constant speed until it just bumps the streambed. It is then raised, also at a constant rate. This transit rate should not exceed 0.4 times the stream velocity. Ideally, the bottle should be between two-thirds and three-quarters full. If it is not full enough, the sampler can be lowered and raised again. If it is too full, the entire sample should be dumped out and the vertical re-sampled. With practice, one can become fairly proficient at judging the speed of the sampler needed to fill the bottle two-thirds full in one pass.

Brakensiek *et al.* (1979) and WMO (1981) describe a sediment sampling method called the *equal transit rate*

*(ETR)* method. A measuring tape is first stretched across the stream, usually in conjunction with measurements of current and depth. The ETR method proceeds as follows:

1. The stream width is divided into six to ten equal-width sections. Fewer verticals will be needed in more uniform cross sections and at lower stages.
2. The same transit rate is used at all verticals, and the person proceeds from one vertical to the next until a sample bottle is sufficiently full (the last bottle used may be less than two-thirds full). The total number of bottles collected will vary from one cross section to another.
3. All of the samples from the cross section are composited to form a single representative discharge-weighted sample.

Methods of computing suspended sediment discharge from these measurements are given in Section 7.5.3.

*Point-integrating samplers* are similar in design to depth-integrating samplers but are equipped with a valve, end flaps or a door, which can be opened and closed by a spring or electronically to 'trap' a sample at a specific depth in the stream. A more basic approach is to simply uncap a bottle or jar at a specific depth and allow it to fill, although the escaping air bubbles may affect the size of sediment entering the bottle. Point-integrating samplers might be used to develop horizontal or vertical profiles of sediment concentration within a cross section

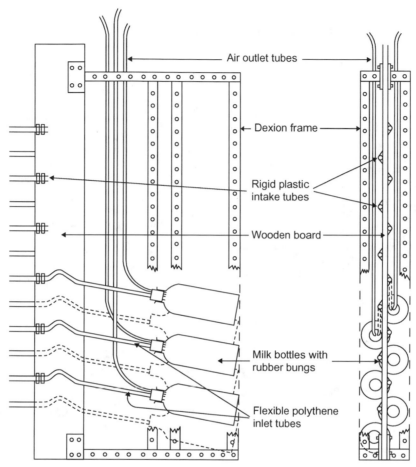

**Figure 5.24.** Rising-stage sampler, home-made design. Reproduced from Finlayson (1981), by permission of the Journal of Geography in Higher Education

or to investigate the conditions experienced by a fish just downstream of a boulder or inside a log.

### Rising-stage Sampler

A rising-stage sampler provides a means of automatically collecting suspended sediment samples from flashy, ephemeral streams. This type of sampler is useful for sampling peak concentrations at remote sites since it is usually difficult to be present when runoff peaks occur.

The rising-stage sampler works on the siphon principle and consists of a number of bottles arranged on top of each other in a frame. Each bottle is fitted with a two-hole stopper, one hole for the sample inlet and one for air exhaust. In the standard US U-59 (Brakensiek *et al.*, 1979), copper tubing of 4.8–6.4 mm inside diameter is recommended for the inlet and exhaust. Finlayson (1981) also gives a design using flexible tubing for the inlet and air outlet. The inlet tube is connected to a piece of rigid tubing, which is mounted on a board and pointed into the flow, as shown in Figure 5.24.

The 'kink' formed in the inlet tubing causes the siphon effect. As the water rises, sample bottles fill when the water level reaches the height of the 'kink'. Thus, samples will be taken just below the water surface. By using more than one sampler, samples can be obtained at several water surface elevations as the water rises.

The rising-stage sampler is most suitable for suspended sediment finer than 0.0625 mm (silts and clays). Larger particles are not sampled adequately with this system since intake velocities are not the same as streamflow velocities. Also, since the sample is taken near the water surface, the coarser particles which are less evenly distributed over the water depth will not be sampled adequately. These limitations should be considered when using concentration data obtained from these samplers.

### Automatic Pumping Samplers

Many different types of automatic pumping samplers are available. Each has its own advantages and disadvantages in terms of mechanical 'vigour', ease and expense of repair, run time on one battery charge and ability to extract a representative sample.

These samplers pump water from the stream, a portion of which is retained in a sample bottle. On most models the intake line is given an initial 'back flush' to minimize cross contamination between samples. The samplers can be set to take samples at a given time interval or attached to a flow rate or flow level meter to sample at a proportionately higher rate during higher flows. A float-controlled switch can also be used to activate the sampler when the water stage reaches some predetermined elevation.

For long-term records automatic pumping samplers are usually located near a streamflow gauging station. At these stations, an 'event marker' can be attached to the water stage recorder to stamp a mark on the chart when a sample is collected.

Bottles should be retrieved after floods or at intervals determined by the time it takes to fill all the sample bottles. Periodic servicing is also required to perform maintenance on the instrument and replace batteries. In areas where temperatures fall below freezing an insulated and/or heated shelter may be required.

Since the sampler intake nozzle is often located near a streambank the samples can give misleading impressions of average stream sediment concentrations. They should be calibrated against concentrations taken across the section and a correction factor applied.

### Turbidity Meters

Attempts have been made to develop methods for continuously monitoring suspended sediment concentrations, which usually rely on the assumption that water with a higher sediment concentration is less transparent. *Turbidity* is a measure of this optical property, which inhibits the transmission of light through a sample due to scattering and absorption. To measure turbidity a sample of the streamflow is usually pumped past a photoelectric cell. Attenuance-type meters measure the amount of light transmitted through the sample while nephelometric-type meters measure the amount of light scattered by the sample. These readings are calibrated against measured concentrations.

Turbidity measurements can be affected by factors other than particle concentration, such as the size distribution, shape and absorptivity of the sediment, and, in attenuance-type meters, the colour of the water. Gippel (1989) investigated both types of turbidimeters for the measurement of suspended sediment concentrations. He concluded that turbidimeters gave satisfactory estimates of storm-event sediment loads if sediment particle size was fairly constant and fell within the range 0.0005–0.01 mm mean diameter (fine silts and clays). An infra-red light source was also found to be superior to a visible light source since it was not affected by water colour or algal growth.

### 5.7.5  Sampling Bedload Sediments

Bedload is the material which generally remains in contact with the streambed and moves by rolling, sliding or hopping. The distinction between bedload and suspended load is discussed in Section 7.1.3. Bedload movement is important because of its relationship to changes in the bars and bends of a stream's morphology. It is also important to benthic stream biota which can be crushed or disturbed by the moving material.

Accurate sampling of bedload is difficult due to the movement of sediments in bars, ripples and dunes, changes in sediment supply, the 'stop and start' nature of sediment movement and the problem of efficiently sampling sediments over a range of sizes. It is important, then, to choose measurement sites where the sampled sediment load is representative of the amount and size of material being carried through the stream. For example, a site downstream from a pond where materials settle out would not be a good location. As with suspended sediments, the highest bedload movement (and in many streams, all of it) will occur during high flows, requiring monitoring during these periods. An adequate bedload sampling program (e.g. to develop rating curves, Section 7.5.4) can be very time consuming and expensive, and estimates made using simple monitoring methods and/or models are more common. Potyondy (1999) recommends a minimum of three years of continuous streamflow records and at least 20 bedload and suspended sediment samples over the range of flows (including those above bankfull) to develop a basis for determining instream flow needs for channel maintenance purposes.

There are several ways of obtaining estimates of the amount of bedload passing though a particular reach of the stream, including (1) bedload samplers, (2) tracer particles, (3) measurements of sediment accumulations behind structures or (4) rule of thumb estimations based on the suspended load and type of channel material.

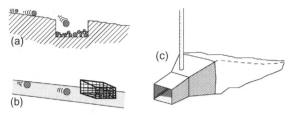

**Figure 5.25.** *Bedload samplers: (a) pit-type, (b) basket-type and (c) Helley Smith pressure-difference sampler*

### Bedload Samplers

Ideally, a bedload sampler should trap large and small particles with the same efficiency yet not alter the natural pattern of flow and sediment movement. An efficiency (ratio of sampled to actual transport rate) of 60–70% can be considered satisfactory (WMO, 1981). Samplers are generally of four types: pit, basket, pan and pressure-difference samplers. Hubbell (1964) reviews and compares several bedload sampler designs.

*Pit-type* samplers (Figure 5.25(a)) are excavated depressions, lined or unlined, for trapping bedload sediments. A classic example is the permanent installation on the East Fork River in Wyoming, described by Leopold and Emmett (1976). At this site, sediments fall into a trench-like pit across the streambed and are transported by conveyor belt to a storage area for weighing. On smaller streams, a box can be placed in the streambed so that its open top is level with the streambed surface. After a period of time, the material is removed from the box and measured for size and weight.

*Basket-type* samplers (Figure 5.25(b)) are made of a mesh which is sized such that suspended material passes through but bed material is retained. An opening on the upstream end allows material to enter the basket. These samplers are relatively inexpensive and simple to use. The mesh size will need to be adjusted depending on the chosen dividing line between suspended load and bedload for a given stream and flood event. Since basket samplers will not function well when nearly full, timing of their removal, as well as care in placing and lifting the samplers, is important. They must also be heavy enough or sufficiently well anchored to keep them aligned into the flow and to prevent the baskets themselves from becoming part of the load moved downstream (Goudie, 1981).

*Pan-type* samplers are usually wedge shaped, with a pan that contains baffles and slots to trap moving sediments. They are set into the stream so that the wedge points into the oncoming current (WMO, 1981).

*Pressure-difference* samplers are designed to produce a pressure drop at the exit of the sampler to ensure that the

entrance velocity equals that of the stream. The Helley-Smith bedload sampler, shown in Figure 5.25(c), has become the standard for bedload measurement. The expanded region causes the streamflow to divert and accelerate around the sampler, leading to separation and a drop in pressure at the exit (see Chapter 6 for theory). A bag is attached to the sampler for collection of sediments. The original design is given by Helley and Smith (1971). Smaller models with a 76 mm orifice can be operated by hand; 152 mm designs can be attached to a backhoe for operation from a bridge. Johnson *et al.* (1977), in measuring bedload transport in a gravel-boulder streambed in Idaho, USA, reported that the standard-size fine mesh (0.2 mm openings) bags of 0.2 m$^2$ were easily clogged by organic debris and fine sediments, and recommended the use of larger 0.6 m$^2$ bags.

Sampling with pressure-difference samplers should be carried out at a uniform cross section as for suspended sediment sampling. A bedrock surface or concrete 'sill', either installed specifically for this purpose or as a road crossing, can act as a platform for sampling sediments travelling over its surface.

Measurements should be taken at three to ten points per cross section (WMO, 1981). The placement of sampling points should reflect the nature of bedload transport, which, except during floods, will generally take place only over a small fraction of the streambed. Samples are collected by lowering the sampler to the streambed, letting it collect sediment over some time period, then removing the sample and bagging it for later analysis (organic material can first be floated off). Care must be taken to keep the sampler correctly aligned and to avoid scooping up excess material from the bed when placing or removing the sampler. The period of time over which material is collected will vary with the flow conditions and bag size. For example, if no sediment is trapped in 5 min, the time period can be extended to 10 or 20 min rather than accepting a zero measurement as indicative of that section. As pointed out by WMO (1981), statistical analyses of field data from up to 100 repetitions have shown that an impracticably large number of samples must be taken at each point to ensure accuracy. A reasonable compromise would be to take two or three samples at each point and combine them. Methods of computing bedload discharge using this technique are given in Section 7.5.3.

### Tracer Particles

Several factors will affect the movement of particles as bedload, including the water velocity and turbulence, and the density, shape, size and compaction of surrounding materials. *Tracer particles* can be used to investigate the types of materials that will move under different conditions. These can be pebbles or boulders which are placed on the streambed and then located again after a period of high flow. Alternatively, pebbles or sand grains of different weights and shapes can be dropped into a stream at various locations to establish which materials can be transported under the existing streamflow conditions.

Tracer particles to be retrieved after a period of high flow should be weighed, measured, numbered and marked with brightly coloured paint. A range of sizes or one representative size such as the median bed material size may be used. The particles are lined up on the streambed and after the spate, an observer notes which sizes have moved and how far—assuming they have not been 'transported' by inquisitive artifact hunters.

Richards (1982) mentions that particles can also be wrapped in aluminium wire and traced with a metal detector in gravel-bed streams. In a large Alaskan gravel-bed river, scientists have used radio transmitters imbedded in large cobbles which were later located with directional antennas (sometimes buried up to 1 m deep) (Emmett, 1989).

Statzner and Muller (1989) propose a quick and simple method for micro-habitat flow characterization using standard hemispheres of 39 mm diameter and varying densities (from 1000 kg/m$^3$ to 10 000 kg/m$^3$). The hemispheres were placed on a level plexiglass plate on the streambed and their movement at differing flows was noted. The authors found that reasonable estimates of shear stress, Froude number and roughness Reynolds number (see Chapter 6 for definitions) could be made by noting the densest hemisphere moved by the flow. Hemispheres could be secured by fishing line for monitoring over a high-flow event.

### Measuring Accumulated Sediments

The pondages behind structures act like 'pit-type samplers' on a larger scale. Thus, repeated surveys of accumulated sediment in reservoirs, ponds, excavated depressions or behind debris dams can give an estimate of the amount of sediment transported by the stream over time. In fact, the rate of sediment deposition should be known before a dam is built because it will affect the useful life of the reservoir.

Surveying techniques, usually employing a grid (Section 5.3.5), are used to obtain an estimate of the sediment volume. Volume is then converted to weight by multiplying by an estimate of the density (Section 5.8.7). Lisle and Hilton (1999) also describe a method for measuring the volume of sediment stored in pools, designed for monitoring changes in the input of fine sediments.

Deposited sediments will actually be composed of bedload and the portion of suspended load which has settled out in the stilled waters. By combining surveys of accumulated sediments with suspended sediment sampling both upstream and downstream, acceptable estimates of bedload and total sediment discharge can be obtained (WMO, 1981).

### Visual Estimation of Bedload

A 'rule of thumb' approach to the estimation of bedload can be taken by using the guidelines given in Table 5.5. If field work is conducted regularly and observations are compared with measured concentrations, a person can become somewhat skilled at estimating suspended load from the colour of the water. For example, a spate of fairly muddy-looking water in what is known to be a gravel-bed stream may be carrying about 10% of its sediment load as

**Table 5.5.** *Estimation of bedload transport. Adapted from Morisawa (1985), after Lane and Borland (1951), by permission of Longman Group, UK*

| Concentration of suspended load (ppm) | Type of material forming the stream channel | Texture of the suspended sediments | Bedload as a percentage of suspended load |
|---|---|---|---|
| Less than 1000 | Sand | Similar to bed material | 25–150 |
| Less than 1000 | Gravel, rock or consolidated clay | Small amount of sand | 5–12 |
| 1000–7500 | Sand | Similar to bed material | 10–35 |
| 1000–7500 | Gravel, rock or consolidated clay | 25% sand or less | 5–12 |
| Over 7500 | Sand | Similar to bed material | 5–15 |
| Over 7500 | Gravel, rock or consolidated clay | 25% sand or less | 2–8 |

bedload. In fact, 10% is a conservative estimate for the middle reaches of many streams (Csermak and Rakoczi, 1987).

### 5.7.6    Erosion and Scour

Changes in channel form can occur as a result of frost action, floods, trampling, bulldozers or other factors. These changes usually fall under the general category of *erosion*, whereas *scour* refers to the movement of bed materials during a flood.

#### *Surveying*

Aggradation and degradation processes can be monitored by repeating cross-sectional surveys at fixed transect sites. These transects should be permanently located with posts, pegs or other marks which are set perpendicular to the channel and far enough away from the streambank to prevent loss during floods. If the channel is easily deformed, cross sections initially lined up perpendicular to the flow may later cut across the stream at odd angles, requiring some 'head scratching' interpretation.

Bank retreat at a specific location can be monitored simply by measuring the distance between a peg or a tree and the edge of a bank. The direction should be established using a compass or by setting out one reference peg and one or more intermediate pegs which are lined up in the direction of measurement.

The line intercept method (Section 5.3.4) can also be used to assess the extent of active erosion along a streambank or the length of streambank protected by vegetation, rock cover or car bodies. Care should be exercised in conducting surveys so the work itself does not cause bank erosion.

#### *Bank Erosion Pins*

For measuring smaller rates of erosion, erosion pins (Figure 5.26) can be installed in the streambank. These are usually metal rods 1 mm or larger in diameter and from 100 to 300 mm in length (Goudie, 1981). They are inserted horizontally into the bank, leaving only a small fraction exposed. Several pins may be used along a vertical profile and surveyed to record their height in relation to some datum. Bank retreat is estimated from the progressive amount of pin exposure, measured with a ruler or callipers.

Some potential problems are that local scour can occur around the base of the exposed pins, or they may catch debris which interferes with the natural rate of bank retreat. Erosion may also be so great that the entire pin is swept away. Despite these drawbacks, erosion pins are a

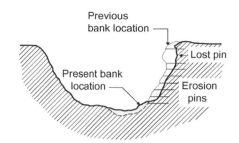

**Figure 5.26.** Using bank erosion pins to monitor the retreat of a streambank

common and inexpensive tool used in the monitoring of bank erosion and gully development.

#### *Scour Chains*

Scour chains can be installed at various points of interest in the channel bed to monitor the amount of scour and fill during floods. These are heavy link-type metal chains, cut to a length corresponding to the expected amount of scour. The chains are anchored onto a metal plate, pin or weight of some type, and buried vertically in the gravels or sands of the bed (Figure 5.27).

When a flood scours away the sediments, the exposed chain falls flat, forming a 'kink'. Subsequent filling reburies the chain. The amount of scour can then be found from the original vertical length as compared to the length below the kink, and the amount of fill found from the depth of sediment deposited above the kink. Obviously, it is essential to know the exact location of the scour chain by surveying so it can be recovered. Scour chains measure the maximum amount of scour during a flood, which may or may not occur at the same time as the flood peak.

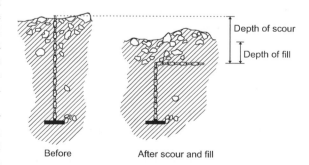

**Figure 5.27.** Measuring scour and fill with a scour chain

## 5.8  Substrates and Sediments: Analysis of Physical Properties

### 5.8.1  General

The fundamental properties of sediment and soil particles are size, shape, mineralogical composition, surface texture and orientation in space. Additionally, bulk properties include colour, average density, porosity and permeability. Some measures are of more interest if a house is to be built on the soil, whereas others are more important if the intention is to grow plants. For the purposes of this text, only those properties relating to the erodibility of the streambed and banks and sediment transport will be treated. The reader is referred to texts on soils, sediment analysis and engineering soil mechanics for more information, such as Brewer (1964), Craig (1983), Guy (1969) and Ward and Harr (1990). Minshall (1984) also provides methods of describing substrate characteristics of biological relevance.

### 5.8.2  Soil Moisture Content, Gravimetric Method

The gravimetric method is a standard method for soil moisture analysis. A 1–100 g sample is placed into a labelled, pre-weighed container and the 'wet weight' of the soil and container is measured using laboratory scales. The sample is then dried at 105 °C for 24 h or until the weight stabilizes at a constant value (Brakensiek, 1979). If a soil can is used the lid should be removed before the sample is dried. Other methods of drying a sample using microwave ovens and heat lamps are given by Klute (1986).

Drying time will depend on sample size, the number of samples in the oven, moisture content of the samples, placement within the oven and type of oven. The samples should be cooled before weighing, preferably in a desiccator to prevent re-absorption of moisture from the air. After weighing, the moisture content (as a percent by mass) is calculated from the formula

$$\text{Moisture content } (\%) = \frac{\text{Wet mass} - \text{Dry mass}}{\text{Wet mass}} \times 100 \tag{5.30}$$

Here, the soil mass is in grams, calculated as

$$\text{Mass of soil} = (\text{Mass of soil} + \text{can}) - (\text{Mass of can}). \tag{5.31}$$

### 5.8.3  Sediment Concentration

#### *Filtration*

In the filtration method a measured volume of well-mixed sediment sample is filtered and the amount trapped on the filter is weighed after drying. The minimum sediment size measured is controlled by the filter size. The choice of filter depends on the type of filter holder available, the particle size, filtration rate and cost. For filtration, samples should contain a sand concentration of less than about 10 000 mg/l and clay concentration less than about 200 mg/l; above these concentrations, filters will plug (Skinner, 2000).

For vacuum filtration using an apparatus similar to the one shown in Figure 5.28(a) cellulose nitrate membranes with a pore size of 0.45 mm are normally used, although glass fibre filters are also available. An apparatus can also be constructed with a Buchner funnel and a glass vacuum flask (Figure 5.28(b)). Glass fibre or paper filters would be used, such as Whatman No. 41, 42 or GF/C. Whatman's website (www.whatman.com) contains a table with information on pore sizes and filtration rates for different types of filters.

If the sediments are predominantly sands, coarser filters can be used and vacuum filtration is not necessary. Generally, paper and glass fibre filters have higher flow rates and are cheaper than membrane filters, although membranes are preferable for precise work because their pore size is more precisely controlled. With any filter the filtration rate will tend to decrease with time as the filter becomes clogged, especially if fine particles are present.

A clean, dry filter is weighed prior to use, preferably using a four figure analytical balance. Filters should only be handled with tweezers. The sediment sample is well agitated and a volume of water-sediment mixture measured with a graduated flask. This is quantitatively transferred to the vacuum funnel using distilled water to wash out any sediment sticking to the flask. The sample is then filtered and the filtrate discarded.

The filter and accumulated sediment should be placed in ceramic crucibles or glass petri dishes and dried overnight at 105 °C. After drying, the filter is cooled in a desiccator to room temperature and then weighed again. The sediment concentration ($c_s$) is expressed in terms of sediment mass per measured volume of water-sediment mix:

$$c_s \ (\text{mg/L})$$
$$= \frac{(\text{Mass of filter} + \text{Sediment}) - (\text{Mass of filter})}{\text{Volume of water-sediment mix}} \times 10^3 \tag{5.32}$$

where mass is in grams and the volume is in litres.

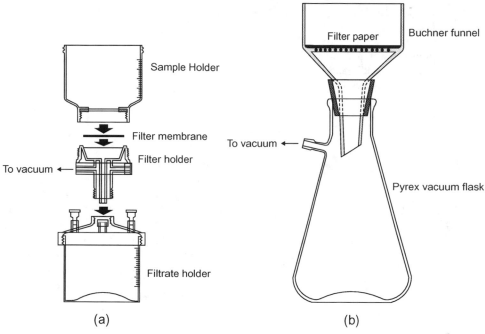

**Figure 5.28.** *Vacuum filtration apparatus for sediment concentration analysis: (a) commercial type for use with membrane filters and (b) Buchner funnel type for use with filter paper. Reproduced from Finlayson (1981), by permission of the Journal of Geography in Higher Education*

A concentration in mg/L is nearly equivalent to ppm (parts per million) by mass for concentrations up to about 8000 ppm (Skinner, 2000). Guy (1969) describes how larger concentrations are converted from mg/L to ppm.

Alternatively, the whole sample can be filtered if the masses of the sample and sample bottle are determined. The concentration would then be expressed as a ratio (with all masses in grams):

$$c_s \text{ (ppm)}$$
$$= \frac{(\text{Mass of filter} + \text{sediment}) - (\text{mass of filter})}{(\text{Mass of sample} + \text{bottle}) - (\text{mass of bottle})} \times 10^6$$

$$(5.33)$$

### Evaporation

A less involved method is to evaporate away the water by pouring a known volume (or mass) of water-sediment mixture into a container and then drying and weighing it. To save drying time, the measured solution should be allowed to settle for at least 12 h so excess sediment-free liquid can be siphoned or poured off. Sediment concentration is calculated as in Eq. (5.32) or Eq. (5.33) after replacing 'mass of filter' with the mass of the container.

With this method, and the method described by Eq. (5.33), the measured 'sediment' concentration will also include dissolved solids. Thus, the technique is best for streams where the salt-mineral content is low in comparison with the suspended load, as in headwater streams or during times of heavy sediment runoff. A method of adjusting the concentration is to separately analyse the weight of dissolved solids in the sampled solution. Alternatively, the electrical conductivity (EC) can be measured and a relationship between EC and dissolved solids used to make the correction.

### 5.8.4 Particle Size

Particle size analysis can be applied to any mixture of sediments, whether collected as suspended or bedload sediment samples, bank materials, samples of floodplain deposits or grit from the crop of a cockatoo. Particle size is a somewhat nebulous length parameter that can be defined by various measures. These include the width of the smallest square mesh through which a particle can pass, the diameter of a circle with an area equal to the maximum projected area of the particle, the diameter of a sphere with a volume or settling velocity which equals that of the particle or simply the longest dimension of the particle (Goudie, 1981).

Size classes have been developed to give a descriptive label to particles grouped within a given size range. Several classifications exist, and standard classes can vary between countries and agencies. Traditionally, the length dimension has been expressed in millimetres, as described by the Wentworth scale in Table 5.6. Sedimentologists typically use the phi ($\phi$) scale, also shown in Table 5.6, where phi is equal to the negative logarithm (in base 2) of the particle size in millimetres. For very small particles this method eliminates the inconvenience of unwieldy numbers. Calculation of base 2 logarithms can be awkward because most calculators provide only base 10 or natural logarithms. However, this is easily remedied by using the formula

$$\log_a n = \frac{\log_b n}{\log_b a} \qquad (5.34)$$

**Table 5.6.** *Grade scales for particle size. Adapted from Brakensiek et al. (1979)*

| Class (Wentworth) | mm | $\phi$ |
|---|---|---|
| Very large boulder | 4096–2048 | −12 to −11 |
| Large boulder | 2048–1024 | −11 to −10 |
| Medium boulder | 1024–512 | −10 to −9 |
| Small boulder | 512–256 | −9 to −8 |
| Large cobble | 256–128 | −8 to −7 |
| Small cobble | 128–64 | −7 to −6 |
| Very coarse gravel | 64–32 | −6 to −5 |
| Coarse gravel | 32–16 | −5 to −4 |
| Medium gravel | 16–8 | −4 to −3 |
| Fine gravel | 8–4 | −3 to −2 |
| Very fine gravel | 4–2 | −2 to −1 |
| Very coarse sand | 2–1 | −1 to 0 |
| Coarse sand | 1–0.5 | 0–1 |
| Medium sand | 0.5–0.25 | 1–2 |
| Fine sand | 0.25–0.125 | 2–3 |
| Very fine sand | 0.125–0.0625 | 3–4 |
| Coarse silt | 0.0625–0.0312 | 4–5 |
| Medium silt | 0.0312–0.0156 | 5–6 |
| Fine silt | 0.0156–0.0078 | 6–7 |
| Very fine silt | 0.0078–0.0039 | 7–8 |
| Coarse clay | 0.0039–0.0020 | 8–9 |
| Medium clay | 0.0020–0.0010 | 9–10 |
| Fine clay | 0.0010–0.0005 | 10–11 |
| Very fine clay | 0.0005–0.00024 | 11–12 |

which, for this method, becomes

$$\log_2 n = \frac{\log_{10} n}{\log_{10} 2} = \frac{\log_{10} n}{0.30103} \qquad (5.35)$$

As an example,

$$\log_2 22 = \frac{1.342}{0.30103} = 4.46$$

The method of particle size analysis should be chosen based on the type and size of material being analysed and the accuracy required. The method will also vary with the amount of sediment available for analysis. For example, many suspended sediment samples will not contain enough sediment for accurate analysis of the larger sizes, and the analysis must be limited to determining the percentage of silts and clays. Methods will be presented in order of their simplicity. For more details on particle size analysis techniques, Allen (1981), BS (1975), Day (1965), Gee and Brauder (1986), Guy (1969) or Lewis (1984) can be consulted.

### Visual Analysis

General classification of surface sediments is often done by eye when assessing the distribution of substrate types in a stream reach. A collapsed version of the grade scale given in Table 5.6 can be used to visually classify sediments as, for example, boulders, cobbles, gravel, sand, silt or clay. Since silt and clay are not easily distinguished, they are sometimes combined into a single class, 'mud'. As with texture triangles used for soil classification, the triangle given in Figure 5.29 can assist in the standardization of substrate classes.

Field identification of sand-sized sediments can be assisted by developing a collection of samples of known size in vials, test tubes, or glued onto plastic slides. A hand lens is useful for identifying smaller grains.

The areal survey methods of Section 5.3 can be used to assess the composition of surface sediments. For most streambed materials, a 1 m$^2$ frame is a functional plot size. Thin wire is strung across the frame every 0.1 or 0.2 m. The grid is placed over the sediment with the top of the grid in the upstream direction. The particle sizes in the surface layer can be measured in the field or a photograph can be taken and percent cover analysed using the grid for scale. Photographs should be shot from a consistent height above the grid, as near to vertical as possible. A card with information on the site and date can be set on a corner of the frame before the picture is taken. Photographs additionally provide a permanent record of substrate

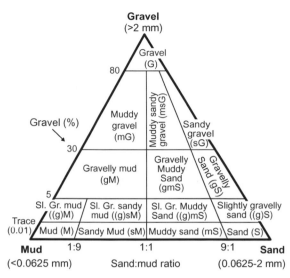

**Figure 5.29.** *Classification of substrates according to gravel, sand and mud composition, sl = slightly. Re-drawn from Folk (1980), by permission of Hemphill Publishing Co. Reproduced by permission of Mr. R.L. Folk*

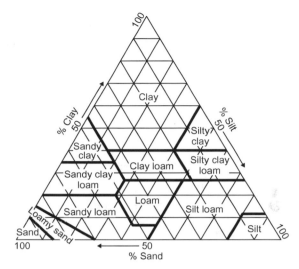

**Figure 5.30.** *Standard US Soil texture triangle. Reprinted from McKnight (1990), © 1990, p. 329, by permission of Prentice-Hall, Inc*

rock type and orientation. Bunte and Abt (2001) give more details on photographic analysis of bed surface sediments.

### *'Hand Texturing' of Soils*

A somewhat subjective but low-effort method of analysing the size composition of soils and finer sediments is by doing a 'texture by feel' analysis. Texture is defined as the relative proportions of sand, silt and clay, as shown in Figure 5.30.

The soil composition is estimated from the feel and malleability of a wetted sample (a 'bolus'). By working the bolus between the thumb and forefinger, a thin 'ribbon' can be created with more coherent soils, its length determined by the soil type. Running the sample under a stream of water can help in detecting the sand content by feel. Table 5.7 is a general key for the textural analysis of soils by observation and feel.

To 'calibrate' a person to the feel of different soil types, samples can be compared to 'type' samples which have been analysed for particle composition by more accurate methods. Such type samples are available in most universities or testing laboratories.

### *Direct Measurement*

Individual boulders, cobbles and large gravels can be measured directly in the field. Callipers or a rule or survey staff can be used, depending on the size. For smaller

particles, graduated eyepieces in microscopes can be employed to make direct measurements.

This method should include multiple measurements along three axes: the *A* or longest axis, the *B* or intermediate axis and the *C* or shortest axis. Each axis is perpendicular to the other two. If only one axis is measured, the *B* axis length will give an adequate estimate of mean diameter for most particles (Briggs, 1977a).

The short axis (*C*) is relatively uncomplicated to identify since it is simply the shortest axis. The *B* axis is defined as the shortest axis of the 'maximum projection plane' (the plane of largest area, perpendicular to the *C* axis). Rather than being the longest axis, the *A* axis is measured perpendicular to the *B* axis. As shown in Figure 5.31 for a 'tabular' pebble, the *A* axis is not the corner-to-corner length.

The value in defining *A* and *B* this way is that it most closely approximates the results from sieving, because it is the *B* axis which determines whether or not an individual pebble falls through a mesh of given size. This is also the reason why if only one measurement is taken, it is made along the *B* axis. A metal template with square holes corresponding to sieve sizes can also be constructed for categorizing particle sizes in the field. Template designs are described by Bunte and Abt (2001, p. 25–27); for example, the US SAH-97 gravelometer is made of 3.2 mm thick aluminium with 14 square holes in units of $0.5\phi$ from $-1\phi$ to $-7.5\phi$. A spreadsheet program for analysing

**Table 5.7.** *Soil texture classification by feel. Adapted in part from Northcote (1979), Reproduced by permission of Rellim Technical Publications*

| Texture class | Behaviour of moist bolus of soil |
|---|---|
| Sand | Crumbles readily; cannot be moulded; single sand grains adhere to fingers |
| Loamy sand | Slight coherence; can be sheared between thumb and forefinger to give minimal ribbon of about 6 mm; discolours fingers with dark organic stain |
| Sandy loam | Bolus just coherent but very sandy to touch; will form a short ribbon; dominant sand grains can be seen, felt or heard |
| Sandy clay loam | Strongly coherent bolus, sandy to touch; medium size sand grains visible in finer matrix; will form a longer ribbon than sandy loam |
| Loam | Coherent and rather spongy bolus; smooth feel when manipulated but with no obvious sandiness or 'silkiness'; may be somewhat greasy to the touch if much organic matter present; will form a short ribbon |
| Silt loam | Coherent bolus, very smooth to silky when manipulated; may form short ribbon |
| Silt | Pure silt will have a smooth, floury or silky feel; bolus can be manipulated without breaking |
| Silty clay loam | Coherent smooth bolus; plastic and silky to the touch; will form longer ribbon than loam |
| Clay loam | Coherent plastic bolus; smooth to manipulate; will form ribbon similar to silty clay loam |
| Sandy clay | Plastic bolus; fine to medium sands can be seen, felt or heard in clayey matrix; will form a thin, long ribbon which breaks easily |
| Silty clay | Plastic bolus; smooth and silky to manipulate; will form long ribbon |
| Clay | Handles like Plasticine, plastic and sticky; will form a long, flexible ribbon of 5 cm or more |

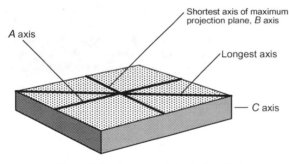

**Figure 5.31.** *Axes on a tabular 'pebble'*

pebble count data is available from the US Forest Service at http://www.stream.fs.fed.us.

### Dry Sieving

Dry sieving is the most commonly used method for the analysis of sand-sized particles. Particles larger than sand are often too bulky and heavy to sieve, and those smaller than sand tend to form aggregates which are not easily sieved in their dry state. If needed, carbonates which cement particles together can be removed using dilute hydrochloric acid, and organic matter can be removed with hydrogen peroxide (Goudie, 1981). Fines should be dispersed prior to drying and sieving with a dispersing agent such as sodium hexametaphosphate (Calgon) or with an ultrasonic probe. These fines are washed through a 0.0625 mm sieve and collected for separate analysis by hydrometer or by drying and weighing the total fraction.

Large samples may require splitting before sieving, either by using commercial sample splitters or by successive sub-sampling. Reducing samples to a manageable size must be done with care, as error can be introduced if a non-representative sub-sample is obtained. One may also wish to separately pick out and weigh all of the larger-sized rocks. Further sub-dividing may be desirable to prevent overloading the sieves when working down to sieve sizes of 2 mm and finer. The relative size of the sub-samples must be used to reconstruct the proportion in each fraction.

Dry sieving requires a completely dry sample. Even if moisture is only 1%, adhesion forces can exceed the weight of grains smaller than 1 mm, preventing them from passing through the smaller sieves. The sample is allowed to air-dry at room temperature by spreading it thinly on trays or newsprint and leaving it for several days. The process can be sped up using a heat lamp or an oven. After drying, aggregates in the sample can be broken up by hand or rolling pin or by grinding, although care should be taken to avoid fragmenting the particles.

A set of sieves of required sizes is stacked together, decreasing in aperture size downwards. Commercially available sieves are sized in a geometric series, normally corresponding to particle size classes (Table 5.6). A regular interval between the sieves should be used. A common convention is to use a $1\phi$ interval with a $4\phi$ (0.0625 mm) sieve as the smallest size and a $-6\phi$ (64 mm) sieve as the largest. This would give a 'nest' of 11 sieves.

A 100–200 g sample of material is weighed out and placed in the top of the coarsest sieve. The nest of sieves is placed on a shaker with a lid on the top and a tray on the bottom, and the sample is shaken for 10–15 min. If, after shaking, more than a few percent of the sample is held within the top sieve or in the bottom tray, the procedure should be repeated to further separate the larger and smaller fractions.

The material trapped on each sieve is transferred to a weighing tray, and the sieves are gently brushed to release sediments stuck in the mesh. The mass of the sediment for each size fraction is recorded. The total of all fractions should be checked with the original weight, although some loss is to be expected. The material passing the 0.0625 mm ($4\phi$) sieve is normally retained for analysis by sedimentation methods such as hydrometer or pipette.

### Wet Sieving

Wet sieving is a good method for sizing coarse particles and aggregated sand-sized particles. If only the sizes of coarser gravels and larger particles are of interest the method is very simple. The coarsest sieve is held over a bucket and a large sample of sediment is poured onto it. The sieve is shaken, with washing as needed, until all particles smaller than the sieve aperture have passed through (typically 10 min). The trapped particles are weighed wet because the mass of the water is considered insignificant compared to the mass of these larger particles. The sediments and wash water passing through the sieve are poured onto the next smallest sieve, and the procedure is repeated until the smallest sieve size is reached.

For particles less than about 8 mm the mass of the water becomes significant and the procedure becomes slightly more involved and time consuming. After each sieving the trapped particles are transferred to a pre-weighed container. Each fraction, including that passing the 0.0625 mm sieve, is separately dried and weighed. The total sample mass is then assumed to be equal to the sum of the masses of the separate fractions. The fraction smaller than 0.0625 mm can be saved for analysis by hydrometer or pipette.

With either wet sieving or dry sieving, the particle shape, sieve opening shape and type and time of shaking will affect the probability of a particle passing through. Standardization of the sieving method is important to ensure reproducibility.

### Sedimentation Methods: Hydrometer and Pipette

Hydrometer and pipette methods are sedimentation methods based on Stoke's law (Section 6.5). These methods are typically used for analysis of the size fraction smaller than 0.0625 mm (silt and clay) or 2 mm (sand, silt and clay). Since, by Stoke's law, larger particles fall more rapidly than smaller ones, an estimate of particle diameter can be obtained by measuring the time taken for them to settle. In the hydrometer method, the density of a sediment-water mixture is measured at specific times as the sediment settles out (Bouyoucos, 1962). The pipette method is a more precise method against which other methods are often compared. In this method, changes in concentration of the settling suspension are found by analysing small subsamples with a pipette at specific settling times.

In both methods, the sediment sample can be pre-treated to remove substances which may affect results such as carbonates, soluble salts, organic matter and iron oxides, and then the sample is oven-dried. It is then mixed with water and sodium hexametaphosphate (Calgon), which reduces flocculation of clays.

Details on these methods are given in Brakensiek *et al.* (1979) and Gee and Brauder (1986). Briggs (1977b) and Foth (1978) provide simpler methods that require less sophisticated equipment and fewer measurements.

Other methods of particle size analysis have used light-scattering principles, such as Coulter counters, laser particle sizers, x-ray attenuation instruments and turbidimeters. Because of their high cost and uncertainties in correction factors, however, they are not routinely used.

### 5.8.5 Presentation of Particle Size Data

Samples of sediment or substrate can be described by the dominant size, by the proportion in each size class, and by statistical measures which describe the distribution of sizes. The results of a particle size analysis, e.g. by sieving, can be plotted as a histogram to illustrate the percentages in each size grade and the skewness and/or bimodality of the size distribution (see Appendix). Sediment particle sizes often tend to follow a log-normal size distribution, with a high proportion of particles in the low-to-middle size class and progressively fewer toward the extremes.

If sediment samples are to be compared, it is more useful to plot results as a cumulative frequency curve. Commonly, a logarithmic $X$ axis is used for the particle size in millimetres, with a linear $Y$ axis indicating the 'percentage finer by mass'. This typically yields an S-shaped curve, as shown in Figure 5.32. The curves give an

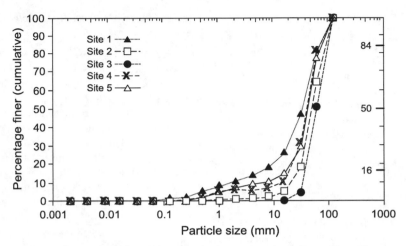

**Figure 5.32.** *Bed material particle size distributions, by mass, for five sites on the Acheron River. Sites correspond to numbered locations in Figure 4.7. Composited samples of surface materials <128 mm collected from several points along a stream reach were analysed*

indication of the spread of sizes present: the larger the range, the flatter the distribution curve, and the more uniform the sample, the more vertical the curve. In Figure 5.32 the sample from Site 3 is the most uniform.

If particle sizes are individually obtained by direct measurement (e.g. pebble counts) it is more appropriate to consider the *number* of particles of a certain size rather than the *mass*. Probability paper is normally used for displaying this *frequency* data, as shown in Figure 5.33.

Particles are ranked from smallest to largest and a plotting position is calculated as $100(m/N)$, where $m$ is the rank ($0 =$ smallest) and $N$ is the number of samples (see Section 8.2.4). This gives the '% finer' value for the $X$ axis. For example, if 30 rocks were measured the largest particle would be plotted at $100(29/30) = 96.7\%$. Because the scale does not include 0% the smallest particle is left off.

It becomes difficult to describe mixtures of very different size classes because the method of measurement

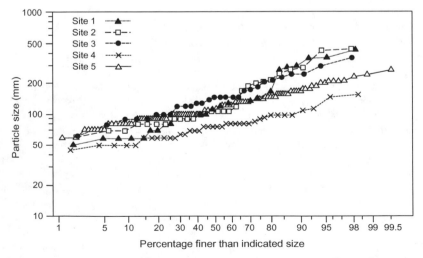

**Figure 5.33.** *Frequency analysis of larger surface bed material sizes for five sites on the Acheron River. Sites correspond to numbered locations in Figure 4.7. Samples were taken by walking up the stream reach and measuring a rock at each step. The particle size is represented by the intermediate B axis*

changes over the range of sizes analysed (e.g. hydrometer, sieving, grid surveys or direct measurement). Techniques for combining these data are given by Bunte and Abt (2001), Church *et al.* (1987) and Griffiths (1967).

Indices for describing the particle size distribution include the mean, median, standard deviation, skewness and other conventional statistical measures (see Appendix). These can be calculated from the data or taken directly from a cumulative frequency plot. Sediment particle sizes are often described by specific percentile values such as the $d_{50}$ or $d_{84}$ or $d_{90}$, which can be easily read from a cumulative frequency curve. The $d_{50}$, for example, is the median particle diameter, meaning that half the sample (by weight or frequency) is larger, half smaller. The $d_{84}$ and the $d_{16}$, the diameters for which 84% and 16% of the particles are smaller, respectively, have particular significance since they represent one standard deviation from the mean.

The median tends to be a more robust measure than the mean for describing sediment size. The problem with using the mean value is that a few particularly large rocks will bias the mean heavily toward the coarse end. This can be avoided by removing unrepresentative particles from the sample or by using a formula which ignores the 'tail ends' of the distribution. However, the preferable approach is to use the mean of the phi ($\phi$) values (called the phi mean) on the assumption that the particle sizes are log-normally distributed. Under this assumption the distribution of particle sizes can be completely described by the phi mean and phi standard deviation.

Formulae for the mean and other statistical measures reflecting the central tendency and shape of a particle size distribution are given in Table 5.8. In natural sediments the parameters tend to be related; i.e. sediments of a larger mean particle size tend to have a larger range of particle sizes and thus a higher standard deviation, whereas finer sediments are more uniform.

### 5.8.6  Particle Shape: Roundness, Sphericity

The form of a particle (i.e. its three-dimensional configuration) can be described in terms of its shape, sphericity, roundness and surface texture. The description of form should be relevant to the field of interest; i.e. whether sedimentological or biological.

As with many measures, an index of form will be dependent on scale. For example, the overall form of a lava rock pockmarked with air cavities may be fairly round, but its surface texture, its ability to be transported by the stream and the number of nooks for benthic organisms would be very different from a more solid

*Table 5.8.* Statistical parameters for describing particle size distribution based on $\phi$ (phi) values. Equation 5.35 can be used to convert from millimetres to $\phi$ and back. After Briggs (1977a)

| Parameter | Method of calculation |
|---|---|
| Median | $\phi 50$ |
| Mean | $\dfrac{(\phi 84 + \phi 16)}{2}$ or $\dfrac{(\phi 16 + \phi 50 + \phi 84)}{3}$ or $\dfrac{(\phi 10 + \phi 20 + \cdots + \phi 90)}{9}$ |
| Standard deviation | $\dfrac{(\phi 84 - \phi 16)}{2}$ |
| Skewness | $\dfrac{(\phi 84 - \phi 50)}{(\phi 84 - \phi 16)} - \dfrac{(\phi 50 - \phi 10)}{(\phi 90 - \phi 10)}$ |
| Kurtosis | $\dfrac{\phi 90 - \phi 10}{1.9(\phi 75 - \phi 25)}$ |

rock of the same roundness. Surface texture of particles is commonly analysed by scanning electron microscope, and methods will not be covered here. Goudie (1981) provides an interesting discussion on particle form.

Two common concepts used in the description of particle shape are sphericity and roundness. The two terms are geometrically distinct; i.e. a pebble with a high sphericity may not necessarily possess a high roundness value. *Sphericity* is a measure of the ratio of a particle's volume to the volume of a sphere which circumscribes it. A sphere is used as a reference form because of the common assumption in many formulae such as Stoke's law that sediment particles are spherical. Sphericity is closely related to the surface area to volume ratio, a fundamental measure of particle shape that reaches a minimum when the particle is spherical. The sphericity of a particle is thus important in controlling lift, settling velocity and sediment transport.

*Roundness* is a measure of 'roughness' or 'angularity' of the particle, and describes the sharpness of the 'corners' on a particle. It is thus strongly affected by abrasion. The roundness of a particle typically changes rapidly during the initial stages of transport, then at a slower rate as the particles continue to be polished. Often, the original shape is reflected in the particle even after a considerable amount of wear.

The techniques given for estimating sphericity and roundness were developed for particles of sizes that are easily handled. The observation of small particles can be assisted using a microscope or an overhead projector to increase the viewed size. Scanning equipment and digital image analysis also have potential for describing particle form. For particles of metre size and larger, surveying equipment may be required.

### Sphericity and Particle Shape

Krumbein (1941) developed a definition of sphericity by using a ratio of the volume of an ellipsoid defined by the *A*, *B* and *C* axes to the volume of a sphere circumscribed around the particle defined by the *A* axis. Sphericity ($\psi$) can thus be expressed as

$$\psi = \sqrt[3]{\frac{BC}{A^2}} \tag{5.43}$$

The *A*, *B* and *C* axes are measured as described in Section 5.8.4. Krumbein (1941) states that sphericity values less than 0.3 rarely occur in nature.

For individual particles the assumption of a triaxial ellipsoid shape may not be valid, but it is approximated when the average of a group of pebbles is taken. Krumbein therefore recommends that 25 or more pebbles should be used in order to obtain agreement with other methods such as Wadell's (1932, 1933), where the actual particle volume is measured.

Zingg (1935) classified pebbles into four basic shapes: disc, spherical, bladed, and rod-like. Brewer (1964) later added three additional classes. Ratios of axis lengths, *B/A* and *C/B* are used to distinguish the classes, as shown in Figure 5.34.

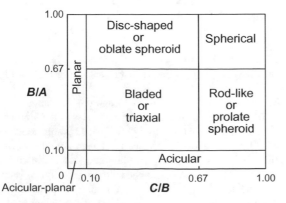

**Figure 5.34.** *Chart for the classification of particle shape. Redrawn from Brewer (1964), adapted from Krumbein (1941), by permission of John Wiley and Sons, Inc. Reproduced by permission of SEPM Society for Sedimentary Geology*

### Roundness

Roundness, a rather vague measure, was defined by Wadell (1932) as the ratio of the curvature of the corners and edges to the average curvature of the particle as a whole. However, Wadell's method of determining particle roundness is very complex. A more practical field method for estimating roundness is to visually estimate it using a chart such as the one shown in Figure 5.35. These images were drawn by Krumbein from pebbles classified by Wadell's method. The original drawings (reduced here) were for pebbles of 16–32 mm diameter. Enlargement or reduction can be used to create charts for any size range.

When using the chart the particle should be viewed such that the maximum projection plane is observed. When estimating a roundness value all the smaller corners and edges should be considered as well as the main ones. Broken pebbles are given a roundness rating for the unbroken half and the value is subsequently halved (in the case of an odd value, e.g. 0.7, the halved value should be rounded down; i.e. 0.3 rather than 0.4).

A set of at least 25 pebbles should be sampled to obtain an average roundness value. It may be advisable to divide pebbles into size classes or even rock type, as there is often a marked change in the roundness of pebbles of different sizes or rock type such as shales and granites. Separation in this manner also makes it easier to compare sediments from different sites, as this can be done on a size or type basis.

### 5.8.7 Particle Arrangement and Other Miscellaneous Bulk Properties

The arrangement of sediment particles affects the degree of 'packing' of grains, which in turn has an effect on the erodibility of substrates and their permeability to air, water and micro-organisms. Packing is a complex factor involving size, grading, orientation and form (Goudie, 1981). Erodibility of soils will also be affected by moisture content and particle size. In this section, measures are presented for bulk density, porosity, aggregate stability, embeddedness and sorting.

### Bulk Density and Porosity

The *bulk density* of a soil or bed material sample is the ratio of its mass to its volume:

$$\text{Bulk density} = \frac{\text{Mass of sample}}{\text{Volume of sample}} \tag{5.44}$$

where bulk density has units of kg/m$^3$.

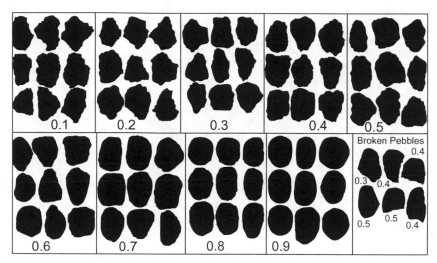

**Figure 5.35.** *Visual comparison chart for the estimation of pebble roundness. Re-drawn from Krumbein (1941), Reproduced by permission of SEPM Society for Sedimentary Geology*

Bulk density is thus an index of the composition and packing of the material (Briggs, 1977b). As the volume of voids between particles increases, the bulk density decreases. It is thus inversely related to *porosity*. If the particles can be assumed to have a density of 2650 kg/m³ (sandy, siliceous particles with little organic matter) then porosity is given as (Briggs, 1977b)

$$\text{Porosity } (\%) = \left(1 - \frac{\text{Bulk density}}{2650}\right) \times 100 \quad (5.45)$$

Bulk density can be measured by taking a sample core of known volume, drying the sample and dividing the dry weight by the core volume. The appropriate volume for a cylindrical core sample is $h(\pi d^2/4)$, where $d$ is the diameter and $h$ is the height of the cylinder. An alternative is the displacement method, which is more applicable for individual rocks, soil clods or other samples of unknown volume, such as those taken by freeze coring. In this method the particle mass is coated with wax or plastic resin (if needed), and the mass is completely submerged in a container filled with water. The volume of water displaced is equal to the volume of the particle mass. The coating is then removed and the sample oven-dried and weighed.

For simply obtaining a 'feel' for the relative compaction of substrates, Gore (1978) and Thorne (1998) suggest observing the penetration of a measuring stick or rod pushed into the sediment. Commercial soil penetrometers can be used for a more standardized version of this technique.

### Aggregate Stability

Aggregate stability is a measure of the erodibility of soil in terms of its tendency to break down upon wetting. A soil with low aggregate stability tends to form a surface crust which impedes infiltration and hinders plant growth.

The method for measuring aggregate stability is described by Briggs (1977b). Loose particles are first gently sifted from a large soil clod through a garden sieve. The aggregates retained on the sieve are poured into a 1000 ml graduated beaker and gently tapped so that the soil settles. The volume of the soil ($\forall_1$) in the beaker is recorded. The beaker is gently filled with water, without damaging the aggregates, and allowed to stand for 30 min. The water is carefully poured off and the remaining volume of soil noted ($\forall_2$). Aggregate stability is calculated as

$$\text{Aggregate stability}(\%) = \left(1 - \frac{(\forall_1) - (\forall_2)}{(\forall_1)}\right) \times 100$$

$$(5.46)$$

### Embeddedness

Embeddedness is an index of the degree to which larger particles (boulders, large cobbles) are surrounded or partially buried by finer sediments. As embeddedness increases, the biotic productivity of the substrate is considered to decrease.

Bunte and Abt (2001) give several methods for describing embeddedness, including the ratio of the total height of a particle to the height buried below the bed surface (the embedded portion). Platts *et al.* (1983) use a rating code to describe the percentage of surface area of the largest size particles covered by finer sediments. A 5-4-3-2-1 rating corresponds with channel embeddednesses of <5%, 5–25%, 25–50%, 50–75% and >75%, respectively.

The Brusven index is a means of describing both sediment size and percent embeddedness (Brusven, 1977). The index is composed of a three-digit number (e.g. 51.5) where the digit in the ten's place represents the largest materials in the sample (called the dominant particle size), the figure in the one's place represents the material surrounding the dominant particles and the decimal place is used to describe the percentage embeddedness. A larger particle completely imbedded in fines is assigned a decimal value of 9.

Brusven's original index was modified slightly by Bovee (1982). Rather than embeddedness, the decimal place describes the percentage of sand and smaller size material in the substrate matrix. The index code for this modified method is given in Table 5.9, and is the method of substrate classification used in the model PHABSIM (see Chapter 9). Vegetation such as rooted macrophytes or algae can also be treated as a form of substrate and included as a decimal in the hundreds place. The numeral might either represent percentage cover or a rating.

As an example, a substrate mixture of medium cobbles (6) surrounded by small gravels (2) and 40% fines (0.4) would have a modified Brusven index of 62.4. One large boulder completely embedded in sand would have an index of 91.9. It should be noted that the index values do not form a continuum from 00.0 to 99.9, because the first number should always be larger than the second; e.g. values of 35.2 or 78.5 would be nonsensical.

**Table 5.9.** *Substrate code for describing size classes in conjunction with the Brusven index method. After Bovee (1982)*

| Code | Substrate description |
| --- | --- |
| 1 | Fines (sand and smaller) |
| 2 | Small gravel (4–25 mm) |
| 3 | Medium gravel (25–50 mm) |
| 4 | Large gravel (50–75 mm) |
| 5 | Small cobble (75–150 mm) |
| 6 | Medium cobble (150–225 mm) |
| 7 | Large cobble (225–300 mm) |
| 8 | Small boulder (300–600 mm) |
| 9 | Large boulder (>600 mm) |

### Sorting

Sorting is a measure of the spread of particle sizes in the substrate. The degree of sorting can be calculated using the equation for standard deviation in Table 5.8. For descriptive purposes the range of values has been divided into five sorting classes as shown in Figure 5.36, which provides a means of visual identification. The degree of sorting can also be estimated by comparison with photographs of known conditions. For smaller sediments, 'standard' vials of sorted sediments can be used.

Andrews (1983) gives an alternative sorting index for bed materials:

$$\text{Sorting index} = \frac{1}{2}\left(\frac{d_{84}}{d_{50}} + \frac{d_{50}}{d_{16}}\right) \qquad (5.47)$$

where $d$ is a particle diameter in mm.

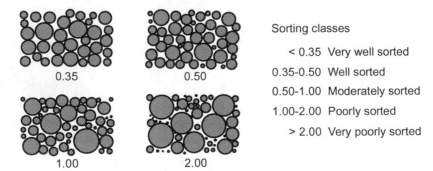

Sorting classes

< 0.35  Very well sorted

0.35–0.50  Well sorted

0.50–1.00  Moderately sorted

1.00–2.00  Poorly sorted

> 2.00  Very poorly sorted

**Figure 5.36.** *Chart for visual estimation of sorting. Re-drawn from Dackombe and Gardiner (1983), adapted from Folk (1980), by permission of Harper Collins Publishers and Hemphill Publishing Co., respectively. Reproduced by permission of Mr R.L. Folk*

## 5.9  Water Quality

Because of its importance to the biota which drink, transpire or live in the water flowing through streams, a large number of attributes lumped under the heading 'water quality' are of interest to water researchers. These might include pH, salinity, temperature, sediment concentration, odour, colour, light penetration, dissolved oxygen, levels of nutrients such as nitrogen or phosphorous, concentrations of pesticides or radionucleotides or heavy metals, or the presence of pathogenic microbes. For example, the UK River Ecosystem Classification system uses an index based on dissolved oxygen, biochemical oxygen demand, ammonia, pH, hardness, dissolved copper and total zinc (The Surface Water Regulations, 1994).

Other than sediment concentration, covered previously, it is beyond the scope of this book to cover all of the aspects of water quality analysis. The interested reader is referred to Clesceri *et al.* (1998), USGS (2003), USEPA (1987) and other standard water quality analysis guides for methods of collecting, preserving and analyzing water samples. Some basic recommendations on sampling are provided here.

Samples can be collected for 'spot' readings or taken over a period of time to monitor diurnal, seasonal or yearly changes. If streams are well mixed, a single 'grab' sample taken from a bridge or streambank may be adequate for determining an instantaneous measure of the stream's average water quality. Fresenius *et al.* (1987) recommend that samples be taken by holding a bottle 30 cm under the water surface. Automatic pumping samplers can be used at well-mixed sites to sample on a regular basis.

In some streams, especially those with slow-moving stagnant pools, layers of different water quality may exist. Specialized sampler designs are available for collecting samples at specific depths, similar to the point-integrating samplers described in Section 5.7.4. Other designs are described in references on water quality analysis and limnology.

Sample bottles can be plastic or glass, depending on the parameter to be tested. They should be rinsed once or twice with stream water before a sample is collected. All bottles should be labelled with the date, time and site (including depth) of sampling.

Field analysis is essential for some water quality parameters such as pH, temperature and dissolved oxygen. Various meters and probes have been developed for field measurement of electrical conductivity (EC), pH, dissolved oxygen (DO) and specific ions. Combined units are also available with dataloggers for automatic data collection, some of which are designed to be left in the field to collect data over longer periods of time.

Field 'test kits' for water chemistry analysis are also commercially available. Colourimetric procedures have been developed for many water quality parameters, where colour intensity or shade is measured with a spectrophotometer or visually compared to a chart or set of standards. 'Test kit' techniques are typically of lower resolution than laboratory methods. However, they may be sufficient for some studies, for example when looking for sudden changes in water quality.

If higher accuracy is needed then there is a stronger argument for using more precise laboratory techniques. For most water quality parameters, samples must be preserved in some manner for transport to the laboratory. The method of preservation depends on the parameters to be measured (for example, samples to be analysed for nitrate should not be preserved with nitric acid). Multiple sample bottles may therefore be required for each sample, each preserved differently. Procedures for the type of sample bottle, sample volumes and methods of preservation are given by Clesceri *et al.* (1998), and should be verified by the designated analytical laboratory.

Samples should be analysed as soon as possible after collection. This will sometimes require shipping the samples to the laboratory by bus, car, train, airplane or 'overnight express'. The bottles should be enclosed in foam containers to prevent breakage and packed in a cooler with ice or dry ice. Keeping samples in the dark and at low temperature (preferably close to 4 °C) retards bacterial growth. Information on the analyses desired, the date and site of collection and the method of preservation should be written on each bottle.

The objective of field sampling should be clearly defined before any samples or measurements are taken. It is easy to become lost in the fog of possibilities for analysis of a stream's physical, chemical and biological features. A study for a high school science experiment may or may not require the same rigorous standards as a study for a court case on setting environmental flows. This chapter has presented some of the standard approaches of field data collection, to serve as a starting point for wetting one's feet with a purpose in mind.

# 6

# Water at Rest and in Motion

## 6.1 General

In contrast to other water bodies, the most noticeable feature of streams is their one-way flow, guided by the influence of gravity. Yet within this unidirectional motion, water molecules follow unpredictable paths seemingly of their own choosing. Like the movement of traffic on a freeway, the general motion can be described in terms of average forward progress. However, the detours, halts and starts of an individual vehicle—or an individual water molecule—defy analysis. Thus one should begin a study of fluid mechanics under the premise that the tools for describing the behaviour of water are only approximations. Nevertheless, by making appropriate simplifying assumptions and generalizations some fascinating traits of the movement of water 'en masse' can be examined and described. There is a large amount of literature on fluid mechanics which is appropriate for engineering applications, and this provides a framework upon which biologists can build to describe the complex interactions of organisms with their flowing environment.

This chapter has been broken into two main streams of study: hydrostatics and hydrodynamics. *Hydrostatics* is the study of water (hydro) at rest (static) and includes the principles of pressure and buoyancy. These principles hold true whether the water is at rest or in motion. However, as soon as a fluid begins to move, viscosity enters the picture, and the study of *hydrodynamics* is therefore more complex.

Hydrodynamics can be further divided into the study of fluid motion in the micro-environment and in the macro-environment. In the *micro-environment* near solid surfaces, viscosity has an important effect on fluid behaviour. Here, the focus is on patterns of viscous action as water passes around a surface, creating lift and drag on sediment particles and affecting the lives of small organisms which live on surfaces within the stream. At this scale, turbulence is manifested in the eddies and velocity fluctuations near solid surfaces. In the *macro-environment*, gross measures are of interest: channel discharge, average velocity and the energy or work required to move the fluid. Here, viscosity is not as important, and turbulence is present over the entire depth of flow as large-scale eddies.

For additional information on biologically relevant topics in fluid dynamics, Vogel (1981, 1988) offers a readable, common-sense introduction to fluid properties and their significance, particularly at the micro-environment level. Newbury (1984) provides a concise, straightforward description of channel-scale flow properties. Other references by Davis and Barmuta (1989), Denny (1988), Denny *et al.* (1985), Hynes (1970), Nowell and Jumars (1984), Silvester and Sleigh (1985), Smith (1975) and Statzner *et al.* (1988) provide additional information from a biological viewpoint. For general information on fluid mechanics, engineering texts include Crowe *et al.* (2000), Douglas *et al.* (2001), Munson *et al.* (2002), Street *et al.* (1995) and White (2000). Some of the newer texts include CD-ROMs with video clips, computer simulations and software.

## 6.2 Hydrostatics: The Restful Nature of Water

### 6.2.1 Pressure

*Pressure* is defined as force per unit area. Pressure can be specified as *absolute*, with zero pressure (complete vacuum) as a reference, or as a *relative* or *gauge* pressure, with respect to the local atmospheric pressure, where

$$\text{Absolute pressure} = \text{Gauge pressure} + \text{Local atmospheric pressure} \quad (6.1)$$

Stream Hydrology: An Introduction for Ecologists, Second Edition.
Nancy D. Gordon, Thomas A. McMahon, Brian L. Finlayson, Christopher J. Gippel, Rory J. Nathan
© 2004 John Wiley & Sons, Ltd ISBNs: 0-470-84357-8 (HB); 0-470-84358-6 (PB)

In this text, gauge pressure will be used. Absolute pressure may also be of biological interest (for example when comparing internal air pressures of organisms at different elevations).

When water pressure acts on a solid surface such as a fish, the side of a levee or the stalk of a water lily, the force acts at right angles to the object's surface, pressing inwards from all sides. Pressure increases with water depth, at a rate of about one atmosphere for each 10.4 m. The relationship between pressure and water depth is linear, and is described by

$$p = \rho g h \qquad (6.2)$$

where $p$ = pressure (Pa or N/m$^2$), $\rho$ = density (kg/m$^3$), $g$ = acceleration due to gravity (m/s$^2$) and $h$ = vertical distance below the water surface (m). This relationship is illustrated in Figure 6.1(a). Constant density is usually assumed for bodies of water. This is not a valid assumption if density changes over depth due to variations in temperature or dissolved solids.

It should be noted that for a liquid of constant density, pressure is dependent only on the depth of water above it. Thus, in Figure 6.1(b), where the two containers both have the same base area and are filled to the same height, the pressure on the base is the same. If force is calculated as (pressure × area) the values are the same for both containers, but if it is calculated as the 'weight' of the water (volume × density × acceleration due to gravity), the values are not the same. This phenomenon is called the *hydrostatic paradox* (Douglas *et al.*, 1983). What is

important to remember is that it is the *depth*, not the weight of the water, that is important. For example, zebra mussels on the submerged outlet gate of a dam experience the same pressure whether the reservoir stretches 10 m or 1000 m upstream.

The measurement of pressure by *manometry* uses this relationship between pressure and water depth. In Figure 6.1(c) a simple manometer attached to a fluid-filled chamber indicates the downward pressure imposed by the weight of the animal. The change in height of the manometer can be measured to obtain the change in pressure from Eq. (6.2). The density, $\rho$, would be that of the manometer fluid (e.g. water, oil, mercury or antifreeze).

By reverse application of the same principle, a measurement of pressure can be used to determine water depth. For example, pressure bulbs or bubbler gauges are often used to measure water depth at stream gauging stations, and 'snow pillows' with manometers are employed for measuring the water content of a snowpack. Pressure transducers can be used to translate the pressure into a voltage, which can then be recorded with a datalogger or telemetered to a receiving station.

*Example 6.1*

Calculate the pressure on the eye of a newt at 1 m depth in 20 °C fresh water. From Table 1.2, $\rho$ = 998.3 kg/m$^3$, so

$$\rho g h = \left(998.3\,\frac{kg}{m^3}\right)\left(9.807\,\frac{m}{s^2}\right)(1\,m)$$

$$= 9790\,N/m^2 = 9.79\,kPa$$

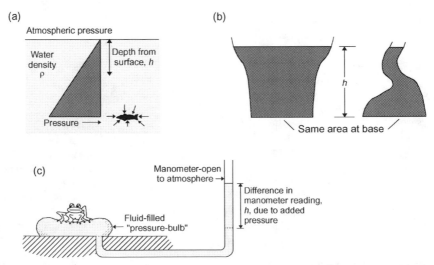

**Figure 6.1.** *Pressure in a body of water: (a) linear variation of pressure with depth, (b) demonstration of the fact that pressure is dependent only on the water depth—the pressure on the base is the same in both containers and (c) the use of manometry for measuring pressure*

## 6.2.2  Buoyancy

The principles for finding the pressures and forces on submerged objects can also be used to determine whether objects sink or float—in a word, their *buoyancy*. Objects can have *positive*, *negative* or *neutral* buoyancy depending on whether they tend to float, sink or remain where they are, respectively.

Submarines can alternate between the three states of buoyancy by pumping water in or out of 'ballast' tanks. People, too, are close to neutral buoyancy and will sink or float depending on whether their lungs are empty or full. Hippopotami must exhale a large volume of air in order to walk along the bottom of rivers and feed on underwater vegetation (Bolemon, 1989). Most fish have swim bladders containing air to regulate their buoyancy; the bladders make up about 7% of the volume of a typical freshwater fish (Bone and Marshall, 1982). The common diving beetle which carries a bubble of air under its wings to breathe from also uses it to change its buoyancy. To sink, the beetle squeezes the trapped air (compressing it and thus increasing its density), and to float back up to the surface, it releases the tension to let the air expand and increase its buoyancy. Buoyancy is not only important to submerged or floating organisms or boats, but many flow phenomena such as the mixing of warm and cold regions are dependent on small buoyant forces (White, 1986).

Archimedes proposed the following two laws of buoyancy in the third century BC:

1. An object totally immersed in a liquid experiences a vertical buoyant force equal to the weight of the liquid displaced;
2. A floating object displaces its own weight of liquid.

These laws yield the following general equation for buoyant force:

$$F = \rho g \forall \qquad (6.3)$$

where $F$ is the buoyant force (Newtons) and $\forall$ is the volume of water displaced ($m^3$). The buoyant force is caused by the differences in water pressure acting on the upper and lower surfaces of an object due to depth (see Figure 6.2(a)).

For example, if the shark in Figure 6.2(a) has a volume of 0.15 $m^3$ and the water is 0 °C, it would experience a buoyant force of

$$\left(1028 \frac{kg}{m^3}\right)\left(9.807 \frac{m}{s}\right)(0.15 m^3) = 1512 \, N$$

For floating bodies such as sitting ducks or icebergs only part of the object is submerged. The volume in Eq. (6.3) is then the volume of water displaced by the underwater portion of the object. Knowing this, the relative proportions of an iceberg sitting above and below the water (Figure 6.2(b)) can be estimated. Using the density of sea water at 0 °C and the density for ice from Table 1.2,

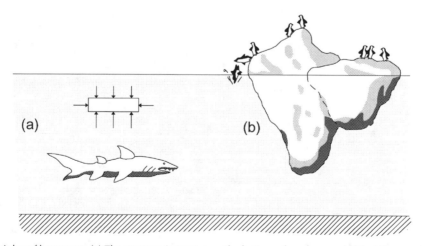

(a)

(b)

**Figure 6.2.** *Principles of buoyancy. (a) The pressure is greater on the bottom of a submerged object than on the top, creating a net upward buoyant force. The buoyant force on the shark is equal to the weight of the water it displaces. (b) For floating objects the buoyant force is equal to the weight of the water displaced by the submerged portion, which equals the weight of the floating object (iceberg + penguins)*

the calculations are performed as follows:

$$\text{Floating body (iceberg) weight} = \left(917\frac{\text{kg}}{\text{m}^3}\right)(g)$$
$$\times \text{(volume of iceberg)}$$

$$\text{Weight of sea water displaced} = \left(1028\frac{\text{kg}}{\text{m}^3}\right)(g)$$
$$\times \text{(displaced volume)}$$

From Archimedes' second principle, the two quantities are equal. By setting the two expressions equal to each other, the ratio (displaced volume/total volume) can be obtained. This is the proportion of ice submerged, 917/1028, or about 89%. The 'tip of the iceberg' would then be only 11% of the total volume. In reality, the ice in icebergs may be less dense than the value used because of air spaces between the crystals of snow and ice.

The two examples given illustrate each of the two laws of buoyancy. Both can be applied in concert to solve the riddle: 'If a person sitting in a boat on a pond tosses a brick overboard, does it raise the level of the pond?' The answer is no, it lowers the level, but the reasoning will be left to the reader.

## 6.3  Studying the Flow of Fluids

### 6.3.1  Steady and Unsteady Flow

The classification of flow as steady or unsteady (Figure 6.3) describes the way it behaves over time. Flow is considered *steady* at a point in space if its depth and velocity do not change over a given time interval. When waves or eddies travel past the point, the water level and/or velocity change from one moment to the next and the flow is said to be *unsteady*.

Turbulence causes the velocity to continuously fluctuate throughout most of the flow. However, in practice, the flow can be considered steady if values fluctuate equally around some constant value (Smith, 1975). An assumption of steady flow is necessary in the solution of many problems concerning water in motion.

### 6.3.2  Streamlines

A *streamline* is a line indicating the direction of fluid movement at a given instant (see Figure 6.6). Figure 6.4 shows isovels (lines of equal velocity). The latter is a common way of presenting data from laser Doppler anemometers. Isovels are interpreted differently from streamlines and should not be confused.

Streamlines represent the paths that would be taken by individual fluid 'particles' if the flow were steady. Convergence of the lines means the flow is accelerating;

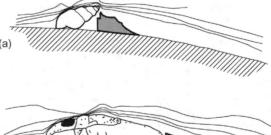

(a)

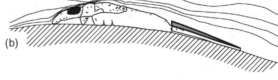

(b)

*Figure 6.4.* Isovels (lines of equal velocity) around (a) water snail (Potamopyrgus jenkinsi) and (b) mayfly nymph (Ecdyonurus cf. venosus), shown as a schematic without legs. Flow is from left to right. Free stream velocity is about 0.18 m/s in both figures, and shaded areas represent regions of zero velocity. Re-drawn from Statzner and Holm (1989, 1982), by permission of Springer-Verlag and B. Statzner

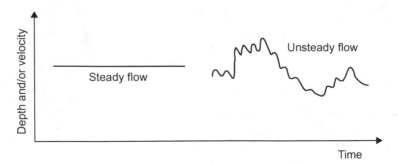

*Figure 6.3.* Classification of flow as steady or unsteady

divergence that the flow is slowing down. A *streamlined* object is shaped such that streamlines remain almost parallel as they pass around the object. For example, the snail in Figure 6.4(a) is less streamlined than the mayfly nymph in Figure 6.4(b).

In three dimensions, a *streamtube* is analogous to a 'bundle' of streamlines. Flow within a streamtube or between two streamlines follows the principle of continuity (Section 6.3.3). Thus fluid does not pass across the boundaries. Moving water is nearly always turbulent, meaning the direction of movement changes from instant to instant and particles do not travel along regular paths. Therefore, streamlines typically represent average patterns of movement, and are visualized with *streaklines* formed by the trails of dye, bubbles or particles released into the flow.

### 6.3.3   Conserving Mass: The Principle of Continuity

By the law of conservation of mass, the mass in any system must remain constant with time. For fluids, this becomes the *principle of continuity*. Strictly applied, the principle of continuity only pertains to incompressible fluids, although in most stream hydrology applications water can be considered incompressible.

In the example of steady flow through an iron pipe, where there is no opportunity for fluids to 'hide' (no storage), the principle of continuity basically states that

$$\text{Outflow} = \text{Inflow} \qquad (6.4)$$

However, in a garden hose with a weak spot that 'balloons' on filling, the equation would be

$$\text{Outflow} = \text{Inflow} \pm \text{Change in storage} \qquad (6.5)$$

where the change in storage would be the changing volume of the ballooning region. Water flowing through a reservoir with a changing water level would follow the more general second relationship (Eq. (6.5)).

Equation (6.4) can be written as follows:

$$Q = A_1 V_1 = A_2 V_2 \qquad (6.6)$$

where $Q$ = discharge (m³/s), $A$ = cross-sectional area (m²), $V$ = average velocity (m/s) and the subscripts refer to sections 1 and 2, where 1 usually represents the inflow point and 2 the outflow point, as illustrated in Figure 6.5(a).

Equation (6.6) is called the *continuity equation*. It was first derived by Leonardo da Vinci in the year 1500 (White, 1986). From the relationship, it can be seen that if

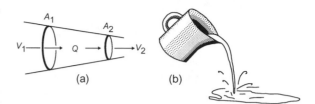

**Figure 6.5.** *The continuity principle applied to (a) a contracting section and (b) a free overfall*

discharge ($Q$) remains constant, but area ($A$) is decreased, the velocity ($V$) will go up. For example, Figure 6.5(a) could represent the tapered nozzle on a firefighter's hose which speeds water towards its destination. In a reverse manner, an increase in velocity can mean a decrease in cross-sectional area. Thus, the flow out of a pitcher contracts as it accelerates from rest in falling towards the earth (Figure 6.5(b)). A counter-intuitive example is that of a stream passing through a constriction, for example where it is partially blocked by sand bars, rock outcrops or bridge piers. At the constriction, the velocity increases and the water depth goes down rather than up as might be expected.

### 6.3.4   Energy Relationships and the Bernoulli Equation

It could be said that water arrives on a catchment with the potential to do great work: to carve smooth pathways across dense bedrock, to carry sediment or simply to rush downstream against the resistance of internal friction. Water contains a certain amount of energy, called *potential energy*, simply due to its vertical location. Water in a prospector's canteen in Death Valley, California, has less potential energy than an equivalent mass of water in a snowfield on Mount Everest. A mass $M$ (kg) of water at a height $h$ (m) above some datum (sea level, a tributary junction or some other reference level) has potential energy (PE) of

$$\text{PE} = Mgh \qquad (6.7)$$

where $g$ is the acceleration due to gravity (m/s²) and PE has units of J (joules).

Potential energy drops as water runs downhill because $h$ decreases. Most of the energy is converted to kinetic energy (KE), the energy of motion. For a mass $M$ (kg) of water, KE is given as

$$\text{KE} = \tfrac{1}{2}MV^2 \qquad (6.8)$$

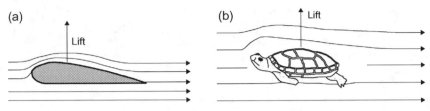

**Figure 6.6.** *Streamlines and lift on (a) an airfoil and (b) a theoretical turtle*

where $V$ is velocity (m/s) and energy and mass have units as before. In a stream, almost 95% of the kinetic energy is consumed as heat loss through turbulent mixing and friction along the bed and banks (Morisawa, 1968). Because of the high thermal capacity of water, and heat transfer from the water to its surroundings, the water temperature only rises a small amount as a result of this internal heat generation. It is also estimated that a small fraction of the energy, about 0.0001%, is converted to the characteristic sounds—the gurgles and roars—of moving water (Hawkins, 1975). The remainder of the kinetic energy is free to run turbines or move mountains a sand grain at a time.

The *Bernoulli equation*, named after Daniel Bernoulli (1700–82), translates the idea of energy conservation into terms applicable to moving fluids. The equation states that energy is conserved along a streamline:

$$KE + PE + \text{pressure energy} = \text{a constant}$$
$$\text{(along a streamline)} \quad (6.9)$$

The terms KE and PE were previously defined. Pressure energy is what people use when they inflate a bicycle tyre with a hand pump, or what archer fish use to squirt water towards aerial prey. In streams it is part of the impetus which motivates a mass of water to continue its work.

Bernoulli expressed the various energy components in terms of *head*, the energy per unit weight of fluid, which gives the terms units of length (m):

$$\frac{p}{\rho g} + \frac{V^2}{2g} + z = \text{constant} \quad (6.10)$$

or

$$\begin{array}{ccc} \text{Pressure} + & \text{velocity} + & \text{elevation} = \text{constant} \\ \text{head} & \text{head} & \text{head} \end{array}$$

Bernoulli's equation can be applied across streamlines if density is constant and the flow is considered 'ideal',

meaning that viscous effects are insignificant. Thus the equation can be applied to small objects in the bulk flow region, away from the surfaces of solids. In general, as the effect of viscosity becomes more important, the principle becomes less applicable. Although it applies to an 'ideal' situation, the Bernoulli equation provides useful approximations in a variety of situations.

Basically, the equation implies that if one term goes up, another must go down. The equation for manometry (Eq. (6.2)) and the equations for Pitot tubes (Eqs. (6.45) and (6.46)) are consistent with the Bernoulli equation. It can also be used to explain why lift occurs on airfoils such as the wings of airplanes or birds. From Figure 6.6(a) it can be seen that streamlines are compressed as they travel over the top of an airfoil, meaning the velocity is higher in that region. From Eq. (6.10) the increase in velocity must be balanced by a decrease in pressure (since the elevation is essentially constant). Thus, the pressure on the upper surface is lower than that on the underside, resulting in a net upwards force on the airfoil. The same principle applies when the fluid is water and the 'hydrofoil' is a penguin's flipper, a flat fish like a flounder or a turtle (Figure 6.6(b)). Lift is also generated when water flows over objects resting on the streambed, such as the organisms in Figure 6.4.

When the bulk flow within a stream reach is considered (e.g. for investigating energy losses over some reach length), an apparently similar but fundamentally different equation is used, the one-dimensional energy equation. This equation is discussed in Section 6.6.4.

## 6.4   Narrowing the Focus: Flow of a Viscous Fluid

### 6.4.1   Laminar and Turbulent Flow

In smoke rising from a lit cigarette the structure of the smoke is at first filamentous, rising straight up. At some point, however, the smooth structure breaks down and

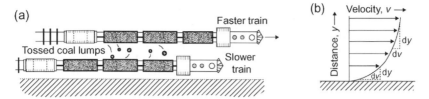

**Figure 6.7.** *Laminar flow: (a) molecular diffusion between 'layers' of fluid, represented by coal lumps tossed between parallel trains, and (b) the velocity profile for laminar flow (see text)*

the smoke curls and swirls and tumbles in erratic motion (Smith, 1975). The smooth pattern, in which all of the smoke rises upward in a uniform column, is called *laminar flow*. In *turbulent flow*, elements of fluid follow irregular, chaotic paths and violent mixing occurs, with eddies continuously forming and breaking down. Practical formulae for dealing with the two types of flow are very different.

### Laminar Flow

In streams, laminar flow may exist as a thin coating over solid surfaces, or where flow moves through the small openings between rocks in a streambed or through dense stands of aquatic weeds. Here, the fluid moves in parallel 'layers' which slide past each other at differing speeds but in the same direction. The fluid layer closest to a solid surface is retarded by the no-slip condition (Section 1.3.2) and decelerates, causing nearby fluid layers to slow empathically.

The process by which one layer encourages its neighbour to slow down is *molecular diffusion*. Binder (1958) likens the process to trains overtaking each other on parallel tracks, with coal tossed from one train to another (Figure 6.7(a)). Higher-speed coal lumps thrown into the slower-moving trains tend to speed them up; conversely, slower-speed lumps thrown into the faster trains slows them down. The resulting velocity profile is illustrated in Figure 6.7(b). As shown, the velocity increases from zero at the solid surface to a maximum value some distance away where the solid boundary no longer influences the movement of the water. Velocity profiles will be discussed further in Section 6.4.2.

If the flow is laminar the amount of force applied to a mass of fluid is directly related to how fast it travels: the more force applied, the greater the speed. This relationship is described by Newton's law of viscosity, stated as

$$\tau = \mu \frac{dv}{dy} \tag{6.11}$$

where $\tau$ is shear stress ($N/m^2$), $\mu$ is dynamic viscosity (Section 1.3.2) and $dv/dy$ is the velocity gradient, representing the rate of change of velocity, $v$, with distance, $y$. Shear stress, as defined in Section 1.2, is a 'sideways' force per unit area. Viscosity is the internal friction which causes the fluid to resist being 'pushed'. The velocity gradient arises as a result of their interaction. Near the solid surface in Figure 6.7(b) it can be seen that the change in velocity, $dv$, with distance $dy$ is relatively large. Shear stress is highest here. Away from the surface, $dv$ gets smaller in relation to $dy$, and thus the shear stress decreases.

Fluids which follow the relationship given by Eq. (6.11) are called Newtonian fluids, after Sir Isaac Newton, who first proposed this law in 1687 (White, 1986). It is perhaps fortuitous that the two most common fluids in nature—air and water—are virtually perfect Newtonian fluids. Non-Newtonian fluids such as mud, blood and paint are treated in a specialized branch of fluid mechanics called rheology.

### Turbulent Flow

Whereas laminar flow can be neatly described by a linear equation, turbulent flow can only be described *statistically*. The motion of an individual water molecule cannot be predicted mathematically, but the harmonized movement of millions of water molecules in turbulent flow can be described by averages. Thus, the velocity measured with a current meter is only a mean velocity, reflecting the average water motion as it fluctuates across the meter. The accurate mathematical modelling of turbulence remains a frontier research topic, and the description of turbulent flow relies heavily on experimentation.

Turbulence exists at all scales, from the swirling motion created when a salmon scoops out a redd to large whirlpools in a river or cyclones in the atmosphere. Larger-scale eddies tend to generate smaller ones, as in Lewis Richardson's poem (as quoted by Gleick, 1987):

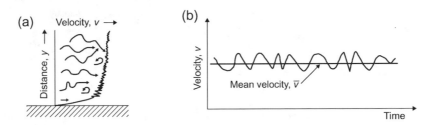

**Figure 6.8.** *Turbulent flow: (a) velocity profile near a solid surface and (b) fluctuations in velocity with time*

> Big whorls have little whorls
> which feed on their velocity;
> And little whorls have lesser whorls
> and so on to viscosity.

The poem itself is a 'spinoff' from an earlier work, *The Fleas*, by Augustus deMorgan (1806–1871).

In turbulent flow, layers of fluid break up into 'globs' which mix with other globs in a chaotic collection of eddies and swirls. It is this turbulent mixing rather than molecular diffusion which is the primary mode of speeding up or slowing down the surrounding fluid. Higher-velocity globs are swept into lower-velocity zones near a solid surface and lower-velocity globs are carried into the higher-velocity zones further away. The behaviour is more like what would occur if the 'laminar' train cars of Figure 6.7(a) were uncoupled at various points and began acting like clusters of carnival bumper cars. Diffusion still continues, but the influence of tossed coal lumps or individual water molecules on the overall motion is much less significant.

In turbulent flow, eddying mixes the higher-velocity fluid into the area closer to the solid surface. This causes the velocity profile (Figure 6.8(a)) to be flatter near the solid boundary than for laminar flow (Figure 6.7(b)). In the turbulent region the velocity is erratic, and is depicted by a 'fuzzy' line in Figure 6.8(a). A small layer of laminar flow may remain in the region near the solid, as will be described in Section 6.4.2.

In turbulent flow, Newton's law (Eq. (6.11)) becomes less relevant. The mixing of fluid 'globs' into zones of higher or lower velocity tends to augment or retard the local velocity. These fluctuations affect the mean velocity in such a way that it seems as if the viscosity has increased. To account for this effect, a term called eddy viscosity, $\varepsilon$, is added to Newton's equation:

$$\tau = (\mu + \varepsilon)\frac{\mathrm{d}v}{\mathrm{d}y} \qquad (6.12)$$

The formula encompasses situations of both laminar and turbulent flow. If the flow is entirely laminar, $\varepsilon$ is zero, and Eq. (6.12) reduces to Eq. (6.11). Alternatively, for fully turbulent flow, effects due to fluid viscosity are negligible ($\varepsilon \gg \mu$), and the equation reduces to

$$\tau = \varepsilon\frac{\mathrm{d}v}{\mathrm{d}y} \qquad (6.13)$$

There are no tables giving values for eddy viscosity, $\varepsilon$. It is dependent on how vigorous the turbulence is, and must be found by experimentation. It is not a fluid property like $\mu$. Equations (6.12) and (6.13) are used in theoretical developments, and are shown here only to demonstrate the difference in the ways laminar and turbulent flow are analysed.

Turbulence can also be described by a measure called *turbulence intensity*. It is useful to think of a local velocity in the turbulent region as composed of two parts: an average value plus a component that represents the fluctuation about the mean (Figure 6.8(b)). Turbulence intensity is a measure of the strength of the turbulent fluctuations. If $N$ instantaneous velocity measurements are made at a point, the turbulence intensity can be expressed as the root mean square of these measured values:

$$\text{Turbulence intensity} = \sqrt{\frac{\sum_{i=1}^{N}(v - \bar{v})^2}{N}} \qquad (6.14)$$

where $\bar{v}$ is the average of the velocity measurements and all variables have units of m/s. Turbulence intensity can also be expressed as a percentage by dividing the result from Eq. (6.14) by $\bar{v}$ and multiplying by 100.

Figure 6.8(b) and Eq. (6.14) apply to one component direction. Usually the component in the direction of flow (normally the horizontal component) is of primary interest. In reality, turbulent fluctuations occur in all directions, and reduce to zero at the solid surface. A common assumption is that turbulence intensities are the same in both horizontal and vertical directions since they arise from the same sets of eddies. In open channels, turbulence intensity is about 0.10 times the local mean velocity, and decreases gradually towards the water surface (Morisawa, 1985).

### The Reynolds Number

By comparing Eqs. (6.11) and (6.13) it can be seen that viscosity is an important factor in laminar flow, but becomes relatively insignificant in turbulent flow. Viscosity tends to dampen turbulence and promote laminar conditions. Acceleration has the opposite effect, promoting instability and turbulence. The resistance of an object or fluid particle to acceleration or deceleration is described by a measure called *inertia*. This is the tendency of an object to maintain its speed along a straight line. It is what keeps a particle of fluid going until it is 'aggressed upon by external authority' (Vogel, 1981, p. 67). Whereas high inertial forces promote turbulence, high viscous forces promote laminar flow. The ratio of inertial forces to viscous forces thus gives an indication of whether the flow is laminar or turbulent.

Late in the nineteenth century a famous professor of engineering, Osborne Reynolds, developed such a ratio by investigating the behaviour of flow in a glass pipe (Reynolds, 1883). A fine stream of dye was introduced into the pipe so that the flow could be visualized. For slower flows, the dye moved as a straight streak, but as the flow rate was increased, the dye stream began wavering. Laminar flow, which produced the straight streak, was termed 'direct' flow by Reynolds, and the turbulent flow which dispersed the dye was termed 'sinuous'. By varying the speed, the diameter of the pipe and density of the liquid, Reynolds tested the significance of a dimensionless number now known as the *Reynolds number, Re*:

$$Re = \frac{VL\rho}{\mu} \quad \text{or} \quad Re = \frac{VL}{\nu} \tag{6.15}$$

with

$V =$ velocity (m/s)
$L =$ some characteristic length (m)
$\rho =$ fluid density (kg/m$^3$)
$\mu =$ dynamic viscosity (N·s/m$^2$)
$\nu =$ kinematic viscosity (m$^2$/s), where $\nu = \mu/\rho$

The terms $\rho$, $\mu$ and $\nu$ are defined in Section 1.3. In Eq. (6.15), the terms in the numerator are related to

inertial forces and those in the denominator are related to viscous forces. Thus, a large value of *Re* indicates turbulence and a small value, laminar flow.

Reynolds also investigated the transition between the two types of flow in his pipe experiment. Starting with turbulent conditions, he found that the flow always became laminar when the velocity was reduced so that *Re* dropped below 2000. This point of transition is called the *critical Reynolds number*. In pipe flow, the transition will not necessarily occur at this value; in fact, laminar flow has been maintained up to $Re \approx 50\,000$ although it is highly unstable and becomes turbulent at the slightest hint of disturbance.

In pipe flow, the 'characteristic length', $L$, used in calculating the Reynolds number is the pipe diameter. Reynolds numbers can be calculated for other situations by substituting an appropriate characteristic length such as the diameter of a sand grain, the length of a fish or the width of a bird's wing. For solids immersed in a flowing fluid the convention is to use the maximum length of the object in the direction of flow. Vogel (1981) and Purcell (1977) give estimates of Reynolds numbers experienced by aquatic organisms, based on their 'typical' values of length and swimming speed, which, in order-of-magnitude terms are:

| | |
|---|---|
| 10 million | Tuna |
| 10 000 | Olympic swimmer |
| 100 | Goldfish or guppy |
| 0.1 | Invertebrate larvae |

For an aquatic organism, both the movement of the fluid and the movement of an organism within it will govern the Reynolds number. In nature, size and speed tend to work together, where 'small' nearly always means 'slow' and 'large' usually means 'fast' (Vogel, 1981). Conditions of both high and low Reynolds number are of interest biologically. Larger creatures such as trout or barramundi live in conditions where viscous forces are less significant and an occasional flip of a tail is sufficient to keep the fish moving through still water. At the low Reynolds numbers experienced by microscopic organisms, however, viscous forces become overwhelming.

Low-*Re* conditions are of little interest to engineers, but are a way of life for bacteria, protozoans or other microscopic organisms which, because of their small 'characteristic lengths', operate at Reynolds numbers in the range $10^{-4}$–$10^{-5}$. Here, inertia is irrelevant in comparison to viscosity, and movement stops immediately when propulsion ceases. Rather than sliding easily through the fluid, the organism essentially carries the fluid along with it, gradually shedding it off to the sides. For a person experiencing the same conditions, it would be like swimming in a pool filled with molasses at a speed of a few metres per week (Purcell, 1977).

The advantage of 'life at low Reynolds numbers' is that the organism is protected from the action of turbulence by a thick 'coating' of highly viscous fluid (as the organism perceives it). However, since mixing is impeded, the transport of energy, nutrients and gases to an organism, and the transport of wastes or current-dispersed gametes away from it, occurs by the slower mode of diffusion. For a sessile species, still water is a hostile environment, and many of these creatures have mechanisms for creating turbulence for improved ventilation and waste dispersal. Mobile micro-organisms may need to 'move to greener pastures' to feed rather than waiting for food to come to them. Purcell (1977) has written a delightful essay on life at low Reynolds numbers, and the reader is referred to this paper and Vogel's (1981, 1988) works for more information on the biological aspects of low-*Re* conditions.

Aquatic invertebrates may experience 'the best of both worlds', both laminar and turbulent. Statzner (1988) points out that these species start life at Reynolds numbers of about 1–10, but when they reach their adult form, they may live in conditions of $Re = 1000$ or higher. For these organisms, he concludes that evolution compromises between life at low and high Reynolds numbers.

For the case of flow in stream channels, a measure called the hydraulic radius forms an appropriate length parameter for Reynolds numbers. For pipe flow, hydraulic radius is given by

$$R = \frac{\text{Area}}{\text{Wetted perimeter}} = \frac{\pi d^2/4}{\pi d} = \frac{d}{4}$$

or

$$d = 4R$$

This suggests that the transitional value of $Re = 2000$ in pipes might be roughly translated to streams if $d$, the pipe diameter, is replaced by $4R$. In wide or rectangular streams the average depth of the water is a good approximation for the hydraulic radius. Hence, in this situation, the second form of Eq. (6.15) becomes

$$Re = \frac{VD}{\nu} \tag{6.16}$$

where $D$ is the average depth. If $Re$ is defined in this way for a stream, the transition from laminar to turbulent would be expected at $Re = 500$ or (2000/4). From experimental data, the transitional range of $Re$ for open channels

is usually considered to be from 500 to 2000, within which the flow can be either laminar or partly turbulent (Chow, 1959). The large variety of shapes and roughnesses of channels make these figures only approximate. A few trial calculations of $Re$ should convince one that the average condition of flow passing through a stream reach would almost never be classified as laminar, except perhaps where the stream is reduced to a thin film of water barely trickling over a smooth bedrock surface. For example, if $D = 2$ mm, $V = 0.1$ m/s and $\nu = 10^{-6}$ m²/s, then $Re = 200$, and the flow would be considered laminar. In the field, laminar conditions are best identified with dyes (e.g. food colouring), injected gently into the flow with a syringe.

### 6.4.2    Flow Past Solid Surfaces: The Boundary Layer

The development of velocity profiles from where a fluid contacts a solid surface to where the flow is no longer effectively influenced by the presence of the surface occurs in what is known as a *boundary layer*. Its outer limit is where the speed of the fluid matches the 'free stream velocity' —the velocity that would exist if the solid were not there. In a stream the boundary layer caused by the presence of the streambed extends to the water surface. Within its depths smaller boundary layers exist on the surfaces of rocks, snags, fish and aquatic insects; in fact, many organisms live within the boundary layers of other organisms.

As Vogel (1981, p. 129) says, 'most biologists seem to have heard of the boundary layer, but they have the fuzzy notion that it is a discrete region rather than the discrete notion that it's a fuzzy region'. Although the boundary layer is somewhat arbitrarily defined, it is a useful concept for explaining interesting phenomena like why a thin layer of dust sticks to a fan or why a sponge works better than pressurized water for washing the last thin layer of grime off a car.

The term 'boundary layer' was originally coined in 1904 by Prandtl, a German engineer. His work on theoretical descriptions of fluid behaviour near solid surfaces formed the basis for many of the engineering formulae still used today. It should be remembered that engineering fluid mechanics methods have been developed to suit the needs of engineers; modifications may be needed for application to problems in stream ecology.

'Life in the boundary layer' usually refers to the organisms which live in the relatively slower-velocity region of flow near solid surfaces such as rocks or the leaves and stems of aquatic plants. Even the most rapid streams have stones covered with 'biofilms' of

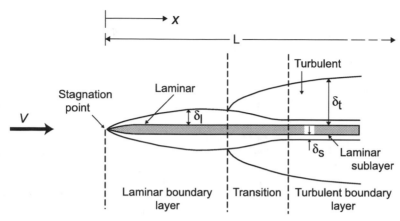

**Figure 6.9.** *Boundary layer formation across the top of a sharp, smooth flat plate (for $Re_L \approx 10^7$, where L is the 'characteristic length', in this case the length of the plate). V is the approach or 'free stream' velocity, x the distance from the leading edge, and δ the thickness of the boundary layer (shown for laminar (l), turbulent (t) and viscous sublayer (s) regions)*

micro-organisms, and mayfly nymphs (*Ephemeroptera*), caddisfly larvae (*Trichoptera*), black fly and midge larvae and pupae (*Diptera*) and others utilize the calmer micro-environments within these swift waters.

The classic approach in most engineering treatments of boundary layer theory is to first discuss the development of boundary layers in the simplest case of flow around a smooth, sharp-nosed, flat plate oriented into the flow. The distributions of velocity and shear stress around the plate are influenced both by the nature of the flow: whether laminar or turbulent, and the nature of the solid: whether rough or smooth. Although flat plates may not have a great deal of ecological significance, the relationships developed are useful for describing the patterns of velocity near surfaces within streams and for calculating skin-friction drag on boats, airfoils and aquatic organisms.

### Flow along a Sharp, Flat Plate

On a sharp, flat plate oriented into the flow, the boundary layer begins at its leading edge. The *stagnation point*, which occurs at this leading edge, is a point where the velocity of the oncoming flow is zero (stagnant) because it has collided with the object. Downstream for some distance, the flow across the plate is *laminar*. As the fluid moves further along the plate, layers of fluid at a greater distance away are slowed and the laminar layer grows. Boundary layer formation across the top surface of a sharp, flat plate is shown in Figure 6.9 for $Re_L \approx 10^7$. The subscript '*L*' refers to the use of the length of the plate, *L*, as the characteristic length for computing *Re* in Eq. (6.15). The velocity used in the equation, *V*, is the

approach velocity, or the 'free stream' velocity that would exist if the plate were not there.

This thickening of the laminar boundary layer continues with distance back from the upstream point until the thickness is so great that the flow becomes unstable and deteriorates into turbulence. The transition point occurs at some critical value of the Reynolds number, given by most authors as

$$Re_x \approx 500\,000$$

where the subscript *x* means that *x*, the distance from the leading edge, is the 'characteristic length' used for computing a 'local' Reynolds number:

$$Re_x = \frac{Vx}{\nu} \tag{6.17}$$

In the *transition region* the flow is both laminar and turbulent. At the transition a large increase in velocity occurs close to the plate as the velocity distribution shifts from a laminar to turbulent velocity profile. (see Figures 6.7(b) and 6.8(a)). This principle can be used to locate the transition region by measuring changes in velocity close to the plate with a device such as a pitot tube (Schlichting, 1961).

In the *turbulent* region further back from the leading edge, the boundary layer continues to grow outward as faster and slower 'globs' of fluid mix together. Since mixing is much more effective than molecular diffusion in encouraging neighbouring globs of fluid to behave in the same way, the turbulent boundary layer grows more rapidly than the laminar layer (see Figure 6.9). In the turbulent region, a very thin layer of laminar flow still exists near the solid surface, protecting it from violations of the no-slip condition. This layer is called the *laminar sublayer* or *viscous sublayer*.

This model of boundary layer phenomena is only valid under specific conditions. At higher Reynolds numbers the whole boundary layer shrinks and becomes turbulent closer to the 'nose' of the plate. The laminar sublayer also becomes thinner. At lower Reynolds numbers the boundary layer is thicker and the laminar region extends further back on the plate. Below $Re_L \approx 1000$, viscosity plays a more important role, turbulence disappears and the whole profile becomes laminar. At very low-$Re$ conditions of about 10, the laminar region extends out in front of the plate (White, 1986), like snow pushed in front of a snowplough.

The described model of boundary layer development is also valid only when the approaching flow is laminar or the plate itself is moving through still water and the plate is smooth. If the oncoming flow is turbulent or the leading edge of the plate is rough, turbulence will set in much sooner.

### Hydraulically Rough and Hydraulically Smooth Surfaces

What if the flat plate is not smooth? In engineering fluid mechanics, 'rough' and 'smooth' have very exact meanings, linked to the definition of the laminar sublayer. In fact, the very existence of the laminar sublayer is dependent upon how rough the surface is. A surface is said to be *hydraulically smooth* (Figure 6.10(a)) if all surface irregularities are so small that they are totally submerged in the laminar sublayer (smooth plastic pipes are hydraulically smooth). If the roughness height extends above the sublayer it will have an effect on the outside flow, and the surface is said to be *hydraulically rough* (Figure 6.10(b)). Since the thickness of the laminar sublayer varies with flow conditions, both flow velocity and roughness height will determine whether a given surface is 'smooth' or 'rough'.

Hydraulically rough conditions will be most prevalent in streams. However, where the surface irregularities become very small in comparison to the water depth, such as in smooth bedrock streams or in deep lowland rivers with streambeds of fine sand, silt and mud, hydraulically smooth flow can occur. On a finer scale, it can also occur along the surfaces of smooth objects submerged in the flow, such as smooth boulders or the leaves of macrophytes.

Much of the original work on roughness was done by Nikuradse, one of Prandtl's students. He studied frictional head losses in pipes which had been coated with sand grains of uniform sizes. Head loss was found to be a function of Reynolds number and *relative roughness*, where the latter was described as the ratio of the sand grain diameter to the pipe diameter. The idea has been extended to open channels, where relative roughness, $R_{rel}$, becomes

$$R_{rel} = \frac{k}{D} \qquad (6.18)$$

where $k$ = some measure of roughness (e.g. particle size) (m) and $D$ = depth of the water (m). In Nikuradse's work,

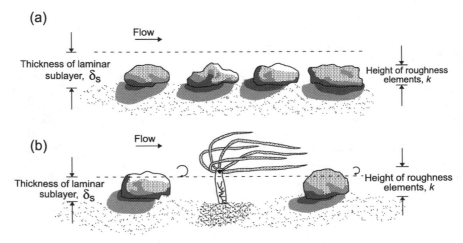

**Figure 6.10.** *Illustration of (a) hydraulically smooth and (b) hydraulically rough surfaces*

**Table 6.1.** *Experimentally derived values of effective roughness height, k. From Chow (1959); reproduced by permission of McGraw-Hill*

| Material | $k$, mm |
|---|---|
| Brass, copper, lead, glass | 0.030–0.90 |
| Galvanized iron | 0.15–4.6 |
| Wood stave | 0.18–0.90 |
| Cement | 0.40–1.2 |
| Concrete | 0.46–3.0 |
| Riveted steel | 0.90–9.0 |
| Natural river bed | 30–900 |

$k$ was defined by the particle diameter, with all grains of the same size. His work provided an important benchmark for determining the 'effective' roughness of surfaces such as concrete, where the surface is more irregular. A few approximate values are given in Table 6.1.

In streams, the roughness height, $k$, is defined by its effect on the flow, and varies not only with the grain size distribution of streambed materials but also with how the roughness elements project into the flow and how they are arranged—including the arrangement of larger 'roughness elements' (e.g. sand dunes and piles of woody debris) and smaller roughness elements (e.g. surface irregularities on individual rocks). Typically, some characteristic diameter of the streambed materials such as the $d_{50}$ or $d_{85}$ (see Section 5.8.5) is used as the roughness height.

A 'roughness' Reynolds number, $Re_*$, can be developed using *shear velocity*, $V_*$ and the roughness height, $k$:

$$Re_* = \frac{V_* k}{\nu} \tag{6.19}$$

(compare with Eq. (6.15)). Here, $V_*$ is a measure of shear stress expressed in velocity units (m/s). It is computed as

$$V_* = \sqrt{\frac{\tau}{\rho}} \tag{6.20}$$

where $\tau$ is the shear stress acting at the surface of a solid (N/m$^2$) and $\rho$ is the density (kg/m$^3$) of the fluid.

A surface is considered hydraulically smooth if $Re_* < 5$, hydraulically rough if $Re_* > 70$ and transitional at $5 < Re_* < 70$ (Schlichting, 1961). Thus, the flow near a solid surface will be disturbed if either (1) the roughness elements increase in height or (2) the velocity increases, causing the laminar sublayer to become smaller than the height of the projections. Davis and Barmuta (1989) state that the roughness Reynolds number appears to be an excellent habitat descriptor since it combines the effects of velocity and substrate type.

### Arranging the Surface Roughness

The way in which objects or organisms are spaced can affect the patterns of flow around them. Morris (1955) explored the concept of roughness spacing in pipes. He proposed that the eddies created between roughness elements would have an effect on flow resistance. Both the longitudinal spacing and height of the roughness elements were considered important. Morris classified flow over rough surfaces into three categories, described by the following conditions (parameters are defined in Figure 6.11).

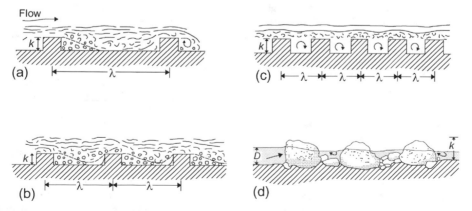

**Figure 6.11.** *The classification of flow near rough surfaces. Diagrammatic illustrations of the flow patterns in (a) isolated roughness flow, (b) wake-interference flow, (c) skimming flow and (d) exposed roughness flow. In (a)–(c), based on the classification of Morris (1955), water depth is fairly large in relation to roughness height; in (d), roughness elements break through the water surface*

1. *Isolated-roughness flow* (Figure 6.11(a)): eddies which form behind each element dissipate before the next element is reached. Isolated-roughness flow will occur when $k/\lambda$ is small.
2. *Wake-interference flow* (Figure 6.11(b)): roughness elements are closer together, and the eddies from the elements interact, causing intense turbulence. Here, roughness height is relatively unimportant compared to the spacing. The depth of flow above the crests of the elements becomes important since it will limit the vertical extent of increased turbulence. Wake interference flow can occur over surfaces of considerable roughness such as corrugated metal (Morris, 1961).
3. *Skimming flow* (Figure 6.11(c)): elements are so close together that the flow 'skims' over the tops of the elements, with low-velocity eddies occurring in the grooves between the elements. The surface acts almost as if it is hydraulically smooth. Skimming flow occurs when $k/\lambda$ is high.

The above three categories are applicable when the water depth is much greater than the height of the roughness elements. Davis and Barmuta (1989) introduced another category for the situation where the roughness element height exceeds one-third of the water depth. An additional category could be included for the situation often found in streams where the roughness elements break through the water surface:

4. *Exposed roughness flow* (Figure 6.11(d)): elements protrude through the water surface. Flow conditions become very complex as water flows over and around these large obstacles, often forming 'whitewater' conditions.

In his study of flow resistance, Morris (1955, 1961) found it convenient to categorize boundary roughness patterns into three broad categories. As suggested by Davis and Barmuta (1989), the classification of boundary roughness patterns may prove to be useful in the study of near-bed environments. It also has relevance in fish-ladder design and the creation of specific flow patterns in stream habitat rehabilitation.

Although Morris used roughness elements of uniform height, his classifications can be extended to surfaces with variable roughness heights and spacing by using average values of the dimensions (Chow, 1959). Gore (1978) and Wetmore *et al.* (1990) describe surface profilers for measuring the local roughness of streambed materials. These instruments are similar to the point frequency frame (Section 5.3.3), where pins are dropped to the surface and the height of each pin recorded. The standard deviation of the pin heights is used as an indication of surface relief.

## A Reality Check

In these somewhat theoretical developments the three important concepts to bear in mind are (1) the no-slip condition, which causes flow to stick to solid surfaces and velocity profiles to develop; (2) the fact that shear stress is highest near solid surfaces such as a rock or streambed and (3) the fact that turbulence is highest near the streambed. In most parts of a stream, flow can be considered turbulent. The velocity profile is actually a continuum—like ecological zones, the division into laminar and turbulent is a convenient way of viewing regions with unique properties that blend together in a transitional zone. The separation is mathematically convenient.

Even under hydraulically smooth conditions, the viscous sublayer, rather than being stable, continuously fluctuates in thickness. For example, when the flow velocity is increased over a smooth bed of sand grains the laminar sublayer shrinks, and the sand grains are eventually exposed to the turbulence which lifts them from the bed. However, this does not occur uniformly over the whole bed surface. Instead, 'patches' of sand are plucked from the bed as the viscous sublayer breaks down in places and the bed is exposed to energetic eddies from the turbulent zone (I. O'Neill, personal communication, 1990). This 'buffeting' effect is no doubt highly significant to organisms living in the high-shear stress, high-turbulence area near the streambed.

Boundary layers are imposed on other boundary layers, and scale determines which one is of interest. For a person picking a spot to sunbathe on a windy beach it pays to find the lee side of a sand dune; a fly landing on the lee side of the person's nose would be similarly sheltered. Within the turbulent boundary layer of a stream, boundary layers form at the surfaces of individual rocks and at the surfaces of snails, mayfly larvae or other organisms sitting on the rocks.

Statzner and Holm (1982) used laser Doppler anemometry to observe flow conditions around *Ecdyonurus* (Mayfly) nymphs as the classic case of 'life in the boundary layer'. From their studies they surmised that it was unlikely that the boundary layer concept could adequately explain the nature of flow around animals. Rather, the flow patterns may have more to do with morphological structures and/or behaviour mechanisms which allow the animals to direct shearing forces to a point where they can be counteracted (e.g. to legs or other 'anchoring' appendages). Vogel (1981) also made the point that the idea of organisms having flattened shapes to reduce drag and help them stay attached to surfaces was a reasonable theory but an oversimplification.

The tools presented in the following sections form a starting point firmly grounded in the field of engineering, from which modifications can be developed for biological applications. Questions which might arise at this point are: 'How thick is the boundary layer?' 'How thick is the laminar sublayer?' 'How fast is the flow at a given point?' and 'What is the drag force on an object within the layer?' Formulae for these quantities are presented in the next section. Since most of the original work was done with pipes, cylinders and flat plates rather than meandering stream channels, irregular rocks and leaves of aquatic plants, the equations should not be applied as hard and fast rules but as approximations. Because flow patterns over and around objects are difficult to describe, simplifying assumptions are made so that problems can be dealt with empirically and theoretically. If budgets, time and equipment permit, the ideal solution is to directly measure velocity profiles, and a few methods are discussed in Section 6.5.6.

## 6.5   The Micro-environment: Flow Near Solid Surfaces

### 6.5.1   General

In the micro-environment the focus is on patterns of viscous action as fluid passes by a solid surface. The boundary layers of interest are those which are produced around objects within the flowing fluid. Finer-scale turbulence such as the small eddies behind objects and velocity fluctuations near the surfaces of solids becomes important.

The patterns of flow within the micro-environment form an important component of the physical habitat for aquatic organisms. As water flows by a solid surface it can generate lift and drag on a sediment particle or a benthic invertebrate. Dead-water zones are created on the downstream sides of boulders which affect the speed of water encountered by migrating fish. Flow patterns can influence the behaviour of an organism clinging to a stone or the stalk of a reed. The distribution of flow patterns in streams has no doubt played a part in the evolution of organisms which are best suited to particular flow environments.

Presented in the next two sections are formulae for describing the distribution of velocity and the thickness of boundary layers for the somewhat idealized case of flow around a sharp, flat plate. From this point, the description of flow around 'bluff bodies' which present a less anorexic profile are developed in Section 6.5.4. Implications in terms of lift and drag are covered in Sections 6.5.4 and 6.5.5, and Section 6.5.6 presents methods of measuring velocity in the microenvironment.

### 6.5.2   Describing the Velocity Profile and Boundary Layer Thickness near a Solid Surface

#### The Laminar Region

As illustrated in Figure 6.7(b), the velocity profile in a laminar boundary layer has a smooth, parabolic shape. Blasius, a student of Prandtl, developed a relationship for describing the way velocity changes with depth in laminar flow over a flat plate. Tables of values for the Blasius velocity profile can be found in White (1986). Although the profile is not described by an exact formula, it can be approximated by a parabolic equation:

$$v \approx V\left(\frac{2y}{\delta_1} - \frac{y^2}{\delta_1^2}\right) \qquad (6.21)$$

where $y$ = distance away from the plate (m), $V$ = free stream velocity (m/s), $v$ = velocity at some distance $x$ along the plate and some distance $y$ above it and $\delta_1$ is the laminar boundary layer thickness at some distance $x$ from the leading edge of the plate (see Figure 6.9). Because there is really no 'outer limit' to the velocity profile (i.e. the retarding effect on the flow continues for large distances), an arbitrary limit was chosen where its effect was considered negligible. Blasius assumed that the edge of the boundary layer occurs at $v = 0.99V$, and developed this equation for boundary layer thickness:

$$\delta_1 = \frac{5.0x}{\sqrt{Re_x}} \qquad (6.22)$$

where $x$ is, again, the distance from the front of the plate and $Re_x$ is the 'local Reynolds number' computed from Eq. (6.17). The Blasius solution has been found to correspond well with experimental values (Roberson and Crowe, 1990).

#### The Turbulent Region

If we 'zoom in' on the turbulent velocity profile of Figure 6.8(a) the average pattern (without the fuzzy lines) will look like either Figure 6.12(a) or 6.12(b), depending on whether the solid surface is hydraulically smooth or rough, respectively. These profiles describe fully developed boundary layers, meaning they are no longer in transition from the laminar conditions at the front of a plate or the entrance to a pipe, flume or channel. In most situations the laminar region is so short in comparison with the total length of the surface that it can be ignored.

The velocity profile in the turbulent zone for either hydraulically smooth or rough surfaces is normally described by a logarithmic equation. Prandtl developed

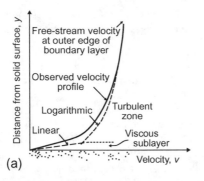
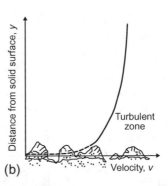

**Figure 6.12.** *Turbulent velocity profiles and their descriptors for (a) a hydraulically smooth surface and (b) a hydraulically rough one. The curves represent averages; in reality, velocity fluctuates about the average profile in the turbulent zone. Adapted from Daily and Harleman (1966), by permission of Harleman*

the *universal velocity-distribution law* for describing turbulent velocity profiles by applying a few mathematical sleights-of-hand such as assuming that 'globs' of fluid exchange momentum (to speed others up or slow them down) over some 'mixing length'. The equation which resulted is of the form

$$\frac{v}{V_*} = \frac{1}{\kappa} \ln\left(\frac{yV_*}{\nu}\right) + B \qquad (6.23)$$

where $v$ is the velocity (m/s) as it varies with distance $y$ (m) away from a solid surface. $V_*$ is the shear velocity (m/s), described in Eq. (6.20) as the square root of shear stress divided by fluid density. Thus, the expressions on both sides of Eq. (6.23) are dimensionless. The factors $\kappa$ and $B$ are empirically derived constants, varying with the type, concentration and size variation of roughness and the shape of the pipe or channel conveying the flow. Equation (6.23) is usually attributed to both Prandtl and von Karman (another student of Prandtl's), and $\kappa$ is sometimes called Karman's universal constant. From experimentation, the nominal value of $\kappa$ is 0.40, but it can vary widely.

For *hydraulically smooth* boundaries (and base 10 logarithms), Eq. (6.23) becomes

$$\frac{v}{V_*} = 5.75 \log\left(\frac{yV_*}{\nu}\right) + 5.5 \qquad (6.24)$$

In practice, the coefficients $1/\kappa$ and $B$ may differ slightly from 5.75 and 5.5, respectively.

For *hydraulically rough* conditions, the equation for the velocity distribution is usually given as

$$\frac{v}{V_*} = 5.75 \log\left(\frac{y}{k}\right) + 8.5$$

or

$$\frac{v}{V_*} = 5.75 \log\left(\frac{30y}{k}\right) \qquad (6.25)$$

where $k$ is a roughness measure (see Eq. (6.18)) and the other terms are as defined for Eq. (6.23). Eq. (6.25) is given in two forms; in the second version, the constant 8.5 has been incorporated into the log term. The velocity profile is normally assumed to apply down to $y = 0.5k$, a depth of one-half the roughness height. If actual velocity measurements are taken, the equation can be fitted to the measured velocity profile to calculate $V_*$, the shear velocity (see Section 6.5.3).

For a smooth, flat plate (or similar objects within the flow), the turbulent boundary layer can be described by the *one-seventh power law*, a simple relationship which provides a good fit when the Reynolds number is between about $10^5$ and $10^7$ (White, 1986):

$$v \approx V\left(\frac{y}{\delta_t}\right)^{1/7} \qquad (6.26)$$

where $V$ is the free-stream velocity (m/s) and $\delta_t$ the thickness of the turbulent boundary layer, commonly given as

$$\delta_t = \frac{0.37x}{Re_x^{1/5}} \qquad (6.27)$$

Here $x$ is, again, the distance from the leading edge of the flat plate (Figure 6.9). By comparing Eqs. (6.22) and (6.27) it can be shown that the turbulent boundary layer will grow more rapidly than the laminar layer with distance along the plate. Example 6.3 illustrates the use of turbulent zone velocity equations.

### The Viscous Sublayer

Perhaps of more interest to organisms living on submerged surfaces is the thickness of the laminar or viscous sublayer, which exists when the surface is hydraulically smooth. In modelling the turbulent velocity profile as a logarithmic function (e.g. Eq. (6.23)) a problem develops as the curve approaches the solid surface because the logarithm of zero makes no sense. To account for this anomaly, Prandtl used a linear velocity profile near the solid surface to 'connect' the logarithmic profile to it. This region was assumed to be the region of laminar flow, in which viscous effects are concentrated.

The velocity profile in this region is described by

$$\frac{v}{V_*} = \frac{V_* y}{\nu} \qquad (6.28)$$

where the terms are defined as before. The thickness of the laminar sublayer, $\delta_s$, is defined by the point where the logarithmic and linear profiles intersect, most often given by

$$\delta_s = \frac{11.8\nu}{V_*} \qquad (6.29)$$

It was formerly a commonly held belief that stream invertebrates which live on substrate surfaces in high-velocity environments have flattened body shapes to allow them to lead a 'sheltered life' in the laminar sublayer. Smith (1975) used Eq. (6.29) and an approximation that the mean stream velocity is twenty times the shear velocity to compute several values of the laminar sublayer thickness (for water at 15 °C):

| Mean velocity, $\bar{v}$(m/s) | 0.01 | 0.05 | 0.10 | 0.50 | 1.00 |
|---|---|---|---|---|---|
| $\delta_s$(mm) | 27 | 5.4 | 2.7 | 0.54 | 0.27 |

He concluded from these calculations that with the mean velocities normally found in streams, 'it seems unlikely that the larger invertebrates can be considered as being sheltered in the laminar sub-layer' (Smith, 1975, p. 36). In fact, their presence may actually cause the surface to be considered hydraulically rough, in which case the laminar sublayer does not exist (Davis and Barmuta, 1989). It is more likely that these organisms have flattened body shapes for other reasons such as ease of movement under and through rocks that provide shelter from the current.

The separation of the velocity profile into zones allows the separate mathematical description of a region in which viscous effects are important and one where they are no longer significant. From a physical and biological perspective, however, it may be preferable to maintain the view of the profile as a continuum, with a lower-velocity region near a solid surface and eddies of varying scale throughout.

### 6.5.3  Shear Stress and Drag Forces

The shape of the velocity profile, particularly in the region closest to a solid surface, will reflect the amount of shear stress that the flow exerts on a surface submerged in the flow. We know from Eqs. (6.11)–(6.13) that shear stress is related to the steepness of the profile. Considering Figure 6.9 again, it can be seen that the boundary layer thickens with distance along the flat plate. At the plate, the velocity is zero, and at the outer edge of the boundary layer the velocity is nearly equal to the free stream velocity. As the boundary layer thickens, this difference in velocity is 'stretched' over increasingly longer distances, and the velocity gradient (the ratio d$v$ to d$y$) decreases. Thus, shear stress generally reduces with distance back from the leading edge, as shown in Figure 6.13, with a jump at the transition from laminar to turbulent.

By integrating shear stress over the surface area, a value can be obtained for the total shearing force. This is the drag force or 'skin friction' force on the plate, symbolized by $F_s$. For one side of the plate, it is

$$F_s = C_f W L \rho \frac{V^2}{2} \qquad (6.30)$$

where $W$ and $L$ are width and length of the plate, respectively, and $C_f$ is a skin friction coefficient which depends on the flow type (e.g. whether laminar or turbulent). To calculate the total drag on both sides of the plate (for example, on both sides of a fish), the 2 in the denominator of Eq. (6.30) would be eliminated.

For a fish swimming within the flow, $F_s$ might represent the amount of force that the fish must exert to maintain the

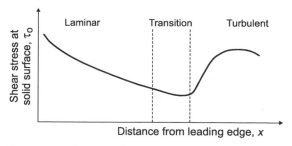

*Figure 6.13. Change in shear stress with distance along a flat plate (compare with Figure 6.9). Adapted from Douglas et al. (1983), by permission of Longman Group, UK*

same speed. For an object on the streambed, such as a flat boulder, a waterlogged leaf or a benthic invertebrate, it denotes the resistance to flow caused by the object, and is thus related to the likelihood with which it will be swept away. Organisms living on surfaces within the stream must balance the benefits of thinner boundary layers (greater mixing, supply of nutrients and gases, and waste removal) with increased shear stresses and problems of remaining attached. Microbes or suspension-feeding organisms located in the thinner boundary layer at the leading edge of a flat plate, for example, would have better supplies of nutrients and gases than those within thicker layers further back, yet the shear stress would be higher (see Figures 6.9 and 6.13).

### The Laminar Region

In laminar flow the local shear stress, $\tau_o$, is given by

$$\tau_o = 0.332 \mu V \frac{Re_x^{1/2}}{x} \qquad (6.31)$$

where $\tau_o$ is in N/m$^2$, $\mu$ is fluid viscosity and the other terms are as previously defined. This equation can be used to obtain the local shear stress at any point along the plate within the laminar region.

The skin friction force or drag force, $F_s$, is calculated from Eq. (6.30), where

$$C_f = \frac{1.33}{Re_L^{1/2}} \qquad (6.32)$$

Here, $Re_L$ is the Reynolds number in which the 'characteristic length' is the plate length, with length measured in the direction of flow.

### The Turbulent Region

Because of turbulent fluctuations, shear stress is increased above what would be predicted from average velocities. The shear stresses caused by turbulence are referred to as *apparent shear stresses* or *Reynolds stresses*. Assuming that the turbulent boundary layer starts at the leading edge, equations can be developed for shear stress and skin-friction drag. Shear stress at the solid surface, $\tau_o$, is given as

$$\tau_o = \rho c_f \frac{V^2}{2} \qquad (6.33)$$

where $c_f$ is the 'local skin friction coefficient' at a distance $x$ from the leading edge. For *hydraulically smooth* conditions,

$$c_f = \frac{0.058}{Re_x^{1/5}} \qquad (6.34)$$

The drag force or overall shear resistance force is equal to the shear stress integrated over the length of the plate. For hydraulically smooth conditions (and for one side of the plate) this is given by Eq. (6.30), with

$$C_f = \frac{0.074}{Re_L^{1/5}} \qquad (6.35)$$

As for Eq. (6.32), $Re_L$ is the Reynolds number based on the plate length. The relationship between $C_f$ and $Re$ for flat plates is included in Figure 6.20 (in the figure, $C_f$ is given as $C_D$, the total drag coefficient). For higher Reynolds numbers, more refined analyses are required. A careful comparison of Eqs. (6.32) and (6.35) will reveal that a strategy of maintaining a laminar boundary layer as far along the plate as possible will reduce drag.

If the smooth plate grows barnacles or is otherwise considered *hydraulically rough*, the resistance coefficients are related to the relative roughness. White (1986) gives these equations for resistance coefficients in the fully rough regime:

$$c_f \approx \left(2.87 + 1.58 \log \frac{x}{k}\right)^{-2.5} \qquad (6.36)$$

$$C_f \approx \left(1.89 + 1.62 \log \frac{L}{k}\right)^{-2.5} \qquad (6.37)$$

where $k$ is a measure of roughness (m) (see Eq. (6.18)), and the other terms are as previously defined.

### The Transition Region

For the transition from laminar to turbulent flow ($500\,000 < Re_L < 10^7$) the boundary layer will be laminar along part of the upstream section of the plate. Prandtl developed the following 'hybridized' equation for the transition region (Streeter and Wylie, 1979):

$$C_f = \frac{0.074}{Re_L^{1/5}} - \frac{1700}{Re_L} \qquad (6.38)$$

*Example 6.2*

(a) Determine the total drag on a smooth, flat plate of length 0.5 m and width 0.23 m, towed through still water (10 °C) at a speed of 2 m/s.

(b) Determine the drag if the plate is rough, with an effective roughness height of 1 mm.

(a) The Reynolds number, $Re_L$ (Eq. (6.15)) is:

$$Re_L = \frac{(2)(0.5)}{(1.308 \times 10^{-6})} = 765\,000$$

thus the flow is transitional. From Equation 6.38

$$C_f = \frac{0.074}{(765\,000)^{1/5}} - \frac{1700}{765\,000} = 0.002\,70$$

and drag on both sides of the plate (Eq. (6.30)) is

$$F_s = (0.00270)(0.23)(0.5)(999.7)(2)^2 = \mathbf{1.24\,N}$$

(b) From Eq. (6.37) (we are assuming the flow is hydraulically rough and fully turbulent),

$$C_f \approx \left[1.89 + 1.62\log\left(\frac{0.5}{0.001}\right)\right]^{-2.5} = 0.0102$$

$F_s = (0.0102)(0.23)(0.5)(999.7)(2)^2 = \mathbf{4.69\,N}$—almost four times the drag force on the smooth plate!

*Example 6.3*

If a bacterium were released into the flow near a stationary, thin, flat piece of wood oriented into the flow, at what velocity would it travel when it reached a point 140 mm back from the leading edge of the solid and 5 mm away from it? Assume the boundary layer is turbulent, the surface is rough, the free stream velocity is 0.6 m/s and the water is 15 °C. Use (a) Eq. (6.25) and (b) the one-seventh power law (Eq. (6.26)).

(a) Not knowing anything about how rough the wood is, we'll start by selecting a value of $k$ within the range given in Table 6.1, say 0.50 mm (0.0005 m). From Eq. (6.36):

$$c_f = \left[2.87 + 1.58\log\left(\frac{0.140}{0.0005}\right)\right]^{-2.5} = 0.008\,49$$

By combining Eqs. (6.20) and (6.33),

$$V_* = \sqrt{\frac{\tau_0}{\rho}} = \sqrt{c_f\frac{V^2}{2}} = \sqrt{0.00849\frac{(0.6)^2}{2}} = 0.0391\ \text{m/s}.$$

Then, from Eq. (6.25),

$$v = (0.0391)\left[5.75\log\left(\frac{30(0.005)}{0.0005}\right)\right] = \mathbf{0.557}\ \text{m/s},$$

almost the free-stream velocity.

(b) From Eq. (6.17),

$$Re_x = \frac{(0.6)(0.140)}{(1.141 \times 10^{-6})} = 73\,600$$

Then, from Eqs. (6.27) and (6.26),

$$\delta_t = \frac{0.37(0.140)}{(73\,600)^{1/5}} = 0.005\,51\ \text{m}$$

$$v \approx 0.6\left(\frac{0.005}{0.00551}\right)^{1/7} = \mathbf{0.592}\ \text{m/s}$$

Although the answers are not identical, they are of the same order of accuracy, and approach (b) is much simpler.

### Calculating Shear Stress from the Velocity Profile

If the velocity profile is measured near a surface and it plots approximately as a straight line on a linear-log plot such as in Figure 6.14(b), then shear stress can be calculated from the slope of the profile. From Eq. (6.20), $\tau_0 = \rho(V_*)^2$, where the equation for $V_*$ along

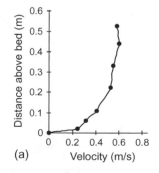
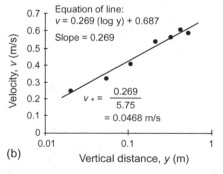

**Figure 6.14.** *Calculation of shear velocity on a hydraulically rough streambed from velocity measurements taken at a vertical at Site 4 on the Acheron River (Figure 4.7): (a) measured velocity profile and (b) regression line (see Appendix) fitted to the relationship between velocity and log(depth). Notice that the axes are reversed in (b) because v is the dependent variable in velocity distribution equations (see Eqs. (6.23)–(6.26))*

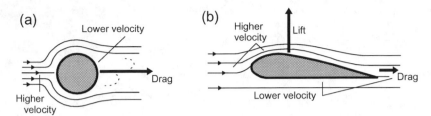

**Figure 6.15.** *Lift and drag on bluff bodies in a viscous fluid*

a hydraulically-rough streambed is obtained from Eq. (6.25) and can be expressed as

$$V_* = \frac{b}{5.75} \qquad (6.39)$$

where $b$ is the slope of the logarithmic velocity profile (velocity versus log(depth)). An example calculation of $V_*$ is shown in Figure 6.14.

### 6.5.4   Flow around Bluff Bodies

The idealized case of flow across a sharp, flat plate is useful in developing a foundation for boundary layer theory. In this section, flow patterns around immersed objects like sediment particles and crocodiles are examined. In engineering terminology these objects are referred to as *bluff bodies* or *blunt bodies*.

Normally, when flow occurs around a bluff body the velocities and pressures on opposite sides of the object will be different, producing a force perpendicular to the oncoming flow (lift) and/or a force in line with the direction of the flow (drag), as shown in Figure 6.15 (see also Figure 7.17). If viscosity did not exist, drag would not occur—a piece of wood dropped into a flowing stream would remain where it had been dropped rather than floating downstream, and a fish (or other neutrally buoyant example) set into motion would have difficulty stopping.

### Separation and Wake

Viscous fluids stick to solids and lose energy to internal friction. Several noticeable features can be observed in the patterns of streamlines around the objects in Figure 6.16. On the upstream side of each object a streamline 'collides' with the object, creating a *stagnation point* (a point of zero velocity). Boundary layer development begins here. As the fluid progresses around the solid, it loses energy to friction, and at some point the boundary layer and the solid part company. This is called the *separation point* and the phenomenon *boundary layer separation*. Analogous to congested traffic, when one particle stops, others stack up behind it or attempt to move around the 'stalled' particle, and the boundary layer thickens. Particles are carried out into the free-stream flow, and the boundary layer as a whole is stretched and drawn away from the solid. A 'gap' is left, and downstream particles are pulled up into the region, producing a flow reversal or *vortex*. Intense turbulence and energy dissipation occur in this region, called the *wake*.

Within the wake, a slow, lazy, retarded flow region occurs just behind the solid, called '*dead water*' or the 'cavity region'. It is here that many aquatic species take refuge from high-speed currents. For objects resembling cylinders (Figure 6.16(b)) a rough rule of thumb is that the length of the dead water region is about one-half of the length of the bluff body. A *stagnation region* of near-zero velocity can also occur in front of a bluff body, such as the flat plate of Figure 6.16(a). In a study of diatom

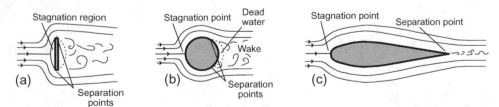

**Figure 6.16.** *Streamlines and flow separation around (a) a flat plate perpendicular to the flow, (b) a cylinder and (c) a streamlined object*

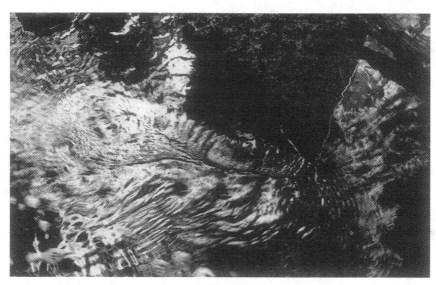

*Figure 6.17.* *Separation and vortex formation at the edge of a rock outcrop. Flow direction is from right to left. Photo: N. Gordon*

colonization on disc-shaped glass 'stubs' placed on a streambed, Korte and Blinn (1983) found increased colonization at the upper leading and trailing edges, and attributed this to interception of drifting organisms by the upstream face and the presence of micro-eddies (separation) downstream.

Although prediction of the separation point has not been resolved mathematically it is easily recognized by observation. The point of separation is dependent on the shape and roughness of the object and boundary layer properties, including whether the flow is turbulent or laminar. Separation tends to take place at sharp breaks in the surface of solids. In streams it occurs not only around bluff bodies like irregular rock outcrops or a stream ecologist's waders but also at the edges of the stream where it widens suddenly and at the downstream side of bends (see Figure 6.17).

In general, separation does not occur when the fluid is speeding up but is favoured by deceleration. The more abrupt the deceleration, the more likely separation is to occur—analogous to the way an abrupt deceleration causes car drivers to be separated from their seats. Streamlining (Figure 6.16(c)) allows flow to decelerate gradually, which delays the onset of separation (Schlichting, 1961). As might be guessed, separation is an important factor in determining drag on an object.

### Vortex Formation and Shedding

The eddying patterns behind an object change in picturesque and dramatic ways with the flow speed. Using the

example of a circular cylinder, separation is not considered to occur at Reynolds numbers below about 5 (the value is not known precisely) because high viscous effects inhibit eddying motion (Roberson and Crowe, 1990). Here, nearly all the drag is due to skin friction. Above $Re \approx 5$, separation starts to occur (Figure 6.18(a)). Initially, eddies in the wake region are symmetrical, as shown in Figure 6.18(b). Downstream of the cylinder the wake begins to oscillate sinusoidally. As the flow speeds up, inevitably one eddy will grow faster than the other until the largest one is 'shed' into the flow and a turbulent, oscillating wake develops behind the cylinder.

Along with conservation laws pertaining to energy and mass, fluids must also conform to the law of *conservation of vorticity*. This states that circulation created in one place must be balanced by an equal and opposite circulation elsewhere (Vogel, 1981). Thus, vortices are shed from alternating sides of an object and each rotates in an opposite direction to the one just before it. This type of wake region with alternating vortices is called a 'von Karman trail' or 'Karman vortex street', after von Karman who gave it a theoretical explanation in 1912. The effect of increasing $Re$ can be seen in Figure 6.19.

The elegant flow patterns of vortices around a cylinder are described as *unsteady laminar flow* at Reynolds numbers below about $10^5$ (with cylinder diameter the 'characteristic length' in Eq. (6.15)). When $Re$ is somewhere between 100 000 and 250 000 the flow becomes turbulent. In the turbulent boundary layer, fluid particles near the solid surface have higher velocities, so a fluid particle will travel further around the cylinder before

(a)    (b)

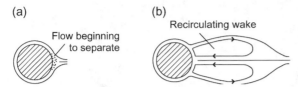

**Figure 6.18.** *Symmetrical eddying around a circular cylinder at (a) Re ≈ 5 and (b) Re ≈ 40*

separation occurs. The wake is thus narrowed and drag is reduced at the critical Reynolds number where the transition from laminar to turbulent flow occurs. This phenomenon will be discussed further in the next sub-section.

Shedding of vortices sets up a rhythmic side-to-side force on the cylinder. If this coincides with a natural vibration frequency of the solid, it can produce resonance—which gives rise to the 'singing' of wires in wind, and which played a part in the dramatic failure in 1940 of the Tacoma Narrows suspension bridge in Washington, USA. A researcher holding a rod in swiftly flowing streams when taking current speed or depth measurements will have a 'feel' for the vibrations caused by vortex shedding.

The frequency of vortex shedding is described by the Strouhal number, *St*, named after a physicist who experimented with singing wires in 1878:

$$St = \frac{\eta d}{V} \qquad (6.40)$$

where $\eta$ = frequency of shedding of vortices from *one* side of the cylinder, in Hz (hertz), $d$ = cylinder diameter

(m) and $V$ = free-stream velocity (m/s). *St* is a function of Reynolds number, with $d$ the characteristic length. Vortex shedding occurs in the range $10^2 < Re < 10^7$. Relationships between *St* and *Re* are given in most fluid mechanics texts. An average value for *St* is about 0.21 (White, 1986).

*Example 6.4*

Calculate the frequency of vortex shedding for the rigid stem of a reed at a point where the stem diameter is 5 mm and the mean flow velocity is 0.3 m/s.

$$St \approx 0.21 = \frac{\eta(0.005\,\mathrm{m})}{0.3\,\mathrm{m/s}}$$

$$\eta = 12.6\ \text{vibrations/second}$$

Vortices can also occur in other situations. For example, when birds fly in formation the vertical vortices shed from the wingtip of one bird gives an added lift to the bird flying just next to it. Whirlpools are vortices (see Figure 6.17), and 'boils' (areas of strong upwelling) can form when vortices detach from the streambed, affecting bed scour and sediment transport (Wohl, 2000). A 'kolk' is an upward-spiralling vortex which can occur in streambed depressions, and produces a strong hydrodynamic lift (Morisawa, 1985). A 'horseshoe vortex' is a region of circulating fluid motion around an obstacle, wrapped around the upstream side with the arms of the 'horseshoe' stretching downstream. It is responsible for the semicircular scour patterns around bridge piers (Bunte and Abt, 2001). On a micro-environment scale, a horseshoe vortex system is present around an isolated organism or other object when its height is close to that of the viscous boundary layer thickness (Davis, 1986). Vortex patterns

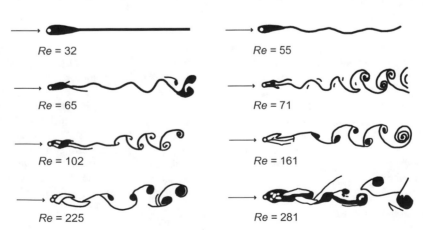

**Figure 6.19.** *The effect of increasing Reynolds number (Re) on flow past a cylinder. Re-drawn from photographs of Schlichting (1961), by permission of Streeter*

may affect patterns of colonization on boulders, and may be used by suspension feeders for concentrating food.

### Drag

The drag on bluff bodies is composed of two parts: the *skin-friction drag* as described for flat plates plus a drag component due to the shape of the object called *pressure* or *form drag*. Form drag is by far the most important component for bodies which are blunt rather than presenting a slender profile to the flow. It is something that has been experienced by anyone who has opened an umbrella on a windy day or waded in knee-deep water along the ocean's edge. Form drag on bluff or 'blunt' bodies has less to do with the 'bluntness' of the front side than with separation and pressure differences from front to back (Schlichting, 1961).

If, for example, the flat, sharp plate of the previous section were turned so that the flat side rather than the sharp nose faced the current, as in Figure 6.16(a), one would intuitively expect the drag force to increase considerably. Not only does the drag force increase but the skin-friction drag becomes insignificant since it acts only at the sharp edges. Thus, the drag on the body is totally form drag, and is calculated from

$$F_D = C_D A_p \rho \frac{V^2}{2} \qquad (6.41)$$

with $A_p$ the area of the plate (width × length), $V$ and $\rho$ the free-stream velocity and fluid density as before, and $C_D$ a coefficient of drag, which will be discussed shortly.

In fact, Eq. (6.41) is applicable for all bluff bodies if $F_D$ is considered the total drag (pressure drag plus skin-friction drag). $A_p$ becomes the projected area of the object as the current 'sees' it. Values of $A_p$ for various shapes are given in Table 6.2.

Other 'area factors' may be used instead of the projected area, such as the 'wetted area' which is the total surface exposed to flow (used for streamlined bodies). $C_D$ is usually determined from experiments with small models in wind tunnels, flumes or towing basins (in which a model is towed through a long body of water). It varies with Reynolds number (based on some characteristic length of the object, e.g. length or diameter), as can be seen in Figure 6.20. The projected area used in Eq. (6.41) should be consistent with the area used in the determination of $C_D$.

$C_D$ is relatively constant for angular bodies at Reynolds numbers over about $10^4$. Extensive tables of $C_D$ for bodies at $Re = 10^4$ and above have been developed; a sampling is given in Table 6.3. The two-dimensional coefficients are used for objects like cylindrical rods or rectangular posts

**Table 6.2.** Projected area, $A_p$, for various shapes ($d$ = diameter)

| Type of object | $A_p$ |
| --- | --- |
| Plate perpendicular to the flow | Width × length |
| Plate parallel to the flow | 0 (not a blunt body—use Eq. (6.30) instead) |
| Cylinder perpendicular to the flow | Diameter ($d$) × cylinder length |
| Cylinder parallel to the flow | $\dfrac{\pi d^2}{4}$ |
| Sphere | $\dfrac{\pi d^2}{4}$ |

which are the same shape over a large distance (the dimension extending into and out of the page). When a cylindrical-like caddis fly larvae case is oriented into the flow, it resembles a three-dimensional 'flat-faced' cylinder, whereas when it is turned sideways to the flow, it is more like a two-dimensional round cylinder. A flounder might be best modelled by a three-dimensional flat ellipsoid.

For circular cylinders and spheres, a characteristic drop in $C_D$ is apparent in Figure 6.20. This drop coincides with the critical Reynolds number where the flow changes from laminar to turbulent. In laminar flow the separation point on a cylinder or sphere occurs close to the halfway mark on the object. In turbulent flow, however, because of the increased mixing, higher-velocity particles are carried further along the solid surface before detaching—at a point about 30° around from the midpoint (White, 1986). The turbulent boundary layer can be 'triggered', for example by adding roughness to the surface as in the case of the dimples on a golf ball. Because separation occurs further back on the sphere, the wake associated with the turbulent boundary layer (golf ball) in Figure 6.21(b) is much smaller than that for the laminar flow conditions (smooth ball) of Figure 6.21(a).

As counter-intuitive as it seems, adding roughness at these near-transitional $Re$ situations *reduces* rather than *increases* the drag. This is the reason why golf balls, which fly in this range of $Re$, are dimpled. Vogel (1981) points out that cylinders and spheres are rather ideal cases, and more irregular biological objects are likely to yield graphs without this characteristic drop. In fact, for a streamlined shape, roughness would *increase* the drag.

At the lower-$Re$ end of the graph in Figure 6.20 (below $Re \approx 1$) the coefficient of drag for a sphere or cylinder varies almost linearly with $Re$. This is the range for

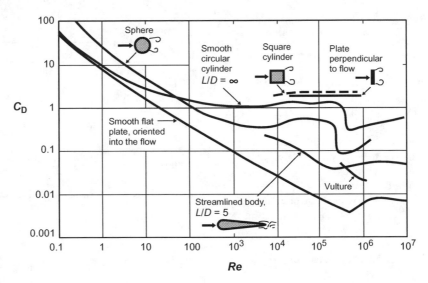

**Figure 6.20.** *The relationship between coefficient of drag ($C_D$) and Reynolds number (Re) for various shapes; L = length, D = diameter. Adapted from White (1986) and Roberson and Crowe (1990), by permission of McGraw-Hill and Houghton Mifflin Company, respectively*

**Table 6.3.** *Drag coefficients ($C_D$) for various shapes at Re > $10^4$. From White (1986) and Roberson and Crowe (1990), by permission of McGraw-Hill and Houghton Mifflin Company, respectively. Drag coefficients are based on the frontal area of the object*

| Two-dimensional | | | Three-dimensional | | |
|---|---|---|---|---|---|
| Shape | $C_D$ | | Shape | $C_D$ | |
| Plate (facing flow) | 1.98 | | | | |
| Rectangular plate (parallel to flow) | | | Flat-faced circular cylinder (parallel to flow) | | |
| $l/b = 1$ | 1.18 | | $l/d = 0$ (disc) | 1.17 | |
| $l/b = 5$ | 1.20 | | $l/d = 1$ | 0.90 | |
| $l/b = 20$ | 1.50 | | $l/d = 8$ | 0.99 | |
| Equilateral triangle | 1.60 | | 60° cone | 0.50 | |
| Half-tube | 2.30 | | Cup, open towards flow | 1.40 | |
| | Laminar | Turbulent | | Laminar | Turbulent |
| Round cylinder | 1.2 | 0.3 | Sphere | 0.47 | 0.20 |
| 2:1 Elliptical cylinders | 0.6 | 0.2 | Ellipsoid | | |
| | | | $l/d = 0.75$ | 0.50 | 0.20 |
| | | | $l/d = 2$ | 0.27 | 0.13 |
| 8:1 | 0.25 | 0.1 | $l/d = 8$ | 0.20 | 0.08 |

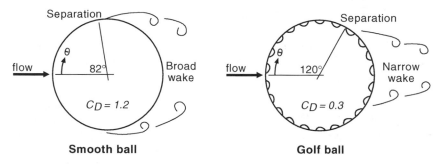

**Figure 6.21.** *Flow of air around (a) a ball with a smooth surface (laminar boundary layer) and (b) a golf ball with dimples on its surface(turbulent boundary layer). θ is the angle at which separation occurs. (Re-drawn from White, 1999)*

droplets of mist floating through the air, the settling of fine sediment particles and the movement of tiny organisms through their highly viscous environment.

### Terminal Velocity

Free descent is common to parachuting baby spiders, wind-borne seeds, falling raindrops and settling sediment particles. An object will at first accelerate downwards under the effect of gravity. However, as its speed increases, drag also increases until it reaches a level where all external forces balance. At this point, acceleration is zero (by Newton's second law, if the net force is zero and the mass is the same, then in the equation $F = Ma$, $a$ must be 0). Thus, velocity is a constant and is called the *terminal velocity*. The maximum speed reached is dependent on the size, shape and density of the object, as well as fluid properties, which must remain constant over the path of descent if terminal velocity is to occur. A human body has a terminal velocity of about 56 m/s (200 km/hour) when falling through the atmosphere.

In 1845 Stokes developed a relationship for the terminal velocity of spheres at very low Reynolds numbers, in the range $Re < 0.1$ (with $Re$ calculated using the particle diameter). In this range the drag coefficient varies linearly with $Re$, and is given by

$$C_D = \frac{24}{Re} \qquad (6.42)$$

The terminal velocity for spheres is then computed by Stokes law:

$$V = \frac{2gr^2}{9\mu}(\rho_s - \rho) \qquad (6.43)$$

in which $\rho$ is the density of the fluid, $\rho_s$ the density of the sphere, $r$ the radius of the sphere and $\mu$ the viscosity of the

fluid (see Table 1.1 for units). Since the equation can be used to determine settling times of particles it is the basis for analysing soil particle size by hydrometer and pipette methods (Section 5.8.4). By assuming that air bubbles in water are spherical, the same equation can be used to approximate the velocity of bubbles rising towards the water surface (although in reality the bubbles will flatten somewhat). In water the law does not apply to particles smaller than 0.0002 mm, since the settling of these minute particles is influenced by Brownian motion.

### Example 6.5

Calculate the terminal velocity for a spherical particle of silt with a diameter of 0.05 mm falling within a column of 25° fresh water. Assume the particle has a density of 2650 kg/m³.

$$V = \frac{2(9.807 \text{ m/s}^2)(0.000025)^2}{9(0.894 \times 10^{-3})}(2650 - 997)$$

$$V = 0.0025 \text{ m/s}$$

Check

$$Re = VD/\nu = \frac{(0.0025)(0.00005)}{(0.897 \times 10^{-6})} = 0.14$$

(This is slightly outside the Stokes range.)

Thus, in a column of 0.4 m it would take about 160 s or 2.7 min for all of this coarse silt to settle out.

For higher-$Re$ conditions where inertial effects are important and $C_D$ is not a linear function of $Re$, calculating terminal velocity is an iterative process (see Roberson and Crowe, 1990). An additional complication arises for disc-like objects because they oscillate while falling, making the values of $C_D$ uncertain (Smith, 1975).

Other limitations on Stoke's law include the assumption that the particle is falling in isolation. If the concentration of particles is high, terminal velocities are reduced. Flocculation of small particles also increases the rate of settling above that predicted, which is why a dispersant is used in particle-size analyses. For the opposite effect, alum is added at water-treatment plants to clarify the water. The alum flocculates small particles and makes them settle out more quickly (Smith, 1975).

### *Reducing Drag*

In aerodynamic and hydrodynamic engineering designs, 'the name of the game is drag reduction' (White, 1986, p. 418). Similarly, for life in a swiftly moving stream it is normally to an organism's advantage to reduce drag and thus decrease the amount of energy needed to either move against the current or maintain position within it. Unlike inanimate objects, organisms usually have some control over how they are oriented with respect to current direction. Vogel (1981) gives the example of ocean limpets which align their longest axes with that of the seaweed leaves to which they are attached. Then, when the seaweed leaf acts like a wind vane blown by the tidal currents, the limpets are also oriented in the proper direction to reduce drag. Other organisms flatten themselves against the substratum; still others have evolved with a sleek, streamlined figure or have appendages which affect drag in complex ways.

Unlike the rigid 'bluff bodies' discussed earlier, most organisms are somewhat flexible. The flapping of leaves or the lateral movements of fish further increases drag over what would be predicted from a rigid shape. The complex hydrodynamics of how organisms interact with flow patterns is a fascinating area for research. Webb (1984) provides an interesting article on how fish swim, describing the different body shapes specially adapted for acceleration, manoeuvring and cruising.

Drag is, again, due to both skin friction and form drag, the latter resulting from flow separation. At *low-Re conditions* essentially all of the drag is skin-friction drag, which acts over the whole surface area of the organism. Here, minimizing drag means minimizing the surface area-to-volume ratio. In general, smaller objects have higher ratios than larger ones, and flatter objects have higher ratios than more spherical ones. However, the advantages of drag reduction must be balanced with the benefits associated with a large surface area-to-volume ratio such as the increased area for absorption of nutrients and gases and the elimination of wastes. Thus, micro-organisms tend to be cylindrical or spherical despite the hydrodynamic disadvantage.

At the higher Reynolds numbers more likely to be encountered by insect nymphs, freshwater mussels and stream ecologists, pressure drag is much more significant than viscous friction. Factors affecting drag at *higher-Re conditions* are:

1. *Shape of the object.* In general, round edges produce less drag than sharp ones. The best way to reduce drag is through streamlining—tapering the trailing edge to a point and moving the separation point further downstream. The girth measurements of trout, for example, correspond almost exactly with the profiles of low-drag airfoils. Maude and Williams (1983) found that when freshwater crayfish were exposed to increases in current they lowered their bodies closer to the substrate and created a 'plough' profile by drawing their front claws together, allowing them to maintain position.
2. *Orientation.* A streamlined shape is best for reducing drag on an object only if the object can orient itself so the streamlining does it some good. If the direction of the current is unpredictable and the organism is attached to its foundation like a limpet or aquatic plant stem, a cylindrical or hemispherical shape may produce the least overall drag.

    Alternatively, drag can influence the orientation of stream-transported objects, such as rocks which line up in a streambed with their long axes pointed downstream (Section 7.4.2).
3. *Modifying the wake size.* The principle of *initiating turbulence* to reduce the wake size and thus form drag was mentioned in regard to the dimples on golf balls. Vanes have also been used to *suppress separation* and wake size, reducing drag. For example, vanes on the sides of trucks or slotted wings on aircraft deflect the airflow closer to the surfaces, keeping the flow accelerated and hindering the onset of separation (Schlichting, 1961; White, 1986). Jets which discharge fluid into the boundary layer to increase acceleration, or suction mechanisms to remove decelerated flow also help prevent separation (Schlichting, 1961). The castor oil fish, *Ruvettus*, for example, has structures on its skin which inject fluid into the boundary layer to reduce drag (Bone and Marshall, 1982).

### 6.5.5  Lift

Organisms use the principles of lift and drag to move horizontally or vertically and to stay put with the minimum amount of effort. Lift is produced because of differences in pressure above and below an object, as demonstrated in the discussion of Bernoulli's equation

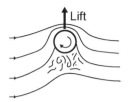

*Figure 6.22. Flow past a rotating cylinder. Redrawn from photograph in Schlichting (1961) by permission of V. Streeter*

(Section 6.3.4). The magnitude of the lift force is computed much the same as drag force, by integrating the pressure difference over the area upon which it acts. In fact, the equation for lift is very similar to Eq. (6.30) for drag:

$$F_L = C_L A_p \rho \frac{V^2}{2} \qquad (6.44)$$

where $A_p$ for an airfoil is the 'planform area' (as seen from above), $\rho$ and $V$ are as previously defined and $C_L$, the lift coefficient, is determined from experimentation. For airfoils it is common to also use the planform area in the calculation of drag because of the relatively large skin-friction drag.

As anyone knows who has experimented with the lift on a hand extended out of a car window, the *angle of attack* (the angle of an airfoil to the direction of fluid motion) has a significant effect on lift and drag. From tests on airfoils, $C_L$ increases almost proportionately with the angle of attack until about 15° to 20° when boundary layer separation occurs. $C_L$ then drops and the airfoil *stalls* (White, 1986). An inverted airfoil has the opposite effect, producing 'negative lift'. An example is the vane on the back of a race car which improves stability and keeps the rear tyres on the ground.

Another means of producing lift is by *rotation* of a cylindrical or spherical object. Note that in Figure 6.22 separation occurs below rather than behind the rotating object. In the illustration the net force is in the direction away from the wake. This is why a cricket ball curves or a tennis ball rises when given an undercut. The phenomenon is called the Magnus effect, named after a German scientist who did studies on rotation-produced lift in the nineteenth century (Roberson and Crowe, 1990).

### 6.5.6 Methods of Microvelocity Measurement

A solution to the problem of estimating velocities in or out of boundary layers is to measure them directly. Muschen-

heim *et al.* (1986) provide details on flume design for benthic ecology work. Field current metering techniques are covered in Section 5.6.3. Here, we will be concerned with techniques for measuring point velocities within the 'micro-environment.' Some of the more common methods are listed in order of complexity.

### *Flow Visualization*

Flow patterns around obstacles or organisms and flow phenomena such as separation or the change from laminar to turbulent flow can be observed by injecting dye (coloured, fluorescent or luminescent) or India ink into the moving fluid. Some of the first studies of the behaviour of boundary layers were made by sprinkling aluminium particles into the flow to create streaklines around objects. Fine seeds (e.g. mustard), egg-white and some eggs of invertebrates can also be used for visualization. The salinity of the water can be adjusted to make some particles neutrally buoyant. Tufts of yarn attached to boundary surfaces will also reveal flow patterns.

Additionally, flow visualization with small, neutrally buoyant particles or hydrogen or air bubbles yields an effective method of estimating flow velocities. By taking photographs with a stroboscope or video camera, point velocities can be determined by dividing the distance travelled by an individual particle, by the time interval between flashes or frames. For this method to be accurate it is important that the fluid motion is truly simulated by the particle motion. Too-large particles may be diverted away from an object by the very boundary layers they are supposed to reveal. For example, when Statzner and Holm (1982) compared their results from laser-Doppler anemometry to previous studies in which velocities were measured using relatively coarse acetyl cellulose particles they found boundary layer thicknesses to be much smaller for equivalent flow conditions.

Since fluid mechanics is such a highly visual subject, much can be gained from qualitative observations. Appendix 2 in Vogel's (1981) book gives several techniques for flow visualization, and films on flow phenomena are available from several sources, i.e.:

1. Encyclopedia Britannica Educational Corp; 425 No. Michigan Avenue, Chicago, IL 60611, USA.
2. Federal Inter-Agency Sedimentation Project, St. Anthony Falls Hydraulic Laboratory, Mississippi River at 3d Avenue SE, Minneapolis, MN 55414, USA.
3. University of Iowa, Media Library, Audiovisual Center, Iowa City, IA 52242.
4. Engineering Societies Library, 345 East 47th Street, New York, NY 10017, USA.

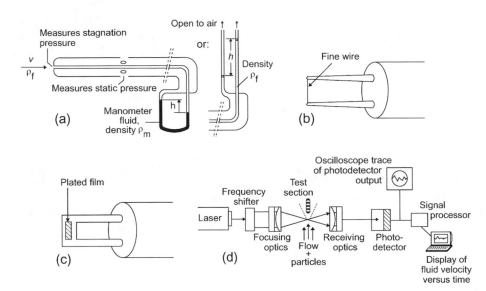

**Figure 6.23.** *Four types of micro-velocity meters: (a) pitot tube, (b) hot wire anemometer, (c) hot-film anemometer and (d) laser-Doppler anemometer. Adapted from White (1986), by permission of McGraw-Hill*

### Pitot Tube

A pitot tube, named after the French engineer who designed it in 1732 (Wilson, 1969), works on the principles of the Bernoulli equation. As shown in Figure 6.23(a), one aperture faces upstream to measure *stagnation pressure* and small holes on the sides of the tube are exposed only to the *static pressure* of the undisturbed fluid (Vogel, 1981). Since, at the stagnation point, all of the velocity head has been changed to pressure head it is a simple matter to subtract the static pressure to obtain the velocity term. By connecting a manometer between the two apertures this difference in pressure can be measured directly. The relevant formula is (Vogel, 1981)

$$V = \sqrt{\frac{2gh(\rho_m - \rho_f)}{\rho_f}} \qquad (6.45)$$

where $g$ is the acceleration due to gravity, $h$ is the difference in manometer readings (m), $\rho_m$ is the density of the fluid in the manometer and $\rho_f$ the density of the flowing fluid. If the fluid in the manometer is the same as the fluid being measured (in which case, the tubes are left open to the air—see second diagram in Figure 6.23(a)), the formula becomes

$$V = \sqrt{2gh} \qquad (6.46)$$

Because these equations are based on Bernoulli's relation they perform best when the flow around the probe is

considered frictionless, at $Re > 1000$ (the probe diameter, $d$, is taken as the characteristic length in Eq. (6.15)).

A disadvantage of Pitot tubes is that they are sensitive to how well they are aligned with the flow direction. The fluid in the manometer acts somewhat like a small version of a 'stilling well', so manometer levels do not reflect small-scale turbulent fluctuations. Because the usual velocities measured produce only a small difference in head, pitot tubes tend to be relatively insensitive. They are also physically intrusive, although they can be made very small (even for measuring blood flow in arteries). Advantages of pitot tubes are that 'point' measurements can be taken, they require no calibration and they do not need a power source (Vogel, 1981; White, 1986). Their greatest usefulness is for measuring high velocities in pipes, chutes or overfalls.

### Hot-wire Anemometer

A hot-wire anemometer (Figure 6.23(b)) consists of a probe with a fine wire heated by an electric current (Boyer, 1964; White, 1986). When the probe is submerged in a flowing fluid the wire is cooled, changing its resistance. This change in resistance is related to the flow velocity. Hot-wire anemometers are available commercially, and Vogel (1981) gives 'do-it-yourself' instructions in his book. Although these anemometers work well for air flow measurements, they are not well suited to measurement of water flow because of hysteresis effects and frailty of the thin wire. It can also be difficult to get a stable calibration. An improvement is the hot-film anemometer (Figure 6.23(c)), which follows the same principles as a hot-wire anemometer but is less fragile and more stable (White, 1986).

### *Laser-doppler Anemometer (LDA)*

A superior device for measuring velocities in water is the laser-Doppler anemometer (LDA). This instrument has the ability to measure point velocities quickly and continuously without disturbing the flow field. With an LDA (Figure 6.23(d)) velocity is measured by detecting the Doppler shift of light scattered by particles moving through the flow (the change in the sound of a train as it moves away is a Doppler effect). Commonly, two laser beams of different frequencies are used, which are focussed on a test section. When particles move through this test section the Doppler shift is detected as the difference frequency between the two beams. The Doppler signal is directly proportional to the velocity of the particle. For an LDA to work there must be a sufficient concentration of particles in the fluid. Normal impurities in water will often suffice, but it is sometimes necessary to add 'seed' particles which are typically 0.0005–0.005 mm. The inertia of the scattering particles is considered negligible so that their movement reflects that of the surrounding fluid.

LDA systems are extremely expensive and not well suited to field work. The advantages are that (1) the flow is not disturbed during measurement, (2) detailed measurements can be taken very close to surfaces to investigate boundary-layer phenomena, (3) there is no need for calibration and (4) the LDA is capable of measuring reverse flows and turbulent fluctuations. LDA systems are available which can measure multiple components (vertical and horizontal) of velocity. A summary of LDA principles can be found at www.dantecmt.com/LDA/Princip/Index.html.

## 6.6   Open-channel Hydraulics: The Macro-environment

### 6.6.1   But First, A Few Definitions

*Depth (D)*.   The vertical distance between the water surface and some point on the streambed.
*Stage (y)*.   The vertical distance from some fixed datum to the water surface. The datum might be the elevation of zero flow (e.g. the thalweg point) or mean sea level.

*Discharge (Q)*.   The volume of water passing through a stream cross section per unit time.
*Top width (W)*.   The width of the stream at the water surface. Except in channels with vertical walls, it will vary with stream depth.
*Cross-sectional area (A)*.   The area of water across a given section of the stream. If a 'slice' is made through the flow at right angles to the flow direction, this is the area exposed.
*Wetted perimeter (P)*.   The distance around the outside edge of the cross-sectional 'slice' where the stream's bed and banks contact the water.
*Hydraulic radius (R)*.   The ratio of the cross-sectional area to the wetted perimeter

$$R = \frac{A}{P} \qquad (6.47)$$

*Hydraulic Depth (D)*.   The ratio of the cross-sectional area to the top width

$$D = \frac{A}{W} \qquad (6.48)$$

In streams which are very wide in relation to their depth (a width-to-depth ratio of about 20:1 or more) the hydraulic radius and hydraulic depth are almost equal and approximate the average depth of the stream.

The symbol *D* will be used interchangeably for depth at a vertical, mean depth across a section, and hydraulic depth. Its interpretation will be explained in the equations in which it appears. Some of the terms which describe cross-sectional dimensions are shown in Figure 6.24.

### 6.6.2   Introduction to Hydraulics

The analysis of bulk flow patterns of water surface shape, velocity, shear stress and discharge through a stream reach comes under the heading of *open channel hydraulics*. Here, the effects of viscosity are not as important as the forces causing the water to move and stay in motion. The vertical distribution of velocity is dependent on the size of roughness elements in the flow path. Turbulence, as large scale eddies and spiralling flow, occurs over the entire depth of flow.

In comparison to flow through pipes, in open channel flow the water surface is exposed to atmospheric pressure. Because the water depth and channel geometry can change in natural stream channels, problems are much more complex. However, many of the formulae for describing open channel flow have been developed from the mathematical description of pipe flow. The description

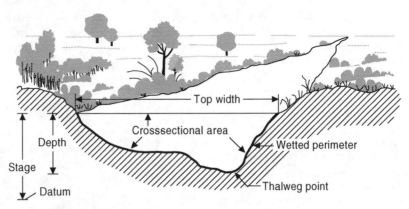

***Figure 6.24.*** *Terminology describing the geometry of open channels*

of pipe flow will not be covered here, although it can occur in natural pipes in the soil and where streams flow through karst areas. For information on pipe flow and further information on open-channel flow, readers can refer to fluid mechanics texts and texts on open-channel flow such as Chow (1959) and Henderson (1966).

Flow in open channels can be classified as

- Steady/unsteady,
- Uniform/varied,
- Laminar/turbulent,
- Supercritical/critical/subcritical.

The classification of steady/unsteady depends on whether the flow depth and velocity change with time at a point, and is discussed in Section 6.3.1. In contrast, the classification uniform/varied is based on whether depth and velocity vary with respect to distance (Figure 6.25). If depth and velocity remain constant over some length of channel of constant cross section and slope (a 'prismatic' channel), then the water surface is parallel to the streambed and the flow is said to be *uniform* in that reach. If the flow in a stream is uniform it is said to be moving at its *normal depth*, an important parameter in

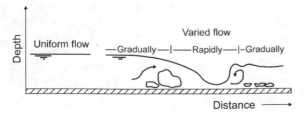

***Figure 6.25.*** *Classification of open channel flow as uniform or varied*

engineering design. The assumption of uniform flow conditions considerably simplifies the analysis of water movement in streams.

In *non-uniform* or *varied* flow the water depth and/or velocity change over distance. Examples are where the flow moves through a bedrock constriction or passes from a pool to a riffle. Varied flow can be sub-divided further into the categories *rapidly varied* and *gradually varied.* If the depth changes abruptly over a relatively short distance, as at a waterfall or wave, the flow is rapidly varied; when changes are more widely spread the flow is gradually varied. Gradually varied flow can be analysed relatively easily (Section 6.6.6), whereas the description of rapidly varying flow (Section 6.6.7) requires experimentation.

Conditions favouring uniform flow in natural streams are rare compared with the well-controlled flow conditions in concrete canals or irrigation ditches. However, uniform flow can be approximated by long, straight runs of constant slope and cross section, and it is under these conditions that the uniform flow equations of Section 6.6.5 provide reasonable estimates.

The third classification, laminar/turbulent, is somewhat irrelevant at the 'macro-environment' level. Although regions of laminar flow can exist near the surfaces of rocks or organisms within the stream, the bulk flow is nearly always turbulent except perhaps in the rare case where it flows slowly as a thin film of a few millimetres. The fourth classification, super-critical/critical/sub-critical, is related to the combined patterns of velocity and depth, and will be discussed in Section 6.6.4.

### 6.6.3  The Variations of Velocity in Natural Channels

Velocity, as the rate of movement of a fluid particle from one place to another, varies with both space and time in a

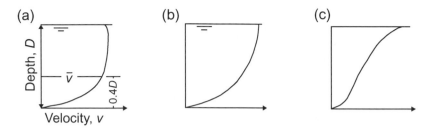

**Figure 6.26.** *Three variations on the vertical velocity profile*

stream. The effects of velocity on stream organisms are discussed in Section 2.1. From continuity (Eq. (6.6)), mean velocity across any section is simply

$$V = Q/A \qquad (6.49)$$

the discharge (m³/s) divided by the cross-sectional area (m²). Flow velocity tends to increase as slope increases and as bed roughness decreases. In the extreme case of a waterfall, where slope is maximal and 'bed roughness' is non-existent, velocities of up to 8.1 m/s have been reported (Hynes, 1970). Water velocities in natural channels tend to remain below about 3.0 m/s. To put speeds into context, a flow velocity of 0.1 m/s is barely perceptible, whereas at 3.0 m/s it is difficult to stand up in knee-deep water.

Because of frictional resistance, the flow is retarded near the streambed and streambanks, and slightly at the water surface. Along with flow turbulence, this resistance causes variations in the distribution of velocity (1) with time; (2) with depth; (3) across a section and (4) longitudinally. Additionally, it leads to spiralling cross-currents which impose another degree of complexity on the main downstream movement of water.

### Variation with Time

At any given point in a stream the flow velocity fluctuates rapidly because of surges and turbulent eddies. As mentioned earlier, measurements of point velocities taken with current meters or probes are actually time-averaged values of velocity. Instantaneous velocities can be much larger or smaller than the average because of the variability in turbulent flow. These fluctuations often appear to have a cyclical or 'pulsing' behaviour, rather than a random trend (Morisawa, 1985). Although merely a statistical nuisance to hydrologists, the nature of turbulence may have profound implications for the organisms exposed to it.

Velocities also change in response to changes in discharge. In general, velocities will increase most where

resistance is least, at the centre of the stream. Even during floods, low-velocity zones will persist near the stream edges. The change in velocity with discharge falls under the topic of hydraulic geometry, which is covered in Section 7.2.6.

### Variation with Depth

If current meter readings are taken over a vertical, a graph of the measured velocities plotted against depth illustrates the vertical velocity profile. Figure 6.26 shows some of the forms that a measured velocity profile could assume. From zero velocity at the bed the velocity increases with vertical distance, at first rapidly but then levelling off as it reaches its maximum value. Channel shape, bed roughness and the intensity of turbulence all influence velocity profiles.

In a 'typical' velocity profile (Figure 6.26(a)) the maximum velocity tends to occur just beneath the water surface because of resistance with the air. The closer the measurements are taken to the streambanks, the deeper is this maximum velocity (Chow, 1959). Slower surface velocities will occur where the flow is retarded by the floating leaves of aquatic plants. In the centre of broad, rapid streams the profile will look more like Figure 6.26(b), with the maximum velocity found right at the free surface. When the depth of rocks, boulders, plants, dead logs, sand dunes and other 'roughness elements' is high in relation to the depth of the water, water velocities within the roughness elements may differ substantially from velocities over the top of the protrusions. For example, Jarrett (1984) mentions that shallow, steep, cobble and boulder-bed streams in mountainous areas can have S-shaped profiles such as in Figure 6.26(c). Thus, the velocity profile may alter in shape as the water level changes.

A logarithmic function is conventionally used for describing vertical velocity profiles, and in most instances it provides a good approximation. If the velocity varies logarithmically with distance from the streambed it can be demonstrated mathematically that the mean value of velocity, $\bar{v}$, occurs at about 0.4 of the water depth (measured upwards from the bed, as shown in Figure 6.26(a)). This is the point at which velocities are measured if only one current-meter reading is taken at a vertical (see Section 5.6.3). For the average velocity at a vertical, Eq. (6.25) for hydraulically rough conditions becomes

$$\bar{v} = 5.75 \, V_* \log \left( \frac{12.3R}{k} \right) \qquad (6.50)$$

from which $V_*$, the shear velocity, can be calculated for a vertical if the mean velocity and relative roughness are known. Smith (1975) indicates that the value of relative roughness, $k/R$, varies from more than 0.2 for a shallow stream flowing over a shingle bed to less than 0.0002 for a deep flow over fine clay sediments. Thus, in rocky streams, the shear velocity is approximately 1/10 of the mean velocity but only about 1/30 of the mean velocity when the streambed is of fine sediments.

Using the information in Figure 6.14, we can back-calculate a value of $k$ from Eq. (6.50) using the water depth as an approximation for $R$. This gives $k \approx 0.09$ m and $k/R \approx 0.17$. This profile was taken near the thalweg of the stream, where the median bed material size (see Figure 5.33) was on the order of 100 mm (0.1 m)—quite similar to the computed value of $k$.

Other velocity profile equations have also been developed for describing velocity distributions over irregular streambeds, based on particle sizes. These equations are discussed by Coleman and Alonso (1983) and Nezu and Rodi (1986).

### Variation across a Section

Velocities tend to increase towards the centre of a stream and decrease towards the perimeter because of frictional resistance at the bed and banks. *Isovels*, lines joining points of equal velocity, can be plotted as a 'map' of a stream cross section, like elevation contours plotted on a topographical map. The distribution of velocity is easily visualized in the plots shown in Figure 6.27. Where isovels are crowded, velocity gradients, and thus shear stresses, are higher. At a bend in the river (Figure 6.27(b))

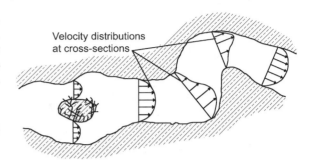

**Figure 6.28.** Longitudinal velocity variations in a stream reach

isovels tend to be closer together on the outer bank where the main current is forced against it by centrifugal force.

### Longitudinal Variation

Patterns of velocity variation within a channel reach can also be shown as a 'bird's eye view', as in Figure 6.28. If velocities (surface or average velocities) are plotted from a number of cross-sections this can give an indication of how close the reach approximates uniform flow conditions, as well as identifying, for example, areas of high bank erosion potential or good fish habitat. Areal 'maps' showing zones of different velocities can be created from a large number of measurements, much like soils or vegetation maps. These can show habitat features such as the sheltered areas next to high velocity zones that are required by salmonids (Maddock, 1999).

### Spiral Flow

For a complete description of velocity distributions in streams, the concept of *spiral, helical* or *secondary flow* should not go without mention. Spiral flow (Figure 6.29) is a consequence of frictional resistance and centrifugal force. When a cup of billy tea is stirred this is the reason why the leaves congregate in the middle rather than at the edges. In a stream, water is hurled against the outside banks at bends, causing the water surface to be 'super-elevated'. The increase in elevation creates a gradient,

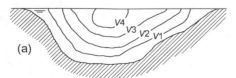
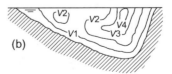

**Figure 6.27.** Velocity contours or isovels at stream cross sections: (a) in a relatively straight section and (b) at a bend. In both diagrams, V4 > V3 > V2 > V1. Adapted from Morisawa (1985), by permission Longman Group, UK

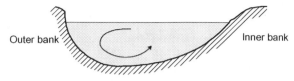

**Figure 6.29.** *Spiral flow at a river bend*

causing flow movement from the outer to the inner bank. A spiralling motion is generated along the general direction of flow (Petts and Foster, 1985). The first observation of spiral flow is credited to Thomson in 1876 (Chow, 1959). Compared to the forward, downstream currents, secondary lateral and vertical currents are relatively small, yet they cause the mainstream current to vary from a predictable course and contribute to energy losses and bank erosion at bends.

On the outside of a bend the rotary motion is from top to bottom, which tends to scour the bank. On the inside the flow is upward and decelerating, and any material carried tends to deposit out, creating point bars (Vennard and Street, 1982). Spiral flow also occurs in straight sections, where the welling of waters near the stream's centre can increase the local surface elevation, resulting in a curved water surface (Leopold *et al.*, 1964). Spiral flow is particularly pronounced where channel boundaries are irregular.

Chow (1959) gives a formula for the strength of a spiral flow as a percentage ratio of the mean kinetic energy of the lateral flow to the total kinetic energy of flow at a cross section. This ratio is relatively high if the Reynolds number of the approach flow is low. It decreases as the channel becomes deeper in relation to its width and as bend angles become less severe.

Circulation patterns in streams are very complex and unpredictable, yet an awareness of their existence can provide a starting point from which to study their effects as well as material for contemplation while watching a fishing line snaking from side to side in a stream. For ecological purposes, velocities near the bottom of a channel or at the water surface will often be of greater significance than averages across a cross section or a reach. Variations of velocity with time, neglected in gross measures of mean velocities, may be critical to the aquatic insects or sand grains buffeted by pockets of turbulence.

### 6.6.4 Energy Relationships in Streams

#### The One-dimensional Energy Equation

The Bernoulli equation (Eq. (6.10)) describes the conservation of energy along streamlines. It can be assumed to apply to objects within a mass of flowing water when viscosity can effectively be ignored. However, in developing a general equation for describing energy relationships in streams the frictional resistance along the channel bed and banks due to viscosity must be taken into account.

For this purpose, we use cross-sectional values of elevation, pressure (depth) and velocity, and the *one-dimensional energy equation*

$$z_1 + D_1 + \frac{V_1^2}{2g} = z_2 + D_2 + \frac{V_2^2}{2g} + h_l \qquad (6.51)$$

where $z$ = elevation (m), $D$ = mean water depth (m), $V$ = mean velocity (m/s), $g$ = acceleration due to gravity (m/s$^2$), and the subscripts 1 and 2 refer to upstream and downstream sections, respectively. As in the Bernoulli equation, all terms have dimensions of length and are called *head* terms (e.g. elevation head, pressure head and velocity head). The *head loss* term, $h_l$, accounts for energy loss in the form of flow separation, turbulence, heat generated as a result of frictional resistance and other forms of internal energy. Empirical formulae have been developed for calculating head loss which take into account the fluid viscosity and boundary roughness effects. Head losses can occur due to local effects such as hydraulic jumps (Section 6.6.7), sudden expansions or contractions, or rapid changes in flow direction at bends.

Water depth, $D$, is the average water depth in rectangular channels, but is measured at the thalweg point (Figure 6.24) in irregular channels. The use of water depth as the pressure head term is valid for slopes less than about 10° (Chow, 1959).

A diagram showing the terms in Eq. (6.51) for non-uniform flow conditions is given in Figure 6.30. The *total energy line* represents the total energy head of the flow, and its slope is called the *energy slope* or *gradient*. In uniform flow the slope of the streambed, water surface and energy line are the same, and thus the head loss is simply equal to the difference in elevations of the streambed or water surface from the top of a reach to the lower end.

#### Specific Energy

*Specific energy* is the energy of a section relative to the streambed; i.e. the distance between the streambed and the total energy line in Figure 6.30. For a channel of small slope the specific energy at a section, $E_s$, is

$$E_s = D + \frac{V^2}{2g} \qquad (6.52)$$

where the terms are the same as in Eq. (6.51). Again, all terms, including $E_s$, have dimensions of length. For a

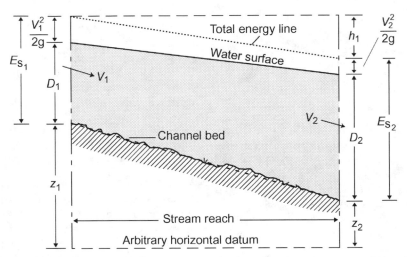

**Figure 6.30.** *Schematic showing the terms in the energy equation. z = elevation above some datum, V = mean velocity, D = water depth, $h_l$ = head loss over a reach and $E_s$ = specific energy, the sum of velocity and pressure head terms. Subscripts 1 and 2 refer to upstream and downstream cross sections at a stream reach*

given channel section and discharge, specific energy is a function of water depth, and the relationship is described by a specific energy curve. Specific energy curves for three different discharges are shown in Figure 6.31(a).

For each value of discharge, there is a minimum depth—the *critical depth*, $D_c$. If the water depth, $D$, is higher than $D_c$, the flow is considered subcritical, and if $D < D_c$, the flow is considered supercritical, as will be discussed later in this section. Theoretically, the critical depth is the minimum depth which can occur as water flows over the top of a boulder or log (Figure 6.32(a)) or

the crest of a waterfall or spillway (Figure 6.32(b)). This minimum depth occurs slightly upstream from the brink because of curvature of the water surface.

If critical flow can be located in a stream (or created with a weir) then the flow rate can be determined from the critical depth. For this application the weir must extend across the section (with no flow underneath or around the edges). At critical flow,

$$E_c = \tfrac{3}{2} D_c \qquad (6.53)$$

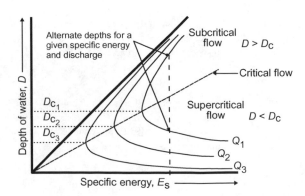

**Figure 6.31.** *Generalized graph of specific energy for different discharges, where $Q_1 > Q_2 > Q_3$. Critical depths $D_{c1}$, $D_{c2}$ and $D_{c3}$ correspond to discharges $Q_1$, $Q_2$ and $Q_3$, respectively*

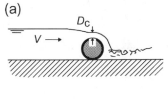

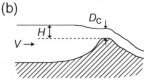

**Figure 6.32.** *Critical depth at overfalls: (a) at a log and (b) at a spillway crest. $D_c$ is the critical depth, V is the mean velocity, and H is the water depth above the crest measured some distance upstream (see also illustration of weir, Figure 5.20)*

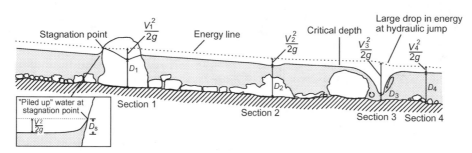

**Figure 6.33.** *Schematic showing the balance of energy in a longitudinal 'slice' through a stream reach. Scale is exaggerated for the velocity head term. The flow is subcritical in Sections 1, 2 and 4, and supercritical in Section 3. Rapidly varied flow is discussed further in Section 6.6.7 and by Chow (1959). Drawn with the assistance of R. Newbury*

and thus from Eq. (6.52)

$$\frac{V_c^2}{2g} \text{(the velocity head)} = \frac{D_c}{2}$$

yielding the equations

$$V_c = \sqrt{gD_c} \qquad (6.54)$$

and

$$Q_c = A\sqrt{gD_c} \qquad (6.55)$$

where $D_c$ is the hydraulic depth at critical flow, $Q_c$. In wide or rectangular channels, $D_c$ can be approximated by the average water depth, but in irregular channels, the definition of hydraulic depth, $A/W$, should be used. Texts on hydraulics should be consulted for a more complete derivation of these equations.

*Example 6.6*

Calculate the discharge flowing over a broad, level concrete sill if the critical depth is 0.31 m and the width of the water surface is 8.2 m. Assume the flow area is rectangular, so $A = (8.2)(0.31) = 2.54 \text{ m}^2$.

*Answer*

$$Q = 2.54\sqrt{9.807(0.31)} = 4.43 \text{ m}^3/\text{s}$$

If the overfalls in Figure 6.32 extend across the whole stream, they function as *broad-crested weirs*. Because flows near critical depth tend to be unstable (from Figure 6.31, it can be seen that a slight change in energy will cause the depth to change markedly), measurements at weirs are usually taken some distance upstream of the

crest. If it is assumed that the velocity head term is negligible at this upstream point in relation to depth, and that the energy is the same as at the crest, then this depth above the crest ($H$) is equal to $E_c$. From Eqs. (6.53) and (6.55), and assuming $A = WD_c$, the following equation for discharge can be derived

$$Q = 1.70 \, W \, H^{3/2} \qquad (6.56)$$

For broad-crested weirs used in stream gauging the coefficient 1.70 is usually somewhat less due to energy losses. Equation (6.56) is often verified in flumes as a laboratory exercise in engineering courses.

Flow velocity and depth 'balance' each other for a given specific energy and discharge. Figure 6.33 illustrates this principle for an idealized 'longitudinal slice' along a stream. Upstream of the boulder at Section 1, a *stagnation point* is created. The velocity increases where the water flows around the boulder, with a corresponding drop in depth. Velocity also increases if the channel contracts or the bed is elevated by cobbles, as at Section 2. As shown, the depth drops rather than increasing. Sections 3 and 4 show a region of rapidly varied flow, accompanied by a large drop in the energy line.

Equations (6.52)–(6.55) are based on the assumption of uniform, steady flow. They are approximately correct for the gradually varied flow shown in Figure 6.33 upstream of Section 3. In the rapidly varied flow of Sections 3 and 4 they are not applicable. It should also be realized that in a stream, the energy line would be determined by cross-sectional values of velocity and depth rather than those at individual verticals.

If a stagnation point occurs in a reach, such as at a boulder, a bridge pier or a tree trunk, it can be used to give a rough indication of the flow velocity. Since the velocity there is zero, the depth of flow is equal to the specific energy. The depth of the 'piled up' water, $D_s$, is thus approximately equal to the velocity head of the approaching flow (see insert in Figure 6.33)

$$D_s = \frac{V^2}{2g}$$

or

$$V = \sqrt{2gD_s} \tag{6.57}$$

The water depth at the stagnation point fluctuates greatly, and a visual average should be made if this depth is measured with a rule or staff gauge.

### Froude Numbers and Flow Classification

The line connecting the critical depths in Figure 6.31 separates the portions of the specific energy curves representing *supercritical* and *subcritical flow*. Thus, except at critical flow, flow of a given specific energy can have one of two depths, called *alternate depths*. This provides a means of classifying the flow on the basis of a dimensionless quantity, the *Froude number*. This number represents the ratio of inertial forces to gravitational forces, where the gravitational forces encourage water to move downhill, and inertial forces reflect the water's compulsion to go along or not. The Reynolds number (Eq. (6.15)) is a better measure of 'internal' conditions whereas the Froude number is a better descriptor of bulk flow characteristics such as surface waves, sand bedforms, and the interaction between flow depth and velocity at a given cross section or between boulders.

Froude number, *Fr*, is defined as

$$Fr = \frac{V}{\sqrt{gD}} \tag{6.58}$$

where *V* is the mean velocity (m/s), *g* the acceleration due to gravity (m/s$^2$), and *D* the hydraulic depth (m). Again,

for rectangular or very wide channels, the hydraulic depth can be replaced by the average water depth. 'Local' Froude numbers can also be calculated, for example, where water flows over or between boulders, and in this case the local depth and velocity at a vertical are used.

At critical flow, Eq. (6.54) can be rearranged to show that $V_c/\sqrt{gD_c} = 1$. Three flow classes can now be designated:

- *Fr* < 1 subcritical (or slow or tranquil) flow
- *Fr* = 1 critical flow
- *Fr* > 1 supercritical (or fast or rapid) flow

The analogy can be drawn between Froude number and Mach number, where a value of 1 separates supersonic from subsonic based on whether the speed of sound is exceeded or not. As with Mach number, the Froude number also deals with speed. The denominator, $\sqrt{gD}$, represents the speed of a small wave on the water surface relative to the speed of the water, called *wave celerity*. At critical flow the wave celerity is equal to the flow velocity. Any disturbance to the surface will remain stationary. In subcritical flow the flow is controlled from a downstream point and any disturbances are transmitted upstream (e.g. a control point for a gauging station, Section 5.6.5). By comparison, supercritical flow is controlled from an upstream point and any disturbances are transmitted downstream.

The direction of wave propagation can be used to locate regions of subcritical, critical and supercritical flow in a stream. A pencil, stick or finger (we will call it a 'Froude indicator') contacting the water surface will generate a 'V' pattern of waves downstream. If the flow is subcritical, waves will appear upstream of this 'indicator' as in Figure 6.34(a), whereas they do not appear when the flow is supercritical (Figure 6.34(b)). By passing the 'indicator' through a region of transition from subcritical to supercritical the location of critical flow can be

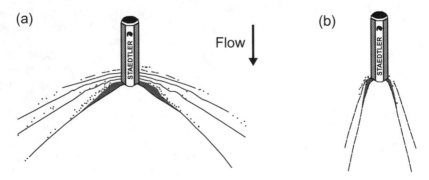

**Figure 6.34.** Detection of (a) subcritical and (b) supercritical flow at the water surface

identified as the point where all upstream waves disappear or the downstream angle of the 'V' is 45°.

In streams most of the flow will be subcritical. Supercritical flow can be found where water passes over and around boulders, and in the spillway chutes of hydraulic structures. Usually, it is accompanied by a quick transition back to subcritical flow (a hydraulic jump—see Section 6.6.7), which appears as a wave on the water surface.

The Froude number is gaining acceptance as an index for the characterization of local-scale habitats. In a study of the distribution of the filter-feeding caddisfly, *Brachycentrus occidentalis* Banks, for example, Wetmore *et al.* (1990) found that the Froude number was a better predictor of larval habitat than depth or velocity taken individually. The larvae preferred regions where streamlines converged over and around boulders and cobbles, exposing them to higher concentrations of food materials. The authors describe a laboratory profiling device modified for the field measurement of local water depths and velocities in shallow riffles. A small current meter with a 10 mm diameter propellor was used for measuring velocities at 0.6 depth (subcritical regions) and 0.5 depth (supercritical regions), as measured upwards from the streambed. For applications where point measurements rather than cross-sectional averages are used, critical flow is not necessarily defined by $Fr = 1$.

### 6.6.5 Shear Stress and the Uniform Flow Equations of Chezy and Manning

Much theoretical and empirical research has been done in an attempt to develop a relationship between velocity and stream channel characteristics. If the flow is uniform, then depth and velocity are constant along the channel. Under these conditions, gravity (as a force causing motion) and friction (as a force opposing it) are in balance.

The frictional force causing flow resistance along the channel boundary can be expressed per unit area as a *shear stress* (see definition, Section 1.2). For channels, shear stress, $\tau$, is given as

$$\tau = \rho g R S \qquad (6.59)$$

where $\tau$ is in N/m², $R$ is hydraulic radius (m) and $S$ is the slope of the energy line, which in the special case of uniform flow is equal to the bed slope and the water surface slope. If the channel is wide or rectangular, $R$ can be replaced by the average flow depth. Shear stress generally declines in the downstream direction since bed slope decreases, and thus organisms are more likely to have adaptations against dislodgement in headwater streams (Townsend, 1980).

By setting the gravitational force in the direction of flow equal to the frictional resistance (assumed approximately proportional to the square of velocity), the following equation can be derived

$$V = \sqrt{\left(\frac{\rho g}{\alpha}\right) R S} \qquad (6.60)$$

where $V$ = mean velocity (m/s), $R$ = hydraulic radius (m), $S$ = energy slope, $\alpha$ = a coefficient that is mainly dependent on boundary roughness, and $\rho$ and $g$ are water density and gravitational acceleration, as previously defined. A French engineer, Antoine Chezy, introduced this equation in 1768 when given the task of designing a canal for water supply in Paris (Henderson, 1966). The term $\sqrt{\rho g/\alpha}$ is normally lumped into a single coefficient, Chezy's '$C$', and the equation written in the form

$$V = C\sqrt{RS} \qquad (6.61)$$

Chezy's $C$ has units of m$^{1/2}$/s, and varies from about 30 in small, rough channels to 90 in large, smooth ones (White, 1986). It is considered to be a function of relative roughness and Reynolds number. A variety of methods for obtaining $C$ have been developed, and details are provided by Chow (1959). The Chezy equation is still popular in many European countries.

Robert Manning, an Irish engineer, first introduced an alternative to Chezy's equation in 1889. It was later simplified to its commonly used form, given in SI units as

$$V = \frac{1}{n} R^{2/3} S^{1/2} \qquad (6.62)$$

or

$$Q = \frac{1}{n} A R^{2/3} S^{1/2} \qquad (6.63)$$

where $V$ is the mean channel velocity (m/s), $Q$ is the discharge (m³/s), $S$ is the slope of the energy line (in uniform reaches, equal to the bed and water surface slopes), $R$ is the hydraulic radius (m) and $n$ is a coefficient referred to as 'Manning's $n$'. Manning's equation was derived from data collected on artificial and natural channels with a range of shapes and boundary roughness. The exponent on $R$ was actually found to vary from 0.6499 to 0.8395, and Manning adopted the value 2/3 as an approximation. Because of its practicality, Manning's equation is probably the most widely used uniform flow equation for open-channel flow calculations in English-speaking countries. A table of Manning's $n$ values is given in Section 5.6.6.

To balance the dimensions of the equation, the $1/n$ term must have units of $m^{1/3}/s$. The normal preference is to leave $n$ dimensionless and attach all remaining units to a coefficient. This has a value of 1 in SI units, as in Eqs. (6.62) and (6.63), but becomes 1.486 in the Imperial system. Thus, a book of $n$ values is applicable in America or Australia or Antarctica. In general, $n$ increases as turbulence and flow retardance effects increase. Sediment load can have a variable effect on flow resistance, as it dampens turbulence which reduces resistance in some circumstances, but the consumption of energy in transporting the load acts like an increase in resistance (Trieste and Jarrett, 1987). Manning's $n$ can thus be thought of as a 'tuning' or 'calibration' factor which integrates the effects of flow resistance caused by bed roughness, the presence (and flexibility) of vegetation, the amount of sediment or debris carried by the flow and other factors (Chow, 1959; Trieste and Jarrett, 1987).

Information on the field estimation of Manning's $n$ is given in Section 5.6.6. Attempts have also been made to give $n$ more of a physical basis. For example, Strickler (1923) developed the following equation for gravel-bed streams with median grain size $d$

$$n = sd^{1/6} \qquad (6.64)$$

where $n = $ Manning's $n$, and $s$ is a coefficient. When $d$ is the median diameter ($d_{50}$, in mm), $s \approx 0.013$ (Henderson, 1966).

Manning's $n$ has also been found to vary with flow depth. Equation (6.64) may be applicable when the roughness elements are very small in comparison to the water depth, but $n$ typically increases as the relative roughness (Eq. (6.18)) increases. According to Chow (1959), Manning's equation is not considered to be applicable when the roughness height exceeds 1/3 of the water depth ($R_{rel} > 1/3$). Relative roughness is included in an equation for estimating Manning's $n$ developed by Limerinos (1970) from data on lower gradient streams with bed material of small gravel to medium-size boulders. The equation is given in SI units as

$$n = \frac{0.1129R^{1/6}}{1.16 + 2.0\log\left(\dfrac{R}{d_{84}}\right)} \qquad (6.65)$$

where $R$ is hydraulic radius (m) and $d_{84}$ is the diameter (m) for which 84% of the streambed particles are smaller. For higher-gradient streams (slope 0.002–0.04), Jarrett (1985) gives

$$n = 0.39\,S^{0.38}R^{-0.16} \qquad (6.66)$$

where the terms are as defined in Eq. (6.63). This equation is applicable to stable channels with hydraulic radii between 0.15 and 2.1 m and similar in character to the Colorado streams from which the equation was derived. This approach is similar to that of Riggs (1976) requiring no estimate of roughness (Eq. (5.28)). For the data used in deriving the equation, the standard error of estimate was 28%.

Equations (6.64) and (6.65) provide options for calculating Manning's $n$ that can be checked with estimates made from other methods (e.g. Cowan's, Section 5.6.6) or used for guidance when adjusting $n$ for different flow levels.

When the stream overtops its banks the water velocity in the main channel is typically much higher than that on the floodplains. For these compound sections Manning's equation is applied to sub-sections (e.g. the main channel and one or more side channels) to obtain sub-section mean velocities and discharges. These discharges are then summed to get the total discharge for the section. The mean velocity for the entire section is equal to the total discharge divided by the total cross-section area (Chow, 1959), although this may have little relevance in compound channels. It is assumed that the water surface is horizontal across the whole section, and that the energy slope is the same in all sub-sections. Example 6.7 below illustrates the procedure. It should also be noted that wetted perimeter is calculated only over the wetted distance; it does not include the dividing lines between the sub-sections.

In natural streams which are irregular in cross section, and which wind around bends, and drop over falls and into pools, bed roughness is not the only source of energy loss. In fact, these uniform flow equations may be inapplicable in very steep streams with pool-step structures, as the flow often passes through critical depth. Jarrett (1985) suggests 0.04 as an upper limit for slope when using the Manning's equation. An understanding of the assumptions behind the equation and the factors which affect Manning's $n$ is helpful in narrowing the error margins around the estimates of velocity or discharge. Although this equation is based on an assumption of uniform flow, it is common practice to use it when describing gradually varied flow. For conditions which deviate substantially from uniform flow, adjustments must be made to account for head losses (see Eq. (6.51)). These procedures are discussed by Chow (1959), Henderson (1966), Jarrett (1985) and Jarrett and Petsch (1985).

As semi-empirical equations, the equations of Chezy and Manning are safely used only within the range over which they were developed—trapezoidal channels with uniform flow and clear water. For practical purposes, flow through a straight stream section of fairly constant cross section, mean velocity and roughness characteristics can be considered approximately uniform, and the equations will perform best if these conditions are met.

*Example 6.7*

Calculate the discharge through a section where the stream has overflowed onto the floodplain and the dimensions of the water area are as shown. For both sub-sections, $S = 0.005$. In sub-section 1, $n = 0.060$, and in sub-section 2, $n = 0.035$.

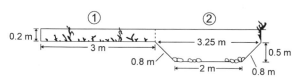

*Sub-section 1*

$$A = 3(0.2) = 0.6\,\text{m}^2$$
$$P = 0.2 + 3 = 3.2\,\text{m}$$
$$R = 0.6/3.2 = 0.188\,\text{m}$$
$$\Rightarrow V_1 = \frac{1}{0.060}(0.188)^{2/3}(0.005)^{1/2}$$
$$= 0.39\ \text{m/s}$$

*Sub-section 2*

$$A = \left(\frac{2 + 3.25}{2}\right)(0.5) + 0.2(3.25) = 1.96\,\text{m}^2$$
$$P = 0.2 + 2(0.8) + 2 = 3.8\,\text{m}$$
$$R = 1.96/3.8 = 0.516\,\text{m}$$
$$\Rightarrow V_2 = \frac{1}{0.035}(0.515)^{2/3}(0.005)^{1/2}$$
$$= 1.30\,\text{m/s}$$

Therefore total discharge for the section is

$$Q = A_1 V_1 + A_2 V_2 = (0.6)(0.39) + (1.96)(1.30)$$
$$= 2.78\,\text{m}^3/\text{s}.$$

### 6.6.6  Water-surface Profiles in Gradually Varied Flow

Since true uniform flow seldom occurs in streams, most problems involve the treatment of gradually varied flow or rapidly varied flow. Gradually varied flow is usually considered to occur over a relatively long length of channel. The longitudinal shape of the water surface within it can be predicted from the 'energy budget' of the flowing water. These profiles are also known as *back-water curves* when they describe subcritical flow conditions.

### Classifying Flow Profiles

The type of water surface profile exhibited (i.e. how the flow depth changes longitudinally) depends on the relationship between the actual water depth ($D$), the normal depth ($D_n$) and the critical depth ($D_c$). The normal depth is the depth that would occur if the flow were uniform and steady, and can be predicted from the equations of Chezy or Manning. Critical depth is the depth at which specific energy is minimum for a particular discharge.

A *critical slope* is one which sustains uniform critical flow ($D_n = D_c$). Because a small difference in energy will cause the flow to fluctuate to either supercritical or subcritical, the flow across a critical slope tends to have an unstable, undulating surface. A *mild slope* is one which is less than the critical slope, the normal depth is subcritical ($D_n > D_c$) and the flow has a downstream control. A *steep slope* is steeper than the critical slope, the normal depth is supercritical ($D_n < D_c$) and the control is upstream. Long steep slopes are uncommon in natural channels; instead, they are usually broken up by pools and drops. It is possible for a given channel slope to be classified as mild, steep or critical at different streamflows since both $D_n$ and $D_c$ vary with discharge.

Based on the relationship between $D$, $D_n$ and $D_c$, the resulting profiles can be categorized. These have conventionally been used for describing the entire flow within a reach, but may have potential use for describing local water surface patterns around and over boulders and branches. Labels for each category consist of a letter designating the type of slope and a number which indicates the relative position of the flow depth:

1. *Type 1 curve.* The actual depth is greater than both $D_c$ and $D_n$. Flow is subcritical.
2. *Type 2 curve.* The actual depth is between $D_c$ and $D_n$. Flow can be either subcritical or supercritical.
3. *Type 3 curve.* The actual depth is smaller than both $D_c$ and $D_n$. Flow is supercritical.

Only mild and steep profiles will be described here. There are other categories for critical, horizontal, and adverse (uphill) slopes, which can be found in Chow (1959) or in chapters on gradually varied flow in fluid mechanics texts. The diagrams shown are drawn with exaggerated slopes since most streambed slopes are very small.

*M profiles.* The M1 profile is one of the more well-known curves, and can be seen where a mild-sloped stream

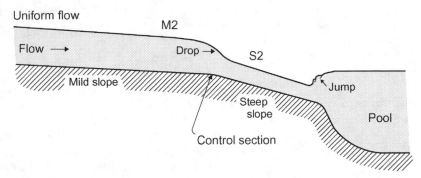

***Figure 6.35.*** *Flow profiles in combination along a reach*

enters a pond. The M2 profile may occur upstream of a sudden enlargement in a channel reach or where the slope becomes steeper. M3 profiles are found where super-critical flow enters a mild-sloped channel such as under a sluice gate (e.g. at a dam) or under a log. They usually end in a hydraulic jump.

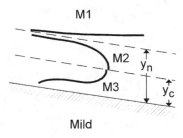

*S profiles.* The S1 profile begins with a rise at the upstream end, then becomes horizontal. It can occur where a steep channel empties into a pool. S2 is called a draw-down curve, and may be found at the downstream end of a channel enlargement. S3 can occur where water enters a steep channel from under a gate. It is considered 'transitional' and is usually very short.

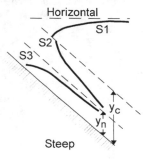

The starting point for water surface profiles is at a *control section*, which can occur at sluice gates and weirs (or their natural equivalent), bedrock constrictions, at sharp changes in channel slope or any location where critical flow occurs. Larger discharges can 'drown out' features which might act as controls at lower stages. Thus, the flow profile is affected by control sections; yet the flow itself determines whether a section is controlling or not (Henderson, 1966). A combination of a few of the flow profile categories is illustrated in Figure 6.35.

### Synthesizing the Shapes of Gradually varied Flow Profiles

In gradually varied flow, depth, area, roughness, and/or slope change slowly along the channel. A mathematical description of the water surface shape can be derived from principles of energy (Section 6.6.4) and continuity (Section 6.3.3). In practice, the *standard step method* is most commonly used. It is applicable to natural channels with their changing slopes and irregular banks, under gradually varied flow conditions. In this method the channel section is divided into short reaches and computations are carried step by step from one end of a reach to another. If the control point is downstream (subcritical flow), calculations are performed in the upstream direction (to derive a backwater curve), whereas calculations are carried downstream if the flow is supercritical.

To apply the standard step method in natural streams a field survey should be conducted to collect data on the cross-sectional geometry, channel slope, roughness and present water surface profile (see Chapter 5 for survey techniques). The computation method requires an iterative solution, which is described by Chow (1959) and Henderson (1966). PHABSIM, a physical habitat simulation model described in Chapter 9 has an option for developing backwater curves, and the HEC-RAS computer program developed by the US Corps of Engineers is commonly used in practice for deriving water surface profiles in natural streams

(HEC-RAS can be downloaded at: www.hec.usace.army. mil/software/hec-ras/hecras-hecras.html). Davidian (1984) provides a comprehensive description of the technique.

By assuming different values of discharge at a cross section, a 'family' of flow profiles can be generated for various conditions of stage and discharge. In this manner, stage-discharge relationships (Section 5.6.5) can be developed for ungauged sections. Other applications include deriving water surface profiles upstream from snags or bridges, or describing the profiles in a tributary stream as the depth in the main stream level fluctuates.

### 6.6.7 Hydraulic Jumps and Drops, Alias Rapidly Varied Flow

Rapidly varied flow occurs over relatively short lengths of channel and it is typically a location of high energy loss. In these situations of intense turbulence a sketch is preferable to an analytical solution for describing the surface profile. Examples are *hydraulic jumps*, where the flow changes from supercritical to subcritical, and *hydraulic drops*, where the reverse occurs. Because of the energy loss, supercritical flow will not 'jump' all the way up to its alternate depth (Figure 6.31) in a hydraulic jump, and subcritical flow will not 'drop' all the way down to its alternate depth in a hydraulic drop.

Hydraulic drops occur where flow accelerates—for example, as it passes over an obstacle, through a chute, or from a mild slope to a steep slope. Hydraulic jumps take place where upstream supercritical flow meets subcritical flow, such as at the downstream side of large boulders, below narrows created by rock outcrops or where the slope changes from steep to mild. Because of the sudden reduction in velocity, hydraulic jumps are associated with highly turbulent conditions, whitewater and large losses of energy. Since they are such effective energy dissipators they are often encouraged in the design of spillway chutes and structures for dissipating the erosive power of water. They are also effective for aeration of water and for the mixing of nutrients in streams. Fish capitalize on the backflow in the standing waves of hydraulic jumps to give them a 'boost' upstream (Hynes, 1970).

The length of flow affected by the hydraulic jump ranges from four to six times the downstream water depth. Its appearance is influenced primarily by the upstream Froude number, with the channel geometry having a secondary effect. Some of the patterns exhibited by hydraulic jumps are illustrated in Figure 6.36. The range of upstream Froude numbers given for each type can serve as a basis for classifying hydraulic jumps.

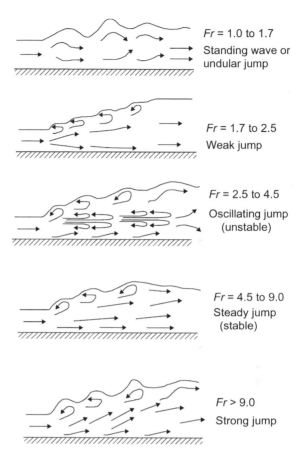

**Figure 6.36.** *Hydraulic jump patterns based on upstream Froude number (Fr). Adapted from White (1986), by permission of McGraw-Hill*

Energy dissipation in hydraulic jumps increases from less than 5% in standing waves to over 70% in strong jumps. Oscillating jumps tend to be unstable, generating large waves that can travel long distances and damage streambanks (White, 1986). It should be noted that hydraulic jumps are not possible if the upstream flow is subcritical (Fr < 1).

The upstream Froude number can be calculated at a hydraulic jump by measuring the upstream and downstream water depths and using the formula (Henderson, 1966)

$$\frac{D_2}{D_1} = \frac{1}{2}\left(\sqrt{1 + 8Fr^2} - 1\right) \qquad (6.67)$$

where the subscripts 1 and 2 refer to upstream and downstream depths, respectively, and *Fr* is the upstream Froude number.

*Example 6.8*

Calculate the upstream Froude number and give the classification for the hydraulic jump where the upstream depth is 0.15 m and the downstream depth is 0.27 m.

*Answer*

$$\frac{0.27}{0.15} = \frac{1}{2}\left(\sqrt{1 + 8Fr^2} - 1\right)$$

$$2(1.8) + 1 = \sqrt{1 + 8Fr^2}$$

$$(4.6)^2 = 1 + 8Fr^2$$

$$Fr = \sqrt{\frac{20.16}{8}} = 1.59$$

This is an undular jump.

Flow in natural channels is typically varied, unsteady, turbulent and subcritical. However, uniform, steady conditions are often assumed in order to simplify the equations that describe flowing water. The various categories are useful for classifying the flow environment experienced by aquatic organisms, and they give insight into the usefulness and limitations of equations which have been based on theoretical definitions of flow conditions. In complex flow environments, flow visualization and mapping of depths and velocities may be more appropriate. It should be kept in mind that the theory of open channel flow assumes flow in prismatic channels (constant cross section and slope). In applying the theory to irregular natural channels we are stretching thin the boundaries of truth, and must interpret results with judgement and caution.

# 7

# It's Sedimentary, Watson!

## 7.1 Introduction to Stream Channels, Streambeds and Transported Materials

### 7.1.1 General

As water works its way downstream some of its energy is expended on the transport and rearrangement of materials in the stream's bed and banks. Its ability to create geometric patterns such as braided deltas, rippled sand bars and regular meanders is as fascinating to the casual observer as it is to the scientist attempting to explain or model the processes involved.

Stream channels display more or less regular downstream changes in width, depth, velocity and sediment load, accompanied by changes in the distribution of stream biota. Streams are considered 'open systems' because they experience continuous inflows and outflows of energy and matter. More importantly, they are dynamic systems, with changes occurring over a range of time scales from instantaneous to geological. Stream levels shift and sediment loads fluctuate, meanders migrate, floods scour and deposit, banks collapse, sand bars grow and the effects of change at one point are reflected elsewhere in the system.

Adjustment of channel form is of interest both to geomorphologists studying the behaviour of natural rivers over long time periods and to hydraulic engineers concerned with shorter-term changes affecting channel stability near bridges, dams and property boundaries. Ecologists, who study the evolution, distribution and interaction of organisms and their adaptation to the environment, are concerned with both long-term changes which form and re-form habitats and the short-term fluctuations which have a more immediate impact.

In comparison to the study of fluid mechanics (Chapter 6), *fluvial geomorphology*, the study of water-shaped landforms, tends to be more qualitative. The focus is on trends and descriptions rather than precise predictions. General texts for further information on the topics discussed in this chapter include Gregory and Walling (1973), Knapp (1979), Knighton (1984, 1998), Leopold *et al.* (1964), Morisawa (1985), Petts and Foster (1985), Richards (1982, 1987), Schumm (1977), Tinkler and Wohl (1998) and Wohl (2000).

### 7.1.2 Making Up a Channel Bed

#### Effects of Geology and Hydrology

In bedrock streams, channels are eroded as a result of mass failure of large rock slabs and the slow chipping and grinding of the channel bed by stream-transported debris. *Potholes* are a common feature in bedrock channels, which result when stones are ground against the bedrock by spiralling flow.

In alluvial streams the flow moves across beds of material deposited as a result of riverine or glacial processes. The distribution, composition and shapes of rocks, pebbles, sands and clays in an alluvial channel bed reflect the ease with which its source materials are broken up and rounded, and the hydraulics of both high and low flows. Weak rocks such as mudstones, shales or sandstone, for example, are easily broken down, whereas quartzites resist erosion. The size of sediments and the distance they have travelled from their sources thus becomes a geologic record of the stream's evolution and hydrology.

In headwater regions, bed materials are usually large, often exceeding the sediment-carrying capability of the flow. These large, often angular rocks disintegrate in the channel near to the original source. Progressing downstream, the mean grain size of substrate materials

Stream Hydrology: An Introduction for Ecologists, Second Edition.
Nancy D. Gordon, Thomas A. McMahon, Brian L. Finlayson, Christopher J. Gippel, Rory J. Nathan
© 2004 John Wiley & Sons, Ltd ISBNs: 0-470-84357-8 (HB); 0-470-84358-6 (PB)

generally decreases as sediments are fragmented and abraded and smaller sediments are sorted out and carried off. This occurs rapidly in the first few kilometres of a stream and more slowly thereafter.

Local flow conditions can also have a sorting effect on sediments. Coarse materials line riffles and other regions of high shear stress. Finer sediments are found in depositional regions such as large pools, between boulders in headwater streams, or in the 'flow shadow' of bends, confluences, tree roots or other obstructions. The patterns of substrate size may only change at high flows, especially in streams with coarser, more heterogeneous materials. Thus, the spatial distribution of bed materials will typically be more closely related to previous flood events than the 'normal' flows carried by the stream, except in streams with large influxes of fine sediment that coat the streambed.

For the Acheron River, the stream system used as an example throughout this text (see map, Figure 4.7), a downstream trend is not apparent in the particle size data of Figures 5.32 and 5.33. The third-order site has the coarsest materials due to armouring (Section 7.4.2), whereas at the fifth-order site the increase in the $d_{50}$ is due to an influx of large materials from rock outcrops along the Little River.

### Flow Resistance

Bed materials offer resistance to the flow, which influences the rate of energy loss along a stream and, in turn, has a strong relationship with channel patterns. Flow resistance is caused by (1) grain or surface resistance and (2) form resistance, somewhat analogous to the categories of surface and form drag described in Section 6.5.4.

*Grain resistance* is the resistance offered by individual grains. In gravel streambeds it tends to be the major component of flow resistance. Grain resistance is considered a function of relative roughness, the ratio of roughness height to water depth (Eq. (6.18)). Thus, its effect diminishes as the roughness elements become submerged at higher discharges. To reflect the influence of the largest particle sizes, roughness height is commonly described in terms of the $d_{84}$ or $d_{90}$ of the bed materials (see Section 5.8.5). The spacing and arrangement of these larger particles can also affect flow resistance (Section 6.4.2).

*Form resistance* is associated with the topography of the channel bed. Bedforms result from the interaction of streamflow patterns and bed sediments, particularly in sand-bed streams (Section 7.3.4). However, form roughness can also refer to the troughs, potholes and plunge pools of bedrock channels and the larger-scale forms of pool-riffle or pool-step sequences (Section 7.3.2).

Progressing downstream, channel topography typically changes from a poorly defined pool-step structure in headwater streams to more developed pool-riffle sequences in gravel-bed reaches, and finally to sand bedforms as the grain sizes become smaller and more uniform. In conjunction with channel patterns, streambed forms seem to regulate resistance in a self-adjusting manner so as to effectively dissipate energy over a wide range of flow and bed material conditions (Morisawa, 1985).

### 7.1.3    What Sort of Debris Is Transported?

*Total load* refers to the amount of dissolved and particulate organic and inorganic material carried by the stream. Although it should be realized that sharp boundaries do not exist, the total load can be divided into three groupings:

- Flotation load
- Dissolved load
- Sediment load

The last category can be further sub-divided on the basis of particle transport rate, size distribution, density, or chemical and mineralogical composition. Sediment is usually considered to be the solid inorganic material, which is separated from particulate organic matter during analysis. Commonly, sediment load is separated into the following categories:

- Washload
- Bed-material load—which can be transported as
    Suspended load or
    Bedload.

The sub-division of the various loads within a stream is illustrated in Figure 7.1.

### Flotation Load

The *flotation load* consists of the logs, leaves, branches and other organic debris which are generally lighter than water (until they become waterlogged). The amount of organic debris supplied to a stream depends on the density and type of vegetation along the banks, the amount of bank failure and tree fall, and the amount of floating debris picked up from the floodplain by flood waters.

The organic debris is the foodstuff of decomposers and some aquatic invertebrates. Large trees, rootwads and

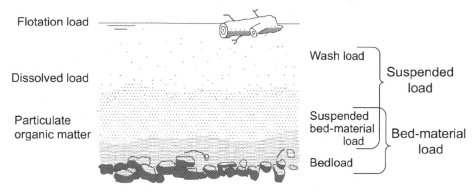

**Figure 7.1.** *Categories of transported materials in a stream*

saplings provide shelter for fish, and form debris dams which trap sediments and modify the channel shape by redirecting swift currents. The large woody debris (LWD) can play an important part in channel stabilization (Section 7.5.1). Removal of riparian vegetation reduces the supply of organic material to the stream, which may have long-term effects on both its ecology and morphology. Flotation load, however, can cause problems when it impinges against bridges or other structures.

### Dissolved Load

A stream's *dissolved load* is the material transported in solution. Local geology, land use and weathering processes affect the amount of dissolved load, which can often exceed the sediment load by total weight. Natural origins of dissolved loads include sea salts dissolved in the rainwater of coastal areas, and the chemical weathering of rocks—sometimes enhanced by organic acids from the decay of vegetation. In general, water originating as groundwater tends to have a higher soluble load than surface-derived runoff. Industrial effluents and agricultural fertilizers and pesticides can also be significant sources of solutes. Dissolved load is therefore associated with chemical water quality (Section 5.9). Although it will vary from river to river, an average of 38% of the total load of the world's rivers is dissolved material (Knighton, 1984).

Because of turbulence, dissolved loads will generally be uniformly distributed over a stream cross section. Exceptions exist in localized areas of saline stratification and in reaches where groundwater of high dissolved load enters the stream. Dissolved loads also change with changes in discharge. During a runoff event, concentrations generally decrease at first as rainwater dilutes the stream water, but later increase as groundwater reaches the stream, bringing in dissolved materials (Hjulstrom, 1939).

### Washload

*Washload* refers to the smaller sediments, primarily clays, silts and fine sands, which are readily carried in suspension by the stream. This load is 'washed' into the stream from the banks and upland areas and carried at essentially the same speed as the water. Only low velocities and minor turbulence are required to keep it in suspension, and it may never settle out. Its concentration is considered constant over the depth of a stream. For practical purposes, the smallest washload grain is considered to be about one-half micron (0.0005 mm), to separate it from dissolved materials. The largest size is usually taken as 0.0625 mm. This division is somewhat arbitrary, but objective.

It is the *rate of supply* from uplands or streambanks which determines the amount of washload transported, rather than the ability of the stream to carry it. According to theory, streams have an almost unlimited capacity for transporting washload. Streams cannot become 'saturated' with sediment as they can with dissolved solids (Hjulstrom, 1939). The washload can constitute a large percentage of the total volume and mass carried by a stream or river.

High washloads may typify streams with banks of high silt-clay content. However, washloads can also be contributed from fire-denuded slopes, the ash from volcanic eruptions or other disturbances in the catchment such as road or dam building and agricultural practices.

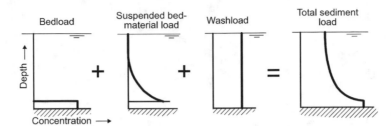

*Figure 7.2.* Schematic diagram of the vertical distribution of sediment load concentrations

### Bed-material Load

*Bed-material load* is the material in motion which has approximately the same size range as streambed particles. In alluvial streams, flow conditions control the amount of bed-material load transported. This load may be further sub-divided according to whether it is transported in suspension or remains in touch with the bed.

*Suspended bed-material load* is the portion that is carried with the washload, remaining in suspension for an appreciable length of time. It is supported by the fluid and kept aloft by turbulent eddies, but will settle out quickly when velocities drop. Over a vertical profile its concentration is highest near the streambed. *Bedload* is that portion which moves by rolling, sliding or 'hopping' (saltation), and is partially supported by the streambed. It is thus only found in a narrow region near the bottom of the stream (see Figure 7.2).

For field measurements, the distinction between sediment load types is commonly based on the method of data collection (see Section 5.7). Suspended sediment samplers will capture both washload and suspended bed-material load, which are often grouped into the single category, *suspended load*. Most depth-integrating suspended sediment samplers (Section 5.7.4) can measure to within about 90 mm of the streambed. The amount of material travelling along in contact with the bed can be measured with bedload samplers or by surveying the amount of sediment accumulated over time in depressions or behind logs (Section 5.7.5).

Another conventional separation of suspended load from bedload is based on the sand particle size threshold at 0.0625 mm, but obviously in a steep, cobble bed stream the sand fraction may be in suspension. Concentration of sand-sized and coarser material tends to peak with discharge, while concentration peaks of the finer material may be out of phase with discharge peaks. Also, material finer than sand tends to be evenly distributed in the cross-section, while coarser material is more concentrated near the bed (Richards, 1982). The division into suspended load and bedload is convenient for the development of sediment transport equations based on the different modes of movement. In reality, a particle of a given size can move in suspension, hop along the streambed, or stay put, depending on the flow rate. Density and shape as well as size can also affect the way in which a particle is transported. On the average, less bedload than suspended load is transported over a year. The ratio of bedload to suspended load is typically in the range 1:5 to 1:50 (Csermak and Rakozki, 1987). This ratio will be higher in headwater streams and during floods, when bedload can often exceed the suspended load.

### 7.1.4  Sediment Distribution and Discharge

Figure 7.2 gives a general illustration of the vertical distribution of sediment. Larger sediments that move only as bedload are the most highly concentrated at the bed, whereas silts and clays are distributed more or less uniformly from bed to water surface. The actual distribution over a particular vertical in a stream is dependent on the particle sizes and the velocity and turbulence intensity (see Eq. (6.14)) of the water.

Considering all particle sizes together, both concentration and mean grain size increase towards the bed. Variations can occur across stream cross-sections and with increases in discharge. The largest quantities of sediment and the coarsest fractions tend to be transported in the path of maximum velocity, which does not always coincide with the thalweg (the path of deepest flow). If velocity and turbulence increase, the larger particles will be distributed more evenly.

*Sediment discharge* is the amount of sediment moving past a cross-section over some period of time. It is usually reported in units of mass per unit time such as kg/s or tonnes per day or year. Vertical and lateral patterns of both sediment concentration and velocity determine the distribution of sediment discharge at a cross-section, as shown in Figure 7.3. This distribution must be taken into account when sampling sediments (Section 5.7). A depth-integrating sediment sampler, for example, takes a velocity-weighted sample, which integrates the distributions

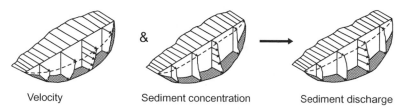

Velocity                Sediment concentration              Sediment discharge

*Figure 7.3.* Graphical representation of sediment discharge at a cross-section as a result of velocity and sediment concentration patterns. Adapted from Nordin and Richardson (1971)

of both velocity and sediment concentration. Additionally, if an automatic sampler intake nozzle is located near the streambank, its samples may be sufficiently representative in a stream which carries mostly washload, but will give misleading information in streams which transport coarser materials.

Patterns of sediment discharge do not necessarily coincide with runoff patterns. At sediment source areas, erosion is usually highest at the beginning of a rainstorm when sediments are more readily available. These are washed into the stream with the first rivulets of overland flow, and as the sediment supply is exhausted, the concentration drops quickly. In these streams, sediment concentrations will ordinarily be greater when the stage is rising, with coarse particles depositing out quickly as the hydrograph starts to fall. Thus, the concentration will be different at the same discharge on the rising and falling limb of a hydrograph (see Figure 7.21). This effect is called *hysteresis*, and complicates the prediction of sediment concentration from streamflow (Section 7.5.4). In headwater areas the sediment concentration will typically peak before discharge does. Downstream, in large catchments, the peak sediment load may match or even lag behind the discharge peak when the site must 'wait' for the sediment to be delivered from upstream.

Patterns of *total* sediment concentration do not give the whole story, however, as the *sizes* of sediments transported will also change with time. Washload fines typically occur in higher concentration at the beginning of an event. Bedload movement and entrainment of bed materials increases as the discharge reaches its peak, and then a reduction in particle size occurs as the hydrograph tapers off. This pattern is dependent on the types of sediment and dissolved materials available from uplands and channel. As shown by Johnson *et al.* (1985), sediment loads also tend to be different for snowmelt and rainfall runoff events.

The amount of sediment carried through the outlet of a catchment depends on two factors:

1. The amount of sediment eroded and transported to the stream from upland sources.

2. The ability of a stream to carry the washed-in sediments and to re-work and transport bed and bank materials.

Streams can therefore be considered either *supply-limited* or *capacity-limited*, depending on whether their ability to carry sediment exceeds the amount available or vice versa. In regard to the type of sediment carried, washload is considered supply-limited and bed-material load capacity-limited. Mountain channels with large streambed materials and pool-step structures are typically supply-limited except after disturbances such as fire or road construction.

The first of the above two factors is dependent on climate, land use and the geology and topography of the catchment. Hillslope erosion can result from raindrop splash, the hydraulic action of overland flow, rill and gully formation, mass movement such as landslides, the scuff of a boot, or the twirl of a tractor tyre. A large amount of effort has gone into the modelling of soil loss and the effects of land management on hillslope erosion. Upland processes are beyond the scope of this book, and the interested reader is referred to Branson *et al.* (1981), Finlayson and Statham (1980) and geomorphology texts.

The second of the above factors is dependent on the hydraulic and hydrological properties of the channel and the erodibility of its bed and banks. Relevant field measurements are covered in Chapter 5 and the processes which cause individual grains to be lifted into the flow are covered in Section 7.4. The larger-scale processes that contribute to the amount of sediment delivered from a catchment are discussed in Section 7.5.

### 7.1.5 Ecological Implications

For many aquatic organisms the channel bed is a 'substrate' to be used as a foothold, as a site to deposit or incubate eggs, as 'grit' for grinding food or as a refuge from floods (Minshall, 1984; Statzner *et al.*, 1988; see also Section 2.1). In engineering practice the channel bed is normally considered the boundary between solid and fluid—a section of the 'wetted perimeter'. For many organisms the boundary is not as distinct. The streambed surface is rich with organic matter trapped in the pits between grains, which provides nutrients for organisms near the base of the food chain. Below the surface, the hyporheic zone forms an interface between stream and groundwater systems. This region can be extremely active biologically. Many organisms reside temporarily in the interstices of the surface materials, whereas some may carry out their whole life cycle deep within stony stream-beds (Hynes, 1970). The streambed acts as a refuge for benthic organisms, providing shelter from floods, drought, and extremes of temperature. Ward and Stanford (1983a) call it a 'faunal reservoir', capable of recolonizing the stream if stream populations are depleted by adverse conditions.

Species differ in their substrate preferences and requirements. The suitability of a substrate for colonization by aquatic flora and fauna depends on its average particle size, its mix of sizes, the size of pore spaces, degree of packing and imbeddedness, and its surface topography. Freshwater crayfish and some aquatic insect species such as dragonfly and stonefly larvae live in the crevices between and beneath rocks. Others, such as the purse-case caddisfly, require unstable fine-grained sands where moss cannot grow. Still others such as midge larvae require mud into which they can burrow. Salmonids require a mix of gravels with small amounts of fine sediments and rubble as an optimum spawning substrate mix (Beschta and Platts, 1986). Algae, mosses and other aquatic plants also have specific substrate requirements. The plants, in turn, provide substrates to shelter or support other organisms.

Thus, the distribution of sediment sizes along a stream will be one of the physical habitat factors influencing the distribution of organisms. In general, the highest productivity and diversity of aquatic invertebrates seems to occur in riffle habitats with medium cobble and gravel substrate (Gore, 1985a). Areas of shifting sands commonly have reduced species abundance and richness (Minshall, 1984).

Biological activity in coarser substrates is dependent upon the maintenance of inter-gravel flow rates for the replenishment of nutrients and oxygen and the removal of metabolic wastes. If excessive fines are washed into a stream, as, for example, from road or dam construction, they can form a 'mat' on top of the coarser bed materials. Fines can also work down between the coarser grains to form a type of 'hardpan' layer. The infilling of gravels with finer sediments can reduce inter-gravel flow rates, suffocate eggs, limit burrowing activity and trap emerging young. Gravel-bed streams which become filled with silt may display a shift in the insect species compositions from mayflies (*Ephemeroptera*) and caddisflies (*Trichop-*

*tera*) towards midgefly larvae (Diptera), which, in turn, can affect fish species compositions (Milhous, 1982). Because of these ecological effects, there has been some interest in estimating the flow required to remove fines from the streambed. Methods for estimating these 'flushing flows' are discussed in Section 7.4.3.

The transport of particulate matter is both bane and benefit to aquatic organisms. Organic particulate matter is a source of food for downstream organisms. However, when the flow quickens, the larger grains can become deadly projectiles. Fine silts and clays clog gills like a particulate 'smog', reduce light needed for photosynthesis and periphyton production, and interfere with the foraging success of sight feeders and filterers. The shifting of whole segments of the streambed uproots and scours away benthic organisms. Heavy metals and other toxic substances can also be adsorbed onto sediment and are thus transported and deposited along with it.

Vogel (1981) implies that abrasion and alterations of form of the stream bottom during floods have more critical impacts on biota than velocity *per se*. Jowett and Richardson (1989) cite a study on rivers in New Zealand where the abundance of trout decreased significantly after a major flood, although the coarsening of substrate, removal of excessive algal growth and deepening of pools improved habitat for future use. Aquatic vegetation such as algal mats and macrophytes will affect bed stability during high flows. After floods, recolonization by bacteria, fungi and algae binds and stabilizes the substrate, improving conditions so that other organisms can come back more quickly.

The movement of sediments and the composition of streambeds will thus have different effects on different species. The ecological advantages and disadvantages of sediment movement and streambed composition should therefore be weighed carefully in studies of instream flow needs, channel changes, or the effects of land use practices (Chapter 9).

## 7.2  Stream-shaping Processes

### 7.2.1  A Note about Stream Power

As described in Section 1.2, power is the amount of work done per unit time, where work and energy have the same units. *Stream power* has a number of definitions, related to the time rate at which either work is done or energy is expended. It is a useful index for describing the erosive capacity of streams, and has been related to the shape of the longitudinal profile, channel pattern, the development of bed forms, and sediment transport.

In studies of sediment transport, Bagnold (1966) originally defined *stream power per unit of streambed area* ($\omega_a$) as

$$\omega_a = \tau_o V \qquad (7.1)$$

where $\tau_o$ is the shear stress at the bed (N/m$^2$) and $V$ is the mean velocity (m/s) in the stream cross section. Thus, $\omega_a$ has units of N/m s (watts/m$^2$).

Perhaps a more useful definition of stream power is the rate of potential energy expenditure over a reach or *stream power per unit of stream length* (this equation can easily be developed from Eq. (1.5) as an exercise)

$$\omega_l = \rho g Q S \qquad (7.2)$$

where $S$ is the energy slope of the reach (see Section 6.6.4), and $\omega_l$ has units of kg m/s$^3$ (watts/m). Another form is *stream power per unit mass* of water (a mass of 1 kg)

$$\omega_m = g V S \qquad (7.3)$$

where $\omega_m$ has units of m$^2$/s$^3$ (watts/kg). Alternatively, it can also be expressed as a *stream power per unit weight* (a weight of 1 N)

$$\omega_w = V S \qquad (7.4)$$

where $\omega_w$ has units of velocity (m/s) or (watts/N). This measure can also be considered the time rate of head loss over a reach, where head is energy per unit weight (Section 6.3.4). Any of the terms $\omega_a$, $\omega_l$, $\omega_m$, or $\omega_w$ can be referred to as a 'unit stream power', because the stream power is expressed per unit area, length, mass or weight, respectively. In the literature, however, $\omega_l$ is commonly given the symbol $\Omega$ and called *total stream power.*

As slopes become steeper and/or velocities increase, stream power goes up and more energy is available for reworking channel materials. In a bedrock stream with no sediment transport all the stream power is spent in the frictional dissipation of energy. In alluvial channels with mobile boundaries part of this stream power is used for transporting sediment. Figure 7.4 compares high and low stream power situations. It can be seen that straightening and clearing a channel would increase its slope and velocity, and thus its stream power. This increases the amount available for erosion and sediment transport. Alterations to the stream power at one point can therefore initiate changes in sediment transport and channel shape elsewhere in the stream.

Stream power also increases as discharge increases. However, even though discharge typically increases in the downstream direction, stream power per unit area ($\omega_a$) typically decreases because slopes decrease. Flash floods in steep ephemeral channels, for example, can generate very large values of stream power. Costa (1987) ascertained that a 1973 flood on a tributary to the Humboldt River, Nevada had a unit stream power ($\omega_a$) of 8160 N/m s, as compared to 12 N/m s for floods on the Mississippi and Amazon Rivers. Jarrett and Malde (1987) estimated a unit stream power ($\omega_a$) of 75 000 N/m s for the prehistoric Bonneville Flood, which catastrophically discharged an enormous volume of water down the Snake River in southern Idaho. The estimated peak discharge of approximately 935 000 m$^3$/s inundated the Snake River Canyon to depths greater than 130 m in places, and deposited large gravel bars and huge boulders in its path.

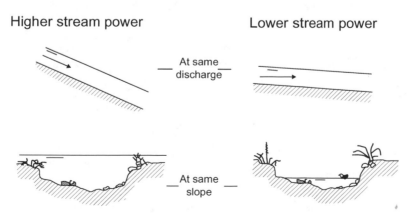

**Figure 7.4.** *Illustration of high and low stream power situations*

The relationship between the channel and its floodplain is important in determining stream power. In a study of alluvial rivers in Victoria, Australia, Brizga and Finlayson (1990) found that rivers which remained within their banks at high flows tended to have high stream power and relatively coarse bed materials. In comparison, rivers which flooded over their banks at high flows had lower stream power, transported finer sediments and had more stable channels.

Langbein and Leopold (1964) proposed that the shape of streams is a compromise between two opposing tendencies: (1) for energy to be expended uniformly over the length of a stream (implying constant stream power), and (2) for the total expenditure of energy to be minimized over the length of a stream. The 'typical' concave shape of the longitudinal profile (Section 4.2.6) may be partially explained by the first theory. In headwater streams, discharge is low and slope is high, whereas in valley streams, discharge is high and slope is low; thus the product QS remains relatively constant over the length of a stream. Based on modelling studies, Knighton (1999) predicted that total stream power peaked at an intermediate distance between drainage divide and mouth, and unit stream power peaked closer to the headwaters. The reader is referred to the fluvial geomorphology texts mentioned in Section 7.1 for further discussion on the use of stream power in the description of stream channel form and formation.

### 7.2.2   Adjustments and Equilibrium

One of the most interesting characteristics of open systems such as streams is their capacity for self-regulation. In terms of channel adjustment a stream has several variables with which to work. It can change local slopes and velocities, rearrange bed materials, transport more or less sediment, and change its channel pattern—within certain constraints. 'Feedback' mechanisms act to stabilize the system so that some degree of equilibrium can be established. This discussion applies primarily to alluvial streams; in bedrock streams, channel adjustment is not quickly achieved.

If a stream is in equilibrium condition it is considered both stable and graded. A stream is described as *stable* if its cross-sectional geometry remains relatively constant over some time scale. It is considered *graded* if its slope is just sufficient to transport all of the material delivered to the stream. If a stream's ability to transport sediments (its sediment transport capacity) is less than that required to move sediments arriving from upstream, then some of the

sediment is deposited, leading to *aggradation*. In contrast, if not enough sediment enters to 'consume' the transport capacity, the flow will erode the river bed and/or banks to pick up sediment, leading to *degradation*.

Aggradation is commonly seen where a stream enters a reservoir or pond. The 'delta' slowly grows downstream, eventually filling the impoundment (see Figure 7.6). In stream channels, aggradation can elevate the channel bed within natural levees, causing adjacent areas to become swamp-like wetlands. Degrading streams, on the other hand, can cause a lowering of the surrounding groundwater level, leaving both native and agricultural plants high and dry. Degradation can occur where the local transport capacity has increased, e.g. from the straightening of a meander. The erosion will often propagate back upstream—a process called *headcutting*. A steep knickpoint usually exists at the upper end of the cut. Erosion is more pronounced in this region, and the swirling action of water in the downstream pool can create an undercut which eventually leads to failure of the overlying material and upstream migration of the knickpoint.

Sites of aggradation and degradation can be revealed by comparing the longitudinal profiles from two different time periods. Over time, aggrading sections tend to become steeper and degrading sections flatter. Because changes in slope affect velocity and sediment transport, 'feedback' can occur to slow or reverse the trend.

A channel may undergo gradual change in form, steepness and sediment transport, which can occur as an average trend or cycle. Oscillations about the average state can also take place. Furthermore, systems may be influenced by *thresholds*; i.e. abrupt, dramatic changes may occur when some critical value is exceeded. As an example of a threshold response, a rock can remain perched on a small pedestal until progressive weathering eats away at the base and the rock suddenly tumbles. In fluvial systems, thresholds may exist for sediment movement, bank collapse or the sudden cutoff of a meander bend. Thus, changes in erosion and deposition may not necessarily be due to external influences like land-use change but may have appeared suddenly when long-term processes reached a threshold.

Schumm (1977) suggests that drainage basin evolution can be considered at four time scales, as illustrated in Figure 7.5. Over a major period of geologic time an uplifted terrain is gradually worn down, and slopes progressively become less steep. Within this time frame the channel is continuously adjusting to changes in discharge and sediment load, causing fluctuations about some average trend. *Dynamic equilibrium* describes this type of behaviour. Abrupt periods of adjustment can also occur within this time frame, and the term *dynamic metastable*

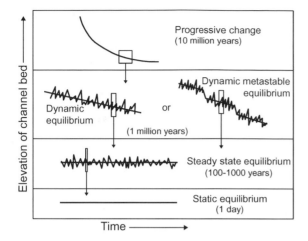

**Figure 7.5.** *Drainage basin evolution at varying time scales. Adapted from Schumm (1977), by permission of John Wiley & Sons, Inc, Reproduced by permission of Stanley Schumm*

*equilibrium* was introduced to allow for the influence of thresholds. *Steady-state equilibrium* refers to a generally stable form, about which seasonal and other short-term fluctuations occur, such as scour and fill during floods. At the time scale of one day the channel form is essentially in a *static equilibrium*, unless a flood or a bulldozer is passing through.

Fluctuations and thresholds can result from the complex interactions of many variables. A major event such as a flood, a bushfire, channelization or tectonic activity can cause a chain reaction of responses that continue for many years. For example, Schumm (1977) describes the complex response of a tributary to the lowering of the elevation at its outlet. This can occur when the stream into which it flows degrades to a lower base level. Predictably, headcutting will progress upstream from the mouth. However, the pattern of erosion, deposition, changes in sediment load and renewed incision within the tributary as it adjusts to the new base level can be extremely complex.

At the time scale appropriate to most stream research a stream system will most likely exist in a state of 'metastability' rather than equilibrium. Like wars interrupting relatively peaceful phases to change the direction of history, channel developments may also be characterized by relatively sudden changes between quiescent periods. Vannote *et al.* (1980) propose that the structure and function of stream communities adjust to changes in physical habitat. It is thus important to consider the type and direction of change when evaluating the condition of a stream channel. As Bovee (1982) mentions, a habitat and

instream flow evaluation based on the assumption that the stream will remain in its present form will be invalid if the stream is not in an equilibrium condition.

### 7.2.3 Balancing Slope, Streamflow, and Sediment Size and Load

The shapes of a stream—its snake-like meanders, its tumultuous drops, its stretches of calm water—are controlled by climate and landscape geology. Over time, channel form is adjusted to accommodate the discharge and sediment loads of a catchment. Complex interrelationships exist between channel dimensions, channel patterns, sediment supply, streambed roughness and steepness of the valley floor and stream channel. Alterations to any of these components, either natural or human-induced, will have an impact on others. Effects are difficult to predict because of the difficulty in isolating the role of a single variable.

A general, qualitative expression for the balance between sediment discharge ($Q_s$), stream discharge ($Q$), particle size ($d_{50}$) and stream slope ($S$) was presented by Lane (1955), and states that

$$Q_s d_{50} \sim QS \qquad (7.5)$$

where $d_{50}$ is the median sediment particle size (see Section 5.8.5). Because this equation is qualitative, no units are given. However, it can be used to obtain a general sense of how a stream will respond to changes. For example, by shifting the relative 'weights' of variables on either side of the equation, it can be seen that

- A channel will remain in equilibrium (neither aggrading nor degrading) if changes in sediment load and particle size are balanced by changes in water discharge and slope.
- A reduction in the sediment load ($Q_s$) can result in a decrease in slope if other factors remain constant. When a dam is installed, accumulations of sediment in the reservoir often results in the release of clearer water downstream (lower $Q_s$ and possibly lower $d_{50}$). This can lead to degradation (lower $S$) downstream, as shown in Figure 7.6.
- An increase in sediment load ($Q_s$) can result in aggradation (higher $S$) if other factors are constant. At the point where a small but highly sediment-laden tributary enters a clearer main stream, for example, aggradation occurs at the confluence and upstream.
- Larger sediment particles can be transported by steeper slopes and/or higher discharges.

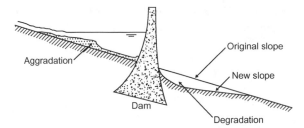

**Figure 7.6.** *Aggradation and degradation at a dam*

Changes in streamflow ($Q$) can result from climatic variations or diversions either into or out of a stream. Inter-basin transfers made for hydropower or mining operations or municipal supplies can affect both the source stream and the stream into which the water is diverted. The large increase in flow in the receiving stream can result in rapid degradation downstream from the transfer point. Sediment load ($Q_s$) can be increased by natural failures of mountain sides or streambanks or by human-caused disturbances. In most situations, however, streamflow and sediment load tend to increase and decrease together rather than in isolation.

### 7.2.4   Floods and Floodplain Formation

To a geomorphologist, a *floodplain* is the relatively flat valley floor formed by floods. To engineers, it may have a more precise definition as the region covered by the 100- or 200-year flood (see Chapter 8). In this discussion it is used more loosely as the valley floor adjacent to the stream channel which becomes inundated at high flows.

*Floods* are commonly defined as those flows which overtop the banks of a stream. They are renowned for their awesome power and disastrous effects on the cities and farmlands which lie in their paths. Their benefits include the ability to replenish topsoil and nutrient supplies on floodplains, to provide water to seedlings and trees requiring periodic inundation, to flush out anoxic or saline waters or deposits of fine sediments, and to permit aquatic animals to migrate to calmer, nutrient-rich shallows in the flooplains to feed and breed. The ecological impacts of floods are further discussed in Chapter 8.

At low flow, the stream may follow a winding path between rocks and around bends and bars. With increasing discharge, the path of travel is shortened as the flow 'shortcuts' across the tops of bars and envelops meanders. This increases the slope of the water, delivering it down-stream more efficiently. As floodwaters rise, they begin to erode the more susceptible portions of the bed and banks. In ephemeral streams and large rivers in semi-arid areas, for example, it is characteristic for the whole width of the bed to be downcut as the stage rises. When the water spills over onto the floodplain its slope approaches that of the surrounding valley. The hydraulics of overbank flow become very complex as the faster-moving water of the main stream interacts with the slower-moving water on the floodplain. 'Rollers'—water spinning around a vertical axis—may develop alongside the channel banks, accelerating erosion in these areas. Bank vegetation may be uprooted and added to the heavy debris load carried by the rushing water. Extremely large floods can leave a lasting imprint on streams; others may have little effect, especially if they have followed on the heels of an even larger flood which has already rearranged the channel.

When floods subside, sediment is deposited on the channel floor, filling in the scoured areas. This *scour and fill* process is a well-known phenomenon at gauged cross-sections, where adjustments must be made to the stage-discharge relationship (Section 5.8.1) to account for its effects. This 'classic' picture will not apply to all sites, however. Even during floods, sediment deposition takes place behind obstructions and other areas of slack water at the channel perimeter. Leopold *et al.* (1964) state that when scour occurs at a pool, there seems to be simultaneous filling on downstream bars or riffles.

Because the water spilling over onto the floodplain moves at a lower velocity it cannot carry as much sediment. Coarser materials tend to deposit out close to the channel rim, and large levees can be naturally constructed by this mechanism. Finer sediments are widely deposited over the floodplain, concentrated in the wake region behind obstacles such as fences or vegetation. Valley floors are thus gradually built up of layers of coarse material from old streambeds and glacial deposits, and finer silts and clays that drop out of suspension onto the floodplain.

The floodplains of braided and meandering streams (Section 7.3.1) form differently. Braided streams migrate widely across the valley floor, leaving isolated bars behind. In meandering streams, bars grow from deposits on the inside of river bends and the meanders migrate outwards and down the valley. Although floodplain formation is considered a long-term process, channel shifts may occur frequently enough to cause changes in habitat for aquatic and wetland species. In a floodplain of the Little Missouri River, for example, a study of the distribution of trees in different age groups led Everitt (1968) to conclude that half of the floodplain had been reworked over a period of only 69 years.

### 7.2.5    Channel-forming Discharges

The concept of a 'channel-forming' or 'dominant' discharge is a convenient one for analytical and conceptual purposes. The *dominant discharge* is considered to be a single discharge equivalent in its effect to the range of discharges which govern the shape and size of the channel. Whereas the gross form of the channel may be shaped by larger, rarer discharges, the maintenance of that form and its smaller-scale features such as gravel bars and bedforms may be more closely related to more frequent discharges (Harvey, 1969). A stream's form is also affected by the relative stability of the stream's bed and banks.

The discharge which just fills the stream to its banks, sometimes termed the *bankfull discharge*, is often assumed to control the form of alluvial channels. Flow resistance reaches a minimum at bankfull stage, and thus the channel operates most efficiently for the transport of water at this level (Petts and Foster, 1985).

In the USA, Leopold *et al.* (1964) report that bankfull discharge occurs approximately every one to two years, although a wide range of values have been found for this average recurrence interval. For 72 rivers in New Zealand, for example, Mosley (1981) reported average recurrence intervals of 1 to 10 years, with a median value of about 1.5 years. In streams with fairly constant flows the bankfull channel dimensions may be controlled by a discharge close to the mean annual flood, but in streams characterized by sharp flood peaks amid long periods of low flows the channel capacity will be related to the higher, less frequent events (Gregory and Walling, 1973). High values of indices such as the coefficient of variation ($C_v$) of annual flows or the index of variation ($I_v$) of annual peak flows (see Appendix) may indicate that a river is less likely to develop a form in equilibrium with average discharge and sediment load conditions. In a study of 15 sand-bed streams in the midwestern USA, for example,

Pizzuto (1986) found that bankfull depth increased with increasing flow variability. Additionally, if the channel has become incised below the floodplain it will not fill to its banks as frequently (Gregory, 1976). In recently incised channels in Australia, Woodyer (1968) used bench levels corresponding to the present floodplain level to derive average recurrence intervals between 1.2 and 2.7 years, fairly close to the interval given by Leopold.

Bankfull stage has been defined in a number of ways. A popular approach is to define it as the height of the floodplain surface. However, the definition of 'bankfull' becomes much more difficult if the section is not well defined, for example: (1) where the bank tops are not of the same elevation, (2) in braided streams, (3) where the break between the channel banks and floodplain is not obvious, such as in steep, rocky canyons, and (4) at complex cross sections where benches or terraces (e.g. former, abandoned floodplains) are present. Slight differences in interpreting bankfull elevation can mean large differences in discharge and thus the associated average recurrence interval. Williams (1978) provides a useful discussion of this concept and also makes a comparison between the recurrence interval as estimated from the annual flood series and that estimated from the partial duration series. Since in most cases the bankfull discharge has a recurrence of less than 10 years, the recurrence interval should be estimated using the partial duration series (see Section 8.2.3).

Various authors have developed criteria for defining bankfull stage. Ridley (1972) used a 'bench index' to find the maximum break in slope on the banks, and Wolman (1955) suggests using the minimum width-to-depth ratio. Another definition is the average banktop elevation. These morphological definitions of bankfull stage at a cross section are shown in Figure 7.7.

Other indicators include the tops of point bars, breaks in slope along the streambank, changes in particle size of bank material, and the top of bank undercuts (Harrelson

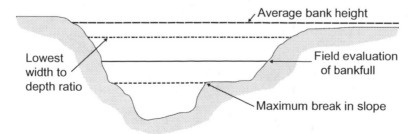

**Figure 7.7.** *Bankfull stage as defined by various indices for a cross-section at Site 4 on the Acheron River (see Figure 4.7). Vertical scale is exaggerated*

et al., 1994). Vegetal clues such as the lower limit of grasses, mosses, liverworts and forbs may also give an indication of bankfull stage. Gregory (1976) suggests the use of lichen limits, which were found to be fairly consistent and associated with less than a 2-year average recurrence interval. Limits may be associated with the sensitivity of plants to inundation or to levels of bedload abrasion. This latter approach may be of more ecological relevance, and was suggested by Newbury (1989) for the evaluation of hydraulic habitats in streams. It may also prove more useful for incised or human-modified channels. Rather than being defined by an interpretation of channel geometry, it is defined by the channel which is regularly maintained by scouring every year or so. Field identification of this *channel maintenance* stage using clues derived from scour lines and vegetation limits was used in the development of hydraulic geometry relationships for the Acheron River (see Figure 7.9). Figure 7.7 shows the interpretation of bankfull using field identification at one cross section. From a field interpretation at the downstream gauging station, the channel maintenance discharge for the Acheron River at Taggerty was about 25.2 m³/s, with a recurrence interval of approximately 0.3 years (partial duration series, see Figure 8.6).

Unless the cross-section has a gauging station it will usually be necessary to estimate bankfull discharge by indirect methods. Chapter 5 gives procedures for measuring bankfull width and depth and for calculating discharge by the slope-area method.

*Effective discharge*, another index of channel-forming flows, is defined as the discharge that transports the most sediment over the long term (see Section 7.5.4). Effective discharge was found to be closely related to bankfull discharge in mountainous streams in Colorado, USA (Gordon, 1995).

### 7.2.6   Fluvial Geometry

A stream can adjust its channel dimensions to accommodate the amount of water and sediment carried. The description of cross-sections and general trends in channel geometry over the length of a stream are valuable for defining patterns of aquatic habitats. The width-to-depth ratio and hydraulic geometry relationships provide methods for quantitatively describing channel shape. Hydraulic geometry can also describe the way in which factors vary with discharge.

#### *Width-to-depth Ratio*

The *width-to-depth ratio* (*W/D*) is often used as an index of cross-sectional shape, where both width and depth are

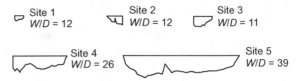

**Figure 7.8.** *Examples of bankfull cross section from Sites 1 to 5 on the Acheron River (see map of Figure 4.7). Cross sections are plotted on the same scale, with the vertical scale exaggerated by a factor of 5*

usually measured at the bankfull level. $D$ is the mean depth across the section. Examples from the Acheron River are shown in Figure 7.8.

The width-to-depth ratio generally increases in the downstream direction. However, it is strongly dependent on the composition of the stream banks. Channels in easily worked sand beds are wider and shallower with sloping sides. In comparison, channels carved into bedrock or silt-clay deposits are narrower and deeper with more vertical banks (Morisawa, 1985). Schumm (1977) used a silt-clay index as a surrogate for the complex influences on bank stability to develop the relationship

$$(W/D) = 255I^{-1.08} \qquad (7.6)$$

where $I$ is the percentage of particles in the channel bed and banks less than 0.074 mm. Both (*W/D*) and $I$ are dimensionless. This relationship was derived from data on channels in the Great Plains of the USA and the Riverine Plains of New South Wales, Australia. Schumm suggests that stable channels should plot close to the line described by Eq. (7.6). The scatter in the original data should be taken into account rather than regarding this as an precise boundary. Aggrading streams would be wider and shallower, plotting above the line, and degrading streams would be deeper and narrower, plotting below the line. For example, at Site 4 on the Acheron River, the silt-clay index from Figure 5.32 is about 5%. From Eq. (7.6), $W/D = 255(5)^{-1.08} = 44.8$. Since the actual *W/D* ratio is 26, this implies that the reach may be slightly degrading.

Both depth and width can respond rapidly to changes in sediment load and/or discharge. Whether a stream erodes downwards or outwards is influenced by both local shear stresses and whether the bed or banks are the most easily eroded. Bank vegetation also increases the resistance to erosion through its binding effects, with erosion decreasing as the percent (by weight) of roots in the soil increases (Richards, 1982), and this leads to narrower channels than would otherwise be expected. The effect of vegetation on channel shape is more pronounced in smaller streams.

## Hydraulic Geometry

*Hydraulic geometry* describes the way in which channel properties change with streamflow. A stream's cross-sectional area, for example, is generally determined by the amount of water it must carry; i.e. headwater streams are smaller than the rivers into which they flow.

In a study on a large sample of rivers in the Great Plains and the Southwest area of the United States, Leopold and Maddock (1953) proposed that mean depth ($D$), width ($W$), mean velocity ($V$) and sediment load ($Q_s$) varied with discharge ($Q$). These inter-relationships, which they termed the hydraulic geometry of streams, are described by the following equations:

$$W = aQ^b \qquad (7.7)$$

$$D = cQ^f \qquad (7.8)$$

$$V = kQ^m \qquad (7.9)$$

$$Q_s = pQ^j \qquad (7.10)$$

The coefficients $a$, $c$, $k$ and $p$, and the exponents $b$, $f$, $m$ and $j$ are empirically derived. The hydraulic geometry equations are power functions. When plotted on log-log paper the slopes of the curves, described by the exponents $b$, $f$, $m$ and $j$, indicate the average change of width, depth, velocity and sediment load, respectively, with changes in discharge, and do not vary with the units used. The coefficients $a$, $c$, $k$ and $p$ do have dimensions.

Since, by the continuity equation (Eq. (6.6)),

$$Q = VA = WDV \text{ (with } D \text{ the hydraulic depth, } A/W)$$

it follows from Eqs. (7.7) to (7.9) that

$$WDV = aQ^b \cdot cQ^f \cdot kQ^m = Q$$

and thus

$$b + f + m = 1, \text{ and}$$
$$ack = 1.$$

Hydraulic geometry relationships can be applied to the description of how variables change with discharge: (1) at a particular location ('at-a-station') or (2) over a drainage basin ('downstream').

*At-a-station* variations are due to the local configuration of the channel and the way in which water flows through the section. For example, velocity will increase more rapidly with discharge in narrow channels constrained by vertical cliff walls than in broad, shallow channels. In general, at higher discharges, meandering or braided streams (Section 7.3.1) spread out (higher $b$) and straight streams speed up (higher $m$). Both cross-sectional shape and flow velocity tend to change abruptly when the stream begins to flow over its banks. At-a-station relationships are of interest particularly if streamflows at that section are critical to some biological behaviour such as fish migration.

Leopold and Maddock (1953) cite average at-a-station coefficients of $b = 0.26$, $f = 0.40$ and $m = 0.34$. Thus, as discharge increases at a cross section, velocity goes up and depth increases faster than width (the width-to-depth ratio drops). These coefficients can be expected to differ considerably from reach to reach.

At a station, Eq. (7.8) is the inverse of the familiar stage-discharge equation (Section 5.6.5) if stage is used rather than mean depth. Equation (7.10) is a rating curve for sediment yield, which will be discussed further in Section 7.5.4.

*Downstream* changes in channel geometry can be investigated by linking information from a number of sites within a stream system. This method is only valid if the discharges used for comparison are of the same average recurrence interval (see Section 8.2.2). Values of mean annual flow or bankfull discharge are commonly used, under the assumption that these levels occur at approximately the same frequency on a large number of rivers.

Using mean annual flows, Leopold and Maddock (1953) found that downstream increases in depth, width and velocity relative to discharge were similar for rivers of varying drainage basin size and setting. Average values of the hydraulic geometry exponents for the rivers studied were $b = 0.5$, $f = 0.4$ and $m = 0.1$. From these exponents it can be seen that large rivers tend to be wider and shallower than smaller streams, and that velocity increases slightly in the downstream direction. The latter conclusion may be somewhat surprising because whitewater mountain streams give the impression of flowing faster than meandering valley streams. Although the headwater streams are steeper, the lower roughness in the valley streams from reduced particle sizes can lead to increases in velocity.

Discharge typically increases with distance downstream because of the increasing area of drainage. Thus, catchment area, as a more readily measured factor, is sometimes used in place of discharge in hydraulic geometry relationships. The downstream changes in width, depth and bankfull discharge with catchment area for the Acheron River system are shown in Figure 7.9, along with data from rivers in the USA.

The patterns of discharge as well as the total amount will have an influence on channel size and shape. 'Flashy'

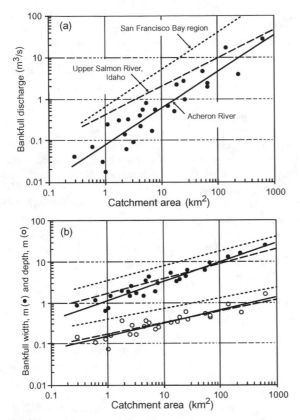

**Figure 7.9.** *Downstream hydraulic geometry relationships for (a) bankfull discharge and (b) bankfull width and depth. Data from the 26 sites on the Acheron River (see map of Figure 4.7) are shown as individual points with a fitted (solid) regression line. Trend lines for the Upper Salmon River are from Emmett (1975), and for the San Francisco Bay region, from Dunne and Leopold (1978). Line patterns are the same in (a) and (b)*

sites and field sites, and the methods used for fitting a line through the data.

Hydraulic geometry relationships have been used for the quantitative description of riverine habitat by Hogan and Church (1989) and Kellerhalls and Church (1989). They can also be useful in studies of how land-use changes affect channel shape and size. For example, increased discharges and stream channel enlargement may accompany the urbanization of a catchment (Morisawa, 1985). Departures from general downstream trends may reveal points of impact, e.g. as a result of dams, channelization or diversions. In stream-rehabilitation work (Chapter 9), the hydraulic geometry relationships from undisturbed areas can be extrapolated downstream or to other basins as a guide for reconstructing the natural geometry of degraded streams (Newbury and Gaboury, 1988). Another implication is that the curves can be used in reverse to estimate discharge from channel geometry. For example, Wolman (personal communication, 1989) states that the mean annual flow can be reasonably estimated for ephemeral stream channels from the height of naturally formed levees.

## 7.3    The Ins and Outs of Channel Topography

### 7.3.1    Channel Patterns

The term 'channel pattern' describes the planimetric form of streams. Channel patterns can be classified as straight, meandering, braided or anastomosing, as illustrated in Figure 7.10. The term 'anabranching' is also applied to

rivers of semi-arid areas with quick, large peak flows may develop wider channels than those in areas where stream-flow is more constant (Gregory and Walling, 1973). Park (1977) summarized and examined worldwide hydraulic geometry data from several studies. Although he found that the at-a-station exponents showed considerable scatter, the downstream exponents were more consistent and clustered near the original values of Leopold and Maddock. In some streams of humid temperate areas, negative values of the velocity exponent, $m$, were found, indicating that mean velocity decreased in the downstream direction. Park also pointed out that differences in the relationships could be due to the flow level used (e.g. bankfull, mean annual flow, etc.), the differences between gauging station

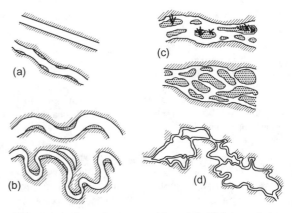

**Figure 7.10.** *Channel patterns: (a) straight, (b) meandering, (c) braided and (d) anastomosing*

anastomosing rivers (Nanson and Huang, 1999). These categories are an arbitrary means of classifying a continuum of forms. Streams can vary from straight to braided due to differences in geologic history, stream slope, discharge patterns and sediment load. Channel patterns can also change with time; for example, streams which are braided at one flow level may meander at higher or lower stages.

In upland areas the channel pattern closely follows that of the incised valley as it threads its way between hillslopes. Downstream, however, where accumulated deposits of alluvial material create wide valleys, the stream's slope is not as strongly influenced by that of the landscape. In general, a braided pattern tends to coincide with high slopes and high stream power and a meandering one with lower slopes and lower stream power. Braided streams are also associated with coarse bed and bank materials and the movement of sediment as bedload. In comparison, meandering streams tend to have more cohesive bed and bank materials and suspended sediment loads.

The channel patterns are distinguished on the basis of channel multiplicity and sinuosity. *Sinuosity* is a measure of the 'wiggliness' of a watercourse, and has a number of definitions. The most commonly used measure is the sinuosity index (*SI*), given as

$$SI = \frac{\text{Channel (thalweg) distance}}{\text{Downvalley distance}} \quad (7.11)$$

The *SI* is normally computed from measurements of stream and valley lengths taken from maps or aerial photos. The reach length should be at least 20 times the average width of the channel (Bell and Vorst, 1981). Stream length is measured by methods described in Section 4.2.3. 'Valley length' presents some difficulty, since, by strict definition, streams tightly confined within V-shaped valleys have the same length as the valley, giving an *SI* of 1 whether or not the streams appear 'straight' to the eye. In practice, straight line segments which follow the broad-scale changes in channel direction can be used as a measure of valley length. For the section of the Acheron River in Figure 3.4 upstream of the junction with the Little Steavenson to the edge of the map, $SI \approx 154$ mm/$115$ mm $= 1.34$. The fractal dimension (Section 4.2.3) may be a more consistent measure of a river's 'crinkliness' as it is independent of map scale; however it is more difficult to compute than *SI*.

In straight streams, $SI = 1$, whereas a value of 4.0 is considered to be highly intricate meandering. Meandering streams are somewhat arbitrarily defined as those with an *SI* value of 1.5 or more. The term *sinuous* is sometimes given to stream patterns which are intermediate between straight and meandering. A description of channel patterns based on their sinuosity, bank characteristics, sediment loads, relative dimensions, bankfull velocity and stream power is given in Table 7.1 and the relationship between channel form, slope and sediment load is shown in Figure 7.11.

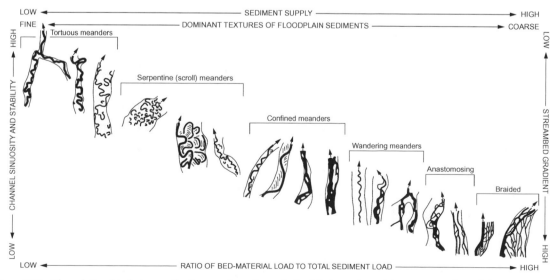

**Figure 7.11.** *The relationship between the form and gradient of alluvial channels and the type, supply and dominant textures (particle sizes) of sediments. Adapted from Figure 10.6 (p. 269) from "Earth's Changing Surface, An Introduction to Geomorphology" by Selby, M.J. 1985. By permission of Oxford University Press, www.oup.com*

**Table 7.1.** Classification of channel patterns. Based in part on Leopold et al. (1964), Morisawa (1985) and Selby (1985)

| Stream type | Description | Width-to-depth ratio | Sinuosity | Bankfull velocity (m/s) | Stream power |
|---|---|---|---|---|---|
| Straight | Single channel with meandering thalweg. Well-defined banks, often containing bedrock. Channel is typically stable, with minor widening or incision. Sediment load is suspended and/or bedload; load is usually small in comparison to transport capacity. Cross sections tend to have a marked central 'hump'. Found in short reaches (also channelized streams). | Low <40 | Low 1.0–1.5 | High >3 | High |
| Meandering | Single winding channel, usually with well-defined banks. Channel shifts mainly due to erosion by undercutting on outside of bends, causing the outward growth and downvalley migration of meanders. Pools form in this region of high velocity and turbulence. 'Cutoffs' can occur across the base of a meander loop, leaving crescent-shaped 'oxbow lakes' (billabongs), which support an ecology different from that of the river. Point bars of sand and gravel form on the inside of bends. Sediment load is mainly suspended load, approximately balanced with transport capacity. Meander wavelength, averages 10 to 14 times channel width. | Low <40 | Moderate to high 1.5–4.0 | Low to moderate 1–3 | Low to moderate |
| Braided | Multiple channels with bars and islands, often with poorly defined banks of non-cohesive materials. Channel and bank erosion follow a fairly random pattern. Flow is concentrated into flanking chutes, increasing velocity and sediment transport capacity. Sediment load is primarily bedload, and load is large relative to transport capacity. Found in glacial streams, alluvial fans and deltas. | High >40 | Low to moderate 1.0–2.0 | Varied, depending on slope and straightness of individual channels. | All ranges, from high in straight streams to low in sinuous streams with islands, channel bars or deltas. |
| Anastomosing | Multiple channels with relatively permanent, stable vegetated islands, in comparison to braided streams where channelways are constantly shifting. Banks are cohesive, and sediment load is primarily suspended load. | Low <10 | Varied | Varied | Moderate |

The channel pattern represents an adjustment of shape in the horizontal plane, and is one of the variables which can be modified by a stream to improve the efficiency with which it conveys water and sediment. Changes in slope, and type and amount of sediment load can lead to changes in channel patterns and thus stream habitats. Thresholds may also exist, meaning that in one reach a large change in slope and/or sediment load may have little effect on the channel pattern, but if the reach is close to some threshold level, a slight change in slope or sediment load can have 'striking repercussions' (Gregory and Walling, 1973). Channels may also change from non-braided to braided in association with a reduction in riparian vegetation along streambanks (Leopold *et al.*, 1964).

### 7.3.2 Pools, Riffles and Steps

The words 'pool' and 'riffle' immediately bring to mind an image of the features they represent, especially to trout anglers, from whom the terms originated. Loosely defined, a *pool* is a region of deeper, slower-moving water with fine bed materials, whereas a *riffle* is a region with coarser bed materials and shallower, faster-moving water, often associated with whitewater. At riffles, the cross-sectional profile tends to be rectangular, whereas pools have more asymmetric profiles. The term *run* is sometimes given to an intermediate category in which the flow is less turbulent than in riffles but moves faster than in pools. Pools and riffles alternate in a 'pseudo-cyclic' manner (Knighton, 1984), with the depth of the pool controlled by the elevation of the riffle just downstream (see Figure 7.12). They can be thought of as 'vertical meanders'. This *pool-riffle* periodicity may be important in the cycling of nutrients along a stream (Goldman and Horne, 1983).

Riffles tend to support higher densities of benthic invertebrates, and are thus important food-producing areas for fish. Due to competition and predation as well as size limitations, young fish and small fish tend to inhabit riffles, whereas deep pools with overhanging banks and vegetation support larger fish. During low flows, riffles may be exposed and pools can become isolated pockets of water which allow the survival of aquatic organisms. Bovee (1974) suggests that riffle-inhabiting species should be used as indicators in determining low-flow requirements of streams (Chapter 9). In terms of physical habitat, the pool-riffle structure provides a great diversity of bedforms, substrate materials and local velocities. The most productive streams have a combination of pool sizes (Hamilton and Bergersen, 1984). Brussock *et al.* (1985) propose that the reason biotic diversity is greatest in the middle reaches of streams is because they typically possess a pool-riffle morphology.

Pools and riffles are fairly easily distinguished by eye at low flow. However, where one begins and the other leaves off becomes a matter of interpretation. A reach-length longitudinal profile (see Figures 4.14 and 5.11) can be divided at sharp breaks in slope to distinguish the forms. An aerial photograph may be helpful in distinguishing changes in bed material between riffles and pools as well as changes in width, since riffle areas tend to be about 12% wider than pools on the average (Richards, 1982). Richards (1976) defines riffles and pools more formally as positive and negative residuals, respectively, from a regression line fitted through the bed profile. Lisle (1987) proposed the use of 'residual depths' to define the extent of pools from a horizontal line through the crest of the downstream riffle. Alternatively, Bren (personal communication, 1990) suggests using Froude numbers to distinguish between pools ($Fr < 0.1$) and non-pools ($Fr > 0.1$).

The alternating pool-riffle bedform is most common in streams with mixed bed materials ranging from pea to watermelon size (2–256 mm) (Knighton, 1984). Jowett and Duncan (1990) demonstrate that it is more pronounced in streams with high flow variability. Pool-riffle sequences are often found in meandering streams, with pools located at meander bends and riffles at crossover

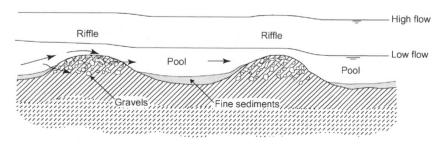

**Figure 7.12.** Pool-riffle sequences at low and high flow

stretches. As would be expected from the average mean-der wavelength (Table 7.1), the deeps and shallows both follow a more or less regular spacing of five to seven channel widths (Leopold *et al.*, 1964). 'Noise' in the rhythmic patterns can result from local controls such as bedrock constrictions or tree roots, or an increased supply of coarse sediments. This semi-regular pattern is also found in straight channels, bedrock streams and the dried remains of semi-arid ephemeral streambeds.

In steep, boulder-bed mountain streams the pool-riffle sequence is replaced by a *pool-step sequence*, where water tumbles over accumulations of boulders and short water-falls plunge into small scour pools. 'Organic' steps can also be created by large fallen trees or debris dams. These debris-created steps also tend to be regularly spaced (Morisawa, 1985).

The pool-riffle bedform is considered a means of self-adjustment in gravel-bed streams to regulate the energy expenditure. In meandering reaches, energy loss is high at channel bends because of the curvature. This may be balanced by energy losses to turbulence in the straight riffled stretches where roughness is greater.

Steps, riffles and pools become more important at low flows when they become more dominant components of channel geometry. At low flows, the sites immediately downstream of riffles are locations of intense energy loss, which becomes concentrated over a very small percentage of the stream's length. For example, the riffles in the Grand Canyon, Arizona which provide thrills and spills for river rafters, concentrate most of the river's vertical drop into segments that average only 10% of the down-stream distance (Leopold *et al.*, 1964).

As kayakers are well aware, the water surface slope, depth of flow and speed of the current become more uniform over the stream reach at high flows (see Figure 7.12). At these times, it becomes questionable whether the terms 'pool' and 'riffle' are even applicable. As discharge increases, velocity and depth rise more rapidly in pools than in riffles, and energy loss becomes more uniform. The shear stress in pools can eventually exceed that in

riffles, which may be part of a sorting mechanism for concentrating coarser materials in riffles (Knighton, 1984). Sediment movement from riffles is postponed until very high flows occur, at which time the coarse particles move from riffle to riffle. Very coarse fragments, however, will still tend to collect in the deepest part of pools.

Of significance to the resident flora and fauna, the pool-riffle bedforms remain relatively fixed in location, unlike sand bars and dunes (Section 7.3.4) which tend to migrate. Pool-riffle structures are usually formed by rare, large historic events, and pool-step systems may relate to even rarer, higher intensity discharges (Petts and Foster, 1985). Thus, the flows which form these structures differ from those that maintain them, and they remain relatively stable under all but extreme flow conditions (Knighton, 1984). Pools may deepen as a result of localized scour during low to moderate flows, especially at bends or the downstream side of logs (Beschta and Platts, 1986). They may also fill with sediment if the sediment supply increases.

### 7.3.3  Bars

Bars are fairly large bedform features created by the deposition of sediments. Whereas submerged sand bars can pose a threat to the unskilled river navigator, exposed bars can become stabilized by vegetation to create island refuges for migrating waterfowl. Like pool-riffle struc-tures, bars also tend to be formed at higher discharges, and then remain in place to define the path of low flows. They have a variety of shapes, and can be composed of a wide range of grain sizes. Bars can be classified by their location in the stream as shown in Figure 7.13 and described by Knighton (1984) as follows:

*Point bars* primarily form on the inner bank of meanders and often create sandy beaches that slope gradually into the water.

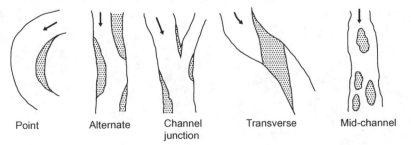

| Point | Alternate | Channel junction | Transverse | Mid-channel |

**Figure 7.13.** *Classification of bars (see text)*

*Alternate bars* occur periodically first along one bank and then along the opposite one, with a winding thalweg running between the bars. These can form in relatively straight sections of sand-bed streams, creating a meandering pattern at low flow.

*Channel junction bars* develop where tributaries enter a main channel.

*Transverse bars* cross the width of the stream (for example at riffles) often at an angle diagonal to the flow. In sand-bed streams these tend to be flat-topped and covered with smaller bedforms such as ripples (Section 7.3.4).

*Mid-channel bars* are characteristic of braided reaches, often existing as diamond- or lozenge-shaped gravel mounds. These are aligned with the flow, separating it into smaller rivulets. Mid-channel bars tend to grow from the downstream end. They typically have coarser materials on the upstream side and finer materials on the downstream one.

Bars are stable features in many locations, with erosion equalling deposition during floods. Even when bars change constantly in form or location, as in braided streams, the amount of area covered can remain fairly constant. The removal of bars may actually lead to instability and an attempt by the river to 'heal' itself. Hooke (1986), for example, observed that bars and shoals removed from the River Dane in England redeveloped in two to three years.

In a gravel-bed stream in Northern California, Lisle (1986) found that bars tended to form three to four bed-widths downstream and one bed-width upstream of large obstructions and bends. The downstream bars typically formed on the same side of the stream as the obstructions. He proposed that 'non-alluvial boundaries' of woody debris, bedrock, fabricated material or root-defended bank promontories actually stabilized the locations of gravel bars and pools by affecting downstream secondary currents and through backwater effects. Thus, the introduction of large obstructions into a stream can lead to changes in the channel's shape both upstream and downstream.

### 7.3.4 Dunes, Ripples and Flat Beds of Sand

Sand-bed streams present a 'Sahara-like' landscape to the benthic invertebrates and aquatic plants seeking a foothold. Because of the smaller, more uniform grain sizes, sand beds are highly mobile and readily moulded into different bedform shapes under the sculpting effects of flow patterns. In turn, the bedforms provide feedback through form resistance to affect local velocities, shear stresses and sediment transport. Local scouring and

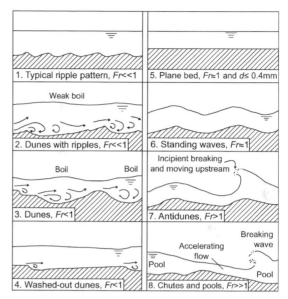

*Figure 7.14. Progressive changes in bedform shape. Adapted from Simons, D.B. and Richardson, E.V. 1961. Forms of bed roughness in alluvial channels. J. Hyd. Div., ASCE* **87**: *87–105. Reproduced by permission of the publisher, ASCE. Fr = Froude number*

deposition within troughs and across crests causes the bedforms to be somewhat self-perpetuating.

From extensive studies in both flumes and natural channels a fairly predictable sequence of bedforms has been found to occur as stream power increases. The forms are divided into two categories, based on a Froude number less than or greater than one: the lower flow and upper flow regimes. In the *lower flow regime* sediment transport is relatively low and flow resistance is high due to the large separation zone behind the crests of dunes or ripples. In the *upper flow regime* resistance is reduced and sediment transport is increased. The progression of bedforms typically follows the sequence given as follows and in Figure 7.14. Forms illustrated in Figure 7.14 but not listed are transition or intermediate phases in which smaller forms can be superimposed on larger ones.

0    *Initial flat bed.* A flat sand bed is an oddity outside of a laboratory environment, and finding one in a natural stream is unlikely. For flumes and theoretical discussions it serves as a starting point. When the water velocity is increased, sand grains begin moving singly at first, then in patches. As the grains move together over the streambed, particles tend to accumulate in clusters. Then, suddenly, the clusters of particles orient in a series of regular waves and hollows longitudinal to the flow (Hjulstrom, 1939).

1-2    *Ripples.* Ripples are small corrugations in the bed with relatively sharp crests. They form under *hydraulically smooth* conditions. The more fine-grained the sediment, the more well-developed the ripple (Hjulstrom, 1939). The distance between ripples is fairly uniform. This ripple 'wavelength' is dependent on particle size and is independent of flow depth (Smith, 1975). It typically ranges from 150 to 450 mm (Simons and Richardson, 1961).

2-3    *Dunes.* If the flow is *hydraulically rough*, dunes will form rather than ripples. Dunes have rounded crests and are larger than ripples. Their dimensions are related to flow depth and are only slightly dependent on particle size (Smith, 1975). Dunes enlarge to a point where further growth is impossible because of high velocities and transport rates at the crest (Leopold *et al.*, 1964). In the Mississippi River, this type of dune is represented by long sand bars up to hundreds of metres long.

Both ripples and dunes migrate *downstream* as sand grains move up the more gradually-sloped upstream side and fall down the steeper downstream face. Sand deposited on the upstream face is closely packed whereas the material on the downstream side is unstable. Researchers who wade in sand-bed streams will have an appreciation for the denseness of the upstream sides of dunes which will support their weight, as compared to the quicksand-like material on the lee side.

4    *Transition zone.* As stream power increases and the Froude number (*Fr*) approaches one, ripples and dunes are washed out. This constitutes the transition from lower to upper regime.

5    *Plane bed.* In beds of finer sediments (<0.4 mm) a plane bed develops. In comparison to the initial flat bed, at this step the bed and fluid have less distinct boundaries and the high sediment transport creates a dust storm-like environment (Leopold *et al.*, 1964). Suspension of sediments further decreases flow resistance by dampening turbulence and thus reducing energy loss (Smith, 1975).

6    *Standing waves.* With larger sediment sizes, standing waves develop rather than a plane bed (Simons and Richardson, 1961). Standing waves are those in which the water surface and bed surface are synchronized and both sand and water waves are stationary. These can begin forming at a Froude number of about 0.84.

7-8    *Antidunes.* At this step, velocities and sediment transport are both high. Waves form on the water surface, accompanied by the formation of rounded bed waves called antidunes. These are extremely unstable and constantly form, disintegrate and re-form. They progressively move *upstream* as sand erodes from the downstream side of a dune and deposits out on the upstream face of the next dune downstream. Antidune wavelengths range from 230 mm to 6 m (Hjulstrom, 1939), and are approximately twice the depth of the water. They have been observed in natural streams with beds of fine sand to coarse gravel (Simons and Richardson, 1961).

8    *Chutes and Pools.* At even higher stream powers, the bed rearranges itself to create a series of hydraulic jumps for energy dissipation. These high-*Fr* situations may occur when flash floods sweep down steep sandy gullies of semi-arid regions. Large quantities of sediment are suspended in the breaking wave. WMO (1980) gives a method for computing discharge from the distance and time between these 'translatory waves'.

Bedform shape acts as another type of self-regulating mechanism in streams, ensuring efficient transport of both water and sediment. In less-uniform sands, armouring (Section 7.4.2) can stabilize bedform size and shape until the flow is great enough to remove the coarse surface layer.

As indicated in Figure 7.15, flow resistance increases through the lower regime until dune formation, then decreases through the transition to a plane bed, increasing again in the upper regime. Thus, in a sand-bed river, Manning's *n* (Sections 5.6.6 and 6.6.5) can vary fourfold simply as a result of changes in bedform (Schumm, 1977). Ripples and dunes slow the increase in velocity until, at higher stages, the dunes wash out, producing a more efficient channel and possibly reducing the height of flood peaks (Schumm, 1977). Bedforms, therefore, can wreak havoc with at-a-station hydraulic geometry relations (Section 7.2.6) such as the relationship between depth and discharge.

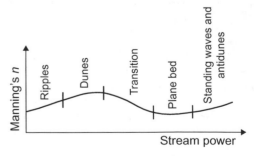

**Figure 7.15.** *The variation of flow resistance with bedform. Source: CSU (1977), based on data in Simons and Richardson (1966), and see also Richards (1982, p. 85)*

Although somewhat unpredictable, the self-adjusting mechanisms of streams have fascinating implications to those studying habitat dynamics and the movement of water, sediments, nutrients and biota through streams.

## 7.4   Sediment Motion

### 7.4.1   Erosion, Transport and Deposition

Sediment motion consists of three stages: initiation of motion, downstream transport, and deposition. At the extremes of particle size, some of the finer sediments may never settle out, and large boulders may never get off the ground. For the large majority of particles, however, motion consists of periods of rest alternating with periods of activity, and any given particle will be eroded, transported and deposited many times *en route* to some accumulation point. The rests, stops and starts of an individual particle have a random nature, but as with fluid motion there are techniques for addressing the overall movement by relating it to easily measured, average flow properties.

Since most of a stream's energy goes into overcoming frictional resistance along its bed and banks, a critical energy level must be reached before a stream can erode or transport channel materials. Thus, the concept of some 'threshold' or 'critical' value is fundamental to studies of particle motion. Several thresholds may exist for a particle of a given size and shape: a bed-erosion threshold at which the particle just begins to move; a 'liftoff' threshold which must be exceeded to suspend the particle; and another 'settling' threshold, below which the particle will drop out of suspension. Threshold values can be specified by a number of hydrodynamic factors such as velocity, shear stress and stream power.

At low discharges, only washload will be transported. When discharge is increased to a certain level, particles on the bed and/or banks will begin to *erode*. Observations of particle movement in both flume and field situations indicate that particles begin to vibrate as the flow intensity is increased. Lighter organic materials such as leaves and bark will be transported first. Initially, a few grains are entrained, and then whole patches of surface material are lifted off into the flow. The forces which 'pluck' a particle from the streambed and keep it aloft are mainly due to upward surges of water—'gusts' from small, turbulent eddies.

*Sediment transport* is considered to occur once the threshold for movement has been crossed. Larger particles tend to roll in the direction of the bed slope, whereas fines usually follow a longer path dictated by the spiralling

movement of the fluid. Eventually, an equilibrium is reached between the amount of sediment and the energy available to carry it. *Flow competence* refers to the maximum particle size that can be carried at a given flow state. *Apparent* competence, or the maximum particle size actually carried by a stream, may be less than the *true* competence if the largest transportable particle sizes are not available.

Numerous approaches to the problem of estimating sediment transport rates have been taken. Most are based on some relationship between the amount of sediment carried and the difference between 'actual' and 'critical' values of some factor such as stream discharge, shear stress, mean flow velocity or stream power. Some equations are physically based, some are probabilistic in an attempt to account for the intermittency of particle movement; but all are empirical to some degree. Because of the extreme complexity of the problem due to channel configurations, velocity variations, and mixtures of particle sizes and packing, the performance of most sediment discharge equations 'does not inspire confidence in results based on them' (Dawdy and Vanoni, 1986, p. 73S). If sediment transport equations are used, they should ideally be calibrated with field data and not used outside the range of calibration. Some of the fluvial geomorphology texts listed in Section 7.1.1, and texts on sediment transport such as Graf (1971), Raudkivi (1967), Thorne *et al.* (1987) and Yalin (1972) will provide further information on the subject.

A stream basically carries its load until it lacks the energy to do so, at which time *deposition* takes place. For bedload materials, deposition occurs when the material stops rolling, sliding or hopping; for suspended material, it occurs when the material settles out of suspension. In still water, a particle will settle out at a rate dependent on its terminal velocity (Section 6.5.4).

Briggs (1977a) describes three major processes which lead to deposition of sediment:

1. *The orientation of the particle in the flow changes, increasing resistance.* The lift and drag on a non-spherical particle will change with its orientation. Like an aircraft wing, a particle can 'stall' at a certain angle. The forces acting on the particle will also affect its orientation on the bed (see *imbrication*, this section).

2. *The competence of the stream decreases*, meaning the energy available for transporting sediments decreases. Suspended sands and gravels settle out when the streamflow drops, and when local velocities decrease, such as at pools, the insides of meander bends, tributary confluences, canyon mouths or other widenings,

or the downstream side of gravel bars or islands, sand dunes, rocks or tree trunks. At ponds and reservoirs, the coarser particles deposit out at the upstream end whereas finer particles move farther downstream. Larger bedload materials may cluster together at riffles or steps.

3. *The quantity or size of the sediment load suddenly increases.* This type of deposition may occur if a stream suddenly receives an influx of sediment from a landslide or collapsing streambank. Particles will remain oriented as deposited until they are re-worked by the stream.

The 'size selectivity' of deposition helps to create and shape large and small bedforms. These, in turn, affect local flow resistance, velocities, and turbulence which encourage further erosion or deposition. Their interaction is part of the self-regulating feedback that enables a stream to adjust to changing discharges and sediment loads.

### 7.4.2   Deviations from 'Ideal'

The picture presented in the previous section is muddied by the great variability and unpredictability of sediment motion. In natural streams, bed materials are made up of a conglomeration of particle sizes, especially in cobble- or gravel-bed streams and rocky headwater reaches. The mixture of particle sizes, their arrangement and the amount in suspension can have significant effects on local velocities and thus sediment transport.

#### Concentration of Suspended Materials

The velocity required to lift and transport sediment can be affected by the amount of sediment already in suspension. Suspended fines tend to reduce the turbulence and thus lessen the eroding power of the flowing water (Hjulstrom, 1939). For example, Bayly and Williams (1973) provide data that show that the flow velocity required to entrain clays in muddy water is about 1.7 times that for clean water. However, because suspended fines also increase the density and viscosity of the water, coarser particles do not settle out as quickly in muddy water as they would in clean water.

In contrast, larger materials such as suspended sands or rock fragments may increase erosion if they are hurled against banks composed of fine materials such as clay, loam or volcanic ash. An additional consideration is the effect of suspended materials on flow resistance. Since turbulence is decreased, flow resistance is reduced and

thus higher velocities and sediment movement are possible. However, Parker *et al.* (1982) found that high concentrations of suspended sand in gravel-bed streams seemed to suppress erosion. The effect, therefore, will depend both on the type and concentration of materials in suspension and the composition of the streambed (see also Table 5.5).

#### Mixtures of Particle Size

The mixture of particle sizes found in natural streams creates interactions which would not occur if materials were of a uniform size. One example is the scouring of sand hollows around the base of boulders by horseshoe vortices (Section 6.5.4). Interactions also affect the ease with which an individual particle will be eroded and transported. The filling of interstices between larger particles can act to 'cement' them in place. Particles that are partly buried or in the wake of larger particles or bedforms will move less readily than isolated particles sitting on top of a flat streambed. Smaller particles can also 'hide' in the lower-velocity region of the boundary layer, whereas larger particles will protrude farther into the velocity profile (Section 6.4.2). For rugged mountain streams with large, immobile rocks, sediment transport may decline with increasing discharge because of sheltering effects (Wohl, 2000).

Research by Parker *et al.* (1982) and Andrews (1983) in self-formed rivers with naturally sorted gravel and cobble bed material indicated that mixed-size particles move over a much narrower range of discharge than expected. This led to the controversial theory of 'equal mobility', which says that nearly all of the grain sizes begin moving at nearly the same discharge. The theory does not imply that the entire bed surface begins moving at the same time. Instead, at any instant, the sediment load may consist of particles over a range of sizes, and the bed 'selectively unravels' from different locations as discharge increases (Prestegaard, 1989).

Thus, the movement of coarse particles probably occurs more frequently than is commonly assumed. Andrews (1983), for example, found that in nine Colorado rivers, particles as large as the $d_{90}$ of the bed material were entrained by bankfull discharges.

#### Armouring

*Armouring* is the development of a surface layer that is coarser than the bed material beneath it (Figure 7.16(a)). It 'protects' the finer materials underneath, which are not mobilized until the armour layer is removed. Armouring may or may not occur in streams; for example, it is less common in braided streams (Bunte and Abt, 2001). If

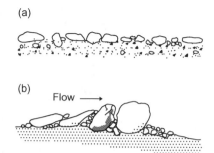

**Figure 7.16.** *Arrangement of surface bed materials: (a) armour layer over finer sub-surface materials and (b) imbrication of disc-shaped particles*

present, it may take place over the whole bed or only in patches where scour is greatest. In some areas, thick, erosion-resistant streambeds can develop which are rearranged only during extreme floods.

Explanations for the mechanics of armouring are subject to controversy. One explanation is that the layer results from the 'winnowing away' of finer materials from the surface layer. On arid land surfaces, the washing of upland slopes by wind and water may act in a similar manner to leave a protective gravel layer called 'desert pavement'. Some armour layers may result from the accumulation of large materials rather than the removal of fines. However, there seems to be an inherent tendency for large particles to find their way to the surface. Leopold *et al.* (1964) give the example of a truck dumping dry gravel, where the largest particles roll across the smaller materials and down the outside of the pile.

Once a channel is armoured, the sub-surface materials are protected from erosion until the armour layer is broken up. Dawdy and Vanoni (1986) cite studies that propose that armouring causes a restructuring of the sediments transported, acting as a type of 'regulator' to enable a stream to transport more of its coarse sediments. The few sites stabilized by armouring may be critical to the stability of the entire channel.

An index of the strength of an armour layer ($A$) can be defined as the ratio of the armour layer grain size to the sub-surface grain size:

$$A = \frac{d_a}{d_{sub}} \quad (7.12)$$

where, commonly, $d_a$ is the $d_{50}$ of the surface, armoured layer, and $d_{sub}$ is the $d_{50}$ of the sub-surface bed materials. The index, $A$, is typically found to be between 1.5 and 3 in gravel-bed streams (Parker *et al.*, 1982). It can reach 1.0 in

streams with a high sediment supply (capacity-limited streams), and drop to less than 1.0 if a fine sediment layer covers the streambed surface (Bunte and Abt, 2001). The influence of armouring should be considered not only in predictions of sediment movement but also in the collection of bed material samples.

### *Imbrication*

The orientation and angle of deposited particles depends on the particle shape and the forces affecting the particles at the time of deposition. If deposition is caused by a sudden fall in the stream's competence, particles tend to be deposited in their position of transport. In fairly steady flow, deposited particles tend to fall into a more stable position. For example, rod-shaped particles will line up with their long axes parallel to the direction of flow. Many gravelly stream deposits, particularly those with disc-shaped pebbles, will exhibit particle *imbrication*, where particles are stacked against each other, nose-down into the oncoming current (Figure 7.16(b)). Briggs (1977a) states that imbrication may be a position of maximum resistance to movement for large bedload sediments.

### 7.4.3 Predicting a Particle's 'Get Up and Go'

In studies of aquatic systems it may be of interest to predict the flow level which will shift sediments, upon or under which benthic organisms reside. For streambeds that have become filled with fines the same principles can be used to calculate flows required to remove sediments from the surface of the bed or from beneath the armour layer. The magnitudes of past floods are also sometimes reconstructed using the maximum size of deposited rocks. Gregory (1983) is a source of information on these 'paleohydrological' techniques.

As mentioned in Section 7.4.1, prediction of sediment entrainment is usually based on some 'critical' state, above which particles begin to move. The initiation of motion of a sediment particle can be described by a variety of factors: (1) lift and drag forces, (2) a critical velocity, or (3) a critical shear stress. Critical stream power has also been used in sediment transport equations by Bagnold (1980) and Yang (1973), but will not be discussed because it is not normally used for the application presented herein. These are all slightly different but related ways of looking at the same phenomenon.

Because of the intermittency with which particles are lifted into the flow, the visual detection and interpretation of the 'critical' state is highly subjective. In some approaches this threshold is treated as a statistical

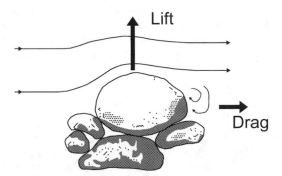

**Figure 7.17.** *Lift and drag forces on a sediment particle*

property, and methods of extrapolating back to 'zero transport' have been employed. Most of the background work has been done in flumes with grains of uniform size. Adjustments must be made to apply the techniques to the mixed particle sizes and varying flow patterns of natural streams.

### Lift and Drag Forces

Lift and drag forces (Chapter 6) act on a sediment particle when differences in pressure and velocity exist from top to bottom or front to back of the grain (Figure 7.17). These forces tend to jostle the particle in place, or, if the forces are strong enough, they can start it rolling or lift it into the flow. The Bernoulli concept of lift (Section 6.3.4), however, probably has less bearing on particle movement than the instantaneous upward velocity components of turbulent flow. Because of difficulty in modelling the effect of turbulence, other approaches using averages are favoured for predicting particle movement.

### Critical Velocity

Hjulstrom (1939) developed the graph shown in Figure 7.18, which relates average velocity to particle size. The curves show the limiting velocities for the three states of erosion, transportation and deposition. As illustrated, fine to medium sand between 0.3 and 0.6 mm is the easiest to erode. For larger particles the erosion velocity increases as particles become larger and more difficult to lift. Finer materials present a smoother profile to the flow and are thus less affected by turbulence; silts and clays are also bound together by a cohesive electrochemical force, making erosion more difficult. Clays tend to erode as aggregates rather than as individual particles. Unconsolidated silts and clays, however, may erode at lower velocities than indicated by the graph.

The Hjulstrom curves demonstrate that the velocity needed to entrain a particle is greater than that required

to keep it moving. Once set into motion, particles continue to be transported until the velocity drops below some speed, indicated by the dotted line in Figure 7.18. Fine materials will remain in suspension at very small velocities as long as there is sufficient turbulence. The difference between erosion and deposition velocities is much less for larger particles; thus the velocities setting them into motion must be maintained or the particles will drop out again. This is consistent with the concept that larger materials tend to travel as bedload.

A limitation of the Hjulstrom curves is that they were developed from data collected on sediments of uniform grain size and streams with depths greater than 1 m (Hjulstrom, 1939). In reality, a large range of particle sizes will move at any given average stream velocity depending on local currents and particle characteristics. For example, Smith (1975) states that the critical velocity of organic matter with a density of 1050 kg/m$^3$ is about 1/6 that of an equivalent-sized mineral particle. The Hjulstrom curves are, however, useful for general estimates of sediment entrainment.

In the development of a model describing the amount of usable habitat in a stream, Jowett (1989) states that the suitability of a streambed for instream flora and fauna depends on its stability. He defines *relative bed stability* (RBS) as the ratio of the critical velocity required to just move a particle ($V_c$) to the actual or predicted water velocity near the bed ($V_b$):

$$\text{Relative bed stability (RBS)} = \frac{V_c}{V_b} \qquad (7.13)$$

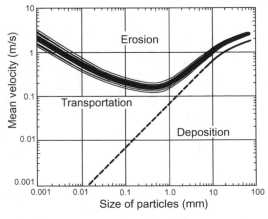

**Figure 7.18.** *Hjulstrom curves showing the limiting velocities required for erosion, transportation and deposition of uniform material. Adapted from Hjulstrom (1939), AAPG ©1939, reprinted by permission of the AAPG whose permission is required for further use*

with both $V_c$ and $V_b$ in m/s. $V_c$ can be obtained from the Hjulstrom curves. Additionally, the US Bureau of Reclamation (USBR, 1977) gives simple equations for both critical velocity and bed velocity, presented in SI units as

$$V_c = 0.155\sqrt{d} \qquad (7.14)$$

where $d$ is the average particle diameter (mm) and

$$V_b = 0.7V \qquad (7.15)$$

with $V$ the mean stream velocity (m/s). Equation (7.14) is considered applicable for uniform particles of sizes greater than 1 mm. In place of Eq. (7.15), Jowett also uses a variation on the logarithmic velocity profile (Section 6.5.2) with the assumption that $V_b$ occurs at a distance of 0.01 m above the bed.

### Example 7.1

Calculate the RBS for Site 3 on the Acheron River (Figure 4.7) at bankfull discharge, if average bankfull velocity is estimated as 0.37 m/s and the relevant particle sizes are those in Figure 5.33.

### Answer

From Figure 5.33, $d_{50} \approx 150$ mm. Thus, from Eq. (7.14),

$$V_c = 0.155\sqrt{150} = 1.90 \text{ m/s}$$

and from Eq. (7.15) for bankfull conditions,

$$V_b = 0.7(0.37) = 0.26 \text{ m/s},$$

and therefore

$$\text{RBS} = \frac{1.90}{0.26} = 7.3$$

This is much higher than 1.0, the value at which particles would be expected to move. Thus, the bed would be considered highly stable at bankfull discharge.

### Critical Shear Stress

An alternative to critical velocity is the concept that a critical shear stress is required to set a particle into motion. With regard to sediment erosion and movement, the term *tractive force* is commonly used as a synonym for shear stress (Dingman, 1984). Shear stress typically increases with discharge, but as with velocity, it is unevenly distributed within a channel.

In deriving an equation for critical shear stress it is assumed that when a particle is just about ready to move ('incipient motion'), the shear force acting to overturn it is balanced with the submerged weight of the particle, which holds it in place. By equating the two forces at the threshold of movement, an equation for critical shear stress ($\tau_c$) can be obtained

$$\tau_c = \theta_c g d(\rho_s - \rho) \qquad (7.16)$$

Here, $d$ is a 'representative' particle size in metres, $\tau_c$ is in N/m$^2$, $g$ is the acceleration due to gravity, and $\rho_s$ and $\rho$ are particle and water densities, respectively, in kg/m$^3$. The dimensionless constant, $\theta_c$, is a function of particle shape, fluid properties and arrangement of the surface particles. It is commonly termed the *dimensionless critical shear stress*.

Equation (7.16) was first proposed by an American engineer, Shields, in 1936. He related $\theta_c$ to another dimensionless factor, the 'grain' or 'shear' Reynolds number, $Re_*$ (Equation (6.19)). The Shields curve, which shows the relationship between $\theta_c$ and $Re_*$ is illustrated in Figure 7.19.

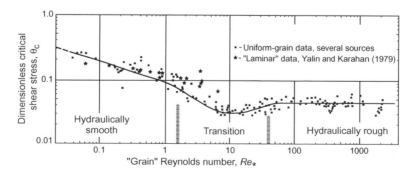

**Figure 7.19.** *'Shields curve', showing the relationship between dimensionless critical shear stress, $\theta_c$, and the grain Reynolds number, $Re_*$. Shaded bars indicate approximate boundary between hydraulically rough, smooth and transition zones. Adapted from Yalin, M.S. and Karahan, E. 1979. Inception of sediment transport. J. Hyd. Div., ASCE, **105**: 1433–1443. Reproduced by permission of the publisher, ASCE*

The value of $Re_*$ gives an indication of whether the flow is considered hydraulically rough or smooth. In reference to Figure 7.19, Yalin and Karahan (1979) give the transitional region as approximately $1.5 \leq Re_* \leq 40$. Under their assumption that the roughness height is approximately $2d_{50}$, this gives boundaries of 3 and 80 for the roughness Reynolds number, which closely corresponds with the range given in Section 6.4.2. Different values for this range will be obtained depending on the representative grain size used (e.g. $d_{65}$, $d_{85}$, etc.) and the interpretation of 'incipient motion'.

Under hydraulically smooth conditions, the particles are enveloped by the laminar sub-layer. Grains in the uppermost layer of the bed are dragged together as a 'grain carpet'. Under hydraulically rough conditions, grains are exposed to the turbulent flow, and grains detach individually and randomly as a result of instantaneous shear stresses. Thus, it takes more 'mobility' in laminar flow for a grain to lift off, as indicated in Figure 7.19 by the left end of the curve. These laminar data were derived by Yalin and Karahan using sand and glycerole.

For larger values of $Re_*$, the Shields curve levels off. Experimenters have established different values for this constant, typically within the range 0.040 to 0.060. Shields (1936) originally found 0.056, and the graph in Figure 7.19 gives an average value of about 0.044. Because of the considerable amount of scatter in the experimental data, it should be stressed that these values are not exact.

For $\theta_c \approx 0.06$, and assuming a sediment density of 2650 kg/m³, Eq. (7.16) becomes

$$\tau_c = 0.06(9.807)(d)(2650 - 1000)$$

or

$$\tau_c = 970d \qquad (7.17)$$

with $d$ in metres, or

$$\tau_c = 0.97d \qquad (7.18)$$

with $d$ in millimetres. Thus, as a 'rule of thumb', the critical shear stress required to move a particle (N/m²) is approximately the same as the particle's diameter in millimetres.

Although the Shields curve may work well for the uniform and fine sediments from which it was developed, its application to mixed gravel-bed materials is more difficult because of the interactions discussed in Section 7.4.2. An implication of the concept of equal mobility (see 'mixtures of particle sizes', Section 7.4.2) is that since all particles move at essentially the same shear

stress, $\tau_c$ can be calculated based on one characteristic grain diameter rather than computing individual values of $\tau_c$ for each size fraction. For streambeds, the median size of the bed materials ($d_{50}$) is often taken as the representative diameter, $d$.

Andrews (1983), for example, recommended a value of 0.020 for gravel-bed streams based on field studies, and suggested that a completely exposed particle may be entrained at values of $\theta_c$ as low as 0.010. In a study of seven steep mountain streams in California, however, Kondolf et al. (1987) used a value of 0.060, but found that gravels did not move at the predicted shear stresses because of the non-uniform flow patterns in the irregular, boulder-cascade channels and the shielding of gravel deposits behind boulders. Church (1978) cites a value of 0.11 for a streambed with imbricated and closely packed materials. Thus, order of magnitude variations in the Shield's parameter can be expected in natural channels. Suggested values are given in Table 7.2.

The effect of armouring on the Shields parameter was addressed by Parker et al. (1982) and Milhous (1986) and Milhous et al. (1989). For armoured streams, both recommend using the $d_{50}$ of the armour layer as the appropriate grain size in Equation 7.16. From research on Oak Creek, an armoured stream in Oregon, USA, Milhous found that if enough shear stress was applied to just move a small portion of the larger particles in the surface layer, the fines deposited among the armour particles were flushed out. He recommends two values of $\theta_c$: (1) 0.021 for flushing fines from surface sediments and (2) 0.035 for movement of 30% of the armour layer and 'deep' flushing of trapped fines. He points out that the parameters are somewhat subjective and applicable only if it is assumed that the Oak Creek results can be extrapolated to other streams.

For a particle to move, the actual shear stress must exceed the critical value. If velocity profiles are available, the local shear stress at a vertical can be calculated from the procedure given in Figure 6.14. Alternatively, the

**Table 7.2.** Suggested values for the Shields parameter, $\theta_c$, for mixed bed sediments. Based on Andrews (1983), Carson and Griffiths (1987) and Church (1978)

| Condition of streambed | $\theta_c$ |
|---|---|
| *Loosely packed.* 'quicksands' and gravels with large voids filled with water | 0.01–0.035 |
| *Normal.* uniform materials or a 'settled' bed with fairly random grain arrangements | 0.035–0.065 |
| *Closely packed.* smaller materials fill the voids between larger particles | 0.065–0.10 |
| *Highly imbricated.* | >0.10 |

shear stress over the whole channel perimeter can be estimated from Eq. (6.59). Andrews (1983) suggests that the actual depth of the zone of maximum bedload transport should be used in Eq. (6.59) rather than the hydraulic radius. However, without bedload measurements, this may be difficult to determine.

*Example 7.2*

Determine the particle size which will be entrained by bankfull flow at Site 3 on the Acheron River using critical shear stress methods (see also Example 7.1). Relevant measures are: channel slope $= 1.2\%$ and bankfull depth $= 0.40$ m. At this site, the cross sections are fairly wide and rectangular. The reach has a slight pool-riffle topography and is heavily armoured. Assume particles are spherical with $\rho_s = 2650$ kg/m$^3$.

*Answer*

From the sieve analysis of the smaller bed materials (Figure 5.32) it can be seen that there are few fines in the surface layer, meaning it is not closely packed. We will choose a value of 0.04 for $\theta_c$, to be consistent with both Milhous's recommendation and the data of Table 7.2.

Letting $\rho = 1000$, $S = 0.012$, and, since the channel is wide and rectangular, $R \approx$ mean depth $= 0.40$ m, then, from Eq. (6.59), the shear stress at bankfull flow is:

$$\tau_{bkf} = 1000(9.807)(0.40)(0.012)$$
$$= 47 \, \text{N/m}^2$$

This value is set equal to $\tau_c$, the critical shear stress for movement. Then, from Eq. (7.16),

$$\tau_c = 47 = 0.04(9.807)(d)(2650 - 1000)$$

giving

$$d = 0.073 \, \text{m} = 73 \, \text{mm}$$

This is approximately equal to the $d_{50}$ of the smaller surface materials (Figure 5.32), but from Figure 5.33 it represents less than 5% of the larger materials sampled from the bed surface. Assuming that the latter is indicative of the armour layer sizes, this would mean that only a small fraction of the bed surface could be mobilized at bankfull flow in this armoured reach.

Additionally, if it is assumed that $\theta_c = 0.010$ for isolated particles (e.g. tracer particles), then the largest isolated particle which will move is four times the above value, or 280 mm. This corresponds to nearly the $d_{95}$ of the larger materials. Thus, large isolated particles may move over the armoured surface at bankfull flows.

## 7.5 Sediment Yield from a Catchment

### 7.5.1 Sediment Sources and Sinks

*Sediment yield* is defined as the total sediment outflow from a catchment over some unit of time, usually one year. It can be calculated from measurements of suspended and bedload sediments (Section 7.5.3) or estimated from relationships between sediment yield and water discharge (Section 7.5.4).

The amount of sediment arriving at a downstream site represents the net balance between the amount of sediment stored in the channel and the amount contributed to it from upland sources, bank erosion and transportation of streambed materials. Schumm (1977) states that the majority of sediment yield is explained by storage and periodic flushing of alluvium. The relative contributions of sediment from channel and non-channel sources varies with basin size and is difficult to assess. In general, the steeper, shorter slopes of upper catchments supply more sediment from hillslopes. Further downstream, the storage of eroded material increases and channel erosion becomes more important (Knighton, 1984).

The relationship between the amount of sediment carried into a stream and the amount measured at some point downstream is described by the *sediment delivery ratio* (SDR), where

$$\text{SDR} = \frac{\begin{array}{c}\text{Sediment yield at a measurement point}\\\text{on the stream}\end{array}}{\begin{array}{c}\text{Total amount of eroded material contributed}\\\text{from slopes above the measurement point}\end{array}}$$

(7.19)

Sediment yield and upland erosion are usually calculated on an annual basis. The value of the SDR is normally less than 1, although values range widely. In general, it is larger for small catchments where drainage dissection is higher and sediment delivery more efficient. The ratio may exceed 1.0 during individual events if hillslope input is minimal and stream hydraulics favour the removal of channel materials (for example, during snowmelt or low-intensity rainfall events when water mainly reaches the channel as interflow). Although the SDR is difficult to evaluate, it forces investigators to consider the amount of sediment stored on the hillslope or in the channel before making generalizations about upland erosion rates from sediment yields measured at some point in the stream.

### *Bank Erosion*

A large number of factors control the rate of bank erosion, including the composition of bed materials, streamflow patterns and amounts, soil moisture, frost action, channel geometry, vegetation type and cover, and activity of burrowing animals (Knighton, 1984). In bedrock, the channel banks erode slowly under the grinding action of swirling rocks or from erosion along fracture lines, which can cause large chunks of rock to fall away. In alluvial channels, banks tend to be composed of finer materials than the streambed. Weak banks of gravel or other unconsolidated alluvium collapse easily, forming wide, shallow channels; whereas banks of more cohesive materials form deep, narrow ones (see Section 7.2.6). In composite banks the erodibility will be controlled by the strength of the weakest layer.

Bank erosion in alluvial channels commonly results from the slumping of saturated soils or the 'pseudo-cyclical' process of undercutting, failure of the overhanging bank and gradual removal of the fallen material (Knighton, 1984, p. 62). Bank undercutting is caused by the combined actions of large-scale eddies, spiral flow, and waves from wind or passing boats, which gradually loosen and remove bank materials. Undercutting may eventually cause trees to topple into the stream, releasing large quantities of sediment. However, the downed trees provide habitat for aquatic flora and fauna, retard the downstream movement of sediments and may protect the bank from further erosion.

Bank erosion is highly variable and episodic, and is usually associated with flood flows. The condition of the bank will affect its susceptibility to erosion. For example, deep saturation, frost action or trampling can cause soils to be more easily washed away. Multi-peaked runoff events may be more effective at eroding banks than single-peaked ones because the bank is saturated when subsequent flood peaks arrive.

*Streambank stability* refers to a bank's resistance to change in shape or position, whether attacked by flood flows or ice floes. Streambank condition and quality of aquatic habitat are closely linked. Stable undercut banks provide shade and cover for fish or burrow sites for river-dwelling mammals such as platypus or muskrats. Swallows and water ouzels may nest higher up on the face of steep banks. A one-metre loss of bank (measured horizontally) from the collapse of an overhang will thus alter the physical habitat differently than a similar loss from erosion of a vertical or sloping bank (Bohn, 1986). Therefore, for ecological purposes, it may be more important to define bank loss in terms of its function rather than the distance it shifts.

The health and composition of riparian vegetation will influence streambank stability. Smith (1976), for example, found that a bank with a 50-mm-thick root mat of 16%–18% root volume afforded 20 000 times more protection from erosion than a comparable bank without vegetation.

Excess trampling of banks removes protective vegetation and reduces bank stability. Platts and Nelson (1985) observed that large floods badly damaged channel banks in heavily grazed sections of a river in Utah, USA, but actually improved bank form in a protected ungrazed enclosure, with more undercut area and lower angles on exposed bank slopes.

Field techniques in Chapter 5 relevant to bank stability include stream surveys, methods of measuring areal extent of vegetation, and methods of analysing soil properties such as moisture content, bulk density, aggregate stability, soil texture and particle size.

### Channel Storage of Sediment

If the amount of sediment leaving a stream system is less than the amount entering, then the difference must be due to storage within the channel. Channel storage can be divided into three categories: (1) temporary storage in the channel bed and bedforms; (2) longer-term storage behind obstructions and (3) very long-term storage in valley flood-plain deposits (Megahan, 1982).

In reservoirs and canals, the trapping of sediments is referred to as *sedimentation*. It normally has a negative connotation since it reduces the life of reservoirs and can clog irrigation ditches. In natural streams, numerous sites for storage exist behind 'mini-dams' created by boulders, rock outcrops, logs, roots or accumulations of debris. These storage sites may be so effective that the impacts of landslide deposits, gravel-mining operations, road construction or forest harvest may not be felt downstream until sediments are removed by a large event—possibly many years later.

In a 6-year study on forested catchments in Idaho, USA, Megahan (1982) concluded that logs were the most important type of obstruction because of their longevity and the large volume of sediment trapped behind them. Streamflow variations were also found to affect the amount of sediment stored. Only large, stable obstructions remained in the channel during a high-flow year, whereas lower flows appeared to favour more obstructions with smaller storage volumes. On average, fifteen times more sediment was stored behind obstructions than was delivered to the drainage outlets. The study illustrates the need to consider all components of the sediment budget both under natural conditions and when monitoring the effects of land-use change.

When organic debris no longer enters a stream the banks become unstable, streamside erosion accelerates and the channel topography can become smoothed from the filling of pools and flattening of riffles. Lisle (1986, p. 46) states that riparian trees and large woody debris should be treated as if they 'belong to the aquatic ecosystem'. Further, Beschta and Platts (1986) point out that many streams are relatively 'starved' of large organic material in regard to channel stability.

### 7.5.2 Sediment Yield Variations

Variability in sediment yield can result from geological and climatic variations, intermittent bank or hillslope collapse, channel incision into alluvium of varying composition, the impact of fire or volcanic activity, change in land use, variable patterns of streamflow, and activity in the channel such as dredging, the cavorting of cattle, or the stirring up of bed sediments by large bottom-feeding organisms. Sediment yield also changes with drainage basin size. As the size of the drainage basin increases, there are more sites for permanent or temporary storage of sediments, and the sediment yield per unit area decreases (Schumm, 1977).

In general, streams in semi-arid areas have more highly variable sediment loads than those in humid regions (Nordin, 1985). It is characteristic of streams, especially those with high-flow variability, for most of the sediment transport to occur during a few days of high flow. Concentrations also tend to be higher at the beginning of the snowmelt or rainfall season when there is more erodible material on catchment surfaces. Subsequent runoff events carry less and less sediment over the runoff season.

Sediment loads also change from year to year. Changes in the amount of sediment delivered during low- and high-flow years can have a considerable impact on the nature of the streambed and its inhabitants. For example, in a pool of the Mississippi River, Bhowmik and Adams (1986) found that the low turbidity during a low-flow year caused a shift in the benthic community from clams to plants.

Regional variation of average annual sediment yield as a function of climate was studied by Langbein and Schumm (1958). They produced the classic curve shown in Figure 7.20 using data from approximately 100 sediment gauging stations in the USA. The 'effective precipitation' is the average annual precipitation, adjusted to a value which produces the same runoff as from regions having a mean annual temperature of 10 °C.

Sediment yield per unit area was found to reach a maximum at an effective precipitation of about 300 mm, which corresponds to areas receiving intense, infrequent rainfalls. Langbein and Schumm reasoned that lower precipitation levels would not produce as much runoff for transporting sediments, whereas higher precipitation encouraged vegetation growth, protecting soils from erosion. Subsequent studies with global data have yielded two more 'peaks', one representing Mediterranean climates with annual precipitation from 1250 to 1350 mm, and another tropical monsoon conditions (>2500 mm) (Walling and Kleo, 1979). Table 7.3 presents average annual sediment yields for some of the world's larger rivers.

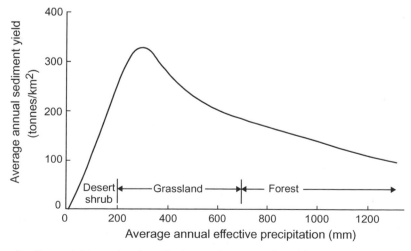

**Figure 7.20.** *Regional sediment yield, as related to effective precipitation. Adapted from Morisawa (1985) after Langbein and Schumm (1958), by permission of Longman Group, UK. 'Effective precipitation' is adjusted to a value producing the same amount of runoff as from regions with a mean annual temperature of 10 °C*

***Table 7.3.*** *Sediment yields of some major rivers, as measured at the mouth (compare with Table 4.2). Adapted from Holeman, J.N. 1968. The sediment yield of major rivers of the world.* Water Resources Research **4**: 737–747. *Copyright 1968 American Geophysical Union. Modified by permission of American Geophysical Union*

| River | Mean sediment yield ($10^6$ tonnes/yr) |
|---|---|
| Amazon, Brazil | 498 |
| Congo, Congo | 65 |
| Orinoco, Venezuela | 87 |
| Yangtze, China | 500 |
| Brahmaputra, Bangladesh | 726 |
| Mississippi, USA | 312 |
| Mekong, Thailand | 170 |
| Parana, Argentina | 82 |
| St. Lawrence, Canada | 4 |
| Ganges, Bangladesh | 1450 |
| Danube, USSR | 19 |
| Nile, Egypt | 111 |
| Murray-Darling, Australia | 32 |

### 7.5.3  Computing Sediment Discharge and Yield from Measured Concentrations

Field methods for the collection and analysis of suspended and bedload sediment samples and dissolved solids were covered in Chapter 5. The 'daily' sediment concentrations published by water authorities are commonly instantaneous suspended sediment values from samples taken at the same time each day, rather than an average concentration for the day. In large rivers where concentrations usually change little over the day, these values are probably adequate for calculating daily suspended sediment discharge as

$$Q_s = 0.0864 \, Q_D c_t \qquad (7.20)$$

where $Q_s$ = suspended sediment discharge (tonnes/day), $Q_D$ = average daily discharge ($m^3$/s) and $c_t$ = daily suspended sediment concentration (mg/L—or ppm if the water-sediment mixture is assumed to have a density of 1000 kg/$m^3$). The 0.0864 converts from seconds to days and milligrams to tonnes. Daily dissolved load can be computed using the same equation.

If a peak event occurs between sampling periods, as will be common in flashy or ephemeral streams, the sample will not be particularly representative of the day's sediment load. Thus, records will generally be biased towards low flows and low estimates of sediment load. In a monitoring programme samples should be taken more often so that a 'sediment hydrograph' can be constructed. Since the analysis of sediment samples is time consuming, various sampling strategies have been developed to minimize the number of samples collected (Nordin, 1985). The general strategy is to collect more samples where the sediment concentration is high, as indicated in Figure 7.21. Some 'smoothing' of the line drawn through the data points may be required because of the large fluctuations in sediment concentration.

Once a hydrograph has been constructed, sediment discharge for individual time steps is calcuated from Eq. (7.20), where time steps can coincide with sampling times or 'breakpoints' where the slope of the sediment or runoff curve changes. Quantities are calculated for each time interval as

$$\left( \frac{(Q_s \text{ at time 1}) + (Q_s \text{ at time 2})}{2} \right) (\text{time 2} - \text{time 1})$$

$$(7.21)$$

Daily sediment discharge totals, in units of tonnes/day can then be calculated by summing up values over individual time intervals within the day. Porterfield (1972) is a source

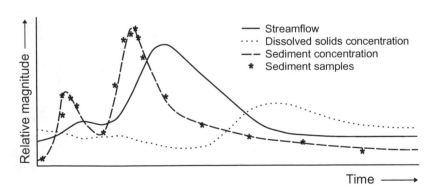

***Figure 7.21.*** *Generalized graphs showing the change in streamflow, sediment load and dissolved load, and spacing of sediment samples during a runoff event*

of more information on computing sediment discharge. Again, the same procedure can be used for daily dissolved load.

Bedload discharge ($Q_{bl}$) is calculated from the amount of sediment sampled over some unit of time. Often, a particle size analysis is done to determine the bedload discharge for each size fraction. For pit-type samplers or reservoir surveys, it is simply calculated from the mass of sediment trapped over some time interval (e.g. one year). If volume is measured, it must first be converted to mass using an estimate of the bulk density of the deposited sediment (Section 5.8.7).

Using Helley-Smith-type samplers (Section 5.7.5), bedload discharge per unit width of stream ($Q_{bl}$ in kg/m·s) is calculated for each measurement point as

$$Q_{bl_i} = \frac{M_i}{Wt} \qquad (7.22)$$

where $t$ is the measurement time interval (s), $M$ is the mass of the sample (kg), $W$ is the width of the sampler entrance (m) and $i$ refers to the $i$-th measurement point. As with the velocity-area method of computing discharge (Section 5.6.3), each measurement is assumed to represent a certain segment of the cross section. Total bedload discharge for a cross section is obtained by multiplying each measurement by the width interval it represents and summing over the width of the stream.

### 7.5.4 The Estimation of Sediment Discharge from Streamflow

Sediment concentrations are often poorly related to discharge. This is partly due to the hysteresis effects discussed in Section 7.1.4 or varying amounts of washload. Therefore, it is always preferable to measure sediment concentrations, rather than attempting to estimate them from streamflow data. However, sediment data will not always be available during particular years, seasons, or events of interest. Estimating sediment concentrations from streamflow data is a widely used approach in hydrological practice.

Sediment concentrations obtained over a range of discharges can be used to develop a *sediment rating curve*, given previously as Eq. (7.10)

$$Q_s = pQ^j \qquad (7.22)$$

where $Q_s$ and $Q$ are sediment and water discharge, as before, and $p$ and $j$ are determined empirically. The value

of $j$, which does not depend on the units used, typically lies between 1.5 and 3.0 (Knighton, 1984). From hydraulic geometry relationships (Section 7.2.6), Leopold and Maddock (1953) found that an increase in sediment discharge is related to an increase in velocity and a reduction in depth.

Depending on the intended use, sediment rating curves can be developed for instantaneous, daily, monthly or yearly data, with the scatter in the relationship decreasing as the time interval increases. In large, stable rivers the relationships will be fairly well defined. However, short-term rating curves for most streams, especially those which are supply limited, will exhibit a considerable amount of scatter. Estimating instantaneous or event concentrations from discharge in these streams is risky at best. Brakensiek *et al.* (1979) further discusses the preparation of sediment rating curves and methods of treating hysteresis effects.

Annual or seasonal rating curves may still be fairly useful for streams where instantaneous, daily, or event rating curves are unacceptable. For example, Nordin (1985) found that the relation between annual sediment load and annual flow was 'reasonably well defined' for the Rio Puerco, an ephemeral tributary of the Rio Grande in New Mexico, USA.

To develop a rating curve, sediment concentrations should be measured over a large range of discharges. Sediment discharge is calculated for individual samples by Eq. (7.20), and a log-log relationship fitted to the sediment discharge and streamflow data. If measurements allow, curves can be developed separately for suspended load and for bed-material load, as shown in Figure 7.22. It may also be appropriate to develop separate curves for each season or for rainfall and snowmelt events.

From the rating curve, a record of daily, monthly or yearly sediment discharges can be generated from streamflow data. These can be used to derive yearly or seasonal sediment yields that may be useful in locating regions of high (or low) sediment production, e.g. for targeting erosion control works. *Effective discharge*, the flow which is associated with the most sediment transport over time, can be computed by combining the rating curve with a flow duration curve (Section 8.3); the peak of the resulting curve is the effective discharge.

Limitations of the data should be taken into account when estimating sediment yields from streamflow data and sediment rating curves. For example, a rating curve developed during an 'average' year would not be applicable to an extremely wet year in which mudflows and bank collapses were common. There is also a danger in extrapolating beyond the data; for example, a rating curve

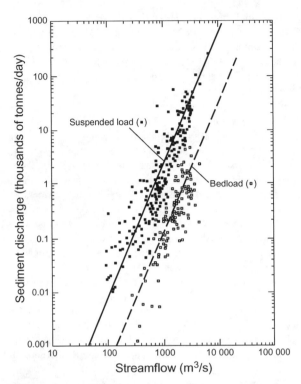

***Figure 7.22.*** *Sediment rating curve for the Snake and Clearwater Rivers in the vicinity of Lewiston, Idaho. Based on data of Emmett (1977); redrawn from Csermak and Rakoczi (1987), Reproduced by permission of Water Resources Publications, LLC*

developed for low flows should not be applied to a streamflow record in which large floods occur. It should be realized that the rating curve is an example of spurious correlation (see Appendix), and should be interpreted only in reference to sediment discharge, not sediment concentration. An evaluation of the error involved should always

be included when estimating sediment discharge using a rating curve.

With a sufficiently long record of either estimated or measured sediment data, the same techniques used for the analysis of streamflow data can be applied; e.g. basic statistics and histograms (Section 4.4), and sediment duration or frequency curves (Section 8.2). However, care should be taken when interpreting the sediment record to infer the future from the past. Gradual trends are common as a result of changes in grazing or forestry practices, or in the amount of cultivated or urbanized area. Sudden changes due to fire, floods, or the construction of reservoirs may impact sediment yields for many years afterwards or permanently. This effectively alters the population from which subsequent samples are drawn (see Section 2.3.1). An example is the impact on the Toutle River from the 1980 eruption of Mount St Helens in Washington, USA. Concentrations of suspended sediment up to 200 000 mg/L were measured shortly after the eruption, and the long-term effect on the estuary of the Columbia River into which the Toutle drains can only be guessed (Nordin, 1985).

This chapter has focused on the dynamic nature of the stream channel, to which stream biota must adapt. The patterns of habitat within stream channels are affected by the geological and hydrological setting of the stream. For example, in mountain streams with stepped formations composed of large cobbles and boulders, extreme flows may be required to rearrange the streambed; however, sands and gravels held in the pools may be mobilized more frequently, as is the case with alluvial rivers. The mobilization of streambed sediments and the relationship between sediment transport and channel form are important considerations in assessing stream health and environmental flows. We will return to these issues in Chapter 9.

# 8

# Dissecting Data with a Statistical Scope

"The importance of information is directly proportional to its improbability"

(Raudkivi, 1979; p. 396).

## 8.1 Introduction

### 8.1.1 General

How often will a river overtop its banks? How long will this drought last? If a stream's flow is such and such today, what is the probability it will be so and so tomorrow? How often will a platypus burrow be inundated? How many days out of the year will the flow be below five cumecs? How large does a reservoir need to be if a town's water supply is to be maintained with a 99% reliability?

These are the types of questions addressed by hydrologists using methods in statistics and probability. Increasingly sophisticated satellites for weather monitoring and complex global computer models are improving the prediction of near-future streamflows. Yet, forecasts of the streamflows that will occur next year or over the next 20 years still cannot be modelled with much certainty. However, if records are available for a sufficient length of time and we assume that these past records are an indication of what will happen in the future, we can estimate with some confidence the average yearly streamflow, the variability of the stream or the probability that a flood of a certain magnitude will occur within the next year.

A streamflow record is a sample out of time of all the flows that will ever course through a stream, past and future. Statistics calculated from this sample are used to predict what might occur in the future under the same conditions. The actual sequence of high and low flows in the past record is not expected to repeat itself exactly—in fact, it would be quite extraordinary if it did. Variations over time occur more or less randomly, although some 'persistence' is seen in streamflow data; e.g. low-flow days tend to follow other low-flow days. Scientists continually look for cyclical trends in streamflow data which might correlate with a measurable phenomenon such as sunspot cycles or the El Niño-Southern Oscillation variation in atmospheric circulation. Success, however, has been limited. Thus, streamflow data will continue to be treated statistically until such time as the weather can be predicted—or controlled—with greater reliability.

In Chapter 2 definitions are given for various statistical terms, and Chapter 4 provides methods of exploratory statistical analysis for describing the pattern, relative magnitude and variability of streamflow. The Appendix includes a summary of basic statistical formulae. In this chapter, methods of analysing probabilities of streamflow magnitudes and durations are presented. The methods given in Sections 8.1–8.4 require streamflow records from gauged sites. However, in many cases it is the stream behaviour at ungauged sites that is of most interest. If gauging stations are located nearby, it may be possible to adjust the records slightly for the ungauged site. Alternatively, a regional analysis can be conducted to develop estimated records for stream sites within the region by using catchment or channel characteristics. Section 8.5 covers regionalization methods and Section 8.6 describes methods of numerical taxonomy which can be used to classify streams and catchments into hydrologically homogeneous groups.

### 8.1.2 Floods and Droughts

In engineering hydrology the emphasis is primarily on the extremes—the floods and droughts—because these are of

---

Stream Hydrology: An Introduction for Ecologists, Second Edition.
Nancy D. Gordon, Thomas A. McMahon, Brian L. Finlayson, Christopher J. Gippel, Rory J. Nathan
© 2004 John Wiley & Sons, Ltd ISBNs: 0-470-84357-8 (HB); 0-470-84358-6 (PB)

the most importance *economically*. These extremes are also important *ecologically* since they affect the populations and distributions of aquatic organisms. Hydrologists are probably most well known for their work on predicting the magnitude of floods of a certain probability (e.g. the '100-year flood'). Knowledge of flood behaviour is needed in the design of dams, bridges, highway culverts, in the development of flood-insurance rates and flood-zoning maps and in the ecological assessment of instream flow requirements and floodplain connectivity.

A flood can be defined simply as a flow which overtops the streambanks. However, this definition is difficult to apply in mountain gorge streams or others which have poorly defined banks (see discussion of bankfull flow, Section 7.2.5). An ecological definition was given by Gray (1981), who considered a flood to be a discharge which scoured substrates and disrupted the biota. Ecologists also use the term 'spate' for high flows.

Floods affect the ecology of a stream by rearranging streambed habitats, scouring away aquatic or riparian plants and increasing the drift of aquatic insects. Most biota avoid high velocities and hurtling particles by sheltering behind rocks or snags, burrowing into the streambed and banks, moving to slower water along the stream's edges and in backwaters, or by having life cycles which are terrestrial or aerial during flood-prone seasons. To some fish, floods act as a cue for spawning, and they may migrate upstream or downstream during flood periods. When floods occur at unusual times the fauna may be severely depleted, requiring many years to recover (Goldman and Horne, 1983). In desert streams which are often subject to sudden, unpredictable floods, aquatic insects have developed specialized coping mechanisms such as rapid development to the adult stage (Gray, 1981).

Peak flows are caused by various factors such as heavy rainfall on saturated or frozen soils or rapid snowmelt from warm Chinook winds. Even though the highest flow in a year or the peak of a given hydrograph may or may not constitute a 'flood' in the strictest sense, these flows are still important in an analysis of how often floods can be expected to occur. Techniques for flood frequency analysis are given in Section 8.2.

The length of time (duration) an area is underwater is also important in a flood analysis. For a riverside cottage the damage may be the same whether it is underwater for 5 minutes or 5 days; however, for a tree, it can mean the difference between renewed vigour and root suffocation. The period of inundation will affect the length of time during which nutrients and biota pass between the stream and floodplain. Inundation for long periods of time can also raise groundwater levels. To assess the duration of floods, techniques such as flow-duration curves

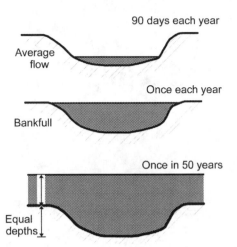

**Figure 8.1.** *Typical river stages at various rates of occurrence. Adapted from Leopold and Langbein (1960), Reproduced by permission of U.S. Geological Survey*

(Section 8.3) or flow-spell analysis (Section 8.4) can be applied.

Flood frequency and flow duration are related to the channel form, as shown in Figure 8.1. In this 'typical' river the average flow is very low, and the stream's level is much less than bankfull level for most of the year. About once each year there is enough runoff to fill the channel. Overflow onto the floodplain accommodates the waters of larger, rarer floods. This pattern will vary from stream to stream and from one cross-section to another on the same stream.

At the opposite extreme from floods are droughts and other periods of low flow. Very low flows and even zero flows are normal in streams during years of low precipitation, particularly if the stream is intermittent or ephemeral. They may also be artificially imposed by diversions or upstream storages. No uniform definition of 'drought period' has yet been established, but it generally refers to long-duration periods of low flow. Like 'flood', 'drought' is usually defined in economic terms, based on water supply needs (McMahon and Finlayson, 2003). Droughts can have immediate and lasting effects on the animal and plant populations of streams. Unless aquatic organisms have developed special adaptations such as long dormant phases, rapid development and prolific reproduction (e.g. the fairy shrimp and toads which live in the temporary lakes of dryland regions), the ecological impacts of drought can be more long-lasting than the effects of floods.

An analysis of low flows is necessary before a stream can be used as a reliable source of water supply (McMahon and Arenas, 1982). Dry spells put pressure on managers who must allocate limited water supplies between irrigators, cities and instream biota. There has been increasing interest in defining 'minimum instream flows' for sustaining aquatic life and/or diluting pollutants. The provision of flows for environmental purposes is covered in detail in Chapter 9.

Various indices have been developed for describing the severity of droughts, including: (1) the lowest flow during the drought period, (2) the mean discharge during the drought, (3) the volume of water 'deficiency' below some flow-threshold level, and (4) the duration of the drought (WMO, 1983). Low-flow frequency analysis (Section 8.2), flow-duration curves (Section 8.3), flow-spell analysis (Section 8.4) and recession-curve analysis (Section 4.3.4) are methods applied to the analysis of low flows.

The discussion this far has been about techniques for analysing the hydrological record for past droughts and low-flow events. In the management of water resources a knowledge of historic low-flow events is useful as it provides information about what can be expected, in a probabilistic sense. In real-time management, how can we tell when a severe low-flow event, or drought, has begun? While significant progress is being made in long term forecasting using large scale circulation anomalies such as the El-Niño/Southern Oscillation, it is also possible for water supply authorities to monitor the state of their systems and respond progressively. Drought Management Plans (Moran and Rhodes, 1991) provide water management authorities with a framework for progressively introducing drought responses using trigger levels. As the volume of stored water declines specific actions are recommended at pre-determined levels. These actions include accessing alternate water supplies (such as groundwater bores) and placing legal restrictions on the use of water, for example, limiting the hours of garden watering.

### 8.1.3  Data Considerations

As mentioned in Section 3.2, hydrological data must be accurate, representative, homogeneous and of sufficient length if they are to be statistically analysed to provide useful answers. To ensure accuracy, the data should first be examined for gross errors and missing data. Representativeness means that the data represent the conditions of interest. Many studies will require natural ('unregulated' or 'virgin') streamflows. If a dam has been installed on a stream, the natural flows must be reconstructed by adjusting the recorded values by the amount of storage, difference in evaporation, and any diversions. In contrast, if regulated flows are of interest, reservoir operating rules

can be applied to past 'unregulated' records to create a record that simulates the presence of a reservoir. *Homogeneity* implies that the streamflows have occurred under uniform conditions; i.e. stream channelization, a modification in the forest density or a change in the gauging station location may lead to a non-homogeneous record. The double-mass curve, described in Section 3.2.2, is a technique for detecting non-homogeneity. Additionally, to use past records to describe the future, the data must be *stationary*, meaning that statistical properties remain constant over time; e.g. climate changes do not occur.

A streamflow record is only a sample in time. The 'population' consists of all streamflows over an indefinitely long time period. A typical record of 20 years in length is usually made up of mostly medium-sized events, with a few large events and a few small ones. From this information inferences can be made about the future and past behaviour of the stream. The value of these inferences depends on how well the sample represents the range of high, low and medium values occurring over time. Since the true population is unknown, it can only be assumed from statistics that more records will bring us closer to the 'truth'. In fact, a rigorous mathematical treatment of very short records may be quite inferior to 'guesstimates' made by old-timers whose families have lived along a river for generations. Analysis of historical documents for accounts of past floods and droughts can help improve knowledge of the behaviour of a river beyond the period of numerical records.

To improve the reliability of short records, streams should not be treated in isolation but with knowledge of the behaviour of similar streams. Several short-term records from stations within hydrologically similar basins can be combined using regionalization methods to increase the effective length of record. These methods are presented in Section 8.5.3.

### 8.1.4  Putting Statistics into a Proper Perspective

In an ideal situation, we would be able to apply uniform methods and obtain consistent results under all situations—short or long records, small or large catchments, stable or flashy streams. However, as Dalrymple (1960, p. 20) cautions, 'A method is not better because it leads to uniform answers if those answers are uniformly wrong'. Therefore, in this chapter several different methods may be presented for the same type of analysis, some more appropriate to specific situations than others.

The treatment of hydrologic data with statistics has led to considerable debate about which techniques are best and the further refinement of the statistical methods used. When reading the vast literature on the testing of these methods one is often left with the impression that researchers have forgotten that the numbers were generated by a catchment rather than a computer. In fact, many validations of methods have been made using synthetic data. It is easy to lose sight of the fact that streams were running at varying rates of flow long before anyone developed statistics, and will continue to do so in spite of our attempts to pin them down numerically. Klemes (1986) has written a refreshing treatise on 'mathematistry' in hydrology which is well worth reading by those needing a healthy dose of reality about the limitations of statistics.

In methods such as flood-frequency analysis and regression analysis where a line is fit through the data there is always the temptation to define the line in its geometric sense—that it has infinite length. However, extrapolation beyond the data should be done cautiously, if at all, with knowledge of the error margins on the extrapolated values. Overmire (1986, p. 35) gives this quote from Mark Twain's *Life on the Mississippi* to illustrate the perils of reckless extrapolation:

> In the space of one hundred and seventy-six years the Lower Mississippi has shortened itself two hundred and forty-two miles. That is an average of a trifle over one mile and a third per year. Therefore, any calm person, who is not blind or idiotic, can see that in the Old Oolitic Silurian Period, just a million years ago next November, the Lower Mississippi River was upward of one million three hundred thousand miles long, and stuck out over the Gulf of Mexico like a fishing rod. . . . There is something fascinating about science. One gets such wholesale returns of conjecture out of such a trifling investment of fact.

For ecological work the consequences of an inaccurate estimate may not be as severe as in engineering. However, the error margin on an estimate and its consequences should still be evaluated as an essential part of any statistical analysis. As Haan (1977, p. 5) says, 'statistics should be regarded as a tool, an aid to understanding, but never as a replacement for useful thought'.

## 8.2   Streamflow Frequency Analysis

### 8.2.1   General Concepts

Frequency analysis is a method for assigning probabilities to events of a given size. The Appendix gives a description of the concepts of frequency and probability, and methods of displaying their distribution. In streamflow frequency analysis we most often use cummulative distributions and speak of probabilities of exceedance. For high flows, 'exceedance' usually means a streamflow value is *greater* than some amount, whereas for low flows, it is usually *less*; e.g. the probability that the flow is greater (or less) than 12.3 m³/s.

In Section 5.8.5 particle size data are analysed with a type of frequency analysis. However, there is a subtle difference in that the particle size frequencies are *percentages* rather than *probabilities*. In sizing sediments the idea is to simply *describe* the bucket-full of sediment which was passed through a sieve. In flood or low-flow frequency analysis the object is to use data from a short time period not only to describe the 'bucket-full' of past flows but also to evaluate the probability of future events.

In a streamflow frequency analysis the basic procedure begins with a ranking of the most extreme events of the past. Probabilities can be determined either by graphical methods or by fitting a theoretical probability distribution to the data, depending on the question being asked. As a general rule graphical methods can be used for cases where the flood or low-flow estimates and the probabilities can be interpolated. However, for estimates larger than a 1-in-$N$/5-year flood or low flow based on $N$ years of data, the fitted curve or distribution should be used to find probabilities associated with events of a given size, within the limits of prudent extrapolation. The techniques will be discussed with a focus on flood-frequency analysis. Low-flow frequency analysis is presented in Section 8.2.6.

With the 20 years of record typically available for most streams, estimation of extreme events rarer than about the 1-in-50-year event should only be considered rough guesses, and extrapolation beyond the data calls for professional judgement by 'experienced individuals who are knowledgeable about the subject and the impacts of their decisions' (Hagen, 1989, p. 18). Regionalization (Section 8.5.3) can help to increase the accuracy of extrapolated values.

### 8.2.2   Probability and Average Recurrence Intervals

In flood-frequency analysis, probability is often expressed as a '1-in-$N$-year' chance. For example, a 1-in-100-year flood is one that would be expected to occur, on average, once in every 100 years. This *average* length of time between two floods *of a given size or larger* is called the *average recurrence interval* or *return period*. The 1-in-100-year flood would have a *probability* of 0.01 or 1% of being equalled or exceeded in any one year. Probability

($P$) and average recurrence interval ($T$) are thus reciprocals:

$$P = \frac{1}{T} \tag{8.1}$$

For example, if $T = 100$ years, $P = 1/100 = 0.01$.

A 1-in-6-year flood occurs with the same frequency as rolling a 7 with two (non-loaded) die. A 1-in-50-year flood occurs with almost the same probability as drawing an ace of spades from a pack of playing cards. And an extremely rare flood (of Noah magnitude or larger) may only occur with the same chance as that of a monkey randomly typing out the word 'Ecology' on a typewriter.

Probability or recurrence interval only tell us how *likely* a flood (or drought of zero flow) is. It says nothing about *when* it will actually come. As someone once said, 'one out of 100 Alaskan bears bite ... the trouble is, they don't come in numerical order'. Thus, a 100-year event might occur this year, next year, several times or not at all during our lifetime.

### 8.2.3   The Data Series

For flood-frequency analysis it is important to have a record which is representative of the total population of floods which are likely to occur on a stream. Records should be relevant to the problem and based on a consistent set of stream, catchment and climatic conditions. Peak flows resulting from different meteorological causes (e.g. snowmelt, rainfall, typhoons, dam failures) can represent different populations, and it may be preferable to analyse the data separately, as suggested by Jarrett (1987).

For probabilistic analysis, streamflows are also required to be *independent*. For example, if a storm produces a double-peaked hydrograph, only the highest peak should be used in a flood-frequency analysis. A problem can also arise when a large event in December spills over into January, becoming the largest 'event' in the two adjacent calendar years. For this reason, water years rather than calendar years are preferred (see Section 4.4.2).

The data series must also be *adequate*. If the sample is too small the probabilities will not be reliable. From Table 8.1 it can be seen that 18 years of data would allow the 10-year flood to be estimated with an error of about 25%, whereas over 100 years of data are needed to estimate either the 50- or 100-year flood within plus or minus 10%. The table was developed from data generated synthetically from a probability distribution, and errors will undoubtedly be higher for real data. These limitations

**Table 8.1.** *Length of record (years) required to estimate floods of various average recurrence intervals with 95% confidence. Adapted from Linsley et al. (1975b), by permission of McGraw-Hill*

| Average recurrence interval (years) | Error | |
|---|---|---|
| | 10% | 25% |
| 10 | 90 | 18 |
| 50 | 110 | 39 |
| 100 | 115 | 48 |

are not meant to be discouraging, but to point out the need for extracting as much information as possible from gauged data and field observations, and for using the most appropriate statistical methods.

Since a longer record (larger sample size) is more likely to be representative of the 'true' population than a short record, some sleuthing may be warranted to obtain information for years outside the published streamflow record. Old newspapers can be searched for historical information on the dates of floods and approximate depths and discharges. Discovering a lack of evidence of historical floods can be just as important. Botanical information from flood-damaged trees such as the number of tree rings in a re-grown branch or a tree scar can also give an indication of both the height of a flood and the year of occurrence. Hupp and Bryan (1989), for example, used tree-ring analysis of flood-damaged vegetation to reduce the standard error of estimation of flash flood frequencies by as much as an order of magnitude. Harrison and Reid (1967) also cited good correspondence between a frequency analysis based on tree scars and a traditional analysis using gauged data. They found that most of the scars were caused by floating ice, and that summer floods which lacked debris did not scar trees. Special procedures must be used to incorporate historical data into a frequency analysis, and the reader is referred to IAC (1982) and IEA (1998).

Flood frequencies can be analysed using data on the stage, volume or discharge of flood events. *Discharge frequency* is analysed most often, and this will be the approach taken in this text. *Stage frequency* may be of value in determining how often a stream will overtop its banks or have sufficient depth for migrating fish. However, a stage frequency analysis only applies to the *actual gauge location* since the shape and stage of the stream can change considerably over a short distance. It may also change over time if the channel adjusts its shape. Stage frequency analysis is best accomplished by analysing discharge frequencies and then converting back to stage

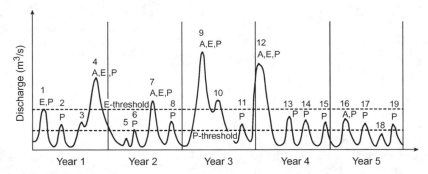

***Figure 8.2.*** *Flood peak data series: annual maximum (A), partial duration (P) and annual exceedance (E). The threshold value for the partial duration series ("P-threshold') is set at a chosen value; for the annual exceedance series ("E-threshold"), it is set to give N peaks in N years of data (here, 5 peaks for 5 years of data)*

with the most recent stage-discharge relationship at a particular site of interest (WMO, 1983). *Volume frequency* has potential use in the prediction of water supply, and can, with caution, be extrapolated to reaches away from the gauging site. Dalrymple (1960) should be consulted for techniques of stage and volume frequency analysis.

Instantaneous flows are normally used for analysis, particularly for flood frequencies. Daily flows may be used, but with the knowledge that they represent an average over the day or for a specific time (e.g. an 8 a.m. reading) rather than the highest streamflow occurring within that day. The difference between the daily flow and the instantaneous flow decreases as catchment size increases.

Flood-frequency analysis is generally performed on a data series composed of the single highest peak flow in each year, the *annual maximum series*. For low flows we can instead speak of an *annual minimum series*, which is comprised of the single lowest flow (instantaneous or daily) from each year. The value of low-flow frequency analysis is often improved by using periods of significant length, e.g. the flow over 7 or 10 consecutive days. The annual minimum series would then be made up of the minimum 7- or 10-day flow from each year of record.

Flood-frequency analysis has been developed mainly within engineering hydrology where much of the interest centres around the characteristics of large infrequent floods because of their relevance for the design of spillways, bridges and similar structures. In ecology and fluvial geomorphology the characteristics of small floods have much more relevance. Channel forming flows, and flows which move the substrate, often occur more frequently than once a year and the annual maximum series does not deal with them appropriately. The *partial-duration series* is a much more appropriate technique for analysing flood frequency in those disciplines. A

partial duration series consists of all flood peaks above a threshold level, chosen for its relevance to the issue for which the analysis is being carried out. A special case of the partial-duration series is the *annual exceedance series*, where the *N* highest peaks within the *N* years of data are chosen, irrespective of when they occur (Shaw, 1988). These concepts are illustrated in Figure 8.2.

The threshold value for a partial duration series can be set arbitrarily, depending on which flow levels are of the most interest. One method is to set the threshold equal to the smallest annual peak flood. Another possibility is to set it to bankfull flow (Section 7.2.5). For calculating average recurrence intervals of one year or less, the base is usually set such that, on average, three to four floods per year are included in the series (Dalrymple, 1960). In Figure 8.2 the threshold for the partial series is set at 3*N*, i.e. there are 15 independent floods making up the partial series out of 19 flood peaks. Numbers 3 and 10 are not considered independent as they occur, respectively on the rising and falling limbs of a hydrograph. Flood peaks 5 and 18 are below the partial duration threshold.

To visualize the difference between the various types of series, one could consider the movements of a cockatoo soaring above a forest. The highest altitude reached each year by the cockatoo would represent the annual maximum series. In the partial-duration series the "threshold level" could be represented by the tree top elevation. Each time the cockatoo flew above the trees the highest point reached would be recorded. This record, collected over some period of time (say 20 years) would constitute the partial-duration series. The 20 highest points reached in that 20 years would represent the annual exceedance series.

From the annual maximum series we could calculate the probability that the cockatoo would fly up to a certain altitude or higher in any given year. From the partial-duration series, we can make a statement some-thing like this: 'on average, the cockatoo flies to at least a height of 500 metres above the ground 4.3 times per year'. Interpretation of the annual maximum, annual exceedance and partial-duration series is different, and is discussed in Section 8.2.9.

### 8.2.4   Graphical Methods: The Probability Plot

In graphical methods of flood-frequency analysis the peak flow data are ranked from highest to lowest and given a *plotting position*. For an annual series this plotting position is a sample estimate of probability. The data and their respective plotting positions are plotted as points on graph paper and a line is drawn to interpret the points. The result is called a *probability plot* and the fitted line a *flood-frequency curve*.

Plotting the data on arithmetic paper will usually result in a S-shaped curve which is difficult to interpret. Instead, the data can be plotted on special *probability paper*. The *Y* axis represents the peak flow value and the *X* axis either the probability of a flow being equalled or exceeded, or the average recurrence interval (see Figure 8.4).

Probability paper is designed to produce a straight-line plot when the appropriate theoretical probability distribution is plotted as a cumulative distribution function (see Appendix). Sample data which plot more or less as a straight line on a specific paper can be judged to be of that distribution type. Commercially available probability paper is most commonly based on the normal distribution, and normally distributed data will plot as a straight line on this paper. The same paper with a logarithmic scale is available for log-normally distributed data.

To plot peak discharge data (either the annual maxima or the peaks above some threshold level), the data values are first ordered by size. The largest value is given a rank of 1, the second largest a rank of 2 and so on until the lowest value has a rank equal to $N$, the total number of data points. If two values are equal, they should still be assigned different ranks.

A plotting position is assigned to each data value using a plotting position formula, three of which are given in Table 8.2. Theoretically, the largest flood should plot at 0 (there would be no chance of it ever being exceeded) and the smallest flood at 1 (every flood would be equal to or greater than this value). However, there is no guarantee that the sample record contains the largest and smallest values; hence, plotting positions should lie between 0 and 1 (or 0% and 100%). The different plotting position formulae tend to give similar values near the middle of the data, but can vary considerably at the tail ends.

**Table 8.2.** *Plotting position formulae (N represents number of years of record, m is the rank of the event, the largest event having rank 1, α denotes a constant, P represents probability and T is the average recurrence interval, years). Reprinted from Journal of Hydrology, Vol. 37, Cunnane, C. 'Unbiased plotting position - a review', pp 105–222, Copyright 1985, with permission from Elsevier*

| Name | Probability of exceedance, $P$ | Average recurrence interval, $T$ |
|---|---|---|
| Weibull | $\dfrac{m}{N+1}$ | $\dfrac{N+1}{m}$ |
| GEV[a] | $\dfrac{m-0.35}{N}$ | $\dfrac{N}{m-0.35}$ |
| Cunnane | $\dfrac{m-\alpha}{[(N+1)-2\alpha]}$ | $\dfrac{[(N+1)-2\alpha]}{(m-\alpha)}$ |

[a] using probability weighted moments to estimate the GEV parameters.

The Weibull plotting position has been commonly applied to both annual flood data and the partial-duration series (Dalrymple, 1960). The choice of a plotting position formula is, in fact, the same as choosing an underlying probability distribution. Hosking *et al.* (1985) recommends the second formula for the general extreme value distribution (Section 8.2.6) using probability-weighted moments (Section 8.2.5). Cunnane (1978) reviewed several plotting position formulae and found that some of the traditional plotting formulae yielded plots which were biased, generally leading to overestimation of flood peaks at high recurrence intervals. He suggested the third formula in Table 8.2 and recommended $\alpha = 0.40$ as a single compromise value for use with all distributions. Other values of $\alpha$ will be given in the discussion of individual probability distributions (Section 8.2.6).

There can be a great amount of uncertainty in the plotting positions assigned to the largest events. The extreme events may plot as 'outliers', far off the line defined by the more frequent events. There are statistical tests available for testing outliers, as given by IAC (1982) and IEA (1998). Data from another stream with a longer record can be used to obtain a better estimate of the average recurrence interval for these larger events, as described by Dalrymple (1960). Regional analysis (Section 8.5.3) is also recommended rather than studying a single site in isolation.

A line is fitted through the plotted data to develop a flood-frequency curve. Linsley *et al.* (1975b) recommend as a general rule of thumb that a curve can be fitted by eye if one intends to use the frequency analysis for information on floods with recurrence intervals less than $N/5$. For larger recurrence intervals a theoretical probability distribution should be fitted to the data to enable more

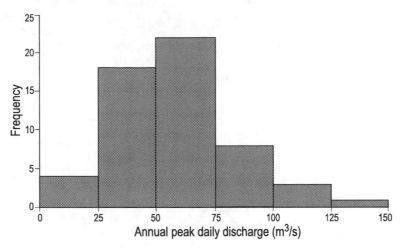

**Figure 8.3.** *Frequency histogram for the annual maximum daily flows on the Acheron River at Taggerty (see Figure 4.7), years 1946–2001 (56 years of data)*

consistent, objective estimates. Straight lines can be drawn between consecutive data points for interpolating values directly from the data, although this should not be attempted for the larger flood flows. More often, a straight line is fitted through the entire data set. Extrapolation beyond the range of data should never be done with an eye-fitted curve. Errors in the estimates of extreme events can be very large using graphical procedures, especially if the eye-fitted line is poorly judged or if the data do not follow a straight line with the chosen probability paper and plotting position.

### 8.2.5 Fitting Probability Distributions

General information on probability distributions and descriptions of statistics such as mean, variance and skewness are given in the Appendix and Section 2.3.1. The Normal or Gaussian distribution is probably familiar to most scientists and engineers. The probability distribution of flood peak data, however, is rarely bell-shaped. One reason for this is that flood data have some finite lower limit. This will be zero in many streams; in others, there will be some theoretical lower limit that represents the smallest expected annual flood or is the threshold value chosen in a partial series analysis. A second reason is that flood data are usually skewed to the right (positively), with a few very high values creating a smoothly tapering 'tail' on the upper side. The flood data for the Acheron River are shown as a frequency histogram in Figure 8.3.

The 'true' underlying mathematical law to which a stream's flood data conform will never be known abso-

lutely. But, we can *assume* that the data closely follow a certain distribution and *test* how well they are described by it. By fitting a theoretical distribution to flood data, statistical methods such as confidence limits and tests of hypotheses can be employed to assist in the interpretation of flood-frequency curves. Inferences can thus be made about how 'good' a flood estimate is. In general, a probability distribution will provide the most accurate estimates near the middle values; less so near the tail ends.

Although no one distribution will fit all flood data, specifying the distribution used and the method of fitting it will allow other researchers to obtain the same results from the same set of data. The procedure is thus much more objective than graphical methods using eye-fitted curves. An additional benefit of using probability distributions is that the estimated parameter values compactly summarize the characteristics of the distribution. Often, these parameters are related to factors such as catchment area, rainfall, topography and other physiographical and meteorological measures (Haan, 1977). The estimated parameter values can be used in a regional analysis (Section 8.5.3) or as factors for stream classification (Chapter 9).

A large variety of distributions has been investigated for application to flood data, and hydrologists constantly debate the relative merits of each. Some distributions will be more appropriate for some streams than for others. In the United States and Australia the *log Pearson Type III* (LPIII) distribution has been selected as a standard by government agencies. Benson (1968), Thomas (1985) and IEA (1998) describe the criteria considered in making the decision. The LPIII distribution includes the *log-normal*

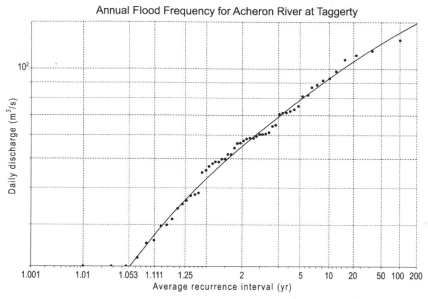

**Figure 8.4.** *Frequency analysis for the annual maximum series (daily average flows), Acheron River at Taggerty, years 1946–2001. A GEV distribution has been fitted to the data and plotted on log-normal probability axes.*

(LN) distribution as a special case. The *general extreme value* (GEV) distribution and its special cases have also been widely adopted (NERC, 1975; Wang, 1997; Rao and Hamed, 2000). The *Weibull* distribution is often used for the description of low-flow frequencies. These distributions are discussed in Section 8.2.6. In Figure 8.4, a GEV distribution has been fitted to flood data.

Probability distributions are defined by their parameters. To fit a distribution to flood peak data one must estimate the 'true' parameter values from information contained in the sample data series. The three approaches most often used for fitting probability distributions are:

- Graphical
- Method of moments (including probability-weighted moments)
- Method of maximum likelihood

In the *graphical method* plotting positions are assigned as described in Section 8.2.4. The plotting position formula and probability paper must be compatible with the distribution to be fitted. If a straight line can be fitted through the data, the parameters can be directly estimated. For example, if normal probability paper is used the 50% mark represents the mean and the 15.9% and 84.1% marks are one standard deviation from the mean.

Probability paper can be developed for particular distributions with specific parameter values by scaling the axes appropriately. For distributions in which skew is a parameter, such as the LPIII or Weibull distributions, a different probability paper must be constructed for each value of skewness. For the interested reader, Haan (1977) gives further information on constructing probability paper.

In the *method of moments*, moments of the 'true' probability distribution are estimated from the data series. An explanation of moments, which are measures describing the location, scale and shape of a distribution, is given in the Appendix. They are derived from sample statistics of the data such as the mean, standard deviation and coefficient of skewness. For hydrological variables, moments greater than third order are not generally computed. Higher-order moments are very unreliable unless extremely large sample sizes are used—much greater than the 20 years or so of streamflow record typically available.

In most frequency analyses using the method of moments, the relation between flood magnitude and probability can be reduced to a simple equation given by Chow (1964b) as:

$$x_T = \bar{x} + K_T s \qquad (8.2)$$

where $x_T$ is the magnitude of an event with an average recurrence interval $T$ years, $\bar{x}$ and $s$ are the mean and standard deviation of sample values, respectively, and $K_T$ is a *frequency factor*. To fit the distribution, a mean and standard deviation are calculated from the sample data and entered into Eq. (8.2) with the appropriate $K_T$ value. If an appropriately scaled probability paper is used only two values of $x_T$ need to be calculated (3 to be safe) to define the distribution by a straight line.

The *method of maximum likelihood* was first introduced in the 1920s by Fisher, a geneticist and statistician (Devore, 1982). For large sample sizes this method is superior to others since the resulting estimators of population parameters are considered to be more efficient and accurate. In this method a likelihood function is derived which indicates how likely the observed sample is, assuming that it is from a certain distribution with a range of possible parameter values. Maximizing this likelihood function yields parameter values which agree most closely with the observed data (Devore, 1982).

Solving the maximum likelihood equations to obtain parameter values normally requires an iterative procedure. Since an efficient estimate will not necessarily exist, a solution may or may not be found.

*Probability weighted moments* (PWMs) are similar to conventional moments in that they summarize and describe the characteristics of a probability distribution. Hosking (1989) defines a type of PWMs called L-moments, in which the moments are linear functions of the data values, hence 'L'. In comparison to conventional moments in which the data values are squared, cubed, etc., with L-moments it is the probabilities which are manipulated. This gives less weight to the very high or very low data values.

The advantage of L-moments is that they are less sensitive to sampling variability and less subject to bias. They are robust in the presence of outliers, meaning they give consistent results even if the extreme values contain measurement errors. For small samples they produce parameter estimates which are sometimes more accurate than even maximum likelihood estimates (Hosking, 1989).

### 8.2.6  A Few Good Probability Distributions

In general, the more parameters a distribution has, the better it will fit a set of data and the more flexibility it has for fitting many different sets of data. However, for the amount of data normally available the reliability of estimates of more than two parameters may be very low. 'Thus a compromise must be made between flexibility of the distribution and reliability of the parameters' (Haan, 1977, p. 146).

The distributions presented are those with the widest use in hydrology, and some of their more common applications will be discussed. NERC (1975) and Rao and Hamed (2000) provide comprehensive summaries for further information.

### Normal Distribution

The normal distribution is a two-parameter distribution, defined by its mean and variance. The $K_T$ value in Eq. (8.2) is simply the standardized normal variate $Z$ tabulated in most statistical texts. In the Cunnane plotting position (Table 8.2) $\alpha = 0.375$ for the normal distribution (Blom, 1958).

The mean of a normal distribution corresponds to a 0.5 probability, meaning that 50% of the values are higher, 50% lower. For an annual series this would correspond to the 2-year average recurrence interval. Haan and Read (1970) found that annual water yield from small watersheds in Kentucky followed a normal distribution.

### Log-normal Distribution (LN)

Like the normal distribution, the two-parameter log-normal distribution is defined by its mean and variance. However, in this case it is the logarithms of the data that follow a normal distribution. The data are usually transformed by taking either natural (base-$e$) or base-10 logarithms. The mean and standard deviation would be calculated using these transformed values, and Eq. (8.2) becomes:

$$\log x_T = \overline{\log x} + K_T s_{\log x} \qquad (8.3)$$

where 'log' can represent either base-10 logarithms ($\log_{10}$) or natural logarithms (ln). The $K_T$ value is the same as for the normal distribution. When a value of $\log x_T$ is computed, the antilog is taken to convert back to original units:

$$x = 10^{\log_{10} x} \qquad \text{(for base 10 logarithms)} \qquad (8.4)$$

or

$$x = e^{\ln x} \qquad \text{(for natural logarithms)} \qquad (8.5)$$

As mentioned in Section 5.8.5, the log-normal distribution is often used in the description of sediment size data; it also applies to the distribution of raindrop sizes (Chow *et al.*, 1988). Flood peak data and low-flow data also often follow a log-normal distribution. In the Cunnane plotting position (Table 8.2), $\alpha = 0.375$ for this distribution.

### Log-Pearson Type III Distribution (LPIII)

The log-Pearson Type III distribution is a three-parameter distribution. It lends itself well to flood data which tend to have a lower limit but no upper limit. It also contains the log-normal distribution as a special case when the skew coefficient of the logarithmic transformed data is zero. These reasons as well as the flexibility of a three-parameter distribution have made the LPIII popular for flood-frequency analysis.

Estimates of the moments are computed from the logarithms of the data series similar to the procedure for the LN distribution. $K_T$ values are provided by Haan (1977), IEA (1998) and Linsley *et al.* (1975b). Equations (8.4) or (8.5) is used to convert back from the log domain. For the Cunnane plotting position, $\alpha = 0.4$ (Srikanthan and McMahon, 1981).

A negative skew value, which is common for Australian flood data, means that the LPIII distribution has an upper bound. This can theoretically cause difficulties if one wants to estimate, say, the 1000-year flood. However, for most applications it is not a problem (IEA, 1998). Since small sample sizes tend to give unreliable estimates of the coefficient of skew, the US Water Resources Council (IAC, 1982) recommends using a regionalized estimate. This regional skew coefficient would be used to obtain the $K_T$ value for Eq. (8.2). Chow *et al.* (1988) give this procedure as well as a US map showing generalized skew coefficients of annual maximum streamflows.

### The Extreme Value (EV) Distribution

The largest or smallest values—the extremes—are of particular importance in hydrology. Examples would be the maximum annual peak flow, the minimum 10-day flow or the minimum monthly flow. The extreme value distribution is based on the idea that these extreme values are taken from the ends of some 'parent' distribution. If the fur thickness on water rats were considered as an analogy, the 'parent' population might be the fur lengths over the whole rat and the 'extreme values' the lengths on just the tails or noses. The distribution of these extreme values is dependent on the sample size as well as the parent distribution.

Three types of extreme value distributions have been developed (Haan, 1977):

*EVI*    distribution has no upper or lower limits;
*EVII*    distribution is bounded on the lower end;
*EVIII*   distribution is bounded on the upper end.

These three forms are special cases of a three-parameter distribution called the *general extreme value* (GEV) distribution (Jenkinson, 1955). This distribution which is fitted using probability-weighted moments is the method of flood-frequency analysis included in AQUAPAK, and a sample output is shown in Figure 8.4. The plotting position for the GEV is given in Table 8.2.

The EVI distribution has a constant skewness of 1.1396; for the EVII, it is higher and for the EVIII, it is

lower. There has been little interest in the EVII distribution in hydrology. The EVI distribution is often used in flood-frequency analysis, and a form of the EVIII distribution is commonly used in the analysis of low flows. Only the EVI and EVIII distributions will be discussed here.

### EVI distribution

An EVI distribution is also called the Gumbel or double-exponential distribution. It is described by two parameters, a scale parameter and a location parameter, where the latter is the mode of the distribution. As mentioned earlier, the coefficient of skew is a constant, 1.1396. The log-normal distribution is a special case of the EVI distribution when the coefficient of variation is 0.364. Also, the average recurrence interval for the mean value is a constant 2.33 years. It is from this relationship that the *mean annual flood* is often defined as a flood with an average recurrence interval of 2.33 years.

The EVI has been found to be satisfactory for describing the distribution of yearly maximum daily discharges (Haan, 1977). Frequency factors ($K_T$ values) for the EVI are given by Haan (1977) and Chow (1964b), and Gumbel paper is commercially available. For the Cunnane plotting position, $\alpha = 0.44$ for the EVI distribution (Gringorten, 1963).

### EVIII Distribution

In hydrology, the EVIII distribution has most often been applied to the analysis of low flows, particularly the annual minimum flows. This distribution is particularly attractive because of its fixed limit. By 'turning over' the curve the limit becomes a lower rather than an upper limit. This is accomplished by fitting the distribution to $(-x)$ rather than $(x)$. For this special case, it is known as the *Weibull* distribution, after Weibull, who first used it in an analysis of the strength of materials.

The Weibull distribution has three parameters: (1) a lower limit (zero or some low finite value), (2) a location parameter and (3) a skewness parameter. The lower limit has been used as the *probable minimum flow*; i.e. the stream is never expected to drop below this value if the same hydrological conditions are maintained. The location parameter defines the low-flow value which will be exceeded (low flows will be higher than this value) about 37% of the time (recurrence interval of 1.58 years). It is sometimes called the *characteristic drought*.

The data used in a low-flow analysis are the annual low flows of some duration (e.g. the lowest 1-day or lowest monthly flow in the year). If seasonal flows are of concern, the low-flow values can be selected from a particular season within each year. Generally, the number of con-secutive low-flow days is set equal to 1, 7, 15, 30, 60, 120, 183 and 284 to cover the range of possible durations. AQUAPAK provides for low-flow analysis using the Weibull distribution in which the parameters are fitted by the method of probability-weighted moments.

As for flood-flow analysis, it is important that the data are independent. For long durations they will inevitably contain serial dependence. Again, a calendar year may not be an appropriate time base because a drought period may cross the December–January boundary, especially in the southern hemisphere. The streamflow data should be divided into climatic years which start at the time of year when the flow is likely to be high and seasonal low-flow periods are not bisected (WMO, 1983). Using a 'climatic year' which is 6 months out of phase with the water year is one approach. The starting month for the year can be specified as an option in AQUAPAK.

The first few (most frequent) values in a low-flow frequency curve often plot as a much steeper line than the remaining values (IH, 1980). It can be assumed that the break in the curve indicates the point where flows are no longer considered 'drought flows', and should be excluded from the frequency analysis. IH (1980) indicates that this break tends to occur at an exceedance probability of around 65%. However, for Victorian streams, Nathan and McMahon (1990c) adopted a value of 80% as being more appropriate. The distribution is then fit only to those flows with exceedance probabilities less than this thresh-old value. In the same manner as for zero flows (Section 8.2.7), a correction is applied to the probabilities to account for the excluded flows. In AQUAPAK the Weibull distribution is fitted only to the lower 80% of flow data.

As with other types of analysis, extrapolation beyond the record should be done with caution. However, Gumbel (1963) found that satisfactory forecasts could be made up to an average recurrence interval of $2N$ on several American rivers.

Some countries base water quality standards on low-flow conditions such as the 7-day, 2-year low flow (McMahon and Mein, 1986). Bovee (1982) also states that the 7-day, 10-year low flow is often considered a 'minimum streamflow' for determining reservoir storage for water supply and in the design of wastewater treatment.

A sample of the output from AQUAPAK is shown in Figure 8.5. From the graph, the 7-day, 2-year low flow (probability 50%) is about 12 m$^3$/s-days (as a 7-day total) or an average daily flow of $12/7 = 1.7$ m$^3$/s. This means that a 7-day period in which the flow averages 1.7 m$^3$/s occurs as a minimum about every 2 years (or conversely, there is a 50% chance of this occurring in any one year).

### 8.2.7   Zeros and How to Treat Them

In hydrology, data often have zero as a lower limit—negative streamflows making little sense except perhaps where water runs upstream in an estuary. However, zeros

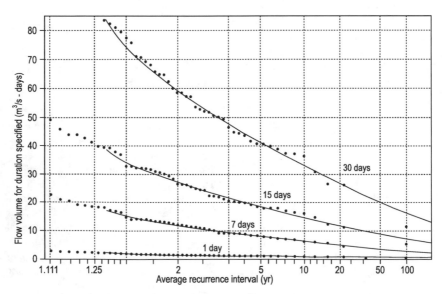

**Figure 8.5.** *Frequency analysis for the annual minima series, Acheron River at Taggerty, years 1946–2001. A Weibull distribution has been fitted to the data and plotted on normal probability axes*

present a problem because they cause the distribution to be truncated; i.e. $P(0)$ is not zero: ⌐‿. One approach which has been taken is to add a small, constant value to all the observations and then make a correction to the final solution. However, this can greatly affect estimates of extreme events, and as a result, the method should not be used.

Another approach uses the theorem of total probability as described by Haan (1977). For non-negative values this theorem simply says that the probability of flows equalling zero plus the probability that they are greater than zero accounts for all of the probability; or, as more concisely stated,

$$P(x = 0) + P(x > 0) = 1 \quad \text{(if all } x \geq 0) \quad (8.6)$$

To account for the proportion of zero values, all non-zero observations are analysed and the resulting probabilities are multiplied by the fraction of non-zero values in the data. For example, if one quarter of the data were zero values, and from the remaining data the probability of a flood being exceeded was calculated as 0.2, the actual probability of exceedance would be $(0.75)(0.2) = 0.15$. This relationship can be written as:

$$P(x) = cP^*(x) \quad (8.7)$$

where $P(x)$ is the probability of exceedance for all values, $c$ is the probability that $x$ is not zero $(P(x \neq 0))$, and $P^*(x)$ is probability of exceedance for the non-zero values.

For those who prefer their water glasses empty rather than full, the probability that $x$ equals 0 can be used instead:

$$P(x) = (1 - c')P^*(x) \quad (8.8)$$

where $c'$ is now the probability that $x$ is equal to zero, $P(x = 0)$. This technique can be easily adapted for threshold values other than zero, as mentioned in regard to low-flow frequency analysis, Section 8.2.6.

All data above zero or some threshold are analysed by graphical or numerical frequency analysis techniques and the calculated exceedance probabilities are adjusted using Eqs. (8.7) or (8.8). A new flood-frequency curve is then re-plotted with the adjusted values. The zero values have the effect of shifting the frequency curve to the right, making large events even more rare and discharges lower for the same probability.

*Example 8.1*

We will consider the effect of zero values on the flood-frequency curve for the Acheron River (Figure 8.4). Let us say that 6 years of very severe drought occurred after the actual period of record, during which the flow was zero.

This would give a total of 62 years of record. For the original observations, we will assume that they are distributed log-normally with $\overline{\ln x} = 3.963$ and $s_{\ln x} = 0.493$. Taking the zero flows into account:

1. What is the probability of an annual peak flow which equals or exceeds 120 m³/s?
2. What is the magnitude of the 10-year flood?

*Answer*

1. The probability of non-zero values is estimated as $c = 56/62 = 0.9032$. Then Eq. (8.7) becomes:

$$P(120) = 0.9032 P^*(120)$$

$P^*(120)$ can be obtained by solving Eq. (8.3) for $K_T$ and then finding the probability from a table of values for a standardized normal distribution.

$$\ln x_T = \overline{\ln x} + K_T s_{\ln x}$$

$$\ln(120) = 3.963 + K_T(0.493)$$

$$0.824 = 0.493 K_T$$

$$K_T = 1.67$$

Using this value in place of the Z values in a standard normal table, this gives a probability of 0.9525. This represents the cumulative area ◣. However, we want the exceedance probability, which is the area ◢. Because the total area is 1, the probability of exceedance for the non-zero values is:

$$P^*(120) = 1 - 0.9525$$
$$= 0.0475 \text{ (the inverse of this is 21, the average recurrence interval in years)}$$

Thus,

$$P(120) = (0.9023)(0.0475) = 0.0429$$

meaning that the probability of a flood of 120 m³/s being exceeded in any year is 4.3%. The recurrence interval is $1/0.0429 = 23$ years, as compared to 21 years for the non-zero data.

2. The probability of a 10-year flood being equalled or exceeded in any year is $1/10 = 0.1$. This value is entered into Eq. (8.7) along with the previous value of $c$ to obtain:

$$P(x) = 0.1 = (0.9032)(P^*(x))$$

giving $P^*(x) = 0.111$. Again, to enter the standard normal tables, a cumulative value is needed: $(1 - 0.111) = 0.889$. This yields a $K_T$ value of about

1.22. Then, from Eq. (8.3), the peak flow associated with a 10-year average recurrence interval is:

$$\ln(x_{10}) = 3.963 + 1.22(0.493) = 4.56$$

and $x_{10} = e^{4.56} = 96$ m$^3$/s (as compared with 99 m$^3$/s for the original data).

### 8.2.8   Goodness, The Distribution Fits!

From a statistical point of view we often want to know how well the chosen distribution fits the observed data. With graphical methods this is done by visually assessing the goodness of fit. Somewhat less subjectively, statistical goodness-of-fit tests are available to test the hypothesis that the observed data are actually from the fitted probability distribution. Two common methods are the chi-square test and the Kolmogorov-Smirnov Test.

According to Haan (1977), many hydrologists discourage the use of these goodness-of-fit tests when fitting distributions to streamflow data. One main reason is that these statistical tests tend to be insensitive in the tails of the distributions—the areas of most importance for prediction of extreme events. Neither of the tests given are very powerful, especially with small samples; i.e. the probability of accepting the hypothesis that the distribution fits, when in fact it does not, is very high. This may be of little comfort to those looking for an objective basis for assessing how well a distribution fits the data. However, the statistics can be of assistance when comparing the relative merit of one distribution or one set of parameters against others.

Readers wishing to apply goodness-of-fit tests should refer to appropriate references—Haan (1977) and Maidment (1993) are good starting points.

### 8.2.9   Interpreting Frequency Curves

Annual maximum frequency curves are analysed and interpreted differently to partial-duration curves. Although more points can be produced from the partial-duration series for defining the flood-frequency relationship, more points does not necessarily guarantee better estimates (NERC, 1975). One drawback of including more floods in the analysis is the increased possibility that flood peaks will be interdependent, since the conditions affecting one flood (i.e. weather patterns or soil moisture conditions) may also affect others occurring within the same season or year.

However, smaller floods will occur more frequently than indicated by the annual maximum series, because several peaks in one year may be higher than the highest flood in others (Figure 8.2). Also, some of the smaller peaks selected as annual maxima may not really be floods. Thus, for these smaller, more frequent events the partial-duration series is more appropriate.

For average recurrence intervals (ARI) of about 10 years or more the difference between the annual maximum series and the partial-duration series becomes very small. As can be seen in Figure 8.6 for the Acheron River,

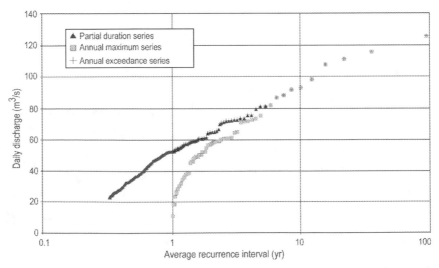

**Figure 8.6.** *Flood data for the Acheron River at Taggerty, years 1946–2001(56 years). Shown are the annual maximum series, annual exceedance series and partial-duration series with threshold = 22.6 m³/s (total of 168 values, 3/year). The data are plotted on semi-logarithmic axes. The average recurrence intervals were computed from the Cunnane plotting position with α = 0.4*

the two series converge at an ARI of about 6 years. It can also be seen that the partial-duration series is simply an extension of the annual exceedance series. IEA (1998) recommends that the partial-duration series should be used for ARI < 10 years and the annual maximum series for ARI ≥ 10 years. For the partial-duration series, with ARI < 10 years, graphical interpolation is considered sufficiently accurate. However, a probability distribution should be fitted for making inferences about the probability of rarer events.

For the annual maximum series the exceedance probability is the probability that a flood of a given size or larger will occur in any one year. The average recurrence interval is the average interval in which a flood of a given size or larger will occur as an annual maximum. Referring to the curve in Figure 8.4, the average recurrence interval of a flood of 100 $m^3/s$ (as a daily average) is about 40 years. This means that on average, the flow will exceed 100 $m^3/s$ once every 40 years. Alternatively, the exceedance probability is $1/40 = 0.025 = 2.5\%$. This is the probability that the highest flood in a year (as an average daily flow) will exceed 100 $m^3/s$.

Probabilities obtained from the annual maximum series can be used to answer questions of the form: 'What is the probability of the 10-year flood being exceeded in the next 20 years'? This is done by simple 'dice-rolling' probability methods. For example, the probability of one or more events exceeding the 10-year flood in the next 20 years is:

$$1 - (0.90)^{20} = 0.88$$

The general formula is given by Chow *et al.* (1988) as:

$$P(X \geq x_T \text{ at least once in } N \text{ years}) = 1 - \left(1 - \frac{1}{T}\right)^N$$

(8.9)

In the partial-duration series, of which the annual exceedance series is a special case, the 'average recurrence interval' is still the average interval between floods exceeding a given size. However, the floods are no longer *annual* maxima, and probability is no longer a useful concept (IEA, 1998). The annual exceedance series ($N$ values in $N$ years of record) can still be plotted on probability paper, but the probability scale should be converted to average recurrence intervals (ARI) by taking the inverse (Eq. (8.1)). For example, in Figure 8.6, the '2-year flood' from the annual exceedance series is about 65 $m^3/s$. Thus, in the 56 years of record used in the analysis this discharge was exceeded $56/2 = 28$ times. We

would expect a flood of this size to be exceeded 50 times (100/2) within a 100-year period.

If the threshold for constructing the partial-duration series is reduced to include more than $N$ values for $N$ years the series can no longer be plotted on probability paper. Plotting positions for the average recurrence interval can still be calculated from the formulae in Table 8.2, with $N$ the number of years (not the number of values). The results are plotted on semi-logarithmic axes with discharge on the linear axis and average recurrence interval on the log scale, as in Figure 8.6. Floods with average recurrence intervals of less than one year can then be evaluated. For example, from the partial-duration series in Figure 8.6 the flow exceeded every 4 months on average (ARI = 0.33 yr) is about 20 $m^3/s$. NERC (1975) provides a more rigorous explanation of the partial-duration series.

The partial-duration flood series may be practical for studying how frequently various levels within a stream channel are inundated. Gregory and Madew (1982) recommend an evaluation of the annual exceedance series when investigating changes in flood frequency due to land-use change. Their reasoning is that the less extreme discharges may show a change in frequency due to land use, whereas the largest events may not be affected.

A frequency curve is interpreted under the assumption that it is a good representation of the population from which the observed flood data were taken. When comparing frequency curves for two different streams it should be kept in mind that the curves may differ because of actual flood-generating differences or simply due to chance, or both. Reliability can be assessed by computing confidence limits and/or standard errors of estimate. Methods will not be given here, but can be found in references such as Chow (1964b), Haan (1977), IEA (1998) and NERC (1975).

## 8.3 Flow-duration Curves

### 8.3.1 General

Flow-duration curves display the relationship between streamflow and the *percentage of time* it is exceeded. They have been used for assessing the percentage of time a streamflow will provide adequate dilution for industrial wastes or sewage, for evaluating the feasibility of hydropower stations and for investigating environmental flow requirements. By evaluating these curves we can ask, for example, 'what percentage of time is the daily flow on the Acheron River above 2 $m^3/s$?' From the daily curve in Figure 8.7 it can be seen that approximately 90% is the correct answer. Thus, in a period of one year, about

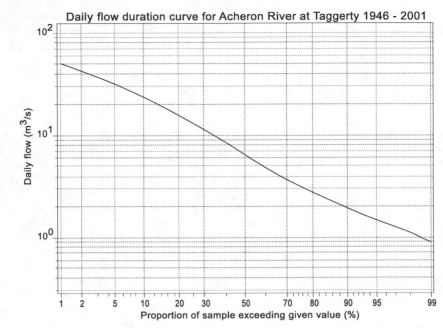

**Figure 8.7.** *Daily flow-duration curve for the Acheron River at Taggerty, based on calendar year data, 1946–2001 as computed by* AQUAPAK

329 days (0.9 × 365.25) will be above 2 m³/s and 36 days below. This discharge is also called the daily $Q_{90}$.

As compared to a flood or low-flow frequency analysis, a flow-duration curve is derived from all the data rather than just the high or low flows. The streamflows analysed are the flows over some period of time or *duration* (e.g. one hour, one day, 10 days, 6 months, or a year). The period chosen will depend on the time interval of the available data as well as the intended use of the duration curves.

Daily curves will show more of the details of variation in the data than monthly or yearly curves, where extremes are smoothed out through averaging. For evaluating the flows over the length of time required for, say, the hatching of mayflies, the migration of ducks, or the germination of sedges, specific durations such as 3 days, 10 days or 2 months may be appropriate.

These data, particularly the shorter-duration data, will be serially correlated (see Appendix) rather than being independent. Therefore, a probability distribution is not fitted to the data. Rather, flow-duration curves are simply a way of representing the historical record, and provide another method for characterizing the pattern of stream-flows.

### 8.3.2  Constructing Flow-duration Curves

Computationally, the technique is much like the sieving of sediment. Flows of a given duration are 'sifted' into class intervals based on size. Searcy (1959) recommends 20–30 intervals. The number of items within each class interval is counted and expressed as a percentage of the total number of items 'sifted' (either daily, monthly, yearly, or *n*-day flows). Beginning with the largest discharge, percentages are summed to obtain cumulative percentages of time above the lower limit of each class. In an alternative computerized method, the flows can be ranked from highest to lowest and assigned a plotting position using the Weibull method (Table 8.2). The highest flow would have rank 1.

When the cumulative frequencies have been computed the data are usually plotted on log-normal probability paper. This tends to linearize the relationship, which helps to define the tail ends of the curves. For water-supply analysis the values are sometimes plotted on arithmetic paper, since the area under the curve is then equal to the total volume of water flowing by the gauging station.

It should be emphasized that the sequence of flows and the time of year in which they occur are not considered. It makes no difference whether July's data are dumped into the 'sieves' before January's or whether the lowest flows occur in one month or are spread across all years. Because flow-duration curves are based on the whole record, the data are essentially averaged. Variability within the time period (e.g. diurnal fluctuations within a day or seasonal fluctuations within a year) is not expressed. It may be desirable to supplement the 'whole record' curve with duration curves for the single wettest and driest years on record to show the extremes. If the data record is of sufficient length it can also be divided into separate 'samples' and flow-duration curves developed for each month or season.

Sometimes the flows are standardized as percentages of the average annual flow. This allows the comparison of curves from streams of different size. Searcy (1959) also suggests standardizing the discharges by dividing by the drainage area and expressing them in units of cubic metres per second per square kilometre. Class intervals are based on 1%, 2%, 5%, 10%, 20%, ..., 80%, 90%, 95%, 98% and 99% exceedances. The number of intervals actually used in the analysis is based on the record length. For example, if a data series has less than 100 values the 1% exceedance (1/100) cannot be evaluated.

### 8.3.3 Interpretation and Indices

Flow-duration curves are a convenient way of portraying the flow characteristics of a stream under natural or regulated conditions. At the high-flow end of the curve they can give an indication of the duration of overbank flows; at the lower end, they can be used to estimate the amount of time a particular log or rock is under water. The shapes of the curves also summarize the flow characteristics of a stream for comparison within or between catchments.

Typically, the curves flatten out as the duration increases from daily to yearly. As with a moving average (Section 4.4.2), increasing the amount of time over which the data are averaged tends to reduce the variability. The slope of the flow-duration curves reflects the catchment's response to precipitation. Stable streams such as the Acheron will tend to have monthly curves that differ little from daily ones in terms of slope. For 'flashy' streams they will differ considerably.

The daily curve for flashy streams will tend to have a steep slope at the high-flow end. A flatter high-flow end is characteristic of streams with large amounts of surface storage in lakes or swamps, or where high flows have mainly resulted from snowmelt. The low-flow end of the curve is valuable for interpreting the effect of geology on low flows. If groundwater contributions are significant, the slope of the curve at the lower end tends to be flattened whereas a steep curve indicates minor baseflows (McMahon, 1976). Streams draining the same geologic formations will tend to have similar low-flow duration curves (Searcy, 1959).

Flow-duration curves can provide a number of indices to characterize the stream for classification or regionalization purposes. If two stream sites are to be compared, it is important that the same period of record is used in deriving the duration curves. The median ($Q_{50}$) value (the flow exceeded 50% of the time) gives an 'average' measure. The $Q_{90}$ or $Q_{95}$ values are commonly used as low-flow indices. The ratio $Q_{90}/Q_{50}$ can be used as an index of baseflow contribution. At the high-flow end of the curve a value such as the $Q_{30}$ or $Q_{10}$ may be of interest in indexing the length of time a floodplain is underwater or as a variable that might correlate with channel dimensions. Harvey (1969), for example, suggests that the duration of floods of a given frequency generally increases in the downstream direction.

Measures of variability can be obtained from the slope of the flow-duration curves. One index, an estimate of the standard deviation of the logarithms of the streamflows ($s_{\log x}$), is given by IH (1980) as

$$s_{\log x} = \frac{\log Q_5 - \log Q_{95}}{3.29} \qquad (8.10)$$

Another measure is Lane's variability index (Lane and Lei, 1950), defined as the standard deviation of the logarithms of the $Q_5$, $Q_{15}$, $Q_{25}$, ..., $Q_{85}$ and $Q_{95}$ values. This index is unsuitable for streams where zero flows comprise 5% or more of all the flows.

The indices apply only to the given duration, whether daily, monthly, annual or $n$-day. For example, the $Q_{90}$ for the annual duration curve is the flow which is lower than 90% of the annual values; the $Q_{90}$ for the 10-day flows is the flow which is less than 90% of the 10-day flows. The standard deviation or Lane's index is also associated with a particular duration curve. Thus, the duration should be stated when citing index values.

Duration curves of sediment, turbidity, water hardness or other water quality characteristic may also be derived if a relationship is established between streamflow and the characteristic of interest (e.g. the sediment rating curve, Section 7.5.4). A flow-duration curve must first be developed and then the discharge values are simply converted to the water quality measure by using the appropriate relationship. The amount of error depends on both the

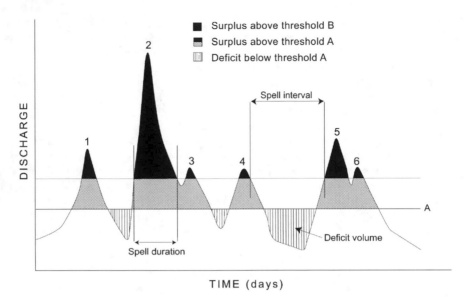

*Figure 8.8.* Generalized diagram of the flow-spell analysis technique

flow-duration analysis and on the correlation between discharge and the water quality measure.

In Section 8.5 methods of estimating flow-duration curves for ungauged stream sites and gauged sites with short record lengths are presented. Further information on the flow-duration curve technique is given by IH (1980) and Searcy (1959).

## 8.4    Flow-spell Analysis

Flow-spell analysis is a procedure developed by the UK Institute of Hydrology (IH, 1980) for the analysis of low-flow periods. Flow-duration curves (Section 8.3) give no information on how the low-flow days are distributed, and streams with similar flow-duration curves may be very different in the way low flows are grouped into long or short periods of time. Flow-spell analysis, in contrast, considers how long a low flow (below some threshold) or a high flow (above some threshold) has been maintained and how large a 'deficit' or 'surplus' has been built up, and thus takes into account the sequencing of flows.

The technique has been used for estimating the amount of storage needed on a catchment to maintain water supplies, and in checking the representativeness of synthetically generated streamflow time series (McMahon and Mein, 1986). In the analysis of flows for ecological and water quality requirements it may be of more value than

flow-duration curves. Donald *et al.* (1999) have identified key components of the flow regime that are considered to be ecologically important and can be defined by spell analysis.

A graphical description of the method is shown in Figure 8.8. Two main measures are obtained from a flow spell analysis: *spell duration* and *deficiency volume*. For low-flow analysis, the spell duration is the length of time that the streamflow is continuously below a threshold and the deficiency volume is the amount of water that would be required to keep the stream at the threshold level. *Annual frequency* refers to the proportion of years in which a deficit volume or spell duration is exceeded. To analyse the spell data using this method, the *longest* spell duration or *largest* deficit volume below a given threshold is found for each year. These are the *annual spell maxima*.

A flow-spell analysis can also be used to consider the length of time and the stage or volume above a given threshold. Such an analysis is particularly useful in establishing the strength of the connectivity between a river reach and its adjacent floodplain or in examining, for example, the effect on the aquatic environment of pumped withdrawals from an unregulated stream.

Figure 8.8 shows that there are six spells above threshold B yet only four above threshold A. This apparent inconsistency can be partly overcome by specifying a minimum spell interval at which adjacent events are considered independent. In Figure 8.8 if the spell intervals between events 2 and 3 and events 5 and 6 were small,

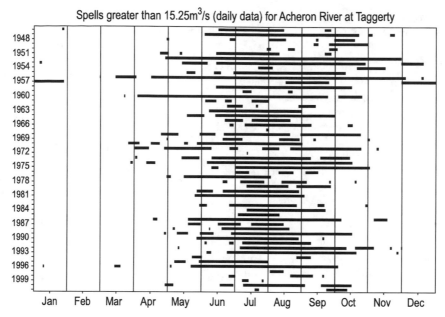

**Figure 8.9.** *Periods of spells greater than 15.25 m³/s for the Acheron River at Taggerty, daily flow 1946–2001*

then depending on the environmental variable being considered the successive events may be determined as dependent. In this situation events 3 and 6 would be considered as part of events 2 and 5, respectively.

In AQUAPAK the threshold discharge can be set at any value and the results of the analysis are plotted as a bar chart. A sample is shown in Figure 8.9. This plot reveals that if, based on environmental considerations, the minimum winter flow (southern hemisphere) threshold was set at 15.25 m³/s (50% greater than the mean daily flow), then potential pumping from the stream would be restricted to the periods shown. Results can also be tabulated as an output file for carrying out a frequency analysis. Nathan and McMahon (1990a) recommend fitting a log-normal distribution to spell maxima.

## 8.5 Extrapolating from the Known to the Unknown

### 8.5.1 General

In practice, gauged streamflow records will often be limited or non-existent at the site of interest, particularly on small streams. The typical problems are therefore (1) the development of streamflow estimates for sites which have no records and (2) the extension of short-term records.

Both problems can be addressed with regression methods (see Appendix). A regression model can be developed which relates a short-term record to the corresponding record from a long-term station to improve estimates of measures such as the mean, variance, 20-year flood or the daily $Q_{90}$. Relationships between discharge and catchment and/or rainfall characteristics can also be developed for a region. This allows the estimation of streamflow measures for ungauged sites within the same region. To be more specific, the region must be *hydrologically homogeneous*, meaning that the catchments should be as similar as possible in terms of their water-producing characteristics (e.g. rainfall patterns, vegetal cover, land use, geology and topography). For example, a rainfall measuring station in the rain shadow of a mountain range should not be used for predicting streamflows on the other side (Searcy, 1959). In some techniques, large and small catchments can be grouped together, but as a rule, they behave very differently hydrologically.

Section 8.5.2 covers methods of extrapolating data from one gauged site to another and Section 8.5.3 covers regional methods. In both sections, general methods for estimating streamflow characteristics will be given first, followed by methods of deriving flood-frequency and flow-duration curves for short-term and ungauged sites. Homogeneous regions can be defined using the methods of numerical taxonomy in Section 8.6 and stream and catchment classification in Chapter 9.

### 8.5.2    Transposing Data from Gauging Stations to Less-endowed Sites

#### Infilling Missing Data and/or Extending a Record

Simple linear regression is one method of extending or infilling missing data using records from another site. The dependent variable ($Y$) represents records from the short-term or intermittent site, and the independent variable ($X$) the long-term or 'index' site. Maps showing correlation coefficients between the short-term site and surrounding stations can be developed to assist in the choice of an index station.

General guidelines for simple log-log regression (see Appendix) should be followed in developing a relationship between the short-term and index station records. However, the relationship will not always be consistent over the entire range of data. A break in the slope may occur at a point where one stream goes dry and the other does not, or where the relative proportions of baseflow and direct runoff shift. In these instances, an 'eye fitted' curve or several line segments may be preferable to a regression line. A line of 1:1 slope is sometimes plotted with the data. For higher flows, the relationship between the streamflows at two sites tends to parallel this line (Searcy, 1960).

The resulting curve can be used to generate data for missing periods or to extend the short-term record to the length of the long-term one. A major assumption is that the relationship for the short concurrent period remains valid over the entire period of estimation. The magnitude of error can be assessed using one or more 'test' sets of data which were not included in the regression analysis.

#### Frequency Curves for Floods and Low Flows

Flood or low-flow frequency curves can also be translated from long-record stations to produce estimates for ungauged sites and improve estimates for short-term stations. If records from several nearby stations are available, regionalization (Section 8.5.3) is a superior method. With only one other station the flood frequency curve developed for the long-term station is transposed to the site of interest using regression methods.

To obtain information at a study site, a recording or crest gauge can be installed and operated for a few seasons or a few years to obtain records for correlation with the long-term station. The slope-area method can also be used to estimate the magnitudes of floods from the recent past (Section 5.6.6).

The steps for transposing frequency curves from long-term (index) stations to short-term ones are described by McMahon and Mein (1986) as follows:

1. For both the short-term station and the index station a frequency curve (annual maximum, partial-duration or low flow frequency) is constructed for the *concurrent* period of record.
2. From these curves, discharges are selected for several average recurrence intervals (e.g. 0.5, 0.8, 1.1, 1.5, 2, 5, 10, 20 and 50 years) for both stations. Discharges for the short-term station are plotted against those for the index station on log-log paper.
3. A curve is drawn through the points, giving additional 'weight' to values in the region of most interest (high, medium or low flows). The line will generally approach a 1:1 slope for the higher values.
4. A frequency curve is developed from the full record for the index station. Using this curve, specific discharges are again selected for several recurrence intervals and adjusted to the short-term site using the log-log relationship.
5. A new frequency curve for the short-term site is drawn through the translated points.

To standardize the discharge values they can first be divided by catchment area. A 1:1 slope in step 3 then indicates that both catchments are producing the same amount of runoff per unit area. Error is dependent on both the accuracy of the index frequency curve and the relationship between the sites.

The regional flood-frequency method described in Section 8.5.3 can also be abbreviated to extend or verify the flood-frequency curve at a particular station. For example, if a curve for a long-term station indicated that the 1965 maximum flood was a 50-year flood, it can also be assumed as a rough guideline that the 1965 flood at the short-term station was also 50 year. A regional flood-frequency curve (see Figure 8.11) can be used to improve the estimates of rarer floods at short-term stations.

#### Flow-duration Curves

Flow-duration curves for short-term stations can be improved using records from long-term stations in the same manner as for flood-frequency curves. Again, the procedure requires a number of concurrent measurements. However, with flow-duration curves the measurements are not restricted to, for example, annual maxima or annual low flows. Spot measurements taken over a period of one or two years and a range of flows may be adequate for general estimates.

The procedure is exactly the same as in the previous section except that in step 2 the discharges are selected for a number of percent duration values (e.g. 10, 20, 30, . . . ,

80, 90). The curve drawn through the points plotted on log-log paper at step 3 will normally be curved in the lower range due to response differences in the catchments from differences in geology and topography. For the higher range of flows the curve often tends towards a 1:1 slope as baseflow becomes insignificant. In step 4 an improved flow-duration curve for the short-term station is obtained by selecting discharges of specific percent duration values from the index curve and adjusting them using the relationship developed in step 3.

To standardize the flow-duration curves, the discharges can first be converted to cubic metres per second per square kilometre by dividing by catchment area. This eliminates differences in mean annual flow due to catchment size. Synthetic duration curves can be approximated as a straight line on log-probability paper by developing estimates for the median flow (the $Q_{50}$) and the standard deviation (which defines the slope). Dingman (1978) is an additional source of information on this procedure.

### 8.5.3   Regionalization

The regionalization of streamflow characteristics is based on the premise that areas of similar geology, vegetation, land use and topography will respond similarly to similar weather patterns; i.e. a weather pattern that produces large floods or droughts in one catchment will likely have the same effect on nearby ones. Regional analysis methods are useful for strengthening estimates of rare events at gauged sites, both short- and long-term, as well as providing a means of estimating flow characteristics at ungauged sites. Additionally, they can be used to identify sites which do not 'belong' in the region and to detect anomalies in gauged streamflow data.

*Rough Estimates of General Parameters*

For ungauged sites, rough estimates of the mean annual flow can be made from nearby gauged sites simply by adjusting for the difference in area, e.g

$$\bar{x}_1 = \bar{x}_2 \left( \frac{A_1}{A_2} \right) \qquad (8.11)$$

where $\bar{x}_1$ = mean annual flow (volume units) for the ungauged site, $\bar{x}_2$ = mean annual flow (volume units) for the gauged site, and $A_1$ and $A_2$ are the areas of the ungauged and gauged catchments, respectively. More generally, this equation becomes

$$\bar{x}_1 = \bar{x}_2 \left( \frac{A_1}{A_2} \right)^a \qquad (8.12)$$

where $a$ is generally less than 1.0. For Australian streams, McMahon (1976) gives a value of 0.65.

The coefficient of variation can be similarly estimated by adjusting for differences in the mean annual runoff

$$C_{V_1} = C_{V_2} \left( \frac{\bar{x}_1/A_1}{\bar{x}_2/A_2} \right)^b \qquad (8.13)$$

where $b$ is generally less than 0.0. McMahon (1976) gives a value of $-0.33$ for Australian streams. It should be noted that in Eqs. (8.11) and (8.12) the ratio of the mean annual *flow* (MAF) is used (total volume over the year), whereas in Eq. (8.13) mean annual *runoff* (MAR) is used (the total flow per unit area). MAF tends to increase with catchment area whereas MAR does not. Regional relationships using gauged records can be developed to obtain values of $a$ and $b$.

*Multiple-regression Approaches*

There have been a multitude of studies in which multiple regression has been applied to the prediction of discharge from catchment, channel and rainfall characteristics. Where no flow records exist at a site, a common approach for estimating streamflow measures is to develop relationships between gauged data and measures obtained from maps or field measurements.

Some of the factors which have been considered include catchment area, channel slope, drainage density, elevation, stream length, mean annual precipitation, mean summer air temperature and percentage of the area covered by forest, swamps, lakes or permeable rock. Many of the factors are intercorrelated, and often the equation is 'boiled down' to a few representative variables. Comparative studies have shown that catchment area, precipitation and geology are typically the most important characteristics.

Regression relationships can be developed between these factors and any of a number of streamflow indices derived from gauged data. These might include the mean annual flow, mean annual flood, annual coefficient of variation, flood peaks of given recurrence intervals, low-flow indices or the $Q_{50}$ or $Q_{90}$ from flow-duration curves. McMahon and Mein (1986) summarize studies using regression methods in low-flow hydrology.

A regression analysis provides (1) insight about which factors have the most influence on flow attributes and (2) an equation which allows the prediction of these flow attributes for ungauged catchments. The coefficients obtained will be specific to the hydrologically homogeneous area from which they were derived. Their reliability for predicting ungauged flow parameters can be tested by collecting a range of actual stream flow measurements at an ungauged site or by reserving the record from one or more gauging stations to act as test data sets.

For example, in a study of 81 catchments smaller than 250 km$^2$ in southeastern Victoria, Australia, Gan *et al.* (1990) found that a regression equation based on catchment area and mean annual rainfall explained 97% of the variance in mean annual streamflows. The equation they established was

$$Q = 9.3 \times 10^{-6} \, A^{0.99} R_{\mathrm{m}}^{1.48} \qquad (8.14)$$

where $Q$ is the mean annual streamflow (million m$^3$), $A$ is the catchment area (km$^2$) and $R_{\mathrm{m}}$ is the mean annual rainfall (mm). Using the Marysville, Victoria precipitation average of 1380 mm (see Appendix), the equation would give a mean annual streamflow of 10.3 million m$^3$ for Site 3 on the Acheron River (catchment area 25.8 km$^2$; see Figure 4.7). This compares to a value of 41.0 million m$^3$ obtained using Eq. (8.12) using data from the gauged site at Taggerty (mean: 323.40 million m$^3$, catchment area: 619 km$^2$) and $a = 0.65$. The lack of agreement illustrates the difficulty in estimating flows for small, ungauged streams.

### Envelope Curves

To examine flood-producing properties of a catchment and to estimate the maximum expected flood on ungauged streams, envelope curves are a useful tool. Within a given region, the *highest observed discharge* at all gauging stations is divided by the corresponding catchment area and then plotted against area using log-log axes. A curve that forms an upper bound to the data is called an *envelope curve*. The graph provides a summary of the flood magnitudes experienced in a region. In Figure 8.10, points have been plotted for several stations in the vicinity of the Acheron River. As can be seen from the graph, the highest

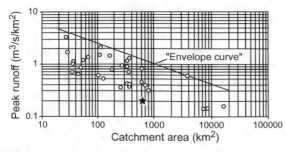

**Figure 8.10.** *Envelope curve for stream gauging stations in the vicinity of the Acheron River. The points represent the highest instantaneous flow on record for each site. The flood for the Acheron River is shown as a star*

flood recorded on the Acheron River is low for its catchment size in comparison with other stations in the region. This may be due to the fact that flood-generating characteristics for the Acheron basin are different, or that none of the floods recorded during the 56 years of record have been rare events (e.g. the 100-year flood).

By labelling each point with a catchment name, trends in the flood data with elevation, aspect, catchment slope and location can be examined. Major floods recorded at longer-term stations will also reveal the effects of short record lengths. As more data are collected the envelope curve will shift upwards. Thus, the method is not satisfactory for accurate flood estimation but can be used for preliminary estimates as well as for checking whether flood estimates obtained by other methods are realistic (IEA, 1998).

### Regional Flood Frequency

Combining data from a number of catchments will help to reduce the uncertainty resulting from short records. In this way, the record is effectively 'lengthened' by extending it across space rather than time (Linsley *et al.*, 1975b). Stations should be from a region that is homogeneous in terms of its flood-producing characteristics and those with very short records or regulated flows should be excluded.

Even for sites with long-term records, a regional flood-frequency analysis is recommended for improving the estimates of rare floods. For ungauged sites or stations with only a few years of record, this is essential. As mentioned previously, parameters based on higher moments such as skewness will not be good estimates of the true population parameters if they are calculated from short records. Regional skew values can be obtained by pooling information from other sites in the region. Generalized maps of skew coefficients for the LPIII distribution have also been published. Chow *et al.* (1988) give a map with regional skew values for the USA.

If a region is relatively homogeneous, flood-frequency curves should have approximately the same slope. Curves from several stations can thus be put on the same scale and superimposed to develop a composite regional curve. From this curve, estimates of extreme events can be made for ungauged and short-record sites. The method presented here is the Index Flood Method, given by Dalrymple (1960). Other variations on this method are given by IEA (1998) and NERC (1975).

The first step is to assemble records from a number of stations and establish a standard record length or base period. Since the common record length will usually be very short, the normal approach is to 'fill in' some of the peak flows. Estimates are obtained from another longer-term station by a regression analysis of annual peak flows. A certain amount of judgement is necessary in deriving the standard base period. For example, if only one station has a record of 100 years, with all others being less than 25 years, not much is gained by using regression to put all stations on a 100-year basis. An analysis of the 100-year station alone would produce as much information.

All flood peak data are divided by an *index flood* discharge, usually the mean annual flood, to make the quantity dimensionless. The mean annual flood is commonly assumed to be the flood with an annual recurrence interval (ARI) of 2.33 years (see EVI distribution, Section 8.2.6). This can be changed to reflect the chosen probability distribution (e.g. for a normal distribution, the mean has an ARI of 2 years). Bankfull discharge (Section 7.2.5) can also be used as the index flood, which allows development of frequency curves for ungauged sites using field estimates of bankfull discharge. This is based on the assumption that bankfull flow is exceeded on different streams with about the same ARI (Leopold *et al.*, 1964).

The values of $Q/Q_{index}$ for each station are plotted against probability on the same graph, and a composite curve is fit through all the points by eye, as shown in Figure 8.11. Alternatively, WMO (1983) recommends using the median value of $Q/Q_{index}$ at several probability levels to establish the curve. For the regional curve, some departure from linearity is acceptable if it appears to improve the fit (NERC, 1975). The line should always

pass through the index flood probability at a $Y$ value of 1.0, as shown.

The procedure is based on the assumption that all streams have the same variance in $Q/Q_{index}$. Floods which plot as 'outliers' on individual curves will still plot as outliers on the composite curve, but regionalization will assist in obtaining better estimates of the average recurrence interval for these floods. Scatter in the data may be due to the variability of short record lengths or actual within-region differences in geology or climate. Individual curves which exhibit slopes noticeably different from the trend of the others may not belong in the region. A homogeneity test given by Dalrymple (1960) can be performed on sets of flood-frequency curves in a region to determine whether they are similar.

To apply the regional frequency curve to ungauged sites, an estimate of the index flood must first be made. If the index flood is the bankfull flow, this is done using field measurements. If it is the mean annual flood, it is estimated by developing a multiple-regression relationship between the mean annual flood at gauged sites and catchment and/or climatic factors. The equation is often just a simple log-log relationship between the mean annual flood and drainage area (NERC, 1975; WMO, 1983). Scatter in the plot may indicate that other factors besides area should be considered or that the stations are from a non-homogeneous region. Once an estimate of the index flood for the ungauged site is obtained, the value is multiplied by the $Q/Q_{index}$ ratios from the regional curve to develop a frequency curve for the ungauged site.

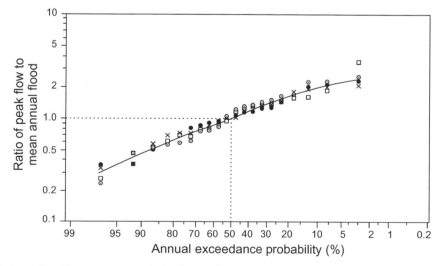

**Figure 8.11.** *Regional flood-frequency curve derived using four stations in the vicinity of the Acheron River with similar catchment sizes. The solid dots represent the Acheron data. The fitted curve must pass through the point (index flood probability, 1.0), indicated by the dotted line. Base period is 1962–81*

### *Regional Flow-duration Curves*

Flow-duration curves from several stations within a region can also be drawn on the same scale and superimposed as for flood-frequency curves. The regional flow-duration curve allows 'smoothing' of local effects such as thunderstorm-produced events that occur in one basin and not another. Again, the discharge values are divided by an 'index' flow such as the median ($Q_{50}$) value, or by the catchment area. The individual curves are then superimposed and averaged. The regional curves will be most reliable towards the middle and less reliable at the extremes of low or high flow.

Depending on how the regional curve was constructed, a duration curve for an ungauged station can be developed by multiplying back by either catchment area or an estimate of the index flow. The index flow would be estimated from catchment, channel and/or climatic factors using multiple regression methods, as for the flood-frequency curves.

### *Regional Flow Spell Curves*

Nathan and McMahon (1990a) developed a method for regionalizing spell duration and deficiency volume frequency curves. Because the data are considered log-normally distributed, the frequency curves plot as straight lines on log-normal probability paper. Thus, to define a curve, only two points need to be estimated. Nathan and McMahon suggest developing regional equations using multiple-regression analysis for the prediction of 2- and 50-year events. These points are plotted on log-normal probability paper and a straight line is drawn through them.

In their study, catchments were first divided into hydrologically homogeneous groups and regression techniques used to select and weight the most important variables. Preliminary groupings were first obtained using cluster analysis (Section 8.6.4), and Andrews curves (Section 8.6.4) were applied to fine-tune the groupings. For the multiple-regression equations, they found that the most important variables were mean annual rainfall and estimated ratio of baseflow to total streamflow.

## 8.6    Numerical Taxonomy: Multivariate Analysis Techniques

### 8.6.1    General

In science, classification is a fundamental principle for imparting order onto the diversity of nature: biologists classify organisms, geographers classify landscapes and psychologists classify personalities. *Numerical taxonomy* is a mathematical procedure for classifying items based on how numerically similar they are to others within the same class and how different they are from items in other classes (e.g. bugs with six legs are insects; those with eight legs are arachnids). These procedures evolved from efforts to separate categories which are *intuitively* recognized as being distinct by using more objective statistical methods. Although the statistical methods add some standardization, the final classification obtained is still largely a matter of preference.

With the increased availability of statistical software, numerical taxonomy techniques have seen wide application in biology, ecology and biogeography and in the analysis of remotely sensed spectral data. There are numerous opportunities for their use in hydrology—for example, in the classification of catchments based on their sensitivity to land modifications or of streams on the basis of their runoff characteristics or biota. However, it has only been recently that these techniques have been applied to the classification of streams and catchments. Stream and catchment classification will be covered in Chapter 9.

To develop the idea of numerical classification, let us consider the classification of ice cream flavours. Although raspberry frozen yoghurt could easily be distinguished from chocolate ice cream, finer criteria would be needed to distinguish raspberry frozen yoghurt from blackberry frozen yoghurt. Qualitative approaches might include classifying flavours according to personal taste preferences or those of a panel of ice cream experts. To numerically classify flavours, some combination of objective measures such as ingredients or water content or melting temperature would be used to distinguish one flavour from another. Attributes can also be 'weighted' according to importance; e.g. the number of walnuts may be less important to flavour than millilitres of mint extract.

The methods that will be discussed are considered forms of *multivariate analysis*, which considers the interrelationships between a number of variables. Whereas multiple regression analysis results in an equation, multivariate analysis in numerical taxonomy results in a 'grouping' of individual observations which have similar characteristics. Multiple regression is sometimes considered a type of multivariate analysis. Tabachnick and Fidell (1989) have written an easily followed text on multivariate analysis, from which much of the information in this section has been gleaned. Other information sources are given in Section 8.6.5.

Primary 'classes' under the heading of numerical taxonomy are those of ordination and classification, which can be jointly or separately applied to data. The techniques can be based on actual measurements, presence/

absence data, rankings (e.g. 1 = poor water quality, 5 = excellent), or statistical abstractions that define the similarity or dissimilarity between individuals.

### 8.6.2 Similarity/Dissimilarity Indices

Any two sets of data in continuous, discrete or attribute form can be represented by *similarity/dissimilarity indices*. A high-similarity measure (low dissimilarity) indicates 'alikeness' whereas a low-similarity measure (high dissimilarity) indicates 'unlikeness'. For example, the amount of fruit might be used to calculate a dissimilarity measure which would indicate that tutti-fruiti ice cream is much more like apricot swirl than it is like mint chocolate chip.

One index is the $r^2$ value from a regression analysis (see Appendix), where high values indicate similarity and low values indicate dissimilarity. Measures based on the geometric distance between points include (SPSS, 1986):

1. *Euclidean distance*

$$\text{Distance } (x, y) = \sqrt{\left[\sum (x_i - y_i)^2\right]} \qquad (8.15)$$

2. *Squared Euclidean distance*

$$\text{Distance } (x, y) = \sum (x_i - y_i)^2 \qquad (8.16)$$

Here, $x$ and $y$ might represent values of different attributes such as catchment shape and water temperature. If a graph of catchment shape versus water temperature were developed, similar sites would plot near each other, whereas sites which were different would plot at a greater distance away. Thus, high values obtained from Eqs. (8.15) or (8.16) indicate dissimilarity. A pattern similarity measure, used by Haines *et al.* (1988) to separate river regimes (Figure 4.20), is given by SPSS (1986) as:

3. *Cosine similarity measure*

$$\text{Similarity} = \frac{\sum (x_i y_i)}{\sqrt{[(\sum x_i^2)(\sum y_i^2)]}} \qquad (8.17)$$

For this measure, higher values indicate similarity and lower values dissimilarity. For ecological data, Faith *et al.* (1987) make explicit recommendations on appropriate measures.

### 8.6.3 Ordination

*Ordination* is a method for investigating the structure in data. It may be based on the original data, such as ice cream density or the presence of fine sediment in a stream, or on some index of similarity or dissimilarity. One aim of ordination is to reduce a large number of characteristics (dimensions) to a lower number of indices which still account for nearly all the variance (Pielou, 1977). The process is somewhat like the editing of technical prose, where the object is to condense the essence of a sentence into as few words as possible while still retaining all of the original meaning. Redundancies in the form of intercorrelations between variables are removed.

For example, if several factors were measured in an assessment of stream physical habitat, one might ask whether they can be condensed into a smaller number of combined characteristics which more effectively or efficiently describe the habitat. Factors such as stream density, relief, bifurcation ratio and catchment shape might all be grouped into one 'catchment morphology' index. The procedure is best used where a number of variables contribute 'overlapping' information but none works well by itself. If the original variables are essentially independent, there is no need to use ordination.

*Factor analysis* (FA) and *principle component analysis* (PCA) are two common methods used in ordination. In both methods, variables that are correlated with each other—but not with other groups of variables—are combined. This combination of variables is called a 'component' in PCA and a 'factor' in FA. The factors or components are considered to reflect underlying processes that might have caused the intercorrelations. For example, stream density, relief, bifurcation ratio and catchment shape may all be related through landscape-evolution processes.

Components and factors are represented by a linear combination of variables, which forms a model much like a regression equation. The model effectively 'weights' each variable according to its importance. This relationship is used to summarize patterns of intercorrelations among the variables, to define some underlying process, or to test a hypothesis about whether an underlying process exists.

For example, the hypothesis could be made that elevation, flow variability, substrate, canopy cover, water temperature and a number of other factors influence the population of mayfly larvae in streams. Ordination techniques can be used to reveal patterns and order in the hodgepodge of measurements. These patterns might indicate that canopy cover is an unnecessary measurement, that there is a need to find an additional measure, or that substrate, water temperature, stream width and depth can be combined into a common factor that may reflect some underlying process.

The choice of whether to use PCA or FA is based on the purpose of the study. PCA is the method of choice for reducing a large number of variables into a smaller number of components, whereas FA is of more use in generating hypotheses about underlying processes. Cattell (1965) considers PCA to be a 'closed model' in that it accounts for all the variance, whereas FA is an 'open model' in which part of the variance is reserved for factors that may yet be included.

### Principal Component Analysis (PCA)

As mentioned previously, PCA is used to develop a smaller set of components that summarize the correlations among the original variables. By collapsing several different measures onto a common axis of ordination a type of 'ranking' is obtained. This ranking can then be used as an index in subsequent analyses. The groupings are also useful for exploring differences between groups; e.g. the components for trout habitat may be made up of a different set of variables than for bass habitat.

Continuing with the ice cream example, an axis representing the amount of coffee flavouring, chocolate content and weight of almonds might provide a basis for ordering (on an axis, not over a counter) the ice cream flavours chocolate chip and Swiss almond mocha. The position of a flavour on the axis tells how 'coffee-chocolatey-nuttey' it is (component 1 in Figure 8.12). Multiple axes would be needed to represent *all* flavours of ice cream; 'chunkiness', 'berriness' or 'creaminess' being other possibilities. Each flavour would then be represented by a position (given by coordinates) in a 'hyperspace' of multiple dimensions. In Figure 8.12, two axes are shown, the second representing 'wateriness'.

The object of PCA is to extract the maximum amount of variance from the data set with each component. The first component extracted is a combination of variables which best separates the individual observations (i.e.

spreads them out along an axis). It accounts for as much of the variation in the original data as possible. The second component extracts the maximum amount of remaining variance that is not correlated with the first component (orthogonal to it). Ideally, the interrelationships will be displayed using as few dimensions as will suffice (Pielou, 1977).

A problem of PCA is that components are dependent on the scale of the variable; e.g. the variable with the highest variance will dominate the first component. It is preferable to first scale variables such that all have a variance near one. Results of a PCA are normally plotted with two axes to represent pairs of components. From these plots, the distance, clustering and direction of the points relative to the axes can be examined. A large distance away from the origin along an axis indicates a close correlation with that component. Optimally, points should cluster near the end of an axis and near the origin; if clustering is not obvious, a component may not be clearly defined. If clusters do not line up on the axes, it may indicate a need to rotate the axes.

*Rotation* of axes can improve the usefulness and ease of interpretation of the solution. Varimax (orthogonal) rotation is most commonly employed, which assumes that the underlying processes influencing each component are independent. Spatially, the axes are rotated so that they more closely pass through the variable clusters, as in Figure 8.12. This allows each cluster or individual to be more easily ranked by distance along an axis.

The resulting components should also be examined in an attempt to understand the underlying, unifying principle. This is usually characterized by assigning the component a name; e.g. 'richness of flavour' or 'landscape form'. In fact, there is no objective measure for testing how 'good' the resulting PCA solution is. The final choice is up to the researcher, based on its interpretability—i.e. does it make sense? Useful solutions ideally have a few

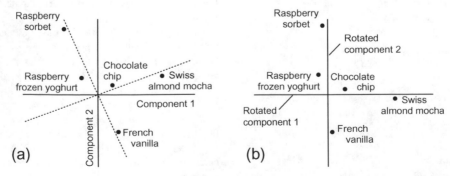

**Figure 8.12.** *A fictitious illustration of the use of rotation in principal component analysis: (a) location of various ice cream flavours after extraction of principal components; and (b) location after rotation of axes. Component 1 might represent 'coffee-chocolate-nuttiness' and component 2, 'wateriness'*

components which are related by some common process and are highly unlike the other components.

Once obtained, principal components can be used as independent variables in a multiple-regression analysis. Because the components are uncorrelated, the regression results tend to be more stable and reliable than regressions made using the original variables (Wallis, 1965).

### Factor Analysis (FA)

The process of factor analysis is very similar to that of principal component analysis. However, FA differs from PCA both in theory and in applicability. In PCA *all* the variance in the observed variables is analysed, whereas in FA only the variance *shared* by the variables is analysed. Thus PCA is an analysis of the total variance whereas FA is an analysis of covariance. In many statistical references the term 'factor analysis' is often used for both factor and principal component analysis.

PCA is normally used for condensing the information contained in the data into indices. Factor analysis, however, might be employed to estimate the amount of variance in the data set which is accounted for by those indices; e.g. 'how effective is catchment morphology at explaining the variability in endemic leech populations?' A low 'score' may highlight the need to measure other variables. Thus, FA may be of use in the exploratory stages of research to identify critical variables and possibly eliminate others which do not add any more information and may be expensive to measure.

Factor analysis is a more formal statistical technique than PCA. Limitations on its use, such as the requirements of independence and normality, can be relaxed for *exploratory* uses of the technique. However, they become more critical in *confirmatory* factor analysis, where the goal is to test an hypothesis about underlying processes (Tabachnick and Fidell, 1989). Factor analysis is most appropriate for generating and testing hypotheses about underlying structures, patterns or processes, whereas PCA is more useful for summarizing the amount of information contained in the data, and is the technique more commonly used in hydrology.

### An Example

Principal component analysis (with Varimax rotation) was used by Hughes and James (1989) to classify streams in Victoria, Australia, into regions according to hydrologic characteristics. The classification procedure produced five distinctive groups which generally corresponded with climatic regions. Sixteen variables were calculated using data from stream-gauging stations. These variables included peak and low-flow variables, mean flows, and coefficients of variation for monthly and annual stream-flow data. It was found that 72.4% of the variation was accounted for by three components.

Figure 8.13 shows how the five groups 'cluster' on a scatter plot in ordination space. The axes represent components 1 (monthly variability) and 2 (annual variability).

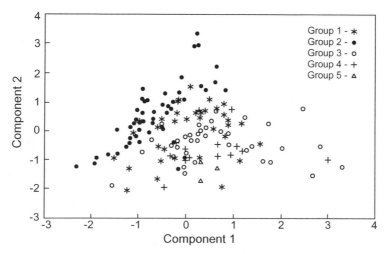

**Figure 8.13.** *Two principal components used in a hydrological regionalization analysis of streams in Victoria, Australia. The five groups of streams shown are hydrologically distinct. Component 1 is an index of monthly variability whereas component 2 reflects annual variability. The Acheron River is in Group 4. Reproduced with permission of CSIRO Publishing, Melbourne, Australia from the Australian Journal of Marine and Freshwater Research vol.* **40***: 303–326 (Hughes, J.M. and James, B. 1989), http://www.publish.csiro.au/nid/126.htm*

### 8.6.4    Classification

Whereas similarity measures and ordination give an indication of how closely individual data points are 'related', *classification* methods group individuals (objects, traits or measures) which are 'alike' into classes. 'Alike' may mean close together in space or time, or close together in terms of having similar characteristics such as chocolate flavouring or catchment size (Sokal and Sneath, 1963). This grouping can be done by dividing the whole collection of data into increasingly finer groups, or by starting with individuals and combining and recombining them to form successively larger groups (Pielou, 1977); i.e. 'splitting' versus 'lumping'.

Two techniques used in classification are discriminant analysis and cluster analysis. In both the aim is to predict membership in a group from a set of variables. These variables might be actual measurements, similarity/dissimilarity indices or ordination scores from a principal component analysis. The two classification techniques are very similar except that in discriminant analysis, class membership is known beforehand, at least for the data that are being analysed. In cluster analysis, classes are 'assembled' based on similarities among individuals.

For example, cluster analysis might be used to classify regions based on vegetation communities. In contrast, discriminant analysis would be preferred for classifying new areas as being likely to contain certain vegetation types, based on physical, meteorological or other measures. Equations would be developed from data that included observations of vegetation type; thus 'class' membership of a particular area (e.g. containing spider orchids and sundews) would be known prior to analysis.

Andrews curves (Andrews, 1972) provide a third technique for classification. This method provides a means of concisely displaying the data so that classes can be grouped visually. With this method, new observations can be assigned to existing groups or identified as being unique.

Another technique, multidimensional scaling, has been recommended for displaying patterns in complex ecological data. This technique will not be discussed here, and the reader is referred to Minchin (1987) for further information.

#### *Cluster Analysis*

Cluster analysis (not to be confused with cluster *sampling*, introduced in Section 2.3.5) is a numerical technique for grouping the data points floating around in hyperspace into clusters of 'like' members. Individual points in different groups should be dissimilar. A common technique of clustering data into similar groups is the nearest-neighbour method. Starting with single points, the groups which are closest to each other, based on a similarity/dissimilarity measure (Section 8.6.2), are combined. Each combination reduces the number of groups by one.

Cluster analysis results in a 'tree diagram' or 'dendrogram' (see Figure 8.14), which shows the sequence in which the groups were combined or divided. The structure of the diagram indicates the similarity between 'stems'. Vertically, the scale gives an indication of the amount of heterogeneity remaining or removed at each step. Different methods of analysis can create dendrograms of different structures.

Continuing the ice cream example, the top of the dendrogram might represent the class encompassing 'all things that taste like ice cream' to the other extreme at the lower end where one class exists for every existing flavour. Where to 'cut' the dendrogram for classification purposes is subject to one's own tastes. The final level of classification can be chosen based on (1) the number of classes to be recognized, (2) the amount of heterogeneity permitted in the final classes or (3) the point at which reduction in heterogeneity from further sub-division becomes too small to be worthwhile (Pielou, 1977).

Probabilistic stopping rules can provide an objective means for deciding on the final level, but it is more often guided by practical judgement. Like PCA, interpretability is a criterion in selection of the final classification. It is best to investigate what the classes represent, and whether they match one's own perceptions and requirements. Romesburg (1984) and several of the other texts on multivariate analysis listed in Section 8.6.5 provide further information on cluster analysis.

#### *Discriminant Analysis*

In discriminant analysis the objective is to predict membership in groups from a set of independent variables. For example, one might ask whether biologists, engineers, environmental scientists and geographers can be distinguished, based on a set of spatial perception and mathematical test scores. A significant difference between groups implies that, given certain test scores, the group in which the individual belongs can be predicted with some degree of certainty. In stream ecology an example might be the separation of 'good trout habitat' from 'good bass habitat' on the basis of variables such as the density of blackfly larvae, the substrate $d_{85}$, the mean summer water temperature, the variability of streamflow, and the percentage of streambank covered by boat docks. The classes would have been previously established by some method such as fish sampling or angler interviews.

Discriminant analysis can also be used to assess the relative contribution of the individual variables to the prediction of group membership. For the ecological example, the main sources of discrimination between trout and bass habitat might be water temperature and substrate, with little predictability contributed by the other factors. When a classification is produced, results can be analysed to evaluate how well individual observations fit into their appropriate classes.

The combination of variables (predictors) which separate groups from each other are called *discriminant functions*. These are much like multiple-regression equations but produce a discriminant function 'score'. This score gives an indication of the group to which an individual or stream or catchment belongs, and whether it falls solidly into one class or into a grey area between two different classes. In the previous example using test scores, the scores might be combined into one discriminant function score, with 'low', 'below average', 'above average' and 'high' groupings reflecting membership in various career 'classes'.

Several functions may be needed to reliably separate groups; e.g. one function may separate bass habitat from trout habitat whereas another might be needed to separate brook trout habitat from brown trout habitat. The second discriminant function would separate groups on the basis of associations not used in the first function. The total number of possible 'dimensions' is either one less than the number of groups, or the number of predictor variables, whichever is smaller.

Analogous to multiple-regression analysis, the predictors can all be entered at once or in a stepwise fashion, with contribution to prediction of group membership evaluated as each predictor is entered. It is typical for only one or two discriminant functions to be chosen. Individual points (or group centroids) can be plotted along axes representing the discriminant function scores. A large distance between groups along an axis indicates that the function is effective in separating them.

The major questions that discriminant analysis is designed to answer are: 'Can group membership be reliably predicted from the discriminant functions?' 'What is the likelihood of mis-classification?' 'If individuals are mis-classified, with what other groups are they most often confused?' As with multiple-regression analysis, the more variables used, the better the relationship; however, after a certain point, the variables are no longer adding much in the way of additional information. Tests of significance can be performed to determine how many discriminant functions are needed and the strength of the relationship between class membership and a set of predictor variables. Williams (1983) discusses uses of

discriminant analysis in ecology, and other references on multivariate analysis listed in Section 8.6.5 will provide further information on the technique.

### Andrews Curves: Displaying Multi-dimensional Data

If only two variables are required for describing the similarity between sites, streams or catchments, then a simple two-dimensional scatter plot is sufficient for displaying groupings. Displaying these data becomes much more difficult with a larger number of variables. A graphical approach presented by Andrews (1972) provides a good method of viewing patterns of similarity or dissimilarity across multiple dimensions. A point in multi-dimensional space is represented by a curve described by the function

$$f(t) = \frac{x_1}{\sqrt{2}} + x_2 \sin(t) + x_3 \cos(t) + x_4 \sin(2t)$$
$$+ x_5 \cos(2t) + \cdots \tag{8.19}$$

where $x_1$, $x_2$, ... are the variables used to characterize a particular site. The function is plotted over the range $-\pi$ to $+\pi$ ($-3.14$ to $+3.14$).

Curves representing points that are located near one another in multi-dimensional space will look similar, whereas points which are distant will produce curves that look different. Results will depend on the order in which the variables are labelled. The first variables will be described by low frequency components (wider 'waves'). These are more readily seen than the higher-frequency components representing the latter variables. Thus, it is more useful to associate the most important variable with $x_1$, the second with $x_2$, and so on (Nathan and McMahon, 1990d). This relative importance can be determined from a stepwise multiple-regression analysis (see Appendix). The values should also be scaled to the same order of magnitude (e.g. by choosing an appropriate unit). One method to standardize is by subtracting the mean and dividing by the standard deviation of all observations of a given variable.

Andrews curves have great potential in stream classification since they provide a method for visual comparison of biological and/or hydrological data. Stream sites or catchments with similar properties would produce a band of similarly shaped curves. If a curve falls outside some margin, the given site can be assigned to a different group. The curves are thus useful in evaluating the results of cluster analysis.

Importantly, group membership can also be determined for a new, unclassified site or catchment. The curve for the new site can simply be compared to bands defined for other site groupings to determine where the new site 'belongs'. Nathan and McMahon (1990d) point out that the technique thus has an advantage over discriminant analysis for regionalization (Section 8.5.3) because it is possible to identify catchments which do not belong to *any* of the existing groups and thus would not be properly described by regional prediction equations.

### Case Examples in Classification: Example 1

Barmuta (1989) analysed patterns of benthic macro-invertebrate occurrence at one site on the Acheron River, Victoria, to test the hypothesis that distinguishable habitat 'patches' would support different communities. Environmental data on velocity, substrate, temperature, DO, EC, pH and food resources as well as invertebrate populations were obtained from four riffle-pool sequences over a period of one year. Community pattern was defined by similarity of species composition among samples. Clustering techniques were first used to find groups of samples with similar faunal characteristics. Then, the discreteness of these groups was assessed using ordination. Finally, faunal patterns were related to environmental data.

The dendrogram from one cluster analysis is shown in Figure 8.14. The nine groups represent clusters of sites with similar faunal patterns. Groups A1–A4 generally contained fauna intolerant of fast-flowing, turbulent water, whereas groups A5–A9 had higher abundances of these organisms (e.g. hydrobiosid caddisflies, baetid mayflies). Thus, the two main 'stems' of the dendrogram were deemed to represent depositional (pool) and erosional (riffle) habitats.

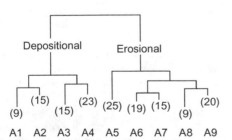

**Figure 8.14.** *Dendrogram from a clustering procedure used by Barmuta (1989) to separate benthic macro-invertebrate samples on the basis of faunal similarity. Numbers in parentheses indicate the size of each sample group. The first 'split' in the dendrogram suggested a major division into 'depositional' (pool) and 'erosional' (riffle) habitats. Environmental variables were insufficient for explaining the remaining divisions. Reproduced by permission of Blackwell Science*

In the ordination analysis the first axis was found to be highly positively correlated with velocity and the presence of aufwuchs ('biofilm'), and negatively correlated with mean particle size, sorting coefficient and organic detritus. This reflected the difference between samples collected in pools and riffles. However, the ordination did not show clear distinctions between groups, indicating that community structure was continuous rather than composed of distinct 'pool' and 'riffle' unit communities. A few individual species preferred distinct habitats, particularly those whose diets were restricted to the aufwuchs on the upper surfaces of rocks, those which required fast, turbulent water for feeding or physiological requirements and conversely, those which could avoid fast-flowing water by burrowing.

The author concluded that environmental variables, at the scales measured, were insufficient to account for the patterns of species co-occurrence, and stated a need for a more precise description of near-bed flows that may be more meaningful for the species studied.

### Case Example 2

In a study on clear New Zealand rivers, Jowett (1990) used both discriminant analysis and classification techniques to relate the distribution and abundance of trout to environmental factors. An experienced team of divers made observations on trout abundance by drift-diving. At each site, instream cover was evaluated and measurements were made of water depth, velocity and substrate along at least two pool-run-riffle sequences. Instream habitat and hydraulic parameters were determined from hydraulic modelling.

Brown trout and rainbow trout populations were classified into six groups based on species, size and biomass (computed from size class and abundance). The primary division was between rainbow and brown trout, with further sub-divisions based on biomass. This classification corresponded well with geographic patterns of trout distribution in New Zealand streams, with rainbow trout occurring only on the North Island.

Much like the study of Example 1, the groups were then examined to determine whether there were significant differences in hydrological, water quality, biological, instream habitat and catchment characteristics between groups. Variables were selected for discriminant analysis if the variance between groups was about twice the within-group variation. Temperature, catchment lithology, hydrological indices, instream habitat and total aquatic invertebrate biomass showed the most significant between-group differences.

A discriminant model to classify sites was developed based on eleven environmental factors. Three discriminant

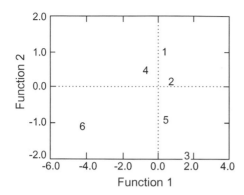

*Figure 8.15. A plot of the first and second discriminant functions from a study by Jowett (1990). Riverine sites were divided into six groups based on trout species, size and abundance. The points on the plot represent group centroids. Group 6, for example, represents streams with rainbow trout. Both functions were found to be related to several environmental factors. Reproduced by permission of The Royal Society of New Zealand*

functions were found to be statistically significant. Variables closely related to the first two functions were minimum annual water temperature, percentage volcanic ash, ratio of mean annual low flow to median flow, square root of river gradient, percentage lake area, and physical habitat. Figure 8.15 shows the group centroids plotted using the first two functions. The resulting discriminant functions correctly predicted group membership for 48 of the 65 sites.

From the analysis the researcher found that lake and spring-fed rivers with stable flow regimes were likely sites for rainbow trout. Examination of mis-classified sites suggested that connection of the stream to lakes might affect species distribution and abundance. Overall, the author concluded that the factors most related to the distribution of trout species were climatic whereas factors

determining abundance were related primarily to instream habitat (both river morphology and streamflow magnitude and variation).

### 8.6.5 Other Sources of Information on Numerical Taxonomy

This brief introduction to numerical taxonomy is designed to introduce various techniques that may be helpful in classifying streams and catchments and to aid in the interpretation of studies in which the techniques have been used. Prior to conducting an analysis, statistical texts should be consulted for information on theoretical limitations and assumptions. A number of computer programs are available for rapid number-crunching, which lowers the 'threshold' for entry into the field of numerical taxonomy. When selecting and using computer software packages it is important to choose the right one, use it correctly and know how to interpret the output. Some software packages provide more options and more readable documentation than others. Cost, reliability, and 'user-friendliness' of the sortware, and the opinions of experienced users should serve as a guide in the selection of an appropriate package.

Pielou (1984), Sneath and Sokal (1973) and Sokal and Sneath (1963) are a few of the many references on the subject of numerical taxonomy. Especially for biologists, an article in *Scientific American* by Sokal (1966) provides interesting reading on the topic. The user's guides to commercial statistical software packages are also good sources of information on the various techniques. Other important references on multivariate analysis include Afifi and Clark (1984), Cattell (1965), Chatfield and Collins (1980), Green (1979), Legendre and Legendre (1983), Romesburg (1984), and Tabachnick and Fidell (1989). Other references and statistical packages are given in the Appendix.

# 9

# "Putting It All Together": Assessing Stream Health, Stream Classification, Environmental Flows and Rehabilitation

## 9.1 Putting Theories into Practice

In previous chapters the interrelationships between ecology and hydrology have been explored at a number of scales: whole river systems, the pools, riffles and meanders of a stream reach, the flow patterns around rocks and the lift and drag on individual organisms. Whereas scientists may study these relationships in an attempt to improve our understanding of streams and their biota, operationalizing research into management recommendations requires a more practical problem-solving approach, as well as placing the scientific considerations within a social, economic and cultural context.

### A Paradigm Shift in Stream Management

Stream management has traditionally been focussed on the utilitarian uses of streams. The main factors of concern were the amount of water available, and the quality of the water with respect to its suitability for agricultural, industrial, domestic or recreational uses. In the endeavour to put freshwater sources to productive use and to tame and control floodwaters and their pathways, the consequences in terms of habitat loss have often been overlooked. The level of environmental awareness has now reached a point where many of the modifications to streams and their catchments have been viewed by a large sector of society as undesirable and in need of rectification. In response to increasing scientific understanding of streams, the well documented decline of stream resources, increased public interest in sustainable resource use, and the need for accountability in resource management, widespread

changes in legislation over the past 10–20 years now requires protection of the inherent ecological resources of streams. Managers have realized that protection of natural ecological processes in streams also helps to protect some of their utilitarian values, although there will still be conflicts over how to best use the resources. The ecosystem concept, which originated in ecology as a research paradigm, has now been transferred to the realm of public policy and stream management methodology (e.g. Grumbine, 1994; Thoms and Sheldon, 2002).

Stream management now involves the input of various stakeholder groups with different and sometimes conflicting interests and priorities regarding how the stream should be managed: for recreation, wildlife habitat, drainage, water supply and residential, industrial and agricultural uses. These interests are shaped by the different ways of valuing a stream's resources, even if these values are not always clearly articulated in the management debate. How we perceive stream values underpins the way we research and manage rivers, so this chapter begins with an exploration of stream values in Section 9.2.

Dams and diversions modify river flows and stream works make channels more hydraulically efficient in order to bring great benefits to society in terms of water supply, hydroelectric power generation and protection of buildings and farmlands on floodplains. River management will always be a balancing act; science cannot provide the 'right' answer. Science can make valuable contributions by refuting false notions and providing predictive models of how various components of the environment work. However, it has no power to establish the relative merit of values that the community assigns to the environment

Stream Hydrology: An Introduction for Ecologists, Second Edition.
Nancy D. Gordon, Thomas A. McMahon, Brian L. Finlayson, Christopher J. Gippel, Rory J. Nathan
© 2004 John Wiley & Sons, Ltd ISBNs: 0-470-84357-8 (HB); 0-470-84358-6 (PB)

(Gippel, 1999). Modern stream management must meet the requirements of humans and the aquatic biota—indeed, to integrate the human component into the natural ecosystem.

The increasing complexity of water-resource problems and the overwhelming amount of information available has created a need for multi-disciplinary teams that include zoologists, botanists, microbiologists, geomorphologists, hydrologists, economists, communicators, hydraulic engineers, chemists, anthropologists and sociologists (King and Brown, 2003). Some of the areas in which interaction is required include the classification of streams for environmental value, the assessment of environmental flows to be protected in unregulated streams and design of new flow regimes for regulated rivers, the design of fishways, the simulation of field hydraulic characteristics in laboratory flumes to study flow patterns around obstacles and organisms, the development of new biologically relevant ways of measuring streams, and design of physical habitat rehabilitation projects. As well as increasing the level of integration among the different scientific and engineering disciplines that investigate streams, the changed emphasis in stream management towards protection and sustainable management of resources has given rise to new fields of endeavour in applied river science.

### Causes of Stream Degradation and Some Solutions

'Disturbance' to the stream ecosystem can result from floods, prolonged droughts, volcanic activity, wildfires as well as anthropogenic factors such as pollution, channel modification, flow modification or direct interference to biota, such as clearing vegetation or introducing alien species. In general, maximum biotic diversity is maintained in streams by a level of disturbance that creates environmental heterogeneity, yet still allows the establishment of communities (Ward and Stanford, 1983a; Resh et al., 1988). For example, a 'patchy' substrate maintained by periodic flooding will support a more diverse biota than a silted-up streambed. Degradation of streams is a worldwide phenomenon. Gippel and Collier (1988) reviewed the causes and extent of degradation of Australasian streams, and this was complemented by a study of human impacts on Australian stream channel morphology by Rutherfurd (2000). Australian streams may vary in some respects from those in other countries (Rutherfurd and Gippel, 2001), and there are cultural differences, but the causes of degradation are reasonably consistent throughout the world. One difference between countries is the length of time that streams have been managed; Australia was a late starter compared to Europe, but managed to catch up in less than two centuries of European settlement. Also, the particular physical and ecological characteristics that operate in various places may act to produce a different type or degree of response to modification.

*Land-use changes* can have an impact on streams by affecting runoff rates and the input rates of sediment, woody debris and chemical pollutants. A well-vegetated catchment with deep soils will absorb rainwater, releasing it slowly. If this vegetation is removed or changed, as by clearing lands for farmland, logging, or (less obviously) by grazing, changes in stream hydrographs can occur. Clearing a large percentage of a catchment for urbanization, agriculture or timber harvest is generally thought to increase flood peak discharges and reduce their duration, and baseflow can also be altered.

Runoff changes due to land clearing can give rise to enlarged and/or entrenched channels. Gregory and Madew (1982), for example, suggest that urbanization can lead to increased peak flows and enlarged channels. Land-use changes can also lead to increased siltation in streams as a result of both channel bank and bed erosion and increased influxes from upland sources. Siltation can affect the survival of fish eggs and benthic invertebrates and can have an influence on aquatic and river-edge plant growth (Reiser et al., 1985).

Changes in runoff can also lead to alterations in the stream biota which have evolved under a different flow regime. For example, a shortened spring runoff can reduce the available period for fish spawning and egg incubation (Newbury and Gaboury, 1988). Changes can also make conditions more favourable for introduced species, as Davies et al. (1988) discovered in a study on a river system in Tasmania, Australia. They found that a near-doubling in the amount of cleared land led to an increase in the mean annual flow but a decrease in the interannual variability. It was found that brown trout, *Salmo trutta L.*, had higher populations and a more stable age structure in the period following clearing. Australian streams are characterized by high variability (see Section 4.4). As brown trout were introduced from England, the change to a less variable flow regime more like that of English streams may have had a positive effect on these fish.

Studies on the impacts of land-use change on runoff, channel morphology and stream biology are difficult because upland changes (e.g. urbanization) are often accompanied by instream changes (e.g. channelization), and because there are few 'control' catchments for comparison. However, it is undeniably a good practice to leave a buffer strip of natural vegetation along streams to reduce the impact of land-use changes in the catchment (Gore and Bryant, 1988). This serves to filter out sediments from surrounding areas, slow overland flow, eliminate bank disturbance and preserve riparian vegetation.

*Channel modifications* directly impact streams. The impacts can occur not only in the modified reach but also in upstream and downstream sections. Channelization is typically carried out to improve drainage or flood-carrying capacity, usually leaving a smooth, trapezoidal channel with improved conveyance and more predictable hydraulic behaviour. In extreme cases the riverbed may be reduced to a concrete channel or a buried conduit.

The straightening of channels and reduction in roughness leads to greater flow velocities and thus higher erosive forces; as a result, the channel will often erode downwards or outwards. This degradation often progresses upstream as headward erosion (Section 7.2). Increases in channel erosion can create increased turbidity, and can cause sedimentation in downstream reaches where the slope decreases or where the stream enters a lake, estuary or ocean. To control erosion, regular maintenance or bank-stabilization measures must be employed which can cause frequent disturbance of stream biota (Lewis and Williams, 1984). However, stabilization works can be designed for biological habitat improvement as well as geomorphological soundness (see Section 9.6.4).

In terms of habitat, channelization reduces the structural diversity of streams through the reduction in meanders, smoothing of pools and riffles and irregular bank boundaries and removal of snags and riparian vegetation. This not only reduces the total amount of stream area and shoreline length for habitation but also eliminates the natural diversity of velocity and substrate patterns. Fish no longer have backwaters, pools or low-velocity regions for refuge during high flows, and fish eggs may be swept downstream by the higher velocities (Newbury and Gaboury, 1988). Changes in hydraulic conditions selectively alter or reduce fish fauna, as increased velocities and shear stresses affect the hydrodynamics of body shape (Scarnecchia, 1988). Riffles, which aerate the flow, are removed, shelter in the form of undercut banks and overhanging vegetation is eliminated, and the substrate is typically more unstable, reducing benthic invertebrate production (Statzner and Higler, 1986). The clearing of riparian vegetation during channelization can reduce food input in the form of leaf litter and affect water temperatures. The removal of snags from within the channel alters stream hydraulics, releases stored sediments and changes the bed topography, leading to alterations in the fish community structure (Sullivan *et al.*, 1987; Gippel, 1995). Channelization also isolates the river from its floodplain, resulting in the loss of marshes, billabongs and their ecological diversity (Petts, 1989). The hydraulic function of the floodplain for storing, releasing and directing waters is also lost, and downstream flood peaks can be accentuated.

By employing channel designs that do not destroy the natural structural diversity or morphological processes, some of the detrimental effects of channelization can be avoided (Shrubsole, 1994). The incorporation of environmental features into the design of flood-control channels was reviewed by Shields (1982). It is important for scientists to be involved during the design phase so that habitat features can be preserved or incorporated within the engineering constraints. Practices might include minimal straightening or shaping, emulating the natural stream morphology and structure, selective removal of trees and snags, single-bank modification, ecologically based bank stabilization and riparian revegetation and stock exclusion. Fish stocking is practiced in many rivers that are unable to support self-staining populations of fish. The use of these practices in the rehabilitation of streams is covered in Sections 9.6.2, 9.6.4 and 9.6.5.

*Dams* typically affect both the hydrology and channel morphology of the regulated stream (Finlayson *et al.*, 1994). They impose an artificial lake environment on the stream, which changes the biota from lotic to lentic and can increase water losses to evaporation and groundwater recharge. The areas downstream of the dam are also markedly affected by changed flow regimes.

Flow regimes can be altered through regulation, in terms of the duration of flows of a given magnitude, the total annual discharge, flow variability or the frequency of flood peaks. Irrigation storage may generate short-term variable flows during peak demands and constant flows otherwise, whereas hydroelectric dams can yield rapidly fluctuating flows (Walker, 1985). These changes can be quantified with standard hydrological statistical methods such as flow-duration curves, flow-spell analysis and flood frequency, monthly histograms and autocorrelation methods (e.g. Gustard, 1992) (Chapters 4 and 8 and Appendix), although specialized hydrological analysis methods that describe variability and flow events are also useful (Gippel, 2001a).

Altered flow regimes can influence oxygen levels, temperature, suspended solids, drift of organisms and cycling of organic matter and other nutrients, as well as having direct impacts on biota. Sudden fluctuations in flow, for example, can wash away deposited eggs or leave fish, crustaceans and molluscs stranded out of water. Regulation can affect community composition by altering 'triggers'; for example, fish activities may be synchronized to periods of low or high flow. Inundation of the floodplain may be reduced, altering the frequency with which floodplain vegetation is 'watered', and reducing opportunities for fish spawning or juvenile growth in flooded backwaters (Fenner *et al.*, 1985; Petts, 1989). Decreased flooding of overbank areas also reduces the amount of food input to the stream (Cadwallader, 1986; Ward, 1989).

The thermal stratification that occurs in most lakes can affect the temperature of released waters, depending on the depth of the reservoir, the location of the outlet and the time of storage. Typically, water is released from the lower depths of reservoirs. In temperate regions this regulated discharge may warm more slowly in spring and remain warmer longer into autumn, which can affect ice formation and melt (Ridley and Steel, 1975). Regulated discharges also tend to have more stable temperatures rather than going through a daily cycle of warming and cooling. The altered temperature regime can result in the elimination of many species of aquatic insects (Ward, 1984). It also affects fish species that require specific temperatures to spawn (Cadwallader, 1986). The release of colder waters from deep-release dams can therefore permit the establishment of fisheries where high temperatures had previously prevented it. However, these waters may also be anoxic as a result of decomposition processes, and may contain reduced chemical compounds that increase the total oxygen demand in the tailwaters and can be toxic to aquatic biota (Ward, 1982). Thus, large releases of these waters can lead to fish kills in downstream reaches. Aeration of the released water can be accomplished by installing air draughts in the water-release ports (Ward, 1984). In contrast, releases of warmer waters from surface layers can encourage the growth of large quantities of algae downstream (Ridley and Steel, 1975). Multiple-level outlets and other structural measures allow the temperature and water quality of released water to be controlled (Section 9.6.3). To improve the temperature regime of released water, Ward (1984) suggests that a certain number of degree days (Section 1.3.4) should be programmed within the annual cycle of releases.

Dams act as barriers to fish migration, and there is little doubt about their effect on the production of Northern Hemisphere anadromous fish such as salmon and sturgeon (Goldman and Horne, 1983). Dams also affect fish in the Southern Hemisphere, as demonstrated by a study of the fish communities upstream and downsteam of Tallowa Dam on the Shoalhaven River, New South Wales, Australia (Gehrke *et al.*, 2002). Species richness was greater downstream of the dam, ten diadromous species were believed to be extinct above the dam because of obstructed fish passage. Four migratory species capable of climbing the wall had reduced abundances upstream. Accumulations of fish, particularly juveniles, were observed directly below the dam for nine species (Gehrke *et al.*, 2002). Options for providing fish passage include the addition of various 'fishways', or in the case of dams that are obsolete, or uneconomical when their environmental

costs are considered, removal may be a feasible option (Section 9.6.3).

The effect of dams on downstream channel morphology will depend on the streambed materials, the amount, size and source of suspended sediment, and the extent of alteration to the natural flow regime. Reservoirs are usually effective sediment traps, releasing clearer water downstream. In some cases the channel downstream has scoured, because the sediment transported from the reach was not replaced by incoming sediment. In some cases channels have silted up and become congested by vegetation because the dam eliminated the frequent floods that formerly removed the sediment input by tributaries. Channel changes can often be detected several kilometres downstream from dams (Gregory and Madew, 1982). Petts (1980) suggested that channel adjustments extend downstream to a point where the reservoir's catchment becomes less than 40% of the total upstream catchment. However, the prediction of channel adjustment is difficult because of the uniqueness of each reservoir-catchment system.

Environmental flows, or instream flows, are the flows that will maintain ecological integrity or a defined level of stream health in a regulated river, or an unregulated river subject to direct diversion of water. There has been considerable growth in research and assessment of environmental flow needs over the past twenty years, although to date there are few cases of implemented environmental flow regimes that have been thoroughly evaluated. The methods of assessment of environmental flows are reviewed in Section 9.5.

In order to manage rivers and streams effectively, a necessary first step is to measure the availability and condition of the resources. Stream condition has traditionally been measured in terms of physico-chemical parameters, because this was appropriate to the emphasis on utilitarian use of the resource. Physico-chemical characteristics are still important, but there has been a paradigm shift in the way stream condition is perceived and measured. We now speak of 'stream health' and measure it in terms of water quality, habitat availability and suitability, energy sources, hydrology and the biota themselves (termed bioassessment). 'Ecohydrological' (SINTEF NHL, 1994; Leclerc *et al.*, 1996) or 'Ecological Health and Integrity' (Scrimgeour and Wicklum, 1996) approaches acknowledge, seek to understand, and make use of the inter-relationships between the biological, chemical and physical nature of streams. Methods of measuring stream health are reviewed in Section 9.3.

Stream classification (Section 9.4) operates at a different scale to stream health assessment, although measures of stream health can and often do form the basis of classification schemes. The main purpose of classification is to simplify the inherent complexity of stream systems. Classification is used as a communication tool that helps to facilitate many aspects of the management process, such as taking an inventory of the resource, prioritizing issues or areas for management action, allowing stakeholders to make trade-offs, and documenting and demonstrating the effectiveness of management to the public. Many different types of classification scheme have been devised, all tailored for a different purpose, geographical location or issue.

## 9.2 Understanding Stream Values

### 9.2.1 What Are Stream Values?

In the search for general principles that may guide the protection and management of biological diversity, Burgman and Lindenmayer (1998) presented several ethical perspectives on the environment. Two main categories of values were identified, each with sub-categories:

1. Utilitarian Value
   - Consumptive use value
   - Productive use value
   - Service value
   - Scientific and educational value
   - Cultural, spiritual, experiential and existence value
   - Aesthetic, recreational and tourist use
2. Intrinsic Value
   - Ecocentric ethic
   - Biocentric ethic

Natural stream resources provide goods in the forms of fish, and other wildlife for harvest and enjoyment, as well as services such as regulation of hydrologic and nutrient cycles, and purification of water. Highly degraded ecosystems are not effective providers of goods and services, so in this way, conservation and economics are inextricably linked (Burgman and Lindenmayer, 1998). This argument also underpins the emerging ecosystem services approach to natural resource management, which values resources in terms of their contribution to sustaining and fulfilling human life (Daily, 1997). This approach is being used to bridge the gap between scientific understanding of environmental, economic and social aspects of river resources and perceptions by members of the public about the tradeoffs they regularly make concerning river resources (Cork et al., 2001; Shelton et al., 2001). Conservation biology provides the tools and the expertise to help maintain and restore the natural resource base, but it is not driven exclusively by economic or utilitarian criteria.

Ethical considerations are an alternative system for the valuation of conservation. Intrinsic values differ from utilitarian perspectives in that the former place value on species and communities, independent of people, while the latter are evaluated with respect to the preferences and needs of people (Burgman and Lindenmayer, 1998). With the ecocentric ethic, the object of concern is the biological community as a whole, with criteria for stream values based on naturalness, representativeness, diversity, rarity or special features (Boon et al., 1998; Dunn, 2000). The biocentric value argues for the value of all individual organisms. It requires a set of rules for human behaviour that includes respect for all entities in the natural environment.

In its purest sense, *ecological potential* is grounded on the ecocentric ethic, while the philosophical stance of practical stream management is strongly conditioned by the wider community's priority for protection of utilitarian values [for a recent example illustrating the persistence of this stance see Ladson and Finlayson (2002)]. While the new ecosystem paradigm may drive public policy, managers at the coalface, i.e. the catchment and reach scales, still have to deal with the reality of trying to balance or resolve competing interests, and stakeholders who hold conflicting convictions. The vagueness of the ecosystem concept is recognized in the research domain, but that is not a barrier to hypothesis testing. However, when operationalized into the management arena, the ecosystem concept can appear nebulous, creating ambiguity and uncertainty, especially with respect to defining terms and objectives (Fitzsimmons, 1996). A policy that elevates, or appears to elevate, environmental protection over other legitimate public goals will likely create division among stakeholders. In this atmosphere of radically altered resource management policy, stream health assessment programs, environmental flow assessments and stream rehabilitation projects will struggle to reach concensus on a definition of ecological potential.

### 9.2.2 Ecological Potential

While Regier (1993) made no real distinction between ecosystem health and integrity, Karr (1996) and Karr and Chu (1999) defined ecosystem health as the preferred state of ecosystems that are modified by human activity, while ecological integrity is an unimpaired condition, reflective of natural, pristine, reference or benchmark ecosystem. The natural condition does not typically exist as an idealized balanced or equilibrium state. Rather it is dynamic, often changing in an indeterminate way (Belovsky, 2002). Thus, the idea of defining a fixed state of ideal stream condition as a reference point from which to grade stream health may have intuitive appeal, but it is far from straightforward in practice.

Pristine condition is usually interpreted to mean the so-called 'primitive' or 'original' state that existed prior to intensive and widespread disturbance by humans. In many places the period of major disturbance would have corresponded with rapid population expansion. However, significant cultural changes that affected the patterns of land use would also have been relevant, as would have occurred during the agricultural and industrial revolutions for example. However, the pre-human disturbance date is just an arbitrary reference point. In much of Europe, river landscapes have been directly and indirectly modified for so long that it may not be possible to determine the pristine condition due to lack of data (Ward *et al.*, 2001). In such cases, the earliest date for which data are available usually determines the reference point. However, data from reference (relatively undisturbed) sites and better-documented comparable sites in other areas can be used to help reconstruct the pre-disturbance conditions. A complicating factor here is that if significant climatic change has occurred, a river would not return to its original or pristine (pre-human disturbance) condition, even if it was fully restored and the whole catchment designated a wilderness area. Also, some disturbances may have ceased and some may still be operating. Even when a disturbing activity has ceased (such as mining) its legacy (sand slugs) may still exist as a major stream disturbance.

Sometimes the pre-disturbance condition may refer to conditions that prevailed prior to a *specific* disturbance, such as widespread catchment deforestation, channel desnagging or flow regulation. However, the terms 'original' and 'pristine' should probably be reserved for conditions that applied prior to widespread and intensive human disturbance. The term 'natural' can mean 'not affected by humans or civilization' but probably is too ambiguous to be of any real value in describing the condition of a river or stream. When referring to ecological potential as it relates to some pre-disturbance condition, it is important to explicitly state the historical time period to which 'pre-disturbance' refers.

It may be better to define ecological potential in terms of what can be achieved under the current situation with respect to factors that are either not easily remedied or for which change is considered undesirable, such as established land use systems, climate, existence of alien species or use of river water for agriculture, recreation, town supply and industry. For example, the expected landscape ecology of a lowland river system would have to include agriculture, as it would be unhelpful to ignore the reality

of human use of river resources. Ecological potential then becomes a question of the expected ecology of a well managed, multiple-use, sustainable system [see Lélé and Norgaard (1996) for a discussion on the difficulty of defining 'sustainability']. The desired future state of a river system, as agreed by the community of stakeholders, is often termed the rehabilitation 'vision' (Gippel, 1999).

Streams also suffer major natural disturbances, and from a narrow management perspective, streams suffering such disturbances could be regarded as being temporarily in less than ideal health. One important difference between human disturbances and natural disturbances is that human disturbances were often undertaken with the intention of altering the stream to a desirable condition that could be maintained, whereas after a natural disturbance the stream might recover to its previous state, or perhaps shift to another condition. These days, natural disturbances are superimposed on human disturbances, and this may change the nature of their impact. For example, catastrophic floods may cause more serious changes in channel morphology if the riparian vegetation has been removed or replaced by less dense or less robust species.

It is clear from the above discussion that ecological potential is not a fixed concept that can be applied in a simple and uniform way across all streams. Stream health can only be assessed relative to arbitrary benchmarks. Establishment of benchmarks involves application of value judgements, usually associated with the normative concepts that biodiversity should be maximized and that ecological systems should be sustainable (which gives rise to ideas such as native species are superior to exotic species, high diversity is better than low diversity, an unregulated flow regime is better than a regulated flow regime and stable channels are more desirable than unstable channels). Lackey (2001) pointed out that in most scientific formulations of ecosystem health, there is a premise that natural systems are healthier than human-altered systems. This is an example of normative science, whereby the results of experiments or surveys are interpreted through the filter of an assumed policy preference, so it is not value-neutral (Lélé and Norgaard, 1996).

### 9.2.3  The Role of Science in the Stream Health Policy Debate

Achievement of stream health is driven by societal preferences and the crux of the policy challenge is deciding which of the diverse set of societal preferences are to be adopted (Lackey, 2001). Scientific information is merely one input to this process, but distinguished from the rest by its ideal of impartiality and objectivity (while other stakeholders unapologetically argue for their particular interests). Suter (1993) argued against the use of 'unreal properties' such as stream health because of the potential for such terms and concepts to obscure the bases for decision making; increase the opportunity for arbitrariness; and decrease the opportunity for informed input by stakeholders. Stream health is actually a normative scientific concept with associated personal values, and is not an impartial contribution to the policy debate (Lackey, 2001). Individuals must decide the preferred state of stream health, because ecosystems have no preferences about their states (Jamieson, 1995). To present ecosystem health as merely a scientific construct is misleading (Lackey, 2001).

The term 'healthy working river' has recently been coined as part of The Living Murray initiative, a process of assessing the environmental flow needs of the River Murray, Australia (MDBC, 2003). A similar concept known as the 'living working river' was applied to the Fraser River estuary in Canada (FREMP, 1994). A healthy working river is one that is managed to provide a compromise, agreed to by the community, between the condition of the river and the level of human use. This definition acknowledges the need for negotiation and compromise between the often-competing values and uses of the river. A key aspect of the healthy working river concept is that the river is managed to sustain an agreed level of work and an agreed state of river health indefinitely. If the level of work reduces the health of the river below what the community desires it is no longer a healthy working river, regardless of the economic gains that are made in the interim (Jones *et al.*, 2002). The 'working' part of the definition refers to the use of water resources for economic gain, water supply or for recreational purposes. Working river objectives are easily quantified using specific measures such as security of supply for irrigation water, navigability, end-of-valley salinity targets, as constrained by channel capacities, capacity of impoundments, legal requirements, minimization of bank erosion, water demands, travel times, infrastructure maintenance requirements and other factors. Some stream health characteristics and objectives can be similarly well defined, such as stream temperatures to match spawning requirements, and fish passage, while others rely on the use of indices (Scientific Reference Panel, 2003).

Suter (1993) and Kapusta and Landis (1998) argued that the potential for misuse of the ecosystem health concept, even if unintentional, warranted its abandonment as a policy goal. The alternative to using ecosystem health would be to state management goals in more simple and specific terms, rather than to rely on metaphor. At the very

least, consensus is required on the exact meaning of ecosystem health (Lackey, 2001). The following section explores the meaning of ecosystem health, as it has been applied to stream environments.

## 9.3 Assessing Stream Health

### 9.3.1 Introduction

The need for cost-effective, rapid and comprehensive environmental monitoring techniques for streams and rivers became apparent in Europe, North America and Australia by the mid-1980s. This realization was a response to community demands for more scientifically informed and transparent stream management decisions, and more reliable, consistent and accessible reporting of the condition of stream resources. Changes in water laws put more emphasis on protection of ecological integrity and biodiversity, and required managers to demonstrate effective management of streams. Also, in many places governments recognized that a large proportion of stream resources were un-assessed. These problems required methods that would allow rapid collection, compilation, analysis and interpretation of environmental data in order to facilitate management decisions and resultant actions for control and/or mitigation of impairment (Barbour *et al.*, 1999). Historically, environmental assessment of streams was based on physico-chemical measures of water quality. However, this approach ignores physical habitat quality, the impacts of alien plant and animal species, the state of fish stocks, alteration of flow regimes, and does not indicate how degraded stream conditions affect the biota.

Various authors, including Metcalfe-Smith (1996), Barbour *et al.* (1999) and ANZECC and ARMCANZ (2000a, b), have reviewed the traditional and more recently developed approaches to environmental monitoring of streams. Bioassessment originally focused on benthic macroinvertebrate assessment, but these approaches now also include periphyton, vegetation, fish and other components of the aquatic system. More recently, the need for even more rapid, inexpensive and comprehensive surveys has led to the concept of 'stream health' (Karr, 1996; Scrimgeour and Wicklum, 1996; Karr, 1999; Boulton, 1999; Norris and Thoms, 1999; Maddock, 1999), the assessment of which is usually based on integrative measures of the quality of the physical habitat structure, hydrological indices and perhaps also including biological and water quality data (e.g. Ladson *et al.*, 1999; Parsons *et al.*, 2002a).

There has been a good deal of argument in the literature about whether it is appropriate to use the 'health' metaphor, which originates from the idea of 'human health', in an ecological context (Karr and Chu, 1993; Scrimgeour and Wicklum, 1996; Gaudet *et al.*, 1997; Lackey, 2001). One argument against the use of the 'health' metaphor is that it is a value-laden concept and therefore inappropriate in science (Wicklum and Davies, 1995; Kaputska and Landis, 1998), and another is that 'health' is not an observable ecological property (health is a property of organisms but not easily defined for ecosystems) (Suter, 1993, Calow, 1992). However, the 'health' metaphor has such public appeal and utility precisely because people have a good understanding of personal health (Ryder, 1990; Karr and Chu, 1993; Meyer, 1997). Stream managers can trade on this universal understanding of the 'health' concept as a means of achieving effective communication between the general public, policy officials and scientists.

Stream health indices do not describe a specific property or entity that can be measured in the normal scientific sense, because 'health' is not an inherent property of a stream ecosystem (Wicklum and Davies, 1995). Rather, these indices combine data collected on a range of variables thought to impart or detract from environmental value, or reflect 'health' or otherwise, of the stream. Indices of stream health are used as classification tools for the purpose of assessing the resource (inventory), prioritizing streams and stream reaches for management actions, establishing a benchmark condition from which change can be assessed, and judging the long-run effectiveness of waterway management programs (Ladson and White, 2000). Stream health is usually measured on a reach scale (with the measured or estimated value applying to a length of stream varying from $10^1$ to $10^3$ m).

Some aspects of stream health can be expressed in absolute terms (e.g. metres of lateral migration per year, or species diversity and abundance). The more common approach is to make a visual evaluation using relative terms (e.g. a site is judged to have less channel erosion, or more dense riparian vegetation, compared with another site), or with respect to a classification scale (e.g. channel erosion is judged to be severe, moderate, slight, none). The alternative but more difficult approach is to measure or model the underlying causative factors that impart stream health, which then enables prediction of stream health under various scenarios.

### 9.3.2 Describing Reach Condition

Stream health assessments usually measure in-stream and riparian vegetation, channel bank and bed stability, water quality, water quantity and aquatic organisms. The description of stream health can refer to present static conditions (i.e. on the day of the survey), or dynamic aspects (i.e. rate and direction of changes). Rutherfurd *et al.* (2000) identified five types of benchmarks against which the health of a reach can be compared:

- The stream's ecological potential [e.g. the German Leitbild concept of restoration (Kern, 1992) uses as a benchmark an ideal solution involving restoration of all natural stream and floodplain properties, ignoring economic and political factors that may act to limit the adoption of this option, even though it is recognized that only in very few cases would this option be feasible].
- A nearby template or reference reach that is undisturbed, with the greater the deviation the poorer is stream health [e.g. the AusRivAS system for describing health of aquatic macroinvertebrate communities in Australia (Simpson and Norris, 2000; Norris and Thoms, 1999) is based on the ratio of observed over expected, where the expected community characteristics are derived from relatively undisturbed reference streams].
- Established criteria, such as those that often exist for water quality [e.g. in Australia, each water quality variable has an associated range that is appropriate for protection of aquatic ecosystems, now regarded as guideline values that can be modified into regional, local or site-specific criteria, or relaxed or tightened depending on the stream's conservation value or level of disturbance (ANZECC and ARMCANZ, 2000c).
- General models of a desirable stream condition [e.g. Young (2001) described a model of a healthy River Murray, Australia, based on scientific knowledge of the important physical, chemical and ecological processes].
- Known empirical relationships [e.g. measures of channel width against catchment area may suggest that the target reach is unusually wide relative to established regional hydraulic geometry relationships (Leopold and Maddock, 1953; Park, 1977)].

Stream condition or health can be assessed across a hierarchy of measurement scales (Table 9.1) and various survey techniques have been devised to capture such data in a systematic way. Methods for surveying stream health vary according to the objectives they were originally designed to satisfy, the level of detail involved (Table 9.1), and the nature of the rivers used in the development of the method. The methods have been thoroughly reviewed in numerous publications, including UNESCO/WHO/UNEP (1992), Metcalfe-Smith (1996), Barbour *et al.* (1999), Ladson and White (1999), Parsons *et al.* (2002a) and ANZECC and ARMCANZ (2000a), and only a selection of these methods will be briefly described here.

**Table 9.1.** *Descriptive variables for assessment of stream health. Source: Rutherfurd et al. (2000). Reproduced by permission of Land and Water Resources Research and Development Corporation and Co-operative Research Centre for Catchment Hydrology*

| Measure | Examples |
| --- | --- |
| Visual description of presence or absence at one point in time | "Native vegetation was absent from the reach, with willows being the only species present" <br> "The water looked turbid" |
| Measured description of presence or absence at one point in time | "Willows were present at density of one tree per 20 m$^2$" <br> "Turbidity was 43 NTU on 26/12/96" |
| Visual comparison with a template reach or original state | "The upstream, uncleared reach had a dense stand of *Eucalyptus camaldulensis*, whereas the target reach had only willows" |
| Measured comparison with template reach | "The template reach had twice as many fish as the target reach when it was sampled" |
| Visual description of change through time | "According to the landholder the head-cut had migrated 200 m since 1974" |
| Measured description of change through time | "Three electro-fishing sweeps, one year apart, showed a statistically significant decline in the number of blackfish present" |

The approaches can be divided into four main groups:

1. physico-chemical assessment
2. habitat assessment
3. hydrological assessment
4. bioassessment

### 9.3.3 Physico-chemical Assessment

Physico-chemical characteristics have long been used as indicators of stream and catchment health, because these variables react to changes in stream flow, land use and riparian conditions (Table 9.2). Physical measurement parameters include flow, temperature, conductivity, suspended solids, turbidity and colour. Chemical measurement parameters include pH, alkalinity, hardness, salinity, biochemical oxygen demand, dissolved oxygen and total organic carbon. Other major controls on water chemistry include specific major anions and cations, and nutrient species (phosphate, nitrate, nitrite, ammonia, silica) (ANZECC and ARMCANZ 2000b).

There have been numerous attempts to integrate various water quality indicators into single indices of water

**Table 9.2.** *General measurement parameters used for assessing aquatic system health. Source: ANZECC and ARMCANZ (2000b, p. 3–19). Reproduced by permission of the Australian and New Zealand Environment and Conservation Council and the Agriculture and Resource Management Council of Australia and New Zealand (both now Natural Resource Management Ministerial Council)*

| Measurement parameter | Input | Potential effects |
| --- | --- | --- |
| Electrical conductivity | Salt | Loss of sensitive biota |
| Total phosphorus | Phosphorus | Eutrophication (nuisance algae) |
| Ratio of total phosphorus to total nitrogen | Phosphorus and nitrogen | Cyanobacterial blooms |
| Biochemical oxygen demand | Carbon in organic material | Asphyxiation of respiring organisms, e.g. fish kills |
| Turbidity | Sediment | Changes in ecosystem habitat <br> Loss of sensitive species <br> Altered light climate that affects productivity and predator–prey relationships |
| Suspended solids | Sediment | Changes in ecosystem habitat, loss of sensitive species |
| Chlorophyll | Nutrients | Eutrophication |
| pH | Acid drainage | Loss of sensitive biota |
| Metals, organic compounds | Toxicants | Loss of sensitive species |

quality (e.g. Brown *et al.*, 1970; Ellis, 1989; Ward *et al.*, 1990, pp. 73–74; Cude, 2001; Hallock, 2002). For non-specialists, raw water quality data can present a challenge for interpretation, so integrative indices are particularly useful for communicating information to the general public. An integrative index provides a single number that can express the relative level of impairment of a given water body and how this has changed through time.

The Oregon Water Quality Index (OWQI) (Cude, 2001) is a single number that integrates eight water quality parameters (temperature, dissolved oxygen, biochemical oxygen demand, pH, ammonia + nitrate nitrogen, total phosphates, total suspended solids, faecal coliform). National Sanitation Foundation (Boulder, Colorado) Water Quality Index (NSFWQI) (Brown *et al.*, 1970; Mitchell and Stapp, 2000) also includes turbidity, while the Washington State Department of Ecology Water Quality Index (DEWQI) (Hallock, 2002) excludes biochemical oxygen demand. These indices were designed to aid in the assessment of water quality for general uses, as a tool to readily communicate water quality information to non-specialists, to provide a basis to evaluate effectiveness of water quality improvement programs, and to assist in establishing priorities for management purposes.

To calculate the OWQI, the raw analytical results for each parameter, having different units of measurement, are first transformed into non-dimensional sub-index values ranging from 10 (worst case) to 100 (ideal) depending on that parameter's contribution to water quality impairment. These sub-indices are then combined to give a single water quality index value ranging from 10 to 100. The unweighted harmonic square mean formula used to combine sub-indices allows the most impacted parameter to impart the greatest influence on the water quality index (Cude, 2001). The NSFWQI is calculated in a similar way, except that the weightings are decided subjectively, on the basis of the perceived importance of each variable to local water quality issues. The DEWQI aggregates turbidity and suspended solids concentration, and considers the TN/TP ratio. Conversion of data to index scores uses formulae that are specific to the location, which depend on the stream class or ecoregion for that location (Hallock, 2002). The heavy weighting of faecal coliform in some integrated water quality indices suggests an orientation towards assessing suitability for contact recreation.

In Latvia, water quality is described using a comprehensive physico-chemical Water Pollution Index (WPI) that incorporates dissolved oxygen concentration, oxygen saturation percentage, pH, BOD, COD, nutrients, salt composition, oil products, heavy metals, pesticides, benthic saprobic index and phytoplankton biomass (Lyulko, 2000). The WPI is calculated as the sum of the ratios of the observed values to the freshwater quality standards in force. Values of WPI <1 are regarded as pure, while values >2 are regarded as polluted. Values of WPI >6 are regarded as heavily impure (Lyulko, 2000).

### 9.3.4    Habitat Assessment

Habitat usually refers to the in-stream and riparian physical and chemical conditions suitable for habitation by biota, so it is specific to biota, and may even vary according to the life cycle of biota (Townsend and Hildrew, 1994; Norris and Thoms, 1999). Habitat quality can be expressed as the presence or absence of suitable habitat (noting the presence of elements that prevent the habitat being accessed), the volume or area available of ideal habitat or a rating of the relative quality of the habitat that is present.

In general, spatial and temporal habitat variability and biological diversity in rivers are closely linked (Petts and Calow, 1996; Downes *et al.*, 1998; Harper and Everard, 1998; Raven *et al.*, 1998). Assuming that water quality and hydrology remains constant over time, the hypothetical relationship between physical habitat quality and biological condition is linear over most of the range (Plafkin *et al.*, 1989) (Figure 9.1). Habitat quality can range from zero to 100% of reference conditions, and can be categorized as non-supporting (<59%), partially supporting (60%–74%), supporting (75%–89%) or comparable (>90%) (Sarver, 2000). The quality of the biological community can range from zero to 100% of the reference, and can be categorized as severely impaired, moderately impaired, slightly impaired or non-impaired (Barbour and Stribling, 1991).

Habitat assessment aims to broadly measure the instream and riparian conditions that influence the structure and function of the aquatic community in a stream. The presence of an altered habitat structure is one of the major stressors of aquatic systems (Karr *et al.*, 1986). Unlike scientific investigations that begin with a hypothesis regarding the likely cause and impact of any disturbance, and then design the measurement scheme accordingly, stream health assessment techniques are expected to be universally applied to a diverse range of problems and environments. Time and resource limitations, the need for expediency, plus the sheer size of the task, have made 'rapid assessment' techniques a popular choice. These methods rely largely on visually assessed or estimated habitat indices, and ignore potentially critical variables because they require instrumental measurement. For example, Gippel and Rhodes (2001) documented cases of headwater streams downstream of diversion

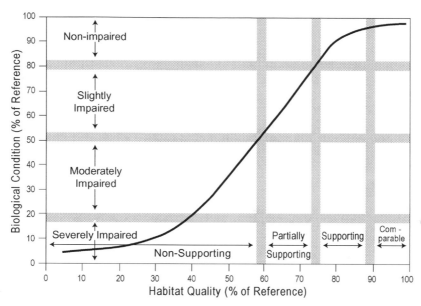

**Figure 9.1.** *Theoretical relationship between physical habitat quality and biological condition. Source: Plafkin et al. (1989)*

weirs where simple visual indices failed to detect degraded stream health (independent macroinvertebrate surveys demonstrated impairment), because the main problem was loss of fast-flowing water in riffles.

The habitat-based approach to impact assessment and resource inventory is well entrenched in the USA. Bain and Hughes (1996) selected and reviewed 50 different methods used by USA agencies for habitat measurement, inventory and reporting. The methods were primarily based on physical habitat, although many also considered water chemistry and the biota. It was common for the methods to require measurement of a long list of variables, even though many of these variables appeared to be redundant (Bain and Hughes, 1996). In general, the analyses incorporated in most methods were simple, with the more sophisticated methods involving synthesis of data using indices, or development of predictive relationships (Bain and Hughes, 1996). This review would apply equally to the methods used in other parts of the world. Some habitat-based stream health assessment methods are summarized below.

### *Habitat Evaluation Procedure and Habitat Suitability Indices (HEP/HSI)*

Habitat Evaluation Procedure (HEP) is often used in the USA to document the quality and quantity of available habitat for selected wildlife species (U.S. Fish and Wildlife Service, 1980a; Brown *et al.*, 2000). Over one hundred

and fifty Habitat Suitability Index (HSI) model reports (for a vast range of species) have been published by the U.S. Fish and Wildlife Service (Raleigh *et al.*, 1984; 1986; Wesche *et al.*, 1987; The National Wetlands Research Center, 2003). HSI models specify hypotheses of habitat relationships that are open for testing and improvement. The original HSI models involved the development of an index from 18 variables that were believed to control the carrying capacity of brown trout in streams (including water temperature, depth, cover, DO, substrate type and baseflow as a percentage of average flow). Indices of habitat suitability are presented on a 0.0 to 1.0 scale, based on the assumption that there is a positive relationship between the index and habitat carrying capacity (U.S. Fish and Wildlife Service, 1981). The HSI measures habitat suitability of a sample plot relative to optimum habitat suitability for a species in regions defined in the individual models.

One drawback with use of individual HSI models in a single stream is that what constitutes suitable habitat for one species may be unsuitable for another. In this situation, a combined, community-based habitat suitability model would be preferable (Hartman, 1999).

### *Rapid Bioassessment Protocol (U.S. E.P.A.) Habitat Assessment*

The United States E.P.A. Rapid Bioassessment Protocol (RBP) (Plafkin *et al.*, 1989; Barbour *et al.*, 1999) is a comprehensive method that covers periphyton, macroin-vertebrates, fish and habitat assessment. The habitat assessment component includes measurement of traditional physico-chemical water quality parameters and quantifiable physical characteristics such as surrounding land use, description of the stream origin and type, summary of the riparian vegetation features, large woody debris density and in-stream parameters such as width, depth, flow and substrate type and size. The RBP also includes a visual-based habitat assessment protocol, similar to the River Habitat Survey methodology used in the United Kingdom (Environment Agency 1997; Raven *et al.*, 1998).

The RBP uses different parameters for low gradient and high gradient streams. All parameters are evaluated over a reach approximately 100 m in length and rated on a numerical scale of 0 to 20 (highest) for each sampling reach. The ratings are then totalled and compared to a reference condition to provide a final habitat ranking. Scores increase as habitat quality increases. To ensure consistency in the evaluation procedure, descriptions of the physical parameters and relative criteria are included in the rating form. The actual habitat assessment process involves rating the 10 parameters as optimal, sub-optimal, marginal or poor based on given criteria. Reference conditions are used to scale the assessment to the 'best attainable' situation. This approach is critical to the assessment because stream characteristics vary dramatically across different regions. The ratio between the score for the test station and the score for the reference condition provides a percent comparability measure for each site. The site of interest is then classified on the basis of its similarity to expected conditions (reference condition), and its apparent potential to support an acceptable level of biological health (Barbour *et al.*, 1999).

### Hydrogeomorphic (HGM) Index of Function Approach for Riverine Wetland Assessment

While rivers and floodplains are assumed to be integral parts of the riverine wetland ecosystem, for practical reasons, the two are separated for the purpose of functional assessment under U.S. Army Corps of Engineers HGM approach (Brinson, 1993; Brinson *et al.*, 1995). This method identifies fifteen functions for riverine wetlands, falling within four categories: hydrologic (5 functions), biogeochemical (4 functions), plant habitat (2 functions) and animal habitat (4 functions). For each of these, an Index of Function (IF) describes the level of functioning relative to reference conditions. Variables are factors that are necessary for functions to occur. To determine an IF, pertinent variables are combined in equations. The reference standards represent the highest level of sustainable functioning in the landscape. These are the conditions used to calibrate the models so that both variables and the IF are set at 1.0 for reference conditions.

To illustrate how the IF is calculated, we will examine the Organic Carbon Export function (within the biogeochemical category). This function requires a source of organic matter in the wetland and appropriate water flows to transport the organic matter downstream. The flow pathway variables include the frequency of overbank flow from the channel ($V_{FREQ}$), flow from subsurface flow ($V_{SUBIN}$) and surface flow ($V_{SURFIN}$), and surface hydraulic connections ($V_{SURFCON}$) with the stream channel. The source of organic matter ($V_{ORGAN}$) is defined as the types and amounts of organic matter in the wetland including leaf litter, coarse woody debris (down and standing), live woody vegetation, live and dead herbaceous plants, organic-rich mineral soils and histosols. The IF equation for Organic Carbon Export is:

$$IF = \{[(V_{FREQ} + V_{SURFIN} + V_{SUBIN} + V_{SURFCON})/4] \times V_{ORGAN}\}^{1/2}$$

If $V_{ORGAN} = 0$, the function is absent.

The equation that models the Organic Carbon Export function is arranged so that the geometric mean of the last variable and the arithmetic mean of the first four can result in the IF being zero if either all hydrologic variables are absent or organic matter is missing from the wetland being assessed. For some variables that are difficult or costly to measure, indicators must be used as surrogate measures for variables.

In the HGM approach, levels of certainty (or uncertainty) are not determined for indicators, variables or functions. The user is only responsible for applying the method to riverine wetlands insofar as the method is capable of detecting thresholds of functioning (Brinson *et al.*, 1995).

### RCE—Riparian, Channel and Environmental Inventory

The RCE was developed in Sweden to assess the physical and biological condition of small streams in the lowland agricultural landscape (Petersen, 1992). The model assumes that disturbance of physical channel and riparian structure is the major cause for a reduction of stream biological structure and function. RCE consists of sixteen characteristics which define the structure of the riparian zone, stream channel morphology and the biological condition in both habitats (Petersen, 1992). The variables are hierarchical and range from the landscape scale to the macrobenthos scale. The RCE score falls into five classes, ranging from excellent to poor. In a study of fifteen Italian streams, the RCE was found to be positively correlated to benthic macroinvertebrate community as measured using the Trent biotic index. One advantage of the RCE is that it is very rapid, requiring only visual observation (Petersen, 1992). Surveys take around 11–20 minutes per site, compared to 45 minutes for the Rapid Bioassessment Protocol of Plafkin *et al.* (1989).

### Index of Stream Condition

The Index of Stream Condition (ISC) is the primary means of reporting stream condition in the State of Victoria, Australia (Ladson and White, 1999; Ladson *et al.*, 1999; Ladson and White, 2000) (Section 4.1.4). The ISC uses a subjective ranking system based on comparing the current conditions with pristine conditions, but also includes some measured physical characteristics of the stream. The ISC is designed to provide a broad, long-term summary of all of the major environmental attributes that affect river health. It may be used for monitoring and to flag potential problems, but it is not useful for scientific hypothesis testing (i.e. a change may be measured through time, but the cause of the change can only be speculated). The ISC is designed to measure long-term changes (i.e. reported every 5 years) over a reach of stream about 10-30 kilometres long. This index provides a basis for reporting the environmental condition of streams to the community and government. The focus of the ISC is on state-wide stream values, and other indicators may be required for local issues (Ladson and White, 1999).

The ISC is compiled by measuring variables (see Section 4.1.4, Table 4.2) and allocating a rating to the measure when compared with the expected 'natural' state (Figure 4.4, Table 9.3). The individual ratings are summed and then scaled so that each sub-index value lies between zero and 10. The final index of stream condition is presented as a sum of the sub-index values, which can have a maximum value of 50 (Table 9.3). Thus, the index shows the relative value of each sub-index, plus the total value. This allows the user to identify the aspect of the stream that is in the worst condition.

The appeal of the ISC is that stream health is based on a core group of variables that are relatively easy to measure. The index also includes water quality and macroinvertebrates, specifying key variables to measure and the acceptable levels or ranges. The ISC can provide a well-structured, rapid assessment of the condition of stream reaches that is comparable between reaches and between streams, and unlike some other rapid assessment methods it includes some basic hydrological, water quality and macro-

**Table 9.3.** *Point scale for unscaled indicator measurements and overall classification for the ISC. Source: Ladson and White (1999)*

| Indicator category (deviation from reference state) | Indicator Rating | Overall stream condition classification | Overall Rating |
|---|---|---|---|
| Very close to reference | 4 | Excellent | 45–50 |
| Minor modification | 3 | Good | 35–44 |
| Moderate modification | 2 | Marginal | 25–34 |
| Major modification | 1 | Poor | 15–24 |
| Extreme modification | 0 | Very poor | <15 |

invertebrate data (see Section 4.1.4 for more details on the ISC).

### Rivercare Approach

The Rivercare methodology is based on the premise that the foundation for stream rehabilitation is a stable, vegetated stream. Raine and Gardiner (1995) summarized the Rivercare approach, with special emphasis on the north coast streams of New South Wales, Australia. Stream reaches are classified by a traffic-light system, where reach vegetation and reach stability (comprising width and alignment) are ranked as being in red (bad), yellow (average) or green (good) condition. Landholders decide on the course of action in the reach (usually a few kilometres of stream) by overlaying clear sheets onto an aerial photograph base. Each layer covers property boundaries, environmental values, geomorphology, permits, management options and a final management plan.

The Rivercare system provides a comprehensive community-based planning tool for managing stream erosion and deposition in short reaches. The emphasis is on producing a stable stream by creating a stable width and stable alignment (both being defined by empirical equations) by clearing inappropriate vegetation from stream channels, using some engineering structures, and planting riparian vegetation.

### State of the Rivers Survey

The State of the Rivers Survey method (Anderson, 1993; Anderson, 1999) was initially devised for streams in Queensland, Australia. The method assesses the 'state of a river in terms of the physical and environmental condition of the rivers and streams throughout the catchment at the time of the survey, relative to the presumed pristine original condition' (Anderson 1993).

Stream sections for assessment are selected using map-based data, and 11 data components are surveyed at each sampling reach. Each data component is composed of different types of variables that characterize the stream channel. Most variables are measured using visual estimation, but some require physical measurement rating on a relative scale. Condition ratings are based on the extent of degradation, ranging from a value of 100% for full pristine condition and 0% for a complete loss of naturalness (Anderson, 1999).

The usefulness of the State of the Rivers Survey method relies heavily on how accurately the operator can determine the 'pristine' condition of the stream. The method does not explain why a reach is in a particular condition. In part this is because it emphasises individual sites rather than the interaction between sites. Also, this method assumes certain cause-effect relationships that in reality may be difficult to determine in the field, such as the relationship between bank erosion and removal of vegetation and/or bar deposition. The method provides an inventory of stream condition, but it lacks a procedure for prioritizing action. Like all rapid snapshot visual-based methods, it does not identify interactions between reaches, and it cannot establish changes over time.

### River Habitat Survey

The River Habitat Survey methodology for rapid assessment of river habitat was developed in the United Kingdom (Environment Agency 1997; Raven et al., 1997; Raven et al., 1998). This method utilizes a large reference site database to predict the physical features of a stream that would occur under unmodified conditions. The field method involves visually assessing a range of features, measuring channel morphology at a cross section, and measuring some other variables from maps. This approach focuses on habitat variables that are of known value to stream organisms. The core of the European standard method (CEN, 2002) is based on the hydromorphological survey system covering channel, river bank and floodplain (IETC, 2002), and is similar to the River Habitat Survey Method.

One difference between the European methodology and the various Australian approaches is that the former is less inclined to apply value-laden terminology to the habitat descriptors. Thus, channel form is described in terms of dimensions, bank material, bank modifications, substrate material and channel modifications, rather than being classified as stable/unstable. The River Habitat Survey does use visual descriptors, but also makes use of simple measurements of channel form that can be more readily checked and evaluated.

### Geomorphic River Styles

The River Styles framework is an Australian classification system based on the hierarchical model of Frissell et al. (1986). The framework characterizes river form and behaviour at four inter-related scales: catchments, landscape units, River Styles (reaches) and geomorphic units. As geomorphic units constitute the basis to assess aquatic habitat availability, and they form the building blocks of river and floodplain systems, the framework assumes that intact reaches of a particular River Style should have similar assemblages of instream and floodplain habitat (Brierley et al., 1996; Brierley, 1999; Thomson et al., 2001). Geomorphic data are measured from maps and in the field, and detailed sediment analysis is also conducted in each geomorphic unit. The information is used to interpret river behaviour, and to map geomorphic process zones. Habitat condition is assessed on the basis of a comparison between current character and behaviour with what would be expected in an undisturbed situation.

A recent study by Thomson et al. (2004) tested the ecological significance of the River Styles framework by comparing the macroinvertebrate assemblages and habitat characteristics of pool and run geomorphic units for three different River Styles on the north coast of New South Wales, Australia. This study found that macroinvertebrate communities within different River Styles were similar, probably because some important large-scale drivers of local habitat conditions are not included in River Styles designations.

### Index of Stream Geomorphology

The Index of Stream Geomorphology is a component of the South African River Health Programme. Geomorphological variables are measured on the assumption that channel morphology provides the physical framework within which the stream biota live (Rowntree and Wadeson, 2000). The geomorphological index has two components: channel classification and channel condition assessment. Information is collected using desktop and field surveys. The desktop exercise includes a reach analysis, which allows assignment of the site to a zone class according to the classification scheme of Rowntree and Wadeson (1999).

Field data collection involves survey of channel transect or cross sections (width-depth ratio and entrenchment ratio). Re-survey of these transects enables monitoring of channel change. The index of channel condition is coupled to a habitat index that describes the diversity of hydraulic habitats at a site in terms of flow hydraulics and substrate conditions, as well as the overhead cover (Rowntree and Wadeson, 2000).

In developing the Index of Stream Geomorphology, Rowntree and Wadeson (2000) combined aspects of the scheme of Rosgen (1996a) for North American rivers, the British River Habitat Survey (Raven *et al.*, 1998), which emphasises the structure of the river, and the Australian Index of Stream Condition (Ladson *et al.*, 1999) which emphasises the condition of the river.

### Habitat Predictive Modelling

Habitat Predictive Modelling uses large-scale catchment features to predict local-scale stream physical habitat features (Davies *et al.*, 2000). Reference site information forms a template against which test sites are compared to assess habitat condition. Reference sites are classified into groups on the basis of similarity of local-scale habitat features, and a predictive model is developed based on large-scale catchment characteristics. The model calculates the probability of occurrence of each habitat feature at a test site, based on the occurrence of that feature within the corresponding reference site groups. The habitat features predicted to occur at a test site are compared against the habitat features that were actually observed at the test site, with the observed:expected ratio being an indication of habitat quality. Despite some analytical limitations of Habitat Predictive Modelling (Davies *et al.*, 2000), the technique has successfully predicted small-scale habitat features in Australia, and represents a promising step forward for habitat assessment (Parsons *et al.*, 2002a).

### 9.3.5   Hydrological Assessment

### Standard Methods of Hydrological Assessment

Dams, water diversions, hydropower generation and change in land use have altered the hydrology of many streams. These factors can affect all aspects of the flow regime, but they particularly reduce the frequency and magnitude of mid-range floods, alter flow seasonality, modify the characteristics of low flows and change the rate of rise and fall. The effect of flow alteration on various aspects of stream health has long been recognized (e.g. Petts, 1980; Nilsson and Dynesius, 1994), and includes impacts on wetlands, channel morphology and substrate, native fish, invertebrates, algae, birds, mammals, riparian vegetation, macrophytes and water quality.

Hydrologists have developed numerous techniques of flow analysis, and these have been widely used to examine the degree of hydrological alteration. The inter-annual coefficient of variation of annual runoff is a fundamental variable in hydrological analysis concerned with evaluation of available water resources, determining hydrological change through time, and comparing the regime of different rivers. Monthly discharge bar charts are commonly used to display intra-annual distribution of flows. The flow-duration curve is one of the most commonly used ways of displaying the distribution of discharges. A series of flow-duration surface curves characterizes the seasonal variability of the full range of flows. The low-flow frequency curve shows the proportion of years when a flow is exceeded, or equivalently, the average interval in years that the discharge falls below a threshold value. The flood frequency curve estimates the return period of floods of various peak magnitudes.

Most of these standard statistical analysis tools used by hydrologists were developed in response to the traditional engineering problems of drought management, flood mitigation or development of water supply systems. The emphasis has been on developing techniques of extrapolating beyond the gauged record, describing runoff in terms of its availability for human use and calculating the hydrological risk posed to humans and their activities and assets. In contrast, consideration of hydrology as part of stream health assessment relies on established or hypothesized relationships between hydrological characteristics of a stream and ecological response. Standard methods of hydrological analysis were not specifically devised to describe flows in terms of their ability to sustain aquatic life. Some standard indices may be appropriate, but it is also necessary to describe discharge records in other ways that relate well to ecological and geomorphological processes.

In assessing flow regimes from an ecological perspective, hydrologists should not feel constrained by the tradition of lumping low flow and high flow data, and concentrating on the characteristics of extreme floods and droughts. Although organisms respond to the flow regime in its entirely, for the purpose of assessing stream health, there is merit in dissecting the flow regime into facets or events that occur across a wide range of time scales.

### Characterizing Flow Variability

Discharge variability is central to sustaining and conserving biodiversity and ecological integrity (Walker *et al.*, 1995; Stanford *et al.*, 1996; Poff *et al.*, 1997; Richter *et al.*, 1997). Flow variability is defined in various ways, covering a wide range of temporal scales. Some studies emphasize the seasonality or timing of certain events (e.g. Tharme and King, 1998), while others emphasize inter-annual variability (Poff *et al.*, 1997). It is useful to simplify the flow regime by reducing it to a number of facets or events that have particular significance for river ecology, geomorphology or water quality:

- Cease-to-flow
- Baseflow
- Flows important for maintenance of water quality
- Small, medium and large flow events of biological significance
- Small, medium and large flow events of geomorphological significance

In assessing these facets of the hydrological regime it is relevant to consider magnitude, frequency, duration, timing and rate of change. These are the equivalent to the five critical flow regime 'components' referred to by Poff *et al.* (1997) when describing stream hydrology as the 'master variable' that regulates ecological integrity. Some limited understanding of the links between biological processes and aspects of flow variability has been achieved, with most progress being made on species of high conservation or commercial value. Puckridge *et al.* (1998) defined twenty-three measures of hydrological variability based on monthly and annual data, but only five were independent. However, they collated evidence that eight of these variables were linked (not necessarily causally, and not universally) to features of fish biology. Jowett and Duncan (1990) classified New Zealand rivers according to seven indices that described inter-annual flow variability, or the range of extremes in the records. All but one of the hydrological indices were highly correlated with each other. This study found strong associations between flow variability and periphyton communities (positive and negative correlations) and rainbow trout abundance (negative correlation), weak associations with benthic invertebrate communities, and no relationship between flow variability and water quality. Jowett and Duncan (1990) concluded that variability in water velocity (associated with discharge variability) was the most important factor influencing river ecology and geomorphology. Clausen and Biggs (1997a, b) further explored the relationship between flow indices and benthic biota in New Zealand. The Range of Variability Approach (RVA) (Richter *et al.*, 1996, 1997), which uses daily data, defines thirty-two parameters that are Indicators of Hydrologic Alteration (IHA). Poff's (1996) hydrological classification scheme uses twelve to fifteen parameters, of which three measure variability.

There is no limit to the number of hydrological parameters that can be contrived to describe flow variability (Ladson and White, 1999; Growns and Marsh, 2000); the

challenge is to find the key parameters that have strong ecological and geomorphological associations, and that are simple enough to include in regular stream health assessment procedures. Olden and Poff (2003) reviewed 171 hydrologic indices (including the IHA) using long-term flow records from 420 sites from across the continental USA. Their results provide a framework from which researchers can identify hydrologic indices that adequately characterize flow regimes in a non-redundant manner.

### Simple Hydrological Index of Flow Alteration

There appears to be no consensus regarding the most ecologically significant hydrological indicators, and there has been little empirical evaluation of indicators, especially in Australia. Stream health assessment aims to identify major deviations from reference or expected conditions and this can be achieved with a simple index. Simple indices will not fully characterize the nature of any hydrological alteration, but they will highlight where change has occurred, and in these cases more detailed hydrological analysis can be conducted as a separate exercise.

Ladson and White (1999) recommended flow indices based on monthly flows, because daily data were less likely to be available universally, and because annual indices cannot shed any light on seasonal variations. One disadvantage of using monthly flows is that information on very low flows and flood peaks is lost. For use in the Index of Stream Condition, Ladson and White (1999) modified the Annual Proportional Flow Deviation (APFD) that Gehrke *et al.* (1995) found was correlated with fish species diversity. The modified index, called the Amended Annual Proportional Flow Deviation (AAPFD), is the sum of the ratio of change in flow (actual—reference) to average monthly flow. The original APFD used reference monthly flow as the denominator, but this is problematic for ephemeral streams where flow can be zero in some months. Higher values of AAPDF indicate a greater degree of hydrological alteration. For a range of regulated New South Wales and Victorian streams examined by Ladson and White (1999, p. 22), the highest AAPFD score was 5, while the lowest rating was 0.24.

For the River Murray, Australia, the AAPFD is highest immediately downstream of the major headwater storages, reaching a value of 4 downstream of Hume Dam, where flows are higher than natural due to inter-basin transfer (Figure 9.2). The AAPFD value is less further downstream at Yarrawonga Weir, although it could be argued that the degree of regulation here is higher, due to diversions. The

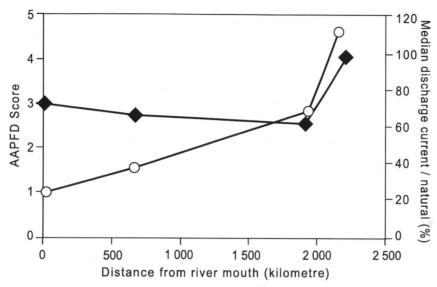

**Figure 9.2.** *Comparison of the AAPFD hydrological alteration index (◆) (Ladson and White, 1999) with change in median annual discharge (○) (Gippel et al., 2002) for the River Murray Australia, for natural versus current flow conditions. Hume Dam is located at 2225 km, while major diversions occur at Yarrawonga Weir, at 1988 km*

AAPFD value then gradually increases towards the mouth, as further diversions are made. This is despite a semblance of natural seasonality returning to the river due to tributary inflows. The AAPFD value is only loosely related to the reduction in annual median discharge, compared to natural conditions (Figure 9.2).

For use in the ISC, the raw AAPFD index value is converted to a rating in the range 0–10, with lower values indicating greater flow alteration. Ladson and White (1999) also included two additional sub-indices: daily flow variation due to change in catchment permeability from urbanization (PERM), and daily flow variation due to peaking hydroelectricity generation (PHE). These indices are rated as present-or absent to give a value of zero or 1. The hydrology index is then calculated as the AAPFD rating, minus PERM and minus PHE, to give a final rating in the range of 0–10. Unregulated streams would be expected to score >8 (some alterations in flow may be apparent due to other causes).

The recent National Land and Water Resources Audit (2002) assessment of river condition in Australia utilized a Hydrological Disturbance sub-index as part of the Environment Index. The hydrological disturbance sub-index is based on comparisons of the current flow regime to the modelled natural or pre-European settlement flow

regime (Young *et al.*, 2001)). The key aspects of flow regime change included in the sub-index were:

- changes in total flow volumes using a mean annual flow index
- changes in flow regime variability using a monthly flow-duration curve difference index (a measure of flood frequency)
- changes in the seasonal pattern of flows using a seasonal periodicity index to assess changes to the seasonal timing of high and low flows, and a seasonal amplitude index to assess changes in the magnitudes of seasonal highs and lows

### 9.3.6  Bioassessment

In many parts of the world, legislation now requires agencies to adopt biologically based monitoring systems for assessment of stream health. For example, since December 2000, European water management has been directed by the EU Water Framework Directive, which defines a methodological framework for assessing the ecological status of rivers in the member countries (Blöch, 1999; Kallis and Butler, 2001; EUROPA, 2003). Bioassessment methods directly measure a biotic

**Table 9.4.** *Summary of bioassessment approaches and measurement parameters (from ANZECC and ARMCANZ 2000b, p. 3–25). Reproduced by permission of the Australian and New Zealand Environment and Conservation Council and the Agriculture and Resource Management Council of Australia and New Zealand (both now Natural Resouce Management Ministerial Council)*

| Approach | Measurement parameters | Advantages | Disadvantages | Overall value |
|---|---|---|---|---|
| Diversity indices | Various | Provide summary of complex data; easy to understand, allow comparisons between sites or times | Ecological significance of indices is unclear; can be affected by sampling and analytical factors | Attractive for their simplicity, but their ecological value is questionable |
| Biotic indices | Principally macroinverte-brates and algae | Simple, easy to interpret summaries of complex data; can provide contaminant-specific response | Detailed knowledge of contaminant tolerance required for diagnostic use | Usefulness limited by baseline; site-specific and contaminant tolerance information needed |
| Stream community metabolism | Benthic flora and fauna | Integrates impact across the entire benthic biota; relatively rapid; provides a simple output | Technique not proved; may be less useful in disturbed catchments; diagnostic capability unclear | Technique has potential, but its sensitivity and diagnostic capacity have not been demonstrated |
| Macroinvertebrate community structure (e.g. AusRivAS) for rapid biological assessment; quantitative methods for site-specific studies | Macroinverte-brates | Integrates over appropriate temporal and spatial scales; much background information available; good diagnostic capability | Relies on complex modelling approach; output not as readily understood as other techniques | Great potential for identification of impacts; reasonable potential for establishing causes of impacts |
| Macrophyte community structure | Macrophytes | Easily sampled, respond to a range of impacts | Gives poor understanding of factors affecting community structure; insensitive to some chemical contaminants | Limited use |
| Fish community structure, biomarkers (biochemical, physiological, immunological or histopathological) | Fish | Readily sampled, taxonomically well known | Gives poor knowledge of population dynamics and water quality factors; temperate fauna are impoverished; biomarker techniques require sophisticated equipment and high level of expertise | Community structure uses more applicable in tropical than temperate waters |
| Algae: biomass and community structure | Algae | Sensitive, taxonomically well known, has diagnostic potential; community structure (AusRivAS-type) approach most promising | Identification requires high level of expertise; community structure approach not well tested | Community structure approach has good potential |
| Bacteria, protozoa and fungi: community structures | Bacteria, protozoa and fungi | Organisms occupy key ecological role so community change can provide valuable key to impacts | May recover too rapidly from impact for moni-toring purposes; taxo-nomy and response to chemical contami-nants poorly known | Limited use at present; would require extensive taxonomic and diagnostic work before they could be useful |

characteristic of the health of a stream. Biassessment methods have also been applied to measuring the health of estuaries (Deeling and Paling, 1999). Alteration of stream habitat, hydrology or water quality can produce a range of effects on aquatic organisms (Friedrich *et al.*, 1992):

- changes in the species composition of aquatic communities
- changes in the dominant groups of organisms in a habitat
- impoverishment of species
- high mortality of sensitive life stages, e.g. eggs, larvae
- mortality in the whole population
- changes in behaviour of the organisms
- changes in physiological metabolism
- histological changes and morphological deformities

The methods of bioassessment measure these effects using a wide range of biotic parameters, or bioindicators, although benthic macroinvertebrates are the most popular choice (e.g. Hellawell, 1977; Friedrich *et al.*, 1992) (Table 9.4). More than one hundred different bioassessment methods exist in Europe, two thirds of which are based on macroinvertebrates (Rosenberg and Resh, 1993; Verdonschot, 2000).

Bioassessment methods often rely on the identification of regionally distinct control or reference sites, where water quality and community composition is assumed to be undisturbed. Other approaches use indexes of diversity on the assumption that high levels of diversity are desirable and equate with high levels of biological integrity (Simon and Lyons, 1995; Barbour *et al.*, 1999). Numerous metrics can be calculated based on presence/absence data or on relative abundances of taxa. Other methods relate the presence or absence of indicator taxa to ranges of water quality (Friedrich *et al.*, 1992; ANZECC and ARMCANZ, 2000a).

### Target Species

A bioindicator is a selected aspect of the environment (a surrogate of a parameter) that represents the parameter of interest, which is either not directly measurable or is too difficult and/or expensive to measure. It is not possible to sample all components of the biotic community, so an alternative is to measure a *target species*, of which there are several types. Species with a high degree of acceptance are known as *flagship species*. They are usually individual, high profile, charismatic animals promoted by conservation campaigns in the hope that the lesser-known

species that coexist with them will also be protected. *Umbrella species* are those that are assumed to represent a system, such that by protecting the umbrella species and its habitat a number of other species that depend on the same habitat are also protected. *Protected species* are those that have special legal protection because of rarity, cultural or historical importance, or because their habitat is under threat.

Species that are easy to catch and identify may be the pragmatic choice of indicator, but a meaningful assessment of stream health can really only be obtained if there is some known process link between the ecosystem and the selected indicator. This means that species whose autecology, life history, habitat needs and population dynamics are reasonably well known are the best candidates as indicators. *Scale species* are those that, because of their body size or ability to move, have a relation to a certain spatial scale or range of mobility. *Keystone species* are the dominant producers or consumers of a system, so that their condition is a determinant, rather than an indicator of the health of the system as a whole.

For assessing stream health, the most promising target species appear to be found among benthic algae, macroinvertebrates and fish. Stream community metabolism is also a useful bioindicator (Davies, 1997). Macrophytes are easily measured but suffer limitations in interpretation. Phytoplankton (Hötzel and Croome, 1998) and zooplankton are other potential bioindicators, but the main difficulty is their inherently high temporal and spatial variability (Chessman and Jones, 2001; ANZECC and ARMCANZ, 2000a). Assessment of the condition of riparian vegetation as an indicator of stream health is not well advanced, but holds some potential (Werren and Arthington, 2002). Bacteria, protozoa, fungi, frogs, reptiles and waterbirds have limited potential as bioindicators (ANZECC and ARMCANZ, 2000a; ANZECC and ARMCANZ, 2000b) (Table 9.4).

### Benthic Algae

Benthic algae (periphyton or phytobenthos) grow attached to surfaces such as rocks, plants and snags. They are primary producers that form the basis of many stream food webs. The characteristics of benthic algae largely reflect the physical, chemical and biological conditions that prevailed during the time in which the assemblage developed. Benthic algae are useful ecological indicators because they are abundant in most streams, because they respond rapidly to changed conditions, are relatively easy to sample and their tolerance to environmental conditions is known for many species (Collins and Weber, 1978;

Lowe and Pan, 1996). Periphyton have been widely used as bioindicators in Europe, where many different approaches have been used for sampling and data analysis (Whitton *et al.*, 1991; Whitton and Rott, 1996). Some phytoplankton and periphytic algae have been shown to have very narrow tolerance ranges of pH, and diatoms have been used to indicate acidification in rivers (Coste *et al.*, 1991). Considerable work has been done in New Zealand relating characteristics of periphyton communities to organic enrichment (Biggs, 1989), and a manual for using periphyton as a means of bioassessment is now available (Biggs, 2000).

Field sampling of periphyton can be done on natural substrate or artificial substrates that are placed in aquatic habitats and colonized over a period of time (Barbour *et al.*, 1999). Rapid, semi-quantitative assessments of benthic algal biomass and taxonomic composition can be made with a viewing bucket marked with a grid and a biomass scoring system. This technique is a survey of the natural substrate performed in the field, but hand picked samples can be later analysed in the laboratory to verify field identification (Barbour *et al.*, 1999).

### Macroinvertebrates

Macroinvertebrates are a group of animals without backbones that are large enough to be seen with the naked eye (>0.5 mm). They may be carnivores, omnivores, herbivores or detritivores. The major groups include worms, molluscs (snails and mussels), crustacea (shrimps and crayfish) and insects (larvae and adults). Macroinvertebrate communities are composed of a number of species, which differ in their tolerances and habitat requirements (Jackson, 1997).

Benthic invertebrates live on, or in, the solid substrates at the bottom of rivers, wetlands and lakes, so they are particularly responsive to changes in flow regime, habitat availability and water quality. Suitable habitat includes wood, aquatic plants, fine organic sediments and inorganic substrata such as sand, gravel and cobbles. These invertebrates cover a large range of sizes, but most are less than 20 mm in length. Macroinvertebrates graze periphyton (and may prevent blooms in some areas), assist in the breakdown of organic matter and cycling of nutrients and, in turn, may become food for predators (e.g. fish). (ANZECC and ARMCANZ, 2000a).

According to Jackson (1997) macroinvertebrates are a widely used indicator of aquatic ecosystem condition because:

- They are a very important part of food chains in aquatic ecosystems
- They are relatively easy to study

- They have potential for monitoring a wide range of river management issues (Arthington, 1993)
- Their lifespan (a few months to several years) and relative lack of mobility (cf. fish) means that changes in communities can sensitively indicate impacts
- Changes in behaviour (e.g. drift) can also indicate impacts and are readily measured

Benthic invertebrates are the organisms most commonly used for biological monitoring of freshwater ecosystems worldwide, including Australia. This is because they are found in most habitats, they have generally limited mobility, they are quite easy to collect by way of well established sampling techniques, and there is a diversity of forms that ensures a wide range of sensitivities to changes in both water quality (of virtually any nature) and habitats (Hellawell, 1986; Abel, 1989). Many forms live in sediment, so they are the most common choice of bioindicator in assessment of sediment toxicity. A number of invertebrate species also live for a sufficiently long time (e.g. molluscs and crustaceans) to be of value as bioaccumulating indicators. Information about an organism's preferred habitat and tolerances to certain types of pollution can be used to interpret habitat quality and degree of water pollution. The responses of different taxa to specific types of chemical contamination has been documented for Northern Hemispheric waters (e.g. Hellawell, 1986). While this information is also available for large parts of Australia and New Zealand, there has been no useful synthesis and review of the scattered reports (ANZECC and ARMCANZ, 2000a).

### Fish

Fish are popular bioindicators because they are known to be sensitive to water quality, they are known to have characteristic habitat preferences, are relatively easy to sample and identify in the field, and they tend to integrate effects of lower trophic levels; thus, fish assemblage structure is reflective of integrated environmental health (Barbour *et al.*, 1999). As they have a relatively large range, fish are best suited to assessing macrohabitat and regional differences. They are long-lived, so fish can integrate the effects of long-term changes in stream health (Simon and Lyons, 1995). Fish populations and communities can respond actively to changes in water quality, but are also strongly influenced by changes in hydrology and physical habitat structure (ANZECC and ARMCANZ, 2000a; Chessman and Jones, 2001). Additionally, fish are highly visible and much valued by the wider community, so fish monitoring usually has strong community approval and interest.

**Table 9.5.** *Condition indicator metrics of fish diversity and abundance used for Swedish watercourses. Source: Swedish Environmental Protection Agency (2002). Reproduced by permission of Swedish Environmental Protection Agency*

| Class | Level | No. of species | Biomass (g/100 m$^2$) | No. of fish | Propotion of anadromous fish | Reproduction ratio | Combined index | Description |
|-------|-------|----------------|----------------------|-------------|------------------------------|--------------------|----------------|-------------|
| I | Very high | >4 | >2200 | >222 | 1 | 1 | <2.0 | Fish communities of great size and species diversity |
| II | High | 3 | 640–2200 | 64–222 | 0.90–1.00 | 0.67–1.00 | 2.0–2.5 | |
| III | Moderately high | 2 | 260–640 | 23–64 | 0.73–0.90 | 0.50–0.67 | 2.5–3.6 | Fish communities of average size and species diversity |
| IV | Low | 1 | 95–260 | 6–23 | 0.16–0.73 | 0.33–0.50 | 3.6–4.0 | |
| V | Very low | 0 | <95 | <6 | <0.16 | <0.33 | >4.0 | Fish communities of small size and diversity |

*Notes on Table:*
- Number of species, biomass and number of fish refer to species that are native to Sweden.
- Proportion of anadromous fish is calculated as the number of such fish caught in relation to the total number of fish caught.
- Reproduction ratio is the number of anadromous fish species represented by offspring in their first year.
- Anadromous fish species are those that migrate up watercourses from the sea in order to breed; they include salmon, brown trout, char and grayling.
- The combined index in the above table is calculated as the average for the various kinds of classifications of fish fauna in a watercourse. Thus, each index rating is based on as many as five separate classifications.

The method for assessing the condition of fish in Sweden utilises a range of metrics based on abundance and diversity (Swedish Environmental Protection Agency, 2002). On this basis, fish condition at the sampling point is graded according to one of five health classes (Table 9.5). The number of species index has a reference value range based on channel size and altitude. For this index the degree of deviation from reference condition is expressed by observed/expected ratio, with five classes of deviation ranging from none (>0.85) to very large (<0.35).

In areas where fish diversity is low or dominated by exotic species or migratory species, fish are less useful as a bioindicator of local stream health. In Australia, the use of fish to monitor stream health is also limited by lack of fundamental biological knowledge of many species (ANZECC and ARMCANZ, 2000a). While methods for surveying fish are well established, the recent development of techniques for sampling of fish larvae (Humphries and Lake, 2000) raises the possibility of assessing fish breeding success rapidly and at smaller spatial scales (Chessman and Jones, 2001).

*Stream Community Metabolism*

Community metabolism refers to the biological transfer of carbon and involves measurement of the basic ecological processes of production (via photosynthesis) and respiration. Community metabolism is sensitive to small changes in water quality and riparian conditions, including light inputs and may enable early detection of an impact before it is manifest in changes in organism assemblages (ANZECC and ARMCANZ, 2000a; Chessman and Jones, 2001).

Metabolism is best measured by monitoring oxygen concentration (e.g. Chessman, 1985; Treadwell *et al.*, 1997; Bunn *et al.*, 1999). Stable isotope ratios have been used to study patterns of nutrient and energy flow through Australian aquatic ecosystems (Bunn and Boon, 1993). The P/R (Gross Primary Production: Respiration) ratio is considered a key biological indicator of system health. Unimpacted streams are typically heterotrophic (i.e. P/R<1) and therefore are a net consumer of carbon. Davies (1997) showed that impacted sites typically have a P/R ratio >1 (autotrophic), indicating a fundamental shift in the energy base of the ecosystem. A shift from heterotrophy to autotrophy can indicate catchment disturbance and/ or nutrient enrichment (ANZECC and ARMCANZ, 2000a).

### Macrophytes

Macrophytes are aquatic plants, growing in or near water that are emergent, submergent or floating. These plants provide cover for fish and habitat for aquatic invertebrates. Reduced numbers of fish and waterbirds can be expected in systems where macrophytes have been lost (Crowder and Painter, 1991). Absence or degradation of macrophytes may indicate high levels of turbidity, salt or toxic substances, while prolific macrophyte growth can result from high nutrient levels (ANZECC and ARMCANZ, 2000a). Advantages of macrophytes as bioindicators are their ease of sampling, the possibility of mapping at broad spatial scales using remote sensing (Chessman and Jones, 2001), and no requirement for laboratory analysis.

Rooted plants predominantly take up nutrients from sediments, while submersed species can take up nutrients through their leaves as well as their roots, making them more responsive to water quality changes (ANZECC and ARMCANZ, 2000a). Outridge and Noller (1991) found that rooted macrophytes were of little use for biomonitoring of sedimentary metals but that free-floating species could be potentially useful for biomonitoring of metals in water. Limited knowledge of macrophyte population dynamics, and how factors other than water quality affect their growth, currently limit the use of macrophytes as bioindicators (ANZECC and ARMCANZ, 2000a). However, aquatic and semi-aquatic plants in south-eastern Australia do show inter-specific differences in their responsiveness to water regime, so they are useful indicators in the context of modified flow regimes (Brock and Casanova, 1997; Chessman and Jones, 2001).

### Riparian Vegetation

Riparian systems have an intimate connection with instream systems and appear to be sensitive indicators of environmental change (Werren and Arthington, 2002). The riparian zone is the link between terrestrial and aquatic systems (Auble *et al.*, 1994), and Tabacchi *et al.*, (1998) argued that this zone is now well integrated into conceptual models of stream ecosystem functioning. A review of existing methods for quantifying riparian vegetation as an indicator of stream health by Werren and Arthington (2002) found that most methods measured structure and failed to consider activity (metabolism or primary productivity), while few considered resilience [i.e. the three system attributes that define health, according to Karr (1999)]. They developed a generalized rapid assessment protocol for riparian vegetation that addressed these shortcomings, but the method awaits testing. A riparian vegetation index used in northeast Spain, named QBR (from its Catalan abbreviation, 'Qualitat del Bosc de Ribera', which translates to 'Riparian Forest Quality'), is based on four components of riparian habitat: total riparian vegetation cover, cover structure, cover quality and channel alterations. It also takes into account differences in the geomorphology of the river from its headwaters to the lower reaches (Munné *et al.*, 2002).

### *Biotic-Water Quality Indices*

There are three principal approaches to biological assessments that utilise taxonomic and pollution tolerance data: saprobic indices, diversity indices and biotic index and scoring systems that use both the saprobic and diversity index approaches to evaluate taxa richness and pollution tolerance (IETC, 2002).

### Saprobic Indices

The Saprobic system is based on observations made over 150 years ago that there was a distinct change in biota downstream of a major source of organic matter pollution (Kolkowitz and Marsson, 1908; Sharma and Moog, 1996). The system was originally developed in central Europe, but it is also popular in eastern European and some other western European countries where the taxonomy of aquatic organisms is well developed, and it is possible to use regional species level relationships. The Saprobic system requires identification at the species level, which is demanding in terms of sample processing effort, but it provides a more precise picture of the water quality.

The Saprobic system is based on four zones of gradual self-purification: the polysaprobic zone (extremely severe pollution), the $\alpha$-mesosaprobic zone (severe pollution), the $\beta$-mesosaprobic zone (moderate pollution), and the oligosaprobic zone (slight or no pollution). These zones are characterized by the presence of characteristic indicator species and chemical properties (Table 9.6). Therefore, comparison of the species list from a specific sampling point with the list of indicator species for the four zones enables surface waters to be classified into quality categories (Friedrich *et al.*, 1992).

Saprobic indices focus on species presence in relation to organic pollution (Pantle and Buck, 1955; Zelinka and Marvan, 1961). The tolerance of an organism is described by the parameters of the preferred Saprobic zone of the

**Table 9.6.** *Traditional Saprobic System Classes and typical biological oxygen demand (BOD) and dissolved oxygen (DO) levels. Water quality class corresponds to the traditional system used in Europe, including Germany, Austria and Denmark.*

| Water Quality Class[a] | Description[a] | Saprobic Class[a] | Saprobic Index[a, b] | BOD (mg/L) typical range[c] | DO (mg/L) typical minimum[c] |
|---|---|---|---|---|---|
| I | no or very little pollution | oligosaprobic | <1.3 | 1 | 8 |
| I to II | slight pollution | oligosaprobic-β-mesosaprobic | 1.4–1.7 | 1–2 | 8 |
| II | moderate pollution | β-mesosaprobic | 1.8–2.1 | 2–6 | 6 |
| II to III | moderate to heavily polluted | β-mesosaprobic-α-mesosaprobic | 2.2–2.5 | 5–10 | 4 |
| III | heavily polluted | α-mesosaprobic | 2.6–3.0 | 7–13 | 2 |
| III to IV | heavily to very heavily polluted | α-mesosaprobic transition zone | 3.1–3.4 | 10–20 | <2 |
| IV | very heavily polluted | polysaprobic | >3.5 | 15 | <2 |

[a] Form Telford (2003), based on information from Eugen Rott (University of Innsbruck, Austria).
[b] Slightly different ranges of Saprobic Index provided by Hosmani (2002) and UNESCO/WHO/UNEP (1992).
[c] Hoffmann (2003).

species, a weighting that depends on the range of tolerance of the species across different quality classes, and species abundance. The Saprobic Index of a site, which ranges from 1 to 4, is calculated as the 'weighted mean' of all individual indices (for all species present) (Table 9.6). Examples include the Saprobic Index (DEV, 1992) and the Saprobic Water Quality Assessment (Moog *et al.*, 1999).

Within the European Union, the EU Water Framework Directive requires that biological assessment be described using only five classes (high, good, moderate, poor and bad), it must be river type specific, and be based on reference conditions that are defined for each stream type separately (EUROPA, 2003). Rolauffs *et al.* (2003) modified the traditional Saprobic system in order to meet these requirements for Germany by defining five 'saprobic quality classes' on the basis of the degree of deviation from the 'saprobic reference condition'. The reference condition was defined as the mean of the lowest 10% of Saprobic Index values for a given river type, minus two standard deviations. The saprobic quality classes were then defined on the basis of degree of deviation from the reference condition (Table 9.7). Unlike the previous fixed saprobic classes, Rolauffs *et al.* (2003) suggested different ranges of saprobic index for each of five saprobic water quality classes as they applied to the 20 German river classes (defined in terms of regional geology and bed material size) of Schmedtje *et al.* (2001).

### Species Richness and Diversity Indices

The idea that high biodiversity is associated with a more stable ecosystem is one of the commonly held assumptions that ground modern environmental management. This

hypothesis, first formalised by MacArthur (1955), basically states that the more complex the ecosystem, the more successfully it can resist a stress (Commoner, 1972). More recently, it has been suggested that ecosystem processes depend more strongly on functional diversity than on species diversity *per se* (e.g. Diaz and Cabido, 2001). While functional diversity and species diversity are often correlated in experimental settings, the relevance of this correlation to natural communities is subject to debate (Hooper *et al.*, 2002).

Species richness is the number of species present, and species diversity is their relative abundance. Many species richness and diversity indices have been used in aquatic ecostems (Washington, 1984; Friedrich *et al.*, 1992) (Table 9.8). The most widely used diversity indices are the Shannon (also called Shannon-Weaver, or Shannon-Wiener) Index (Spellerberg and Fedor, 2003), Simpson Index (Simpson, 1949), and Margalef Index. Each index

**Table 9.7.** *Saprobic Quality Classes that conform with the EU Water Framework Directive. Source: Rolauffs et al. (2003)*

| Status | Saprobic quality class |
|---|---|
| High | ≤5% deviation from the reference condition |
| Good | >5% to ≤25% deviation from the reference condition |
| Moderate | >25% to ≤50% deviation from the reference condition |
| Poor | >50% to ≤75% deviation from the reference condition |
| Bad | >75% deviation from the reference condition |

**Table 9.8.** *Some measures of diversity sed in aquatic systems. Source: Friedrich et al. (1992). Reproduced by permission of World Health Organisation*

| Index | Calculation |
|-------|-------------|
| Simpson Index $D$ | $D = \dfrac{\sum_{i=1}^{S} n_i(n_i - 1)}{n(n-1)}$ |
| Species deficit according to Kothé | $\dfrac{A_1 - A_x}{A_1} \times 100$ |
| Margalef Index $D$ | $D = \dfrac{S-1}{\ln N}$ |
| Shannon Index $H'$ | $H' = \sum_{i=1}^{S} \dfrac{n_i}{n} \ln \dfrac{n_i}{n}$ |
| Shannon Eveness $E$ | $E = \dfrac{H'}{H'\text{max}}$ |

$S$ the number of species in either a sample or a population
$A_1$ the number of species in a control sample
$A_x$ the number of species in the sample of interest
$N$ the number of individuals in a population or community
$n$ the number of individuals in a sample from a population
$n_i$ the number of individuals of species $i$ in a sample from a population

has its strengths and weaknesses (Boyle *et al.*, 1990). For example, the Margalef Index is easy to calculate, but it is not sensitive to species evenness, Simpson's Index is a measure of the dominance in a sample and is insensitive to rare species. Patil and Taillie (1979) pointed out that in a diverse community, the typical species is rare, so they suggested defining diversity as the average rarity of species within a community.

Species diversity can increase as a result of low levels of nutrient pollution, although this may not be considered ecologically desirable. Diversity can be very low where it is naturally limited by the conditions of the habitat, such as in small springs and headwaters (Friedrich *et al.*, 1992). Diversity indices are probably best applied to situations of toxic or physical pollution which impose general stress (Hawkes, 1977). Benthic organisms are the most suitable for diversity indices, as more mobile organisms, such as plankton, may reflect the situation elsewhere in the water body rather than at the monitoring site (Friedrich *et al.*, 1992).

### Pollution Tolerance Biotic Indices and Scores

There are hundreds of methods for biological water quality assessment. Sharma and Moog (1996) differentiated between table-based methods, the origin of which is the Trent Biotic Index (Woodiwiss, 1964), and score-based methods, the origin of which is the Chandler's

Biotic Score (Chandler, 1970) (although the Chandler Biotic Score is actually a later variation of the Trent Biotic Index). The Trent Biotic Index is based on the number of defined taxa of benthic invertebrates in relation to the presence of six key organisms found in the fauna of the sample site.

Pipan (2000) found that for the Reka River (Slovenia), the Chandler Biotic score, the Shannon-Weaver Index and the Saprobic Index were correlated. For the Woluwe River in Brussels and Flanders (Belgium), the diatom Saprobic Index and the Ellenberg macrophyte index were strongly correlated (Triest *et al.*, 2001). Both groups showed strong correlations with phosphate, ammonium and chemical oxygen demand. The Belgian Biotic Index (De Pauw and Vanhooren, 1983) showed lower correlations with the nutrient variables, but was slightly better correlated to chemical oxygen demand, chloride and dissolved oxygen. None of the indices showed a correlation with nitrate. Triest *et al.* (2001) concluded that the three methods were complementary, because the indices based on the primary producers were more indicative for the trophic status, whereas the Belgian Biotic Index showed a broader relationship to the general degree of pollution.

Some other methods that have found application are BMWP Average Score Per Taxon (ASPT) and family-level BMWP number of taxa Scores (Armitage *et al.*, 1983), the Indice Biologique Globale Normalisé (AFN, 1985), Danish Fauna Index (Skriver *et al.*, 2000), Acidity Index (Henrikson and Medin, 1986), O/C Index (density of oligochaetes/density of both oligochaetes and sediment-dwelling chironomids) (Wiederholm, 1980) and the sequential comparison index (Cairns *et al.*, 1968). These methods are based on the presence or absence of certain 'indicator' groups, and/or 'indicator' species and are best suited to use in waters polluted with organic matter, as the indicator organisms are usually sensitive to decreases in oxygen concentrations (Friedrich *et al.*, 1992). The advantages and disadvantages of biotic indices were discussed by Washington (1984). While most indices are applicable to water polluted by organic matter, a biological monitoring approach has recently been developed for acidified streams, using an Acidification Index based on the tolerance of invertebrates to acidity (Raddum *et al.*, 1988; Fjellheim and Raddum, 1990).

The ECOSTRIMED Index (ECOlogical STatus RIver MEDiterranean) was developed for use in Spain to assess the ecological status of rivers in accordance with the EU Water Framework Directive (Prat *et al.*, 2004). This protocol (known as ECOBILL) includes two indexes of biological quality based on macroinvertebrates, the FBILL and IBMWP. The FBILL is a simplification of the original BILL index and keeps the taxonomic level to family level.

**Table 9.9.** *Calculation of the Spanish ECOSTRIMED index of ecological status on the basis of two macroinvertebrate indices (FBILL and IBMWP) and a riparian vegetation condition index (QBR). The five classes of ecological status are those proposed in the EU Water Framework Directive. Source: Prat et al. (2004)*

| FBILL | IBMWP | QBR | | |
|-------|-------|------|------|------|
|       |       | >75  | 45–75 | <45 |
| 8–10  | >100  | HIGH | GOOD | FAIR |
| 6–7   | 61–100 | GOOD | FAIR | POOR |
| 4–5   | 36–60 | FAIR | POOR | POOR |
| 0–3   | <36   | POOR | BAD  | BAD |

The IBMWP index is a Spanish version of the BMWP index used in the UK. The ECOSTRIMED Index also incorporates the QBR index of riparian environment quality. The habitat index IHF (from the Spanish 'Indice de Habitat Fluvial') is also measured, but is not used in the calculation of the final index. The main factor determining ecological status is the biotic index, because even when riparian habitat is healthy, highly polluted (Table 9.9) waters prevent the functioning of freshwater ecosystems (Prat *et al.*, 2004).

The method for assessing the condition of benthic invertebrate fauna in Sweden utilises a range of biotic indices (Swedish Environmental Protection Agency, 2002). On this basis, water quality at the sampling point is graded according to one of five health classes (Table 9.10). Each index has a range of reference values defined for the six defined geographical regions of Sweden. The degree of deviation from reference condition is expressed by observed/expected ratio, with five classes of deviation ranging from none (>0.9) to very large (<0.3).

The macroinvertebrate-based SIGNAL index has been developed for use in eastern Australia by Chessman (1995). This index is a modification of the British BMWP score system developed by Armitage *et al.* (1983). Each type of macroinvertebrate has a 'grade number' (Chessman *et al.*, 1997; Chessman, 2003a, b). A low grade number means that the macroinvertebrate is tolerant of a range of environmental conditions, including common forms of water pollution. A high number means that the macroinvertebrate is sensitive to most forms of pollution. The higher the number, the greater the average sensitivity. The SIGNAL index for a site is calculated by summing the grades for all the families present, the total is then divided by the number of families at the site, which gives the average grade per family. The resulting value of SIGNAL can vary from 1 to 10, and reflects the site's status with respect to organic pollution (Table 9.11). The latest version of SIGNAL has versions to suit both family and order-class-phylum identification (Chessman, 2003a, b).

Rivers with high SIGNAL scores are likely to have low levels of salinity, turbidity and nutrients such as nitrogen and phosphorus. They are also likely to be high in dissolved oxygen. When considered together with macroinvertebrate richness, SIGNAL can provide indications of the types of pollution and other physical and chemical factors that are affecting the macroinvertebrate community. SIGNAL scores can be calculated with or without abundance weighting. The scores are plotted against number of orders-classes-phyla or families observed, and the distribution of points is divided into quadrants (Chessman, 2003b). Quadrant 1 is indicative of favourable habitat and chemically dilute waters, Quadrant 2 is indicative of high salinity or nutrient levels (may be natural), Quadrant 3 is indicative of toxic pollution or harsh physical conditions, and Quadrant 4 is indicative of urban (e.g. sewerage treatment plants), industrial or agricultural pollution or sites downstream of dams (Figure 9.3).

**Table 9.10.** *Indicator metrics of benthic macroinvertebrate diversity and distribution used for Swedish watercourses (riffle sites). Indicators explained in text. Source: Swedish Environmental Protection Agency (2002). Reproduced by permission of Swedish Environmental Protection Agency*

| Class | Index value | Shannon diversity index | ASPT index | Danish fauna index | Acidity index |
|-------|-------------|-------------------------|------------|--------------------|---------------|
| I     | Very high        | >3.71      | >6.9    | 7   | >10  |
| II    | High             | 2.97–3.71  | 6.1–6.9 | 6   | 6–10 |
| III   | Moderately high  | 2.22–2.97  | 5.3–6.1 | 5   | 4.6  |
| IV    | Low              | 1.48–2.22  | 4.5–5.3 | 4   | 2.4  |
| V     | Very low         | <1.48      | <4.5    | 1–3 | <2   |

**Table 9.11.** *Interpretation of SIGNAL value with respect to reference organic water quality pollution. Source: Chessman (1995) and Ladson and White (1999)*

| Class | SIGNAL value (upland reaches) | SIGNAL value (lowland reaches) | Water Quality |
|---|---|---|---|
| I | >7 | >6 | Excellent |
| II | 6–7 | 5–6 | Clean water |
| III | 5–6 | 4–5 | Doubtful, mild organic pollution |
| IV | 4–5 | 3–4 | Moderate organic pollution |
| V | <4 | <3 | Severe organic pollution |

### Multi-metric Techniques

*Indices of Biotic Integrity*

Biological integrity is defined as 'the ability to support and maintain a balanced, integrated, adaptive biological system having the full range of elements (genes, species and assemblages) and processes (mutation, demography, biotic interactions, nutrient and energy dynamics and metapopulation processes) expected in the natural habitat of a region' (Karr and Dudley, 1981; Karr 1996). Biological integrity is equated with stable systems that have been minimally disturbed. The biotic integrity concept provides a system-specific framework in which species assemblage data can be ranked on a qualitative scale. This approach to measuring stream health is, therefore, more ecologically relevant than traditional measures of diversity (Brooks *et al.*, 1998; Karr, 1999).

Numerous metrics can be calculated based on presence/absence data or on relative abundances of taxa. These metrics can be combined into an Index of Biotic Integrity

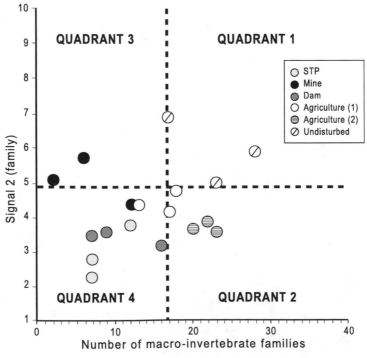

**Figure 9.3.** *Example of a biplot of SIGNAL 2 (family) scores from three samples from each of 6 sites in the upper Macquarie River catchment, New South Wales, Australia. The dotted lines represent possible quadrant boundaries. Source: Chessman (2003b)*

(IBI). The concept was first developed by Karr (1981) for warmwater fish communities in small wadeable streams in central Illinois and Indiana, and since that time the IBI has been widely used in the USA and elsewhere (e.g. U.S. EPA, 1990; Karr, 1991; Goldstein *et al.*, 1994; Simon and Lyons, 1995; Karr, 1997, 1999; Yoder and Kulik, 2003). The core principle of multimetric IBI is to detect divergence from biological integrity due to human actions (Karr, 1999). For small, wadeable headwater streams, sufficient and suitable reference analogs usually exist, so reference condition can be empirically derived. However, for large lowland rivers, such analogs are either rare or do not exist, so reference state must be recreated through historical knowledge (Yoder and Kulik, 2003).

The original IBI compares 12 measures (called metrics) of the structural and functional components of the fish community with those from minimally disturbed reference communities from ecologically similar areas. The metrics are concerned with fish species composition, trophic composition, reproductive behaviour and abundance and condition (Karr, 1997). To determine an IBI for a stream, each metric is assigned a value of 5, 3 or 1 depending on whether the condition is comparable to, deviates somewhat from, or deviates strongly from the reference condition. Metric scores are then summed to yield an index (Karr, 1997).

Modifications of the original metrics have been made by various researchers to account for regional or local differences in fish communities (e.g. Goldstein *et al.*, 1994; Lyons *et al.*, 1995; Simon and Lyons, 1995; Bowen *et al.*, 1996; Roth *et al.*, 2000; Yoder and Kulik, 2003) to include macroinvertebrate indices (e.g. Kerans and Karr 1994; LaViolette, 1998; Karr, 1999; Llansó *et al.*, 2003), and to apply to vascular plants (Mack, 2001) and amphibians (Micacchion, 2002) in wetlands.

### Simultaneous Multi-metric Measure of Health and Conservation Value—PBH Method

The Pressure-Biota-Habitat (PBH) framework was developed for the simultaneous assessment of the environmental conservation value and health of rivers in New South Wales, Australia (Chessman, 2002). The method uses six criteria: physical diversity, biological diversity, vigour, resilience, rarity and risk factors. The risk factor criterion evaluates threats to existing ecological values and impediments to the natural recovery of lost values. Each criterion can be considered in relation to selected ecosystem components, such as fish, diatoms and macroinvertebrates. The framework groups the components into three classes:

- *Pressure* components that describe human-generated disturbances of the stream ecosystem
- *Biota* components that describe important properties of the native plants and animals using the stream (alien organisms introduced by human agency are considered as pressure components)
- *Habitat* components that determine the suitability of the stream to support particular types of plants and animals, and often mediate ecosystem responses to human pressure

To carry out a PBH assessment, raw data on a series of pressure, biotic and habitat components are obtained for the study river, and organized into attributes that align with the six criteria. Conservation assets and potential problems are evaluated by comparing attribute values with appropriate thresholds. Attribute values have to be standardized by dividing the actual value by the value expected for an average site of the same catchment area, elevation and slope [to give standardized attribute values (SAVs)]. The thresholds are value judgments, informed by scientific information, about what constitutes good health, conservation value and potential problems. One way of reporting results of the PBH assessment is the number or proportion of relevant SAVs above and below conservation and problem thresholds. The principal uses of the PBH method are the identification of special assets and potential problems as input to planning procedures, the establishment of a baseline of current conditions and the monitoring of management performance (Chessman, 2002).

### Stressor-specific Multi-metric Approach Based on Macroinvertebrates—European AQEM Method

The EU Water Framework Directive requires a stream type-specific approach to measuring stream health. At the European-wide scale this is inevitable. AQEM is a Europe-wide (involving eight European countries) method to put the EU Water Framework Directive into practice (Buffangi *et al.*, 2001). This method uses the idea of reference condition sites (least modified), and compares these with sites impacted by the three main perceived degradation types present in Europe: acidification (e.g. Northern Sweden), water (organic) pollution (e.g. Southern Europe) and morphological alteration (e.g. Central Europe). In some cases more than one stressor is separately assessed and the results of the individual steps are then combined to a final assessment result. For each stream type various metrics are calculated and these are combined in a multi-metric index that is converted into the final score ranging from 5 (high quality) to 1 (bad quality). The software and manuals for AQEM can be downloaded from the AQEM Project website (http://www.aqemde/start.htm). In Italy, the AQEM method has been extended to enhance the 'information potential' of the field surveys, by conducting a River Habitat Survey at the same time. This complies with the EU Water Framework Directive requirement for the hydromorphological assessment of rivers, to collect data that will yield a better understanding of biological and chemical data (Buffangi *et al.*, 2001).

One important objective of the EU Water Framework Directive is to be able to easily compare river quality across all of Europe. Hence, assessment must be performed in a standardized way or, as a minimum requirement, intercalibration of the different assessment methods used in member countries must be achieved. This is the main focus of the European Union funded STAR (STAndardisation of River classifications) project (http://www.eu-star.at/) (Hering and Strackbein, 2002).

### Multivariate Analysis Techniques

Multivariate statistical techniques allow detection of patterns of variability within groups of taxa and/or between groups of taxa and environmental variables. The British RIVPACS (River InVertebrate Prediction And Classification Scheme) (Wright *et al.*, 1993, 1984, 1998) system uses multivariate analysis techniques to classify unpolluted running waters, and to predict community types from environmental data. This model calculates the probability of occurrence of an expected taxon from the weighted reference site group. Comparing the observed fauna (at the species or family level) with the expected or target predicted fauna, yields a measure of site quality. The same technique was used for development of the Canadian BEAST (BEnthic Assessment of SedimenT) predictive models for rivers and lakes (Reynoldson *et al.*, 1997; Bailey *et al.*, 1998; Reynoldson *et al.*, 2000; Rosenberg *et al.*, 2000).

Australia has adopted a nationally standardized approach to biological assessment of stream condition using macroinvertebrates (Coysh *et al.*, 2000; Simpson and Norris, 2000). The system, known as The Australian River Assessment System (AusRivAS) is based on RIVPACS. Following field sampling of sites, a ratio of the observed number of macroinvertebrate families to the expected number of families (the O/E score) can be calculated for each test site. The value of the O/E index can range from zero (none found) to around one (all families which were expected were found). By comparing the O/E scores to categories (bands) representing different levels of biological condition (Table 9.12), an assessment of the level of

**Table 9.12.** *Description of classes for AusRivAS observed/expected (O/E) Scores. Source: Gray (2003), based on Barmuta et al. (2002)*

| Band label | AusRivAS O/E score | Band Name | Description |
|---|---|---|---|
| X | >1.15 | Richer than reference | More families found then expected<br>Potential biodiversity 'hot spot'<br>Possible mild organic enrichment |
| A | 0.85 to 1.15 | Reference condition[a] | Most or all of the expected families were found at the site<br>Water quality and/or habitat condition roughly equivalent to reference sites<br>Impact on water quality and habitat condition does not result in a loss of macroinvertebrate biodiversity<br>AusRivAS score within the range of the central 80% of all reference sites |
| B | 0.55 to 0.84 | Significantly impaired | Several expected families not found<br>Water quality and/or habitat condition significantly impaired<br>Ecologically and statistically significant loss of macroinvertebrate biodiversity |
| C | 0.25 to 0.54 | Severely impaired | Many expected families not found<br>Water quality and/or habitat condition severely impaired<br>Severe loss of macroinvertebrate biodiversity |
| D | <0.25 | Extremely impaired | Extremely few of the expected macroinvertebrate families found<br>Extremely poor water and/or habitat quality<br>Extreme loss of macroinvertebrate biodiversity<br>Highly degraded |

[a] Reference sites used in the construction of the AusRivAS models have an O/E score of approximately 1.0. These sites are described as undisturbed/natural/pristine, or least disturbed or impacted.

impact on the site can be made. AusRivAS also calculates the SIGNAL score (Chessman, 2003b). The recent National Land and Water Resources Audit (2002) assessment of river condition in Australia used AusRivAs O/E scores as the basis of the aquatic biota index.

Successful AusRivAS urban models were constructed using both family and lowest taxonomic level data for hinterland sites by Breen *et al.* (2000). O/E scores were negatively correlated to catchment imperviousness and biochemical oxygen demand, which are considered to be indicators of urban density and efficiency of pollutant delivery to streams.

Along with bioassessment, AusRivAS also has a physical assessment protocol. This is a stand alone method of physical and geomorphological assessment, but it is intended to complement the biological assessments of stream condition (Parsons *et al.*, 2002b). Assessment of stream physical condition involves the collection of local scale and large-scale physical, chemical and habitat information from test sites. This information is then entered into the predictive models and an O/E ratio is derived by comparing the features expected to occur at a reference site against the features that were actually observed at the test site. The deviation between the two is an indication of physical stream condition (Parsons *et al.*, 2002b).

### Large-scale Bioassessment Based on Ecological Traits

Bioassessment methods based on taxonomic metrics do not provide insight into the causal mechanisms resulting in stream health degradation. Indices that focus on biological traits of macroinvertebrates (e.g. body size, descendants per cycle, numbers of reproductive cycles per year, life expectancy of adults, general mobility, regeneration potential, intensity of attachment to substrate, body flexibility, occurrence of resistant or resting stages, intensity of respiration, and active/passive food/feeding characteristics), and based upon the functional diversity of biocommunities, have been proposed for use in Europe (Statzner *et al.*, 1994, 2001; Townsend and Hildrew, 1994; IETC, 2002). These indices provide an approach to defining the mechanisms that control stream macroinvertebrate community assemblages. This is achieved through linking large geographic-scale environmental conditions with evolutionary adaptations among taxa that give rise to the formation of distinctive biocommunities (IETC, 2002).

### 9.3.7    Types of Stream Health Monitoring Program

Stream health assessment is increasing being standardized and coordinated under catchment, state or national-scale programs. The EU Water Framework Directive (Blöch, 1999; Kallis and Butler, 2001; EUROPA, 2003) distinguishes between *surveillance monitoring* for providing a coherent and comprehensive overview of current health status, and assessment of long-term changes, at the scale of large river basin districts; *operational monitoring* for those water bodies (down to 10 $km^2$) that are identified to be at risk of failing to meet environmental objectives, and for monitoring changes in the status after rehabilitation measures, and; *investigative monitoring* for those water bodies that have failed, or are likely to fail, to meet the environmental objectives for unknown reasons or to ascertain the magnitude and impacts of accidental pollution.

*Campaign monitoring* is a special case of operational monitoring, used to detect and measure environmental response to specific management interventions. The basis of this type of monitoring is hypothesis testing, and the selection of response variables and the appropriate spatial and temporal scales at which to monitor will be governed by the hypothesis tested. This type of monitoring, which may have particular statistical requirements (Sit and Taylor, 1998), is applicable to the adaptive management approach (e.g. Holling 1978; Walters 1997).

### 9.3.8    Selecting a Method for Measuring Stream Health

Measurement of physico-chemical characteristics is a traditional, widely applied and robust approach to assessing stream health. This has proved a valuable and necessary approach (especially for relatively straightforward issues such as effluent compliance monitoring, drinking water standards and contact recreation standards), but it does not explicitly measure the effects of concentrations of pollutants on organisms. Physico-chemical profiling may identify situations where biotic communities are at risk, but provides no information on the actual damage, if any, to the biota. Even where the tolerances of organisms to individual pollutant concentrations are known, this information may have limited application to the common situation of streams being affected by multiple pollutants, as well as habitat disturbance, and hydrological alteration.

Habitat assessment methods are often rapid, visually based and integrative (perhaps incorporating physico-chemical and bioassessment). This approach relies on assumptions about what constitutes unimpaired habitat for organisms of interest, and stream health is rated against this ideal, or against some other reference condition. Observations of changing habitat conditions through time can provide an early warning of possible future biotic responses. Habitat assessment methods, even those that employ quantitative measures, are not usually adequate to identify the causes of degraded stream health, but once degradation has been noted, a targeted process study can be undertaken.

Hydrological indices are easy to calculate but depend on the availability of an adequate hydrological record—a serious limitation in many cases. Even where a discharge record is available, complex modelling may be required to derive the unimpaired discharge record. While some recent progress has been made on relating hydrological indices to ecological processes, this remains a major weakness of using hydrological alteration indices as measures of stream health. Like the physico-chemical approach, hydrological assessment and habitat assessment may identify the condition of the factors known to threaten or enhance biota, but it says nothing about the actual condition of the biota.

Rather than measuring the physical and chemical factors that give rise to stream health, bioassessment methods directly measure the condition of the aquatic biota. Compared with full physico-chemical characterization, targeted bioassessment requires less equipment and a large area can be surveyed intensively in a short time. The weakness of bioassessment is that it provides no information on the causes of the observed biological effects (e.g. Ravera, 1998). There is an obvious advantage for managers in combining chemical (possible causes) and biological (effects) approaches to stream health monitoring (Cairns, 1995). The condition of biotic and abiotic factors also partially reflects habitat factors and hydrological regime, so assessment of these variables will also contribute to the understanding of stream health.

Hunsaker and Levine (1995) raised the question of whether local or catchment-wide factors have more of an impact on biotic integrity of streams. Lammert and Allen (1999) found that local land use and habitat predicted biotic integrity while regional land use showed no relationship. In contrast, Roth *et al.* (1996) in a study of the same catchment (but covering a larger area with greater contrast between sub-catchments) found that regional land use factors were relevant to biotic integrity. It appears that measurement of catchment variables may need consideration. The recent National Land and Water Resources Audit (2002) assessment of river condition in Australia used an Aquatic Biota Index as well as an Environment Index. The Environment Index combined the cumulative effects of catchment-scale features (woody vegetation loss and development of infrastructure) and local features including habitat (bed deposition, riparian vegetation cover and connectivity), hydrology, and nutrients and suspended sediment loads.

Macroinvertebrates are the most popular choice for use in bioassessment of stream health. Compared with other groups of organisms and parameters, they can be more easily and reliably collected, handled and identified. In addition, there is often more ecological information available for such taxonomic groups. Another factor which makes macroinvertebrates the most broadly applicable group is that there are very few stressors to which macroinvertebrate community structure is unlikely to respond (ANZECC and ARMCANZ, 2000a). However, macroinvertebrates are generally not as sensitive as algae to herbicides or to nutrients such as nitrogen and phosphorus, nor do they provide an indication of the presence of microorganisms that can cause disease in humans. They are not generally as sensitive to altered river flows as fish (Harris, 1995).

The condition of the biota indicates whether maintenance or rehabilitation action is required, while understanding the likely causes of observed stream health enables management to act in an informed way. A combined approach to stream health assessment will provide this information. Ravera (2001) called this approach 'ecological monitoring', whereby biotic and abiotic variables that relate to the characteristics of the ecosystem and the goals of management are chosen for assessment.

## 9.4  The Use of Stream Classification in Management

### 9.4.1  Introduction

Historically, most fields of science have undergone a phase of classification during their early stages of development, with the objective of ordering observations and descriptions. As advancements are made, classification gives way to development of empirical relations, and then to theoretical understanding of fundamental processes (Goodwin, 1999). Classifications only label objects; they do not of themselves produce any information (Cowardin, 1982). Leopold and Langbein (1963) recognized that the main limitation of description and classification was its poor predictive power. O'Keeffe *et al.* (1994) argued that classification helps us organise, and thereby understand complex objects, systems and ideas, with the main problem being that most of the criteria used for grouping were, in fact, continuous in nature. However, O'Keeffe *et al.* (1994) saw the point of classification as not to identify all of the distinctive features of groups of rivers, but to clarify their differences. They realized that the criteria had to be chosen with the end user in mind if the scheme was to be useful. Thus, classification should have a purpose.

For river and stream management, classification is used to simplify what would otherwise be impractical tasks, including taking an inventory of the resource, periodically reviewing the impacts of human actions (including management actions), judging stream condition against criteria or legislative requirements, describing the resource in simple and common terms so that stakeholders can debate trade-offs and separating streams into management classes that have different objectives. Some classification models attempt to rate the likelihood of being able to restore a disturbed channel form to a previous or alternative state. Scientists utilize classification to select representative reaches when undertaking stream condition assessments.

When selecting criteria to set up classes a fundamental principle is that the causes of class difference are a better basis for classification than the effects that the differences produce (Lotspeich, 1980). For example, there is some value in classifying streams on the basis of the presence/absence of a fish species of interest, but a model that classified the streams on the basis of the factors that caused the fish to be present would better serve management if the goal was to protect or extend the distribution of that species. Scientists develop such process-based classification models as a framework for generating hypotheses to test, in order to improve their understanding of river processes.

Using the nomenclature of Smith and Medin (1981), Goodwin (1999) identified three approaches to fluvial classification. In the *classical approach* all entities having a given collection of common properties form a category; these properties are both necessary and sufficient to form a category. No member of the category has special status or is more representative of the class than other members. The three class (straight/meandering/braided) geomorphic river classification system of Leopold and Wolman (1957) and the hierarchical Rosgen (1994) scheme are examples of the classical approach. *Probabilistic or prototype theory* does not require all properties to be true of all class members (Smith and Medin, 1981). Class members are assumed to vary in degree both to which they share the class and to which they represent the class. An individual is considered to be a member of a class if the sum of its weighted values exceeds some critical value. The geomorphic classification systems of Montgomery and Buffington (1997) and Nanson and Croke (1992) implement prototypes to some degree (Goodwin, 1999). The British Columbia Forest Practices Code Fish and Habitat Inventory stream classification method uses a probabilistic approach to compute the probability of each stream class variable for each reach (BC Fisheries

Information Services Branch, 2000). In the *exemplar view*, specific exemplar members are used to define the class (Smith and Medin, 1981). Miall's (1996) sedimentological approach is close to the exemplar view (Goodwin, 1999). Sixteen fluvial styles are named in terms of predominant characteristics, and some are named after particular exemplar streams (e.g. Platte type, Bijou Creek type).

Wiley and Seelbach (1997) recognized four fundamental ways of viewing river systems:

- a landscape-scale system,
- a hydrologic system integral to regional water cycling,
- a geomorphic system that shapes the landscape, including and its own channel,
- an ecological system that supports a diverse and highly adapted biota.

These different ways of conceptualizing rivers has influenced the development river classification models. Models have been devised to suit different landscape scales, and while some emphasize the physical aspects of rivers, others classify ecological characteristics or processes. More recently, models have been developed that integrate ecological and physical aspects of rivers across hierarchically arranged spatial scales.

### 9.4.2 Ecological Classification Models Based on Energetics, Structure, Function and Dynamics of Rivers

Ecologists have developed numerous models that describe basic aquatic ecosystem functioning. These models necessarily involve simplification, a major part of which is classifying organisms and processes according to differences and similarities. Stream organisms derive energy through the chemical process of *respiration*, which involves breakdown of organic compounds, obtained through either the *autotrophic* pathway (i.e. through photosynthesis by primary producers such as plants, algae and phytoplankton) or *heterotrophic* pathway (i.e. through consumption of primary producers, detritus or other organisms by secondary producers such as insects, crustacean, fish and birds). The organic matter present in a stream comes from either *autochthonous* (internal) or *allochthonous* (external) sources. *Primary production* by instream autotrophs is an important source of autochthonous material. Many in-stream heterotrophs have evolved to consume particular types of food. This is particularly true of aquatic invertebrates, which can be classified into feeding guilds: shredders, gougers, filterer collectors, collector gatherers, grazers, macrophyte piercers, predators and parasites.

Organisms require nutrients in order to synthesise new organic material, and these nutrients may be obtained through consumption of other organisms, from their roots or from the water column. Nutrients occur in various forms in streams, and their cycling is an important ecosystem process. The longitudinal and lateral movement of water in streams creates a *nutrient spiral*. Flow rate and level of biological activity determine the nutrient levels at any one location, but where a system has reached its capacity to recycle nutrients, there will be a net downstream transport of nutrients. *Trophic status* is determined by nutrient concentration and level of productivity. Waterbodies with extremely low nutrient status that are unable to support life are termed *dystrophic*. Those with low nutrient concentrations and low productivity are termed *oligotrophic*, moderately enriched systems are termed *mesotrophic*, while those with high nutrient concentration and high productivity are termed *eutrophic*. Eutrophication is a natural process, but is often accelerated through excessive inputs from human activities. The trophic system is used mainly to classify lakes, but the criteria apply equally well to streams and rivers. Vollenweider (1968) classified trophic status on the basis of mean concentrations of epilimnetic total phosphorus ($\mu$g/L):

| | |
|---|---|
| Ultra-oligotrophic | <5 |
| Oligo-mesotrophic | 5–10 |
| Meso-eutrophic | 10–30 |
| Eutrophic | 30–100 |
| Hypereutrophic | >100 |

Where primary production is not limited by light or nutrients, instream photosynthesis exceeds respiration (P/R > 1) and the system is described as autotrophic. If respiration exceeds photosynthesis (P/R < 1) then the system is defined as being heterotrophic. Forested, shaded upland streams are often heterotrophic, with energy transfer driven through the input of allochthonous organic material (i.e. leaves). In autotrophic streams, energy transfer is driven through autochthonous primary production in the form of growth of in-stream plants and algae. Chlorophyll *a* concentration is an indicator of primary production, and Reckhow and Chapra (1983) used mean chlorophyll *a* concentrations ($\mu$g/L) to define trophic status as:

| | |
|---|---|
| Oligotrophic | <4 |
| Mesotrophic | 4–10 |
| Eutrophic | 10–25 |
| Hypereutrophic | >25 |

The simple trophic status classes can be used to aid management. For example, Carroll *et al.* (1996) used trophic state values to indicate that Broken Bow Lake in Oklahoma shifted from oligotrophic to mesotrophic between 1987 and 1991, with some areas nearing eutrophy. Harris (2001) argued that application of catchment-scale biogeochemical models to management questions requires better understanding of the forms, fluxes and transformations of nitrogen and phosphorus in catchments. In Australia, dissolved inorganic nitrogen (DIN) and dissolved inorganic phosphorus (DIP) are better indicators of trophic status than the more commonly measured total nitrogen and total phosphorus, and the ratio of DIN to DIP is a good predictor of algal blooms.

Although Hynes (1975) proposed that 'every stream is likely to be individual', various structural and functional models have been devised in an attempt to explain the distributions of organisms and process in streams. A simple descriptive model based on stream appearance and temperature is that of Illies (1961, 1962):

- *Crenon* (spring), subdivided into eucrenon (spring proper) and hypocrenon (spring run).
- *Rhithron* (stony stream), subdivided into epirhithron (upper), eurhithron (true) and hyporhithron (lower).
- *Potamon* (river), subdivided into epipotamon, eupotamon and hypopotamon (equivalent to the estuary).

Each zone represents a region of the river with distinctive habitat conditions and similar faunal composition. The point on a stream separating rithron and potamon is defined as the location where the monthly mean temperature is 20 °C. For tropical rivers, Illies and Botosaneau (1963) modified the criterion to a summer maximum monthly mean of 25 °C. The dividing point will lie at different altitudes depending on the general climate, with the rithron extending to lower altitudes at latitudes closer to the poles. Because a variety of factors may influence where the boundary occurs, it should not be considered precise, or fixed. Additionally, some streams will not fit the scheme, such as those that never have a monthly mean temperature exceeding 20 °C.

Huet's (1949, 1954) longitudinal zonation uses a combination of channel gradient and width to classify reaches according to fish communities characterized by individual species (Figure 9.4). This simple model is difficult to apply to rich faunal assemblages covering various climatic zones and it also shows inconsistencies with observed data (Cowx, 2001). Despite these problems, the model is still used for classifying river type in some parts of Europe (e.g. Goethals and De Pauw, 2001).

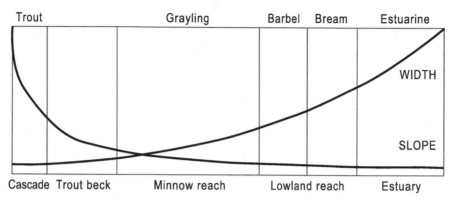

**Figure 9.4.** *Classification of fish zones for European rivers based on stream width and slope. Source: Huet (1949, 1954, 1959). Reproduced by permission of Birkhäuser Verlag AG*

The most influential structural and functional ecological fluvial model is the *River Continuum Concept* (RCC) (Vannote *et al.*, 1980), which predicts changes in physical, chemical and biological characteristics along the length of a stream (Figure 9.5). For example, the RCC predicts that the number of species will increase and the proportion of shredders (organisms that consume leaf materials) will decrease from headwater streams to larger rivers. In mid-sized rivers there is a shift to grazer communities, and in the lowland reaches the collectors dominate. The RCC explains the distribution of biological characteristics in terms of an interaction with geomorphological factors, which also vary downstream. For example, as the size of a river increases from a headwater stream to a mid-sized river, the influence of the surrounding riparian forest decreases due to the change in the dominant biological community. The RCC divides a river system into three major groups comprising headwater streams, medium-sized streams and large rivers.

A weakness of the RCC is that it only applies to perennial streams, and the model does not account for disturbances that interrupt the natural pattern, such as dams and water diversions. Another shortcoming of the longitudinally focused RCC is its lack of consideration of movement of water onto floodplains during flood events. Ward and Stanford (1983b) introduced the *Serial Discontinuity Concept* (SDC) to account for disruptions such as dams, and geomorphic features that disrupt the continuity of the RCC predicted habitats. The *Hyporheic Corridor Concept* (HCC) (Stanford and Ward, 1993) recognized the importance of lateral and vertical connectivity. The *Flood Pulse Concept* (FPC) (Junk *et al.*, 1989; Bayley, 1990; Tockner *et al.*, 2000) suggests that the primary source of productivity in lowland rivers comes from the nutrients

and particulate material derived from the lateral exchange between the floodplain and the channel. Another complementary model for large floodplain rivers, the *Riverine Productivity Model* (RPM) (Thorp and Delong, 1994), stresses the importance of local autochthonous production and allochthonous inputs to the food webs of large rivers. This locally sourced material is thought to be more easily assimilated than refractory carbon from the floodplains and tributaries. Also, allochthonous material that accumulates in slow moving, shallow river edge areas is an important energy source because it is available over the long periods between flood pulses.

The *Ecological Niche Concept*, as proposed by Hutchinson (1957), was an early attempt to provide a stronger basis for understanding biological community structure. Niche theory predicts that communities will establish an equilibrium structure with microhabitat and resource boundaries determined by biological interactions. Community structure was believed to reflect the variable ability of species to dominate other species under variable environmental conditions and the distribution of resources. Many early studies of the interaction between flow and stream organisms adopted this equilibrium model of niche theory and examined spatial patterns in flow and species distribution (Gorman and Karr, 1978; Statzner and Higler, 1986; Grossman and Freeman, 1987). This alternative to the FPC and RPM energetic models, termed the *hydraulic model*, proposes that the physical characteristics of flow (both flow regimes and more local hydraulic habitat) are major determinants of stream organization.

One limitation of the ecological niche concept is the lack of consideration given to dynamic aspects of the abiotic environment. Environmental conditions are rarely constant and there may not be time for a competitively

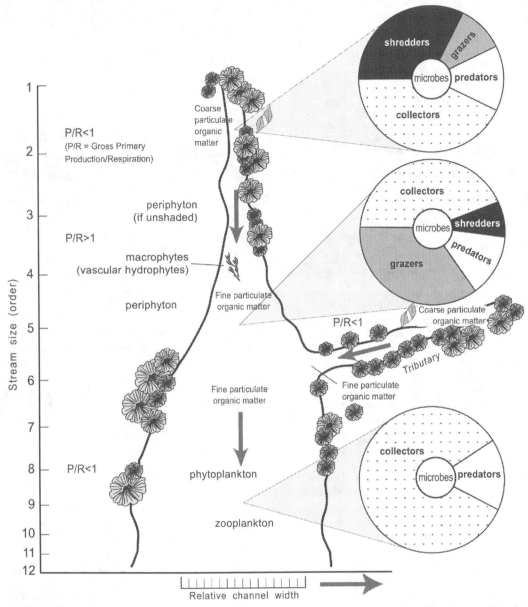

**Figure 9.5.** *River Continuum Concept. Source: Vannote et al. (1980). Reproduced by permission of NRC Research Press*

dominant species to exclude a competitively inferior species (Townsend, 1991). The consequence of this is that equilibrium conditions may be rare in ecology (Wu and Loucks, 1995). Increasing attention has been given to the importance of *temporal environmental variability and disturbances* that force communities away from a static or near-equilibrium condition by creating gaps for colonization by new organisms (Karr and Freemark, 1985; Levin and Paine, 1974). Recent studies provide evidence of the importance of disturbance associated with floods or droughts (Ward and Stanford, 1983a; Resh *et al.*, 1988; Lake, 1995; Poff and Allan, 1995; Poff and Ward, 1989) and cyclic geomorphic perturbation (Shields *et al.*, 2000a) in regulating stream community structure.

Poole (2002) recently proposed a general framework for *fluvial landscape ecology*. Landscape ecology is concerned with integration of pattern, process, hierarchy, scale, directionality and connectivity in order to derive relationships between system structure and system function (Forman and Godron, 1986; Wiens, 2002). Poole's (2002) hierarchical framework acknowledges that rivers form a patchy discontinuum from headwaters to mouth, and attempts to integrate the ecological relevance of the discontinuum by highlighting the importance of *uniqueness* (Hynes, 1975) in fluvial landscapes.

Conceptual models of the ecology of particular rivers often draw on various models of energetics, structure, function and dynamics. For example, Young *et al.* (2001) developed a conceptual model of the River Murray Australia, and Water and Rivers Commission (2002) proposed a different conceptual model to suit the rivers of Western Australia.

A plethora of ecological classification methods exist (Hawkes, 1975; Pielou, 1977). Aquatic ecosystems can be classified based on dominant species, indicator species, assemblages of organisms, or on the inputs and cycling of energy and nutrients. Most of the bioassessment schemes involve classification of waterways according to how close they are to an ideal or reference state. Such models do not classify rivers according to type, but to relative condition. The concept of river type mostly refers to physical characteristics, such as its position in the catchment, bed material and hydrology. These factors come within the realm of fluvial geomorphology.

### 9.4.3 Geomorphological Classification Models Based on River Process and Structure

The aim of fluvial geomorphology is to describe and analyse landform features created by flowing water, and to develop an understanding of the ways in which surface processes operate and control the development of these landforms. Fluvial geomorphology grew from a tradition of explaining long-term landscape evolution. William Morris Davis, one of the first influential American geomorphologists, described landscape evolution in terms of youth, maturity and old age using his now discarded 'geographical cycle' (Davis, 1899). Although Davis was a mathematician he did not feel any need to express his ideas quantitatively (King, 1966, p. 5). He had little interest in the mechanisms involved in the formation of landscapes and promoted his model more as a tool for communication than for scientific explanation. In contrast to Davis and twenty years prior, Grove Karl Gilbert approached landscapes from a quantitative process perspective using the scientific method (i.e. hypothesis testing) in order to explain the mechanisms of formation of landscapes (Gilbert, 1877).

Although Gilbert's work was respected, the Davisian approach dominated American geomorphology for the first half of the twentieth century. At the time it was fashionable to apply Darwinian analogies to other fields, the language of the geographical cycle was easy to understand, and being professor at Harvard University, Davis was able to exert a strong influence on many students who further promoted his ideas (Sack, 1992). Davis travelled widely, and although his views were enthusiastically adopted in Australia, New Zealand and South Africa, he received lukewarm acceptance elsewhere. Although Davis was acknowledged in Britain, his ideas did not take root there, and his method of simplistic generalization did not appeal to German geomorphologists (King, 1966, p. 5).

A shift in approach to geomorphologic investigation was signalled mid-way through the twentieth century by Strahler (1952), who argued that progress required adoption of a quantitative process-based approach, as used by the until then unfashionable Gilbert. Since that time, investigation of process has overwhelmed the interest in historical geomorphology (e.g. Richards, 1982). Most process-based research has been empirical in nature, resulting in numerous generalized models, such as relationships between aspects of channel morphology and discharge indices (Richards, 1982, pp. 1–12). Others have attempted to understand causal mechanisms through physically based process models, such as the way knowledge of the physics of particle dynamics, hydraulics and fluid mechanics has assisted explanations for observed sediment transport (Richards, 1982, p. 7).

Although research geomorphology has largely moved on from classification and other simple descriptions of landforms, the same cannot be said for some branches of applied geomorphology. River classification has become a popular method in the USA for designing river restoration works. This approach is intended to help 'predict a river's behavior from its appearance', and proposes that, by assigning a classification to a reach of stream, it is possible to assess the channel's stability, trend if left undisturbed, and prospects for change under restoration (Rosgen, 1994); this is reminiscent of the Davisian approach (Doyle *et al.*, 1999).

Almost since their inception, classification schemes have been roundly criticized as a design tool by research geomorphologists, who argue that they oversimplify the complexity of the fluvial system and can produce misleading results, which may carry significant economic and public safety implications (Kondolf, 1995; Miller and

Ritter, 1996). This has not prevented growing interest in application of the procedure, especially in the USA, where classification methods are commonly endorsed by government agencies, such as the Forest Service (Stream Systems Technology Center, 2001; U.S. Fish and Wildlife Service, 2001). Doyle *et al.* (1999) noticed that the reasons for the appeal of classification were similar to those that made Davisian geomorphology seem so attractive more than 100 years ago: non-specialists find the simple language and concepts relatively easy to understand; the method is widely taught through short courses; and there is no need to have a good understanding of geomorphology, hydraulic engineering or hydrology because there is an implication that the river is 'known' once the river is classified (Kondolf, 1995).

The geomorphology of rivers is mainly concerned with sediment erosion, transport and deposition. Although there has been a shift towards understanding process, most fluvial geomorphology in the twentieth century has used an empirical and descriptive methodology (e.g. Leopold *et al.*, 1964; Gregory and Walling, 1973; Schumm, 1977). This work has been adapted to river management in the form of regime and tractive force equations (Ministry of Natural Resources, 1994), reference reaches (Newbury and Gaboury, 1993a, 1993b; Rosgen, 1998), hydraulic geometry relations (Leopold and Maddock, 1953) and combinations of these approaches (Gillilan, 1996). Stratigraphic analysis, including dating, is an important tool for placing sediment processes into a historical perspective (e.g. Kenyon and Rutherfurd, 1999; Grayson *et al.*, 1998). Also, the value of fluid mechanics to help explain sediment processes in rivers has not gone unnoticed (e.g. Richards, 1982), and numerous models are now available for analysis of hydraulics, sediment transport and bank stability (Doyle *et al.*, 1999; Shields *et al.*, 2003a). Robust physically based process models of channels are under development. For example, CONCEPTS (CONservation Channel Evolution and Pollutant Transport System) is a process based dynamic computer model that simulates open channel hydraulics, sediment transport and channel morphology, including width adjustment and streambank mechanics (Langendoen, 2000; Langendoen *et al.*, 2001).

There has been a convergence of interest in the fields of fluvial geomorphology and river engineering (Gilvear, 1999) and modern applied fluvial geomorphology makes use of hydraulics, geology, geotechnical slope stability, sediment transport and hydrology (FISRWG, 1998; Doyle *et al.*, 1999; Gippel *et al.*, 2001; Shields *et al.*, 2003a). The roles of geomorphology in stream management are to link the local site management concerns to the wider catchment and channel processes, define an acceptable level of

instability within the system, lengthen the time scales of concern by management, identify system thresholds and provide a conceptual and communicative link between engineering and ecology (Gilvear, 1999). Under this approach, classification is only the beginning of the analysis process.

Geomorphic classification schemes are used in the environmental field with the assumption that the physical characteristics define the likely biological characteristics (Lotspeich and Platts. 1982; Frissell *et al.*, 1986; Poff, 1997; Naiman, 1998), even though this has not been well tested (Naiman, 1998) and biologists do not universally agree with this assumption (O'Keeffe *et al.*, 1994). Most classification schemes use structural rather than functional (process) characteristics as their criteria for similarity. Reviews of fluvial classification schemes have been conducted by Mosley (1987) Naiman *et al.* (1992), Rosgen (1994), Miers (2001), Downs (1995), Kondolf (1995), Thorne (1997), Naiman (1998) and Goodwin (1999), and only a few examples will be discussed here.

At a simple network level streams can be classified in terms of stream order. For example, Smith and Lyle (1978) counted the total number of streams of different order in Great Britain. Stream order can act as a surrogate for river habitat elements only at the very simplest level, so such a crude scheme has limited application (Newson, 1994, p. 124). Geomorphic classification systems have traditionally been based on river shape. The best known is the division of streams into straight, meandering and braided (Leopold and Wolman, 1957) (see Section 7.3.1). Schumm (1977) envisaged a broad-scale system of three-channel zones based on sediment transport: an upper zone of sediment production (source), a middle zone (transfer) essentially in equilibrium and a lower zone (sink or depositional area) (Figure 9.6). Kellerhals *et al.* (1976) proposed that channels generally become straighter with an increased discharge of sediment and/or water. Later, Schumm (1985) and Selby (1985) combined morphological and sediment models to create a six-class model of channel forms (Figure 7.11). In a similar model, Church (1992) separated large channels into three phases (a bed material supply-dominated phase, a transitional phase and a wash material supply-dominated phase). Each phase may have a range of morphological characteristics, depending on sediment supply.

Rosgen (1994, 1996a) used stream channel geometry to define eight primary stream channel types (Figure 9.7). The alphabetical categories roughly indicate decreasing channel slope from A to G. Rosgen's scheme relies on measurement, so it should be reproducible, i.e. the information collected by different observers should lead to the same classification. The core set of measured attributes is:

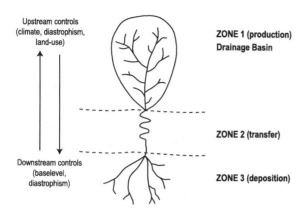

Upstream controls
(climate, diastrophism,
land-use)

**ZONE 1 (production)**
**Drainage Basin**

**ZONE 2 (transfer)**

Downstream controls
(baselevel,
diastrophism)

**ZONE 3 (deposition)**

**Figure 9.6.** *Zones of an 'ideal' fluvial system. Modified from Schumm (1977), by permission of John Wiley & Sons, Inc. Reproduced by permission of Stanley Schumm*

entrenchment ratio, width-depth ratio, dominant channel materials (from the cross section) slope, bed features (from the long profile), and sinuosity and meander width ratio (from the plan-form). There are four levels in

Rosgen's classification hierarchy: geomorphic characterization (Level 1), morphological description (Level 2), stream condition assessment (Level 3) and validation and monitoring (Level 4). Each level rests on the information derived from the previous level. Definitions used in the first two levels are mainly with respect to the conditions during bankfull discharge. Rosgen (1994) included management interpretations for each stream type, in terms of sensitivity to disturbance, recovery potential, sediment supply, streambank erosion potential and vegetation controlling influence on the width/depth stability. Rosgen (1996a) also listed the suitability of each stream type for application of various fish habitat improvement devices.

The Rosgen scheme provides detailed descriptions of the reach within the stream network, but there is no link to the hillslopes. This weakness contradicts the accepted view that the structure and dynamics of a stream are determined by the surrounding catchment (Wadeson and Rowntree, 1994). Such classification schemes do not independently identify channel stability and Rosgen's (1994) method has been criticized by Miller and Ritter (1996) for being subjective, and failing to properly identify terms such as 'channel stability'. Rosgen (1996b)

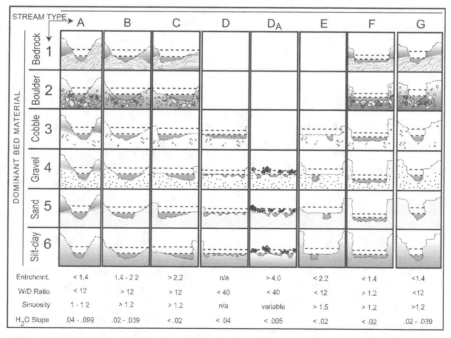

| STREAM TYPE | A | B | C | D | D$_A$ | E | F | G |
|---|---|---|---|---|---|---|---|---|
| Entrchmnt. | < 1.4 | 1.4 - 2.2 | > 2.2 | n/a | > 4.0 | < 2.2 | < 1.4 | <1.4 |
| W/D Ratio | < 12 | > 12 | > 12 | < 40 | < 40 | < 12 | > 1.2 | <12 |
| Sinuosity | 1 - 1.2 | > 1.2 | > 1.2 | n/a | variable | > 1.5 | > 1.2 | >1.2 |
| H$_2$O Slope | .04 - .099 | .02 - .039 | < .02 | < .04 | < .005 | < .02 | < .02 | .02 - .039 |

**Figure 9.7.** *Level I and II of the Rosgen classification scheme for natural rivers. Further subclasses can be defined on the basis of channel slope classes. Revised version of Fig. 4 in Rosgen (1994). Source: Wildland Hydrology, Pagosa Springs, CO. Reproduced by permission of Wildland Hydrology*

attempted to answer these criticisms, and encouraged further development of the model. Ashmore (1999) illustrated with an example that grain size and slope (along with discharge) are the primary variables for channel width design and that stream type is irrelevant; in the Rosgen system the emphasis is the other way around. Further limitations of the Rosgen classification approach were recently highlighted by Juracek and Fitzpatrick (2003) as: time dependence; uncertain applicability across physical environments; difficulty in identification of a true equilibrium condition; potential for incorrect determination of bankfull elevation; and uncertain process significance of classification criteria.

Several approaches are being taken to stream classification in South Africa (Uys 1994). Wadeson and Rowntree (1994), Rowntree and Wadeson (1998), Rowntree and Wadeson (1999) and Rowntree and Wadeson (2000) developed a hierarchical geomorphological classification model, based on a modification of the model of Frissell *et al.* (1986), which provides a scale-based framework that can be applied to any river system. In declining order of scale the hierarchy is catchment, segment, geomorphological zones, reach, morphological unit and hydraulic biotope. The model has nine classes of geomorphic zones, each having an associated range of channel slope. The main value of the methodology was perceived to be its ability to highlight areas of potential disturbance, and to focus attention, in an objective way, on the various aspects of the fluvial system at a number of different scales (Van Niekerk *et al.*, 1995).

Downs (1995) reviewed a number of classification schemes that interpret channel processes specifically for management purposes. Such models are known as channel evolution models (CEM). Two CEMs that have received wide acceptance in the USA for streams with cohesive banks (FISRWG, 1998, pp. 7–32) are those of Schumm (Schumm *et al.*, 1984) and Simon (Simon and Hupp, 1986; Simon, 1989; Simon and Downs, 1995). CEMs begin with a pre-disturbance condition, then identify phases of degradation associated with increased stream power, aggradation and widening, then achievement of a new equilibrium form located at a lower elevation than the original channel. CEMs help management by establishing the current direction of change, they can help to prioritize rehabilitation actions (by directing attention to the more stable reaches), and they can help to match solutions to the problems, and help to set rehabilitation goals (FISRWG, 1998, pp. 7-34 to 7-36; Watson *et al.*, 2002). An example of the application of Simon's CEM to planning stream management actions can be found in Simon and Thomas (2002).

The River Styles classification framework (Brierley *et al.*, 1996; Brierley, 1999) (Section 9.3.4) is another model that is based on the hierarchical model of Frissell *et al.* (1986). The context of development of the model was the observation that some coastal rivers in southeastern Australia had gone through a disturbance phase so catastrophic that they would stabilize at a new incised form, rather than return to the pre-disturbance condition. Like the method of Rosgen (1994), River Styles includes post-classification stages, the first being assessment of river condition to predict likely future river form, and the second being prioritorization of catchment management issues and identification of suitable river structures for Rivercare (Raine and Gardiner, 1995) planning (Brierley *et al.*, 1996). The scheme is strongly evolutionary (i.e. CEM model), and it provides a common geomorphic language with which to describe the fluvial characteristics of rivers and predict their recovery potential.

River Styles involves collection of geomorphic data from maps and from field surveys. Links are deduced between geomorphic units and formative processes, such as a sediment-choked channel, with dissected bars, suggests bed aggradation and channel contraction is reflected by bench formation and in-channel sedimentation. Reaches are amalgamated into Schumm's (1977) source, transfer, throughput and accumulation process zones (Figure 9.6) based on the assemblage of geomorphic units (such as pools, runs, riffles) and associated sediment relations along reaches (Brierley *et al.*, 1996). More recently, Thomson *et al.* (2001) described an extension of the River Styles framework to the smaller-scale of hydraulic units—patches of uniform flow and substrate. The River Styles framework assesses the evolution of the river in a historical context, with a close examination of the relationship between channel form and changing riparian vegetation condition. In a series of articles, Fryirs and Brierley (1998), Brierley *et al.* (1999) and Fryirs (1999) demonstrated the application of the River Styles approach to describing post-European settlement channel changes for some NSW coastal rivers, and determining their recovery potential. The premise is that those sections of the catchment displaying high recovery potential have the greatest likelihood of rehabilitation success.

Brussock *et al.* (1985) proposed a longitudinal model of stream classes based on channel form, considered within three sedimentological settings relevant to biota: cobble and boulder bed, gravel bed and sand bed. Relief, lithology and runoff were selected as the main factors that control other important parameters, such as temperature, depth, velocity and substrate. Whiting and Bradley (1993) proposed a process-based classification scheme for headwater streams. The forty-two classes were process

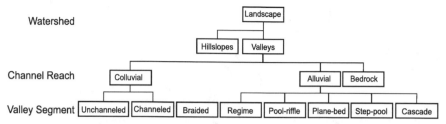

**Figure 9.8.** *Hierarchical subdivision of watersheds into valley segments and stream reaches. Source: Bisson and Montgomery (1996). Reprinted from Methods in Stream Ecology, Hauer, F.R. and Lamberti, G.A. (eds), pp 23–52, Copyright 1996, with permission from Elsevier*

interpretations of dimensional properties of morphological features, which included channel gradient, channel width, valley width and median sediment size. Bisson and Montgomery (1996) proposed a hierarchical classification model with the reach scale distinguishing colluvial, alluvial and bedrock valleys (Figure 9.8).

Montgomery and Buffington's (1997) process-based classification scheme defined seven channel reach types based on overall geomorphic character (Figure 9.9). This model couples reach-level processes with their downstream spatial arrangement, their links to hillslope processes, and external forcing by confinement, riparian

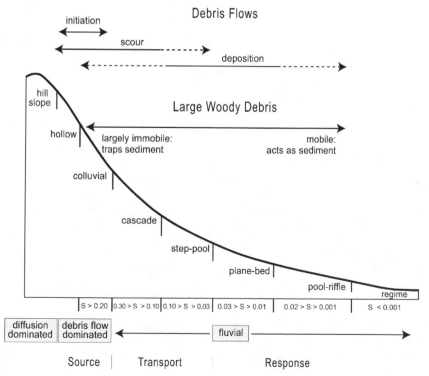

**Figure 9.9.** *Idealized long profile showing distribution of alluvial channel types and controls on channel processes in mountain drainage basins. Source: Montgomery and Buffington (1997). Reproduced by permission of Geological Society of America*

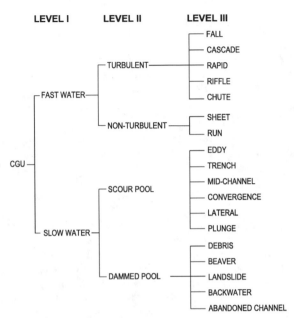

**Figure 9.10.** *Channel geomorphic units (CGU) classified with three increasing levels of resolution. Source: Hawkins et al. (1993). Reproduced by permission of American Fisheries Society*

vegetation and woody debris. The model also provides a framework from which to assess channel condition for the reach types and their response potential to moderate changes in sediment supply and discharge. The three-level hierarchical channel unit model of Hawkins *et al.* (1993) goes a step further down in scale to describe hydraulic conditions within channel geomorphic units, with hydraulic conditions being a reflection of geomorphic channel features (Figure 9.10). Thoms and Sheldon (2002) applied a geomorphic-based, hierarchical classification scheme to a dryland river in Australia. They attempted to relate physical factors to ecosystem structures and processes across a range of spatial scales, but much of this was hypothetical.

While many existing river classification methods successfully predict physical habitats based on catchment characteristics they lack an explicit underlying process model that links the occurrence of habitats to catchment processes (Young *et al.*, 2002). Nanson and Croke (1992) and Croke (1996) developed a floodplain classification scheme based on the energy:resistance concept. Energy was defined by stream power, while resistance was defined in terms of bank cohesivity. Three main system types were identified, and within those, thirteen orders and suborders. This is a process-based, scale-independent scheme. It can

be used to predict the impact of management actions that affect either energy or resistance.

Young *et al.* (2002) hypothesized that sediment transport capacity and sediment supply determine channel type, and that channel type in turn affects the provision of hydraulic habitat and the diversity of bed habitats within river channels. For 36 sites within the upper Murrumbidgee catchment, Australia, they demonstrated that geomorphic stream type [using the classification of Montgomery and Buffington (1997)], bed particle size, and the occurrence of sand slugs varied in the expected way with sediment transport capacity and maximum shear stress (based on digital elevation models and hydrological regionalizations). Young *et al.* (2002) then applied the model across the entire catchment to predict the location of sites degraded by sand slugs. Assuming that higher energy channel types offer the best potential for restoration, they were able to identify degraded sites that were the best candidates for restoration.

Elaborating on the recommendations of Naiman *et al.* (1992), Goodwin (1999) made ten recommendations to improve geomorphic classification schemes (see below). Perhaps the greatest hurdle to overcome is that of unrealistic expectations by potential users. Classification is an exercise in data organisation, which can be a useful tool in aiding decision making, but classification is not equivalent to decision making (O'Keeffe *et al.* 1994).

Goodwin's (1999) recommendations to improve geomorphic classification schemes are:

1. *Base classifications on natural kinds.* This assumes that natural kinds do exist, and that they should be related to some fundamental laws of channel development.
2. *Base classifications upon processes or controlling variables.* Although channel form variables are easily measurable, they may be non-unique end products of the complex dynamic system. There is usually a lack of direct correspondence between fluvial processes and form, suggesting that measurement of processes may be more productive.
3. *Base classifications upon temporal change and thresholds.* Rivers should be viewed as processes acting through time, not just space. Concepts of relaxation time, thresholds and equilibrium will be required.
4. *Base classifications on theory.* There is sufficient information available to guide observations.
5. *Base classifications on a probabilistic view.* Traditional classification models require that for a river to fall into a class that it satisfies certain criteria.

However, probabilistic theory does not require all properties to be true of all class members.

6. *Calibrate and verify classifications for prediction.* If classification schemes are to be used to make management predictions, then an empirical approach to classification can be taken. However, the models will require calibration, verification and updating.

7. *Incorporate size factors in classification schemes.* Most classification schemes ignore scale. Many absolute measurements are converted to dimensionless variables (e.g. width/depth ratio and sinuosity). However, scale issues are important with respect to processes.

8. *Use nomenclature that improves communication.* The scheme will benefit from the use of simple, unambiguous terms.

9. *Treat classifications as hypotheses not paradigms.* As evidence is gathered regarding a scheme's explanatory or predictive capacity, modifications can be made, and if necessary, the scheme can be replaced by a new system. Accepting a classification scheme as a paradigm will slow development of the science.

10. *Ultimately, ignore classification.* The desire to classify rivers may be partly explained by the strong history of field studies, description and classification in geography and geology—the root disciplines of fluvial geomorphology. In many other scientific disciplines classification plays little or no part. The goal should be to develop geomorphic models that have much greater predictive power than classification schemes.

### 9.4.4 Hydrologically Based Classification

Streamflow is a useful measure for classification purposes because it integrates the influences of most landscape features into a single measureable 'characteristic' (Likens *et al.*, 1977, p. 27). Hydrology is regarded by many aquatic ecologists to be the key driver of river and floodplain wetland ecosystems (Harris *et al.*, 2000; Bunn and Arthington, 2002). For example, in Michigan, USA, hydrologically flashy rivers can frequently and unpredictably disturb important periods of fish reproduction (Wiley and Seelbach, 1997). Stable groundwater fed rivers, on the other hand, are relatively free of such disturbances. The hydrological classes, which also have characteristic temperature regimes, give rise to typical dominant fish communities (Table 9.13).

Classification of alpine streams is traditionally based principally on water temperature. Such models do not take into account spatial and temporal variations in water source contributions unless associated temperature changes occur. Using data from the French Pyrénées, Brown *et al.* (2003) proposed a classification system that better describes spatial and temporal variability in glacial, snowmelt and groundwater inputs to alpine streams, based upon the mix of proportions of water contributed from each of these sources.

Poff *et al.* (1997) described stream hydrology as the 'master variable' that regulates ecological integrity. Bunn and Arthington (2002) posed four guiding principles regarding the influence of flow regimes on aquatic biodiversity (Figure 9.11). Not surprisingly, researchers have attempted to find links between flow indices and biological indices (e.g. Puckridge *et al.*, 1998; Jowett and Duncan, 1990; Clausen and Biggs, 1997a, b; King *et al.*, 1995).

With an interest in developing objective, reproducible methods for delimiting community structure in unpolluted streams, Jones and Peters (1977) used a flow-regime classification. They reasoned that changes in regime could be used to predict alterations in invertebrate

*Table 9.13.* Ecological correlates of streamflow sources in Michigan rivers. Source: Wiley and Seelbach (1997)

| Dominant source of streamflow | Degree of flood and drought disturbance | Summer stream characteristics | Dominant fishes {community type} families |
|---|---|---|---|
| Runoff | High | Warm temperatures (max >26 °C) after with large diel flux, sluggish flows, shallow depths, silt deposition on riffles | {warm-water fishes} suckers, sunfishes, catfishes, minnows and mudminnows |
| Throughflow or mixed sources | Moderate | Cool temperatures (max 22–26 °C), modest currents, shallow to moderate depth, little silt deposition in riffles | {cool-water fishes} suckers, sunfishes, pikes, perches, minnows |
| Groundwater | Low | Cold (max <22 °C) and stable temperatures, swift flows with good depth, clean coarse substrates | {cold-water fishes} trouts and salmons, sculpins |

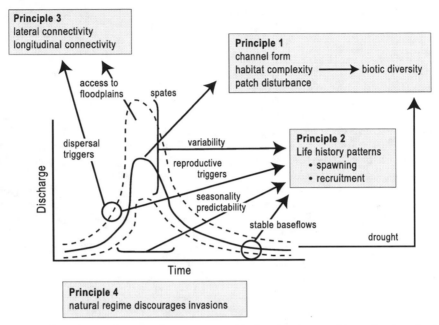

**Figure 9.11.** *The natural flow regime of a river influences aquatic biodiversity via several interrelated mechanisms that operate over different spatial and temporal scales. There are four guiding principles regarding the influence of flow regimes on aquatic biodiversity. Source: Bunn, S.E. and Arthington, A.H. 2002. Basic principles and ecological consequences of altered flow regimes for aquatic biodiversity. Environmental Management* **30**(4): 492–507 ©*Springer-Verlag GmbH & Co. KG. Reproduced by permission*

communities. Regime was obtained by visually inspecting graphs of mean, maximum and minimum monthly flows (both mean and standard deviation) and ranking the regimes from 'stable' to 'spatey'.

Harris *et al.* (2000) applied a classification procedure to four British rivers and found that the typical regimes for each of the four catchments were composite features produced by a small number of clearly defined annual types that reflected interannual variability in hydroclimatological conditions. Three seasonality classes and one magnitude class dominated annual discharge patterns. Annual patterns of air temperature were classified evenly into three seasonality and four magnitude classes. Harris *et al.* (2000) argued that this variety of flow-temperature patterns is important for sustaining ecosystem integrity and for establishing benchmark flow regimes and associated frequencies to aid river management.

Poff and Ward (1989) used detailed analysis of fifteen streamflow variables to group 78 streams from across the continental USA into nine stream types. Highly variable and/or unpredictable flow regimes were considered to have a dominant effect on ecological patterns, whereas in more predictable flow environments the patterns were

thought to be influenced more strongly by biotic interactions such as competition and predation. Poff and Ward (1989) developed a conceptual stream-classification model, which was based on a hierarchical ranking of four components of flow regime (intermittency, flood frequency, flood predictability and overall flow predictability) (Figure 9.12). Poff and Ward (1989) provided a discussion on the implications of the results in terms of ecological patterns. For example, highly intermittent streams would tend to support fish with small body size and invertebrates with resting stages and increased dispersal capacity. In contrast, streams with high flow predictability would tend to support larger, more specialized fish and invertebrates and more long-lived species. The discussion was based on observations from a number of other studies because of a lack of data on stream organisms for the gauged sites.

The regionalization of streamflow characteristics is based on the premise that catchments with similar climate, geology, topography, vegetation and soils would normally have similar streamflow responses (Smakhtin, 2001). Classification of catchments into hydrological groups may be based on flow characteristics estimated from

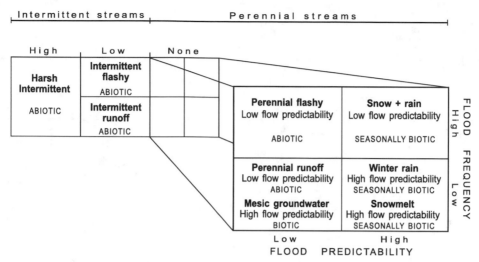

**Figure 9.12.** A conceptual model of stream classification based on discharge regime. The relative effects of abiotic (flow regime) and biotic (e.g. predation, competition) processes on community structure differ between the various classes, as indicated. Redrawn from Poff and Ward (1989). Reproduced by permission of NRC Research Press

observed or simulated flow records. Hughes and James (1989), for example, used streamflow indices to classify Victorian streams for 'hydrobiological' purposes (see Section 8.6.3). The regime-classification method of Haines *et al.* (1988) (see Section 4.4.3) is based on average monthly flows. Alternatively, the hydrological regions are delineated using physiographic and climatic parameters obtained from maps and hydrometeorologial data (e.g. Acreman and Sinclair, 1986).

Joubert and Hurley (1994) attempted to group South African rivers by hydrological characteristics in order to generalize about their ecological flow requirements, with the hope that classification would provide guidelines for situations where catchments were ungauged. They selected two sets of hydrological indices, one based on the variables used by Haines *et al.* (1988) and the other based on variables used by Poff and Ward (1989). Joubert and Hurley (1994) found that there were only two or three hydrologically distinct regions in South Africa, and only one of these had a seasonally predictable flow pattern. Interestingly, it was the only region that had a seasonally predictable cycling of the invertebrate community. Pegg and Pierce (2002) used multivariate analysis to group flow gauges into hydrologically similar units in the Missouri and lower Yellowstone Rivers. The three most influential variables were flow per unit drainage area, coefficient of variation of mean annual flow and flow constancy. One surprising result was the relative similarity of flow regimes between the two uppermost and three lowermost

gauges, despite large differences in magnitude of flow and separation by roughly 3000 km.

Growns and Marsh (2000) examined the hydrology of 107 gauges on eastern Australian rivers by calculating 333 descriptor variables (in seven major categories) that had some basis for having ecological relevance. There was a strong gradient between intermittent and permanent flows, not just with respect to intermittency, but other low flow, seasonality, rates of rise and fall and monthly variation indices. The indices also strongly distinguished between permanent streams in arid zones and those in permanent warm temperate zones. In the warm temperate climatic group there was a separation between regulated and unregulated streams. Arthington and Pusey (1994) found that coastal rivers in Queensland, Australia fell into four distinct biophysical regions, with fish diversity and assemblage structure being related to flow regimes.

### 9.4.5  Water Quality-based Classification

Physico-chemical criteria for waterways can form the basis for stream classification. Classification schemes applied by water quality regulatory agencies categorize streams based on a comparison of physico-chemical characteristics according to established criteria. A shortcoming of many such classifications, from the perspective of holistic stream management, is their narrow focus only

**Table 9.14.** *Some of the variables used in the Scottish river classification scheme. Other variables used in the classification are soluble reactive phosphorus, Fe, pH, bankside ecological index scores, aesthetic condition and toxic substances. Source: Scottish Environment Protection Agency (2000)*

| Class | DO[a] (% sat.) 10th percentile | BOD[b] (mg/L) 90th percentile | NH4-N[c] (mg/L) 90th percentile | BMWP[d] ASPT O/E | BMWP[d] No. Taxa O/E | Ecological description |
|---|---|---|---|---|---|---|
| A1 Excellent | >80 | ≤2.5 | ≤0.25 | ≥1.0 | ≥0.85 | Sustainable salmonid fish population. Natural ecosystem |
| A2 Good | >70 | ≤4 | ≤0.6 | ≥0.9 | ≥0.70 | Sustainable salmonid fish population Ecosystem may be modified by human activity |
| B Fair | >60 | ≤6 | ≤1.3 | ≥0.77 | ≥0.55 | Sustainable coarse fish population. Salmonids may be present. Impacted ecosystem. |
| C Poor | >20 | ≤15 | ≤9.0 | ≥0.50 | ≥0.30 | Fish sporadically present. Impoverished ecosystem. |
| D Seriously Polluted | <20 | >15 | >9.0 | <0.50 | <0.30 | Cause of nuisance. Fauna absent or seriously restricted |

[a] Dissolved oxygen
[b] Biological oxygen demand.
[c] Ammonia.
[d] Biological Monitoring Working Party (BMWP) biotic score system using RIVPACS assessment based on data for 1 year. ASPT is average score per taxon. O/E is observed/expected ratio.

on the properties of the system that fall under the legal jurisdiction of the regulatory agency (Naiman *et al.*, 1992).

The five-class water quality classification used for rivers in Scotland (Scottish Environment Protection Agency, 2000) is derived from a broad-based assessment of selected physico-chemical variables, nutrient (phosphorus), biotic indices (BMWP), aesthetic condition and toxic substances. Stream reaches are defined hydrologically (i.e. on the basis of segments between tributaries) or on the basis of points of entry of known pollutant sources likely to effect a class change. It is a 'default based' system, i.e. the overall class of a watercourse at a particular sampling point defaults to the poorest class determined from the various indices. The water quality classes closely relate to their suitability for certain fish types (Table 9.14).

The Environment Agency recently issued guidelines for General Quality Assessment (GQA) of rivers in the United Kingdom, for chemistry, biology, nutrients, aesthetics and estuaries (e.g. The Environment Agency, 2003; Environment and Heritage Service, 2001). Under the GQA, the same biotic indices, and similar chemical parameters, are used to classify streams as used in Scotland (Table 9.14), but the GQA uses six classes. Also, stream reaches are classified separately for each aspect, so a reach may fall into more than one quality class.

### 9.4.6  Combined Physical-chemical-ecological Classification Models

There are many methods available for assessment and classification of river health. They often combine physical, chemical and ecological metrics to provide an overall rating for the sampled site, defined by five or six classes from 'extremely modified' to 'reference condition' or 'poor (bad)' to 'excellent'. However, these stream health assessment procedures usually fall short of specifically categorizing rivers, or river reaches, according to a management class. The recently devised PBH (Pressure-Biota-Habitat) method of Chessman (2002) attempts to prioritize stream reaches in terms of conservation value, but at this stage of development, the method does not specifically classify reaches.

The so-called 'World-wide stream classification' of Pennak (1971) assumes that ecological character is predicted by abiotic factors. The model uses thirteen parameters, including chemical variables, hydraulic variables, morphological variables and the biotic variables rooted aquatic plants and streamside vegetation. Savage and Rabe (1979) proposed a system for predicting stream biotic classes for small natural streams based on stream order and gradient measured from maps and substrate and flow characteristics measured in the field. Lichthardt (1997) used this scheme to evaluate the representativeness of streams in a protected reference area of the Clearwater National Forest in Idaho.

In Europe, the EU Water Framework Directive (WFD) (EUROPA, 2003) takes a holistic view and calls for integration of physical structure (habitat), biological (water quality) and water quantity assessment in river basin plans and associated monitoring programmes. The WFD river type classification schemes are primarily concerned with hydromorphological and physico-chemical status of rivers. As such, the schemes use purely abiotic criteria to define river types. However, lack of knowledge about the links between abiotic factors and distributions of organisms means that research will be required in order to achieve predictive models (Noble and Cowx, 2002). Although the WFD provides the guidelines for monitoring water bodies, member governments still have a lot of freedom regarding their practical implementation. For example, Goethals and De Pauw (2001) recommended an approach to classifying river type in Flanders, Belgium that combined abiotic and biotic factors at different scales, with differentiation into ecological region and stream and river types an essential starting point (Verdonschot, 2000).

Historically the statutory and operational drivers in the United Kingdom have lead to the development of distinctly different and isolated approaches to the management of water quality, water resources (quantity) and physical habitat structure. Assessment of water quality has been based on information from the survey of benthic macroinvertebrates using the BMWP score and RIVPACS, physical habitat has been assessed using the River Habitat Survey, while water quantity needs have been based on assessment of macroinvertebrate community sensitivity to the hydrological regime (Extence *et al.*, 1999). Logan (2001) encouraged integration of these different approaches.

Amoros *et al.* (1987) developed a classification model that considered fluvial hydrosystems as interactive ecosystems, and applied it to the Rhone River, France, to predict the impact of various management scenarios on channel morphology and ecology (see Section 4.1.2 and Figure 4.2). In Germany, the Leitbild concept provies guidance for ecologically sound stream management (Kern, 1992). Leitbild is a model of the ecological potential of a stream. Bostelmann *et al.* (1998) developed a hierarchical physical-ecological-chemical classification for natural streams in south-west Germany to identify a stream-type dependent Leitbild. Schmedtje *et al.* (2001) developed regional stream typologies for the 'most biologically relevant stream types' in Germany on the basis of ecoregion, altitude, stream and floodplain morphology, water geochemistry (as determined by regional geology), hydrology and macroinvertebrates. All geomorphological stream types, biologically relevant stream types and the relevant longitudinal zonation types were considered. The final twenty defined stream classes were equivalent to geomorphologically and geochemically homogenous landscape units, in which specific water types have their prevalent occurrence.

### 9.4.7  Ecoregions and Multi-scale Classification

Most large countries have a wide diversity of aquatic ecosystems. This diversity creates difficulties in managing these ecosystems at a national level. For example, guidelines for managing water quality and river health may not apply equally well to all ecosystems present. Therefore, a classification that groups similar aquatic ecosystems is a way of accommodating this natural diversity. Ecoregions are defined as areas of relative homogeneity in ecological systems and their components (Woods *et al.*, 1996). Omemik (1987) delineated aquatic ecoregions for the USA as a means of classifying waterbodies for use in more effective aquatic ecosystem management. Other uses for ecoregions include inventorying resources, selection of regional reference sites, and establishment of recovery criteria for impacted aquatic ecosystems (Wells and Newall, 1997).

The ecoregion approach has been found to work well in some areas in terms of predicting certain water quality and ecological attributes. However, in Victoria, Australia, an ecoregion classification based on rainfall, altitude, landform and pre-European vegetation structure and composition regionalizations was not effective in representing aquatic ecosystem patterns (macroinvertebrates and water quality) (Wells and Newall, 1997).

Grouping of similar rivers is a prerequisite to following the river-type specific approach of the EU Water Framework Directive (EUROPA, 2003). The WFD has dictated the classification of river types based on geographical and abiotic criteria. As the first stage in establishing the ecological status of surface waters, the WFD requires that rivers be placed into one of the ecoregions described by Illies (1978). Natural variations within ecoregions are accounted for by dividing rivers into types based on physico-chemical descriptors (Noble and Cowx, 2002). Several stream classification systems used in North America have been based on a hierarchical perspective that links large regional-scale ecoregions with small microhabitat scales. Examples are Cowardin (1982), Lotspeich and Platts (1982) and Larsen *et al.* (1986). This approach has been successful for predicting fish species distribution, water chemistry and physical habitat features (Naiman, 1998). Demarchi (1993) divided British Columbia, Canada, into ecoregions based on physiography and climate. The classification had five levels of resolution. Miers (1994) proposed that these ecoregion units be used as a first level in an ecologically based aquatic habitat hierarchical (multi-scale) framework (Figure 9.13).

Due to the limited availability of biological information on the freshwater ecosystems of Greater Addo Elephant National Park (GAENP) in South Africa, Roux *et al.*

(2002) used a desktop approach, supplemented by aerial and land surveys, to devise a new river classification typology. The model was used to determine priorities in biodiversity conservation. Landscape attributes were used as surrogates for biodiversity patterns, resulting in defined physical 'signatures' for each river type. Riverine biodiversity was considered to be conserved by including rivers of each type as defined by the respective signatures (Roux *et al.*, 2002).

New Zealand's rivers have been classified using a GIS-based tool known as the River Environment Classification (REC) (Snelder and Biggs, 2002; Snelder *et al.*, 2003). The REC is a tool for ecosystem-based inventories of river resources, effects assessment, policy development, developing monitoring programmes and interpretation of monitoring data, and state-of-environment reporting. REC has been used to classify all the rivers of New Zealand at a 1:50 000 mapping scale. The area classified comprises 267 000 km$^2$ and 426 000 km of river network. REC is an advancement on most multi-scale ecoregion approaches in that it is more scalable than existing regionalizations, delineating patterns at a range of scales from approximately 104 km$^2$ to 1 km$^2$, and it is based on a network of 'sections' that are associated with their upstream catchments (Snelder and Biggs, 2002).

The REC is a hierarchical classification where each level is defined by a component of the environment, the 'controlling factor', that is the 'cause' of ecological variation at a characteristic scale. Classification is carried out for individual sections of the network and is based on four abiotic factors that characterize the upstream catchment (climate, topography, geology and land cover) and two abiotic factors that characterize the section itself (network position and valley landform) (Figure 9.14). When mapped, the resulting classification has the form of a 'linear mosaic' of poly lines rather than polygons such as for ecoregions. Classes can change in the downstream direction, so tributary streams can have different classifications to the main stems they meet, and tributaries may collectively change the classification of the main stem (Snelder *et al.*, 2003).

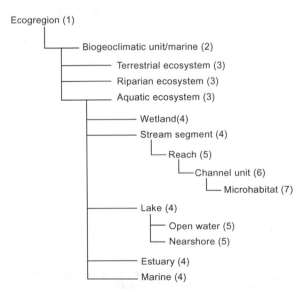

*Figure 9.13. Framework for ecological classification of aquatic habitats in British Columbia, suggested by Miers (1994). Hierarchical levels shown in brackets. Note that channel unit (level 6) corresponds to the channel unit used in the classification of Hawkins et al. (1993)—see Figure 9.10. Microhabitat (level 7) is similar to the hydraulic unit of Thomson et al. (2001)*

### 9.4.8 Wetland Classification

Various systems of definition and classification have been developed for the purpose of inventorying wetlands. The U.S. Fish and Wildlife Service adopted a definition that emphasizes three key wetland attributes: hydrophytic vegetation; hydrology (flooding during the growing season); and hydric soils (periodically inundated and/or saturated) (Burke *et al.*, 1988; Welsch *et al.*, 1995). The

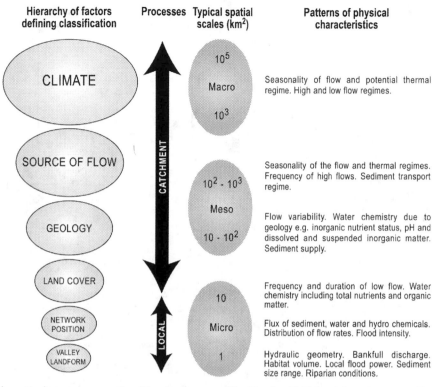

**Figure 9.14.** *Schematic diagram representing River Environment Classification (REC) levels based on controlling factors and the patterns of physical characteristics at typical spatial scales. Source: Snelder et al. (2003). Reproduced by permission of American Water Resources Association*

most important aspect of this definition is that soils should be inundated during the growing season. Legislation to protect wetlands in the USA requires that the existence of a wetland be defined in precise hydrological terms. It appears that a minimum of between 14 and 28 days of saturation near the surface are required to induce anaerobic conditions and hydric soil morphology. Thus, proposed criteria for minimum wetland definition require that the water table is at or less than a critical depth from the surface for a minimum number of consecutive days during the growing season (Skaggs *et al.*, 1994).

The systems used for wetland classification in the USA and Canada have hierarchical structures that at the lowest level rely mainly on vegetation characteristics (Carter, 1986). Some attempts have been made at classifying wetlands on the basis of hydrological characteristics alone, but data and knowledge gaps limit their application (Carter, 1986). Brinson (1993) proposed a simple hydrological classification of main wetland types, based on the relative contributions of groundwater, surface water and precipitation to the water budget (Figure 9.15).

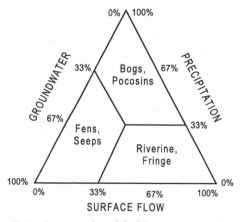

**Figure 9.15.** *Conceptual model of the relative contribution of three water sources to the main wetland types defined for the USA. Boundaries drawn between classes should not be interpreted as distinct; rather, gradients between classes are continuous. Source: Brinson (1993)*

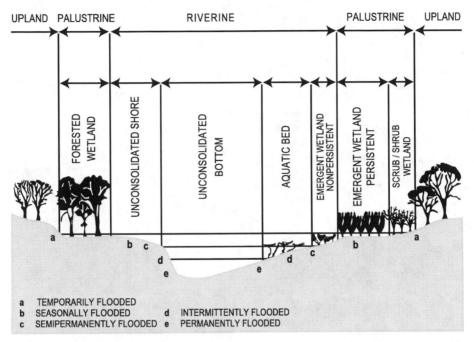

**Figure 9.16.** *Distinguishing features and examples of wetland habitats in the Riverine System of the Cowardin et al. (1979) classification system. Source: Cowardin et al. (1979)*

Wetlands in the USA are classified using the hierarchical system of Cowardin (Cowardin *et al.*, 1979; Cowardin, 1982). Five systems are defined at the highest level of the classification hierarchy: marine, estuarine, riverine, lacustrine and palustrine. The systems have subsystems, and within those, classes are based on substrate material and flooding regime or on vegetative life form. The dominance type, which is named for the dominant plant or animal forms, is the lowest level of the classification hierarchy. Dominance types must be developed by individual users of the classification. Figure 9.16 shows distinguishing features and examples of habitats in the Riverine System.

The Canadian Wetland Classification System contains three hierarchical levels: class, form and type (National Wetlands Working Group, 1997). Five classes are recognized (bogs, fens, swamps, marshes and shallow water) on the basis of the overall genetic origin of wetland ecosystems and the nature of the wetland habitat. Forms are differentiated on the basis of surface morphology, surface pattern, water type and morphology of underlying mineral soil. Types are classified according to vegetation physiognomy. Many of the wetland forms apply to more than one wetland class and some can be further subdivided into subforms. Wetland types are classified according to vegetation physiognomy.

Green (1992) trialed the scheme of Cowardin *et al.* (1979) on wetlands in the Gwydir Valley, New South Wales, Australia, and found that the classification resulted in too many wetland groups, some of which were not significantly different from others. It was also dependent on some threshold values for wetland size and vegetation cover, which were found to have questionable value for management purposes. Green (1997) developed a hierarchical classification scheme for wetlands in New South Wales that grouped wetlands with similar management problems.

In Victoria, Australia, Corrick (1981, 1982) classified wetlands initially on the basis of vegetation, water regime (permanence), depth, salinity and area, and then allocated to subcategories on the basis of vegetation important in determining use by waterbirds. Tunbridge and Glenane (1982) based assessments of wetland value on fish populations. The Wetlands Resource Assessment Package (WRAP) (ABRG, no date) places most emphasis on the presence of rare or endangered species and has been applied in some areas of Victoria. For wetlands in the Kerang area, Lugg *et al.* (1989) used a greater range of criteria to allocate wetland value as high, moderate or low. High value appears to derive mainly from consideration of ecological characteristics. Carter (1986) argued that high value should be assigned to wetlands if they perform a significant role in basin-wide hydrological processes.

### 9.4.9  Estuary Classification

A review of estuary classification schemes by Ferguson (1996) grouped the approaches according to the criteria on which they were based: geomorphology, evolutionary stage, hydrology, climate, water quality, habitat, land use, and aesthetic values. Most schemes focussed on geomorphology and/or hydrological conditions without considering ecological factors, and few included any reference to the evolutionary stage of the estuary. Some estuarine classification schemes are briefly outlined below.

Pritchard (1952a) defined an estuary as positive where freshwater input (river flow and precipitation) exceeds losses due to evaporation. Thus the surface salinities in a positive estuary are lower than in the open ocean. A negative estuary is one where freshwater input is less than losses due to evaporation. Pritchard (1952b) also suggested a four-group topographic classification. Simmons (1955) used a simple classification based on determining the quantity of freshwater flowing into the estuary over a tidal cycle in relation to the tidal prism, which strongly conditioned the salinity structure. Hansen and Rattray's (1966) system described four estuary types based on degree of mixing. Estuaries in Tasmania, Australia are classified into five types based on geomorphology (Edgar et al., 1999). Digby et al. (1999) proposed a classification scheme for Australian estuaries based on easily quantifiable physical characteristics that predicted the proportion of mangrove and saltmarsh habitat.

The Australian Geological Survey Organisation developed a process-based classification covering physical forces (wave, tide and river energies) driving the form

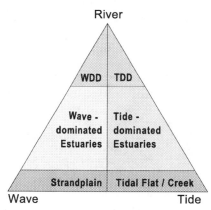

**Figure 9.17.** *Classification of coastal systems divided into six subclasses based on the relative influence of wave, tide and river energies. WDD = wave-dominated delta; TDD = tide-dominated delta. Estuaries, from a geomorphologic perspective, refer to those located in the middle of the triangle. Source: National Land and Water Resources Audit (2002)*

and function of Australian estuaries and coastal waterways (National Land and Water Resources Audit, 2002). Australian coastal systems were classified into six subclasses according to the wave-, tide- and river-energies that shape them and their overall geomorphology (Figure 9.17). Only two of these groups were true estuaries. A seventh subclass grouped 'other' categories: drowned river valleys, embayments and very small coastal lakes, lagoons and creeks. This classification scheme linked estuary type to management considerations through the dominant physical processes operating in the different estuary types (Table 9.15).

**Table 9.15.** *Estuary processes and functions and considerations for management. Source: National Land and Water Resources Audit (2002)*

| Type of coastal environment | Sediment trapping efficiency | Turbidity | Circulation | Risk of habitat loss due to sedimentation |
| --- | --- | --- | --- | --- |
| Tide-dominated delta | Low | Naturally high | Well mixed | Low risk |
| Wave-dominated delta | Low | Naturally low | Salt wedge/Partially mixed | Low risk |
| Tide-dominated estuary | Moderate | Naturally high | Well mixed | Some risk |
| Wave-dominated estuary | High | Naturally low | Salt wedge/Partially mixed | High risk |
| Tidal flats | Low | Naturally high | Well mixed | Low risk |
| Strand plains | Low | Naturally low | Negative/Salt wedge/Partially mixed | Low risk |

### 9.4.10   Classification of Conservation Value of Rivers

Assessment of the value of rivers for nature conservation generally rests on the criteria of diversity, rarity, naturalness, representativeness and the presence of otherwise special features. Each of these criteria can be applied to various physical and biotic components of river ecosystems. Some schemes of conservation evaluation combine assessments for various components and criteria into a single number (e.g. O'Keeffe, 1997).

In the USA, the Wild and Scenic River designation (approved by an Act in 1968) allows for certain rivers possessing outstanding scenic, recreation, geologic, fish and wildlife, historic, or cultural values to be preserved in a free-flowing state (National Park Service, 1982). If a river is found to be eligible a recommendation is made that it be placed into one or more of three classes: wild, scenic or recreational (Table 9.16). The Wild and Scenic Rivers Act does not generally lock up a river like a wilderness designation. The idea is not to halt development and use of a river; instead, the goal is to preserve the character of a river. Uses compatible with the management goals of a particular river are allowed; change is expected to happen (National Park Service, 1982).

The System for Evaluating Rivers for CONservation (SERCON) assesses conservation value of UK rivers using six conservation criteria and an impacts criterion (Boon *et al.*, 1997; Boon *et al.*, 1998; Boon, 2000). The six conservation criteria are: physical diversity, naturalness, representativeness, rarity, species richness and special features. These criteria have been 'designed so that evaluation can be related to the wider field of nature conservation assessment, which is achieved by fitting each attribute into a framework of generally accepted conservation criteria' (Boon *et al.*, 1997, p. 308). A SERCON evaluation comprises three stages. The first stage is a field survey using an extended form of the River Habitat Survey (RHS) (Section 9.3.4) (Boon *et al.*, 2002). In the second stage a wide range of other relevant data from available sources are collected. The third stage involves transforming the data into scores on a scale from zero to five, which are then weighted and combined to provide separate indices of conservation value for each of the six conservation criteria (Boon *et al.*, 1998). The final indices are presented in the form of a five class (A to E) assessment of conservation quality. SERCON has recently been redesigned (named SERCON 2), with one important new feature being closer links with RHS.

In Australia, the Register of the National Estate is the national inventory of natural and cultural heritage places which are worth keeping for the future. The Register of the National Estate deals with place-based localities and cannot readily accommodate values such as the importance for maintaining downstream habitats in rivers. Some parts of rivers are protected by legislation under Ramsar listing, but this system is oriented towards wetland environments, and is also place based (Ramsar Convention Bureau, 1996). The States also have their own legislation that is relevant to conservation classification. For example, the Heritage Rivers Act (1989) makes provision for the protection of particular parts of rivers and river catchment areas in Victoria that have significant nature conservation, recreation, scenic or cultural heritage attributes (Parliament of Victoria, 1997). Dunn (2000) made recommendations regarding a national framework for assessment and classification of ecological values of rivers to be based on five

**Table 9.16.** *Key criteria assessed for the United States Wild and Scenic Rivers Act and the standards applied for each category. Source: National Park Service (1982). Reproduced by permission of the U.S. Government*

| Attribute | Wild | Scenic | Recreational |
| --- | --- | --- | --- |
| Water resources development | Free of impoundment | Free of impoundment | Some existing impoundment or diversion |
| Shoreline development | Essentially primitive. Little or no evidence of human activity | Largely primitive. No substantial evidence of human activity | Some development. Substantial evidence of human activity |
| Accessibility | Generally inaccessible except by trail | Accessible in places by road | Readily accessible by road or railroad |
| Water quality | Meets or exceeds federal criteria or federally approved state standards for aesthetics and propagation of fish and wildlife normally adapted to the habitat of the river | No criteria prescribed by the Wild and Scenic Rivers Act | No criteria prescribed by the Wild and Scenic Rivers Act |

criteria: naturalness, representativeness, diversity and richness, rarity and special features (i.e. the same as the SERCON model).

Frissell and Bayles (1996), Brierley (1999) and Rutherfurd *et al.* (1999; 2000) emphasized the importance of prioritizing reaches for restoration, with least disturbed reaches being the main target for restoration and highly degraded reaches receiving the lowest priority for action. Another issue to consider is the presence of threats to the maintenance of conservation values. Together, value, threat and 'protectability' (the practical feasibility of implementing protection measures) form a basis for conservation priority (Collier and McColl, 1992). This system directs higher priority for protection or restoration to an ecosystem of medium value that is under threat than to a higher value ecosystem that is not threatened (Chessman, 2002).

The Stressed Rivers classification of unregulated sub-catchments in New South Wales, Australia gave high management priority to highly stressed systems and low priority to lightly impacted systems, although systems with high conservation value were also high priority (Department of Land and Water Conservation, 1998). Generally, for unregulated streams, water extraction has the most significant impact on the level, frequency and duration of low flows, so stresses were assessed on the basis of current water usage. Possible future level of hydrological stress was also considered where there were a substantial number of undeveloped water entitlements. The assessment also considered environmental health measures. Environmental stress was not formally defined in this exercise but was inferred from data such as broad-scale vegetation and erosion mapping, water quality surveys, fish kill reports and known barriers to fish migration. The resulting matrix of stress classifications and management categories is shown in Table 9.17. The classification process also attempted to identify all sub-catchments with special conservation value. This included not only low-stress rivers, but also some stressed rivers that had remnant habitats or species of significance. The conservation value was assessed in terms of significant aquatic fauna and flora (including threatened species), dependent wetlands and significant native fish species (including threatened species) or habitat. Sub-catchments classified as high stress and/or high conservation value were given priority for the preparation of river management plans (Table 9.17).

### 9.4.11 Designated Use Classification

Most river classification systems are based around ecosystem processes and values, i.e. they categorize rivers or

**Table 9.17.** Matrix of stress classifications and management priority categories for unregulated sub-catchments in NSW, Australia, showing percentage falling into each category (22.5% unresolved). Source: Department of Land and Water Conservation (1998). Reproduced by permission of NSW Department of Infrastructure, Planning and Natural Resources

| Proportion of water extracted[a] | Environmental stress[b] | | |
|---|---|---|---|
| | Low | Medium | High |
| High | Category U1<br>Medium priority[c]<br>2.2% | Category S3<br>Highest priority<br>8.4% | Category S1<br>Highest priority<br>10.6% |
| Medium | Category U2<br>Low priority[c]<br>1.8% | Category S4<br>Medium priority[d]<br>4.0% | Category S2<br>Highest priority<br>5.4% |
| Low | Category U4<br>Low priority[c]<br>17.8% | Category U3<br>Low priority[c]<br>15.4% | Category S5<br>Medium priority[d]<br>11.9% |

[a] Hydrologic stress indicator was estimated from the proportion of the 80th percentile daily flow extracted during month of peak demand. Stress was classed as low (0–30% extraction of flow), medium (40–60% extraction) or high (70–100% extraction).

[b] Measures or indicators of environmental stress (e.g. riparian clearing, bank erosion, fish barriers, and reduced water quality) were estimated by a rapid desktop assessment method. Measures for each indicator were ranked into low, medium or high stress levels and then combined to create a single high, medium or low environmental stress index.

[c] High priority for management plan if stream has high conservation value.

[d] High priority for management plan if stream has high conservation value or where full water resources development may put sub-catchment into high stress category.

river reaches in terms of some physical, chemical, or ecological aspect or aspects of river health, usually with respect to a reference condition. While such a system is not entirely free of human values, it is centred on intrinsic environmental values (Section 9.2.2). One limitation of using such classifications in the river management arena is that protecting the intrinsic values is just one part of the public policy debate. The challenge of the debate is to balance protection of the intrinsic value with making the best use of the utilitarian values of river resources (Section 9.2.2). This has been recognized in many modern water quality guidelines. For example, The National Water Quality Management Strategy (NWQMS) for Australia and New Zealand (ANZECC and ARMCANZ, 2000c) recognizes six classes of river use: aquatic ecosystems, primary industries, recreation and aesthetics, drinking water, industrial water, and cultural and spiritual values.

The health of waterways in the USA is protected by the Federal Water Pollution Control Act (1972), also known as the Clean Water Act. The purpose of this legislation is to maintain clean waters ('drinkable'), to make sure that game fish are safe for human consumption ('fishable'), and that waterways are safe for recreation ('swimmable'). One of the original interim goals of the Act was to achieve 'fishable/swimmable' standards throughout the nation by 1985 [although a survey in 2000 showed that 40% of rivers had still not achieved this goal (U.S. Environmental Protection Agency, 2002)]. A classification system was developed to help with the implementation of this law. Each river is assigned to a 'designated use' or 'beneficial use'. Designated uses are human uses and ecological conditions that are officially recognized and protected.

States must designate one or more uses for each water body. The designated uses for water bodies must include existing and desired uses that require good-to-excellent water quality. These normally include the uses of fish consumption, shellfish harvesting, drinking water supply, primary contact recreation, secondary contact recreation and protection of aquatic life. Some States list agriculture, industry and navigation among their designated uses. However, these uses require water quality less than that for aquatic life and contact recreation, so this designation does not help attainment of the swimmable/fishable goal. Some States use the term 'designated uses' for contact recreation, aquatic life and drinking water classes, and 'general uses' for other classes of use. The States determine their own classes of designated use, so there is some variation in the classification schemes used throughout the country. For example, some designated uses are 'natural trout waters' and 'exceptional warm water habitat'. The criteria applied to designated uses are usually based on physico-chemical parameters, but some States make use of bioassessment.

Thailand also uses the term 'beneficial uses' to describe national water use classes based on water quality (Pollution Control Department, 2002) (Table 9.18). Unlike the USA classification system, Thailand's is not driven by the objective to achieve swimmable/fishable standards in all waterways. In Japan, for the purpose of administration, all rivers are designated as either Class A (major rivers for conservation or economic purposes) or Class B (tributaries or other lesser rivers) (Kawasaki, 1994). Japan also has a national system for classifying rivers based on water quality aimed at protection of human health and conservation of the living environment (Ministry of the

**Table 9.18.** *Water use classes in Thailand. Source: slightly modified from Pollution Control Department (2002)*

| Class | Objectives/Condition and Beneficial Usage |
|---|---|
| Class 1 | *Extra clean fresh surface water resources used for:*<br>1. conservation, not necessary to pass through ordinary water treatment process, requires only process for pathogenic destruction<br>2. ecosystem conservation where basic organisms can breed naturally |
| Class 2 | *Very clean fresh surface water resources used for:*<br>1. consumption, but passing through an ordinary water treatment process before use<br>2. aquatic organism conservation<br>3. fisheries<br>4. recreation |
| Class 3 | *Medium clean fresh surface water resources used for:*<br>1. consumption, but passing through an ordinary treatment process before use<br>2. agriculture |
| Class 4 | *Fairly clean fresh surface water resources used for:*<br>1. consumption, but requires special water treatment process before use<br>2. industry |
| Class 5 | *The sources which are not classification in class 1–4 and used for navigation* |

Environment Government of Japan, 2003). The criteria are based on pH, dissolved oxygen, biological oxygen demand, suspended solids and total coliform, although faecal coliform, chemical oxygen demand, oil films and transparency criteria are also used for the three highest classes, where swimming is one of the uses (Ministry of the Environment Government of Japan, 2003).

In the Republic of South Africa it has been observed that ecosystems have an inherent capacity to recover from natural disturbances, so it is accepted that ecosystems can function at a number of different levels of health (Palmer, 1999). This principle was used to formulate a river classification that categorizes acceptable levels of human use and impact, and the risk of degradation that a particular use involves. The ecological management categories range from Class A systems, which are near to pristine, to Class D systems, which are judged to be sustainable despite use for abstraction, effluent disposal, and having undergone structural change (Table 9.19). The system also includes categories E and F, which are degraded and/or degrading. The policy requires

these systems to be managed with the goal of achieving at least Class D status (Palmer, 1999). Classes are identified on the basis of physical, chemical and biotic parameters.

The South African system allows for different levels of river health, such that river uses that degrade habitat and impair flow regimes are allowed in resilient systems. Although human use classes are implicit, it remains an ecologically based classification. The examples of water quality classification schemes given above (i.e. USA, Thailand and Japan) are driven by the existing or designated use of the river. Such a scheme could be adapted to consider a range of possible river health conditions corresponding to the major uses (like the South African scheme), rather than simply aiming for the relatively narrowly defined goal of swimmable/fishable. In a classification system that is driven by the designated uses of the river (as determined by stakeholders), the role of scientists and river managers would be to work with the community to achieve the best possible river health outcomes (Jones, 2003).

**Table 9.19.** *Ecological management classes for aquatic ecosystems in South Africa. Adapted from Palmer (1999) and Kleynhans (1996)*

| Ecological management class | Ecological condition description | Management guidelines |
|---|---|---|
| Class A | *Highly sensitive systems.*<br>Biota close to reference conditions | *Allow no human induced hazards.*<br>Remain within the target water quality for all constituents. Allow minimal modification of hydrology, instream and riparian habitat. |
| Class B | *Sensitive systems.*<br>Biota may be slightly modified from reference conditions. Especially intolerant biota may be reduced in numbers or extent of distribution. | *Small risk allowed.*<br>Set water quality objectives which pose only slight risk to intolerant organisms. Set instream flows that allow only slight risk to intolerant organisms. Allow slight modification from natural conditions for instream and riparian habitat. |
| Class C | *Moderately sensitive systems.*<br>Biota may be moderately modified from reference conditions. Intolerant organisms may be absent from some locations. | *Moderate risk allowed.*<br>Set water quality objectives which allow moderate risk only to intolerant biota. Set instream flows that allow moderate risk only to intolerant biota. Allow moderate modification of instream and riparian habitat from natural conditions. |
| Class D | *Resilient systems.*<br>Biota may be highly modified from reference conditions. Intolerant biota unlikely to be present. | *Large risk allowed.*<br>Set water quality objectives that may result in high risk to intolerant biota. Set instream flows that may result in high risk of the loss of intolerant biota. Allow a high degree of modification of instream and riparian habitat from natural conditions |
| Class E | *Seriously modified*<br>Degrading and/or degraded | *Improve to at least Class D* |
| Class F | *Critically modified*<br>Degrading and/or degraded | *Improve to at least Class D* |

## 9.4.12  Summary

Classification is a subjective procedure, dependent upon its purpose and the type of data available. Clearly, there is a need for an 'all-purpose' stream/catchment classification system that encompasses landscape, geology, soil, topography, temperature, climate, runoff, water quality, fauna, vegetation, aesthetics and economic and societal characteristics. A major limitation to the development of a broad classification scheme is the lack of availability of both physical and biological data at the stream reach level, and the lack of agreement on which attributes are most important for classification. Another consideration is the dynamic nature of landscapes and streams: a frog will maintain its identity for its lifetime, whereas changes in a stream reach may cause it to slip into a different class (e.g. braided versus sinuous). In the classification of riparian and wetland areas, Gebhardt (1989, 1990) recognized the potential for change in channels, water table levels, and associated vegetation, and introduced the concept of 'state' to allow process-oriented descriptions of classes.

As stream-classification methods continue to be applied and tested, progress will continue towards a universal, standardized stream-classification system. Multi-disciplinary efforts are needed to determine which variables are the most critical and universal. 'Lumpers' who tend to coalesce related classes and 'splitters' who tend to divide them will need to reach a compromise on the actual format of the classification system. As Pielou (1977) stated in reference to work in ecology, much of the literature is on testing and comparison of methods, and if the classifications are to be used, ecologists must choose one or a few methods and employ them consistently.

As in biology, classifications can be expected to change constantly within a general framework as 'new species' are encountered and new observations and measurements reveal deficiencies in older systems. There appears to be a consensus developing on the fundamental attributes of an enduring ecosystem-based classification system. Such a system would have the ability to encompass broad spatial and temporal scales, to integrate structural and functional characteristics under various disturbance regimes, to convey information about underlying mechanisms controlling in-stream features, and to accomplish this at low cost and ensuring good understanding among stakeholders (Naiman *et al.*, 1992, Hawkins *et al.*, 1993; Naiman, 1998). Most current systems lack sufficient definition of process to be able to reliably predict channel responses to specific disturbances. Despite their shortcomings, hierarchical classification systems have been effective in communicating to managers that stream types are diverse and they require a variety of management prescriptions for habitat protection and conservation (Naiman, 1998).

Most stream classification systems are based on the processes that occur in natural, reference systems as the ideal, and the resulting classes of stream health reflect that philosophy (i.e. a stream is either healthy or not healthy). This can create frustration in the stream management arena, where the goal is usually to seek a balance or compromise between protection of intrinsic stream values and the utilitarian use of stream resources. Perhaps a new generation of classification schemes based on agreed human uses is required, where the role of scientists and managers is to work out how to achieve the highest possible level of stream health for the designated class of stream use.

## 9.5  Assessing Instream-environmental Flows

### 9.5.1  Introduction

Dams, reservoirs, diversions and direct pumping modify the natural patterns of stream flow. Flood-mitigation dams, for example, reduce the peak discharges that would normally overflow the riverbanks and spill onto the floodplain (Petts, 1989). Reservoirs operated for irrigation water supply or hydropower production modify the natural flow regime through storage of water during high-runoff periods for later release when demands are highest. Peaking power hydroelectric production may also impose an on-off pattern on the natural flow regime as turbines are quickly brought on-line to supplement daily electricity requirements during peak-demand periods (Gore *et al.*, 1989).

Many studies conducted over the past two decades throughout the world have cited flow regulation as a major cause of degradation across the spectrum of river health (e.g. Trotsky and Gregory, 1974; Ward and Stanford, 1979; King and Tyler, 1982; Petts, 1984; Cushman, 1985; Mohanty and Mathew, 1987; Bain *et al.*, 1988; Boon, 1998; Klimas, 1988; Petr and Mitrofanov, 1988; Benke, 1990; O'Keeffe *et al.*, 1990; Cambray, 1991; Finlayson *et al.*, 1994; Bain and Travnichek, 1996; Gippel and Collier, 1998; Brizga and Finlayson, 2000; Jansson *et al.*, 2000). It is often difficult to isolate observed changes in the aquatic environment that are due to flow regulation from those that are due to changes in other factors such as catchment land use, fishing pressure, introduced species, riparian vegetation cover, large woody debris distribution and natural variations in flow regime (i.e. the pattern of floods, droughts and seasonal flow variability). The difficulty arises from the fact that changes to these factors can produce similar ecological responses, and the changes in these factors are often overlapping or simultaneous with flow regulation. Another difficulty for interpretation of historical data is that when changes in the controlling factors occur together, the effects are not necessarily a simple additive function of the impacts of each factor. Changes to some controlling factors may take a period of time before they manifest as degradation of the aquatic environment. Some impacts are marked and easily detected, while other more subtle impacts may be very difficult or expensive to measure. Another major problem in determining the environmental impacts of regulation in large river systems is that regulation has often occurred progressively over a long period of time, as water resources were developed to meet demands. Data are often available to quantify the hydrological impact of regulation, but in most situations, regulation began well before ecological monitoring programs were initiated.

Reviews of the impacts of flow regulation on the health of the River Murray, Australia (Gehrke, 1995; Norris *et al.*, 2001; Gippel and Blackham, 2002; Gehrke *et al.*, 2003) found that regulation of flow has played the major role in the overall degradation of the river, albeit in many cases the evidence is circumstantial and there are other potential contributing causes of degradation. The same could be said for the Missouri River, USA (Water Science and Technology Board, 2002). A survey by U.S. Environmental Protection Agency (2002) ranked changes to naturally flowing water as the second main threat to the water quality of USA rivers after agricultural pollution. A recent survey of the health of the uMngeni River and neighbouring rivers and streams in South Africa cited flow regulation as one of the primary causes of degradation in all resource units (River Health Programme, 2002).

Identification of flow regulation as one of the major causes of stream health degradation has led river managers to seek ways of reversing this effect, to the extent that this is currently one of the priority issues in stream management throughout the world. For example, a workshop of experts convened during the development of the Murray-Darling Native Fish Strategy (MDBMC, 2002) ranked flow management as the top intervention that would lead to improved native fish populations (abundance), with habitat restoration a close second (note that the strategy recommended a holistic approach that combined various interventions, including fishways, control of introduced fish species, correcting cold water pollution and declaring reserves). With respect to rehabilitation of the River Murray in general, Jensen (1998) ranked development of an environmental water policy and associated flow management strategy as the top environmental policy

concern. Modifying regulated flow regimes and returning diverted water back to rivers to benefit wildlife is a highly contentious and controversial issue. Conflicts over water use are common, especially where water supplies are limited, as in arid and semi-arid areas, where demands for water supplies are increasing, and where there are other instream uses such as navigation and recreation. It is necessary then to determine a satisfactory balance between competing uses. The process of achieving this balance is the topic of this section.

### 9.5.2   Instream and Environmental Flows Defined

The term *instream flow* is used, particularly in the USA, to refer to a specific stream flow that is identified for the purposes of planning or management of a stream or river. The instream flow is usually defined as a flow that is adequate to meet specific needs or management objectives for the river. In the USA, instream flows are established in legal form, typically through adoption of a State rule (Geller, 2003). Federal laws are also relevant in the USA through the Clean Water Act and the Endangered Species Act (Rushton, 2000). Instream flows are those that are retained in their natural setting, as opposed to those waters that are diverted for offstream users such as industry, agriculture and town water supply. Flows influence adjacent ground water levels, as well as the hydrological status of foodplain wetlands. The level of flow in a stream also conditions aesthetic and scenic values. Navigation is affected by flows. For example, kayakers in mountain streams require high flows, while in large rivers like the Columbia River in Washington State, if flows are below a certain level, the river becomes impassable to commercial barges, tugs and other watercraft because of the lack of draft (Rushton, 2000). Instream flows are also used for the production of hydroelectricity and waste disposal. In some places indigenous people live close to the river and their livelihoods depend on the river's resources, both clean water and biological resources (Quinn, 1991). Some rivers have valuable commercial and recreational fishing industries that depend on instream flows. The intrinsic wildlife values of rivers are also sustained by instream flows. The flows required to maintain or rehabilitate the habitat for riparian and aquatic life are the most difficult to quantify, and this is the main focus of the rest of this section. This subset of instream flows is known, particularly in Australia and southern Africa, as *environmental flows*, although in recent times this term has been used more broadly to also include other related uses.

Brown and King's (2003) definition of environmental flows encompassed aesthetic, recreational and cultural values as well as biophysical ones, and they also placed social and economic issues within the realm of environmental flows. King and Brown (2003) saw the process of environmental flow assessment as a means of describing the potential trade-offs between development gains (such as increased access to water for agriculture or industrial use) and environmental losses (such as reduced habitat for waterbirds or reductions in the quality of life of subsistence users of the river). Under this definition, environmental flows are not just 'flows for nature' (King and Brown, 2003). Instream flows for hydropower releases, irrigation releases, navigation, dilution of pollution, release of wastewater, and inter-basin transfers were excluded from Brown and King's (2003) definition of environmental flows, because these uses traditionally work against achievement of their defined ecological and human needs goals. It should be pointed out that the vast majority of instream flow effort in the USA has focused on assessing the flows required to sustain riparian and aquatic life [sometimes known there as conservation flows (Denslinger *et al.*, 1998)]. The issues of honouring the needs of existing instream users, and consideration of the impacts of instream flows on offstream users are dealt with as a related but separate exercise. So, for the remainder of this section we will use the term environmental flows to refer to instream flows required to sustain riparian and aquatic life in a healthy condition, plus those human uses that rely on, and do not compromise, stream health values.

### 9.5.3   Three Basic Assumptions of Environmental Flows Assessment

*Assessments Are Generally Grounded on the Natural Flow Paradigm, Even Though the Final Regime Is Likely to Be a Compromise on This*

Environmental flow objectives can be expressed as general principles. Regulated rivers are often managed to flow at relatively constant levels for long periods of time, and this also applies to so-called minimum flows for the environment (also called conservation, passing or compensation flows). This conflicts with the currently popular set of principles based around the natural flow paradigm, which states that discharge variability is central to sustaining and conserving biodiversity and ecological integrity (e.g. Walker *et al.*, 1995; Stanford *et al.*, 1996; Power *et al.*, 1996; Poff *et al.*, 1997; Richter *et al.*, 1997; Environment Protection Authority, 1997; Puckridge *et al.*, 1998; Tharme and King, 1998). Some limited understanding of the links between biological processes and aspects of flow

variability has been achieved, with most progress being made on species of high conservation or commercial value. However, given this currently limited understanding, and the improbability of ever being able to fully define the needs of the whole biological community, the conservative alternative is to assume that the natural flow regime is the best indicator of environmental needs. In Australia and South Africa, which are known for their highly variable rivers (Finlayson and McMahon, 1988), this concept has been quickly converted into environmental flow policy objectives. The concept is known as mimicking the natural system, or *physiomimesis* (Katopodis, 2003) and appears as an objective in numerous publications in the environmental flow literature (e.g. Environment Protection Authority of NSW, 1997; Arthington, 1998; Tharme and King, 1998; Snowy Water Inquiry, 1998).

Ecologists define flow variability in several ways, with some studies emphasizing maintenance of seasonality or timing of certain events that are related to known requirements of certain species' or assemblages of aquatic organisms (Tharme and King, 1998), and others emphasizing maintenance of inter-annual flow variability for its role in disturbing or re-setting ecological processes (Poff *et al.*, 1997). Hydrological processes alone do not sustain aquatic life. As well as factors such as adequate water quality, food supply and colonisation sources, aquatic organisms require diverse and abundant in-stream, riparian and floodplain habitats. Understanding the creation and maintenance of these physical habitats within a variable flow regime is the main geomorphological issue in environmental flow assessment (Petts and Maddock, 1996).

Scaling environmental flows as a proportion of the natural flows [i.e. the 'translucent dam' principle (Arthington, 1998)] maintains hydrological variability in a statistical sense. However, this approach fails to mimic natural flow variability at the scale of habitat hydraulics (temporal and spatial distribution of depth and velocity), because the channel morphology, which also controls the pattern of depth and velocity, remains unaltered. Also, the response of hydraulic, geomorphic and ecological characteristics and processes to changes in flow may often be non-linear and discontinuous. For example, half of the peak discharge will not move half of the sediment, and half of an overbank flow will not inundate half of the floodplain (Poff *et al.*, 1997). It is important then, when applying the flow variability paradigm, to consider process thresholds as well as hydrological variability *per se*. The concept of thresholds is well known to fluvial geomorphologists, because the process of sediment mobilization and deposition operates as a threshold phenomenon.

The geomorphological extension of the natural flow paradigm is the concept of river attributes. These are basic hydrogeomorphic characteristics of natural streams thought to be necessary for maintaining ecosystem integrity in rivers. McBain & Trush (1997) devised an attribute system for cobble and gravel-bedded alluvial rivers in California's Central Valley. Their ten attributes really describe processes, and are sufficiently general that they can be applied to most alluvial rivers (rivers with adjustable bed and banks). Five of the attributes explicitly deal with in-channel geomorphological processes: spatially complex channel morphology, frequently mobilized channel bed sediments, periodic channel bed scour and fill, balanced fine and coarse sediment budgets and periodic channel migration. Although not provided by McBain & Trush (1997), the ecological significance of these geomorphic attributes has some support in the literature. It has been argued that the attributes can be applied to both regulated and unregulated rivers, with some attributes achievable without requiring the entire unregulated flow regime (McBain and Trush, 1997).

The natural flow regime paradigm has merit, but translating it into practical recommendations for environmental flows can be problematic. In practice, the process of determining environmental flows does not involve attempting to devise a regulated flow regime that has a statistically defined variability (across all time scales) identical to that of the natural flow regime. For example, all of the environmental flow projects listed by Poff *et al.* (1997) involved only partial restoration of the natural flow regime. Under conditions of limited water resources, competing demands and constraints on flow control imposed by river structures, some flow variability targets will be low priority, and others will be impossible to implement.

### Less Than All the Natural Flow Will Maintain Stream Integrity (Stream Health)

Environmental flow assessments have long been applied to streams and rivers with existing or proposed large flow regulating structures such as dams and weirs (termed regulated streams and rivers). More recently such assessments have been applied to other streams, the hydrology of which is, or in the future may be, altered through direct pumping (abstraction), capture by farm dams, groundwater pumping or changed land use (e.g. urbanization or deforestation). This group of streams has been termed unregulated (Arthington and Zalucki, 1998), even though technically they may have highly modified flow regimes and/or poor water quality. There is an implicit understanding that environmental flow assessment is not about

recommending the full natural flow regime as the ideal. Rather, it is about determining what lesser amount will maintain stream health in the desired condition.

For regulated streams, the objective of environmental flow assessment is to set maintenance or rehabilitation targets (Gippel *et al.*, 2002), while for unregulated streams the objective is to set a sustainable limit to diversions (Nathan *et al.*, 2002). For minimally altered streams it would be prudent to err on the side of caution and recommend that the majority of the stream flow should be retained for environmental purposes. This caution can be justified by the unequivocal evidence of widespread decline in stream health through over-regulation, and the impossibility or high cost of restoration in these cases (Edwards and Crisp, 1982; Gippel and Collier, 1998).

The biotic and abiotic components of stream ecosystems can range in their sensitivity to water resource development (Figure 9.18). Sensitivity (or fragility) refers an ecosystem's resistance and resilience in the face of disturbance (Kleynhans and O'Keeffe, 2000). *Resistance* is the capacity of a system to resist change in structure or function, while *resilience* is the capacity of a disturbed system to return to its previous state once the source of the disturbance has been removed (Milner, 1994; Resh *et al.*, 1988; Peterson *et al.*, 1998; Carpenter *et al.*, 2001; Gunderson and Holling, 2002). For example, a wide, shallow stream will be more sensitive to drying out from flow reductions than a narrow, deep stream; some species of invertebrates, plants and fish are more sensitive to flow or water quality changes than others. Kleynhans and O'Keeffe (2000) developed a methodology for assessing ecological sensitivity in environmental flow studies.

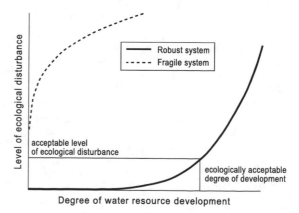

**Figure 9.18.** *Conceptual diagram of the relation between the degree of water resource development and the level of disturbance. Source: Stewardson and Gippel (1997). Reproduced by permission of Hydro Tasmania*

Estimates of how much flow must remain in a river before its health is impaired range from about 65% to 95% of natural flow, provided the natural pattern of flow is also retained (Dyson, 2003). Studies in Queensland, Australia, suggest that around 80% to 92% of natural mean annual flow (and other ecologically relevant hydrological indicators) may be needed to maintain a low risk of environmental degradation (Arthington and Pusey, 2003). An Expert Reference Panel advising on environmental flows for the River Murray (Jones *et al.*, 2002) used benchmarking to conclude that for a range of key hydrological indicators, assessed across several Australian river systems, the limit for an increased risk of unacceptable environmental degradation (environmental flow limit) generally lies within the range of 65% to 75% of natural. Greater levels of flow disturbance resulted in a range of different river conditions. The range of values for an acceptable level of hydrological modification given in these studies partly reflects the enormous variation in sensitivity and importance of the systems that have been assessed.

There is a good deal of uncertainty associated with the environmental flows assessment process. Much of this stems from uncertainty in ecological and geomorphic models, and inherent indeterminacy within the fluvial system. One way of addressing this uncertainty is to use a risk-based approach to environmental flows assessment. For example, Jones *et al.* (2002) presented their recommendations for the River Murray, Australia as a range of flow options, each with an associated likelihood or probability (low, moderate or high) of achieving a healthy working river, as defined in terms of a set of system-wide ecological objectives.

Tunbridge and Glenane (1988) took a different approach to risk in specifying environmental flows for the Gellibrand River and estuary in Victoria, Australia. They recommended higher (less uncertain) *optimum environmental flows* to allow the full production of fish especially for recovery after a period of stress (e.g. drought, overfishing), a *minimum environmental flow* level which would result in little or no reduction in numbers of fish, to apply in average rainfall years, and a *survival environmental flow* level which may cause a reduction in numbers of fish but no loss of species, to apply in low-rainfall years.

### Stream Health Exists on a Continuum

When stream health is defined as ecosystem integrity (Karr and Dudley, 1981) (see Section 9.3) it could be construed as an absolute term that cannot have degrees, i.e. ecosystem integrity is intact (stream is healthy) or it is not (stream is not healthy). As described in Section 9.2.3,

a healthy working river is one that is managed to provide a compromise, agreed to by the community, between the condition of the river and the level of human use. This concept is implicit in stream health assessment methodologies that rate stream condition against a reference condition, but along a grade, such as expressed by a range of observed/expected scores, or as a range of index scores compared against the highest possible score (see Section 9.3.6). Also, grades of stream health are allocated to rivers in classification schemes (see Section 9.4). In situations where the vision for the river, and hence the management goal, is a departure from the reference condition, the environmental flows assessment process seeks to find the balance between the desired ecosystem condition and other social and economic needs for water (Dyson, 2003). The flows allocated to achieve the chosen condition are the environmental flows.

### 9.5.4    Forms of Environmental Flow Assessment

The earliest phase of environmental flow assessment occurred from the end of the 1940s to the 1960s in the northwestern USA using professional judgment (Nestler *et al.*, 1989). During the 1970s, interest in instream flows expanded dramatically in this part of the world. Salmon fisheries of significant commercial value were being threatened by an intense phase of dam building, but at the same time, legislation was being enacted to protect fisheries. Since that time, development and application of environmental flow methodologies has expanded in scope and level of sophistication and has spread to many countries (Tharme, 2003).

Environmental flow assessments vary across a wide range of complexity and depth, as dictated by the level of funding, availability of data, technical capacity, time frame, priority of the site, or expected level of controversy. Environmental flows may be specified at several levels of resolution, from a single annual flow volume or a minimum flow limit below which diversions are not permitted, through to a comprehensive flow regime which specifies the distribution of a range of flows throughout the year. A comprehensive study might specify flows necessary to allow the passage of fish, flows to provide sufficient 'living space' for biota or to ensure acceptable levels of temperature, dissolved oxygen or salinity and higher flows (flushing flows or channel maintenance flows) to remove fine materials from the streambed, scour out encroaching vegetation, or flush anoxic or highly saline waters from stratified pools. The spatial scale of environmental flows assessment also varies widely, from whole of catchment to the river reach.

Methodologies range from relatively simplistic, reconnaissance-level approaches to resource intensive methodologies for detailed studies.

There have been numerous reviews and evaluations of environmental flows methodologies including Stalnaker and Arnette (1976), Prewitt and Carlson (1980), Wesche and Rechard (1980), Mosley (1983), Estes and Orsborn (1986), Loar *et al.* (1986), Richardson (1986), Reiser *et al.* (1985; 1989a; 1989b), Arthington and Pusey (1993), Karim *et al.* (1995), King *et al.* (1999), Tharme (2000), Jowett (1997), Stewardson and Gippel (1997), Dunbar *et al.* (1998), Arthington (1998), Arthington and Zalucki (1998), Espegren (1998), King *et al.* (1999), Annear *et al.* (2002), Dyson *et al.* (2003), Caissie and El-Jabi (2003) and Scatena (2004). Tharme's (2003) recent global review of the present status of environmental flow methodologies revealed the existence of some 207 individual methodologies, recorded for 44 countries within six world regions. A database of these methodologies is available online at International Water Management Institute (2003) (www.lk.iwmi.org/ehdb/EFM/efm.asp).

Tharme (2003) grouped environmental flow methodological types into four main categories, namely, hydrological, hydraulic rating, habitat simulation (or rating) and holistic methodologies. Tharme (2003) also recognized two minor classes of methodologies: 'combination' (or hybrid) approaches which had characteristics of more than one of the four basic types, and a group termed 'other' which comprised techniques not specifically designed for environmental flow assessment, but which had been adapted, or which had potential, to be used for that purpose. Assessment of flood flows necessary for channel formation or maintenance is an important aspect of many environmental flow assessments. There exists a group of geomorphic techniques that are used for this purpose in association with some of the environmental flow methodologies. Brown and King (2003) classified environmental flow methodologies into two categories—prescriptive and interactive—from the perspective of their usefulness as a tool for negotiating trade-offs between stakeholders. The following section illustrates the main characteristics of each methodological type by referring to examples of some of the better-known methods that are currently in use.

### 9.5.5 Hydrological Methods

*Tennant Method*

Hydrological methodologies use simple rules based on flow duration or mean discharge to scale down the natural

**Table 9.20.** Critical minimum flows required for fish, wildlife, recreation in streams identified by Tennant (1976)

| Description of flows | % of mean annual flow | |
|---|---|---|
| | dry season | wet season |
| Flushing or maximum | 200% of the mean annual flow | |
| Optimum range | 60%–100% of the mean annual flow | |
| Outstanding | 40 | 60 |
| Excellent | 30 | 50 |
| Good | 20 | 40 |
| Fair or degrading | 10 | 30 |
| Poor or minimum | 10 | 10 |
| Severe degradation | 0–10% of the mean annual flow | |

flow regime. The Tennant method (Tennant, 1976) also referred to as the 'Montana' method, is the most commonly applied hydrological methodology worldwide (Tharme, 2003). Recommended minimum flows are based on percentages of the average annual flow, with different percentages for winter and summer months (Table 9.20). The recommended levels are based on Tennant's observations of how stream width, depth and velocity varied with discharge on 11 streams in Montana, Wyoming and Nebraska. At 10% of the average flow (the mean daily flow, averaged over all years of record), fish were crowded into the deeper pools, riffles were too shallow for larger fish to pass, and water temperature could become a limiting factor. A flow of 30% of the average flow was found to maintain satisfactory widths, depths and velocities. The choice of a maximum flow was based on the theory that prolonged large releases would result in severe bank erosion and degradation of the downstream aquatic environment. The method was designed for application to streams of all sizes, cold and warm water fish species, as well as for recreation, wildlife and other environmental resources.

One main limitation of Tennant's method is that application of the technique to other streams requires that they be morphologically similar to those for which the method was developed. The required criteria, however, are not given by Tennant, making direct transfer of the technique difficult. Field observation of the stream at the various base flow levels is recommended for verification. Also, since the method is based on the average flow it does not account for daily, seasonal or yearly flow variations. Comparison of the recommended flows with the average 10- and 30-day natural low-flow values is advisable to determine whether the flows are available naturally during low-flow periods (Wesche and Rechard, 1980). Prewitt and Carlson (1977) also recommend the examination of mean monthly flow data to check the validity of the method.

### Texas Consensus Three-zone Concept

The Texas consensus-based state water planning process joins the three primary State water or natural resource agencies, the Texas Water Development Board, the Texas Natural Resource Conservation Commission and Texas Parks and Wildlife Department with other stakeholders to update the State Water Plan (Texas Parks and Wildlife Department, 2003). This effort addresses the long-range, multi-purpose water needs of Texas through broad-based involvement, negotiation, and consensus-building among key parties. The ecological objectives were aimed at the long-term health of the aquatic environment, realizing that periodic dry conditions are a natural part of the climate, hydrology and ecosystem processes in Texas. Water supply goals were also identified. To acknowledge the priority of human needs during dry periods and drought, the relative share of water provided for the environment is successively reduced to protect water supplies. This philosophy led to development of the three-zone concept; three classes of hydrological conditions having different primary objectives:

*Zone 1.* During normal or higher flow periods, the objective is to promote the long-term health of the natural environment.

*Zone 2.* During drier periods, the objective is to provide for minimum ecological maintenance where the aquatic species are impacted by lower flows, but can survive for a short period.

*Zone 3.* During severe drought conditions, the objective is to protect water quality.

Criteria were established for managing instream flows within the three different hydrological 'zones' for new developments downstream of reservoirs, and for new developments of direct diversions from channels (Table 9.21). In all hydrological zones the intention is that flows passed for instream purposes also contribute to meeting the ecological needs of the associated bay and estuary system. In addition to passage of environmental flows, adequate flows are passed to protect downstream water rights. The method was developed from the original work of Bounds and Lyons (1979) [cited in Texas Parks and Wildlife Department *et al.* (2003)] and Mathews and Bao (1991). Note that the three hydrological 'zones' of the Texas consensus concept are similar to the optimum, minimum and survival flow categories identified by Tunbridge and Glenane (1998) for the Gellibrand River, Victoria, Australia.

**Table 9.21.** *Flow criteria for setting environmental flows for new developments using the three hydrological zones of the Texas Consensus Concept. The streamflow values which trigger different zonal operations are calculated from naturalized daily streamflow estimates. Source: Texas Parks and Wildlife Department (2003)*

| | Direct diversion from channel | | Downstream of reservoir | |
|---|---|---|---|---|
| Zone | Criterion | Environmental flow | Criterion | Environmental flow |
| Zone 1 | Streamflow > natural monthly median | Minimum flows passed downstream in amounts up to the natural monthly median | Reservoir water levels >80% of storage capacity | Inflows passed downstream in amounts up to the natural monthly median |
| Zone 2 | Natural monthly 25th %ile ≤ streamflow ≤ natural monthly median | Minimum flows passed will be in amounts up to the natural monthly 25th %ile flows | Reservoir levels 50% to 80% of storage capacity | Inflows passed downstream in amounts up to the natural monthly 25th %ile flows |
| Zone 3 | Streamflow ≤ natural monthly 25th %ile | Minimum flows passed will be the greater of (i) flow necessary to maintain water quality or (ii) a continuous-flow threshold (e.g., natural monthly 15th %ile), such that the diversion does not dry up the stream | Reservoir levels <50% of storage capacity | In lieu of any site-specific data, the 7Q2 low-flow value is the default criterion for flows to pass downstream. |

### Hoppe Method

The Hoppe (1975) regional rule-of-thumb method utilises flow duration curves ideally based on daily discharge. It is based on environmental flow recommendations in the Frying Pan River, Colorado for Rocky Mountain trout streams. The recommended flows are flow exceeded 17% of the time for flushing, 40% of the time for spawning and 80% of the time for rearing. The flushing flow is a flow maintained for 48 hours to flush fines from gravels. An application of the Hoppe method (including a comparison with other hydrological methods) on the Klamath River, California, can be found in Institute for Natural Systems Engineering (1999).

### Range of Variability Approach

The Range of Variability Approach (RVA) (Richter *et al.*, 1997) defines 32 parameters that are Indicators of Hydrologic Alteration (IHA). The method can be used to identify annual river management targets based on a range of variation (e.g. ±1 standard deviation from the mean, or 25th to 75th percentile range) in each of 32 parameters. The method prescribes that environmental flow regime characteristics should lie within the targets for the same percentage of time as they did prior to regulation. This method has been applied in 30 environmental flow related studies (although mainly to examine degree of flow alteration) in North America and has attracted interest in three other countries (Tharme, 2003).

### Conclusion—Hydrological Methods

The main attraction of hydrological techniques is the fact that an answer can be obtained rapidly if gauged records are available, eliminating the time and cost of field data collection. Some methods require modelling of naturalized streamflow records. The techniques tend to be site- and species-specific, and for them to be applicable in other situations the relationship between habitat and discharge must be similar. As given, the methods do not require a detailed understanding of the ecosystem; a limitation which can be significant if, for example, results from snowmelt-dominated streams are applied to the flashy streams in semi-arid lands. It may be possible to develop hydrological methods for other species or streams if a relationship between habitat and discharge can be derived from the experiences of local researchers or anglers, or special field studies.

Orth and Leonard (1990), Dunbar *et al.* (1998) and Tharme (2003) concluded that simple hydrological methods requiring little or no field work are most appropriate for basin-wide planning purposes, or for providing preliminary estimates in low controversy situations. A number of studies have compared hydrological methods with more sophisticated approaches (e.g. Orth and Maughan, 1981; Orth and Leonard, 1990; Caissie *et al.*, 1998; Bureau of Reclamation, 1999). These comparisons were reviewed by Stewardson and Gippel (1997), who concluded that there was no consistent or reliable pattern in the relative magnitude of flows recommended by these methods.

### 9.5.6  Hydraulic Rating Methods

Hydraulic rating methods utilise a quantifiable relationship between the quantity and quality of an instream resource, such as fishery habitat, and discharge, to calculate flow recommendations. Most emphasis has been placed on the passage, spawning, rearing and other flow-related maintenance requirements of individual, economically or recreationally important fish species. This approach is sometimes known as a 'transect method' or 'wetted perimeter method' because it involves measuring and interpolating changes in simple hydraulic variables, such as wetted perimeter or maximum depth, usually measured across single river cross sections, as a surrogate for habitat factors known or assumed to be limiting to target biota. Commonly, shallow riffled reaches are chosen for analysis as these areas are the first to be affected by flow alterations, and because it is reasoned that maintenance of suitable riffles will also maintain suitable pool conditions. There is an implicit assumption in this method that there is a threshold value of the selected hydraulic parameter that if maintained in the regulated flow regime will maintain stream health.

### Washington Method

The Toe-Width Method was developed by the Washington Department of Fisheries, the Department of Game and the U.S. Geological Survey in the 1970s in response to the need to determine minimum instream flows for fish and is still in use (Rushton, 2000; Department of Ecology, 2003). Data were collected over a 9-year period on water depths and velocities at 8–10 different flow levels along transects on 28 streams over known spawning areas. Criteria for spawning and rearing depths and velocities for each fish species and life stage were used to calculate the area of habitat at each measured flow (Collings, 1972; Collings, 1974) (Figure 9.19). These points of habitat quantity at different flows were connected to create a fish habitat versus streamflow relationship. A similar approach is taken in Colorado with the R-2 Cross methodology

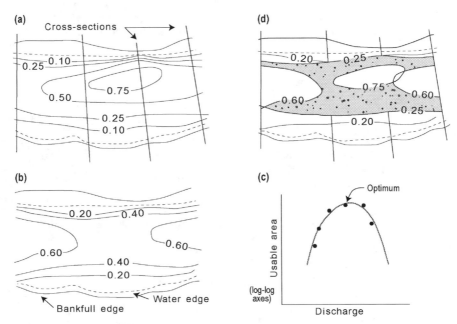

**Figure 9.19.** *Washington method for determining preferred discharge. For this theoretical example, 'preferred habitat' is: depth 0.25–0.75 m and velocity 0.20–0.60 m/s. Figures display: (a) depth contours in metres, (b) velocity contours in metre per second, (c) a combination of maps (a) and (b) with shaded region showing usable area, and (d) trend-fitted curve derived from measurements (point shown) taken at several discharges. The 'optimum' discharge corresponds with the greatest usable area. Adapted from Collings (1972)*

which, despite being developed more than 25 years ago, is still in use (Espegren, 1998).

The established fish habitat relationships for Washington rivers were compared to catchment characteristics to determine if there was an easy way of predicting spawning or rearing flow for a certain fish species. The toe-width was the only variable found to have a high correlation. The toe-width is the distance from the toe of one stream bank to the toe of the other stream bank (i.e. bed width). Power function equations are used to derive the flow needed for spawning and rearing salmon and steelhead based simply on a stream width measurement (Collings, 1974; Swift, 1976; Swift, 1979; Rushton, 2000; Department of Ecology, 2003). A similar approach was taken by O'Shea (1995) to develop predictive relationships for streams in Minnesota.

### Wetted Perimeter Methods

Simple wetted perimeter methods (Nelson, 1980) have been widely applied for many years (Tharme, 2003), and are still being used to make important environmental flow

determinations (e.g. McCarthy, 2003). In the State of Washington, if fish habitat criteria are not available, then a wetted-perimeter method is used (Collings, 1972; Department of Ecology, 2003). When applied in its simplest form, transects are located in several representative riffles and measurements of depth and velocity are taken over a range of different flows. A plot of wetted perimeter against discharge is drawn and the first break in slope in the curve is taken as an indication of the optimum, adequate or minimum discharge for the biota of interest (Figure 9.20). Increases in discharge above this point produce smaller changes in wetted perimeter. In many USA applications, the technique is based on the premise that the slope breakpoint represents the quantity of water preferred by salmon, but there are few studies that have investigated the ecological significance of the breakpoint in any detail. Gippel and Stewardson (1998) provided an application and evaluation of the method for headwater streams in Victoria, Australia, and compared the results with those of hydrological methods, and macroinvertebrate and fish data. In that case the breakpoints occurred at flows considered too low to adequately

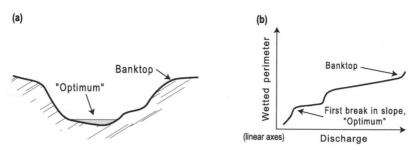

**Figure 9.20.** *Wetted-perimeter method: (a) hypothetical channel cross-section and (b) graph of wetted perimeter versus discharge. The first breakpoint in slope is used as an index of optimum available water*

protect the biota. One long-standing problem with the wetted perimeter method is the subjectivity involved in selecting the breakpoint in the curve of wetted perimeter versus discharge. Gippel and Stewardson (1998) solved this problem by developing a systematic way of selecting the breakpoint. Reinfelds *et al.* (2004) refined this methodology and compared the predictions for a field site in New South Wales, Australia, with those of commonly used hydrological methods.

For unregulated perennial streams in New South Wales, the cease to pump limit is initially set using a hydrological method based on flow duration (80th or 95th percentile flows are protected) (Environmental Protection Authority, 1997). There is provision for revising these limits, which normally requires a field verification based on a transect method that uses breakpoints apparent on relationships between measures of wetted perimeter and discharge to assist in the determination of minimum environmental flows. The method uses detailed topographic riffle survey data, coupled with a GIS, to model the change in area of physical habitat (wetted perimeter) with discharge (Reinfelds *et al.*, 2004). Compared with the traditional transect survey method, this model provides more information about the spatial distribution of available habitat, and it is a useful visual aid for demonstrating the impacts of various low-flow scenarios.

### Conclusion—Hydraulic Rating Methods

As compared with rule-of-thumb methods, transect techniques take into consideration the habitat requirements of biota and the availability of habitat at various discharge levels. The need for field data, however, makes the techniques more time-consuming and costly. Hydraulic rating methodologies were the precursors of more sophisticated habitat rating or simulation methodologies, also referred to as microhabitat or habitat modelling methodologies (Tharme, 2003).

### 9.5.7    Habitat Rating Methods

Habitat rating methods consider not only how physical habitat changes with stream flow but combine this information with the habitat preferences of a given species to determine the amount of habitat available over a range of stream flows. Results are normally in the form of a curve showing the relationship between available habitat area and stream discharge. From this curve, the optimum stream flows for a number of individual species can be ascertained and the results used as a guide for recommending environmental flows.

### Instream Flow Incremental Methodology (IFIM) and PHABSIM

The most well known of the habitat rating methods is the Instream Flow Incremental Methodology (IFIM) (Bovee, 1982). IFIM is the most commonly used flow assessment method worldwide (Tharme, 2003, King and Brown, 2003). IFIM was originally devised in the late 1970s by the then Co-operative Instream Flow Service Group of the U.S. Fish and Wildlife Service, Colorado, with a focus on optimizing habitats for a single important species of fish, such as salmon (Reiser *et al.*, 1989b). In the USA, environmental flows (usually expressed as a minimum flow) are legally recognized through various State flow and pollution laws and Federal laws (Clean Water Act and Endangered Species Act) and are usually described and established in a formal legal document, typically an adopted State rule (Rushton, 2000). Despite the method receiving a good deal of criticism over the years (Tharme, 2003), in many States of the USA, IFIM is regarded as the most scientifically and legally defensible method, so it has become the standard for large rivers, with hydraulic rating methods used on smaller streams (Espegren, 1998). There has been considerable development of IFIM since its introduction over twenty years ago (Bovee, 1982), with some of the earlier concepts becoming redundant. King and Tharme (1993) presented an updated step-by-step guide to applying the IFIM, including a review of the process.

IFIM is much more than a habitat rating method. It is a problem-solving tool made up of a collection of analytical procedures and computer models. It was designed as a communication link between fishery biologists, hydrologists and hydraulic engineers. The aim of IFIM is normally to determine the effect of some activity such as irrigation withdrawals, dam construction or channel modification on aquatic habitat. As each riverine system will have different sets of flora, fauna, hydrological and hydraulic characteristics, types of disturbances and regulating agencies, the IFIM allows the development of a different approach for each situation.

The IFIM is a complete thought process that begins with the structuring of a study design and the description of the present condition, and carries through to the final negotiation of a solution. 'Incremental' means the slight or incremental modification of the problem or the perspective or view of the problem until a solution is found. It also refers to the ability to look at the effects of incremental changes in a variable (e.g. discharge) on available habitat. Since there will be a number of perspectives on management of water for instream uses, incrementalism is a valuable approach to problem-solving in this field. Rather than generating a single answer, the methodology produces a range of solutions that permit the evaluation of different alternatives. Two identical applications of the method can lead to different solutions simply due to different management goals.

The IFIM is grounded in the ecological niche concept (hydraulic model) of explaining community distribution, incorporating both macrohabitat and microhabitat concepts. Macrohabitat applies to large-scale longitudinal gradients in habitat characteristics along streams, and microhabitat is the precise location where an individual species is normally found (Table 9.22). The microhabitat approach was justified by reference to two studies showing that competition between fish species was reduced by physical habitat isolation (Bovee and Milhous, 1978). The Weighted Useable Area (WUA) microhabitat model is the core habitat rating component of the IFIM. PHABSIM (Physical HABitat SIMulation) is a collection of computer programs used to model WUA. The PHABSIM software and manual (Waddle, 2001) can be downloaded free of charge from the U.S. Geological Survey website (http://www.fort.usgs.gov/). PHABSIM has traditionally been used to simulate a relationship between streamflow and physical habitat for various life stages of a species of fish (or a recreational activity), although Gore *et al.* (2001) point out that there is a greater ability to predict macroinvertebrate distribution and diversity without complex population models. Individual organisms will tend to select the most favourable instream conditions, but will also use less favourable ones, with preference decreasing as conditions become less favourable (Stalnaker, 1979).

The IFIM habitat modelling process is conceptualized in Figure 9.21. The final relationship models the change in WUA with discharge. WUA is an indicator of the net

*Table 9.22. Scales used for physical habitat assessment in streams. Source: Stewardson and Gippel (1997). Reproduced by permission of Hydro Tasmania*

| Name | Common length scale (channel widths) | Distinguishing hydraulic features | Description |
|---|---|---|---|
| Segment (macrohabitat) | >100 | Hydrology, mean channel geometry | Long stretch of a stream often starting and ending at a major confluence |
| Reach | 10–100 | Pool-riffle and meander cycles | A length of stream which includes one or more pool-riffle or meander cycles |
| Morphological unit | 0.1–10 | Bed topography | Topographic high or low points of the longitudinal bed profile (eg. pool, riffle and step) |
| Hydraulic biotope (meso-habitat) | 0.01–1 | Surface flow condition, substrate, velocity, depth, location in channel | A region of the stream environment populated by a characteristic biota |
| Physical microhabitat | 0.001–1 | Velocity, depth, bed shear stress, substrate, turbulence, boundary layer hydraulics | A precise location where an individual species is normally found |

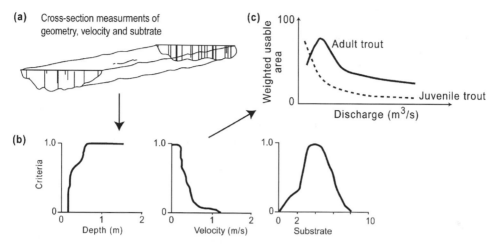

**Figure 9.21.** *A conceptualization of the procedures in* PHABSIM: *a stream reach is selected and surveyed at a particular discharge (a), and the information is combined with habitat Suitability Index (SI) curves defined for a particular species and life stage (b). The procedure is repeated for several discharges (using measured or modelled data) to obtain a curve describing the suitability of use of that reach as a function of discharge (c). Adapted from Gore and Nestler (1988). Reproduced by permission of John Wiley & Sons Ltd.*

suitability of use of a given reach by a certain life stage of a certain species. At a particular stream discharge the pattern of distribution of physical habitat (depth, velocity, cover and substrate) is evaluated over the stream reach [Figure 9.21(a)]. This is combined with the habitat Suitability Index (SI) curves [Figure 9.21(b)] (see 'Habitat Evaluation Procedure and Habitat Suitability Indices (HEP/HSI)' in Section 9.3.4) to determine the WUA for that discharge. The physical habitat is redefined at each discharge and the computations repeated to obtain WUA as a function of discharge [Figure 9.21(c)].

The habitat-discharge functions from PHABSIM may be combined with flow data to obtain monthly or daily habitat time series and habitat-duration curves using a set of programs, TSLIB (Time Series LIBrary), also developed by the USA Fish and Wildlife Service (Milhous *et al.*, 1990). The habitat time series is useful for comparing pre- and post-project habitat availability. Habitat duration curves can be developed for different water allocation rules to help in selection of the best alternative. Periods critical to species survival during a given life stage can also be identified, as can the limiting habitat availability (i.e. physical carrying capacity) for each species and life stage. This information is particularly useful for evaluating potential changes in species composition since

different species will react differently to changes in hydraulic characteristics (Milhous *et al.*, 1989; Stalnaker, 1979).

A survey of IFIM users (who reported 616 applications of IFIM) in the USA by Armour and Taylor (1991) found that application of IFIM stumbled in some cases because users considered the methodology was either technically too simplistic or too complex to apply. Armour and Taylor (1991) concurred with the survey respondents that the emphasis for research should be on addressing technical concerns about SI curves and WUA. The IFIM/PHABSIM approach is not without its critics (Castleberry *et al.*, 1996; Van Winkel *et al.*, 1997), with claims that its misuse has actually hastened the decline of rivers in the western USA, particularly the large alluvial rivers (Woo, 1999). Two major criticisms of PHABSIM are that it promotes the application of flows that are too low for sediment mobilization, and that it promotes relatively constant flows. Much of the criticism relates to the earlier applications of IFIM, when the focus was on a single species of fish, or on studies where PHABSIM was enthusiastically applied outside of the more holistic IFIM framework, and channel forming flows were ignored (Barinaga, 1996). It is crucial that PHABSIM be used within the wider IFIM framework (or other holistic framework), as originally intended, so that inappropriate low flow options generated by PHABSIM can be critically evaluated, and higher channel maintenance flows can be included (Woo, 1999).

### Alternative Hydraulic-habitat Rating Models

A number of alternative software systems with the same or similar components as PHABSIM have been developed for use with IFIM. These include EVHA (Pouilly *et al.*, 1995), which is an adaptation of PHABSIM for use in French streams, RHABSIM (Payne, 1994) (http://www.northcoast.-com/~trpa/), which is a commercial version of PHABSIM, and RYHABSIM (Jowett, 1989), which is a New Zealand adaptation of PHABSIM with a simpler interface and fewer options.

The River System Simulator is a computer-based simulation system, developed in Norway for multi-purpose planning and operation of river systems, with special emphasis on hydropower and its environmental effects (Killingtveit and Harby, 1994; Killingtveit and Fossdal 1994; Alfredsen *et al.*, 1995; Alfredson, 1998) (http://www.sintef.no). The simulation software includes 14 models representing input from watershed into the river system; hydropower system simulation models; physical, chemical and biological processes in rivers and lakes and the consequences for humans and ecosystems.

PHABSIM, like most other environmental flow habitat rating methods, uses a basic 1-D (one-dimensional) hydraulic model for predicting changes in water surface elevation and velocity with discharge. Until the mid-1990s, 1-D hydraulic models like HEC-RAS, WSPRO or SWMM were used almost exclusively to model channel hydraulics for instream flow studies (Texas Parks and Wildlife Department *et al.*, 2003). However, rivers typically have spatially complex hydraulic habitat, including across-channel velocity variations, and 1-D models may be inadequate for resolving detailed channel hydraulics (Leclerc *et al.*, 1995; Moyle, 1998; Railsback, 1999; Crowder and Diplas, 2000). 1-D models are probably adequate for perennial, low gradient streams but may be inaccurate if applied to hydro-peaking streams, ephemeral streams, mountain streams or streams under flood conditions (Gan and McMahon, 1990a). Multi-dimensional models offer expanded options for instream habitat analysis (Bovee, 1996; Hardy, 1998).

2-D hydraulic models predict across-channel variations in flow, but only 3-D models capture both horizontal and vertical velocity variations, and they are appropriate where strong vertical velocity gradients exist (Texas Parks and Wildlife Department *et al.*, 2003). The 2-D, depth-averaged, finite element hydrodynamic model RIVER2D has been utilized for several study sites in Canada and the

USA (Katopodis, 2003), and can be downloaded from www.river2d.ca. 2-D habitat models allow examination of patterns of habitat availability and use at the meso-habitat scale (Ghanem *et al.*, 1996). However, these models need biological validation, user-friendly interfaces, good documentation and training if they are to replace the familiar 1-D models (Bovee, 1996). A comparison of 1-D and 2-D hydrodynamic models by Waddle *et al.* (1996) found that the two methods predicted water surface elevations and mean velocity equally well, but the 2-D model was able to capture the complex flow situations where significant transverse flow was present. The application of 3-D models is currently limited by the difficulty of field verification and the extremely small grid scale that is necessary to resolve variations around local objects such as large woody debris (Texas Parks and Wildlife Department *et al.*, 2003). Hardy and Addley (2003) demonstrated how incorporating behavioural decision rules regarding fish community structure and dynamics could extend the capabilities of 2-D and 3-D modelling of flow hydraulics. Further discussion on the merits and pitfalls of 2-D and 3-D modelling can be found in SINTEF NHL (1994) and Leclerc *et al.*, (1996).

A critical limitation on the use of habitat simulation models is the lack of well-defined habitat-suitability curves. Since these curves are essentially empirical correlations, some authors (e.g. Nestler *et al.*, 1989) state that the curves may not be transferable from one stream to another, and indicate that site-specific curves may be preferable. However, the development of these curves is costly (Bovee, 1986). The strongest criticism of PHABSIM has centred on the ecological interpretation of the WUA index. Gore and Nestler (1988) reviewed and commented on the criticisms put forward by a number of authors. The assumption is made in instream flow uses of PHABSIM that if the habitat is maintained, the fish population will be maintained. Some studies found evidence to support this assumption (Orth and Maughan, 1982; Stalnaker, 1979), while others did not (Irvine *et al.*, 1987; Mathur *et al.*, 1985; Scott and Shirvell, 1987). Other factors such as food supply, biological interactions (e.g. competition, predation), nutrients, dissolved oxygen, presence of ice cover, temperature and flow regime (including the effect of floods) may be of greater importance than physical habitat in limiting species biomass or abundance.

Gore and Nestler (1988) found that the relationship between fish biomass and WUA was better in coldwater streams, which have simpler ecosystems and more predictable hydrological regimes in comparison to more hydrologically and ecologically complex cool- or warm-water streams. Scott and Shirvell (1987) also observed that a better relationship was obtained when usable space was a limiting factor, such as for older age classes of fish and when the biomass was near the system's carrying capacity. If a river is carrying the maximum possible number of individuals, then reducing the amount of habitat will cause a near-immediate reduction in numbers or size of fish, whereas the same reduction at less than carrying capacity may have little effect (Tunbridge and Glennane, 1988).

The large number of options in PHABSIM may diminish the objectivity of the results. In an evaluation of PHABSIM, Gan and McMahon (1990b) found up to an elevenfold difference in computed WUA, depending on the options selected. RHYHABSIM, in comparison, has few options. Thus different analysts will get similar results, but the program does not provide as much flexibility.

Orth and Maughan (1982) conclude that IFIM is a useful framework for managing streams with altered flow regimes but it is not a panacea. Biological expertise is still needed in the interpretation of results. However, the model can provide a basis upon which a biologist may apply professional judgement (Mosley and Jowett, 1985) and a methodology for comparing the relative effects of different management decisions.

### *The Flow Events Method*

Stewardson and Gippel (2003) noted that because IFIM in its original form is grounded in ecological niche theory, it does not adequately account for disturbance, which is an important determinant of community structure (see Section 9.4.2). The Flow Events Method, developed by the Cooperative Research Centre for Catchment Hydrology, Australia (www.catchment.crc.org.au), extends the hydrological variability RVA idea of Richter *et al.* (1997) to also consider temporal changes in physical habitat conditions at the appropriate spatial scales. The method characterizes flow variations for environmental flow studies using knowledge of the influence of flow events on biological and geomorphic processes. This approach has the advantage that ecological benefits of the environmental flow are clearly articulated, available knowledge is included in the development of flow recommendations, and the method accounts for the natural dynamism in flow-related ecosystem processes by using the natural flow regime as a template for the environmental flow regime (Stewardson and Gippel, 2003).

A flow event might be considered a discrete period of time during which the flow conditions influence a geomorphic or biological process. These events are selected using collective multi-disciplinary expert knowledge. An example of a flow event might be drying of the streambed, which can be characterized by the area of wetted peri-

meter of the channel. Hydraulic models are used to relate the selected habitat parameters to discharge, and then time series' of the parameters are created from the flow records (usually the current, natural and proposed future regime scenarios are compared). The time series' are used to generate statistics regarding the temporal distribution of the events (e.g. Stewardson and Cottingham, 2002; Stewardson and Gippel, 2003). The forms of the analyses undertaken are akin to the analyses of flow records to characterize flood frequencies using the annual and partial duration series, or to generate flow duration curves. The Flow Events Method is not a new framework, but an analytical tool that can be used within any framework, such as IFIM.

### 9.5.8 Holistic Methods

Holistic approaches are essentially frameworks for organizing and using flow-related data and knowledge (Brown and King, 2003). This approach is not constrained by the analytical tools, and it is not unusual to make use of several different methodologies. While the hydrological, hydraulic rating and habitat rating methods usually focus on key sport fish such as salmonids, or fish with a very high conservation value, the holistic approach attempts to consider the entire ecosystem, using all available information, much of which may be little more than working hypotheses. The problem of uncertainty is overcome by adopting a conservative 'precautionary principle', and by recommending ongoing monitoring and adaptive management (e.g. Poff *et al.*, 1997; Richter *et al.*, 1997; Environment Protection Authority of NSW, 1997; Richter *et al.*, 1998). Holistic approaches cover a wide range of methodologies, including the Expert Panel Assessment Method (Swales and Harris, 1995), the River Babingley (Wissey) method (Petts, 1996; Petts *et al.*, 1999), the Scientific Panel Assessment Method (Thoms *et al.*, 1996; Thoms *et al.*, 2000), the Building Block Methodology (Tharme and King, 1998; King *et al.*, 2000), the Holistic Approach (Arthington *et al.*, 1992), the Flow Restoration Methodology (Arthington, 1998; Arthington *et al.*, 2000a), the Benchmarking methodology (Arthington, 1998) and the DRIFT process (Brown and King, 2000; King *et al.*, 2003; Arthington *et al.*, 2003).

Proponents of the holistic method take the view that where possible, management strategies for rivers should be aimed at maintaining or rehabilitating as much as possible of the original, functional aquatic ecosystem (Arthington and Pusey, 1993). The holistic method is based on the natural hydrological regime and is intended to provide water required for the complete ecosystem including the river channel, riparian zone, floodplain, groundwater, wetlands and estuary. Explicit numerical models that relate discharge to aspects of the river's geomorphology, water quality or ecology may be available, or even developed through the course of the investigation, but they are usually used as an aid to decision making, rather than as a numerical solution to the problem of defining a suitable regulated flow regime.

Holistic methods operate across a range of spatial scales. An important advance in holistic methodology has been the establishment, in Queensland, Australia, of the benchmarking methodology, which undertakes basin-scale evaluation of the potential environmental impacts of future scenarios of water resource management (Arthington, 1998). Another significant development in holistic methodologies has been incorporation of social, cultural and economic components into environmental flow assessments. For example, the DRIFT method developed for use in southern Africa (Brown and King, 2000; King *et al.*, 2003) evaluates links between flows and social consequences for subsistence users alongside ecological and geomorphological ones, and economic implications in terms of mitigation and compensation. Another example is The Living Murray initiative (www.thelivingmurray.com.au), which is an ongoing process that is examining the ecological, cultural, social and economic consequences of flow scenarios for the entire 2530 km long River Murray, Australia (Gippel *et al.*, 2002).

### Bottom-up (BBM) and Top-down (DRIFT) Holistic Approaches

The holistic approach may adopt a bottom-up philosophy (Arthington, 1998), whereby the regime is built up from hydrological facets (usually a pattern of baseflows combined with certain floods) that is assumed from experience or research to be the minimum combination required to achieve ecological sustainability. The BBM (Building Block Method) approach builds the regime from three main groups of flow facets (Figure 9.22). The strength of the BBM method lies in its ability to incorporate any relevant knowledge, and to be used in both data-rich and data-poor situations (Brown and King, 2003). The alternative is the top-down philosophy, whereby it is assumed that the entire natural flow regime is ecologically important, but that some facets of it can be modified or omitted without threatening ecological sustainability (Arthington, 1998). The top-down South African DRIFT (Downstream Response to Imposed Flow Transformations) process (Brown and King, 2000; King *et al.*, 2003), which emerged from the foundations of the BBM, was developed for the assessment of environmental flows for the Lesotho High-

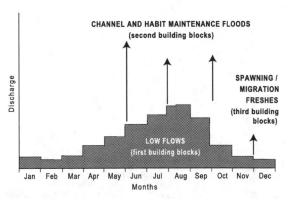

**Figure 9.22.** *The 'building blocks' of a hypothetical environmental flow regime created using the BBM approach. Source: Tharme and King (1998)*

lands Water Project (King and Brown, 2003; Arthington *et al.*, 2003).

DRIFT consists of four modules (Figure 9.23). In the first, or biophysical module, models are developed to predict how the river ecosystem changes with discharge. Tools commonly used in other approaches, such as wetted perimeter or PHABSIM could be utilized in this module. In the second, or socio-economic module, the objective is to develop predictive capacity of how different flows would impact the lives of riparian people who are subsistence users of river resources, known as populations at risk (PAR). In the third module, scenarios are built of potential future flows and the impacts of these on the river ecosystem and the riparian people. The fourth, or economic module, lists compensation and mitigation costs (King *et al.*, 2003).

Both bottom-up and top-down holistic approaches share four main assumptions regarding achievement, or maintenance, of ecological sustainability:

- some facets of the natural flow regime cannot be scaled down, and must be retained in their entirety
- some facets of the natural flow regime can be scaled-down
- some facets of the natural flow regime can be omitted altogether
- variability of the regulated flow regime should mimic that of the natural flow regime, in certain respects

These assumptions arise from the notion that high- and low-flow events are more important than in-between

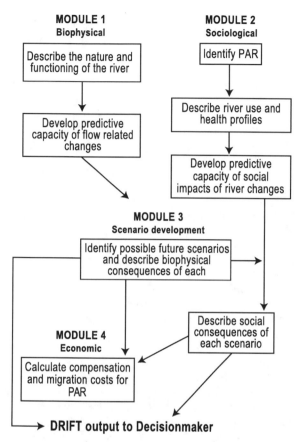

**MODULE 1**
**Biophysical**

Describe the nature and functioning of the river

Develop predictive capacity of flow related changes

**MODULE 2**
**Sociological**

Identify PAR

Describe river use and health profiles

Develop predictive capacity of social impacts of river changes

**MODULE 3**
**Scenario development**

Identify possible future scenarios and describe biophysical consequences of each

**MODULE 4**
**Economic**

Describe social consequences of each scenario

Calculate compensation and migration costs for PAR

➤ **DRIFT output to Decisionmaker**

*Figure 9.23. The four modules used in a DRIFT assessment. PAR is 'populations at risk'. Source: King and Brown (2003), © John Wiley & Sons Ltd. Reproduced with permission.*

conditions because of the stresses and opportunities they present to the biota (Poff *et al.*, 1997). Also, many geomorphic and ecological processes show non-linear responses to flow, requiring a threshold to be exceeded before the process is activated (i.e. they cannot be scaled down) (Poff *et al.*, 1997). This approach attributes lower ecological importance to medium to high level baseflows, and repeated small- and medium-sized flood events, so they are sacrificed in the environmental flow regime. Drought events are normally retained in their entirety by adding the proviso that prevailing natural flows (i.e. as defined by dam inlet discharge) are applied when they fall below the baseflow level set for that month. Very large floods, although their ecological and geomorphological importance has been acknowledged, are not usually con-

sidered in environmental flow assessment because in most situations they cannot be controlled.

### *Expert Panels*

Central to the holistic approach of addressing biophysical aspects is the idea of assembling a team of specialists from various disciplines to combine their expert knowledge to recommend a flow regime that satisfies objectives for the particular site. At one or more stages in the process, the experts are drawn together in a workshop situation with the aim of expediting an outcome. Apart from procedural differences, the main factor that differentiates the way individual expert panels operate is the level of effort applied to obtain information about relations between flow and habitat, geomorphology, water quality and biota. At its simplest level, a visual inspection of the river may suffice, while more sophisticated studies may incorporate aspects of IFIM, or commission a targeted research component.

When applied as a rapid and relatively cheap method to evaluate environmental conditions, there are a number of concerns associated with the expert panel process. The outcomes of expert panel deliberations are conditioned by interpersonal dynamics, the potential for groupings to arise within the panel, the impact of a single dominant personality, and the possibility that consensus can be a product of collective bias (rather than independent assessments).

Bishop (1996) statistically analysed the results of a study of two expert panels that independently rated a range of flows for their suitability for native fish, invertebrates and habitat quality at the same six sites (Swales and Harris, 1995). Of the 18 assessments made, the two panels agreed on only one of them. It was suggested that the differences in the ratings were derived from differences in expertise, from the subjective nature of the assessment, and conflicts between the experts' knowledge of a system and the supplied data (Bishop, 1996; Pusey, 1998). While it is true that there was considerable variation in the absolute rankings of the suitability of the flows between the two panels, the panels gave similar patterns of rankings (Swales and Harris, 1995). The overall result of the study was not surprising; in general, the panels rated flows similar to the natural flow regime as having the highest ranking for provision of suitable ecological conditions.

Cottingham *et al.* (2002) reviewed the use of scientific panels in environmental flow assessment studies in Australia and found that they were prolific, spurred on by new Federal and State initiatives that demanded assessments be completed within a relatively short time-frame (6–12 months). However, these panels have been consistently confronted by the same limitations and knowledge gaps, and Cottingham *et al.* (2002) feared that unless investment was forthcoming to close those gaps, their usefulness might decline. The performance of expert panels would also be improved by developing guidelines for selecting panel members and protocols for guiding the conduct of the assessment process (Cottingham *et al.*, 2002).

### 9.5.9    Other Approaches

Tharme (2003) found that a reasonable proportion of all methodologies surveyed could not be easily classified into one of the four main groups. Some were hybrids of established methods, and some involved use of rapid stream health assessment tools, such as RIVPACS, to recommend environmental flows. Tharme (2003) distinguished one particular approach—the use of managed high flows, or experimental flow releases. Techniques for designing and implementing managed high flows are not a separate class of methodology. Rather, they comprise a set of tools that can be used within other methods or frameworks, principally the holistic methodologies, which address overall river health and therefore recognize the need to consider the full spectrum of flows. There are two main categories of high-flows that are of interest: flows for floodplain and wetland inundation (Acreman, 2003) and flows for performing geomorphic work (Gippel, 2002a). Two other special groups of instream flow studies apply to specific environments not traditionally covered in environmental flow assessments: groundwater dependent ecosystems and estuarine ecosystems. A final special group of instream flow studies concerns assessment of flows for recreation.

### *Managed Floods for Floodplain and Wetland Inundation*

Acreman (2003a) described the concept of reinstatement of flood releases from reservoirs for floodplain inundation. The main issues are the general appropriateness and feasibility of introducing managed floods; the design of appropriate flood flows, including the optimum magnitude, duration, frequency and timing of managed floods (given the economic, social and environmental feedbacks and tradeoffs); and implementation of the flood. Acreman (2003a) emphasized the need to ensure that managed floods are compatible with the livelihood strategies of the floodplain communities. Objectives of managed floods

can include flood recession agriculture, fishing, animal husbandry, groundwater recharge or conservation of biodiversity and bioproductivity. One of the main limitations of managed floods is technical feasibility. Many dams do not have adequate outlet structures or sufficient storage capacity, and water quality can be poor due to stratification. Despite these difficulties, managed floods have been demonstrated to result in stimulation of fish breeding, provided they occur during the normal flood period (Welcomme, 1989).

While the flood pulse concept (Section 9.4.2) emphasizes the ecological importance of floods, in practice it may be difficult to identify thresholds of flooding below which floodplain ecosystems cease to function adequately (Welcomme, 1976). Acreman (2003a) suggested that the alternative to attempting to recommend how much water the floodplain ecosystem requires is to set managed flood objectives in terms of specific targets, such as fish production, for which sufficient data may be available to develop a tentative predictive relationship. In other cases, flows necessary to maintain specific floodplain and riparian plant and animal communities are reasonably well known (e.g. Stromberg and Richter, 1991; Blanch *et al.*, 1999; Roberts *et al.*, 2000; Leslie, 2001; Hughes *et al.*, 2001; Pettit *et al.*, 2001).

In some cases, lack of flood flows for plant recruitment is not the problem; rather it is lack of flows to keep plant growth in check. For example, a 14-year study of tree reproduction and survival in the Platte River, Nebraska by Johnson (1999) found that Populus and Salix recruitment is controlled largely by stream flow in June. Floods produce seedlings in most years, but there is a 90% attrition rate within the year due to summer stream flow pulses from thunderstorms that erode or bury new germinants, river bed restructuring by moving ice in winter and to a lesser extent, desiccation during summer droughts. This kind of information can be used to design environmental flows at key times of the year to raise seedling mortality rates in order to maintain or widen overgrown channels.

Most river management agencies are committed to protection of floodplain ecosystems, particularly wetlands. For example The N.S.W. Wetlands Management Policy (Department of Land and Water Conservation, 1996) specifies as the first of their nine principles that 'Water regimes needed to maintain or restore the physical, chemical and biological processes of wetlands will have formal recognition in water allocation and management plans'. The Australian Federal government has also made a strong commitment to wetland protection; recently publishing a framework for determining the water needs of wetlands of National and International Importance (Davis *et al.*, 2001).

One reality is that the practice of managed flood releases for wetland flooding will mostly involve enhancement (or boosting) of the peak magnitude or duration of naturally occurring floods, rather than generating the entire flood from a dam release. This is simply a matter of the large volume of water required. The appropriateness of managed flood releases will depend on the relative merits of using water to benefit floodplain ecosystems and people who depend on healthy floodplain resources, compared to storing it in the reservoir for utilitarian uses. This decision can only be made on a case-by-base basis (Acreman, 2003a). An example of a managed flood release on the The Waza-Logone floodplain in Cameroon was described by Acreman (2003a). The objectives were to restore biodiversity and the livelihoods of communities—who depend on natural resources of the floodplain—through managed flood releases, while retaining sufficient water in the reservoir for intensive rice irrigation. Pilot flood releases were conducted in 1994 and 1997 to determine the response of the ecosystem and of local communities (Acreman, 2003a).

Research has clearly linked declining health of the Ramsar listed Barmah-Millewa Forest on the River Murray, Australia, to reduced frequency and duration of medium-sized floods due to regulation (Gippel and Blackham, 2002). In late 2000 the opportunity arose to enhance the duration of forest flooding by supplementing a naturally occurring 1 in 5 year flood event that peaked at around 90 000 ML (approx. 1000 m³/s). A total allocation of 341 GL (341 000 m³) was released from a headwater dam in three parcels (Barmah-Millewa Forum, 2001; Leslie and Ward, 2002) (Figure 9.24). At the time, this enhancement represented the largest single-event allocation of environmental water in Australia (Leslie and Ward, 2002). Despite the large volume of this allocation, it comprised only about 8% of the total flow passing and entering the Forest between September 2000 and January 2001. The objective of the enhancement was to 'in-fill' the flood recession limbs so that water levels in the Forest would not drop below levels where birds would abandon nests. Modelling demonstrated that under unregulated conditions, during the recession between the two flood peaks the water levels would have remained well above the level where nest-abandonment might occur [17 200 ML/day or 200 m³/s (Blanch, 1999)] (Figure 9.24). Without enhancement of the recession between the two main flood peaks, a nest-abandonment event would almost certainly have occurred (Leslie and Ward, 2002). By enhancing the flood recessions with the flow releases, this was avoided, and the duration of conditions suitable for bird breeding on the flood recession limb were extended by more than a month (Figure 9.24).

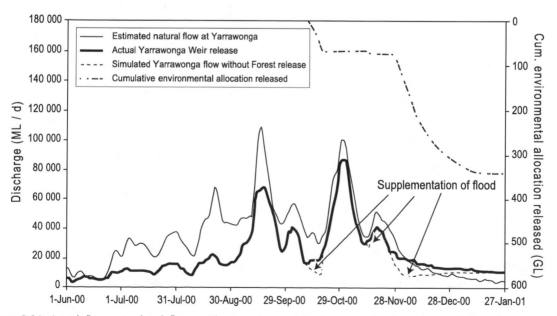

**Figure 9.24.** *Actual flows, simulated flows without environmental releases, and estimated natural flow downstream of Yarrawonga Weir, and cummulative use of the environmental flow account, River Murray, Australia. 100 000 ML/d is approx. 1200 m³/s. Source: Gippel (2003). Reproduced by permission of the Murray-Darling Basin Commission*

### Managed Flows for Channel Formation

Channel forming flows include any flows that have a role in shaping the physical form of the channel, maintaining habitat forms, prevention of vegetation encroachment and removal of fine sediment and detritus from the surface of the substrate (Reiser *et al.*, 1985; 1989a; 1989b; Brooks, 1995). In environmental flow literature the terms 'flushing flows' and 'channel maintenance flows' are also used to describe some of these processes (Reiser *et al.*, 1985; 1989a; 1989b; Kondolf and Wilcock, 1996; Milhous, 1996; 1998; Texas Parks and Wildlife Department *et al.*, 2003). Channel forming flows can be specified in terms of magnitude, frequency, duration and timing. Like many aspects of environmental flow assessment, the idea of providing special flows that maintain channel morphology has attracted some criticism. This in part reflects gaps in the knowledge of the way channels form, regional and site specific differences in the way channels form, and indeterminacy of channel form (Gippel, 2002a).

Given the difficulty of making theoretical predictions about channel form (Hey, 1978; Richards, 1982; Phillips, 1990), one way of establishing a case for retaining channel forming flows is to examine the geomorphic effects of dam regulation, where hydrological modifications have altered the nature of channel forming flows in the river downstream. In the mid-1970s, the USA Federal government filed an application for rights to water for the purpose of channel maintenance flows for streams in the Water Division 1 region in Colorado. There were many opponents to this application, the main one being the State of Colorado, which was concerned about the impact of this application on existing water rights and on the development of future water supply projects (Gordon, 1995). The case presented an opportunity for some of the most prominent geomorphologists in the USA to argue scientific points in a court setting. However, uncertainty in the geomorphic models of the way channels form (uncertainty that geomorphologists acknowledge and expect), raised sufficient doubt and concern that the judge ruled against the Federal government. Given the complexity of the problem, it is not surprising that the geomorphic responses to even a relatively simple impact like dam regulation have included channel widening, narrowing and no change (Kellerhals, 1982; Williams and Wolman, 1984).

Most dam-regulation and water diversion projects decrease the capacity of a stream to transport sediment. The net effect is that sediment delivered to the stream by tributaries, channel banks or side slopes accumulates on the surface and can work its way into the subsurface, rather than being flushed away during flood events. The connections between substrate quality and fish and macro-invertebrate populations are well documented in the literature (Reiser *et al.*, 1985; Brooks, 1995). Channel bed degradation below dams is reportedly a common phenomenon (e.g. Rasid, 1979; Germanoski and Ritter, 1988; Galay, 1993). Channel enlargement can only occur if the post-impoundment flows have the capacity to mobilize the bed and/or bank sediments. Regulation of rivers by large impoundments usually involves a profound reduction in the frequency of small- to medium-sized floods. Because these floods include channel forming flows, it is not surprising that channel narrowing has commonly been observed below dams. Reduction in channel migration rates has also been observed below dams (Shields *et al.*, 2000b). One review suggested that braided rivers were more prone to narrow, while meandering rivers were more prone to slow their migration rate (Friedman *et al.*, 1998). The channel adjustment process can be rapid, with significant deposition observed within two years, while other studies report significant channel narrowing within periods ranging from five to thirty years after dam closure (see Gippel, 2002a). Channel contraction below a dam is normally limited in its downstream extent to the junction of a major unregulated tributary, which reinstates channel forming floods (Carling, 1988). In some cases, coarse sediment delivered by downstream tributaries forms large gravel bars at junctions, because the regulated main stream is incompetent to transport it further downstream (Brierley and Fitchett, 2000). Channel responses to changes in controlling variables are complex, variable and hard to predict, and some systems are geomorphologically insensitive to the effects of dam operation (Phillips, 2003). In an attempt to unify the variety of observed geomorphic responses to dams, Grant *et al.* (2003) developed an analytical framework based on two dimensionless variables that predicts geomorphic responses to dams. The two variables were the ratio of sediment supply below to that above the dam and the fractional change in frequency of bed sediment-transporting flows.

Floodplain maintenance flows are large floods that shape floodplain features through lateral erosion, meander cut-off, avulsion and overbank sediment deposition. These floods are important for maintaining habitat diversity over longer-time scales. Only very large dams interfere significantly with floodplain maintenance floods, so they are not usually considered in environmental flow assessments.

Where flows for geomorphic processes have been recommended (e.g. Gippel and Stewardson, 1995; Wilcock *et al.*, 1996; Snowy Water Inquiry, 1998; Kondolf, 1998a), they invariably represent only a small part of the natural medium- and high-flow regime. Many of the important instream habitat features of rivers are also medium-scale geomorphic features, such as bars, undercut banks, pools and riffles. Geomorphic features like rock bars, sand slugs and cascades can limit fish passage under low regulated flow conditions. There is currently only limited understanding of how specific bed and bank features form and are maintained; the exception is the existence of predictive models for dynamic plane, ripple, dune and anti-dune bedforms in sand bed rivers (Vanoni, 1984). Low flows can transport sand-sized sediment and cause in-fill of scour pools, so there is a need to also consider the impacts of the duration of low-flow events on river geomorphology (Brizga, 1998). It is these lower flows that are the primary geomorphic concern when assessing the limits of sustainable extractions on unregulated rivers.

There is overwhelming evidence in the literature from Australia and elsewhere that marked alteration of channel forming flow processes is associated with declining ecological health, or degradation of the physical channel attributes required for normal ecological functioning (e.g. Gippel and Collier, 1998; Tharme and King, 1998; Petts and Maddock., 1996; Gippel, 2000; Reiser et al., 1989a, 1989b; Rutherfurd, 2000). For example, Bain et al. (1988) compared fish community and habitat in two rivers: one with a natural flow regime and one that was regulated with frequent discharge fluctuations. They found that those species requiring stream margin habitat with shallow depths and slow water velocities were reduced in abundance in the regulated river. Although models of channel formation contain uncertainty, this does not discount the importance of channel forming flows, nor does it prevent quantitative consideration of channel forming flows in environmental flow assessment. Floods that serve a channel forming purpose probably also perform some important direct ecological role, such as cueing fish migration or spawning, or regenerating riparian vegetation. So, when assessing the need for flood flows it is necessary to consider their multiple roles.

Channel form is a complex function of flood frequency, flood duration, sediment transport and boundary conditions (resistance of bed and banks). It has long been thought that the process of channel formation was fundamentally associated with bankfull discharge, or the flow that just fills the channel to the top of the banks (banktop) (Wolman and Leopold, 1957) (Section 7.2.5). Studies from many areas of the world suggest that on average, this banktop discharge occurs every one to two years. However, so much variation in this frequency has been observed that geomorphologists are now wary of applying this as a general rule (Gippel, 2002a). While bankfull flow is defined in terms of channel geometry (i.e. the flow that fills the channel to the bank top), the dominant, or

effective, discharge is the flow that carries the majority of the sediment load over a long period of time (Gordon, 1995). It is the medium-sized flows, which carry an intermediate amount of sediment but occur relatively frequently, that transport most of the sediment in the long term. Some studies have found that the dominant/effective discharge of Australian rivers crosses a broad range, suggesting that channels are naturally adjusted to a wide range of channel forming flows (Gippel, 2002a). This wide band of effective discharge possibly explains the common existence of complex channel morphologies in Australian rivers (Rutherfurd and Gippel, 2001).

A review of the nature of channel forming processes by Gippel (2002a) suggested that geomorphologists should be prepared to apply a range of methods to the problem of understanding channel forming flows, and not expect a simple and consistent explanation to emerge from the analysis. Environmental flow regimes that focus on a single discharge, or limited range of discharges, to perform the channel forming role could result in simpler channel morphology. While short-tem processes are important, channel formation also operates over a long-term time scale. Environmental flows (which are generally designed for and managed over the relatively short time-scale) should be placed within this longer-term context of changing channel form.

Reiser et al. (1985; 1989a; 1989b), Milhous (1986), Brizga (1998) and Gippel (2002a) reviewed the literature on channel forming flows as a component of environmental flow assessment. The approaches generally fell into three categories:

1. *Hydrological*—These methods use an index obtained from runoff records, such as the Tennant and Hoppe Methods. Hydrological records (or regionalized estimates) are needed at the site of interest.
2. *Morphological*—Channel characteristics such as some percentage of bankfull flow (Section 7.2.5) are used to obtain a flushing flow. This discharge can be estimated at the site using the slope-area method (Section 5.8.2) or more elaborate backwater curve methods (Section 6.6.6).
3. *Sedimentological*—A channel forming flow is determined by using a sediment-transport equation to find the flow level needed to move particles of a given size (Section 7.4). Knowledge of channel dimensions, slope, and substrate composition is required.

In the hydrological and morphological methods the selected flow level corresponds with flow levels at which satisfactory channel forming processes take place, as determined from observations at the actual site, in a flume study or from the literature. It is reasonable to assume that statistics derived from one study site can be applied to others within a hydrologically homogeneous area (Kondolf *et al.*, 1987). If operation rules permit, it may be possible to perform experiments downstream of a dam site. Stillwater Sciences (2003) presented a methodological framework for undertaking and assessing managed geomorphic high-flow events. Some experimental managed geomorphic floods have been planned or conducted (Gippel, 2002a). Planned large-scale releases in the Trinity River (Petts, 1984, pp. 260–262; Kondolf and Wilcock, 1993; McBain & Trush 1997), and Glen Canyon (National Research Council, 1991) have not been implemented (Stillwater Sciences, 2003), but a reasonably large flood was released in the Grand Canyon in 1996 (Collier *et al.*, 1997; Schmidt, 1999, Valdez *et al.*, 1999; Webb *et al.*, 1999).

Although Hoppe's (1975) hydrological method gives a recommended duration of 48 hours the other hydrological methods only indicate peak-flow values. The duration of the flow and the shape of the hydrograph (Section 4.3) should be part of the design for both ecological and geomorphological reasons. For example, if the flow is dropped too quickly it can lead to sloughing of saturated stream banks or stranding of fish or other stream-margin biota. The quality of the released water should also be considered; for example, releasing large quantities of anoxic water from the lower levels of a reservoir may kill off much of the downstream aquatic life, somewhat defeating the purpose of habitat improvement. The time of release of flushing flows should preferably coincide with historical periods of high flow, since the biota will be adapted to this regime.

One general rule-of-thumb for channel maintenance that has been used in gravel-bed streams is that sediment begins to be mobilized at a flow depth just greater than 80% of the bankfull flow depth (Parker, 1979), although Andrews and Nankervis (1995) found that 60% of bankfull flow was sufficient to initiate motion of some gravels (noting also that general motion of the bed surface was exceedingly rare). Rosgen *et al.* (1986) give flushing flow recommendations developed by the U.S. Forest Service for snowmelt-dominated, perennial, alluvial streams. Their method uses bankfull discharge as the peak flow of a synthetic flood hydrograph, with the rising limb approximating the natural hydrograph shape. The recession flows are dropped somewhat more quickly than natural floods under the assumption that most of the sediment transport occurs on the rising limb. The authors stated that adjustments to the basic method should be made based on local conditions and professional judgement.

A study to determine the flushing flow requirements to maintain the spawning habitat for Colorado River squawfish used a combination of fieldwork and laboratory tests (O'Brien, 1984). In this case the river was incised, so effective discharge was much lower than bankfull discharge. Effective discharge was recommended for microscale sediment and vegetation reorganization, while bankfull flows were recommended for maintenance of channel form. Andrews and Nankervis (1995) analysed effective discharge of Rocky Mountains Streams and found that 80% of bed material load was transported by flows between 0.8 and 1.6 times bankfull discharge, and averaged 15.6 days per year. Preserving all flows sufficient to initiate bed particle movement required 35% of the mean annual flow. A comprehensive range of environmental flows recommended by Hill *et al.*, (1991) for fishery flows, channel maintenance flows, riparian flows and valley maintenance flows, assumed that bankfull discharge was the important determinant of channel form, with valley slope maintenance assumed to be related to the flood with a recurrence interval of 25 years.

The sedimentological method is physically based, derived from sediment-entrainment theories (Section 7.4). It is normally assumed that fines will be flushed out when the threshold of motion for some percentage of the particles is reached. One simple relationship based on Lane's data equates tractive force (depth of flow in metre $\times$ slope $\times 10^3$) (kgm$^{-2}$) to incipient particle diameter in centimetres. This applies to round particles, with flat-shaped particles requiring half the tractive force to initiate movement (Newbury and Gaboury, 1993a). The amount of shielding, packing or imbrication, or armouring must be taken into account as well as the particle size to be mobilized. The Shield's Eq. (7.16; see Example 7.2) is commonly used to predict sediment mobilization (e.g. Reiser *et al.*, 1985; Pitlick, 1994). However, as mentioned in Section 7.4, the difficulty lies in selection of a proper coefficient since the method was developed for uniform sands.

Milhous (1996) used numerical techniques to calculate the flushing flow requirements of the Colorado River squawfish in the Gunnison River. Three different sediment disturbance processes were identified (each corresponding to a different median particle size). The Meyer-Peter and Müller equation was used to predict the incipient motion of sediment of a particular size for streams in southern Wyoming (Water and Environmental Consultants, 1980). The stream power thresholds of Brookes (1990) could be used to design a channel forming flow regime to suit the desired channel type. Fundamental changes in stream character could result if an environmental flow regime caused the stream to cross any of these stream power thresholds.

The best approach, then, would be to develop flushing-flow estimates using several methods, include ecological requirements for flooding such as a specified level or period of inundation, adopt a conservative figure, test it out at the site and monitor the results to determine its effectiveness (Stillwater Sciences, 2003). Monitoring sediment movement can be done (1) during the flushing event with a bedload sampler or (2) after the event, using scour chains, cross-sectional surveys, bed-sediment sampling, tracer particles of various sizes or photography (see Section 5.9). Biological monitoring should also be conducted to evaluate whether the changes in streambed composition and/or flooding of the floodplains have the desired effects on the aquatic ecosystem.

Natural flow regimes are composed of numerous facets, or components, occurring as a complex time series, not discrete, predictable and independent events. Scaled-down flow regimes, or regimes with certain components culled from the regime, should not be expected to produce the same morphology as a natural flow regime. Environmental (regulated) flows will nearly always be simpler and less variable than natural flow regimes, so the resulting geomorphology will probably be less diverse (Gippel, 2002a). It is likely that in many cases environmental flow assessments will be complicated by the existence of geomorphic disturbances additional to those caused by flow alteration. The problem of flow regulation or modification should be viewed as one of a group of potential causes of physical channel disturbance. This also applies to unregulated streams when the issue is determining sustainable limits to diversion. The first stage of any assessment for channel forming flows should be to place the significance of flow regulation as an agent of geomorphic change within this wider context (Gippel, 2002a). This will provide a realistic perspective on the potential for environmental flows to either maintain or improve channel functioning.

### Assessing Water Requirements of Groundwater Dependent Ecosystems

There are six main types of groundwater dependent ecosystems: terrestrial vegetation, terrestrial fauna, river baseflow systems, aquifer and cave ecosystems, wetlands and estuarine and near-shore marine ecosystems. Nearly all rivers have some level of groundwater interaction. King *et al.* (2000) recognized the need to consider groundwater when doing surface water environmental flow assessments in South Africa, especially for non-perennial rivers, although only a few studies have done so. Cook and Lamontagne (2002) noted that while methods are available for assessing the groundwater dependency of ecosystems, water requirements are much more difficult to determine. This explained why so few water allocation plans in South Australia had adequately assessed the needs of groundwater dependent ecosystems.

One important aspect of rivers that is rarely considered in assessment of environmental flow requirements is the phreatic zone, an area of saturated sediments located beneath and adjacent to streams and rivers (Triska *et al.* 1989). This zone has been recognized as providing a fundamental link between the active channel and floodplain areas (Stanford *et al.*, 1994; Edwards, 1998), with intense, microbially mediated biogeochemical processing regulating nutrient transformations and fluxes to surface aquatic ecosystems (Findlay, 1995, Triska *et al.*, 1993). The hyporheic zone also provides important habitat for aquatic macroinvertebrates, especially as a refuge area (Stanford and Ward, 1988; Stanford *et al.*, 1994; Smock *et al.*, 1992; Cooling and Boulton, 1993; Coe, 2001).

Recognition of the need to consider the water requirements of groundwater dependent ecosystems is relatively recent (Hatton and Evans, 1998). Clifton and Evans (2001) identified three management approaches to meeting the environmental needs of groundwater dependent ecosystems. The first is to ignore the environmental needs and make no specific provision. The second is to provide a fixed water provision, such as fixed percentage of average annual groundwater recharge. For example, a blanket 30% of recharge is being considered in New South Wales, Australia (Department of Land and Water Conservation, 2000). The third is to consider the needs of the environment through a formal assessment process, informed by an understanding of the economic, social and environmental costs and benefits.

Clifton and Evans (2001) developed a conceptual framework for assessing the water requirements for groundwater dependent ecosystems (although to date it has not been applied). The framework is divided into three main components. The first concerns determining the level of groundwater dependency of the ecosystem, which involves identifying the dependent elements of the ecosystem. Next, the key groundwater elements are described in terms of flux (rate of surface or subsurface discharge), level (depth of water table) pressure (applies to confined aquifers) and quality (salinity, nutrients and contaminants). These environments vary in their degree of dependence on groundwater, with Hatton and Evans (1998) recognizing five classes, ranging from those entirely dependent on groundwater to those which make only limited or opportunistic use of groundwater. The second main component of the framework is to assess the water regime in which the dependency operates. Like surface water assessments, this step involves studies of the ecology, habitat, consumptive use and hydrological characterization in terms of timing, frequency and duration. These models are used in the third component of the framework, which is to determine the environmental water requirements. The relationships allow prediction of responses to change. Only by setting a target level of ecosystem health can water requirements be determined.

### Assessing Water Requirements of Estuarine Ecosystems

Most of the research and application of environmental flows has been limited to the freshwater reaches of rivers. However, there is a reasonable basis of understanding of the role of freshwater in maintaining estuarine ecosystems. The 'order of effects' model suggested by Hart and Finelli (1999) distinguishes between the direct and indirect effects of flow on ecological processes. Direct effects are things like changes in salinity or erosion while indirect effects alter some component of the system (biotic or abiotic), which in turn affects organisms. The conceptual models of Kimmerer (2002a) describe the consequences of freshwater inflows to estuaries in terms of physical, chemical and biological components, or resources. The model of Sklar and Browder (1998) also included the effects of landscape modifications, tidal actions and solar activity.

In the USA, a National Symposium in 1980 (Cross and Williams, 1981) was followed by numerous detailed ecological studies of freshwater inflows to estuaries (e.g. Longley, 1994; Powell and Matsumoto, 1994). Freshwater inflow is known to be a strong determinant of fish abundance, and numerical modelling tools for maximizing estuarine fish harvest have been devised on this basis (Bao and Mays, 1994a; 1994b). Salinity is an important variable used to define estuarine processes. The limits of oceanic saline influence have traditionally been defined by the 1 ppt (parts per thousand) salinity isohaline, because this is the upper limit for many agricultural and industrial applications, but a higher limit of 2 ppt (termed 'X2'), measured 1 m from the bottom of the water column, is more useful as an ecological indicator (e.g. Kimmerer *et al.*, 1998; Kimmerer and Schubel, 1994).

A study of flow and water quality in the Upper Derwent River Estuary, Tasmania by Davies and Kalish (1994) developed a model of salt wedge location as a function of river discharge, and was able to specify the duration and magnitude of the flood required to flush oxygen depleted water from the estuary. In the lower Snowy River, Victoria, reduced flows under regulation can allow penetration of the salt wedge further upstream, effectively extending the limit of the estuarine section of the river,

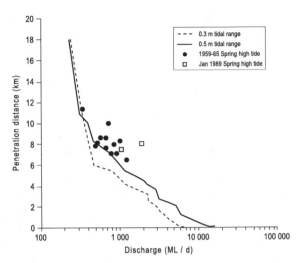

***Figure 9.25.*** *Model of position of the upstream extent of the salt wedge (2 ppt isohaline) as a function of discharge for two tidal ranges, lower Snowy River, Victoria, Australia. Plotted data are observed positions of salt wedge toe under Spring High Tide (1.4–1.6 m) conditions. Source: Gippel (2002b) using data adapted from Webster and McLellan (1965) and Coastal Environmental Consultants et al. (1989). Reproduced by permission of East Gippsland Catchment Management Authority*

inconveniencing landholders wishing to pump from the lower reaches of the river, and reportedly increasing erosion by degrading bank side vegetation (Department of Conservation and Environment, 1992). Gippel (2002b) used a flow-salt wedge (X2) model (Figure 9.25) to demonstrate that regulation of the river caused the salt wedge to shift upstream by about 2–3 km for most of the time (Figure 9.26). This type of model could be used to recommend flows required to move the salt wedge to the desired downstream position at critical times of the year.

In the San Francisco Bay and Sacramento-San Joaquin Delta estuary, about half of the natural freshwater inflow is diverted upstream for irrigation and other uses. An expert panel chose X2 as the key indicator to set flow standards for estuarine management (Jassby *et al.*, 1995). This index was found to be inversely proportional to the logarithm of the net freshwater outflow from the delta. Some other advantages of this index were its ease of measurement, it was meaningful to non-scientists and therefore a good communication tool, it could be directly related to management actions, a historical record was available, and it was correlated with habitat conditions and ecosystem responses (Jassby *et al.*, 1995). The X2 isohaline coincided with the turbidity maximum (or entrapment zone), a zone of particle flocculation with elevated levels of

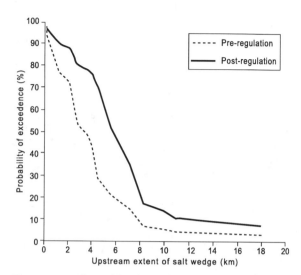

**Figure 9.26.** *Effect of regulation on lower Snowy River salt wedge penetration expressed as percentage of time the up-stream extent of the salt wedge from the mouth is exceeded. Tidal range at mouth in this example is 0.5 m. Source: Gippel (2002b). Reproduced by permission of East Gippsland Catchment Management Authority*

biological activity. The historical value of X2 was statistically correlated with ecosystem variables across all trophic levels, such as input of organic carbon, and abundance and survival of shrimp, fish and macroinvertebrates (Jassby *et al.*, 1995; Kimmerer *et al.*, 1998; Kimmerer *et al.*, 2001; Kimmerer, 2002a). These relationships to flow may be due to several potential mechanisms, but no mechanism has been conclusively shown to underlie the flow relationship of any species (Kimmerer, 2002a). Positive relationships between freshwater inflows and biological response are thought to operate mainly through stimulation of primary production with effects propagating up the food web. Evidence from the San Francisco Bay Estuary found that the increase in abundance or survival of organisms in higher trophic levels with increasing freshwater inflow was not explained by increased primary productivity. There is evidence to suggest that the relationship between fish survival and abundance and freshwater inflow may be explained by attributes of physical habitat that vary with flow (Kimmerer, 2002b; Coastal CRC, 2002). This finding emphasizes the need to protect physical habitats as well as inflows.

In some coastal rivers in Queensland, Australia, a close correlation has been established between prawn and fish catches and freshwater inflows to estuaries, expressed as annual catchment rainfall, summer discharge and annual discharge (Loneragan *et al.*, 1998; Bunn *et al.*, 1998; Loneragan and Bunn, 1999). Other Queensland studies have found direct evidence linking large cohorts of estuarine fish populations in the Fitzroy River and Port Curtis region with major flood events in 1973, 1975, 1991 and 1996 (Coastal CRC, 2002). Coastal fisheries production appears to have strong potential as an indicator of freshwater flows and ecosystem health in estuaries (ANZECC, 2000, p. 61). It is considered important by the general public, it has tangible economic value in the form of recreational and commercial fisheries, and time series' of data are usually available (i.e. through commercial catch or landing records). Even though habitat availability and fishing pressure also affect fish catch, providing environmental flows to sustain fisheries production have become a key feature of many water management plans in Queensland (Arthington *et al.*, 2000b).

The Murray mouth, Coorong and lower lakes are at the downstream end of the River Murray System. The Coorong and lower lakes area is classified as a Ramsar wetland. Strong evidence exists for declining stream health (across a range of ecosystem components) related to the high level of flow regulation; the median annual flow is 27% of what it was under natural conditions (Gippel and Blackham, 2002). There is considerable concern, both locally and nationally, about the increased risk of mouth closure, which potentially carries high economic and ecological costs.

Walker (2002) quantified the degree of opening of the River Murray mouth on the basis of the ratio of tidal energy in the estuary inside the mouth to that in the sea outside the mouth. The more closed the mouth becomes, the greater is the difference in tidal energy, because the closure prevents seawater moving into the estuary. So, the more constricted the mouth becomes, the lower is the value of relative tidal energy. The relative tidal energy at any time is a function of the relative tidal energy in the previous month, and the flow in the river over the previous two months. This model was used by Close (2002) to assess a number of options for modifying the flow regime to reduce the risk of mouth closure. The degree of mouth opening was expressed by Close (2002) in terms of the 'mouth opening index' (MOI), which is another name for the relative tidal energy as described above. The model of Walker (2002) only explains about half of the variation in relative tidal energy, and because low values of the MOI do not necessarily mean that the mouth will close (the mouth has only closed once in historical times), the MOI is best seen as a measure of the risk of mouth closure, with the risk being significant when MOI < 0.05. By providing an increased baseflow each month, the flow in the river would tip the balance in the mouth to a net outward flow that would assist in preventing sediment entering the inlet during a rising tide, and assist in flushing sediment during an ebb tide (Walker, 2002). The model of Close (2002) predicted that a steady flow of at least 2000 ML/d (23 m³/s) reduced the percentage of years with risk of mouth closure from the current level of 31.5% of years to 7.4% of years.

Development of frameworks for assessing the freshwater flow needs of estuaries began in the mid-1970s. One of the first frameworks was devised by Fruh and Lambert (1976) and Lambert and Fruh (1978) for the Corpus Cristi Bay, Texas. In the UK Binnie, Black and Veath Engineering Consultants (1998) provided a systematic procedure for assessing the freshwater flow requirement for estuaries based on: environmental assessment to identify the severity of impacts; risk assessment based on the likelihood of an estuarine standard being derogated; and, simple, intermediate or complicated computational modelling depending on the level of risk. A review of the status of environmental flows in Australian estuaries by Gippel (2002c) found that estuaries have had little attention compared to rivers when it comes to environmental flow assessments. One problem is that there is often a jurisdictional boundary at the tidal limit. More attention could be directed to estuaries if they were made a compliance point in riverine environmental flow assessments. However, it is also recognized that the characteristics of estuaries are sufficiently different to rivers that a different approach is required there. A general approach to determining the flow requirements of estuaries (based on a risk framework) has been proposed by Peirson *et al.* (2002).

In the Republic of South Africa, consideration of the freshwater flow needs of estuaries is well advanced (Department of Water Affairs and Forestry, 1999). The recommended framework is based on the same approach used for freshwater parts of rivers, beginning with determination of current ecological condition (with respect to reference condition) and designation of ecological management class. After an initial desktop estimate is performed, three levels of flow assessment are made, depending on the conservation status and resource use, and consequences of further development: rapid determination, intermediate determination and comprehensive determination (Department of Water Affairs and Forestry, 1999). The most limiting factor in estimating the water quantity requirements of estuaries is the lack of data on the role of floods and seasonal high flows (magnitude and frequency) in maintaining the sediment erosion/deposition equilibrium, so a generic guideline for rapid assessment is that flows >1.5 year average recurrence interval are protected. Another generic guideline states that if an estuary is permanently open, its sensitivity to reduction in seasonal base flows is assumed to be high,

and therefore a reduction in seasonal base flows should not be considered (Department of Water Affairs and Forestry, 1999).

### Assessing Flow Needs for Recreation

Rivers are used for a range of water-based recreational activities, including fishing, kayaking/canoeing/rafting, boating/water skiing and associated land-based activities including sightseeing, walking/jogging, camping, picnicking, hunting and bicycle riding. The peak in recreational activities usually occurs in the warmer months, which coincides with the period of irrigation demand, potentially creating conflicts in resource management. Another management issue is that an environmental flow regime to benefit wildlife will not necessarily maintain or optimize recreational opportunities.

The importance of recreational use of rivers should not be underestimated; for example, in North Dakota, 42% of adults participated in some type of river recreation in 1996, with 80% of the activity occurring in the warmer months (Bureau of Reclamation, 1999). Each recreation activity has different requirements, and recreation is an experience, not just an activity (Merrill and O'Laughlin, 1993). Thus, recreational users desire protection of natural stream values (namely, water quality, riparian vegetation, natural channel features, adjacent wetlands and the opportunity to see and hear moving water), as well as provision of the hydraulic conditions that facilitate the particular activities of interest (Merrill and O'Laughlin, 1993).

Whittaker *et al.* (1993) evaluated the relationship between stream flows and recreational values, and concluded that one of the most effective methods for evaluating flows for recreation was surveys of users. Brown and Daniel (1991) investigated the relationship between flow levels and public perception of scenic quality along the Cache La Poudre River in northern Colorado, which is classified as a 'wild and scenic river'. Positive reactions to scenic beauty increased with discharge up to a point, and then decreased with further flow increases. Suitability for paddling can be assessed by consulting river guides or seeking the opinion of recognized local expert kayakers and rafters. Alternatively, the rivers can be assessed by a dedicated survey crew of paddlers (Rood and Tymensen, 2001; Rood *et al.*, 2003).

Tennant (1976) and Corbett (1990) suggested particular proportions of mean annual flow that would offer suitable conditions for recreational boating and other uses, but this method is relatively crude (Whittaker *et al.*, 1993; Burley, 1990). For kayaking/canoeing/rafting, the single transect method requires identification of the particular shallow location that will initially limit paddling and subsequently, the flow providing the sufficient 10–20 cm depth at that point is determined. Finding the hydraulic control point can be time consuming, so Rood and Tymensen (2001) recommended a general depth criterion using gauging station sites. For rivers in North Dakota, the Bureau of Reclamation (1999) suggested four classes of suitability of flow depth for general recreational uses: 0–45 cm (poor); 45–75 cm (fair); 75–105 cm (good) and 105–135 cm (excellent). For kayaking/canoeing/rafting, Rood and Tymensen (2001) suggested a minimum depth of 60 cm was required to immerse a paddle blade, and depths of 75 cm to 100 cm progressively improved the appeal of many hydraulic features, reduced the chances of hitting rocks, permitted the kayaker to perform a roll, and provided less obstructed conditions for a paddler who swims following a capsize.

Rood and Tymensen (2001) and Rood *et al.* (2003) surveyed the suitability of flows for kayaing/canoeing/rafting in 13 reaches of the Oldman River Basin and the Milk and South Saskatchewan, southern Alberta. The rapids in the reaches ranged from Grade I ('not difficult') to III+ (between 'difficult' and 'very difficult') (in accordance with the International Canoe Federation difficulty classification scale which ranges from I to VI–unrunnable). Recreational flow values were determined for minimal flows, low flows that still provide a worthwhile paddling opportunity, and preferred or sufficient flows that represent the low end of the favoured flow range (Figure 9.27). Four methods were used: a mail-in survey

of general users, expert judgment from paddling guides, survey of all sites by the authors and rating curves from gauging stations to determine flows that would satisfy 60 cm (minimal) and 75 cm (sufficient) depth criteria. There was a good deal of scatter in plots of flow depth suitability versus discharge, as judged by river users (Figure 9.27). The survey by the authors rated the suitability of paddling conditions as 'reasonably favourable' once the minimal depth criterion was exceeded, with small further increases in discharge producing 'optimal' conditions, and higher discharges causing a decline in the suitability of conditions for paddling (Figure 9.28). All four approaches produced generally consistent results for minimal and sufficient discharges, indicating that all methods were valid for this purpose. There was a close correlation between mean discharge (average daily flow) and minimal flow (although absolute mean discharge was slightly higher), allowing an initial estimate of recreational flow to be made for similar rivers based simply on the flow record (Rood *et al.*, 2003).

### Prescriptive Versus Interactive Approaches

Brown and King (2003) placed all environmental flow assessment methods into two categories: prescriptive and interactive (Table 9.23). The prescriptive approach involves the study team spending considerable effort in

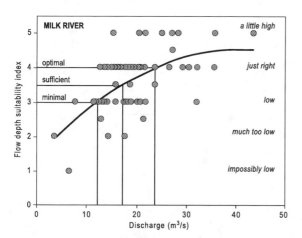

**Figure 9.27.** User survey ratings of suitability of flows for kayaking/canoeing/rafting on the Milk River, Alberta (Grade I—not difficult). Adapted from Rood and Tymensen (2001)

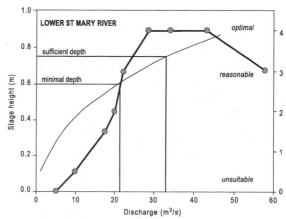

**Figure 9.28.** Suitability of flows for kayaking/canoeing/rafting as rated by competent survey team for the Lower St Mary River, Alberta (Grade I/III+—not difficult to difficult). Graph also shows rating curve from nearby gauging station, with discharges corresponding to minimal and sufficient depth classes (based on paddle blade length and performance requirements). Adapted from Rood and Tymensen (2001)

**Table 9.23.** *Relative data and time requirements of selected flow assessment methods. Source: King and Brown (2003)*

| Output | Method | Data and time requirements | Approximate duration of assessment | Relative confidence in output |
|---|---|---|---|---|
| Prescriptive | Tennant Method | Moderate to low | Two weeks | Low |
| | Wetted-perimeter Method | Moderate | 2–4 months | Low |
| | Expert Panel | Moderate to low | 1–2 months | Medium |
| | Holistic Method | Moderate to high | 6–18 months | Medium |
| Interactive | IFIM | Very high | 2–5 years | High |
| | DRIFT | High to very high | 1–3 years | High |

justifying their derivation of the 'best answer' to specific project objectives, they do not necessarily examine alternative scenarios, and the outputs do not include functions that inform the process of compromise. Interactive approaches, on the other hand, develop relationships between changes in river flow and one or more aspects of the river, allowing multiple interpretations of possible future river health. Interactive approaches generally take longer and have higher data demands, but they produce more confident results (Table 9.23). Most of the documented generic methods were classed by Brown and King (2003) as prescriptive, although some individual studies may have undertaken analysis of scenarios and developed continuous functions of flow dependent variables. Brown and King (2003) illustrated the characteristics of the interactive approach by comparing the IFIM and DRIFT methods. Both are essentially problem-solving tools with similar approaches (Table 9.24). The output is a set of options, termed 'alternatives' in IFIM terminology, and 'scenarios' in DRIFT terminology.

Both IFIM and DRIFT quantitatively describe relationships between possible future flow regimes and stream health characteristics, the effects on water yield as it affects security of supply for offstream users, and economic costs and benefits. DRIFT also considers social costs and benefits, principally for downstream riparian users (population at risk) (Brown and King, 2003). Neither method produces a specific flow recommendation. Rather, they provide information to assist managers and stakeholders decide how to share the available water and to quantify the likely consequences of their decisions.

### 9.5.10  Implementation and Evaluation of Environmental Flows

Methods of environmental flow assessment are at best indicative of the flow required to meet the environmental need. The best way to test hypotheses regarding environmental flows is to trial full-scale implementation. Acreman (2003b) recommended that three types of response be monitored: the river flow, the response of the ecosystem, and the social response to ecosystem change. The number of studies worldwide that report environmental flow recommendations greatly outweighs the number of studies that report scientific evaluation of the results of implementation of these recommendations. In fact, very few such studies can be found in the international literature (Annear *et al.*, 2002, p. 309). This is partly explained by the fact that most environmental flows have been implemented relatively recently, and there has been insufficient time to evaluate their effect. Another reality is that some implemented flows are not well monitored, or the results are not published in a readily accessible form.

Although adaptive management has been used in resource management for around 30 years, it is now being widely promoted as a suitable model for management of river resources (e.g. Boyd, 1999; Annear *et al.*, 2002, p. 309; Ontario Ministry of Natural Resources and Watershed Science Centre, 2001; Newson, 2002). Because it is scientifically based, includes stakeholders, requires attention to documentation, and is based on the idea of implementing interventions as field-scale experiments that generate knowledge useful to management, adaptive management offers a possible framework for implementing and evaluating environmental flows (e.g. Castleberry *et al.*, 1996; McBain & Trush, 1997; Barmah-Millewa Forum, 2000; Reid and Brooks, 2000; U.S. Fish and Wildlife Service, 2000; Irwin and Freeman, 2002; Freeman, 2002; Arthington *et al.*, 2003; Poff *et al.*, 2003).

### *Adaptive Management and Environmental Flows*

Adaptive management has its roots in 'adaptive environmental assessment and management' (AEAM), a struc-

**Table 9.24.** *Phases of IFIM and DRIFT interactive methods compared. Source: King and Brown (2003)*

| PHASES | IFIM | DRIFT |
|---|---|---|
| Problem identification or issues assessment | Identification of interested and affected parties, their concerns, information needs and relative influence or power.<br>Identification of the broad study area, and the extent of probable impacts. | Identification of the main components of the project and the interested and affected parties.<br>Identification of the population at risk.<br>Identification of the broad study area, and the extent of probable impacts.<br>Identification of social concerns (local, national, and international) to be addressed in the biophysical studies. |
| Study planning | *Both approaches require:*<br>Assessment of existing biophysical, social and economic data, and evaluation of the need for further data.<br>Selection of representative river reaches.<br>Design of data collection procedures.<br>Identification of key data collection sites.<br>Interdisciplinary Integration of site selection and data collection avoids overlaps and gaps, and maximizes the usefulness of the data. | Addition of social considerations in selection of study area and sites. In particular, compatibility ensured between biophysical data (collected at river sites) and social data (collected in rural villages). |
| Study implementation | Collection of hydraulic and biotic data.<br>Calibration of habitat model. | Collection of hydraulic, chemical, geomorphological, thermal and biotic data, and analyses and develop predictive capacity on how flow changes will affect each.<br>Multi-disciplinary workshop to compile a database of biophysical consequences of a range of flow manipulations. |
| Options analysis | *Both approaches require:*<br>Development of environmental flow alternatives or scenarios, each describing a possible future flow regime and the resulting river condition.<br>Yield analysis of water available for development with each scenario.<br><br>Determination of the direct costs and benefits of the alternatives. | Determination of the direct costs and benefits of each scenario.<br>Additionally, for each scenario, determination of the social impacts and costs to the population at risk of changing river condition. |
| Problem resolution | *Both approaches require:*<br>Assessment of the bigger picture (for example, data on other costs/benefits of the water-resource development).<br>Negotiation with offstream water users.<br>Public participation.<br>Transparent decisionmaking processes. | |

tured approach to the evaluation and solution of environmental management problems developed by Holling and several colleagues at the University of British Columbia's Institute of Resource Ecology, Vancouver, Canada in the late 1960s. Some of the principles and methods of adaptive management are similar to the 'Deming' management method, and the related 'Total Quality Management' method, which are used in the business world (Forest Service, British Columbia, 1997; Walton, 1986).

Holling (1978) defined adaptive management as:

"a process which integrates environmental with economic and social understanding at the very beginning of the design process, in a sequence of steps during the design phase and after implementation"

Some river managers might argue that they have always practiced adaptive management, through implementing works and then trying different approaches if problems arise. However, this is a superficial understanding of adaptive management. Adaptive management explicitly recognizes uncertainty, so policies (and resultant on-ground actions) are seen as experiments that, if carefully designed and monitored, can provide information about the system. Under an adaptive management philosophy, management is involved in development of best practice, rather than passively adopting new technologies when they become available. The procedure for adaptive management can be conceptualized as comprising several interrelated components that Nyberg (1999a, b) depicted as a cycle (Figure 9.29).

The active adaptive process involves experimentation with multiple management practices, and has rarely been applied to river management. Lee (1993, p. 57) stated that:

'Although virtually all policy designs take into account feedback from action, the idea of using a deliberately experimental design, paying attention to the choice of controls and the statistical power needed to test hypotheses, is rarely articulated and still more rarely implemented".

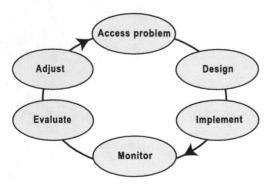

**Figure 9.29.** *Adaptive management cycle. Source: Nyberg (1999a, b)*

For reasons of scale, expense and practical limitations, experiments in adaptive management will not always include controls, replication, multiple treatments, randomization or other features commonly expected of traditional experimental design and frequentist (classical) statistical methods, but alternative statistical methods are available (Sit and Taylor, 1998).

Management decisions may well be made with awareness of the uncertainty of the science, but this is rarely explicitly acknowledged. There is a simplistic appeal in the idea that if managers are going to act then they might as well act on the 'best' available advice. However, in many cases the idea of a single 'best' piece of advice may be an illusion, with the reality being that there exists one or more alternative ideas that may have an effectively equal chance of being the 'best'. A focus on policies and actions as hypotheses subject to critical analysis is the hallmark of an adaptive approach. In Gunderson's (1999) view, an adaptive approach views policies as 'questions masquerading as answers'.

Arthington *et al.* (2003) warned that the holistic DRIFT method should only be applied within an adaptive management framework where there is a genuine commitment to the generation and use of new knowledge derived from monitoring and research. Unfortunately, experience has demonstrated that without a concerted effort to assist its adoption and avoid its pitfalls, adaptive management is likely to fail (Mclain and Lee, 1996; Stankey, 2003; Ladson and Argent, 2002). Ladson and Argent (2002) reviewed adaptive management instream flow projects in three large North American rivers. The three case studies were the Columbia, Colorado and Mississippi Rivers. These rivers have all been developed for irrigation, hydropower or navigation, and adaptive management was used to explore flow changes proposed for environmental reasons while recognising the trade-off with production values. Ladson and Argent (2002) found that only the Colorado River case could be regarded as a success. Annear *et al.* (2002) made the point that adaptive management is not appropriate in every situation. Some environmental flow assessments can be made in a short period of time, with little controversy and a high degree of confidence that the flow will improve stream health. In other situations, some stakeholders may be unwilling to participate in a long-term adaptive management process (Annear *et al.*, 2002). Some other problems with implementing active adaptive management on rivers are the lack of sites to act as controls (each river being unique), and the difficulty of running more than one experiment at the same time to test different hypotheses (experimental sites located along the river may influence each other).

Gippel (2003) undertook a comprehensive review of all implemented environmental flow related actions on the River Murray, Australia. Nineteen projects were reviewed, covering a wide spectrum of actions including artificial wetland watering, enhanced flooding, variable in-channel flow releases, flows for algal bloom suppression, weir level drawdown and raising and opening weir gates for fish passage. The review found ample evidence that environmental flow implementation has been based on science and that it had lead to improvements in river health (although it was too early to be able to assess system-wide and long-term improvements). The actions generally involved monitoring, although it was undertaken using various levels of scientific rigour. Overall there was a high level of awareness of adaptive management, but no examples could be found where a formal adaptive management framework [as described by Holling (1978), Walters (1997), Walters and Holling (1990) and others] (Figure 9.29) had been explicitly followed from inception to implementation and adjustment. This is not to say that the river and floodplain management work reviewed by Gippel (2003) did not fall within the realm of adaptive management. The Barmah-Millewa Forest Water Management Strategy (Barmah-Millewa Forum, 2000) appears to be a working example of adaptive management, and so too does the work on managing wetlands in South Australia (Jensen, 2002a, b). The success of these projects as exercises in adaptive management, and how well they achieve their environmental objectives, awaits future evaluation.

## Some Case Studies of Evaluated Environmental Flows

### Large-scale Monitoring Programs—Australia

A few large-scale environmental flow monitoring projects are underway in Australia. The Federal, Victorian and New South Wales State governments agreed in October 2000 to implement an environmental flow to the Snowy River, increasing up to 21% mean annual natural flow over a ten year period, with a further increase of 7% mean annual natural flow reliant on cost savings by irrigators. The first environmental flow release was made on 28 August 2002 (Rose and Bevitt, 2003). The Snowy Benchmarking Project is examining the full length of the Snowy River. Near Jindabyne Dam the environmental flow will increase annual discharge from 1% of natural mean flow and is expected to vastly improve the condition of the river. In the lower reaches, the flow increase will be around 11% of mean natural flow, and the physical and ecological responses to environmental flows may be difficult to detect, especially in the short term. The

first group of reports examined background condition and establishing hypotheses (Rose and Bevitt, 2003), which is a critical step in evaluation of the effectiveness of the environmental flow. A similar large-scale environmental flow trial is underway on the Campaspe River in Victoria (Humphries and Lake, 1996), but to date, the flow has not been implemented. One of the most important outcomes expected from these trials is the development of appropriate methodologies and indices for measuring the effects of implementing environmental flows.

The New South Wales Government has established a major scientific program to monitor the effectiveness of new flow rules using hypothesis-driven research, known as the Integrated Monitoring of Environmental Flows (IMEF). The IMEF involves the collection of biological, physical and/or chemical data from river sites and wetland sites within sections of the river that are affected by the flow rules. The IMEF program has published two reports that examine river condition prior to implementation of the environmental water allocation (Department of Land and Water Conservation, 2001a, b), but information on how the river and wetland ecosystems are responding to the environmental flow rules has not yet been published.

### Opportunistic Flood Enhancement—River Murray, Australia

In late 2000, a naturally occurring 1 in 5 year flood event in the Ramsar listed Barmah-Millewa Forest on the River Murray, Australia, was supplemented by three parcels of water released from a headwater dam (Barmah-Millewa Forum, 2001; Leslie and Ward, 2002) (Figure 9.24). The flood in this area peaked at close to 90 000 ML/d (approx. 1000 m$^3$/s). The environmental releases (341 ML or 341 000 m$^3$) comprised only about 8% of the total flood discharge, and some ecological response would have occurred even if the releases were not made. Nevertheless, the release effectively filled a gap between two flood peaks and extended the recession of the second flood peak, mainly to maintain conditions suitable for bird breeding. The critical aspect of this managed flood release was its timing. A report on the impact of the flood (Barmah-Millewa Forum, 2001) revealed that bird breeding success was ranked at least a 1 in 10 year event in terms of number of species, and total bird numbers. The Great, Intermediate and Little Egrets (all endangered species in Victoria) bred in the Forest for the first time since 1993, 1992 and 1975, respectively (Leslie and Ward, 2002). Because the breeding events reached completion (at least 15 000 breeding pairs of 20 or more species), a large population of birds became available for future breeding events. Without the environmental flow, Leslie and Ward (2002) estimated that about 3000 nests would have been abandoned by mid-December.

During the same late-2000 flood event on the River Murray, but much further downstream, the flow to South Australia was enhanced by an appropriately timed release of 9700 ML/d (112 m$^3$/s) from a nearby storage (Department of Water, Land and Biodiversity Conservation, 2002), so that the October flood peak of 32 000 ML/d (370 m$^3$/s) was increased to 42 050 ML/d (487 m$^3$/s). Approximately 100 GL (100 000 m$^3$) of water was released for flood peak enhancement, which was intended to increase the extent of floodplain watering. More widespread inundation was achieved by surcharging a weir by about 0.5 m using stop logs. A monitoring program investigated the environmental response of this flood enhancement in terms of hydrology, groundwater, salinity, fish movement and plants. The second, larger December flood peak in this event was also enhanced by flow releases from the storage, but weir surcharging was not carried out, and the ecological response of this second enhancment was not monitored.

The October environmental flow release, plus the weir surcharging produced a river stage that was equivalent to about 70 000 ML/d (810 m$^3$/s) (Department of Water, Land and Biodiversity Conservation, 2002). The area of floodplain inundation was increased by about 10%. There were no negative impacts on groundwater, there was a small increase in salinity, and native fish were not observed to move. Unfortunately the flow enhancement coincided with the local major annual Carp spawning event (Carp are considered a pest species in Australian rivers), and the raised water levels appear to have improved their breeding potential. While some new vegetative growth did establish, recruitment of individuals from new seed was not observed. The success of an enhanced flood event such as this rests upon recruitment of new individuals, and not merely the survival, growth and reproduction of established plants. With respect to vegetation, the value of these small, occasional interventions in environmental flow management may be to maintain existing communities rather than restore degraded ones (Siebentritt et al., 2004).

### Modified Hydro-peaking Releases from Hydroelectric Stations

By the early 1990s the lower reaches of the Bregenzerach River in western Austria had been subjected to the adverse effects of heavy hydro-peaking for several decades, and both fish and benthic invertebrate biomass were less than 5% of the expected values (Parasiewicz et al., 1996). A new hydroelectric station that was constructed on the river in 1992 included a re-regulation reservoir, allowing the amplitude of the flow fluctuations to be reduced, although their frequency remained the same. Using reference sites upstream, it was demonstrated that after the flow change, benthic invertebrate biomass increased to about 60% of the expected value. There was no significant change in the benthic community composition and no change in fish biomass (Parasiewicz et al., 1996).

Several environmental flow regimes have been implemented and evaluated in the USA downstream of dams, often expedited by the need for hydroelectric station pre-licensing or re-licensing negotiations which can produce an atmosphere of cooperation and mutual compromise (King and Brown, 2003). On the Tallapoosa River, near Tallassee, Alabama, a minimum environmental flow from Thurlow Dam was implemented as part of the re-licensing agreement of the hydroelectric project. The implemented flow regime was evaluated using a hypothesis testing approach (Travnichek et al., 1995). Before 1991, river discharge below the dam fluctuated widely and frequently. The environmental flow regime that began in 1991 did not eliminate flow fluctuations but reduced their severity. Fish species richness more than doubled and the mean abundance of fish per sample increased 500%. The shoreline fish species in the river were macrohabitat generalists prior to the environmental flow regime, but one year afterwards, the fish assemblage below the dam was composed of both generalists and fluvial specialists, with the latter accounting for over 70% of the fish collected (Travnichek et al., 1995; Bain and Travnichek, 1996). These findings confirmed the predicted benefits of the environmental flow.

A study on the Susquehanna River downstream of the hydroelectric station on Conowingo Dam, Maryland, evaluated three environmental flow release regimes: a continuous minimum flow; an intermittent minimum flow; and no release (Versar Inc., 1998; Weisberg et al., 1990). Monitoring of aquatic macroinvertebrate communities found that the no flow condition was detrimental while the minimum flow regimes were beneficial. Environmental flows were also an outcome of hydroelectricity station re-licencing on the Skagit River system, Washington, one of the few Puget Sound basins in which salmon are managed on a natural stock basis. The original license expired in 1977, but tribes and agencies concerned about fisheries and other environmental issues opposed the re-licensing application. Operating under an annual licence, a programme of research and implementation of environmental flow management measures was implemented throughout the 1980s, and this coincided with increased numbers of pink and chum salmon (Federal Energy Regulatory Commission, 1996, p. 43). In 1991 a comprehensive Settlement Agreement was signed and in 1995 a new 30-year operating license was granted.

### Modified Flow Release from Dams

In North America, cottonwood is an important riparian tree species, with a lifecycle that is strongly dependent on the pattern of flooding (Fenner et al., 1985; Everitt, 1995). Following the completion of the St. Mary Dam on the St. Mary River, Alberta, in 1951, the riparian cottonwood populations downstream failed to recruit, due to insufficient summer flows and abrupt flow declines in the late spring (Rood and Mahoney, 2000). Changes to the pattern of dam releases began in 1991, when the minimum flow was tripled. Gradual flood recessions were introduced in 1994. Unlike previous floods that occurred after regulation, a 1-in-50-year flood event in June 1995 was managed to have a gradual flow recession. This provided an opportunity to test the 'recruitment box' cottonwood recruitment model of Mahoney and Rood (1998). The model was supported, with the managed flood resulting in extensive recruitment of cottonwoods and willows. By 1999, seedling densities had decreased by about 95% but seedling growth resulted in almost complete leaf area cover.

When the Upper Salmon Hydroelectric Development in central Newfoundland, Canada, was constructed in the early 1980s, an environmental flow release was developed using the Tennant Method, providing flows in Tennant's 'good' category (Table 9.20) (Scruton and LeDrew, 1996). Post-regulation studies have investigated the geomorphology and juvenile fish populations. Initially the river bed accumulated fines, but this appeared to have no effect on spawning and egg incubation, as densities of Atlantic salmon fry were higher after regulation than prior to development. However, numbers of older age class fish declined, which was thought to be due to flows in winter being too low. A retrospective IFIM assessment confirmed that winter flows were too low, and this process also recommended inclusion of a flushing flow (Scruton and LeDrew, 1996).

The Grand Canyon controlled flood flow of 1274 m$^3$/s released from Glen Canyon Dam over a seven-day period in March–April 1996, known as the '96 Beach Habitat Building Flow', was very closely investigated (Collier et al., 1997; Konieczki et al., 1997; Smith, 1999; Schmidt,

1999, Valdez et al., 1999; Webb et al., 1999). The Grand Canyon experiment produced mildly beneficial ecological changes and no adverse impacts (Valedez et al., 1999). The primary geomorphic objective of the Grand Canyon flood was to mobilize sediment in fan-eddy bars (Schmidt and Rubin, 1995) (formed from sediment delivered for years to the main channel by unregulated tributaries) and redeposit it on downstream beaches as the flood slowly receded (Schmidt, 1999; Freeman, 2002). There was widespread deposition of formerly in-channel sand at high stage locations, resulting in an increase in available beach-bar area (Schmidt, 1999). The flow was not strong enough to flush exotic plants (e.g. tamarisk) that had invaded the channel (FISRWG, 1998, p. 3–9). One important finding was that most of the effects were realized in the first 48 hours of the event (FISRWG, 1998, p. 3–9). The wide spatial variation in the patterns of sediment deposition and aggradation meant that it was difficult to judge the 'success' of the project (Stillwater Sciences, 2003). Unfortunately, for legal reasons it was necessary to release the artificial flood in a high water year, and the big new beaches built following the 1996 flood were quickly lost to high flows through the Canyon (Freeman, 2002). A number of other experiments have been conducted based on the findings from the original '96 flood. In 1997, a 878 m$^3$/s habitat maintenance flow was tested to evaluate its ability to conserve sediments recently delivered by silty tributaries. In 2000, low steady summer flows were used to simulate the seasonal pattern of the natural hydrograph to benefit native fish (Freeman, 2002).

Dartmouth Dam is a large storage that has altered the pattern of low flows, flood flows, as well as seasonal and daily flow variations in the Mitta Mitta River, a headwater tributary of the River Murray, Australia. Long periods of relatively constant regulated flow appear to be the main cause of accelerated channel erosion. After a long period of relatively constant flow at 4000 ML/d (46 m$^3$/s), from mid-November 2001 a small variation in flow was introduced to simulate the response to a rainfall event (Gippel, 2003). The variation was based on a fortnightly cycle, with a short rise over 2 days followed by a recession over 12 days. Flow was varied over the range 3200 ML/d to 4800 ML/d (37–56 m$^3$/s), with a mean flow rate of 4000 ML/d. This corresponded to a variation in river stage of 250 mm. A suite of 14 water quality, river productivity and invertebrate indicators were measured, with a reference site used as a control. In general, there was an overall improvement in ecological health of the Mitta Mitta River following the variable flow release pattern, while the data suggested that during the 37-day period of constant flows that followed the simulated event, river health had begun to decline (Sutherland et al., 2002).

*Trinity River—Fifty Years of Environmental Flow Studies*

The Trinity River, located in northwest California, is the largest tributary to the Klamath River. For thousands of years, members of the Hoopa, Yurok and other tribes depended on the rich Chinook, Coho salmon and Steelhead fishery. The Trinity River Diversion (TRD) began full operation in 1964. Petts (1984, p. 256–262) reviewed some of the issues involved in this case, but there has been some notable progress since that time. Over the first ten years of operation 88% of the mean annual flow (MAF) was diverted. An early study recommended 9% of MAF be released at rates of between 4.3 m$^3$/s and 7.1 m$^3$/s to provide fish habitat for spawning chinook (Moffett and Smith, 1950). Despite the implementation of this flow regime, channel deterioration was noticed within one year of completion of the TRD, and within a decade salmonid populations had noticeably decreased (Hubbel, 1973). An environmental impact study in 1980 (U.S. Fish and Wildlife Service, 1980b) determined that an 80% decline in Chinook salmon and a 60% decline in Steelhead populations had occurred since the commencement of TRD operations. The study further estimated the total salmonid habitat loss in the Trinity River Basin to be 80–90%. Insufficient streamflow was identified as the main factor responsible for the decline in fishery resources.

In 1981 four different flow regimes (representing 10%, 16%, 21% and 24% of MAF) were proposed for evaluation. The highest of these environmental flow regimes was implemented in 1991, but this flow was equal to the third driest year on record, so the river was effectively in permanent drought. Modest annual high flow releases (170 m$^3$/s) were insufficient to remove vegetation that had invaded the channel (Barinaga, 1996). Accounting for hatchery releases, fish survey data indicated that by 1999 Chinook salmon abundance had declined by 68% compared to pre-TRD abundance, Steelhead by 53% and Coho salmon by 96% (U.S. Fish and Wildlife Service and Hoopa Valley Tribe, 1999; U.S. Fish and Wildlife Service, 2000). Trinity River Coho is listed as threatened under the Federal Endangered Species Act and Steelhead is listed as a candidate species. The final evaluation of the scenarios that were suggested in 1981, which was not completed until 1999 (U.S. Fish and Wildlife Service and Hoopa Valley Tribe, 1999), concluded that the scenarios would meet known fishery restoration criteria to varying degrees, although all criteria were not fully met even with the greatest volume; the fisheries resource was predicted to continue to decline. One of the main problems identified was that the channel had contracted and required reshaping.

The most recent environmental studies (U.S. Fish and Wildlife Service and Hoopa Valley Tribe, 1999; U.S. Fish and Wildlife Service, 2000) recommended greater variability in the environmental flow regime, and inclusion of channel forming flows ranging from 43 m$^3$/s in critically dry years to 312 m$^3$/s in extremely wet years [based on recommendations by McBain and Trush (1997)]. The recommendations represented a range of 26% MAF (dry years) to 58% MAF (wet years). A 1999 study estimated that economic benefits for instream flows were USD803 million per annum for the highest release scenario, a value that greatly exceeded the social cost estimate (Douglas and Taylor, 1999). In December 2000 the Interior Secretary announced that an environmental flow package of 52% MAF would be returned to the Trinity River, with 43% MAF allocated for low flows.

### 9.5.11 Summary

All instream flow methods will yield 'answers'. The choice of a method will be based on time and budget allocations and the level of competition between instream and offstream water users. With any technique, the dictum of computer modelling, 'GIGO' (for garbage in–garbage out) is relevant. Answers will only be as good as the quality of information examined and the skill of the user in applying the model and interpreting the results. The major trend in instream-environmental flow assessment over the past 30 years has been a shift from narrow studies that catered for a single fish species at one critical life phase to a holistic approach that aims to restore natural river processes.

Proponents of the habitat-simulation methods feel that these are more attuned to biological principles than the hydrological index and hydraulic rating techniques, and are thus superior for evaluating biological impacts resulting from flow alterations. Whatever tools are used, it would appear that they are best applied within a holistic framework. None of the methods can, as yet, directly address the potential changes in biomass or populations resulting from altered flow regimes. There remains a need for more research on the effects of flow regimes on biota, due to resultant changes in physical habitat as well as changes in water quality and biological interactions.

Also, as Stalnaker (1979) recommended, flow requirements should be considered dynamic. Rather than applying a set 'minimum' or 'optimum' flows throughout the year, every year, considerations should be given to seasonal and annual variations, with different regimes for 'wet years' and 'dry years'. The natural regime of streams should be considered in establishing flow requirements since the resident populations have adapted to it. Especially in ephemeral streams or drought periods, plans are needed to balance short supplies between users. Periods of above-average water supply would perhaps be the most acceptable time to allocate water for artificial floods.

It should also be recognized that plants may be able to tap groundwater reserves, frogs can bury into the mud to enter a dormant phase, fish can migrate and aquatic insects may have adult forms that are not as dependent on water for survival. Thus, an allocation of 'cease-to-flow' may be appropriate for streams that historically dry up each year.

The field of instream flow recommendation is still a dynamic, evolving area, with heated controversy and much testing and refinement of methods. The well established IFIM and the new DRIFT holistic frameworks can be considered the present 'state-of-the-art' methods for decision making on large-scale, complex and controversial projects, but other methods are more than adequate for simpler situations, especially where resources and time are limited. Tharme (2003) noted a widespread move towards hierarchical application of environmental flow methodologies, with at least two stages to the framework: (1) reconnaissance level assessment, primarily using hydrological methodologies; and (2) comprehensive assessment, using either habitat simulation or holistic methodologies. Holistic methodologies are particularly appropriate in developing countries, where the focus is on protection of the resource at an ecosystem scale, as well as the strong livelihood dependencies on the goods and services provided by aquatic ecosystems (Tharme, 2003).

The search for an 'ideal' universal flow allocation technique may be fruitless, as the practical reality is that environmental flow problems are highly diverse, in terms of the characteristics of the environment, the funds available, the time available, the scope of the study and the potential cost of making an error in the allocation. Thus, while the importance of developing methods and testing hypotheses cannot be understated, there is a need to ensure that the industry develops the capacity to make prudent use of the outcomes of environmental flows research (Gippel, 2000). At the same time, it is vital that the stakeholders and wider community are informed of, and involved in, developments in the environmental flows field (this has not been a strong feature of progress to date). Lack of understanding disempowers community advisors and stakeholders—a problem that could threaten successful implementation of environmental flows (Gippel, 2001b). The weakest area of environmental flows endeavour is not the science of assessment, but in the transformation of recommendations into applied flow

regimes. It took 50 years of research and negotiation to agree on a flow regime that would restore fish populations in the Trinity River. Implemented environmental flows require performance evaluation, ideally within an adaptive management framework, which encourages refinement of the flow recommendations and the flow assessment methodologies.

## 9.6   Stream Rehabilitation

### 9.6.1   The Basis of Stream Rehabilitation

In most agricultural areas of the world, removal of vegetation from stream banks has been widespread. The vegetation was removed, and is maintained in that state, because it offers ease of access to the river for stock watering. The removal of riparian vegetation also frees up some fertile land for cultivation or grazing, and allows farmers to more easily control fires and local pests, such as foxes in Australia. Riparian zones are maintained in a cleared state mostly through uncontrolled grazing pressure. One consequence of clearance of rural riparian zones, in combination with other factors, was widespread channel instability. In urban areas, as development spread, not only were stream banks cleared, but dramatically increasing runoff coefficients led to more erosive streamflows, causing widespread channel expansion and instability. The response to this channel instability was to establish river management authorities, whose responsibility was primarily to prevent channels from eroding and shifting their location, and for minimizing flooding.

River management has a tradition rooted in civil and hydraulic engineering. Most of the work was grounded in well-established theory of stable channel design. This led to an emphasis on control of flow and structure using embankments, re-shaping channels to trapezoidal cross-sections, clearing snags and riparian vegetation, rock beaching of banks and construction of training structures. The inherently dynamic nature of rivers was seen as an annoyance that should be controlled, or if structures failed, as a catastrophic and unusual event.

This conventional paradigm is now falling out of favour, where it is recognized that a level of channel instability is desirable from an ecological perspective, and that a high level of channel stability is difficult and expensive to attain. Even in the more progressive organizations, there is considerable room to shift the emphasis of stream management. To highlight a typical case from Australia, the 1996/97 waterway management expenditure figures for the managing authority for

Melbourne's waterways (Melbourne Water, 1997) showed that 63% of the budget was spent on waterway stabilization works—addressing a problem that threatened only 5% of the total waterway length. At the time there were no specific programs to address loss of instream habitat, even though it had been established through surveys that this problem threatened 40% of waterway length. This imbalance has been recognized, and Melbourne's waterways are undergoing extensive assessment for rehabilitation, and in some cases active rehabilitation with channel modifications (Gippel *et al.*, 2001).

Riverine biota and ecosystems have evolved in the context of natural channel instabilities, so that in many cases the processes associated with the instability (e.g. channel scour), or the landforms produced by instability (e.g. bare sand and gravel bars, undercut banks), are required for ecosystem maintenance or the survival of particular species (Petts and Calow, 1996). For example, Petts *et al.* (1992) found that lack of geomorphic instability (in regulated rivers in this case) led to succession of vegetation units to mature stages with the loss of pioneer and early successional units. Some examples of dynamic geomorphic features that provide habitat in rivers are:

- undercut banks as fish habitat
- undercut banks required for animal burrows
- use of vertical banks by birds for nesting
- sand and gravel bars used by reptiles for basking
- disturbance resulting from erosion and sedimentation processes provides habitat diversity (patches)
- periodic mobilization of gravel bed material is required for the maintenance of substrate and interstitial void habitats
- meander cutoffs provide important new floodplain wetland habitats.

Natural dynamically stable geomorphic function involves gradual erosion and accretion of stream forms (e.g. Hooke, 1979; Odgaard, 1987; Elmore and Beschta, 1987). Miller (1999) argued that true stream rehabilitation should strive to provide a level of channel mobility. In highly modified systems the rates of geomorphic adjustment are often much higher (channels erode too fast) or lower (channels silt up) than in undisturbed systems, so rehabilitation projects that wish to allow for channel mobility still have to do it in a managed way. Miller (1999) termed this idea the 'deformable stream bank'. The question that arises is: how much mobility is enough? This may be limited by practical considerations, such as the width of the available corridor. In other areas the degree of channel

mobility can be controlled through choice of construction materials and riparian vegetation plantings (Skidmore and Miller, 1998; Miller, 1999).

Water laws now require authorities to maintain or improve the environmental condition of streams. For example, recent developments in European legislation (Habitats Directive and the Water Framework Directive) should give further impetus to river restoration across EU member states, as this legislation places greater emphasis on the hydrological and geomorphological processes that support river ecology (Clarke *et al.*, 2003; Duel *et al.*, 2003). The WFD requires that ecological functioning and quality of rivers achieve 'good or high quality status'. At the same time, public pressure to improve streams appears to be increasing. The growing enthusiasm for restoration of disturbed environments (Brookes and Shields, 1996) is hardly surprising.

### Forms of Stream Rehabilitation

Although Brookes and Shields (1996) acknowledged the practical reality that restoration projects were often associated with other utilitarian goals (such as erosion control or flood defence), the focus of their book is clearly on environmental restoration. They adopt the premise that if natural hydrology and morphology are restored, then there is a strong probability that ecological recovery will follow (Brookes and Shields, 1996, p. xvii). The reality is that stream *restoration*, defined by Brookes and Shields (1996, p. 4) as the complete structural and functional return to a pre-disturbance state, is an ideal that is rarely practiced, and even then, ecological recovery following physical restoration is not guaranteed. Gore (1985a) defined restoration as 'recovery enhancement', or a means of increasing the rate of recovery of a disturbed ecosystem. (Here, disturbance refers to major events that lie beyond the range of predictable, high frequency disturbances.) There are various pathways, or trajectories, that a biological community might follow after major disturbance, such that recovery may result in an endpoint that is different to the pre-disturbance state (Figure 9.30). For a good example of a system shifting to undesirable endpoints see Moss *et al.* (2002). Restoration can increase the probability that the community will return to its pre-disturbance state, but had the community not been disturbed, it may have moved on to a different state anyway, through natural successional processes (Figure 9.30). Predicting what this state would have been is a particularly difficult problem for managers aiming for the restoration ideal (Cairns, 1990; Milner, 1996). The situation is further complicated by the manner in which

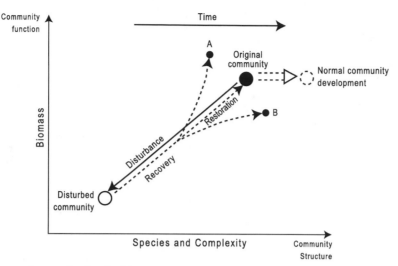

**Figure 9.30.** *Recovery pathways of a disturbed biotic community. A and B are examples of alternative endpoints of recovery. Source: Milner (1994, p. 206), after Bradshaw (1988)*

streams respond geomorphologically to disturbance (Figure 9.31). Stream morphology is the foundation of physical habitat availability, and therefore is a major determinant of the potential for ecological recovery. The threshold model (Schumm, 1977) predicts that channel state changes when a sediment supply or transport threshold is crossed. Indeed, some channels may never achieve geomorphic equilibrium over management time scales [for a discussion of the instability of Australian streams see Brizga and Finlayson (2000) and Gippel (2002a)]. It seems appropriate then that restoration objectives should focus on the desired characteristics for the system in the future rather than what these were in the past, and the targets should be based on dynamic rather than static attributes (Dobson and Cariss, 1999; Hobbs and Harris, 2001).

The outlook for widespread restoration of hydrological regimes, channel morphologies and biotic communities would appear to be bleak when most rivers are already regulated and/or altered physically and/or chemically beyond the point where morphological, hydrological and ecological recovery is possible within management time scales (Rapport, 1999). Stream restoration activities are often focused on highly modified urban landscapes where the chances of achieving ecological restoration are extremely slim, and where the community may not desire a wild (i.e. restored) streamscape. In connection with this point, there is a need for scientists to accept as valid,

alternative visions for streams that include strong human interactions with other natural elements of the stream environment. The engineering profession has responded to the stream restoration movement by softening their traditional view of an ideal stream, and in doing so exploited new opportunities that still require engineering expertise. Similarly, the scientific view of what constitutes a healthy functioning stream ecosystem may not be universally shared or so highly valued by other members of the wider community, and some adaptation may be necessary.

The expectation of stream *rehabilitation* is partial return from a degraded state towards a reference condition, or perhaps return to a different but sustainable condition. *Enhancement* is any improvement in environmental condition, and is not with respect to a previous condition. *Creation* is the intentional development of a new type of ecosystem that never previously existed at the site. *Naturalization* incorporates aspects of both enhancement and creation (for a range of definitions see National Research Council, 1992; Brookes and Shields, 1996, pp. 2–5; Bradshaw, 1996; FISRWG, 1998, pp. 1–3). The concept of naturalization accepts human utilization of natural resources as an integral component of what we understand to comprise the 'natural' environment (see Section 9.2). Return to a condition that pre-dated significant human disturbance is not sought. Rather, efforts to achieve environmental improvements explicitly consider

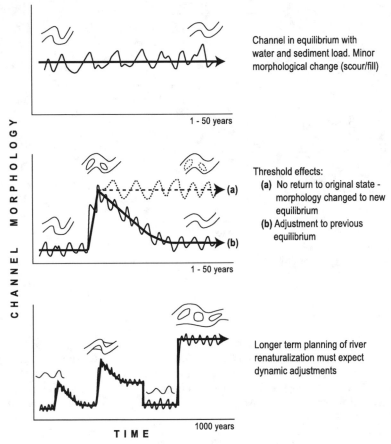

**Figure 9.31.** *Geomorphic dynamic equilibrium concept, and the effect of threshold-crossing disturbances on channel stability. Source: Sear (1996, p. 156). Reproduced by permission of John Wiley & Sons Ltd.*

current human utilization of resources, and they are designed to be compatible with current (usually disturbed) fluvial processes (Brookes and Shields, 1996, p. 4). Under the current paradigm (Section 9.1) almost all stream management activities include ecological objectives, so the terms stream rehabilitation/restoration/reclamation/enhancement/naturalization are now synonymous with stream management. For convenience, in this chapter we use the most inclusive and realistic descriptor 'rehabilitation' to cover all of these categories of stream management.

This chapter is limited to a brief discussion of some basic rehabilitation principles, and focuses on physical channel habitat rehabilitation design at the reach-scale. Our intention is not to encourage localized gung-ho structural 'tinkering' in stream channels as the primary mode of stream rehabilitation. In fact, we agree with

FISRWG (1998) that rehabilitation should follow the principles of:

1. seeking the most minimally intrusive solution to achieve the desired result (FISRWG, 1998, pp. 1–5);
2. managing causes rather than treating symptoms (FISRWG, 1998, pp. 5–17); and
3. nesting reach-scale activities within an understanding of catchment- or corridor-scale processes (FISRWG, 1998, pp. 8-1 to 8-2).

We recognize that there are many important aspects of stream rehabilitation that we have omitted here, such as planning, legislation, resourcing the project (technical and financial), community involvement, setting the vision or desired future state, prioritization, objective setting, risk assessment, cost-benefit analysis, monitoring, wetland rehabilitation, floodplain rehabilitation, water quality amelioration, catchment-scale rehabilitation and urban runoff management. Readers are encouraged to seek out alternative sources for information on these and other related topics. Some relevant material has been included in preceding sections. Hydrological rehabilitation was covered in Section 9.5. All stream rehabilitation projects will have to confront the issue of identifying stream values (Section 9.2) during the planning phase. Sections 9.3 and 9.4 explained how stream health assessment and stream classification is used in stream management/rehabilitation.

Although the focus of this section is on rehabilitation, the same concepts can be applied to new river engineering works and entirely artificial waterways (Ernegger *et al.*, 1998) to produce a "managed river which behaves as much like a natural one as possible, which looks as aesthetically pleasing as possible and which minimizes the morphological and ecological disruption" (Brookes and Gregory, 1988, p. 158). In some rehabilitation situations, localized 'band-aid' solutions may be adequate, such as where a streambed has been dredged for gold or a channel re-routed away from a highway. If whole-river systems are of concern, stream rehabilitation is best undertaken within a total catchment management strategy. Stream improvements should not be used as a substitute for dealing with more complex causes of stream degradation such as grazing, logging, road construction and other impacts to upland areas (Elmore and Beschta, 1987). As Platts and Rinne (1985, p. 127) state, "artificial stream improvement must not be substituted for vigorous, responsible stewardship of the surrounding watershed".

A multi-disciplinary approach is required in the rehabilitation of streams, as modifications must be both biologically and hydraulically sound. Digging out a hole in the streambed may create a nice pool environment for fish refuge at low flows but it will be of little benefit if the stream's geomorphology is such that the pool will fill up with sediment during the first flood. Conversely, planting woody vegetation along streams may improve bank stability and riparian habitat, but it can also increase local flooding hazards if the channel capacity is too low to handle the effects of additional roughness.

Stream rehabilitation can be either:

1. *passive*, in which the factors that prevent recovery are dealt with and the stream is left to heal undisturbed, or
2. *active*, in which specific repair procedures are applied.

In its strictest sense, passive rehabilitation is non-interventionist (as in wilderness areas), but in most areas the stream will still be affected by human activity, so it is not an entirely 'natural' process. An example of passive restoration would be declaration of a conservation area or reserve, where trees are allowed to fall into the river and bank erosion is allowed to proceed. Active rehabilitation occurs when the stream is managed for the benefit of humans and non-humans, and it is usually a 'positive' activity. An example of positive-active rehabilitation is the placement of boulders or other structural devices in the stream to rapidly increase habitat area, or planting riparian vegetation. Most practical passive rehabilitation (see Ebersole *et al.*, 1997) is actually a form of negative-active rehabilitation (preventative maintenance), because ameliorating or removing degrading processes or factors that are preventing recovery is usually a pre-requisite activity. An example is the protection of streambanks from grazing pressure to permit natural revegetation.

A passive 'leave it alone' approach is quite possibly the most difficult one to implement from the social perspective. People seem to have an innate tendency to tinker with things and to get impatient about seeing instant results from investment of considerable time or resources. In severely degraded streams, natural morphological and biological recovery may be a very slow process, typically requiring 10–100 years (Brookes and Gregory, 1989). Telfer and Miller (2001) documented the case of Toorumbee Creek, New South Wales, Australia, which was in a highly degraded condition in the 1960s. In 1998 it was classified as a high conservation value stream, yet in the intervening period there was virtually no stream management intervention. The stream's natural recovery was explained largely in terms of a shifting socio-economic situation: land use changed from intensive dairy to beef grazing and timber harvesting, stocking rates were reduced and corn fodder was no longer grown to the stream edge. Since the 1970s land use intensity further declined due to settlement of 'lifestyle' owners, many of who were absentee landholders. Toorumbee Creek was also geomorphologically resilient, with bedrock controls and a supply of sediment to facilitate structural channel recovery.

'Removing a problem source' might require controlling point pollution sources by introducing wastewater treatment plants, or controlling non-point pollution sources by reducing chemical usage and erosion on range and agricultural lands. Alternatively, it could involve controlling the spread of pest plant or animal species. For example, in many areas, introduced game fish have displaced native species, and vigorous rehabilitation plans might include the elimination of exotic species such as carp. In heavily grazed areas the solution may be to exclude riparian areas to domestic livestock during all or certain times of the year through management or fencing. The fencing of corridors along the stream can protect riparian plants from grazing pressure and banksides from trampling. After the problem source is removed or controlled, natural rehabilitation can be assisted through revegetation and re-introduction of faunal species. Ideally, river management should progress to a point where a preventive maintenance approach can be used, with potential problems foreseen and eliminated before they become a burden.

There is an enduring debate over the merits of passive versus active stream rehabilitation. On the one hand, passive rehabilitation can be less expensive, but it takes a long time to see results. It can also be a lower-risk option; embracement of uncertainty avoids the problem of having fixed expectations about the exact form of the restored ecosystem, which we know cannot easily be predicted (Van Rijen, 1998). However, it may be more difficult to get landholders interested in projects that do not produce instant, predictable results that will immediately justify the capital expense, or opportunity costs. In some cases streams are too altered to return to the desired condition, so active rehabilitation is the only option. While stream health (Sections 9.2 and 9.3) reflects a wide range of human activities and natural processes, there has been a tendency for active rehabilitation projects to focus on single-factor solutions.

Active restoration is a higher-risk activity, because there is no guarantee that the physical form and/or biota will respond in the desired way, and within the expected time frame. In any case, the risk of failure for active rehabilitation projects is certainly reduced by halting the degrading factors through an accompanying passive rehabilitation programme (Kauffman *et al.*, 1993).

Askey-Doran (1999) viewed rehabilitation interventions as a means to overcome system thresholds by either reinstating dynamic processes (process-driven thresholds) or overcoming gross changes to the physical environment or to the range of species available (component-driven thresholds). They defined three classes of rehabilitation intervention:

1. *unassisted regeneration threshold*—the system cannot self-heal without some intervention, but it is still resilient and can return to its pre-existing state with a "kick-start";
2. *assisted regeneration threshold*—organisms or substrate critical for recovery are depleted and need to be restored in order to return to a pre-existing state; and
3. *reconstruction threshold*—component-driven thresholds have been crossed and recovery of the pre-existing system is not possible, so creation of a new state is the only option.

These three intervention classes based on system thresholds are roughly equivalent to the three approaches to rehabilitation identified by FISRWG (1998, p. I-3): *non-intervention and undisturbed recovery*, where active restoration is unnecessary and can be detrimental; *partial intervention for assisted recovery*, where action may increase the rate of natural recovery processes and *substantial intervention for managed recovery*, where recovery of desired functions are beyond the repair capacity of the system.

For our discussion purposes, we group rehabilitation activities under the headings of those that address biotic factors, and those that address physical factors. We illustrate the former group with a discussion of riparian zone stock exclusion and re-vegetation (passive/active partial/assisted intervention) and re-introduction of fauna (active substantial intervention) (Section 9.6.2). Physical rehabilitation is illustrated with a discussion of correcting limiting physical factors by providing fish passage, mitigating the effect of cold water discharges from dams and removing obsolete or uneconomical dams (passive substantial intervention) (Section 9.6.3), and in-channel measures in the form of gross modifications to the channel itself (active substantial intervention) (Section 9.6.4) and modifications within the channel in the form of habitat improvement structures (active partial/assisted intervention) (Section 9.6.5).

### How Close to 'Natural' Can We Go?

Much of the environmentally driven rehabilitation work conducted in highly modified urban streams is more appropriately termed enhancement or naturalization. Indeed, almost the entire rehabilitation effort of some regions and countries, although enthusiastically applied, has been highly constrained by the imperative to protect floodplain land from flooding and channel instability. Japan is a case in point; flat land is rare in this mountainous country, so floodplain land is highly valuable, intensively developed and generally unavailable for inclusion in rehabilitation plans (Gippel and Fukutome, 1998). In such cases it is not feasible to correct catchment disturbances, and channel incision or channelization is too extensive to expect a return of any semblance of the original channel form and function within management time scales. Here it would be prudent to encourage natural river features that suit the modified channel form or modified hydrological and sediment regime. These features may never have existed at this particular location, but would be found on other natural or semi-natural systems. In some cases, the ecology may show minimal change after habitat enhancement, but the stream may have vastly improved natural values, measured in terms of visual and recreational amenity for the public.

From the public's perspective, a stream can still possess a strong sense of naturalness for much of the time, even if the ecology is depauperate, and even if the stream has very poor water quality and intolerable hydraulic conditions during flood events. Creation of this sense of naturalness in urban streams may help to foster growing community–wide acceptance of (and perhaps demand for) expenditure on more ambitious protection and rehabilitation of ecologically important streams in rural and forested areas.

River management authorities are probably best served by adopting flexible definitions of both river rehabilitation and channel stability. Rehabilitation and channel stabilization can be compatible, but it would be safe to say that, until very recently, most channel stabilization works conducted by river management authorities have not been specifically designed to rehabilitate aquatic habitats. Many urban streams are too severely altered to allow rehabilitation in its strictest sense. However, there is much potential to enhance and naturalize these stream channels, by creating new appropriate stream forms that have an appearance and function that is compatible with the surrounding landscape values. Enhancement and naturalization has a strong potential to accommodate both the objectives of improving environmental and amenity values, and protecting public and private assets through stabilization. Enhancement and naturalization is more than engineering a channel for stability or hydraulic efficiency and then applying a cosmetic layer of landscape gardening. Scientific and landscape architectural expertize should be used to help design the works in order to maximize the ecological and amenity values, and to ensure their acceptance by the community.

Individuals in the community must value stream works or they will not be maintained and will not endure (Schauman and Salisbury, 1998). Even though in general people like 'scruffy' natural scenes more than those with buildings and structures, members of streamside communities in the Puget Sound area of Washington State have demonstrated a clear preference for stream enhancement works with a 'refined' appearance in the urban regional context (Schauman and Salisbury, 1998). This survey of scenic quality showed that scruffy is not a preferred attribute of home landscapes. It also demonstrated that expert opinion does not always correlate with layperson choices.

Enhancement/naturalization potential cannot be evaluated in the same way as rehabilitation potential. The reason is that enhancement/naturalization is a much more flexible concept. Enhancement/naturalization can be applied to any stream, while rehabilitation has much tighter constraints. It is probably safe to assume that where protection of cultural assets demands close-to-absolute channel stability, then the scope for rehabilitation is diminished, and enhancement/naturalization will be the best option (Gippel *et al.*, 2001).

### Planning Ahead: 'Preventive' Measures

In the planning phase of water resource development projects it is a routine procedure to evaluate several alternatives in terms of cost, effectiveness and environmental impact. The pro-active inclusion of passive rehabilitation may reduce the need for later, expensive, active rehabilitation works—much like a 'preventive medicine' approach. As an example, thanks to extraordinary foresight by early resource planners, 90% of the water supply for Melbourne, Australia, is drawn from closed catchments (i.e. uninhabited and not accessible to the public) (Melbourne Water, 2003). This has maintained a high water quality, reducing the need for expensive purification. Also, the streams in these headwater catchments are relatively pristine (fish passage is not completely open, and exotic species are present), and they supply high quality water and a source of colonization to the downstream waterways.

If feasible, alternative solutions can replace the need for engineering works such as flood-mitigation dams and channelized streams. With proper planning, lands in the floodplain can be set aside for restricted activities (e.g. parks, golf courses, car parks), where adjustments of the channel and overbank flows do not pose serious problems.

In streams where channel modifications are required for flood mitigation, measures can be taken to reduce the impact on the aquatic environment. Brookes (1989), Brookes and Gregory (1988) and Shields (1982) discuss environmental features for flood channels. These include reducing the amount of channel straightening or shaping, selectively leaving mature trees and other bank-stabilizing vegetation and minimizing the removal of large rocks, snags and aquatic plants from the stream. Retaining the existing features is preferable to 'fixing' the stream after disturbance.

### Sources of Information on Stream Rehabilitation

The past two decades have seen tremendous growth in the publication of manuals, guidelines and research papers on the topic of stream rehabilitation, which reflects the high level of interest and activity in this field. A new international journal *Restoration Ecology* and two new online journals, *Conservation Ecology* and *Conservation in*

*Practice* regularly contain articles on stream rehabilitation, and articles and special editions on the topic are well represented in multi-disciplinary journals such as *Rivers Research and Applications, Aquatic Conservation: Marine and Freshwater Ecosystems, Hydrological Processes, Hydrobiologia, Environmental Management, Ecological Engineering* and *Journal of Environmental Hydrology.*

Some important and influential publications on river rehabilitation have been published recently, including books by Gore (1985b), Petts and Calow (1996), Brookes and Shields (1996), DeWaal *et al.* (1998), FISRWG (1998) and Williams *et al.* (1997); chapters in Gore and Petts (1989), Harper and Ferguson (1995) and Thorne *et al.* (1997); conference proceedings such as International Trout Stream Habitat Improvement Workshops held every two years since 1978 (e.g. Alberta Environmental Protection, Fish and Wildlife *et al.*, 1994), the STREAMS Conferences (e.g. STREAMS, 2003), International Symposiums on Habitat Hydraulics (e.g. SINTEF, 1994; Leclerc *et al.*, 1996), Australian Stream Management Conferences (Rutherfurd *et al.*, 1999; Rutherfurd *et al.*, 2001); Hayes (1998, 2001), Hancock (1993), Collier (1994), Shrubsole (1994), Van de Kraats (1994), Watershed Science Center, Trent University (1999) and Gregory and Staley (2004); frameworks for stream rehabilitation by Koehn *et al.* (2001) and Phillips *et al.* (2001); stream channel and riparian zone rehabilitation design manuals by House (1988), Katsantoni (1990), Johnson and Stypula (1993), Newbury and Gaboury (1993a), Ontario Ministry of Natural Resources (1994), Lowe (1996), Kapitzke *et al.* (1998), The River Restoration Centre (1999, 2002), Rutherfurd *et al.* (2000), Water and Rivers Commission (2002) and Ontario Streams (2003); guidelines for urban stream rehabilitation such as ID&A Pty Ltd (1996), Conrick and Ribi (1987) and Booth *et al.* (2001); books specific to rehabilitation in particular regions such as Bustard (1984), Eiseltová and Biggs (1995) and Montgomery *et al.* (2003); monographs on specialized rehabilitation topics such as large woody debris reintroduction by V.A. Poulin & Associates (1991), Klassen (1991), Oregon Department of Fish and Wildlife (1995) and Millar (1997), riparian zone rehabilitation by Raine and Gardiner (1995), fishway design by Rajaratnam *et al.* (1989), Komura (1990), Katopodis (1992), Clay (1995), Bates (1997), Odeh (2000) and Wildman *et al.* (2002), and rehabilitation for fish by Hunter (1991), Slaney and Zaldokas (1997) and Cowx and Welcomme (1998); critical reviews of rehabilitation works by Hunt (1988), Holmes (1998) and National Research Council (1992); pictorial essays such as Blake *et al.* (2000); citizen science guides such as Washington Department of Fish and Wildlife (1995), Riley (1998) and Lindloff (2000); a children's

book on restoration of wetlands on the lower Mekong delta in Vietnam (Keller, 1994), as well as numerous brochures published by river management agencies [e.g. Australian large woody debris management guidelines: Water and Rivers Commission (2000); NSW Fisheries (2001); Cottingham *et al.* (2003)].

A large number of websites are now dedicated to stream rehabilitation, many of them community-based. Just a few examples are Partners for Fish & Wildlife, Wildfish Habitat Initiative, Montana Water Center, URL: http://water.montana.edu/wildfish/default.asp; Stream Restoration, Ecology & Aquatic Management Solutions (STREAMS), URL: http://www.ag.ohio-state.edu/~streams/index.html; Salmon River Restoration Council, URL: http://www.srrc.org/; San Lorenzo River Restoration Institute, URL: http://members.cruzio.com/~slriver/; Kissimmee River Restoration Project, URL: http://www.sfwmd.gov/org/erd/krr/; Romanian Centre for River Restoration, URL: http://rcrr.mobius.ro/; European Centre for River Restoration, URL: http://www.minvenw.nl/rws/riza/home/ecrr/; American Rivers, URL: http://www.amrivers.org/; Jordan River Restoration Project, URL: http://mountain-prairie.fws.gov/jordan/; River Restoration Northwest, URL: http://rrnw.org/; Klamath Resource Information System, URL: http://www.krisweb.com/krisweb_kt/index.htm; Ontario's Stream Rehabilitation Manual, URL: http://www.ontariostreams.on.ca/OSRM/toc.htm and Upper Saugeen Habitat Restoration Association, URL: http://www.canadianangling.com/index.html.

Add to the above lists the vast number of primary journal articles that have been published and there is clearly a rich and imposing source of literature and information available of stream rehabilitation. Two monographs stand out as being suitable both as a starting point for those new to the topic, and as a comprehensive reference source for the more experienced: FISRWG (Federal Interagency Stream Restoration Working Group) (1998), *Stream Corridor Restoration: Principles, Processes and Practices* and Brookes and Shields (1996), *River Channel Restoration: Guiding Principles for Sustainable Projects.*

### 9.6.2 Addressing Biotic Factors

#### *Riparian Zone Revegetation*

##### *Processes*

In studies on streams in the western USA, Platts and Nelson (1985) found that streams in heavily grazed areas tended to widen and become shallower. In a later survey,

Belksy *et al.* (1999) found that livestock grazing had altered approximately 80% of stream and riparian ecosystems in the western USA. Livestock seek out water, forage and shade in riparian areas, and in the process the stream banks are tramped and overgrazed. The consequences are soil erosion, loss of bank stability, declining water quality and increased stream temperatures (Trimble and Mendel, 1995; Belskey *et al.*, 1999), although the severity of the impacts is highly variable (Kauffman and Krueger, 1984; Myers and Swanson, 1992; Quinn *et al.*, 1992). In turn, these changes have reduced the habitat available for native plant, fish and other wildlife, causing many native species to decline in abundance or go locally extinct (Bowers *et al.*, 1979; Quinn *et al.*, 1992; Williamson *et al.*, 1992; Douglas and Pouliot, 1998). Thus, the main degrading factor with respect to riparian zone vegetation is usually uncontrolled grazing, which can be rectified through fencing (passive rehabilitation) (ID&A Pty Ltd., 2002). Re-vegetating the fenced-off area adjacent to the stream (active rehabilitation) can then assist the rehabilitation process.

Natural riparian vegetation provides bank stabilization, nutrient regulation, filtering of sediments, shading, nesting areas for birds, cover for fish and is the primary source of large woody debris to channels (Gregory *et al.*, 1991; Kapitzke *et al.*, 1998, pp. 170-171; Askey-Doran, 1999; Webb and Erskine, 2003). Where the surrounding land is cleared, as in forestry areas in headwater zones, or in farmland on lowland floodplain streams, the section of remnant or rehabilitated riparian vegetation adjacent to the stream is usually referred to as a buffer strip, or buffer zone (U.S. Army Corps of Engineers, 1991; Large and Petts, 1992; Osborne and Kovacic, 1993; Woodfull *et al.*, 1993; Abernethy and Rutherfurd, 1999).

Vegetation cover has a modulating effect on stream temperatures, keeping the water cooler in summer and insulating it in winter (e.g. Winegar, 1977; Teti, 1998, 2000). Quinn *et al.* (1997) found that New Zealand streams without riparian vegetation were on average 2.2 °C warmer than streams with intact riparian vegetation. The influence of riparian vegetation on stream temperature is one process that can be reasonably well predicted using numerical models (e.g. Rutherford *et al.*, 1997). Without streamside vegetation, creeks can also dry up more readily. Vegetative buffer strips are known to filter sediment and nutrients from non-point sources (Barling and Moore, 1994), but no models have yet been designed specifically for predicting the performance of riparian vegetation zones (Dosskey, 2001). Studies in Europe and the USA demonstrate high levels of nutrient uptake in riparian vegetation zones (Osborne and Kovacic, 1993; Mander, 1995). Bennett *et al.* (2002) conducted a flume experiment with simulated emergent vegetation that demonstrated that flow velocity can be markedly reduced within and near the vegetation zones, that flow can be diverted towards the opposite bank, and that vegetation density controls the magnitude of these effects.

Growns *et al.* (2003) examined fish assemblages in treed and grassed reaches in the Hawkesbury-Nepean River in New South Wales and turned up what may seem like a surprising result—the habitats adjacent to grassed banks supported more individuals and more fish species than well-treed banks. The differences in the distributions of fish species appeared to be related to the greater abundance of aquatic macrophytes near grassed banks, while the shaded treed banks reduced macrophyte growth. Read and Barmuta (1999) found that macroinvertebrate diversity and density was lower in willowed reaches than in reaches with native riparian vegetation. This difference appeared to be due to changes in shading, water quality and the quantity of habitat.

As bankside protection, plants absorb the impact of ice chunks and waves, but they are not structurally 'fixed'. In contrast to inert materials (e.g. rocks and concrete), plants are self-regenerating, allowing 'structural failures'. Riparian vegetation and aquatic macrophytes act to increase flow resistance, and are considered crucial in reducing fluvial scour and bank failure (Prosser *et al.*, 1999). The cohesive strength of riparian vegetation has been shown to stabilize banks (Abernethy and Rutherfurd, 2000). Beeson and Doyle (1995) assessed 748 stream bends for stream erosion after large floods and found that the vegetated banks showed much less erosion that those with semi- or un-vegetated banks. Vegetation on the banks also reduces the velocity of water flowing through it, encouraging sediment accumulation (Lewis and Williams, 1984). It follows that re-vegetation of point bars at meander bends (depositional zones) is a good strategy for building up banks (Platts and Rinne, 1985).

There is considerable debate in the literature, which has also spilled into the management realm, regarding whether trees or grass are superior for stabilizing stream banks and reducing nutrients. With respect to nutrients, the evidence seems to suggest that riparian forests are more efficient at removing nitrate-N in shallow subsurface water than are grass (Osborne and Kovacic, 1993). Studies of gravel-bed rivers in Colorado by Andrews (1984), in Britain (Hey and Thorne, 1986) and in New South Wales, Australia (Huang and Nanson, 1997) revealed a consistent pattern of narrower channels in forested riparian zones. A study of sand-bed channels in New South Wales by Huang and Nanson (1997) found that streams with trees and shrubs growing on the banks, but also in the bed, were double the width of streams with trees on the banks only.

Trimble (1997), in a study in Wisconsin, found that streams running through forest were significantly wider than those in pasture. This was confirmed by a New Zealand study by Davies-Colley (1997). However, this latter study found that width was independent of vegetation cover for catchments draining an area greater than 30 km². It was postulated that as stream power increased with basin area, the protective influence of grassy vegetation became less important. Riparian trees can shade banks, eliminating grass cover and exposing bank surfaces to erosion. The process that explains channels being wider under forest is woody debris and in-channel trees (especially willows) deflecting flows onto the banks and eroding them.

Trimble (1997) warned that restoration of riparian zones with trees might not be good public policy, although this attitude was qualified in a later paper (Lyons et al., 2000). Montgomery (1997) pointed out that the contradictory findings on this topic are due to the highly complex nature of the interaction of stream variables— "This smorgasbord of influences means that simple guidelines and blanket generalizations rarely provide a sound basis for the management of rivers and streams" (p. 328). Abernethy and Rutherfurd (1996) and Abernethy and Rutherfurd (1998) proposed that the stabilizing effect of bank vegetation varied throughout the stream network, generally decreasing downstream. Consideration of scale is important in analysing these situations. For example, woody debris can cause local bank erosion within a channel reach, yet be responsible for storage of vast quantities of material in the bed (Gippel et al. 1992; Montgomery 1997). So, even if forest channels are wider than grassland channels, they probably store more sediment. The literature suggests that the influence of vegetation on channel morphology is highly variable, and so too is the impact of grazing on stream health (ID&A Pty Ltd., 2002), making it difficult to predict the outcome of rehabilitation through fencing and revegetation.

*Methods*

The first step in any revegetation program carried out in a rural area is to exclude stock from the stream buffer area through fencing, and provide offstream watering. Alternatively, stock access points can be located where the animals will do the least damage to the stream, and rock-lined fords can be used to reduce streambed disturbance. The main approaches to stock management in the riparian zone are full fencing, partial fencing (priority zones are fenced) and time-controlled grazing. Making paddocks more attractive to stock for grazing by providing watering points (Miner et al., 1992) and shade can reduce the time stock spend in the riparian zone. Audio electrical stimulation collars stimulated by a transmitter are effective in restricting stock movement (Quigley et al., 1990). Bell and Priestly (1999) provide advice on practical riparian zone fencing techniques.

Vegetation management should aim to maintain a stable, self-sustaining cover of native vegetation. As a general recommendation, indigenous species should be used wherever possible, since they are adapted to the local conditions and utilized by the local wildlife (Friedmann et al., 1995; Carr et al., 1999; Webb and Erskine, 2003). Plants can also be selected for their value to fish, invertebrate, waterfowl or wildlife habitat, their rapid regrowth attributes, their successional standing (e.g. as a primary colonizer), their aesthetic quality or their rareness (e.g. to re-introduce endangered species). The various 'zones' of the stream (i.e. channel edges versus mid-channel; see Figure 2.1) should also be kept in mind during the planning of revegetation work (FISRWG, 1998; pp. 8–12). Site preparation, plant propagation and planting procedures are best addressed on a local basis. Guidelines for riparian vegetation rehabilitation can be found in Lewis and Williams (1984), Cairns (1998), Risser and Harris (1989), Johnson and Stypula (1993), Raine and Gardiner (1995), Abernethy and Rutherfurd (1999) and Webb and Erskine (2003).

The suitable width for a riparian strip depends on the nature of the stream, and the purpose of the rehabilitation. Mander (1995) presented an equation based on soil properties, roughness and slope that predicted widths of between 5 and 50 m. Abernethy and Rutherfurd (1999) suggested that the appropriate width of vegetation for maintaining bank stability into the future should be a basic width (suggested as 5 m) plus the height of the stabilized bank (measured vertically from the toe of the bank to the crest). They stressed the need to stage the vegetation plantings, moving from fast growing plants closer to the bank edge, to slow-growing and long-lived plants further from the bank edge. If areas are to be revegetated as part of an environmental objective, the width of the riparian zone should be at least 30 m (Schueler, 1995; Dignan et al., 1996). This width has been shown to provide the majority of in-stream processes (leaf fall, shading, woody debris, etc.) at near natural levels. Webb and Erskine (2003) found that widths narrower than 30 m contributed to extensive weed invasion and low plant survival rates. Howell et al. (1994) recommended that a 50 m wide corridor of vegetation should be maintained on each bank of the large Hawkesbury-Nepean River, New South Wales, but in particularly active rivers, even wider buffer strips may be needed (Webb and Erskine, 2003).

Willows (Salix sp.) have often been used to protect banksides because they can be established easily, are quick-growing and resilient, can withstand inundation and are dense enough to promote sediment deposition (Petersen, 1986). Strom (1962) advocated pruning willows, or partly cutting the stems and bending them over, to keep the trees short and bushy. This improves water resistance and prevents them from becoming top-heavy. He also recommended planting them by burying cuttings horizontally in the soil to generate a mat of roots and a thicket of stems. A recent study by Schaff *et al.* (2002) found that willow posts subjected to a 10 day soaking prior to planting had greater growth, and this treatment doubled the survival rate. Willow logs can be used to create 'growing' groynes (Section 9.6.4) to encourage silt deposition. They can be grown from slips and cuttings; in fact, this regrowth ability may cause them to be a nuisance weed species when broken branches are washed downstream. In regions where they are not native they should be discouraged. River management agencies in Australasia are currently burdened by expensive willow removal programs (West, 1994; Ladson *et al.*, 1997; Cremer, 1999). Exotic species in general can introduce foreign organic matter into the ecosystem and change the timing and rate of processing of the material (Campbell *et al.*, 1989). It is common for exotics to out-compete and displace native riparian vegetation (Olson and Knopf, 1986).

Revegetation of riparian zones can be done by allowing natural regeneration, which relies on the existence of a viable seedbank and requires control of weeds and protection from grazing, or planting from tubestock, which is expensive and slow, but stands a better chance of success against weeds, dry soil conditions and stock grazing. Burston and Brown (1996) recommended direct seeding, which is fast and inexpensive, but the plants have to be protected from weeds, grazing and insect attack. Protection and maintenance are important components of revegetation work. Seeds and/or cuttings may require watering until their roots reach the water source. Losses should be anticipated; thus allowances should be made for over-planting or re-planting of the sites. The sites should be maintained to reduce weed growth, with mulching or cultivation preferred over herbicide application. Protection of the new plants with stock- and vermin-proof fencing and/or tree guards is also essential.

In recent years bioengineered or 'soft armour' streambank protection with geotextimes has been used as an alternative for rock rip-rap. This approach ultimately relies on vegetation to provide stability, and there is a need for vegetal shear stress information in the design process. A review of literature on the topic by Hoitsma

and Payson (1998) found a wide range of values, most likely related to vegetation type and cover, and soil properties. Unless more appropriate local information is available, the conservative values of Chen and Cotton (1988) should be used. One difference in the vegetative approach to bank stability compared to armouring (hardlining) (Section 9.6.4) is that plants are weak when first installed, but get progressively stronger over time, whereas armouring methods are strong when first installed, but deteriorate and weaken with time.

Consideration must be given to the hazards caused by the introduced vegetation, especially when the stream passes through private lands. The retarding effects of vegetation within a channel will slow the water and encourage silt build-up, which may be desirable from an ecological viewpoint but in extreme circumstances it can exacerbate the extent of local flooding (Gippel *et al.*, 1992). If landholders living next to incised or channelized streams have become accustomed to a reduced threat from floods, they may be reluctant to allow tree planting on the streambanks for erosion control. When vegetation becomes dislodged it can lead to log jams or weed jams at bridges or other constrictions. Fencing can trap debris and act like a levee. Even in a riparian zone, the dense vegetation can pose a fire hazard during dry conditions. Controlled grazing to reduce fuel levels may be feasible in some areas during seasons when the stream banks and plants are least sensitive to destruction by grazing. It may also be a method of removing undesirable species (e.g. the use of tethered goats for blackberry control). Alternatively, mechanical cutting, thinning and removal of aquatic and bankside vegetation may be required. The re-establishment of vegetation should, therefore, be combined with good management practices both during the rehabilitation stage and afterwards.

### Reintroduction of Fauna

Although revegetation has received much of the emphasis in stream rehabilitation efforts, faunal species are also part of a healthy riparian ecosystem. As with re-vegetation, it may be possible to speed up the natural rehabilitation process by re-introducing animals at the appropriate stage of recovery. Again, to guide rehabilitation efforts, it will be necessary to acquire knowledge of the requirements of each species and the typical population densities found in undisturbed areas.

Recolonization of invertebrates will occur rapidly if substrate and nutrient requirements are met, but for some species with limited migration capabilities, 'seeding' larvae from undisturbed sites may also be feasible. Fish stocking has a long history in areas which are heavily fished or which cannot support self-sustaining populations, and an extensive literature is available on game fish species (e.g. Dodge and Mack, 1996). In the re-establishment of native fish species considerations must include competition between species, predators (which may include introduced game species) and food and physical habitat requirements. Bayley and Osborne (1993) found that in small streams in Illinois, restoration programmes did not require fish stocking, provided the streams were connected to permanently flowing streams that contained the full complement of desired species. However, many streams undergoing rehabilitation are not connected to source areas, and stocking can complement physical habitat improvements (e.g. Jutila, 1992).

Supplemental stocking is the practice of stocking fish where the existing self-sustaining population cannot meet the demand for fishing. In places where freshwater fish are an important source of protein, or have high economic significance, introductions are done to increase yields. Stocks of fish in some large commercially important rivers are maintained almost entirely by stocking with artificially propagated fry, and in many cases the species would disappear without such programmes (Welcomme, 1989).

A decision making model for fish introduction proposals developed by Kohler and Stanley (1984) and adapted by Dodge and Mack (1996) has a series of information review steps, each followed by critical questions. Following the process right through without answering 'reject' leads to a decision to approve the proposal to stock fish. A precautionary approach to fish introductions was recommended by Bartley and Minchin (1995).

Among the higher animals, the beaver has had an active role in stream rehabilitation in North America. The trade in beaver fur decimated the population of these animals before the turn of the 20th century, but reintroduction over the past 50 years has now restored the USA population to 6–12 million, compared to the pre-European level of 60–400 million (Naiman *et al.*, 1986). Beaver dams, which can have extreme longevity (Johnston and Naiman, 1990), positively influence riparian function in many ways (Olsen and Hubert, 1994). They can trap sediment, help prevent bank erosion, play a role in nutrient uptake, modify hydrology and the ponds they form also provide important habitat for birds and fish (FISRWG, 1998, pp. 8–26).

Dahm *et al.* (1989) documented the degradation of a stream in New Mexico, USA, following more than a century of heavy grazing, logging and beaver trapping. Expansive meadows, most likely formed and maintained by beaver activity, were greatly reduced in extent, and the vegetation adjacent to the stream became dominated by xeric (dry-adapted) species as the water table dropped. Since acquisition of the land by the U.S. Forest Service,

stream reaches have been rehabilitated through riparian plant re-introduction, bank stabilization and the addition of large woody debris. Beavers naturally recolonized an upper tributary, stabilizing the upper catchment with a network of dams and ponds. On another tributary, beavers were introduced as part of a stream ecology study.

### 9.6.3   Correction of Physical Limiting Factors

#### *Fish Passage*

Dams affect instream biota in three main ways: by imposing unnatural flow regimes (Section 9.5), cold water releases (see below) and by acting as barriers to migration. Fishways are structures that enable fish to pass barriers on streams and rivers. Fishways have been built in connection with dams in Europe for several hundred years, and the Belgian engineer G. Denil designed the first fishways based on hydraulic experiments in the early 1900s (Straub, 1934).

The Columbia-Snake River system is a good example to illustrate the issues associated with fish passage. The total run of Pacific Salmon to the Columbia River in the Pacific Northwest in the 19th century was in the order of 10–16 million, while today it is in the order of 1–3 million, 75% of which are hatchery origin (Park, 1990; Buchanan *et al.*, 1997; Johnson *et al.*, 1997; Thurow *et al.*, 2000; Knudson *et al.*, 2000; U.S. Army Corps of Engineers, 2002a). Wild salmon are now extinct in many areas that were part of their historic range (Stouder *et al.*, 1997; Dunham *et al.*, 2001). The decline has been oscillatory, and is explained by a number of anthropogenic and natural factors (Lichatowich and Mobrand, 1995; U.S. Army Corps of Engineers, 2002a): overharvesting in the late 1800s into the early 1900s; effects on habitat from farming, cattle grazing, mining, logging, road construction and industrial pollution; introduced species (Thurow *et al.*, 2000); altered water temperature (Dunham *et al.*, 2001); the complex of tributary and mainstem dams (Ebel *et al.*, 1989, Wissmar *et al.*, 1994); and fluctuations in response to changing climate and ocean conditions (Anderson, 2000). It has been estimated that wild salmon runs were reduced to less than 5% of their historical levels by the late 1950s, with some recovery achieved through a widespread hatchery enhancement program beginning in the 1960s (Johnson *et al.*, 1991).

Although populations of salmon on the Columbia River system had already declined before the main phase of dam building (beginning with the construction of Rock Island Dam in the early 1930s and finishing with Lower Granite Dam in 1984), dams have clearly played a significant role in continuing the decline and preventing recovery (Buchanan *et al.*, 1997; U.S. Army Corps of Engineers *et al.*, 1999; U.S. Army Corps of Engineers *et al.*, 2002a, b). There are now 27 dams on the Columbia and Snake River mainstems. The dams cause mortality by killing juvenile salmon migrating downstream (when they pass through turbines or over spillways) and by restricting their passage as returning adults. The problem of fish passage was well understood before dam construction began (Baker and Gilroy, 1933), with the Bonneville Dam (60 m high wall, completed in 1938) fishways built as an integral part of the dam at a cost of USD7 million, or 8.5% of the total cost of the dam (Clay, 1951). The fish passage facilities at McNary Dam (67 m high wall, completed in 1953), which were based on the Bonneville Dam facilities, cost USD28 million (Von Gunten *et al.*, 1956). As dam construction continued on the Columbia-Snake River system, considerable effort was made to refine and improve fishway technology (Anon., 1951; Clay, 1951; Anon., 1958; Deurer, 1960; MacLean, 1961; Martin, 1964; Loder and Erho, 1971). Thirteen of the mainstem dams have fishways, which includes all the lower river dams (Northwest Power and Conservation Council, 1994).

Many different types of fishway are used around the world. They include pool-type such as the vertical-slot fishway; Denil fishways, also built on a sloping channel but utilizing U-shaped baffles without intervening pools; lock fishways (either existing boat locks or locks built specifically for lifting fish); trap and transport methods in which fish are trapped in a holding area at the base of a dam and taken upstream to a release site; rock-ramp or 'nature-like' fishways which are particularly suited to smaller streams and small and juvenile fish; bypass fishways, also called nature-like fishways, which are meandering channels that skirt around the barrier; and eel and elver fishways (Petts, 1989; Thorncraft and Harris, 2000). Locks and transport methods may have an advantage for the transport of very small fish that would not be able to tolerate the higher velocities in fish ladders. A major factor in the usefulness of any of these methods is the provision of hydraulic conditions and/or fencing to guide the fish to an entryway. Other design considerations are the fish burst speed, fish endurance and number of fish that will use the structure.

A detailed discussion of fishway design is beyond the scope of this text; however, Clay (1995), Larinier (1987), Rajaratnam and Katopodis (1984, 1988) and Thorncraft and Harris (2000) can provide additional information, and some examples of nature-like fishways can be found in Wildman *et al.* (2002). A book on designing water intake structures for fish protection has also been published by ASCE (1982). One important point to remember is that fishways need to be designed to suit the swimming abilities and preferences of the fish for which it is intended. Only a few Australian fish species jump like salmon in the Northern Hemisphere (Mallen-Cooper, 1994). Australian fish use differing tactics to pass obstacles, with some capable of climbing wet vertical surfaces. Design considerations for rock ramp fishways to suit Australian streams were provided by Lewis *et al.* (1999), who noted that this type of fishway was considerably cheaper to build than a vertical-slot type. White and O'Brien (1999) described two examples of recently constructed successful fishways in Australia, a vertical slot fishway on the River Murray and a rock ramp fishway on the Barwon River, Victoria. Almost half of the native fish species in the River Murray migrate as part of their life cycle. Provision of fish passage along the length of the River Murray from the sea to Hume Dam is considered a high priority for management, and AUD25 million has recently been allocated to achieve this goal (MDBC, 2002).

### Mitigating Thermal Effects of Dams

Cold water pollution is a major problem with older large dams that lack selective withdrawal capabilities and release water from deep within the hypolimnion layer where the temperature of the water is much colder than the river downstream. These colder temperatures can prevent spawning of native fish, and can favour alien species. Some examples from the Murray-Darling Basin in Australia illustrate this phenomenon. Water released from Dartmouth Dam (160 m deep at wall) on the Mitta Mitta River, a tributary of the River Murray, is fairly constant at around 9–11 °C throughout the year, and the effect persists for 100 km downstream, where the river flows into Hume Dam Reservoir (Walker, 1985). Native fish species now comprise less than 7% of the fish biomass in the Mitta Mitta River, but the cold water is favourable for trout, an introduced species (Koehn *et al.*, 1995). Downstream of Hume Dam (41 m deep at wall) water temperature can be reduced by up to 4–6 °C for a distance of around 200 km (Walker *et al.*, 1978). This thermal shock effect is implicated in the circumstantial disappearance or severe depletion of native aquatic faunal species within this zone and dominance of introduced species such as trout and carp (ID&A Pty Ltd, 2001, p. 16). Eildon Dam on the Goulburn River, Victoria, a major tributary of the River Murray, reduces the spring/summer temperature by up to 7 °C, with the effect persisting for 200 km. These conditions are unsuitable for the spawning of threatened native fish species, explaining their absence from the river

(Gippel and Finlayson, 1993). Field experiments conducted by Astles *et al.* (2003) downstream of Burrendong Dam on the Macquarie River in New South Wales found that juvenile native fish were affected by cold water pollution in the areas of growth, survival, distribution and activity. In thermal gradient experiments offering a choice of warm and cold water, the fish exhibited a strong preference for warm water. This work clearly demonstrated the potential benefits of mitigating cold water releases from dams.

The options for mitigating cold water pollution from dams fall into two categories: exploit the stratification of the reservoir by selective withdrawal, or break up the stratification artificially. Sherman (2000) conducted a review of seven different means of mitigating thermal pollution: retrofitting of dams with multi-level outlet structures so water can be selectively withdrawn; destratification using bubble plumes or mechanical mixers; trunnions (floating intakes); surface pumps, and draft tube mixers that pump warm water near the surface into the cold water layer; submerged weirs or curtains suspended at various depths to provide a barrier to the passage of water and stilling basin, which is a shallow pond through which releases pass, delaying water so thermal equilibrium may be reached with the atmosphere.

The capital costs of retrofitting dams with multi-level offtakes may be prohibitive. Destratification systems have high initial costs, and also incur high operational costs. Surface pumps are relatively inexpensive, flexible and effective. The simplicity of a submerged curtain, with no moving parts, is attractive, but this device does not allow release of cold water if necessary (e.g. to prevent release of blue-green algae). Trunnions do not have a large capacity, so they are not suitable for dams that release flows for irrigation. Stilling basins are usually not feasible due to lack of suitable space. The comparison of technologies by Sherman (2000) for the case of Burrendong Dam found that draft tube mixers were the most appropriate, because they were relatively inexpensive, had operational flexibility and were effective in temperature mitigation. The choice of technology depends on the situation, and has to be evaluated on a case-by case basis. Sherman (2000) found adequate evidence in the literature that, if properly installed and operated, these technologies can largely mitigate the problem of thermal pollution.

### Dam Removal

Dam removal is currently a high profile issue in river management (Grant, 2001), particularly in the USA where 57 dams in 15 states and the District of Columbia were scheduled for removal in 2003 (American Rivers, 2003a). Although controversial, dam removal is not a new practice. In the USA, where most dams are privately owned, over 500 obsolete dams were removed in the last two decades (Stanley and Doyle, 2003). Despite growing interest, there are only a few reports of dams being removed in the rest of the world (International Rivers Network, 2001; Epple, 2000; Stanley and Doyle, 2003). In Australia, a proposal in the early 1990s to decommission a dam on on the Serpentine River, Tasmania, that drowned Lake Pedder, formerly a small glacial lake with high conservation value, has been strongly resisted (Gippel and Collier, 1998). In New South Wales there are 3,328 licenced dams and weirs on streams that are barriers to fish (NSW Government, 2001), and a program has been established to review the operation of these structures (NSW Government, 1997). A recent survey recommended the removal of 88 of these structures (NSW Government, 2001). The New South Wales Weir Removal Program has successfully negotiated the decommissioning of 14 barriers in recent years, including the long-obsolete, 100 year old, 15 m high Wellington Dam, dismantled in 2002.

More than 114 dams have been removed in the USA since the highly publicized breaching in 1999 of the 162 year old, 7.5 m high Edwards Dam located at Augusta, near the tidal limit of Maine's Kennebec River (American Rivers, 2003a). The breaching of the Edwards Dam lowered the water level by 3 m at the dam site, and opened 27 km of the lower Kennebec River (further fish passage is prevented by the Lockwood Dam at Waterville). This dam removal was significant because it was a direct consequence of the precedent-setting 1997 decision by the Federal Energy Regulatory Commission (Bryant, 1999) that the environmental costs of the dam outweighed the value of the electricity it produced (the dam produced only a small amount of electricity and performed no irrigation or flood control function). Faced with costs of USD9.9 million to build a fish ladder and carry out environmental remediation, the dam's owner agreed to allow the dam to be removed. The USD7.25 million cost of removal was met by owners of other upstream dams, as well as downstream shipbuilders Bath Iron Works, as mitigation for the impacts on fish of their own projects.

The issue of dam removal has come into prominence for several reasons: increasing awareness of the negative effects of dams on stream health; many dams now reaching their life expectancy; older dams posing a risk to public safety; older dams being costly to maintain; growth of the stream restoration movement; and emergence of the associated natural flow paradigm philosophy (Lindloff, 2000; Grant, 2001; Doyle *et al.*, 2003a). By 2020, 85% of all government owned USA dams will be near the end of their operational lives, which typically ranges from 60 to 120 years (American Society of Engineers, 1997; Federal Emergency Management Agency, 1999). More than 400 dams failed in the USA between 1985 and 1994, causing serious environmental damage and devastating losses of property and human lives (Graham, 1998). Deteriorating structures must eventually be removed, repaired or replaced to avoid these outcomes (Stanley and Doyle, 2003). Dam removal is also emerging as an option in hydroelectric relicensing proceedings. More than 500 licenses will expire in the USA in the next decade (International Rivers Network, 2001).

There are three basic options for dam removal: complete removal, partial removal, and staged breaching. ASCE Task Committee (1997) provided guidelines for dam removal projects that (1) identify types of data options to be considered; (2) describe available engineering, environmental and economic methods for assessing, quantifying and implementing dam retirement and (3) identify types of techniques for comparing and evaluating retirement costs and benefits. Lindloff (2000) provided a citizen's guide for mounting a grass-roots campaign for dam removal. Doyle et al. (2003a), following Pejchar and Warner (2001), recommended the development and adoption of a prioritization scheme for what constitutes an important dam removal, and the establishment of minimum levels of analysis required prior to decision-making about a dam removal. Doyle et al. (2003a, b) argued for an adaptive management approach, with well-studied small dam removal projects used to learn about the processes and impacts surrounding removal, before large-scale removals with potentially much greater environmental consequences are contemplated.

Despite the current enthusiasm, the decision to remove a dam is complex, controversial and may have both positive and negative effects (Stanley and Doyle, 2003). Management and disposal of the trapped sediment is a major issue. This is not just a local issue near the dam, because released sediment will deposit further downstream, and for a period of time increased turbidity can be expected. This creates higher water treatment costs, and can impact some sensitive biota (FISRWG, 1998, pp. 8–78; Stanley and Doyle, 2003; Doyle et al., 2003a, b). Following removal of the 9 m high Fort Edwards Dam on New York's Hudson River in 1973, large quantities of stored polychlorinated biphenyls were suddenly exposed in the old riverbed or flushed downstream (Shuman, 1995). Dam removal lowers the base level of upstream tributaries, which can cause rejuvenation and result in bed and bank instability. Removal of the dam pond results in loss of wetland habitat (FISRWG, 1998, pp. 8–78). In some instances, dam removal will mean loss of a valuable

nutrient sink, which may conflict with nutrient management strategies (Stanley and Doyle 2002). Mortality rates of virtually all reservoir populations, except fish, will be extremely high and can be expected to approach 100% if dewatering is rapid (Stanley and Doyle, 2003). Dam removal represents disturbance of a stable, adjusted system (Grant, 2001), and the loss of resident flora and fauna and the disruption of ecosystem processes should be expected (Stanley and Doyle, 2003). The environmental impacts of dam removal are not necessarily large or lasting. Removal of a small dam (approx. 3 m high) on the Baraboo River in Wisconsin caused relatively small and transient geomorphic and ecological changes in downstream reaches, explained by the relatively large channel size and the small volume of stored sediment available for transport (Stanley et al., 2002).

Most dams removed to date in the USA have been less than 8 m high, with only a few exceeding 20 m high (Poff and Hart, 2002), and this trend is likely to continue (Grant, 2001; Doyle et al., 2003b). The reason for that is the sheer number of small dams and their age, expertize in removal and estimating the economic costs and benefits is available, and the smaller scale is easier to manage (Lindloff, 2000). Issues surrounding small dam removals are thus the most critical focus for new science and policy (Doyle et al., 2003b). There is also strong interest in removing some larger dams (International Rivers Network, 2001; Kareiva, 2000; Stanley and Doyle, 2003; The Pacific Coast Federation of Fishermen's Associations, 2003), e.g. dams between 30 m and 38 m high on the Elwa and the White Salmon Rivers, Washington, and the lower Snake River Idaho, plus the 82 m high Glines Canyon Dam on the Elwa River.

Dam removal costs can be less than the estimated long-term expenditure required for long-term safety and environmental compliance, repair and maintenance costs. Removing the 40 m high Condit Dam in Washington is predicted to cost USD15 million, while estimated repair costs are estimated at twice that amount. The full cost of purchasing and removing the two Elwha River dams is expected to exceed USD200 million over a 20-year period, but paying reparations to local affected people could cost much more (International Rivers Network, 2001).

### 9.6.4  Rehabilitation of Channel Form

For many streams, passive methods will be the only possible solution simply on economic grounds. Without help, however, some degraded streams may take centuries to recover. The rehabilitation process can often be expedited through active intervention. Active restoration methods may be more popular with the public (and thus more likely to be funded and maintained) simply because the effects are more immediate and more visible—they give the impression of 'doing something'.

In comparison to passive methods, active methods for rehabilitating streams are more immediate but will usually be much more costly. The purpose of channel modification and instream structures is to restore diversity of physical habitat to streams that have been simplified through past management practices.

Meanders, pools and riffles, islands, billabongs and side channels contribute to biological richness by providing a diversity of channel habitat: turbulent and still water, shade and sun, sand and mud, eroding cliffs and point bars (Section 9.6.1). Channelization has reduced the total length of many waterways, and removed much of the irregularity of the stream margins. Restoration of a more natural meandering, rough-boundary form increases the surface area and total amount of habitat available for biota. In other cases the sediment supply and/or transport regime has been altered (Bravard *et al.*, 1999).

Active channel rehabilitation, which we refer to here as 'Natural Channel Design' (NCD) might be applied to streams that have been straightened, desnagged and otherwise channelized; streams where the natural channel structure has been totally destroyed by mining or dredging or newly created diversions around highways or strip mines. An idealized application of this approach to rehabilitating a straight channel is illustrated in Figure 9.32.

### The Basis of 'Natural Channel Design' (NCD)

'Natural channel design philosophy' (NCDP) is used in the USA, principally in the Rocky Mountain States, the Southwest and California, to describe the approach to channel rehabilitation that follows the prescriptive, channel-based classification method of Rosgen (1996a) (Section 9.4.3). This template-based method relies heavily on empirical relationships centred on bankfull discharge, reference reaches, regional curves and physiomimesis (mimicking natural processes). The term is widely used among river managers in these States (e.g. Schmetterling *et al.*, 1999; Brown *et al.*, 2003), who learn about the method in popular workshops run by Dave Rosgen. However, the natural channel design concept was not conceived by Rosgen, and has a broader and more general meaning, not tied to a specific method. To "emulate nature in designing channel form" was one of Nunnally's (1978) three principles for channel 'renovation'. Gerdes (1994), a participant at the Canadian conference *'Natural' Channel Design: Perspectives and Practice* (Shrubsole, 1994) wrote that natural channel design (NCD) was based on the 'theory' that

"...by emulating natural form and process, a stream can be assisted in achieving a physically and ecologically stable configuration (Brookes, 1987). 'Natural' channel designs provide more benefits by facilitating river stabilization to reduce erosion, and using natural materials and form instead of more traditional hard treatments. In addition to treating the erosion site itself, the erosion/sedimentation balance of the river is achieved and leads to a greater ability of the river to adjust to future changes; improved water quality; increased aquatic habitat diversity; and enhanced aquatic production. Further, aesthetics and recreational opportunities can be significantly improved by natural channel design. These features all contribute to ecological stability and to land use opportunities...An understanding of river process, floodplain process, and their connection with watershed process is critical to a successful undertaking of 'natural' channel design." (p. 14)

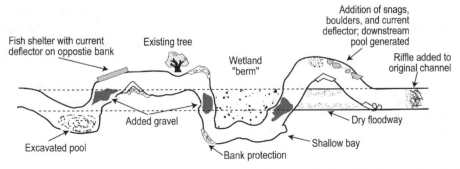

**Figure 9.32.** *A theoretical example of how a straightened, channelized reach (dotted line) might be rehabilitated using techniques described in this chapter*

Skidmore *et al.* (2001) also used NCD in its broad sense, and classified the tools used in NCD into three groups:

1. *Analogue approach* uses templates from adjacent or historical channels and assumes equilibrium between channel form, hydrology and sediment transport.
2. *Empirical approach* uses established equations that predict characteristics on the basis of regional data sets and also assumes equilibrium conditions.
3. *Analytical approach* makes use of hydraulic models and sediment transport functions to derive equilibrium conditions, and is useful when historic or current channel conditions are not in equilibrium with the sediment and hydrological regime.

The Rosgen (1996a) 'reference reach approach' is an example of an analogue approach. Analogue approaches do not require quantification of dependent variables. Emplacement of instream habitat devices is often based on this approach of copying a natural stream form. Similarly, re-instating meanders is usually based on copying a previously existing form (e.g. The River Restoration Centre, 1999), although flume studies and numerical models can also be applied to this problem [e.g. project site 'Blochinger Sandwinkel' on the Danube River, Germany (Kern, 1992)]. Empirical approaches rely mostly on the idea of dominant discharge (Section 7.2.5), which is also known as the 'hydraulic geometry method' (Section 7.2.6) (FISRWG, 1998; Fripp *et al.*, 2001). Williams (1978), Williams (1986) and Wharton (1995) provided reviews of the foundations of this concept; the general conclusions being that there are problems with defining bankfull discharge, the relationships only apply to the range of conditions from which they were derived, and even then, the confidence intervals for predictions are extremely wide. Bob Newbury's 'art of stream restoration' method mixes analogue and empirical approaches (Newbury and Gaboury, 1998; Newbury and Gaboury, 1993a, b; Newbury, 1995). The analytical approach quantifies variables and predicts outcomes with numerical models, and perhaps physical flume models; it is a process-based approach (Section 9.4.3) (Soar *et al.*, 2001). The analytical 'design for stream restoration' method of Shields *et al.* (2003) emphasizes the need for robust hydraulic analysis of flow resistance and sediment transport, and supervision by hydraulic engineers during implementation. The analytical approach can be limited by its high data demands, and the need for a high level of expertize. It is also the most expensive method of design. Outputs from these models do not normally incorporate natural channel variability. Soar *et al.* (2001) recommended ways that channel designers could use all three

approaches in combination, as appropriate for the particular problem at hand. The USACE method also makes use of various approaches (FISRWG, 1998, pp. 8–32; Fripp *et al.*, 2001). Schoor and Sorber (1999) described application of a combined methodology to rehabilitation of channels in the Netherlands.

Here we use NCD to cover all approaches to design. The central theme is that NCD is the holistic alternative to the traditional approach of seeking absolute channel stability at the expense of geomorphological and ecological function. The approach of Rosgen (1996a) is one method that can be used in NCD, but there are others. The equations used to design 'hard-lined' and stable channels are also useful for NCD, but they may require some adaptation. For example, most standard engineering channel designs are based on design floods with return frequencies of 10–50 years; NCD recognizes that geomorphic channel processes are associated with a range of event frequencies, but often focussed on those that occur much more frequently, perhaps every one or two years.

Stabilized, smooth channels with trapezoidal sides and uniform slopes act much more predictably than natural ones. It is thus much easier to describe the behaviour of water flows with simple backwater curves and Manning's equation in channelized reaches. In natural streams, there is much more variability, and simplifying assumptions or modifications must be made in order to apply techniques that were developed for uniform conditions. Some modifications to standard predictive hydraulic equations that account for natural channel features such as large woody debris (e.g. Shields and Gippel, 1995) are already available. However, there remains great opportunity for advances to be made in the modelling of water behaviour in natural streams.

The discussion of designs for both channel modification and instream structures relate primarily to low- and middle-order streams. On larger streams such works become major engineering projects. In all cases, interaction between biologists and engineers during the design, construction and monitoring phases can lead to improved designs that are effective from both standpoints. For example, channels designed for flood conveyance can be modified to include the 'biological component' by allowing for the resistance effects of vegetation. Successful demonstration sites can help to 'sell' rehabilitation ideas to both private landholders and public agency decision makers. Techniques in this section include the establishment of natural channel dimensions, bank-stabilization measures, the re-establishment of natural pool-riffle sequences and/or meander patterns and the reconstruction of floodways.

### Channel Modifications

#### Setting Channel Dimensions

If a streambed has become widened or entrenched it will change the frequency of overbank flows, modify water table levels, influence water temperatures and affect riparian vegetation growth. One goal of stream rehabilitation may be to re-establish more natural stream geometry.

Determining the stable channel dimensions can be guided by the natural characteristics of streams in hydrologically similar settings, such as the width:depth ratio. For example, on the Mink River, an alluvial stream in Manitoba, Canada, Newbury and Gaboury (1988) suggest that although narrow, deep channels provide the best hydraulic radius for conducting flood flows, natural cross sections tended to be wider and shallower with a general width-to-depth ratio of about 15:1. If a reference reach is used, it should be evaluated to make sure that it is stable, and it should be similar to the project reach in terms of sediment transport, bed and bank material and hydrological regime (FISRWG, 1998, pp. 8–32).

Measurements should be taken in both pools and riffles. If there are no suitable, unaltered streams for comparison, hydraulic geometry relationships (Section 7.2.6) obtained from upstream, unaltered tributaries can be used to estimate the 'unaffected' downstream dimensions (FISRWG, 1998, pp. 8–36). For example, a reach of the Steavenson River, a tributary of the Acheron River (see Figure 4.7), was diverted into a straightened channel with a width of approximately 3.5 m. The catchment area for this site is about 140 km$^2$, and from the hydraulic geometry relationships (Figure 7.9) it can be seen that the width should be closer to 10 m. The hydraulic geometry method is most reliable for width, less reliable for depth, and least reliable for slope. FISRWG (1998, pp. 8–37) recommend this method only for the preliminary design stage, or for selecting a starting point for numerical modelling scenarios.

Analytical approaches can be used to design channel geometry, but the problem is that there are more unknowns than there are equations, so the system is indeterminate. This can be solved by computing the unknowns using empirical relations, making assumptions or ignoring some variables by simplifying the channel system. Tractive stress or tractive force analysis is relatively simple and can be applied only in very stable systems, where no bed movement is expected (see Section 7.4.3). In the more common situation of NCD, where a degree of bed mobility is desirable, other analytical approaches are available (see FISRWG, 1998, pp. 8-38 to 8-40). In some situations, because of the complexity of the problem, the high level of controversy or the potentially high cost of failure of the design, numerical modelling can be complemented by physical modelling (flume studies).

Under natural conditions, shorelines are seldom smooth. The natural variability in width, depth and bed topography should therefore be worked into the design (Iverson *et al.*, 1993; FISRWG, 1998, pp. 8–44). Small-scale diversity should also be a design consideration, with efforts made to keep channel edges 'rough'. Lewis and Williams (1984), for example, discuss the addition of shallow bays to the outside of meanders (see Figure 9.32). These create pond-like conditions that support a rich population of algae, plankton, water beetles and pond skaters, and may protect fish fry from predation by larger fish. Shallow bays can also function as stock watering points.

#### Reinstating Meanders

Meandering channel forms are described in Section 7.3.1. In comparison to a straight channel, the presence of meanders reduces the slope of the stream and thus its velocity and sediment-transport capacity. The reinstatement of meanders can accompany stream-rehabilitation works to increase both the quality and quantity of habitat. The slope and bed materials of the rehabilitated channel should be compatible with those of upstream and downstream reaches. For example, if roughness is too low or the channel too straight, water will exit the reach with excess energy, which can cause scour downstream and headcutting upstream [see discussion of Eq. (7.5)].

General relationships between meander parameters and other hydrological or geomorphic measures have been developed (Hasfurther, 1985; Morisawa, 1985). In re-establishing meanders a general rule of thumb is to simply set them five to seven stream-widths apart, one-half of the meander wavelength (Leopold *et al.*, 1964; see Table 7.1). It should be realized, however, that this rule might not be applicable to all streams. It is preferable to develop relationships for a particular region by studying the meander patterns on aerial photographs of undisturbed sections.

It should also be realized that it is normal for meander spacings to have a large amount of variability. There are few streams that look like perfect sine waves; instead, waves are superimposed on waves at different scales, with short, tight meanders and irregular shorelines superimposed on a larger snake-like form. This variability should be worked into a design by taking into account the location of trees, boulders and variations in soil or substrate.

Hasfurther (1985) and FISRWG (1998, pp. 8-34 to 8-35) summarized techniques for using meander parameters in channel design. Overall there appears to be seven methods of meander design:

1. *The carbon-copy technique*: The meander is reconstructed exactly to its pre-disturbance form. This assumes other factors affecting stream patterns (e.g. discharge, bed materials) remain the same. A variation on this technique is to 'carbon copy' the pattern of a similar reach in an undisturbed section.
2. *Established empirical relationships*: A number of relationships have been developed that relate meander wavelength, radius of curvature, meander amplitude and channel slope to discharge or width (FISRWG, 1998, pp. 8–34). Meander wavelength is determined on the basis of channel width or discharge.
3. *Regionally established empirical relationships*: Meander parameters are based on those observed for fully adjustable stream sites within the local region and applied to the affected site.
4. *Use nearby undisturbed reaches*: Hunt and Graham (1975, cited in FISRWG, 1998, pp. 8–35) and Brookes (1990) described the use of nearby undisturbed channels as templates for the project reach.
5. *'Natural' approach*: The stream is allowed to seek its own path. The disadvantage of this method is that it may take a long time for the stream to reach a stable form, and high erosion rates and sediment movement may occur.
6. *Slope first method*: Hey (1994a) recommended designing meanders by first selecting a mean channel slope based on hydraulic geometry.
7. *A 'systems' approach*: This combined approach was advocated by Hasfurther (1985), and includes an analysis of undisturbed meanders, an evaluation of the geomorphology of the disturbed area and consideration of the interaction between the stream and the surrounding areas.

Equation (7.5) should be used as a guideline for modifying the design if streambed conditions in the new channel (e.g. sediment size) are different from those in the reaches used for guidance on meander design. Some self-adjustment of the channel after construction should be expected; however, bank stabilization should be considered in areas where excessive adjustment of the channel is undesirable.

In river engineering work for improvement of flood conveyance, meanders are often eliminated. Lewis and Williams (1984) suggest an alternative method of cutting a bypass channel across the meander neck and leaving the meanders as refuge areas for wildlife. The original meandering channel continues to carry some portion of the flow, with the bypass channel conveying the peak discharges (Figure 9.33). This improves flood conveyance and minimizes erosion within the meander. The bypass channel can be designed to carry a small proportion of the flow year round or to remain dry until floods occur. An example of this approach to meander reinstatement can be found in Kern (1992). Dry bypass channels may require maintenance by grazing or mowing to keep plant cover under control, unless the resistance effects of vegetation are included in the design. Engineering input is needed in selecting the stage at which water enters a dry bypass channel, as poor design may lead to scour or silt deposition at the entryway. Bank protection at the entryway and exit may thus be required. As shown in Figure 9.33(b), weirs or headgates can also be built for a more precise control of flow into the bypass or into the meander.

Horner and Welch (1982) provided a case study from the Pilchuck River in Washington, USA, in which a segment of the stream channel was reconstructed to bypass a new section of highway. The original bends in the river reach were replaced with an S-shaped meander that retained the original stream length. The bases of the banks were stabilized with rock, and grasses and trees were planted along the banks and tops of the slopes. Substrate material of a variety of sizes was placed on the bed of the new channel to create habitat diversity, and large rocks were used for directing the streamflow to scour out pools.

The new channel was left dry for about a year before the streamflow was gradually diverted into it during the following summer. There was a relatively rapid recovery of the bed topography and substrate to that of the original channel, accompanied by a rapid development of extensive and diverse macroinvertebrate and fish populations. Monitoring of fish and invertebrates was conducted over a 5-year period, and showed no deterioration in size, diversity or quantity in the reconstructed channel (Horner and Welch, 1982). On private lands, especially where the stream divides two properties, changing the stream course may meet with resistance because it can mean the loss of land to one landholder and a gain to the other. Thus, the reinstatement of meanders may require negotiation and sufficient verification of potential benefits.

### Re-establishing Pool-riffle Morphology

A meander design would be incomplete without considering the 'vertical meanders'—the pool-riffle sequences.

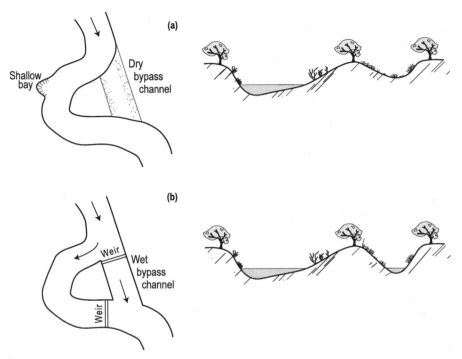

**Figure 9.33.** *Meander modifications, showing plan and cross-sectional views: (a) shallow bay and dry bypass channel and (b) wet bypass channel with optional weirs for control of flows. Adapted from Lewis and Williams (1984). Reproduced by permission of the RSPB*

Section 7.3.2 describes these features and their value to the instream biota. Energy can be dissipated in the turbulent flow between and over the larger rocks of riffles. Thus, pool-riffle structures can be added to straightened reaches to dissipate the energy that would have been lost over a longer meandering reach.

As with meander re-establishment, historical or regional patterns or rules of thumb (e.g. five to seven channel widths) (Keller and Melhorn, 1978; Hey and Thorne, 1986) can guide the placement of riffles, but Roy and Abrahams (1980) warn that pool-riffle spacing is highly variable within any given channel. Average riffle spacing is often (but not always) half the meander length. Leopold *et al.* (1964) cite a study where a stream was dredged by a dragline, but the operator was instructed to leave piles of gravel at about five to seven stream widths apart. These 'riffles' smoothed out over a few flood seasons and then remained stable. A line through the crests of the riffles should follow the general slope of the channel. As these

will act as hydraulic 'control sections' (see Sections 5.6 and 6.6), their effect on flood flows should be considered when deciding on an appropriate height. The slope of the downstream end of the riffle, the 'tailrace', should be similar to that in natural channels, and allow fish passage. For example, on the third-, fourth- and fifth-order streams surveyed in the Acheron River basin (Figure 4.7) the average riffle slope was 2.3%, or about 1:43.

Pools and riffles can be established by (1) excavation, (2) placement of gravel, cobbles and/or boulders to form riffles and an upstream pool, (3) building low weirs or debris dams or (4) strategically placing instream structures so the stream will 'do the work' to scour out pools and create gravel bars. Again, knowledge of morphologic processes is necessary to prevent excavated pools from filling up on the next flood event or riffles from becoming buried in mud. Following natural pool-riffle patterns in the design, with allowances for local variation such as the presence of large boulders or snags and massive roots of bankside trees, can help to ensure stability and longevity of the rehabilitated form.

The first three alternatives will produce immediate effects and may thus be looked upon more favourably by landholders and funding agencies than the fourth. With the first method it may be possible to use material excavated from a pool to build a riffle if the particle sizes are not too small for riffle-living invertebrates; otherwise, it might be employed to create levees or earthen current deflectors (Section 9.6.5). Allowing the stream to do its own excavation work, however, has its advantages in cost as well as in the distribution of substrate materials. The stream will be more selective about what it transports than a scoop shovel. Outhet *et al.* (2001) used an undisturbed river as a template for designing and installing log sills (low energy reaches), and log pin-ramps and rock sills (high energy reaches) to stabilize riffles in gravel bed streams in New South Wales, Australia.

Lewis and Williams (1984) give an example of reinstatement works performed on a channelized stream in Wales which had previously been a salmonid spawning and rearing area. Spawning riffles were created by removing the fine gravel substrate to a depth of 0.3 m and replacing it with a suitable rubble mixture. Stabilizing boulders were placed at the lower end of the riffle to prevent downstream movement of the gravel. The riffles were combined with groynes, instream placement of scattered rocks, revegetation and bank stabilization.

The use of natural channel characteristics in stream-rehabilitation works was described by Newbury and Gaboury (1988). They provided a case study from the Mink River in Manitoba, Canada, in which pool-riffle structures were used to improve stability and fish habitat in a channelized stream. The stream had been channelized and straightened from a tightly meandering form to increase the channel flood capacity. The channelized section was narrower and deeper than the natural channel, and continued to degrade over a period of 30 years, causing the banks to slump and steepen. The degraded reaches were selected for rehabilitation in 1985–86. A series of riffles was constructed in each reach using cobbles and boulders collected from surrounding fields (Figure 9.34). The downstream tailrace was designed to dissipate energy in the reach; its low slope was consistent with natural riffles which did not obstruct upstream fish passage. A 'V'-shaped crest concentrated low flows in the centre of the channel. Riffles were spaced approximately six times the predevelopment bankfull width, creating pools which overlapped part of the next upstream riffle. After a bankfull flood, none of the riffles were displaced. A scour hole formed in the pool downstream of each riffle. In comparison, bank erosion and slumping were severe in the un-rehabilitated sections. Walleye, a game fish species, were observed using the eddies created both upstream and downstream of the reconstructed riffles.

Pool-riffle structures can also be used to dissipate energy and control headward erosion as a 'soft engineering' solution, with environmental benefits. For example, on the Bunyip River in Southern Victoria, Australia, channelization resulted in headward erosion which contributed excessive sediment to Western Port Bay and threatened an upstream water supply pipeline. An approach involving the use of five rock drop structures and floodplain remoulding was chosen over other options, including a single large concrete drop, because it caused minimal adverse impact, provided fish passage and resulted in cost savings. Angular granite rocks of varying sizes up to 1.5 m were used to create the structures, which each had a 1:3 (33%) upstream slope and a 1:12 (8.3%)

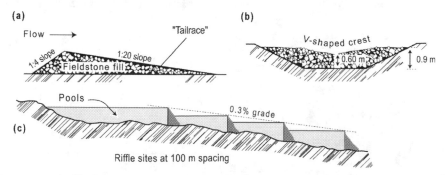

**Figure 9.34.** *Reconstructed riffle design used by Newbury and Gaboury (1988) on the Mink River in Manitoba, Canada: (a) cross-sectional view through riffle, (b) front view and (c) riffle placement, showing pools formed behind each riffle. Modified from Newbury and Gaboury (1988). Reproduced by permission of Canadian Water Resources Association*

downstream slope. Together, the five structures of varying heights took out 8 m of head over about 700 m of stream length (S. Brizga and Associates, 1998).

### Pool Cascade Structures (Artificial Weirs)

The pool cascade forms of high mountain streams (Montgomery and Buffington, 1997) can also serve as a template for rehabilitation designs. In these streams a 'plunge pool' absorbs the energy of the cascading water, also oxygenating it through turbulent action. Weirs are a low profile dam that crosses the channel (Wesche, 1985). They create a pool above the structure, and also downstream of the structure, due to scour. The stable surface can form a site for colonization by mosses, lichens and encrusting algae, which in turn supply food or habitat for organisms higher in the food chain.

Weirs are probably best used in environments where debris dams or large piles of boulders would have occurred naturally; i.e. small, steep-gradient headwater streams. Design of the weir should be consistent with the stream's biology and hydrology. For example, it should not form a barrier to migrating fish species during critical times of the year. One option is to design the weir such that it is underwater during high flow to permit upstream migration, with a notch to concentrate low flows over a 'spillway' into a downstream pool that can provide refuge from summer droughts. Shields *et al.* (1992) depicted four different arrangements, full width, upstream-V, downstream-V and partial width, all creating different pool formations. Weirs may be made of:

1. *Stone walls*: Natural stones with crevices and rough surfaces for invertebrate shelter are preferable to smooth bricks, and are aesthetically more pleasing. The stones can simply be stacked and held in place by stakes of metal or wood, or mortared together for a more permanent structure.

2. *Gabions*: Gabions are wire mesh baskets filled with cobbles and gravel from the streambed. They are relatively low-cost and do not require heavy equipment. Their best application may be in lowland gravel-bed streams where larger rocks are uncommon. Their initial artificial appearance is improved when they become silted and overgrown. However, they can become aesthetically unattractive and a hazard when the wire mesh rusts and begins to break up.

3. *Logs*: As shown in Figure 9.35, different configurations can be used depending on the log size, stream size and desired weir height. More elaborate structures can be fashioned log-cabin style to create boxes or 'cribs' which are then filled with stones (Strom, 1962). Logs will preferably be obtained from local downed trees, although planks and other commercial timbers such as railroad sleepers ('ties') can be used. Weirs can also be treated using some of the 'wicker' structures of woven branches as described in the section on bank stabilization.

A review of a number of successes and failures of these structures is given by Wesche (1985). He recommends that a good location for weir placement is in a straight, narrow reach at the lower end of a sharp break in gradient. Both ends of the dam should be anchored into stable banks and the base sunk into the streambed. Weirs may require bank-stabilization works to prevent erosion around the plunge pool and end-cutting around the weir. The addition of rocks below the overfall can help to dissipate some of the erosive energy. In larger streams, weir design should be done with care to avoid creating 'roller' currents downstream which are hazardous to swimmers and boaters who can be pulled under by the strong currents.

Wesche (1985) recommends that the low-flow spillway should be located near the thalweg line of the channel to preserve the natural flow path. Substrate materials can also be packed against the upstream face of the structure to

Single-log weir

Three-log weir

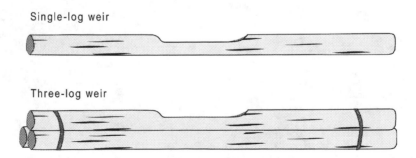

***Figure 9.35.*** *Log weir designs. Redrawn from Wesche (1985), by permission of the Butterworth-Heinemann*

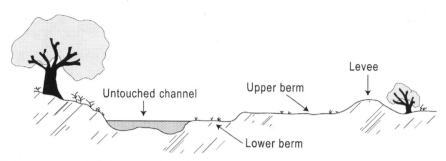

**Figure 9.36.** *Multi-stage channel design. Berms have been excavated and the material used to form a levee. The lower berm would be inundated several times each year, whereas the upper one might only be inundated once every 3–5 years. Modified from Lewis and Williams (1984). Reproduced by permission of the RSPB*

create an upward tapering 'ramp' (as in a beaver dam or riffle) to help to stabilize the weir and to direct flows smoothly over the top. Deposition of sediment upstream of the structure in subsequent years will further stabilize the weir. The gradual loss of the upstream pool should be considered part of the natural evolution of the streambed.

### Floodbanks, Floodways and Floodplains

Sandbags filled and placed in front of buildings, the levees alongside irrigation channels, and the dikes in Holland are all methods of preventing water from spreading into areas where it is not wanted. Levees or floodbanks are probably one of the oldest methods of stopping rivers from flooding adjacent lands. For protection against a flood of a given magnitude, the height of the floodbanks will depend on the distance they are set away from the stream, as this defines the amount of area available for floodwaters: high banks are needed close to the stream; lower banks further away. Levees have disadvantages because they disrupt the natural function of the floodplain in directing waters into the main channel, and in large floods when the levees are overtopped, they can actually prolong the flooding duration. If they are built close to the channel they can also result in the loss of wetland areas of ecological value (Ward, 1989).

From an ecological standpoint it is preferable to set levees further apart to preserve more of the floodplain habitat and to provide a buffer strip to reduce chemical and sediment input from lands which are farmed, inhabited, mined or logged. On the areas between the levees, land uses should be adapted to the flood-prone conditions, perhaps by being expendable or intermittent-use (e.g. cycle paths or seasonal grazing or agriculture).

Traditionally, materials for the construction of levees have been dredged from the stream itself, sometimes as part of a design to create a deeper, straighter flood channel. The use of smaller levees further from the stream would reduce construction costs as less material is required and would reduce impacts to the stream by discouraging the use of stream materials.

Multi-stage channels (Figure 9.36) are a common engineering practice. Usually, the 'normal' lower flows are contained within a relatively narrow channel, with the higher flows carried by the wider, leveed floodplain. Concrete-lined examples of this design are common in metropolitan areas. In a more natural environment the equivalent practice is to leave the stream undisturbed and cut 'berms' adjacent to the stream to increase the flood capacity of the immediate floodplain. Land use, geology and amount of flood mitigation required will determine the height, width and location of the berm(s). Lewis and Williams (1984) recommend cutting berms on one side only (sides may be alternated) to reduce disturbance of bank vegetation.

With this method, not only is most of the original channel habitat preserved, but also the low-lying berm may function as a wetland or damp meadow habitat. Vegetation on the berms may require periodic pruning unless the resistance effect of plants has been included in the design.

For streams that have already been converted to a trapezoidal floodway Bovee (1982) offered some habitat improvements that can be made without affecting the flood-carrying capacity. Even in these constricted channels, meandering and reworking of the bed materials will still occur, creating microhabitat patterns. Figure 9.37 illustrates the addition of cutout areas, boulders and vegetation to provide sites of lower velocity during flood

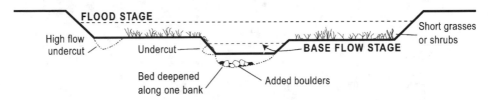

**Figure 9.37.** *Modifications of flood channels for habitat improvement. Adapted from Bovee (1982)*

flows. Bovee (1982) states that it is important to leave some unaltered trapezoidal sections to act as hydraulic controls.

*Bank Stabilization*

Bank slope and stability have an influence on channel form, vegetation growth and habitat for bank-living species such as otters, platypus and kingfishers. Bank stabilization is a traditional and common engineering practice, with rock rip-rap probably the most well-used technique. Artificial bank stabilization might be undertaken in situations where the rate of bank erosion has accelerated beyond the normal range (due to catchment-scale or more local disturbance), or the rate of bank erosion is considered undesirable because of the perceived economic consequences of losing land or public or private assets. Landholders are often particularly sensitive to 'loss' of productive land, even if this land is technically public land under their stewardship. Avulsion, where the stream changes its course by breaking out through the banks and cutting through the floodplain, is another dreaded phenomenon, because it can split properties and cause havoc for access. Stock exclusion and re-vegetation (Section 9.6.2) is a passive bank stabilization measure; however, this form of rehabilitation is often seen as an adjunct to harder forms of protection. A common justification for combining armoring with re-vegetation is that the bank protection will allow time for the vegetation to become well established (FISRWG, 1998, pp. 8-61). Different forms of bank protection materials offer different levels of resistance to erosion, thereby allowing control over the desired rate of mobility for 'deformable stream bank' (Section 9.6.1). Skidmore and Miller (1988) provide a methodology for designing deformable banks. We agree with Miller (1999) that real natural channel design should incorporate deformable banks, but there are many instances, such as near highly valuable assets, or within urban catchments, where absolute stability is an unavoidable requirement.

The methods included in this section are those that predominantly use natural materials, although artificial materials may also be acceptable. These applications are sometimes termed 'soft engineering' or 'biotechnical engineering', as opposed to 'hard engineering' which uses rip-rap, stone or concrete (Lewis and Williams, 1984). Soft engineering techniques can be less costly; however, they are more labour-intensive and may be shorter lived than more conventional engineering works. They also require more monitoring and maintenance. Anchoring of the works is important because if they drift away there is a small possibility that they will rack-up and create flooding or erosion hazards elsewhere, or endanger bridges. The benefit of soft engineering comes in the improvement of bankside habitat and water quality.

An important element in the design of stabilization works for habitat rehabilitation is non-uniformity. This preserves the natural variability, providing different surfaces and textures for colonization by a diverse array of organisms. Bank protection should ideally use local materials for the same reason, and to reduce costs. Flexible structures (e.g. plants) are preferable in some cases as they will absorb the erosive energy rather than reflect it to some other location; although in other cases this will be the intention (see current deflectors, below).

Bank battering or shaping to some standard slope (e.g. 1 in 2) is often a part of the channelization process. In severely entrenched streams this technique is performed to smooth down canyon-like banksides, improving access to the river for stock and anglers and facilitating revegetation efforts. Material from the upper part of the bank is piled at the foot of the cliff to form the slope. Tamping or battering the soil can help to prevent later slumping or erosion of saturated, loose soils. If lower soil layers are of low fertility it is advisable to set the topsoil aside until the slope is formed, then spread it across the surface. This is primarily an engineering practice, and expertise is required to determine the proper bank angle and degree of compaction.

Protection of the re-formed bank with mulch or cut brush weighted down with rocks will help to prevent rainfall-induced erosion until the vegetation takes hold. The toe of the slope may also require protection with

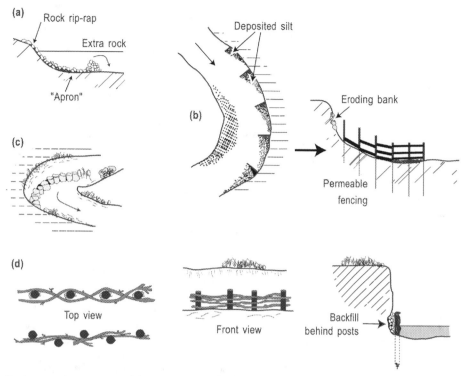

**Figure 9.38.** *Bank-protection measures: (a) stone 'rip-rap', (b) dikes or groynes, (c) vertical timber posts and (d) wicker spiling. Adapted from Lewis and Williams (1984). Reproduced by permission of the RSPB*

rocks or logs to prevent scour by the stream. Again, diversity is important in restoring 'naturalness'. Creative 'sculpting' of the channel in harmony with geomorphic processes might leave undercut banks and some cliff faces, preserve existing trees or shrubs, and even include a few flat ledges on the slopes for angler and wildlife access.

Bank-protection works are applied where bank erosion is likely to be highest, as on the outside of meander bends, on the bank opposite a current deflector, or on newly constructed bank or levee slopes. The selective use of bank protection only where it is most needed can provide stability at a lower cost as well as maintaining habitat diversity. Some of the discussed bank protection designs are shown in Figure 9.38. The longevity of the works will depend on the kind of materials that are used; for example, Australian eucalyptus timbers last a lot longer than their North American counterparts, but Australia lacks the versatile equivalent of the willow (Rutherfurd and Gippel, 2001). Texts on river engineering such as Petersen (1986) can be consulted for additional information. Design guide-

lines for bank protection are provided by U.S. Department of Agriculture (1996), and guidelines for including environmental features in bank protection works were provided by Henderson and Shields (1984). Hey *et al.* (1991) developed expert systems, reprinted in Hey (1994), to guide the selection of bank protection methods based on (1) bankfull velocity, slope and depth criteria, (2) expected bank failure type, soil type and drawdown rate and (3) wave and tidal controls.

### Stone 'Rip-rap' or 'Beaching'

Lining the banks with stone is feasible if a local source of rock is available and equipment is available—for transporting and depositing it. The pits in the stones and the nooks and crannies between them increase the amount of area for colonization by aquatic insects. Plants can take root between the rocks and rock ledges can be used as resting and roosting sites by waterfowl. Stonework is a relatively permanent option if the stones used are sufficiently large that they do not wash away during high flows. More angular stones will interlock and are thus more stable. The methods of Section 7.4 can be used as a guideline for determining the appropriate rock size. Placing a layer of fine gravel or stone under the larger rip-rap material will help to prevent erosion of finer bank materials from beneath the rocks. To avoid undermining at the toe of the rip-rap, Strom (1962) described the use of stone 'aprons' which extend well out into the riverbed. An extra pile of rocks can be added to the edge of the apron—these later fall in and continue protecting the bed if it degrades (Figure 9.38(a)). It is no longer considered necessary to line the full height of the bank with rock, as the most common source of failure begins with scour at the bank-toe. The finished job allows for easier revegetation of the top of the bank, the rock is often not visible, and the costs are much lower.

### Dikes, Groynes or Revetments

These structures are built in the stream bed, projecting from the bank into the flow [Figure 9.38(b)]. The difference between retards and groynes is that groynes are taller and extend a shorter distance into the stream. The primary function of a groyne is to protect an eroding bank. Groynes interrupt flow lines adjacent to the bank, creating a zone of slow flow near the eroding bank where sediment will deposit. Vegetation that establishes there will further enhance stability (Drummond *et al.*, 1995). Groynes may be permeable or impermeable. They can be constructed from the same materials used for constructing retards, except that impermeable groynes are made of quarry rock, concrete blocks or solid timber or concrete walls.

With proper design, silt will be deposited in the quieter water between the groynes. Vegetation can be planted in these silt beds which will eventually form floodplain benches. Normally, a series of groynes are used along a bank. Guidelines for determining the effective length of deflectors (see section below) also apply to retards. As a rule of thumb, a groyne will protect three to five times its own length of bank (Strom, 1962).

The angle of the retard to the flow has minimal effect on the relative velocity achieved. According to Drummond *et al.* (1995) there is a general belief among practitioners that retards direct flow to the opposite bank, and that this was partially related to retard angle. However, flume tests by Drummond *et al.* (1995) demonstrated that this was a water surface feature only, and there was no streaming (lateral deflection) of flow from a retard. The effect of a retard is not to deflect the flow to the opposite bank and create a scour hole. Rather the effect is to create a zone of deposition. If this zone occupies a large percentage of the channel a zone of scour will also be created in another part of the channel.

### Vertical Timber Posts

Wooden posts or piles driven vertically into the streambed can be used to protect the lower part of a bank, for example where a vertical cliff on the outside of a meander bend is to be preserved (Figure 9.38). Some bird species use these vertical cliffs for perching and nesting. Lewis and Williams (1984) recommend that at least half of the post should be driven into the ground against the base of the cliff and local materials banked against the area behind the posts to prevent washout.

### Woody Debris on the Bank

In areas where tree pruning, snag removal and brush cutting must be carried out these materials can be 'recycled' to the stream in the form of bank protection. Debris can also be used to slow the water, trap silt and sediment and permit the growth of river-edge plants. Cut logs and branches can be secured against a bank for protection or for filling scour holes. Large trees can be tethered to the bank with the root wad facing upstream, which is the most common natural position (Gippel *et al.*, 1996a). Alternatively the trunk can be partially buried into the bank by digging a trench, with the root wad facing upstream angled into the flow, and this seems to offer good bank protection while at the same time maintaining the stability of the log (Johnson and Stypula, 1993). Stakes driven into the streambed or heavy-gauge wire or cables looped around the debris and attached to firmly embedded stakes on the banktops may be necessary to hold the materials in place until anchored by siltation and vegetation growth. This is less of a problem in Australian rivers, because the native eucalyptus trees have a high specific gravity, and their low, sturdy branches act like anchors once buried into the bed material (Rutherfurd and Gippel, 2001).

### 'Wicker' Spiling and Hurdles

Smaller materials can be 'woven' together for bank protection. Spiling [Figure 9.38(d)] can be used to protect the base of steep banks and to create dikes on smaller streams. Stakes are driven into the ground and branches woven between them. Alternatively, posts can be staggered in two rows and brush piled between them. The distance between the posts should be consistent with the length and thickness of the 'weaving' materials. As with timber posts (above), the area behind the spiling should be backfilled. If appropriate for the region the use of willow stakes can create a semi-permanent, living form of bank protection.

Hurdles are woven of smaller branches and pegged in place against banks of milder slopes. These function as a 'mulch' through which plants grow, and degrade at about the time that the plants become established. Hazel, which will not grow from cuttings, is the usual construction material in the UK (Lewis and Williams, 1984). Hurdles are only suitable for small streams or low-turbulence reaches.

*Geotextile Systems*

Geotextiles are a well-established choice for erosion control on road embankments, usually in combination with seeding, or seedlings planted in slits in the fabric. The same technology can be applied to river banks, with the best choice of material being biodegradable jute or coconut fibre (Johnson and Stypula, 1993). 'Fibre-schines' are a cylindrical fibre bundle that can be staked to the bank with cuttings or rooted plants inserted through or into the material (FISRWG, 1998, pp. 8–65).

### 9.6.5  Instream Habitat Improvement Structures

Like channel modifications, instream structures are intended to create physical diversity in the stream. The preferred approach to habitat recovery is to restore the natural processes that create instream diversity, such as healthy riparian zone, and stable sediment regime. However, artificially emplaced structures can provide 'instant' habitat where the expected time period for natural recovery is considered too long. Considerations include biological requirements and the geomorphological and hydrological character of the stream. For example, there is little point in providing instream habitat if the biota are limited by unrelated factors such as water quality, temperature, lack of source for colonization, food supply, or passage. Rosgen (1996a) provides a table listing his classification scheme's suitability of 16 instream structures for the 17 channel types.

*Current Deflectors*

Groynes are known to produce scour holes, so they are used in river rehabilitation designs where an increase in hydraulic diversity is required. When applied in river rehabilitation, these structures are commonly known as 'deflectors' (or current deflectors) and they are often constructed of rock (Figure 9.39). In the USA deflectors are also known as spurs or spur dykes. Deflectors are probably the most widely used structure for habitat

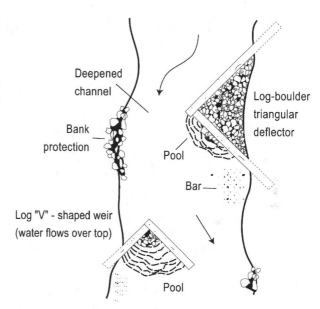

**Figure 9.39.** *Current deflector designs, adapted from Koehn (1987) and Wesche (1985), by permission of Butterworth-Heinemann*

improvement, and there is ample evidence in the literature that they can increase abundance, and perhaps diversity, of fish and macroinvertebrates (e.g. Brookes *et al.*, 1996; Shields *et al.*, 1998; Shields *et al.*, 2000c). Johnson and Stypula (1993) listed recommendations and options for the design variables for deflectors, but noted that there are no strict rules for designing this type of structure.

Deflectors are used to increase hydraulic diversity (depth and velocity), to deepen the channel, to develop erosion areas and to create pools (Brookes *et al.*, 1996). When constructed of rock the coarse material offers cover for benthic macroinvertebrates and fish (Yrjänä, 1998). Wesche (1985) found that the length of a deflector has to be at least half the width of the river to be successful in habitat creation. Brookes (1988) concluded that deflectors are only successful if the current velocity (presumably the mean) exceeds 0.6–0.9 m/s.

Shields *et al.* (1995) designed a series of deflectors for restoration of a 1 km reach of the Hotophia Creek in Mississippi based on general guidelines, previous experience, knowledge of the pre-disturbance channel, hydrological characteristics and biological requirements. Shields *et al.* (1995) did not use more sophisticated design procedures (i.e. numerical simulation) due to the lack of prototype data for model calibration.

Bovee (1982) mentioned that the U.S. Federal Highway Administration recommends that deflectors should extend no more than halfway across a channel and be no higher than 0.5 m. Wesche (1985) recommended a height of no more than 0.15–0.3 m above the low-flow elevation. Deflectors should be no higher than the top of the bank and slope downward to the nose to prevent undermining (Franco, 1967). The most common angle of deflector orientation is 45° to the flow. Johnson and Stypula (1993) reported that deflectors oriented upstream create deeper and larger scour holes than deflectors oriented perpendicular or downstream. However, Garde *et al.* (1961) contend that perpendicular deflectors create the deepest scour. When overtopped, upstream angled deflectors direct water away from the bank (towards the centre of the channel). When perpendicular and downstream oriented deflectors are overtopped, they can direct flow towards the bank and cause erosion (Hey, 1994b). Wing deflectors are triangular in planform and have characteristics of both up and downstream angled deflectors (Figure 9.39). Wing deflectors require a greater quantity of rock for construction than angled deflectors. The greater the channel constriction, the greater the scouring potential.

Recommendations in the literature for constriction range from 25% to 80% of the channel width (Wesche, 1985), with up to 75% required for overwidened reaches (Hey, 1994b). Garde *et al.* (1961) found that in straight reaches, the depth of scour was 0.2 to 0.5 times the effective length of the deflector.

Most spacing recommendations are for deflectors designed to direct the thalweg away from an eroding bank. When closely spaced, a buffer zone of eddies is produced between the deflectors. This flow structure is created when the spacing is 2–4 times the effective length, and it decreases as the deflector length increases beyond 20% of channel width (Klingeman *et al.*, 1984). Spacing deflectors so that their flow pattern interacts creates more scour and diversity of habitat than do individual deflectors (Brookes *et al.*, 1996). A spacing of between five and seven channel widths, which corresponds to the natural pool riffle spacing in rivers, was suggested by Brookes *et al.* (1996) to be appropriate for deflector spacing. In unconsolidated bed materials, it is recommended that rock deflectors be embedded for a distance equal to the height of the deflector. FISRWG (1998) warn that deflectors placed in sand bed streams may settle or fail due to undermining, and that in these applications a filter layer or geotextile might be needed beneath the deflector.

### Boulder Placement

The placement and arrangement of boulders in streams (Figure 9.40) can create diverse flow conditions, shelter for fish, amphibians, invertebrates and mammals, and 'patchiness' of streambed substrates. In smaller disturbed stream sites such as power-line crossings or disused road crossings, rocks can be added or rearranged to improve fish migration through the reach at low flows. The streambed topography is altered not only by the objects but also by scour, which can occur around the sides or under these boulders, and by deposition downstream. As obstructions, boulders cause energy to be dissipated through turbulence, and can be used to slow the total velocity through a reach (e.g. in a fishway).

Size of boulders depends on stream size, flow characteristics and bed stability, but some guidelines were provided by Industry Canada (no date). Typically, stones are 600–1000 mm in diameter and placed 1.5–2.5 m apart, adjacent to the thalweg. Twenty-five times the average diameter of the riffle substrate is a good rule of thumb for determining the size of stone required. Avoid adding boulders that cover more than 1% of the channel bed area as this will increase hydraulic resistance, cause sediment deposition and provoke river shifting. Diamond shaped boulder clusters are commonly used, but they can be arranged to create specific local flow variations preferred by individual organisms (Nowell and Jumars, 1984). With some 'playing in the water', the wake region behind one rock can be connected with that of the next rock downstream, to provide a continuous region of low velocity as shown in Figure 9.40.

In larger streams, boulders should be carefully placed, preferably based on the combined judgement of a biologist and a geomorphologist or engineer. Another consideration is whether the stream is to be used for boating (Wesche, 1985). If the placements are to be relatively permanent an estimate should be made of the size of rock that will move at high flows (see Section 7.4). Embedding the boulders in the streambed will reduce their tendency to roll away. If the object is to re-create a natural boulder pattern, natural arrangements should first be surveyed by observation or by formal methods (Section 5.2) to determine the distribution of boulder spacings. A design should be worked out in advance, especially where larger boulders will need to be moved with heavy equipment.

If boulders are not readily available it may be possible to 'mould' them out of a mixture of concrete and local gravels or cobbles. The wet mixture can be poured into burlap (hessian) bags, shaped and left to dry. Colonization of the surface by algae should soon eliminate the artificial look. Submerged log sections are another alternative.

### Fish Shelters

Fish shelters (Figure 9.41) are substitutes for natural overhanging banks. These provide shade, shelter from

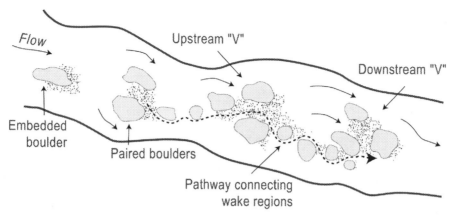

**Figure 9.40.** *Boulder placement patterns. Wake regions created by individual boulders can be connected to provide a continuous path of low velocity between boulder clusters. Adapted from Wesche (1985), by permission of Butterworth-Heinemann*

swift flows and protection from predators. Lewis and Williams (1984) stated that these shelters can be built above or below water level or made to float on the water surface. Covers of brush, either tied to the bank or woven like the hurdles mentioned in the previous section, are a simple form. A platform of split logs or planks projecting from the bank and secured to the streambed by posts or reinforcing rods ('rebar') is another design. Bovee (1982) recommended that pilings (support columns) should be about 2 m long for low-gradient streams. Concrete footings can be added to the pilings for more permanence. Earth, rock and sod can be added to the top for additional weight, to improve aesthetics and to provide opportunities for overhanging vegetation to become established.

Bovee (1982) stated that fish shelters should be located in the erosional parts of the channel such as the outsides of meander bends, otherwise they may be lost to siltation in depositional areas. However, a close look at small, undamaged streams will reveal that undercuts can occur almost anywhere along the length of a stream reach.

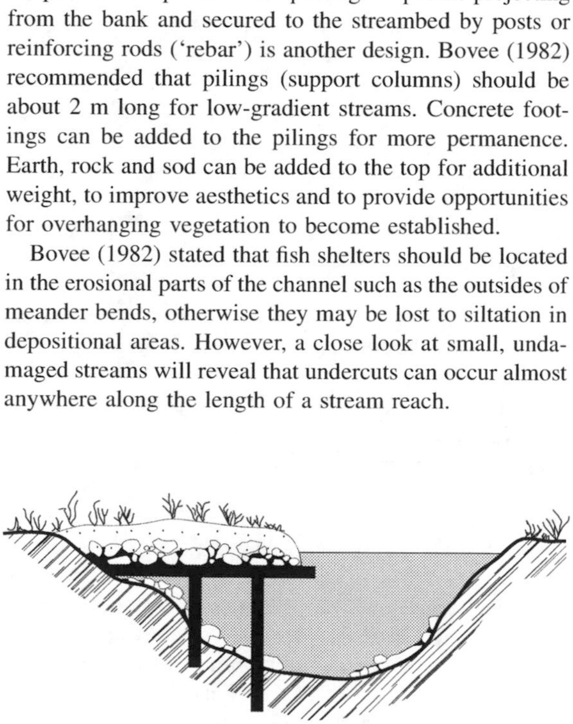

Fish shelters should be installed when the flow is low, both to facilitate construction and to ensure that the sheltered region contains water. Boulders can be placed under the overhang to enhance cover and protect the bank against erosion (Wesche, 1985). Fish shelters can also be attached to other structures such as current deflectors or weirs. Alternatively, current deflectors can be installed on the opposite bank to improve circulation of water under the shelter and lessen sediment deposition.

Although a fish shelter will form an obstruction at high flow, the narrow width should not increase flood hazards significantly. The main concern is their tendency to float or wash away during high flows. Because of this susceptibility they should perhaps be designed to break up during floods so that they do not create problems downstream. As Bovee (1982) mentioned, the failed structure itself may not form a large part of the debris load moving down a river, but if blockage problems occur, the structure can receive a disproportionate part of the blame!

### Large Woody Debris (LWD) Structures

LWD is generally considered to be any logs, stumps, rootwads and branches larger than 0.1 or 0.2 m diameter, and longer than 1 m located within a river channel (Gippel, 1995). LWD is recognized as an important structural and ecological component of many stream environments (Harmon *et al.*, 1986). Woody debris provides hard surfaces for attachment and growth of aquatic plants and invertebrates and also habitat for fish (Harmon *et al.*, 1986; Gippel, 1995). It has recently been argued that LWD (along with riparian vegetation) should assume a place

**Figure 9.41.** *Fish shelter design. Redrawn from Bovee (1982), after White and Brynildson (1967)*

beside the sediment and discharge regime as a primary control on the morphology and dynamics of river systems (Montgomery, 2003).

Because many streams have been de-snagged, or denied a source of woody debris through degradation or removal of riparian timber, reintroduction of large woody debris is often an important aspect of stream rehabilitation. As well as devising riparian management strategies to ensure an ongoing supply of wood to streams, many agencies are attempting to speed the recovery of woody debris loadings by direct reintroduction of wood and wood structures into channels. Debris is reintroduced to rivers in four main forms: as individual logs; as log arches; as log jams; and, as instream structures such as current deflectors, notched weirs, cover devices and revetments along banks (see above).

Debris is most commonly reintroduced as a series of individual logs. Cross-stream logs are normally placed in small to moderate sized streams and rivers (5–20 m wide) for the purpose of providing habitat for fish and macro-invertebrates, both directly, and indirectly through the bed structures that are created by local hydraulic interactions.

Pendants or sweepers are whole trees anchored by the base to a stump, steel post or deadman on the bank using cables. The single point of attachment allows the sweeper to move up and down with flow variations. Naturally occurring sweepers (sometimes termed submerged brush shelters) are common in the upper reaches of Ontario streams (Ontario Streams, 2003). Whole trees provide a wide range of hydraulic habitat, and the branches tend to collect fine organic debris, further adding to the habitat complexity. These structures are intended to attract juvenile fish by providing dense cover and a rich food supply.

Log arches are V-shaped structures comprising two logs angled diagonally to the flow, with the upstream ends cabled together in the centre of the channel and the two downstream ends anchored on opposite banks. A cross-brace can be added to the downstream end of the structure if effective anchoring cannot be achieved. The apex of the structure is excavated into the stream bed flush with the surface of the bed, with the downstream butt ends buried halfway into the stream bed. These ends are further secured by cabling them to stumps, trees or deadmen on the banks. A deadman is buried 2 m deep into the bed to anchor the apex of the arch (Poulin & Associates, 1991).

Abbe and Montgomery (1996) showed that natural accumulations of debris in log jams in large alluvial channels could be extremely stable, with life expectancies exceeding those of many river engineering projects. Abbe *et al.* (1997) used the characteristics of naturally occurring stable log jams, combined with geomorphic and engineering principles, to develop design guidelines for engineered

log jams (ELJ). Some ELJs constructed in the USA are quite large structures. ELJs are appropriate in large lowland rivers, and are designed to provide habitat and to protect banks from erosion. ELJs are best placed next to pools on the outside bends of meanders and along the banks of deep runs (Ontario Streams, 2003). Guidelines for designing ELJ structures for incised streams were provided by Shields *et al.* (2001) (Figure 9.42).

There is considerable literature on in-stream structures that are manufactured from wood, such as crib walls, check dams, deflectors and log weirs (e.g. Swales and O'Hara, 1980; Wesche, 1985; Gore and Shields, 1995; Charbonneau and Resh, 1992). These structures certainly increase the volume of large wood in the stream, but they are not expected to provide the hydraulic, geomorphic and habitat functions of natural large woody debris formations.

LWD occurs naturally at loadings of between 0.001 and 0.1 $m^3/m$ in relatively undisturbed streams and rivers in various countries, including Australia (Gippel, 1995, Gippel *et al.*, 1996a). The spacing of debris is variable, ranging from one log per kilometre to more than one log per metre in lowland rivers (Gippel *et al.*, 1996a). Ontario Streams (2003) noted that a density of 12–15 logs per 100 m was not uncommon. Poulin & Associates (1991) reintroduced debris at a spacing of one structure for every two bankfull channel widths, based on Hogan's (1986) observations of debris density in small undisturbed streams in the Queen Charlotte Islands, Canada.

There are four main philosophies for placement of debris within the channel: random log placement (Booth *et al.*, 1996; Hilderbrand *et al.*, 1997); habitat manipulation (Poulin & Associates, 1991; Hilderbrand *et al.*, 1997); mimic natural analogs (Booth *et al.*, 1996), and; minimizing the hydraulic impact on the stream with respect to the magnitude of the flood afflux generated (Gippel *et al.*, 1996b). Random placement is carried out with the expectation that the debris will be relocated by subsequent high flows. Creation of specific habitat features (such as pools) through strategic debris placement is not well understood, and trials report variable results (e.g. Poulin & Associates, 1991; Hilderbrand *et al.*, 1997).

Debris is most commonly reintroduced as a series of individual logs. There are two basic reintroduction philosophies: anchored (Ontario Streams, 2003) and unanchored (Booth *et al.*, 1996). Placement of anchored individual debris might be the preferred method in areas where there is concern over the potential for negative impacts of log jams or debris accumulations. The common concerns are increased flooding, impediment to fish passage, catastrophic failure during floods that causes severe channel erosion (DeBano and Heede, 1987), or drift accumulations on bridge piers (Diehl, 1997).

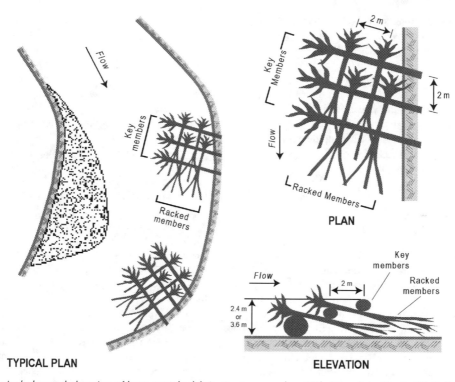

**Figure 9.42.** *Typical plan and elevation of large woody debris structures used on Little Toposhaw Creek, Mississippi. Source: Shields et al. (2001)*

Ontario Streams (2003) recommended the use of aircraft cable and buried T-bar posts as anchors. Their preferred configuration was to allow the log to float above the bed. Millar (1997), Slaney *et al.* (1997) and D'Aoust (1998) provide design guidelines for the mass of ballast required when anchoring debris of various diameters and lengths with boulders. ELJs are designed to be stable under at least moderate-sized floods. The literature on ELJs has not indicated a need for anchoring, but this should be reviewed in situations where the bed is highly mobile during floods.

### 9.6.6 Evaluation of a Range of Stream Rehabilitation Projects

Most rehabilitation projects are not rigorously evaluated to exacting scientific standards. It would be normal to make casual observations of fish or bird presence, or report general approval or otherwise from the local community, but this does not constitute scientific evaluation. Many projects are driven by practical, local enthusiasm, not scientific curiosity, so the budgets do not generally allow for careful long-term assessment. In any case, it would be an inefficient use of research resources to scientifically assess every project—better to concentrate resources on some key projects or processes that are likely to reveal important new information. Some projects are set up as adaptive management experiments, but for reasons of scale, expense and practical limitations, they do not always include controls, replication, multiple treatments, randomization, or other features commonly expected of traditional experimental design and frequentist (classical) statistical methods. In large rivers it may be impossible to find controls, and replicates are problematic given that potential sites along a river are connected longitudinally. Despite these limitations, practicality needs to be balanced with statistical rigour. Statistical analysis allow managers to detect change due to management, and to distinguish it from background variation and sampling errors. The statistical methods to be used in analysis of data must be agreed as part of the experimental design process, i.e. before any measurements are made (Nyberg, 1998).

One limitation of classical statistical methods such as ANOVA and regression analysis in adaptive management experiments is that they are not designed to answer management questions, such as "what is the probability of a 50% increase in fish abundance after installing five large woody debris strutures?' When classical methods are not appropriate, there are alternative options for statistical analysis (Sit and Taylor, 1998). Bayesian statistics directly analyse the probability of a hypothesis being true, allowing scientists to formally update their beliefs in a variety of experimental and non-experimental situations. Bayesian analysis can assign intermediate degrees of belief or probability to hypotheses, unlike the all-or-nothing inferences 'reject/retain the hypothesis' (Anderson, 1998). The Bayesian approach provides a way of incorporating prior knowledge with new knowledge (Bergerud and Reed, 1998).

### Riparian Vegetation Management

Evaluations of riparian rehabilitation projects have examined the effects of revegetation and/or exclusion of livestock. On ten streams in Oregon, prudent riparian management by the U.S. Bureau of Land Management transformed intermittent streams into perennial ones (Stuebner, 1988). The improvement in vegetation made the streams act more like reservoirs, retaining the runoff and then releasing it more slowly. Historically, plants were routinely used for stabilizing banks and deflecting flows, with regular maintenance to monitor and modify their growth as needed. A good example is the Tennessee Valley Authority floodplain restoration projects undertaken in the Southern USA during the 1940s (FISRWG, 1998, pp. 8.18 to 8.19). Recent surveys demonstrated that within a period of 50 years the planted communities had developed structural and understory conditions that resembled those of natural forest stands (Shear *et al*, 1996).

Green and Kauffman (1995) documented significant increases in the mean height of woody plant species ten years after grazing ceased. A long-term study of grazed and fenced riparian areas in Colorado by Schulz and Leininger (1990) found that after 30 years the total vascular vegetation and shrub cover was greater in the exclosures compared to the grazed areas, while forb cover was similar between treatments. Exclosures had nearly two times the litter cover of grazed areas, which in turn had four times more bare ground. Another long-term study in New Zealand (Howard-Williams and Pickmere, 1994) documented a continuous vegetation successional change, with increasing biomass and generally increasing diversity through time. Webb and Erskine (2003) evaluated 10 trial sites of in-channel bench revegetation on streams in the Hunter Valley, New South Wales, Australia. Regeneration of planted species was generally poor, and this was attributed to intense weed growth and consequent low levels of light at ground level. The intense weed growth, in turn, was due to a number of factors, such as 'edge-effects' (Fox and Adamson, 1986) created by the narrow planting widths of the sites, frequent flood disturbance (Tickner *et al.*, 2001) of the riparian zone, and a lack of competition from native plants owing to the sparse 4-m spacing of planted trees and the low survival rates of many species.

Williamson *et al.* (1996) visually mapped an area of eroding streambank of the Ngongotaha Stream, New Zealand, before and after riparian restoration and the exclusion of stock access. Prior to riparian restoration, approximately 30% of the area of streambank surveyed was eroding. Fourteen years after riparian restoration, this had declined to 3.7%. Elimination of grazing from four streams in Oregon resulted in bankfull widths becoming 10%–20% narrower (Magilligan and McDowell, 1997), which is consistent with the findings of Platts (1991). Myers and Swanson (1996) observed increased stream stability over a 14-year period of livestock exclusion on a stream in northwest Nevada. Increased stability also occurred on a similar stream where periodic grazing was allowed, but this stream was more susceptible to erosion during floods. A study of eleven fenced sites with controls in the Blue Mountains of eastern Oregon (McDowell and Mowry, 2002) found that the treatment reach was narrower, deeper, and/or had more pool area than the grazed reach. The response of maximum pool depth and residual pool depth was mixed, with pools deeper in the treatment reach at about half the sites, but equal or shallower at the other sites. The variability of the results was not easily explained. In contrast to the above studies, Kondolf (1993) observed no channel narrowing in areas that had been fenced from cattle for a 24-year period in California, and Williamson *et al.* (1992) found no width change for New Zealand streams that had been fenced for two years. Murgatroyd and Ternan (1983) compared the erosion rates of stream banks in forested and moorland areas in a catchment in England. The forested areas, which had been revegetated in the 1930s, were significantly wider and had much higher erosion rates compared to the moorland streams. This was attributed to the process of suppression of protective grass through shading of the forest and creation of debris dams that caused scour.

Van Velson (1979) found average summer water temperatures dropped from 24 °C to 22 °C one year after stock exclusion. Kauffman and Krueger (1984) reported data that showed a temperature differential up to 7 °C and a lower daily range in an area that was rested for four years and then grazed for 3 months, versus an area grazed for 6 months. Large reductions in suspended solids loads of 40%–90% have been reported after stock exclusion, with the effect attributed to increased bank stability (Kauffman and Krueger, 1984; Owens *et al.*, 1996; Line *et al.*, 2000; McKergow *et al.*, 2001; Parkyn *et al.*, 2003). Stock exclusion and revegetation is less effective in reducing nutrient exports (McKergow *et al.*, 2001; Parkyn *et al.*, 2003). Howard-Williams and Pickmere (1994) found that nutrient removal increased for the first 12 years after revegetation and fencing, but then shading changed the species composition and reduced the nutrient uptake.

Behnke (1977) and Platts (1991) believed that the best opportunity for increasing populations of resident fish species in western North America was to fence and improve riparian habitats that have been adversely modified by livestock grazing. Garcia de Jalon (1995) considered the intermediate stage in plant recolonization (dominated by grasses and low bushes) the most beneficial for fisheries. Bowers *et al.* (1979) reported an average increase in fish production of 184% from five independent studies where livestock grazing was eliminated, and Howard-Williams and Pickmere (1994) reported increased trout spawning within six years of fencing and revegetation. Parkyn *et al.* (2003) assessed nine riparian buffer zone schemes in North Island, New Zealand that had been fenced and planted (age range from 2 to 24 years) and compared them with control reaches upstream or nearby. Significant changes in macroinvertebrate communities toward 'clean water' or native forest communities did not occur at most of the study sites, despite water quality improvements and more stable banks. It seemed to Parkyn *et al.* (2003) that restoration of in-stream communities requires canopy closure, with long buffer lengths, and protection of headwater tributaries.

Thexton (1999) described a revegetation project on the over-widened, sand-bedded Genoa River, Victoria, Australia. The objective of the project was to stabilize the bed of the river using natural regeneration of plants through erosion control and fencing (termed Assisted Regeneration). A moderate-sized flood removed a large proportion of the regenerated plants from the regeneration demonstration site, while those out of the main flow survived. Observations suggested that another cycle of regeneration began quite soon after the flood. Persistence in weed control was critical for success of the Assisted Regeneration approach, because the rivers of temperate Australia are now among the most weed rich of any environment. In creating conditions suitable for the regeneration of indigenous species, conditions are also created for the regeneration of weed species. Some weeds can out-compete the indigenous species.

## Fish Reintroduction

The practice of introduction and relocation of fish is widespread, although success rates are generally poor, and it can negatively impact the native fauna (Holcik, 1991; Billington and Herbert, 1991). Many introduced species are more adaptable to a flow regulated, or physically and/or chemically modified system, and they can out-compete the resident native fish populations. Dodge and Mack (1996) viewed supplemental fish stocking as potentially damaging to the wild populations, which can lose fitness through adverse effects on the genetic diversity of the native gene pool (Martin *et al.*, 1992). For example, the Glomma River, the largest river system in Norway, is stocked with hatchery fish reared from native brown trout to compensate for inefficient performance of fishways. A study by Linløkken *et al.* (1999) found that interbreeding had genetic effects on wild trout populations.

## Fishways

The first 44 fishways built in New South Wales, Australia between 1913 and 1985, did not perform effectively, usually because of design failures (many were based on North American designs) (Thorncraft and Harris, 2000). Mallen-Cooper *et al.* (1995) used some fishway data to illustrate an important difference between what they termed 'assessment' (understanding process using well-structured experimental design and analysis) and 'monitoring' (counting or describing). The first fishway in the River Murray, Australia, was installed as an experiment at Euston Weir in 1938, using a North American design intended for salmon. A monitoring survey at the fishway between 1938 and 1942 (55 months) counted 16 000 native fish at the top of the fishway, which was later reported by Petts (1984) in an internationally recognized text as 'large numbers' in the context of an effective fishway. However, a later assessment of the fishway by Mallen-Cooper and Brand (1992) found that there were 100 times more golden perch at the bottom of the fishway compared to the top. This assessment of the fishway was able to put the 'large numbers' observed through monitoring into perspective, and exposed the performance of the fishway as inadequate.

Australian fishways are now designed to be compatible with the swimming abilities and behaviour of Australian native fish. This is an example of management responding to new information that could only be provided by proper scientific investigation. The vertical-slot type Torrumbarry Weir fishway on the River Murray, built in 1989–1991, was designed on the basis of the burst swimming ability of adult golden perch and silver perch observed in a laboratory study (Mallen-Cooper *et al.*, 1997). A study of the performance of this fishway using pre- and post-fishway surveys and daily monitoring of the number, size class and species that passed found that the fishway was highly effective, except for Australian smelt (Mallen-Cooper *et al.*, 1995). The study also found that the burst swimming capabilities of adult fish had been underestimated by the laboratory study, so the fishway design criteria were conservative. This was serendipitous; because another surprise finding was that the majority of fish using the fishway were immature fish, which have less swimming ability than adults.

Stuart and Mallen-Cooper (1999) assessed the effectiveness of converting a salmonid-type pool-and-weir fishway into a vertical-slot design on a tidal barrage on the subtropical Fitzroy River, in Queensland, north-eastern Australia. The study showed much greater potential for success with the vertical-slot fishway, with its lower water velocities and turbulence, as relatively few fish negotiated the original pool-and-weir design. Despite the improved performance, juvenile crabs and long-finned elvers did not ascend the vertical-slot fishway, and the passage of smaller size classes of immature fish was restricted. Some fish took a long time to ascend the fishway, or remained there, causing crowding and reducing its capacity.

Jensen (2001) pointed out that while vertical slot fishways are generally regarded as a success in Australia, the criteria for measuring success are quite different to those applying in the mid-lower Mekong River basin (shared by Vietnam, Cambodia, Laos and Thailand). In Australia, achievement of any degree of fish passage with a new fishway is often regarded as a success because it is being compared against the previous situation where the dam (with no fishway or an inefficient fishway) allowed zero or low rates of fish migration. In the lower Mekong, migrating fish play an important role in the food security of the people living in the area, and the rate of fish migration can be extremely high. Jensen (2001) was concerned that even a well functioning vertical-slot fishway would not have the capacity to operate with 100% efficiency. He warned that building a dam on the main stem of a free-flowing river could have serious social and economic consequences, because it would almost certainly cause deterioration in fish numbers.

Early data comparing pre-dam migration data with post-dam counts of fish ascending fishways on the Columbia-Snake River system suggested that the facilities were very effective (e.g. Loder and Erho, 1971). However, the numbers of fish migrating naturally fluctuates from year to year, so comparisons of pre- and post-dam annual migration data over short periods reveal little about fishway efficiency. Later surveys demonstrated that problems with operation and design could result in delayed passage, inefficient passage and mortality (Basham, 1985; Northwest Power and Conservation Council, 1994; Bjornn *et al.*, 1999; U.S. Army Corps of Engineers, 2002b; U.S. Army Corps of Engineers, 2003). Also, the fishways were designed for anadromous fish, not resident fish such as bull trout. There is a possibility that the dams have isolated populations of resident bull trout from migratory bull trout, potentially causing a loss of genetic diversity (U.S. Army Corps of Engineers *et al.*, 2000). Screen sytems have been installed to guide fish away from turbines, which can cause mortality. However, not all fish are diverted to the bypass channels, with the percentage efficiency ranging from 30% for fall chinook salmon to 80%–90% for steelhead salmon (U.S. Army Corps of Engineers 2002b). Agencies have combined to attempt to overcome these problems and improve fish passage in the Columbia-Snake River system (Federal Caucus, 2003). Baltic Salmon have suffered a similar fate to their Pacific cousins, with dams blocking or reducing access to their spawning grounds through inefficient fishways (Rivinoja, 2001).

Takahashi (2000) found that most evaluations of the effectiveness of fishways in Japan were based on experiments using stocked fish in the fishway to examine the possibility of their upstream movement. However, it was found that some fishways that appeared to be effective in trials performed poorly in practice. This was explained in terms of external factors that affect performance in the field, but which are not accounted for in narrowly defined evaluations of effectiveness. Nakamura's (1994) hydraulic studies explained the poor field performance of the bending-slope (fan-shaped) fishways used in Japan, and he recommended that they be supplemented with an adjacent alternative fishway of a different design. Oldani and Baigún (2002) monitored the operational performance of an elevator lift system on the Yacyretá Dam on the Paraná River, Argentina-Paraguay. One problem was that the system was mechanically inoperative for 30%–38% of the time during the period of greatest fish migration. They estimated a fish passage efficiency of 1.88% for all species and 0.62% for target species, which was regarded as inadequate to maintain populations of target species.

The reproductive strategies of neotropical potamodromous fish of the large Plata River basin (shared by Argentina, Bolivia, Brazil, Paraguay and Uruguay) involve upstream migration and passive downstream movements of eggs and larvae for dozens of kilometres

(Agostinho *et al.*, 2002). The reservoirs created by dams delay the passive drift of eggs and larvae, and expose them to intense predation and/or promote their settling toward deeper water, where the conditions of oxygenation are generally critical. For larvae that reach the dam wall, most of them cannot be attracted to ladders (elevators are designed only for upstream movements) or they suffer mortality if they pass through turbines or a spillway. For these rivers, fishways are an ineffective tool for fish migration (Agostinho *et al.*, 2002). A study by Fièvet (2000) of passage facilities for diadromous freshwater shrimps in the Bananier River, Guadeloupe, West Indies, found them selective and inefficient because they did not take biological criteria into account. Fièvet (2000) concluded that the benefits of such installations for shrimps were doubtful, and simple waterfalls flowing over spillways would be at least as efficient, if not more so.

### *Dam Removal*

There are only a few published studies of the impacts of dam removal (Kanehl *et al.*, 1997; Stanley *et al.*, 2002; Doyle *et al.*, 2003c). One of the major ecological justifications for dam removal is the elimination of barriers to fish migration. Stanley and Doyle (2003) noted the lack of published refereed articles documenting changes in population sizes of migratory species following dam removal, but movement of fish into formerly inaccessible reaches has been reported (American Rivers *et al.*, 1999; International Rivers Network, 2001). Following the removal of the Edwards Dam in Maine's Kennebeck River, USA, striped bass, alewife, shad, Atlantic salmon and sturgeon all traveled past the former dam site (International Rivers Network, 2001; American Rivers, 2003b). In France, the 1998 removal of two dams (12 m and 4 m high) on tributaries of the Loire River reportedly revitalized native shad, lamprey and salmon populations (Epple, 2000; International Rivers Network, 2001).

Fish and macroinvertebrates adapted to slow-moving water and silty sediments gave way to riverine taxa within a year of removal of two small dams in Wisconsin, USA: the Woolen Mills Dam on the Milwaukee River (Kanehl *et al.* 1997) and the Baraboo River dams (Stanley *et al.*, 2002). In both studies, the recovery of riverine taxa reflected both recolonization of individuals that had previously resided upstream or downstream from the dam and successful reproduction within this newly created habitat (Stanley *et al.*, 2002). Fish diversity in the former impoundment of the Baraboo River more than doubled, from 11 to 24 species, eighteen months after dam removal in 1997 (American Rivers *et al.*, 1999). However, Stanley

*et al.* (2002) noted that one year following Baraboo River dam removal, downstream fish assemblages experienced decreases in abundance and species number.

The nutrient-rich sediment that becomes available after dam removal offers opportunities for revegetation, but it is also prime habitat for invasion of weedy and exotic species (Shafroth *et al.* 2002). Stanley and Doyle (2003) observed abundant weed invasion at several Wisconsin dam-removal sites.

### *Bank Protection, Channel Works, Instream Habitat Structures and Combined Projects*

Most reviews of rehabilitation works are after-the-fact evaluations of small-scale projects. Twenty years ago, reviews of rehabilitation efforts in the Northwest USA (Reeves and Roelofs, 1982; Hall and Baker, 1982) recognized the general failure to document and evaluate the work. They also noted the probable bias in the published literature because of administrative or editorial decisions against publication of inconclusive or unfavorable results. Despite continued encouragement to integrate scientific assessment into rehabilitation works, little has changed since that time. This is not just the result of oversight, inadequate communication of the message by the scientific community or lack of funding. There is a pervading view in the community that stream rehabilitation is a purely practical, not an academic exercise, and the current level of scientific understanding is in most cases sufficient to enable rehabilitation works to proceed with confidence. Many reports of rehabilitation 'success' do not cite specific evidence, or use anecdotal accounts to support conclusions, rarely mentioning 'failure'. This is understandable, when the objectives of such reports is to muster community support, create enthusiasm and maintain funding and resources. Published scientific papers and reviews tell quite a different story, indicating that there is still plenty to learn about river rehabilitation.

The habitat-based approach to rehabilitation has been widely applied in the USA, on the assumption that this was the most important factor limiting the success of the biota (usually fish). The creation of instream habitat has become a major industry in North America, with most of the enhancements involving placement of boulder arrays or large woody debris. However, in many cases this approach has been shown to be unsuccessful. For example, Andrus (1991) reviewed over 1200 stream restoration projects undertaken in Oregon, many of which concentrated on creating pools for summer fish habitat. Recent research has shown that for many Oregon streams, pool habitat is not necessarily the factor limiting fish productivity and the focus for restoration of streams in Oregon is now the provision of cover as refuge for young fish during high winter flows (Andrus, 1991). Some activities undertaken in the context of rehabilitation, such as rock rip-rapping of banks and removal of log-jams, may have been harmful to fish (Andrus, 1991).

Beschta *et al.* (1994) provided a critical review of artificial habitat manipulation projects conducted in the USA, and Richardson and Hinch (1998) produced a similar review for British Columbia, Canada. Haltiner *et al.* (1996) and Kondolf (1998b) reviewed projects undertaken in California, and the review of Frissel and Ralph (1999) focused on the Pacific Northwest area of the USA. There have been a few poorly documented reach-scale projects, and very few catchment-scale efforts (Frissel and Ralph, 1999). The above authors of rehabilitation reviews were highly critical of the idea of structural habitat enhancement on the grounds that the projects typically focussed on physical habitat components in isolation rather than considering the entire ecosystem, and they failed to address the causes of degradation. Structural projects usually have short-term objectives, but fall far short of the long-term ecological requirements of habitat restoration. Structures often create features that do not integrate well with other landscape elements, and they do not emulate the riparian/aquatic functions associated with intact riparian plant communities. High failure rates and escalating costs of maintenance have been reported for many rehabilitation projects in catchments where high erosion and sedimentation rates, high peak flows, or other watershed alterations or stresses are pervasive (Frissel and Ralph, 1999).

Beschta *et al.* (1994) and Richardson and Hinch (1998) argued that without removal or significant reduction of anthropogenic activities that are adversely affecting the entire basin, the restoration of local aquatic habitats for fisheries and other organisms cannot be expected. These authors drew conclusions that are relevant to any planned stream rehabilitation project:

- Abusive land management practices cannot be mitigated by structural works.
- The restoration of healthy riparian vegetation is a necessary requirement for rehabilitation of aquatic habitats.
- It is important to maintain natural disturbance patterns (i.e. retain some geomorphic instability).
- Restoration can take a long time.
- Reliance on in-stream structures can sever linkages between terrestrial, riparian and aquatic ecosystems, and because of their size and permanence, cause a shift in channel morphology and loss of in-stream functions.

- Little is known about the impact of structural rehabilitation because of the lack of scientific evaluation.
- Existing monitoring of fish populations has generally failed to provide evidence that structural approaches are achieving desired goals.
- Non-natural materials should be eliminated from in-stream rehabilitation projects aimed at improving fish habitat.

Pretty *et al.* (2003) assessed the benfits of instream structures on fish populations in 13 lowland rivers in the UK. In general, rehabilitation measures increased depth and flow heterogeneity. However, total fish abundance, species richness, diversity and equitability were not significantly different between rehabilitated and control reaches. There were few significant relationships between the fish fauna and physical variables, indicating that increasing physical (habitat) heterogeneity does not necessarily lead to higher biological diversity. Pretty *et al.* (2003) cautioned against what is the very common use of physical responses to rehabilitation as a surrogate or assumed predictor of ecological response. Given the poor response from in-channel devices, the authors recommended that resources would be better devoted to promoting the development of lateral and off-channel habitats within the river corridor. Shult (1996) measured the effects of log deflectors, rock weirs, rock islands, stumps and revetments placed in small streams in Washington. Physical habitat diversity increased, and so too did trout abundance. In a southeast Minnesota stream, Thorn and Anderson (2001) found that artificially increasing cover and woody debris both increased trout abundance, but the effect was limited to smaller-sized fish. Wesche (1985) provided details on the construction of deflectors for fish habitat enhancement and a number of examples where deflectors were applied both successfully and unsuccessfully. An increase in production of benthic invertebrates and fish was reported in several cases. Failures were mostly due to flood damage and/or improper design.

Frissell and Nawa (1992) evaluated rates and causes of physical impairment or failure for 161 fish habitat structures in 15 streams in southwest Oregon and southwest Washington following a flood of magnitude that occurs every 2–10 years. Results suggested that structural works are often inappropriate and counterproductive in streams with high or elevated sediment loads, high peak flows or highly erodible bank materials. Burial of structures by bedload, or loss of function due to bar or channel shift were the most common failure mechanisms. They advised that for large alluvial rivers, rehabilitation requires re-establishment of natural watershed and riparian processes over the long term. O'Neil and Fitch (1992) also reported that many of the aquatic enhancement projects they evaluated in Alberta, Canada, were ineffective or in some cases even detrimental to the aquatic environments.

Frissel and Ralph (1999) described the remodelling of a 4.3 km long reach of the Blanco River in southwestern Colorado from a braided morphology to a pool-riffle morphology. While the treatments were anecdotally reported as successful (Rosgen, 1988), the results have not been published in a critical way in the scientific literature. Similarly, a detailed description of morphological changes after channel reshaping of a 1.8 km reach of the Lower Rio Blanco River can be found in Kurz and Rosgen (2002). The River Restoration Centre (1999) described four cases of restoring meanders to straightened rivers. After some initial adjustment the meandering streams appeared to be stable, but biological responses were not reported. Restoration of secondary channels adjacent to the main stem of the Lower River Rhine in The Netherlands provided suitable habitat that is currently lacking for a broad range of rheophilic macroinvertebrate and fish species (Simons et al., 2001). Similarly, constructed shallow zones along navigation canals in The Netherlands functioned as a habitat for helophyte communities and contributed to a higher aquatic biodiversity than is associated with traditional banks along navigation canals (Boedeltje et al., 2001).

Restoration activities are a form of disturbance, and an initial decline in stream health should be expected. It took 2 years for riparian vegetation to recover after a restoration project on the River Gudenå, Denmark, but the new community was more diverse (note: the instream plant community had still not recovered after 2 years) (Baattrup-Pedersen, 2000). Restoration of rivers in Finland that had been dredged to facilitate timber floating began in the 1970s (when floating ceased to be a priority). By 1995 around 180 projects had been completed, although very few were monitored (Yrjänä, 1998). The main rehabilitation method was installation of boulder dams to replace the rapids that had been dredged. In a project that involved extensive channel remodelling, the disturbance from construction caused a significant decrease in macroinvertebrate abundance and diversity, and the recovery period of the community was on the scale of 1–3 years (Laasonen et al., 1998). In projects that required minimal physical disturbance macroinvertebrate community composition and abundance returned to pre-disturbance condition within a period of two weeks (Tikkanen et al., 1994; Yrjänä, 1998). Long-term monitoring of restoration projects in Finland found that while hydraulic habitat heterogeneity increased, macroinvertebrate and mussel abundances did not increase (Yrjänä, 1998). Restoration did not improve the capacity of the river reaches to maintain stocked 0-year old trout or grayling, but the biomass and density of stocked 1-year-old trout was significantly greater than in unrestored control reaches. Population densities of other common fish were lower in the restored reaches (Yrjänä, 1998).

Rehabilitation of an incised channel in Mississippi, USA, with grade control structures caused a change in hydraulic habitat variables, but neither fish species richness, evenness, nor assemblage structure differed between rehabilitated and channelized segments, with both exhibiting less richness and different assemblage structures than the unaltered segment (Raborn et al., 2003). Walsh and Breen (2001) found that reinstatement of rocks to create a pool-riffle form in an urban stream in Melbourne, Australia, resulted in little change to macroinvertebrate community composition. They hypothesized that, in such urban sites, quality and hydrology of urban stormwater are the primary determinants of community composition, not channel form.

Shields et al. (2000e) compared the performance of three types of bank protection applied on an incised warmwater stream in Mississippi: longitudinal stone toe; stone spurs; and dormant willow posts. Patterns of fish abundance and diversity did not differ between the treated reaches, but the untreated control site was distinctly different. The reaches stabilized with stone spurs supported large sized fish compared to reaches with other treatments. In a study of macroinvertebrate utilization of different bank treatments, Armitage et al. (2001) found that the shallow-sloping vegetated site supported a total of 115 taxa, in contrast to the artificial iron revetment, on which only 32 taxa were recorded. Total abundances were five to six times greater in the shallow vegetated sites compared with the steeply sloped and artificial banks. Also, seasonal fluctuations in both abundance and number of taxa were least on the reveted bank. Examples of the application of a range of 'soft' bank protection techniques are described favourably in River Restoration Centre (1999), but biological responses were not reported. Doyle (1992) compared the durability of alternative methods of bank protection with that of traditional toe-trenched angular rock rip-rap at seven typical bank erosion sites in British Columbia. On these steep gravel-bed rivers, the gravel dykes did not endure; tree revetments required constant maintenance and did not endure large floods unless extremely well constructed; and toe aprons were not as reliable as toe trenches for the same volume of rocks. Only well-placed, large semi-round rock performed as well as the toe-trenched angular rock rip-rap.

In a study on the rural Ovens River in Victoria, Australia, Koehn (1987) reported a ninefold increase in the number of native blackfish (*Gadopsis bispinosis*) after the addition of artificial habitat. The treatment consisted of a low (1 m height) *V*-shaped weir composed of six telegraph poles installed near the edge of the stream and a 24 m long reach downstream was 'seeded' with large rocks (0.35–1.1 m in diameter). Velocity became more diverse, and the log weir produced a pool 1.4 m deep. Fine sediment accumulated around the large rocks, aggrading the streambed and reducing the amount of fish cover as compared with the initial condition.

Mowry and McDowell (2003) selected seven active restoration sites from the Columbia River Basin and measured riparian vegetation and channel morphology. All sites were gravel-bedded alluvial channels with drainage areas of 25–150 km$^2$. Restoration techniques included instream log weirs, boulder weirs and emplaced large woody debris. These sites were considered typical of active restoration in this environment. In low gradient meadow reaches that would naturally have sinuous pool-riffle channels, pools created by structures are often smaller than naturally formed pools. In steeper reaches, instream structures added pools but also caused stream widening, which reduced fish habitat. Where sediment loads were high, fixed weir structures trapped sediment, filling the pools, and they often failed to create fish cover. In these sites, addition of unanchored natural LWD mimicked natural habitat formation more closely than fixed structures. At other sites, anchoring of structures was recommended.

In evaluating the success of installation of LWD structures, Shields *et al.* (2000d) noted that the structures were vulnerable after they were installed until sufficient sediment had deposited within the structure to counteract buoyant forces. Another hazardous period occurred when the structures decomposed and disintegrated if colonization of the sediment deposits by woody vegetation was not rapid. The long-term outlook for such structures improves if accompanied by a basin-wide strategy to manage sediment sources. Studies by Trotter (1990), Poulin & Associates Ltd (1991), Hunter (1991), Kondolf (1996), Abbe *et al.* (1997) and Borg *et al.* (2003) showed higher morphological diversity, particularly creation of pools, after introduction of LWD. Lemly and Hilderbrand (2000) also observed creation of pools, but there was no change in the structure of insect functional feeding groups. Cederholm *et al.* (1997) found that increased pool habitat was associated with LWD reintroduction in a small stream, and while coho salmon numbers increased, steelhead did not. Reinstatement of LWD in Douglas River, Cork, Ireland, increased hydraulic and geomorphic diversity, and also increased trout density and biomass (Lehane *et al.*, 2002). In Fish Creek, north central Oregon (catchment area 173 km$^2$), despite an

extensive in-stream habitat rehabilitation effort that included a 200% increase in the number of pieces of LWD in the channel, the numbers of anadromous fish did not increase significantly (Beschta *et al.*, 1992).

Shields *et al.* (2003b) recently completed a study aimed at restoring stream habitat along a 2 km incised section of Little Topashaw Creek, Yalobusha River, north central Mississippi (catchment area 37 km$^2$). The project consisted of placing 72 LWD structures along eroding concave banks and planting 4000 willow cuttings in sand bars. Hydraulic diversity increased as a result of the introduction of LWD and there was also an increase in pool habitat and number of fish species. However, progressive failure of structures and renewed bank erosion was observed during the second year after construction. Brooks *et al.* (2001) installed LWD structures on a 1100 m long reach of the River Williams River, NSW, Australia, a previously desnagged, high energy gravel-bed river (slope 0.0025). Twenty separate log structures were constructed, incorporating a total of 430 logs placed without any artificial anchoring. Early results indicate an increase in geomorphic complexity and increase in fish abundance (Brooks *et al.*, 2001). Re-snagging has been undertaken at a number of sites along the River Murray, Australia, with an increase in fish abundance observed. Comparing different sites, the fish abundance was positively correlated with the density of wood (Nicol *et al.*, 2002).

Brown *et al.* (1999) and Stewardson *et al.* (1999) reported a detailed evaluation of the biological and physical impacts of rehabilitation of Ryans Creek and the Broken River in northeast Victoria, Australia, using the addition of LWD and boulders, the replacement of willows with native vegetation, and control of erosion using rock beaching and stock exclusion fencing. This work found that rehabilitation produced the expected results in only some respects, and interpretation of the data was confounded by many interactions in the system.

Hicks and Reeves (1994) conducted a thorough review of projects in western North America that have attempted to restore fish habitat through placement of large wood or boulders in the low-flow stream channel. Performance of the projects has been mixed. Indeed, demonstration of the benefits to fish has proved elusive. In the majority of cases little effort has been expended in determining the response of fish to habitat enhancement, and in some cases fish showed no significant change (e.g. Gowan and Fausch, 1996). The common explanation for failure of the structural approach is that the structures were not put into the appropriate ecological context (Hicks and Reeves, 1994). In-stream structures were the sole component of most programmes reviewed by Hicks and Reeves (1994), whereas in most cases the habitat was altered by inappropriate land management activities. There has been a shift in recent times towards viewing structural rehabilitation as a part of a more comprehensive rehabilitation programme. Structures are viewed as a catalyst that may facilitate recovery of the habitat in the short term, while other components of the rehabilitation are being carried out. These other components are principally removal of the degrading influences, and re-establishment of the riparian vegetation community. Agencies and organisations need to make sure that the wider community are aware that restoration may not produce immediate results. There is no guarantee that the rehabilitation will work, and in some cases the damage simply cannot be reversed (Hicks and Reeves, 1994).

Heller *et al.* (2001) and Roper *et al.* (1998) investigated the durability (defined as the degree to which the structure remained at its original location) of 3946 instream structures in 94 streams located in the Pacific Northwest. These structures had all experienced floods with return intervals exceeding 5 years. Less than 20% of the structures sampled left the placement site and those that did move were often found to re-deposit downstream to create accumulations of superior complexity and stability to the original structure. These high durability figures are reason for some optimism regarding the stability of contemporary instream structures.

### 9.6.7  Some Reflections on Stream Rehabilitation

After a river rehabilitation conference held in North America in 1994, Ken Owen concluded:

> "So now, after two days of intensive study, everyone ... knows for sure that designing and building natural channels, whatever that means, is the solution to our problems. Well, do not be too disappointed if in 10 years you are doing something else to deal with these problems and in 30 years, young whippersnapper engineers and biologists are laughing and pointing their fingers at you as you walk down the street. Do not get me wrong—we have to keep looking for solutions to our problems. ... But if we have learned anything in the last 30 years, I really hope that it is that we need to monitor the changes our solutions create to make sure our hypotheses were correct or, if not, how we can do things better". (Owen, 1994).

Stream management has undergone a paradigm shift, from an emphasis on control over nature, response following flood catastrophe and narrow decision making by a few individuals, to working with nature, planning ahead and community involvement. River management is now called river rehabilitation The advances made in stream management certainly hold the promise of healthier streams for the future, but evaluations of projects to date suggest some lingering doubts and uncertainties.

Fencing out a stream and encouraging re-vegetation has such intuitive appeal, and seems so straightforward, that many practitioners would be surprised to learn that, when held up to scientific scrutiny, many such projects have failed to achieve any real improvement in ecological health. Case studies from around the world demonstrate a wide range of responses, often unexpected and disappointing, and it appears far from certain that this form of rehabilitation will achieve the desired goals. However, the range of responses to fencing and re-vegetation merely reflects the natural variability and unpredictability of the physical and biological aspects of streams, and the significance of scale in determining the dominant processes. Also, every stream reach lies within a particular catchment setting, and undertaking local riparian rehabilitation in isolation from, and in ignorance of, catchment-scale and other local-scale factors can be little more than a guessing game. Regulated rivers that are used to convey irrigation water present particular difficulties for re-vegetation with native plants, because many of them cannot tolerate having 'wet feet' in cold water for long periods throughout the summer.

Managers and communities seeking specific rehabilitation goals should not expect to put up a fence, plant some seedlings and then assume that letting nature take its course will produce the desired results with no adverse or undesirable reactions. It would be better to plan the actions to suit the local processes, and then monitor the changes (including the controlling forces) to improve understanding of the system so that management can be refined. In some situations revegetation will achieve outstanding results, while in others the results will be at best modest. Re-vegetation programmes create a dynamic, organic system, so the outcomes, whether measured in terms of physical factors, water quality, aesthetics or biota, will change through time.

The field of dam removal is gaining momentum, and although it is costly and presents technical difficulties, it appears to be a viable approach to deal with ageing obsolete dams that present a barrier to fish passage. Removing a dam is in itself a disturbance, and some adverse reactions should be expected. The long-term effectiveness of dam removal on stream health has not yet been determined. Active rehabilitation of channels and instream habitat has met with mixed success. It appears that we cannot assume that providing ideal habitat will necessarily result in a biotic response. A critical issue for stream managers is to identify and then address the limiting factors. There is little point in undertaking expensive channel remodelling or instream habitat creation if the biota is limited by water quality. Similarly, there is little point in removing a dam if there are no spawning grounds available, or if the flows required to cue fish movement, or bring a food supply are absent from the regulated regime.

Case studies in conference proceedings often report successful achievement of rehabilitation goals. Sometimes the goals are unclear, or more related to a narrow cultural perception of 'success', rather than in terms of ecological criteria. Admissions of failure are rare, despite the fact that many of the successes actually depend on the experience of numerous failures. Such reports contrast with the scholarly reviews of rehabilitation projects, most of which have concluded that rehabilitation more often than not fails to achieve the desired ecological response. Is it possible that some practioners have grown so used to action without knowledge (or with misconceptions) that they don't realise they lack it? Remember, it was this attitude that failed us in previous generations of river management. In many cases, on-ground 'action' is the end goal of rehabilitation, and the success or failure of the project is measured by whether or not a dramatic and rapid physical change has been achieved. We see this as potentially futile, and urge practitioners to think very carefully about rehabilitation goals, and how to evaluate them.

Poorly thought out monitoring programs are unlikely to demonstrate any statistically significant 'improvement', and even well planned programs can take many years to produce results. Even so, monitoring only tells us if there has been a change, not what caused the change, and it gives us few insights into the processes involved in creating that change. It would be more valuable to measure post-project changes using proper scientific methodology designed to test hypotheses. In this way, the rehabilitation projects will advance scientific understanding of how river systems respond to management actions. This knowledge can be used to continually improve rehabilitation methodologies (i.e. the adaptive management cycle, see Figure 9.29). If this simple message from the scientific community is heeded, the money spent on monitoring and evaluating will pay great dividends by improving the reliability of future rehabilitation efforts.

Stream rehabilitation is still a relatively new field in which there is endless scope for experimentation and refinement of methods. Rigorous method, and thorough and effectively communicated documentation is needed. A well-studied rehabilitation failure is a success for the advancement of scientific knowledge. As stressed throughout this section, each stream site should be considered unique. The sediment transport, hydrology, native plants and animals, bank soil types and channel bed topography must all be taken into account. The continuing interaction of biologists, geomorphologists and engineers is needed to develop guidelines for stream-rehabilitation works that are sufficiently versatile to accommodate a wide range of situations, and flexible enough to incorporate improvements suggested by research and experimentation.

# Appendix

# Basic Statistics

## A.1 Summary Statistics

Summary statistics describe properties of sample data. For this summary, $x_1$, $x_2$, $x_3$, ..., $x_n$ represent sample data values, where $n$ values have been selected from a population of all possible values of $x$. The symbols $\sum$ and $\Pi$ indicate summation and multiplication, respectively, of values from $i = 1$ to $n$. Many of the given statistics can be calculated using AQUAPAK.

### Measures of Central Tendency

*Arithmetic Mean or Average, $\bar{x}$*

$$\bar{x} = \frac{\sum x_i}{n} = \frac{x_1 + x_2 + \cdots + x_n}{n} \tag{A.1}$$

*Geometric Mean, $\bar{x}_g$*

$$\bar{x}_g = (\Pi x_i)^{1/n} = \sqrt[n]{x_1 \times x_2 \times \cdots \times x_n} \tag{A.2}$$

For calculating $\bar{x}_g$, Armour *et al.* (1983) recommend taking the mean of the logarithms of the sample values, then taking the antilog of the mean:

$$\bar{x}_g = \text{antilog} \left( \frac{\sum (\log x_i)}{n} \right) \tag{A.3}$$

*Median*

If a data set is ranked in order of magnitude, the *median* is the middle value; e.g. for the values 1, 3, 4, 6, 8, the median is 4.

*Mode*

The *mode* is the sample value which occurs most often in the data set; e.g. for the values 1, 1, 2, 2, 3, 3, 3, 4, 5, 5, the mode is 3. If a sample has two modes, it is *bimodal*; if more, it is *multimodal*.

### Measures of Spread

*Range*

$$\text{Range} = (x_{\max} - x_{\min}) \tag{A.4}$$

where $x_{\max}$ is the highest sample value and $x_{\min}$ the lowest.

*Root Mean Square (RMS)*

$$\text{RMS} = \sqrt{\frac{\sum (x_i - \bar{x})^2}{n}} \tag{A.5}$$

The RMS is used as a measure of turbulence intensity with $x = \text{velocity}$ (Section 6.4.1).

*Variance, $s^2$*

$$s^2 = \frac{\sum (x_i - \bar{x})^2}{n - 1}$$

or

$$\frac{\sum (x_i^2) - \left[ \frac{\left( \sum x_i \right)^2}{n} \right]}{n - 1} \tag{A.6}$$

The second version is preferred for computational purposes.

Stream Hydrology: An Introduction for Ecologists, Second Edition.
Nancy D. Gordon, Thomas A. McMahon, Brian L. Finlayson, Christopher J. Gippel, Rory J. Nathan
© 2004 John Wiley & Sons, Ltd ISBNs: 0-470-84357-8 (HB); 0-470-84358-6 (PB)

*Standard Deviation, s*

The standard deviation, $s$, is the positive square root of the variance.

*Coefficient of Variation, $C_v$*

The coefficient of variation is a dimensionless measure of variability commonly used in hydrology:

$$C_v = s/\bar{x} \qquad (A.7)$$

*Index of Variability, $I_v$*

$$I_v = \sqrt{\frac{\sum (\log x_i - \overline{\log x})^2}{n-1}} \qquad (A.8)$$

where $\overline{\log x}$ is the mean of the logarithms of the $x$ values. $I_v$ is often used to characterize the year-to-year variability of peak flood flows. Baker (1977) also refers to this as the 'flash flood index'. Streams with high values of $I_v$ are more likely to have flash flood behaviour, and may also have lower species diversity and abundance than streams with only moderately variable peak regimes (Ward and Stanford, 1983a).

**Measures of Shape**

*Skewness ($C_s$)*

$$C_s = \frac{n \sum (x_i - \bar{x})^3}{(n-1)(n-2)(s^3)} \quad \text{or} \quad \frac{n(\sum x_i^3 - 3\bar{x} \sum x_i^2 + 2n\bar{x}^3)}{(n-1)(n-2)(s^3)}$$

$$(A.9)$$

This measure describes the lack of symmetry in a set of data. A positive value of $C_s$ means that the data are skewed to the right, and a negative value means that the data are skewed to the left, . Streamflow data typically show a positive skewness, with the skewness decreasing as the time interval increases from daily to annual.

*Kurtosis ($C_k$)*

$$C_k = \frac{n^2 \sum (x_i - \bar{x})^4}{(n-1)(n-2)(n-3)(s^4)} \qquad (A.10)$$

Kurtosis describes the 'peakedness' or 'flatness' of the data distribution. For a normal distribution (Section A.2), the kurtosis is 3.

## A.2　Describing Distributions of Data

Measurement variables can be either *continuous* or *discrete*. Data which can take on any value—or any value within a certain range (e.g. 0–100)—are considered continuous. Examples of continuous variables are channel width and depth, stream temperature, streamflow and substrate particle diameters. Discrete data, in contrast, have a limited number of possible values. Examples are counts of rocks, plants or fish or quality ratings (e.g. 5 = excellent, 1 = poor). Continuous data are presented in discrete form due to measurement limitations (e.g. measured to the nearest 1 mm).

If large amounts of data are collected (either continuous or discrete), the values can be summarized by a *frequency distribution*. To accomplish this, the samples are ranked from smallest to largest and then grouped into classes. *Frequency* refers to the number in each class. Frequency data can be plotted as a *histogram* (Figure A.1(a)), where the height of each column represents the frequency and the class intervals are given on the $x$-axis.

If the frequencies are divided by a factor such that the area under the curve equals one, then the values become *relative frequencies* (Figure A.1(b)). These can also be expressed as *percentages* if multiplied by 100. By progressively summing up these values, a *cumulative frequency distribution* can be developed, as shown in Figure A.1(c). The sediment data of Figure 5.32 are presented in this form.

When the number of values increases and the size of the class interval decreases, the profile of the frequency distribution becomes smoother. At the limit, where the number of values approaches infinity and the class interval approaches zero, the resulting curve becomes a *probability distribution* or *probability density function* (*pdf*), as shown in Figure A.1(d). The function describing the curve is denoted $f(x)$, or 'function of $x$'. Probability density functions for standard theoretical distributions (e.g. normal, log-normal, etc.) are usually defined by mathematical equations. In a pdf, the function $f(x)$ must be $\geq 0$ for all $x$, and the area under the graph of $f(x)$ is equal to 1. For the probability distribution of Figure A.1(d) the probability that $x$ takes on a value between $a$ and $b$ (written '$P(a < x < b)$') is equal to the shaded area under the curve. A probability of zero means 'impossible' whereas a probability of 1 means 'certain'. The probability that $x$ is *exactly a* or *exactly b* is zero; thus, we usually use the symbols $>$ and $<$ rather than $\leq$ and $\geq$.

Another, more useful form of the probability distribution is the *cumulative distribution function* (*cdf*). The probabilities are cumulated, and cumulative values are plotted against their associated $x$ values, as shown in

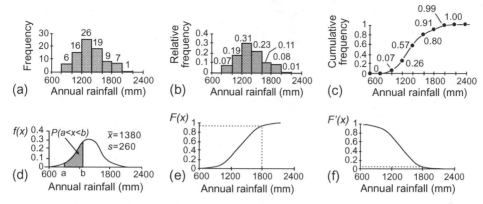

**Figure A.1.** *Marysville, Victoria, annual rainfall, presented as (a) frequency histogram, (b) relative frequency histogram, (c) cumulative frequency distribution, (d) fitted probability distribution (normal), (e) cumulative distribution function and (f) inverse cumulative distribution function. f(x) = probability, F(x) = probability of non-exceedance, and F'(x) = probability of exceedance*

Figures A.1(e) and A1.1(f). The probability becomes the *probability of non-exceedance*, $F(x)$, or the *probability of exceedance*, $F'(x)$, depending on the direction in which the values are cumulated and the intended use of the curves. $F'(x)$ is sometimes called the *inverse cumulative distribution function* (*icdf*). Thus, in Figure A.1(e) the probability that $x < 1800$ mm ('$P(x < 1800)$') is about 0.95, whereas in Figure A.1(f) the probability that $x \geq 1800$ mm ('$P(x \geq 1800)$') is about 0.05. As is apparent, $F(x) = 1 - F'(x)$.

Cumulative distributions are commonly plotted with the $x$ and $y$ axes reversed, using a probability scale based on a theoretical probability distribution (e.g. normal). Data which are described by this distribution will then plot as a straight line. Commercial probability papers are available for a number of distributions. The large sediment data of Figure 5.33, the flood-frequency data of Figure 8.4 and the flow-duration data of Figure 8.7 are plotted in the cumulative form on log-normal probability axes. It should be noted that Figures 5.33 and 8.7 are cumulative frequency plots, whereas in Figure 8.4 a theoretical probability distribution has been fitted to the data. The first summarizes the data; the second allows inferences to be made about the underlying population.

A basic assumption in statistics is that sample data can be described by probability distributions. For example, in a flood-frequency analysis, a distribution is fitted to the data to improve estimates of, say, 100-year floods. If the fitted distribution does not truly represent the population of all possible sample values, then results can be misleading.

Distributions used for describing discrete data include the binomial, negative binomial and Poisson distributions. These will not be discussed here, but are described in many basic statistics texts. The *normal probability distribution* ('bell-curve') is one of the most widely used models for both discrete and continuous variables. Other distributions are described in Section 8.2.6. The normal distribution is defined by its parameters $\mu$ (mean) and $\sigma^2$ (variance). In this distribution, 68.3% of the values lie within $\pm 1\sigma$, 95.4% within $\pm 2\sigma$, and 99.7% within $\pm 3\sigma$. Normal probability tables give values of the standard normal distribution, for which $\mu = 0$ and $\sigma^2 = 1$. To use the table, sample values are standardized with the transformation

$$Z = \frac{x - \bar{x}}{s} \qquad (A.11)$$

where $Z$ is the standardized variate. Thus, if the data of Figure A.1 are assumed to be represented by a normal distribution, then we would expect the annual precipitation to be between 860 and 1900 mm about 95% of the time (solving for $x$ with $\bar{x} = 1380$, $s = 260$ and $Z = \pm 2$).

The normal distribution is particularly important because a number of statistical methods require that the data are normally distributed, e.g. *t*-tests, analysis of variance and correlation analysis. Sometimes *transformations* can be applied to the data (e.g. log, square root, etc.) to satisfy this requirement. Even when variables are not normally distributed, sums and averages of the variables may be. According to the *central limit theorem*, if samples are selected randomly from a given distribution then the distribution of the sample means tends towards the normal distribution. The more the values included in the sample means, the better the approximation. Thus, many variables have distributions that can be fitted closely with a normal distribution, such as measurement errors, heights, weights and other physical characteristics. For rainfall and streamflow data, for example, the distributions tend towards normality as the time base is increased; e.g. average monthly rainfall and yearly runoff may be approximately normally distributed. As a rule of thumb, if $n > 30$, the central limit theorem can be applied (conveniently similar to the number of values in monthly totals).

## A.3   Just a Moment

*Moments* provide a way of quantifying location, spread and shape of probability distributions. 'Moment' is a term from the bending and twisting branch of engineering mechanics, defined as distance times force. For example, if a bag of 500 'Newton's apples' of 1 N each were placed on the end of a 3 m diving board, the applied moment would be 1500 N m. A related concept, torque, explains why less force is needed to budge a locknut if a longer-handled wrench is used.

In statistics, infinitely small strips called 'elemental areas' and distances along the $x$ axis replace the physical measures of force and distance (see Figure A.2). For continuous variables, moments are computed from calculus formulae. 'Distances' are measured in relation to either the origin ($x = 0$) or the mean ($x = \mu$). The two types of moments are called 'moments about the origin' and 'moments about the mean' ('central moments'), respectively.

The mean, $\mu$, is the same as the first moment *about the origin*. If the $x$ axis were 'balanced' on a wedge as shown in Figure A.2 the downward 'forces' on either side would be in balance and the point of balance would be the mean. The first moment *about the mean* is therefore zero.

The second moment is calculated in the same manner except that the distances are *squared*. For the third moment, the distances are cubed, and so on. To generalize

the situation, the $r$th moment about the mean ($m_r$) for a continuous distribution can be described by

$$m_r = \int_{-\infty}^{\infty} (x - \mu)^r f(x)\mathrm{d}x \qquad (A.12)$$

Here, $x$ is the distance along the $x$-axis from the origin, $\mu$ is the population mean and $f(x)$ is the probability density function. To compute moments about the origin ($m'_r$), the term $\mu$ is deleted from Eq. (A.12).

For sample data, Eq. (A.12) becomes

$$M_r = \frac{\sum (x_i - \bar{x})^r}{n} \qquad (A.13)$$

where $\bar{x}$ is now the sample mean. To calculate the $r$th sample moment about the origin ($M'_r$), $\bar{x}$ is deleted from Eq. (A.13).

Probability distributions can be fitted to data by approximating the moments of the distribution by sample moments. This is called the *method of moments*. Sample moments are related to the sample statistics, mean, variance, skewness and kurtosis, as follows:

$$\bar{x} = M'_1 \text{ (the first moment about the origin)} \quad (A.14)$$

$$s^2 = \frac{n}{n-1}(M_2) \qquad (A.15)$$

$$C_s = \frac{n^2 M_3}{(n-1)(n-2)s^3} \qquad (A.16)$$

$$C_k = \frac{n^3 M_4}{(n-1)(n-2)(n-3)s^4} \qquad (A.17)$$

## A.4   Correlation and Regression: How Closely Is This Related to That?

### A.4.1   General

Correlation and regression are well-worn and proven methods used in hydrology and ecology, and a multitude of examples can be found in the literature. For example, regression relationships can be developed to extend the record at one gauging station from data obtained at another, to predict streamflow from catchment characteristics and precipitation or to determine which physical habitat factors are most closely related to the abundance of organisms.

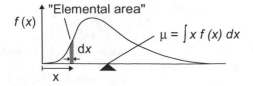

**Figure A.2.** *Illustration of the first moment about the origin, the mean ($\mu$), for a continuous probability distribution. One half of the area under the curve lies to either side of the mean. The sum of all 'elemental areas' is equal to one*

The objective may be (1) to simply determine whether certain variables are related or (2) to develop a relationship so that one variable can be predicted from one or more other variables. The first process is one of correlation and the second, regression. The two have important distinctions although the methods of analysis are much the same. Haan (1977) and Holder (1985) are useful references on the use of these statistical methods in hydrology.

### Correlation

This is a means of defining the strength of the relationship between two variables, e.g. substrate size and benthic populations of blackfly larvae. A sample *correlation coefficient* is a quantitative measure of this 'strength', based on sample data. For this example the sample data might be the substrate $d_{50}$ and numbers of larvae per square metre. A close association between two variables does not imply cause-and-effect or a common influencing factor. The fact that substrate size and numbers of blackfly larvae are correlated does not mean that changing the number of blackfly larvae on the streambed will affect the substrate size or vice versa.

The sample correlation coefficient ($r$) for $n$ pairs of ($x, y$) data is calculated as follows:

$$r = \frac{n \sum x_i y_i - (\sum x_i)(\sum y_i)}{\sqrt{n \sum x_i^2 - (\sum x_i)^2} \sqrt{n \sum y_i^2 - (\sum y_i)^2}} \quad (A.18)$$

The value of $r$ lies between $-1$ and $+1$, inclusive. As a rule of thumb, the correlation is weak if $0 \le |r| \le 0.5$, strong if $0.8 \le |r| \le 1$ and moderate otherwise (Devore, 1982).

To illustrate how $r$ is interpreted, let the graphs of Figure A.3 represent paired measurements of air temperature ($x$-axis) and frequency of frog croaking ($y$-axis). A value of $r$ near $+1$ means that frogs croak more frequently at higher temperatures and less often at low ones. Conversely, a value of $r$ near $-1$ means that frogs croak more often as the temperature goes down. An $r$ value of zero would imply that the correlation between the two variables is non-existent or not linear. Thus, $r$ is a measure of the degree of linearity of the relationship. Clearly, in the fourth graph, there is a relationship (a reasonable one for this example) even though $r \approx 0$.

A high correlation between two variables when in fact none exists is called *spurious correlation*. For example, if the data form clusters of two or more groups such as in the last graph of Figure A.3, a large correlation may result solely due to the heterogeneity of the data. Another potential source of spurious correlation is the use of ratios with a common element, e.g. the correlation of $X_1/X_2$ with $X_1$, or $X_1/X_3$ with $X_2/X_3$. Plots of these variables may show a well-defined relationship with a narrow scatter of points due only to the presence of the common element. This type of spurious correlation is quite common, and often goes undetected. Examples in this text include the $C_v$ versus MAR relationship (Figure 4.19(b)), where mean annual flow is a common factor, and the sediment rating curve (Figure 7.22), where discharge is a common factor. As Benson (1965) points out, the relationship is not invalid as long as interpretations are made in terms of the ratios and not the original factors. For example, the sediment-rating curve should be used only to estimate sediment discharge, not sediment concentration.

If sample values for the two variables are assumed to be selected from a bivariate ('three-dimensional') normal distribution, and free of measurement error, then one can perform tests of hypotheses about whether the population correlation coefficient is significantly different from zero, and construct confidence limits. Neglecting these requirements can bias results. Correlation coefficients can have very large confidence limits for sample sizes of 30 or less. Generally, correlation coefficients are only reported to two significant figures.

### Regression

In correlation, the 'end product' is the correlation coefficient. In *regression*, the 'end product' is a *regression equation* that describes one variable, the dependent ($Y$) variable, as a function of one or more independent ($X$) variables. A coefficient is also obtained, which in this case describes how well the equation fits the data. The term 'regression' was first used by Francis Galton to explain the phenomenon whereby men's heights 'regressed' back towards the average. From a relationship developed between the heights of fathers and sons, he found that if a father was taller than average, his son would also be expected to be taller than average, but not by as much.

**Figure A.3** *Interpretation of the correlation coefficient, r*

Similarly, the son of a shorter-than-average father would be expected to be shorter than average, but taller than his father (Devore, 1982).

The deviations of observed values $(y_i)$ from values predicted using the regression equation $(\hat{y}_i)$ are the error terms or *residuals*, $\varepsilon_i = y_i - \hat{y}_i$. If the regression equation is to be accompanied by confidence intervals or tests of significance, the required assumptions about the residuals are that they

- Are random and independent
- Are normally distributed
- Have constant variance ('homoscedastic')

Plots of the residual values provide a means of visually assessing whether the assumptions are met, and how well the equation fits the data. The recommended procedure is to develop a *scatter plot* of $\varepsilon_i$ versus $y_i$. Plotting the predicted values, $\hat{y}_i$, against $y_i$ is also advisable. Statistical computer packages normally have an option for plotting residuals, and users' manuals will provide information on interpreting the patterns. Various statistical tests of hypothesis are also available for testing the validity of the assumptions. Details will not be given here, but can be found in statistical texts (see Section A.5).

Regression techniques can be used despite violations of the above assumptions, but the resulting equation should be applied only over the range of data to which the equation has been fitted. The correlation coefficient will change over the range of data, making it unsuitable as a measure of adequacy of the regression, and unsuitable for extrapolation.

When nothing is known about the relationship between variables, a 'first-try' approach is to simply assume that the relationship is linear. A linear relationship works well for some variables such as the relationship between annual rainfall and annual runoff. Data are often *transformed* to linearize the relationship or to achieve constant variance or normality of the residuals. Transforming the data by taking logarithms is by far the most common approach in natural sciences. Examples are the stream-order relationships in Figure 4.12, the hydraulic geometry relationships in Figure 7.9 and the relationship of fish height to length (Armour *et al.*, 1983). Relationships based on transformed data are considered *non-linear*.

It should be noted that the correlation coefficient from an analysis of transformed data applies *only* to the transformed data. Therefore, it is not 'legal' to compare correlation coefficients from regressions on original, log-transformed and otherwise-transformed data to determine which has the 'best fit'. This should, instead, be

judged visually or with the assistance of various tests to ascertain whether the transformation does in fact improve the linearity of the model and/or the residual patterns.

Regression and correlation are powerful methods for examining relationships between variables. As with any procedure, it is important to understand the assumptions and limitations. A general caution about the use of regression methods is well summed-up in a phrase by Haan (1977, p. 218):

> The ready availability of digital computers and library regression programs has led many to collect data with little thought, throw it into the computer, and hope for a model. This temptation must be avoided.

## A1.4.2   Simple Linear Regression and Least Squares Fitting

In simple linear regression the fitted line is described by an equation of the form:

$$Y = aX + b \qquad (A.19)$$

or

$$Y = \beta_0 + \beta_1 X + \varepsilon \qquad (A.20)$$

depending on whether one's preferences are algebraic or statistical. The second form will be used in this discussion. In Eq. (A.20), $Y$ is the dependent variable (the one which is to be predicted) and $X$ is the independent variable (the 'predictor'). The coefficients $\beta_0$ and $\beta_1$ are called *regression coefficients*. In this model, $\beta_1$ represents the slope of the line and $\beta_0$ the point where the line crosses the $Y$-axis (the '$Y$ intercept'). A scatter plot of the observed data ($y_i$ versus $x_i$) is always advisable as a first step to investigate whether $Y$ increases or decreases with $X$, whether the relationship is strong, moderate or nonexistent, its linearity, the presence of 'outliers' (points which considerably depart from the general trend), and the general range of $X$ and $Y$ (Haan, 1977).

When fitting a line through the data the objective is to make the deviation between the observed points and the fitted line small. This is usually accomplished by using the principle of *least squares*. A least squares fit of the equation to the data minimizes the sum of the squares of the deviations between the observed and predicted values of $Y$. This deviation is the 'error' or 'residual' term, and is illustrated for one point in Figure A.4.

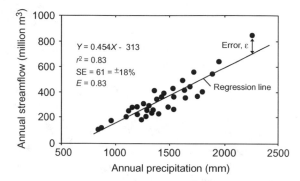

**Figure A.4.** *Simple linear regression analysis of annual streamflow data from the Acheron River at Taggerty versus annual precipitation data recorded at Marysville (see map of Figure 4.7). Shown are the scatter plot of the actual data, the fitted regression line, regression equation, and values of $r^2$, SE (standard error) and E (coefficient of efficiency)*

By using calculus, equations are derived for $\beta_0$ and $\beta_1$ such that the sum of the squares of the errors becomes a minimum. These equations are

$$\beta_1 = \frac{n \sum x_i y_i - (\sum x_i)(\sum y_i)}{n \sum x_i^2 - (\sum x_i)^2} \tag{A.21}$$

and

$$\beta_0 = \bar{y} - \beta_1 \bar{x} \tag{A.22}$$

where $x_i$ and $y_i$ are now actual observations, $n$ is the number of data pairs and $\bar{x}$ and $\bar{y}$ are mean values. From Eq. (A.22), it can be seen that this solution must pass through the point $(\bar{x}, \bar{y})$.

In Figure A.4, 40 years of annual runoff data from the Acheron River at Taggerty (the dependent variable) have been plotted against the annual precipitation data at Marysville (the independent variable). A regression line fitted by least squares has been drawn through the points. The resulting equation allows runoff to be predicted if rainfall is known.

The total sum of the squares of the $Y$ values (SST), $\sum y_i^2$, can be partitioned into three components:

Sum of squares due to the regression (SSR),
    $\sum (\hat{y}_i - \bar{y})^2$;
Sum of squares due to the mean (SSM),
    $(\sum y_i)^2 / n = n\bar{y}^2$;
Sum of squares due to the error or residual terms (SSE),
    $\sum (y_i - \hat{y}_i)^2$.

Thus, $\text{SST} = \text{SSM} + \text{SSR} + \text{SSE}$. A derivation can be found in Haan (1977).

The SSE is a measure of the variability of the $y$ values about the fitted line. It is what is minimized in the least squares procedure. The ratio of the SSR to the total sum of squares adjusted for the mean (SST − SSM) is commonly termed the *coefficient of determination* or $r$-squared ($r^2$) value:

$$r^2 = \frac{\text{SSR}}{\text{SST} - \text{SSM}} = \frac{\sum (\hat{y}_i - \bar{y})^2}{\sum (y_i - \bar{y})^2} \tag{A.23}$$

The coefficient of determination is interpreted as the proportion of variation in $Y$ explained by the regression equation. If the model were a 'perfect fit' to the data, the $r^2$ value would be 1, whereas if none of the variation in the $Y$ values were explained by the regression equation, the $r^2$ value would be 0. Thus, the closer the value of $r^2$ is to 1, the better the fit. The square root of $r^2$, $r$, is a type of correlation coefficient which describes the correlation between the values of $y_i$ and $\hat{y}_i$.

The *standard error* or *standard error of estimate* (SE) is preferred over $r^2$ as a measure of reliability of the regression equation. It is equal to the standard deviation of the residuals about the regression line. The spread is assumed to be normally distributed and of constant variance over the range of data. An unbiased estimate for this SE term is (Haan, 1977)

$$\text{SE} = \sqrt{\frac{\sum (y_i - \hat{y}_i)^2}{(n - 2)}} \tag{A.24}$$

The $(n - 2)$ term indicates that two degrees of freedom are 'lost' by estimating $\beta_0$ and $\beta_1$ (Holder, 1985). If the standard error is squared, the term $\text{SE}^2$ is often called the *mean squared error* (MSE). If the regression equation explains a large part of the variation in $Y$, then SE will be much less than the standard deviation of the $y_i$ values. Because SE has the same units as $Y$, it is sometimes expressed as a percentage of $\bar{y}$, as in Figure A.4. When data are log-transformed, SE can be converted back to the original scale to give a more useful description of the spread about the regression line. For log-log or semi-log regressions (with $y$ the logarithmic variable), these percentages are calculated as

$$\text{SE (lower)} = \left( \frac{1}{\text{antilog}(\text{SE}')} - 1 \right) \times 100 \tag{A.25}$$

and

$$\text{SE (upper)} = (\text{antilog}(\text{SE}') - 1) \times 100 \tag{A.26}$$

where SE′ is the standard error calculated on the log-transformed scale (this is the value normally given on a computer output). For example, if a log-log (or semi-log) regression (base 10) gives $SE' = 0.135$, then antilog $(SE') = 1.36$, giving $SE = -27\%$, $+36\%$. Both SE and $r^2$ are usually listed with the regression equation when reporting the results of a regression analysis.

Another useful measure is the *coefficient of efficiency*, *E*. The coefficient of efficiency describes the degree to which observed and predicted values agree (Aitken, 1973). The term is computed from the equation

$$E = \frac{\sum (y_i - \bar{y})^2 - \sum (y_i - \hat{y}_i)^2}{\sum (y_i - \bar{y})^2} \qquad (A.27)$$

The value of the coefficient of efficiency is always less than or equal to 1.

In obtaining a least squares estimate, no assumptions about the distribution of error terms are required. They enter the picture, however, when confidence intervals and tests of hypothesis are of interest. Haan (1977) and Holder (1985) and most statistics texts offer procedures for calculating confidence intervals about the regression line and for testing the significance of the parameters $\beta_0$ and $\beta_1$. Procedures are also available for testing whether two regression lines are significantly different. For example, in the double-mass curve analysis (Section 3.2.2) a test can be performed to determine whether an observed change in slope is significant. Confidence bands expand at both the high and low ends of the range of data. Also, the relation between *X* and *Y* (or their transformed values) may be linear only over the range studied. For these reasons, extrapolation of a regression equation beyond the range of *X* is discouraged.

It is also worth noting that the regression line for *Y* versus *X* is not the same as that for *X* versus *Y*. For example, in Figure A.4 the inverse relationship could *not* be found by solving the regression equation for *X*. A new relationship must be developed by regressing precipitation versus runoff (which, in this example, makes little physical sense). The $r^2$ value will be identical but the equation will be different. However, the equation will still pass through the point defined by the means of the two variables.

As Holder (1985) points out, the least squares method is an impartial, objective method of fitting a line through the data. For some purposes, a critical assessment by eye which gives more weight to some values (e.g. measurements which were known to be of high quality) and less to others may yield a line which is more useful than a least squares fit. Therefore, the intended use of the equation and the consequences of poor predictions should always be considered when using regression methods.

## A.4.3    Multiple Regression

The 'simple' in simple linear regression simply means that only one independent variable, *X*, is involved. Often, a phenomenon is 'dependent' on several factors. For example, the number of fish in a stream reach may depend on the velocity of the stream, area of bank overhang, substrate type, amount of aquatic vegetation, populations of aquatic insects and other edible organisms, amount of shade, temperature, flow duration and other factors which may or may not be measurable. A regression model for predicting the number or biomass of fish might include some or all of these influencing factors.

The general form of the multiple regression model is

$$Y = \beta_0 + \beta_1 X_1 + \beta_2 X_2 + \cdots + \beta_p X_p + \varepsilon \qquad (A.28)$$

where *Y* is the dependent variable, $X_1$ to $X_p$ are independent variables, $\beta_0$ through $\beta_p$ are parameters of the model and $\varepsilon$ is the error term, as described for Eq. (A.20). Data may consist of actual measurements (e.g. stream width)— or 'dummy' variables can be included to represent presence/absence criteria or a qualitative '1 to 10' rating, as described by Holder (1985).

Equation (A.20) describes a straight line. In multiple regression, the equation describes a 'hyperplane' of *p* dimensions, which becomes more difficult to visualize or plot when $p > 2$. Andrews (1972) presents a method of displaying the observations of a number of variables on a single graph. This allows visual comparison of the sets of observations and provides a concise summary of the data. Andrews curves are described in Section 8.6.4. Two-dimensional scatter plots can still be developed for evaluating residual patterns.

To obtain the parameters $\beta_0$ to $\beta_p$ a number of observations are made of all the variables. This number must be at least as great as the number of parameters, $p + 1$, and preferably three to four times as large or more (Haan, 1977). For example, if one were trying to predict the density of Chironomid larvae in stream sediments from (1) the average size of the sediment grains, (2) percentage organic matter, (3) stream velocity and (4) depth, then at least five concurrent observations of each variable (and preferably 20 or more) would be needed to solve for the parameters $\beta_0$ to $\beta_4$.

This can be done using a least squares procedure, which in this case, minimizes the deviation of points from the 'hyperplane'. The procedure requires the manipulation of matrices containing the observations, and is best carried out with the assistance of commercial statistical software packages. A method of hand calculation is given by Riggs (1968). A large number of significant figures are carried throughout the computations to prevent large roundoff errors. However, in reporting results, the number of significant figures should be reduced to three or four.

The *multiple coefficient of determination* ($r^2$) obtained from a multiple regression analysis is the proportion of the total sum of squares (corrected for the mean) explained by the regression equation. This $r^2$ value also has a range of 0 to 1, as does its positive square root, the *multiple correlation coefficient* ($r$). The coefficient of efficiency can be calculated with Eq. (A.27) and interpreted as for simple linear regression.

An unbiased estimate of the variance of the error terms is normally given in the computer output as the 'expected mean square'. The positive square root of this value is the SE of the multiple regression equation.

The $r^2$ value will never decrease as more values are added; thus a high $r^2$ may be obtained simply by including a large number of independent variables of dubious value. In many studies, these variables are often intercorrelated. If two people always give the same advice then nothing is gained by listening to both of them. By the same token, there is no need to include two intercorrelated variables in an equation when one will do. The final choice of variables may be based on which are the easiest and least expensive to measure.

The $r^2$ value is not necessarily the best indicator of how well the regression equation fits the data. It is preferable to observe the effect of additional variables on SE, since the equation with the smallest SE will have the narrowest confidence intervals (Haan, 1977). If the addition of variables causes SE to increase, this should be an 'alert' signal that these variables are not adding any more information to the equation and can be left out. The values of SE and $r^2$ can be plotted—graphically or mentally— and the number of variables selected by noting where the graph reaches a 'point of diminishing returns', as shown in Figure A.5.

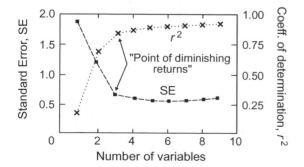

**Figure A.5.** *Illustration of the effect of additional variables on $r^2$ and SE in a multiple regression equation*

One of the first steps in developing an equation is to look over the *correlation matrix*, a table giving the correlation coefficients ($r$ values) between each variable and every other variable. A correlation matrix is normally available as an option in multiple regression programs. Variables should be selected which are the most correlated with the dependent variable and the least correlated with each other. The equation can then be modified by adding or subtracting variables by trial and error, assisted by judgement about which variables make the most practical sense.

A formal, systematic procedure is provided by 'all possible regressions' and 'stepwise regression' methods, which will be discussed. Statistical texts will yield additional approaches. With any approach, the results should always be tempered with knowledge about the phenomena of interest.

In *all-possible regressions*, regression equations are developed with every possible combination of the independent variables. If all regression equations are of the form of Eq. (A.28) then $2^p$ regressions would need to be developed. This may be a simple matter if, for example, only three independent variables are considered, requiring only eight equations. A final equation would be selected based on the $r^2$, SE and relevance of the independent variables. As $p$ increases, however, the number of possibilities increases exponentially; e.g. for $p = 10$, 1024 regressions would be needed. Thus, this method is not practical for a large number of variables.

*Stepwise regression* is one of the more common approaches. The regression equation is 'built,' one step—one variable—at a time. The variable added at each step is the one that explains the largest amount of the remaining unexplained variation (Haan, 1977). Thus, at the first step the variable added is the one with the highest correlation with the dependent variable. After each step, all variables which have been included are re-examined to see if they are still significant. Non-significant variables can be discarded. The procedure is repeated until all variables have been 'tested' for possible contribution to the equation. The final equation should be composed of a 'best' set of significant variables and the variables left out should be insignificant. To make this an objective procedure, the level of significance is set before the regression procedure is carried out. Statistical references should be consulted for methods of testing significance.

After an equation is selected, a scatter plot of the observed and predicted values should be examined to check for numerical errors, outliers, violations of assumptions about the error terms, and the appropriateness of the model. These points should ideally plot evenly about a 1:1

line. Residuals can also be plotted against $Y$ or any of the $X$'s or against time to examine trends.

In some cases, a transformation is needed to linearize a model so that a least squares approach can be taken. Scatter plots of the residuals against individual $X$ variables can identify which variables need transformation. In some cases it may be possible to improve the model by including cross-product terms or ratios (e.g. $X_1X_2$ or $X_1/X_2$). Care should be taken to avoid introducing spurious correlation (Section A.4.1) (for example, by including both the ratio $X_1/X_2$ and $X_1$ or $X_2$ in the relationship). Assumptions about the distribution of errors are made relative to the transformed scale; i.e. the errors in the transformed domain should be of constant variance, random and normally distributed.

The resulting equation is empirical; i.e. not necessarily based on physical reality. In most cases the units on either side of the equation are different. A little 'applied common sense' (Holder, 1985) is valuable when investigating whether the equation correctly mimics the real situation. Regression equations will also change slightly as more data are added. For example, snowmelt forecast equations are often updated each year after both winter snowpack and spring runoff measurements have been obtained.

In deriving a multiple regression equation, simplicity is the goal. The final equation should contain just a few meaningful variables that explain most of the variability in the data. They should all contribute to the regression unless there is some other logical or intuitive reason for retaining them. It is also helpful to evaluate the usefulness of the final equation with a test set of observed values that were not used in its development.

### A.4.4    Autocorrelation and Intercorrelation

#### *Autocorrelation*

A look at time-series plots of daily streamflow data (e.g. Figure 3.1) will reveal that runoff does not occur in a totally random fashion. High flows tend to follow other high flows and low flows also tend to group together. Seasonal precipitation patterns, the carryover effects of soil moisture, and storage of water in reservoirs or in the channel can all create this *persistence* in the record. Persistence has the effect of *increasing* the variability of the mean over the period of record, meaning that more years of data are required to accurately estimate the mean than if the data were truly random.

*Autocorrelation* or *serial correlation* refers to the correlation of data series against the same data series 'shifted' by some time interval. Correlations are computed for a given *lag time*, $k$, to obtain an *autocorrelation coefficient*, $r(k)$. For example, the correlation between today's streamflow and yesterday's (or between this month's and last month's) would be a lag-1 autocorrelation coefficient, $r(1)$.

Figure A.6(a) diagrammatically illustrates a lag-1 autocorrelation, with the arrows indicating the data 'pairs' used in the correlation analysis. The number of pairs of observations decreases as $k$ increases, as can be seen by comparing Figures A.6(a) and A.6(b). Autocorrelations can also be computed in a 'circular' manner, where the shifted data are wrapped around to start over at the beginning, as in Figure A.6(c). The correlation coefficient should only be computed with a value of $k$ much less than

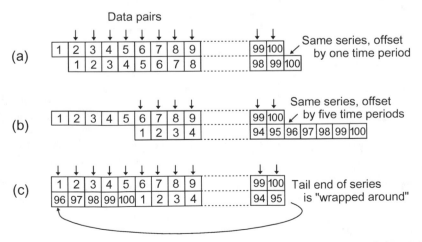

**Figure A.6** *Diagrammatic illustration of (a) lag-1, (b) lag-5 and (c) circular lag-5 autocorrelations, with* n = 100

the number of items in the data series, $n$. The upper limit of $k$ is usually taken as $n/4$. For a series of length 100, this would mean that only lag-1 through lag-25 autocorrelations would be performed.

The values of $r(k)$ can range from $-1$ (high values correlate with low values) to $+1$ (high values correlate with high values and low with low). The value of $r(0)$ will always be $+1$; i.e. an observation is perfectly correlated with itself. A value of $r(k)$ significantly different from zero means that part of the information contained in each observation is already known through its correlation with a preceding observation (Haan, 1977).

Lag-1 autocorrelation coefficients are usually of the most interest in hydrology and are considered a measure of persistence. For daily streamflows, a high positive value of $r(1)$ implies that streamflows are similar to those of the previous day. For annual streamflows, $r(1)$ will not be significantly different than zero for most streams. For monthly flows, however, it will nearly always be positive and significant (McMahon and Mein, 1986). Large, stable streams will have a high correlation between consecutive days and thus a high daily $r(1)$ as compared to flashy streams. Jones and Peters (1977), for example, used the average of the 1-, 2-, 3-, 4- and 5-day lag autocorrelation coefficients as a measure of the instability of the flow regime in stream classification. Negative lag-1 coefficients are unusual and difficult to explain in hydrology, implying that a large flow is followed by a small one and vice versa.

The lag-$k$ autocorrelation coefficient can be calculated from the equation (Wallis and O'Connell, 1972):

$$r(k) = \frac{\sum_{i=1}^{n-k}(x_i - \bar{x})(x_{i+k} - \bar{x})}{\sum_{i=1}^{n}(x_i - \bar{x})^2} \qquad (A.29)$$

If the autocorrelation coefficient is calculated for a range of $k$ values, a graph can be constructed by plotting $r(k)$ against $k$. This graph is called a *correlogram*. Examples are shown in Figure A.7. A data series that is entirely random will yield a graph similar to Figure A.7(a), with the autocorrelation decreasing quickly from 1.0 to near-zero. Figure A.7(b) represents a data series with a cyclical or periodic component. Figure A.7(c) shows the correlogram for daily streamflow data from the Acheron River, $k = 0$ to $k = 730$ (2 years).

Assumptions required in computing a serial correlation coefficient are that observations are equally spaced in time and that the statistical properties do not change with time. Correlation methods are also based on the assumption that the data are independent and composed of random observations (Haan, 1977). This is paradoxical since, by definition, correlated data are not random. Therefore, statistical tests are used to evaluate whether the deviation of the coefficient from zero is too large to have occurred by chance. Raudkivi (1979) gives a simple approximation for the confidence limits on the autocorrelation coefficient as $\pm 2/\sqrt{n}$ (at the 0.05 significance level). If the computed value of $r(k)$ lies outside these limits, it is considered significantly different than zero, indicating persistence in the data. A more precise equation can be found in statistical texts.

### Intercorrelation

*Intercorrelation* or *cross-correlation* was mentioned as a consideration in reducing the number of significant variables in a multiple regression equation. The implication is that if variables are intercorrelated, they do not add as much information as they would if they were totally independent. Thus, a few, carefully selected independent variables provide as much information as several intercorrelated ones.

For the same reason, a few independent measurements of rainfall or streamflow or stream length from different sites will provide as much information about the 'true' mean and variance as several intercorrelated measurements. Raudkivi (1979) gives this equation for computing a regional mean intercorrelation coefficient ($\bar{r}$)

$$\bar{r} = \frac{2}{n(n-1)} \sum_{j=1}^{n-1} \sum_{i=j+1}^{n} r_{ij} \qquad (A.30)$$

where $r_{ij}$ represents the correlation coefficient calculated between stations $i$ and $j$, and $n$ is the total number of

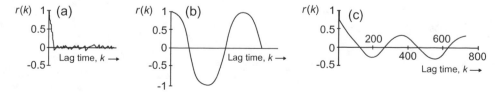

**Figure A.7.** *Correlograms showing the relationship between the lag-k autocorrelation coefficient, r(k) and the lag period, k, for (a) a random data series, (b) a cyclical data series and (c) daily streamflow data from the Acheron River at Taggerty (lag-0 to lag-730)*

stations. Alexander (1954) suggests that the number of independent stations ($n'$) required to give the same variance as some number $n$ of intercorrelated stations is given by

$$n' = \frac{n}{[1 + \bar{r}(n - 1)]} \qquad (A.31)$$

where $n$ is the number of stations having a mean intercorrelation coefficient of $\bar{r}$ and $n'$ is the number of equivalent independent stations which would provide as much information about the variance. It can be shown that as $n$ becomes large, $n'$ approaches $1/\bar{r}$. As an example, if 100 stations have an average correlation of 0.25, $n'$ is 3.9. Thus, only about 4 independent stations would be needed to provide as much information as the original 100. A logical conclusion is that only a few independent stations should be installed rather than many correlated ones. However, 'overlap' of stations is often desirable for other reasons such as the estimation of missing data.

Thus, data from the same station may be serially correlated and those between stations intercorrelated. Intercorrelation actually has the effect of *improving* the amount of information about the mean. For example, if many rainfall gauges in an area report similar values, it strengthens the evidence that mean values are correct. Serial correlation, in comparison, tends to *decrease* the amount of information about the mean (Haan, 1977). For

example, if streamflows occurred randomly about a mean value the mean daily flow for that year could be estimated after only a few months, but because of persistence in the data due to seasonality, we must wait until the end of the year.

## A.5 Further Information on Statistics and Sampling Design

Statistical texts that can be consulted for more information include Cochran and Cox (1957), Devore (1982), Dunteman (1984), Elliot (1977), Green (1979), Haan (1977), Holder (1985), Mann (1995), Snedecor and Cochran (1989), Sokal and Rohlf (1969), Steel and Torrie (1960), Tabachnick and Fidell (1989), Wardlaw (1985), Yevjevich (1972) and Zar (1974).

Commercial computer packages for statistical analysis include BMDP (Dixon, 1984), MINITAB (Ryan *et al.*, 1985), SAS (SAS, 1987), SPSS (SPSS, 1986) and SYSTAT (Wilkinson, 1986). A text written by Kvanli (1988) describes the use of SPSS, SAS and MINITAB, and Afifi and Clark (1984) provide a reference on multivariate analysis using the packages BMDP, SAS and SPSS-X. Statistical software suppliers can now be contacted via the Internet; for example, a demonstration version of SYSTAT can be downloaded from www.systat.com/downloads/.

# References

Abbe, T.B. and Montgomery, D.R. 1996. Large woody debris jams, channel hydraulics and habitat formation in large rivers. *Regulated Rivers: Research & Management* **12**: 201–221.

Abbe, T.B., Montgomery, D.R. and Petroff, C. 1997. Design of stable in-channel wood debris structures for bank protection and habitat restoration: an example from the Cowlitz River, WA. In Wang, S.S.Y., Langendeon, E.J. and Shields, F.D. Jr (eds) *Proceeding of Conference on Management of Landscapes Disturbed by Channel Incision*, pp. 809–815.

Abel, P.D. 1989. *Water Pollution Biology*. Ellis Horwood Limited, Chichester.

Abernethy, B. and Rutherfurd, I.D. 1996. Vegetation and bank stability in relation to changing channel scale, In Rutherfurd, I. and Walker, M. (eds) *Proceedings of the First National Conference on Stream Management in Australia*, Merrijig 19–23 Feb, Cooperative Research Centre for Catchment Hydrology, Monash University, Clayton, pp. 213–219.

Abernethy, B. and Rutherfurd, I.D. 1998. Where along a river's length will vegetation most effectively stabilise stream banks? *Geomorphology* **23**: 55–75.

Abernethy, B. and Rutherfurd, I.D. 1999. Guidelines for stabilising streambanks with riparian vegetation. *Technical Report 99/10.* Cooperative Research Centre for Catchment Hydrology, Monash University, Clayton, Victoria.

Abernethy, B. and Rutherfurd, I.D. 2000. The effect of riparian tree roots on the mass-stability of riverbanks. *Earth Surface Processes and Landforms*, **25**: 921–937.

ABRG (Australian Biological Research Group) (no date). Conservation Values of Lakes and Wetlands in South-Western Region Victoria—Technical Report. Report to Department of Water Resources Victoria, Conservation Forests and Lands and Ministry for Planning and Environment, East Melbourne.

Abt, S.R., Wittler, R.J., Taylor, A. and Love, D.J. 1989. Human stability in a high flood hazard zone. *Water Resour. Bull.* **25**: 881–890.

Ackers, P., White, W.R., Perkins, J.A. and Harrison, A.J.M. 1978. *Weirs and Flumes for Flow Measurement*. John Wiley, Chichester (as cited by Shaw, 1988).

Acreman, M.C. 2003a. *Environmental Flows: Flood Flows. Water Resources and Environment Technical Note C3,* Davis, R. and Hirji, R. (eds), The World Bank, Washington, DC. URL: http://lnweb18.worldbank.org/ESSD/ardext.nsf/18ByDocName/EnvironmentalFlowAssessment-NOTEC3Environmental-FlowsFloodFlows/$FILE/NoteC3EnvironmentalFlowAssessment2003.pdf (accessed 1st Nov 2003).

Acreman, M.C. 2003b. Defining water requirements. In Dyson, M., Bergkamp, G. and Scanlon, J. (eds) *Flow: The Essentials of Environmental Flows*. Water and Nature Initiative, International Union for Conservation of Nature and Natural Resources, Gland, Switzerland and Cambridge, UK, pp. 11–29. URL: http://www.waterandnature.org/pub/FLOW.pdf (accessed 1st Nov 2003).

Acreman, M.C. and Sinclair, C.D. 1986. Classification of drainage basins according to their physical characteristics: an application for flood frequency analysis in Scotland. *Journal of Hydrology* **84**: 365–380.

Afifi, A.A. and Clark, V. 1984. *Computer-aided Multivariate Analysis,* Lifetime Learning Publications, Belmont, CA.

AFN (Association Francaise de Normalisation). 1985. Essais des eaux. Détermination de l'indice biologique global (IBGN), NF T90–350, Paris.

Agostinho, A.A., Gome, L.C., Fernandez, D.R. and Suzuki, H.I. 2002. Efficiency of fish ladders for neotropical ichthyofauna. *River Research and Applications* **18**(3): 299–306.

Aitken, A.P. 1973. Assessing systematic errors in rainfall-run-off models. *J. of Hydrol.* **20**: 121–36.

Alberta Environmental Protection Fish and Wildlife, Trout Unlimited Canada, Pisces Environmental Consulting Services Ltd. and Department of Fisheries and Oceans. 1994. *Proceedings, 9th International Trout Stream Habitat Improvement Workshop*, 6–9 Sep, Calgary, AB.

Alexander, G.N. 1954. Some aspects of time series in hydrology. *J. Inst. of Engineers* (Australia), Sep 196 (as cited by Haan, 1977).

Alfredsen K. 1998. Habitat modelling in Norway—an overview of projects and future developments. In Blažková, Š., Stalnaker, C. and Novický, O. (eds) *Hydroecological Modelling. Research, Practice, Legislation and Decision-making*. Report by U.S. Geological Survey, Biological Research Division and Water Research Institute, Fort Collins, and Water Research Institute, Praha, Czech Republic, VUV, Praha, pp. 33–35.

Alfredsen, K., Bakken, T.H. and Killingtveit, A. (eds) 1995. *The River System Simulator, User's Manual*. SINTEF NHL Report, Trondheim, Norway.

Allan, J.D. 1984. Hypothesis testing in ecological studies of aquatic insects. In Resh, V.H. and Rosenberg, D.M. (eds) *The Ecology of Aquatic Insects*. Praeges, New York, pp. 484–507.

Allan J.D. 2001, *Stream Ecology: Structure and Function of Running Waters*, Kluwer, Dordrecht, pp. 388.

Allen, T. 1981. *Particle Size Measurement*, 3rd edn, Chapman and Hall, New York (as cited by Klute, 1986).

American Rivers 2003a. 57 Dams in 16 States to be removed in 2003. American Rivers, Washington, DC, August. URL: http://amrivers.org/pressrelease/damremoval8.19.03.htm (accessed 1st Nov 2003).

American Rivers. 2003b. Edwards Dam case study. Dam Removal Toolkit. American Rivers, Washington, DC. URL: www.amrivers.org/ damremovaltoolkit/edwardsdam.htm, (accessed 1st Nov 2003).

American Rivers, Friends of the Earth and Trout Unlimited, 1999. Dam removal success stories restoring rivers through selective removal of dams that don't make sense. Washington, DC, December. URL: http://www.amrivers.org/docs/SuccessStoriesReport.pdf (accessed 1st Nov 2003).

Amoros, C., Roux, A.L., Reygrobellet, J.L., Bravard, J.P. and Pautou, G. 1987. A method for applied ecological studies of fluvial hydrosystems. *Regulated rivers* 1: 17–36.

Anderson, J.J. 2000. Decadal climate cycles and declining Columbia River salmon. In Knudsen, E.E, Stewart, C.R., MacDonald, D.D., Williams, J.E. and Reiser, D.W. (eds) *Sustainable Fisheries Management: Pacific Salmon*. Lewis Publishers, New York.

Anderson, J.L. 1998. Embracing uncertainty: the interface of Bayesian statistics and cognitive psychology. *Conservational ecology* [online] 2(1): 2. URL: http://www.consecol.org/vol2/iss1/art2 (accessed 1st Nov 2003).

Anderson, J.R. 1993. State of the Rivers Project. Report 1. Development and Validation of the Methodology. Department of Primary Industries, Queensland.

Anderson, J.R. 1999. Basic Decision Support System for Management of Urban Streams. Report A: Development of the Classification System for Urban Streams. National River Health Program, Urban Sub Program. *Occasional Paper 8/99*. Land and Water Resources Research and Development Corporation, Canberra, Australian Capital Territory.

Andrews, D.F. 1972. Plots of high dimensional data. *Biometrics* **28**: 125–36.

Andrews, E.D. 1983. Entrainment of gravel from naturally sorted riverbed material. *Geol. Soc. of Amer. Bull.* **94**: 1225–1231.

Andrews, E.D. 1984. Bed-material entrainment and hydraulic geometry of gravel-bed rivers in Colorado. *Bulletin of the Geological Society of America* **95**: 371–378.

Andrews, E.D. and Nankervis, J.M. 1995. Effective discharge and the design of channel maintenance flows for gravel-bed rivers. In Costa, J.E., Miller, A.J., Potter, K.W. and Wilcock, P.R. (eds) *Natural and Anthropological Influences in Fluvial Geomorphology: The Wolman Volume*. Geophysical Monograph 89, American Geophysical Union, Washington, DC, pp. 151–164.

Andrus, C. 1991. Improving Streams and Watersheds in Oregon. Inventory and evaluation of efforts to improve the condition of Oregon's streams and watersheds from 1985 to 1990. Water Resources Department, Salem, Oregon.

Annear, T., Chisholm, I., Beecher, H., Locke, A., Aarrestad, P., Burkardt, N., Coomer, C., Estes, C., Hunt, J., Jacobson, R., Jobsis, G., Kauffman, J., Marshall, J., Mayes, K., Stalnaker, C. and Wentworth, R. 2002. *Instream Flows for Riverine Resource Stewardship*. Instream Flow Council, Cheyenne, WY.

Anon. 1951. A fish-pass problem. *Water Power* September: 342–347.

Anon. 1958. A $5 million gamble to build a better fish trap. *Engineering News-Record* August **14**: 46–48.

ANZECC. 2000. *Core Environmental Indicators for Reporting on State of the Environment*. Australian and New Zealand Environment and Conservation Council, State of the Environment Reporting Task Force 2000, Environment Australia, Canberra, Australian Capital Territory. URL: http://www.deh.gov.au/soe/publications/coreindicators.html (accessed 25th Mar 2004).

ANZECC and ARMCANZ. 2000a. Australian and New Zealand Guidelines for Fresh and Marine Water Quality. Volume 2, Aquatic Ecosystems — Rationale and Background Information (Chapter 8), *National Water Quality Management Strategy Paper No. 4*. Australian and New Zealand Environment and Conservation Council, Agriculture and Resource Management Council of Australia and New Zealand, Canberra, Australian Capital Territory. URL: http://www.deh.gov.au/water/quality/nwqms/volume2.html (accessed 25th Mar 2004).

ANZECC and ARMCANZ 2000b. Australian Guidelines for Water Quality Monitoring and Reporting. *National Water Quality Management Strategy No. 7*. Australian and New Zealand Environment and Conservation Council, Agriculture and Resource Management Council of Australia and New Zealand, Canberra, Australian Capital Territory. URL: http://www.deh.gov.au/water/quality/nwqms/monitoring.html (accessed 25th Mar 2004).

ANZECC and ARMCANZ. 2000c. An Introduction to the Australian and New Zealand Guidelines for Fresh and Marine Water Quality. *National Water Quality Management Strategy Paper No. 4a*. Australian and New Zealand Environment and Conservation Council, Agriculture and Resource Management Council of Australia and New Zealand, Canberra, Australian Capital Territory. URL: http://www.deh.gov.au/water/quality/nwqms/introduction/ (accessed 25th Mar 2004).

Armitage, P.D., Lattmann, K., Kneebone, N. and Harris, I. 2001. Bank profile and structure as determinants of macroinvertebrate assemblages—seasonal changes and management. *Regulated Rivers: Research & Management* 17(4-5): 543–556.

Armitage, P.D., Moss, D., Wright, J.F. and Furse, M.T. 1983. The performance of a new biological water quality score system based on macroinvertebrates over a wide range of unpolluted running water sites. *Water Research* 17(3): 333–347.

Armour, C.L., Burnham, K.P. and Platts, W.S. 1983. Field methods and statistical analyses for monitoring small salmonid streams. *FWS/OBS-83/33*, U.S. Fish and Wildlife Service, Washington, DC.

Armour, C.L. and Taylor, J.G. 1991. Evaluation of the Instream Flow Incremental Methodology by U.S. Fish and Wildlife Service field users. *Fisheries* **16**: 36–43.

Arthington, A.H. 1998. Comparative evaluation of environmental flow assessment techniques: review of holistic methodologies. *LWRRDC Occasional Paper No. 26/98*. Land and Water

Resources Research & Development Corporation, Canberra, Australian Capital Territory.

Arthington, A.H. and Zalucki, J.M. (eds) 1998. Comparative evaluation of environmental flow assessment techniques: review of methods. *LWRRDC Occasional Paper No. 27/98*, Land and Water Resources Research & Development Corporation, Canberra, Australian Capital Territory.

Arthington, A.H. and Pusey, B.J. 1993. In-stream flow management in Australia: methods, deficiencies and future directions. *Australian Biology* **6**: 52–60.

Arthington, A.H. and Pusey, B.J. 1994. River health assessment and classification based on the relationships of flow regime, habitat anf fish assemblage structure. In Uys, M. (ed) Classification of Rivers, and Environmental Health Indicators. Proceedings of a joint South African Australian Workshop, Cape Town, Feb 7–11, *Water Research Commission Report No. TT 63/94*, Pretoria, Republic South Africa, pp. 191–204.

Arthington, A.H. and Pusey, B.J. 2003. Flow restoration and protection in Australian rivers. *River Research and Applications* **19**(5-6): 377–395.

Arthington, A.H., Brizga, S.O., Choy, S.C., Kennard, M.J., Mackay, S.J., McCosker, R.O., Ruffini, J.L. and Zalucki, J.M. 2000a. Environmental flow requirements of the Brisbane River downstream of Wivenhoe Dam. South East Queensland Water Corporation, and Centre for Catchment and In-Stream Research, Griffith University, Brisbane.

Arthington, A.H., Brizga, Bunn, S.E. and Loneragan, N.R. 2000b. Burnett Basin WAMP, Current Environmental Conditions and Impacts of Existing Water Resource Development, Appendix J, Estuarine and Marine Ecosystems, The State of Queensland, Department of Natural Resources, June. URL: http://www.nrm.qld.gov.au/wrp/pdf/burnett/vol2/AppJ.pdf (accessed 1st Nov 2003).

Arthington, A.H., King, J.M., O'Keeffe, J.H., Bunn, S.E., Day, J.A., Pusey, B.J., Bludhorn, D.R. and Tharme, R. 1992. Development of an holistic approach for assessing environmental flow requirements of riverine ecosystems. In Pigram, J.J. and Hooper, B.P. (eds) *Water Allocation for the Environment, Proceedings of an International Seminar and Workshop.* Centre for Water Policy Research, University of New England Armidale, pp. 69–76.

Arthington, A.H., Rall, J.L., Kennard, M.J. and Pusey, B.J. 2003. Environmental flow requirements of fish in Lesotho rivers using the DRIFT methodology. *River Research and Applications* **19**(5-6): 641–666.

ASCE. 1982. *Design of Water Intake Structures for Fish Protection.* American Society of Civil Engineers, New York.

ASCE, Task Committee on Guidelines for Retirement of Dams and Hydroelectric Facilities of the Hydropower Committee of the Energy Division. 1997. *Guidelines for Retirement of Dams and Hydroelectric Facilities.* American Society of Civil Engineers, New York, p. 243.

Ashmore, P. 1999. What would we do without Rosgen? Rational regime relations and natural channels. In *Proceedings of the Second International Conference on Natural Channel Systems*, Session 9—Design Applications (CD-ROM), 1–4 Mar, Niagara Falls, Canada.

Askey-Doran, M., Pettit, N., Robins, L. and McDonald, T., 1999. The role of vegetation in riparian management. In Lovett. S. and Price, P. (eds) *Riparian Land Management Technical Guidelines, Volume One: Principles of Sound Management.* Land and Water Resources Research and Development Corporation, Canberra, Australian Capital Territory, pp. 97–120. URL: http://www.rivers.gov.au/acrobat/techguidelines/tech_guide_vol1a_chap08.pdf (accessed 1st Nov 2003).

Astles, K.L., Winstanley, R.K., Harris, J.H. and Gehrke, P.C. 2003. Experimental study of the effects of cold water pollution on native fish. Regulated Rivers and Fisheries Restoration Project. *NSW Fisheries Final Report Series No. 44.* NSW Fisheries, Office of Conservation, NSW Fisheries Research Institute, Cronulla, NSW, April. URL: http://www.fisheries.nsw.gov.au/sci/outputs/ooc/pdf/317_Cold.pdf (accessed 1st Nov 2003).

Auble, G.T., Friedman, J.M. and Scott, M.L. 1994. Relating riparian vegetation to present and future streamflows. *Ecological Applications* **4**: 544–554.

Avery, T.E. and Berlin, G.L. 1992. *Fundamentals of Remote Sensing and Airphoto Interpretation*, 5th edn, Prentice-Hall, Englewood Cliffs, NJ.

AWRC. 1984. Workshop on surface water resources data, *AWRC Conference Series No. 10*, Canberra, 23–25 No. 1983, Australian Water Resources Council, Canberra.

Baattrup-Pedersen, A., Riis, T., Ole Hansen, H. and Friberg, N. 2000. Restoration of a Danish headwater stream: short-term changes in plant species abundance and composition. *Aquatic Conservation: Marine and Freshwater Ecosystems* **10**(1): 13–23.

Bagnold, R.A. 1966. An approach to the sediment transport problem from general physics, *US Geol. Survey Prof. Paper 442–I*, USGS, Washington, DC.

Bagnold, R.A. 1980. An empirical correlation of bedload transport rates in flumes and natural rivers. *Proc. Roy. Soc. London.* **A372**: 453–73.

Bailey, R.C., Kennedy, M.G., Dervish, M.Z. and Taylor, R.M. 1998. Biological assessment of freshwater ecosystems using a reference condition approach: comparing predicted and actual benthic invertebrate communities in Yukon streams. *Freshwater Biology* **39**(4): 765–774.

Bain, M.B. and Hughes, T. 1996. Aquatic habitat inventory and analysis methods used in fishery and environmental management. Report by the Cooperative Fish and Wildlife Research Unit, Cornell University to the American Fisheries Society, Bethesda, Maryland, New York. August.

Bain, M.B. and Travnichek, V.H. 1996. Assessing impacts and predicting restoration benefits of flow alterations in rivers developed for hydroelectric power production. In Leclerc, M. Capra, H., Valentin, S., Boudreault, A. and Côté, Y. (eds) *Proceedings of the 2nd International Symposium on Habitat Hydraulics, Ecohydraulics 2000.* Institute National de la Recherche Scientifique-Eau, co-published with FQSA, IAHR/AIRH, Ste-Foy, Québec, Canada, pp. B543–B552. URL: http://www.dnr.cornell.edu/hydro2/flalt.htm (accessed 1st Nov 2003).

Baker, S. and Gilroy, U.B. 1933. Problems of fishway construction. *Civil Engineering* **3**(12): 671–675.

Bain, M.B, Finn, J.T. and Booke, H.E. 1988. Streamflow regulation and fish community structure. *Ecology* **69**: 382–392.

Baker, V.R. 1977. Stream-channel response to floods, with examples from central Texas. *Geol. Soc. of Amer. Bull.* **88**, 1057–1071.

Bao, Y. and Mays, L.W. 1994a. New methodology for optimisation of freshwater inflows to estuaries. *Journal of Water Resources Planning and Management*, ASCE **102**(2): 199–217.

Bao, Y. and Mays, L.W. 1994b. Optimization of freshwater inflows to Lavaca-tres Palacios, Texas, Estuary. *Journal of Water Resources Planning and Management*, ASCE **102**(2): 218–236.

Barbour, M.T. and Stribling, J.B. 1991. Use of habitat assessment in evaluating the biological integrity of stream communities. In Gibson, G. (ed.) *Biological Criteria: Research and Regulation. Proceedings of a Symposium,* 12–13 Dec 1990, Arlington, Virginia, EPA-440-5-91-005. Office of Water, U.S. Environmental Protection Agency, Washington, DC.

Barbour, M.T., Gerritsen, J., Snyder, B.D. and Stribling, J.B. 1999. *Rapid Bioassessment Protocols for Use in Streams and Wadeable Rivers: Periphyton, Benthic Macroinvertebrates and Fish*, 2nd edn, EPA 841-B-99-002. U.S. Environmental Protection Agency, Office of Water, Washington, DC.

Barinaga, M. 1996. A recipe for river recovery? *Science* **273**: 1648–1650.

Barling, R.D. and Moore, I.D. 1994. Role of buffer strips in management of waterway pollution: a review. *Environmental Management* **18**(4): 543–558.

Barmah Millewa Forum. 2001, Report on Barmah-Millewa Forest Flood of Spring 2000 and the second use of Barmah-Millewa Forest Environmental Water Allocation, Spring Summer 2000/2001. Canberra, Australian Capital Territory.

Barmah-Millewa Forum. 2000. *Barmah-Millewa Forest Water Management Strategy.* Murray-Darling Basin Commission, Canberra, Australian Capital Territory, June.

Barmuta, L.A. 1989. Habitat patchiness and macrobenthic community structure in an upland stream in temperate Victoria, Australia. *Freshwater Biol.* **21**, 223–236.

Barmuta, L.A., Chessman, B.C. and Hart, B.T. 2002. Interpretation of the outputs from AusRivAS. Second Milestone Report of Project UTA3 to the Land and Water Resources Research and Development Corporation. *Monitoring River Health Initiative Technical Report No. 24*, Environment Australia, Canberra, Australian Capital Territory.

Barnes, H.H., Jr 1967. Roughness characteristics of natural channels. *US Geol. Survey Water-Supply Paper 1849*, USGS, Washington, DC.

Barnes, R.K. and Mann, K.H. 1980. *Fundamentals of Aquatic Ecosystems*. Blackwell Scientific, Oxford.

Barnett, V. 1982. *Elements of Sampling Theory*. Hodder and Stoughton, London.

Barrett, E.C. and Curtis, L.F. 1992. *Introduction to Environmental Remote Sensing*, 3rd edn, Chapman and Hall, London (also 2nd edn, 1982).

Bartley, D.M. and Minchin, D. 1995. Precautionary approach to the introduction and transfer of aquatic species. In *Precautionary Approach to Fisheries Part 2. FAO Fisheries Technical Paper No. 350 Part 2*, Food and Agriculture Organization of the United Nations, Rome, pp. 159–189.

Basham, L.R. 1985. Adult fishway inspections on the Columbia and Snake Rivers 1984. Water Budget Center, Portland, Oregon.

URL: http://www.efw.bpa.gov/Environment/EW/EWP/DOCS/REPORTS/UPSTRM/U11797–3.pdf (accessed 1st Nov 2003).

Bates, K. 1997. *Fishway design guidelines for Pacific Salmon.* U.S. Department of Fish and Wildlife, Olympia, Washington.

Bayley, P.B. 1990. The flood pulse advantage and the restoration of river-floodplain systems. *Regulated Rivers: Research & Management* **6**: 75–86.

Bayley, P.B. and Osborne, L.L. 1993. Natural rehabilitation of stream fish populations in an Illinois catchment. *Freshwater Biology* **29**: 295–300.

Bayly, I.A.E. and Williams, W.D. 1973. *Inland Waters and Their Ecology.* Longman Australia, Victoria.

BC Fisheries Information Services Branch, 2000. Reconnaissance (1:20,000) Fish and Fish Habitat Inventory: Users Guide to the Fish and Fish Habitat Assessment Tool (FHAT20). Prepared by BC Fisheries Information Services Branch for the Resources Inventory Committee, Ministry of Sustainable Resource Management, Government of British Columbia, May. URL: http://srmwww.gov.bc.ca/risc/pubs/aquatic/fhat20/index.htm (accessed 1st Nov 2003).

Beaumont, P. 1975. Hydrology. In Whitton, B.A (ed.) *River Ecology.* Blackwell Scientific, Oxford, pp. 1–38.

Behnke, R.J. 1977. Fish faunal changes associated with land-use and water development. *Great Plains Rocky Mountain Geographical Journal* **6**:133–136.

Bell, F.C. and Vorst, P.C. 1981. Geomorphic parameters of representative basins and their hydrologic significance. Australian *Water Resources Council Technical Paper No. 58*, Australian Government Publishing Service, Canberra.

Bell, I. and Priestly, T. 1999. Management of stock access to the riparian zone. In Rutherfurd, I.D. and Bartley, R. (eds) *Proceedings of the Second Australian Stream Management Conference*, Adelaide, 8–11 Feb, Cooperative Research Centre for Catchment Hydrology, Monash University, Clayton, pp. 51–56.

Belovsky, G.E. 2002. Ecological stability: reality, misconceptions, and implications for risk assessment. *Journal of Human and Ecological Risk Assessment* **8**(1): 99–108.

Belsky, A.J., Matzke, A. and Uselman, S. 1999. Survey of livestock influences on stream and riparian ecosystems in the western United States. *Journal of Soil and Water Conservation* **54**(1): 419–431.

Bencala, K.E. 1984. Interactions of solutes and streambed sediment. 2. A dynamic analysis of coupled hydrologic and chemical processes that determine solute transport. *Water Resour. Res.* **20**, 1804–1814.

Bencala, K.E., Kennedy, V.C., Zellweger, G.W., Jackman, A.P., and Avanzino, R.J. 1984. Interactions of solutes and streambed sediment. An experimental analysis of cation and anion transport in a mountain stream. *Water Resour. Res.* **20**, 1797–1803.

Benke, A.C. 1990. A perspective on America's vanishing streams. *Journal of the North American Benthological Society* **9**: 77–88.

Bennett, S.J., Pirim, T. and Barkdoll, B.D. 2002. Using simulated emergent vegetation to alter stream flow direction within a straight experimental channel. *Geomorphology* **44**: 115–126.

Benson, M.A. 1965. Spurious correlation in hydraulics and hydrology. *J. Hyd. Div., Proc. ASCE* **HY4**: 35–42.

Benson, M.A. 1968. Uniform flood-frequency estimating methods for federal agencies. *Water Resour. Res.* **4**: 891–908.

Bergerud, W.A. and Reed, W.J. 1998. Bayesian statistical methods. In Sit, V. and Taylor, B. (eds) *Statistical Methods for Adaptive Management Studies*, Chapter 7. Land Management Handbook No. 42. Research Branch. B.C. Ministry of Forests Forestry Division Services Branch, Victoria, BC, pp. 89–104.

Beschta, R.L. 1983. Channel changes following storm-induced hillslope erosion in the Upper Kowai Basin, Torlesse Range, New Zealand, *J. of Hydrol. (NZ)* **22**: 93–111.

Beschta, R.L. and Platts, W.S. 1986. Morphological features of small streams: significance and function. *Water Resour. Bull.* **22**: 369–79.

Beschta, R.L., Platts, W.S. and Kauffman, J.B. 1992. Field review of fish habitat improvement projects in the Grande Ronde and John Day River basins of eastern Oregon. USDE. Bonneville Power Administration, DOE/BP-21493-1, Portland, Oregon.

Beschta, R.L., Platts, W.S., Kauffman, J.B. and Hill, M.T. 1994. Artificial stream restoration—money well spent or an expensive failure? In *Environmental Restoration, Proceedings UCOWR 1994 Annual Meeting*, Big Sky Montana, The Universities Council on Water Resources, pp. 76–102.

Best, G.A. and Ross, S.L. 1977. *River Pollution Studies*. Liverpool University Press, Liverpool.

Beven, K.J. and Callen, J.L. 1979. HYDRODAT—a system of FORTRAN computer programs for the preparation and analysis of hydrological data from charts. *Technical Bulletin No. 23*, British Geomorphological Research Group, Geo Abstracts Ltd, Norwich.

Bevenger, G.S. and King, R.M. 1995. A pebble count procedure for assessing watershed cumulative effects. *USDA Forest Service Research Paper RM-RP-319*, US Forest Service, Rocky Mountain Forest and Range Experiment Station, Fort Collins, Colorado.

Bhowmik, N.G. and Adams, J.R. 1986. The hydrologic environment of Pool 19 of the Mississippi River. *Hydrobiologia* **136**: 21–30.

Biggs, B.J.F. 1989. Biomonitoring of organic pollution using periphyton, South Branch, Canterbury, New Zealand. *New Zealand Journal of Marine and Freshwater Research* **23**: 263–274.

Biggs, B.J.F. 2000. *New Zealand Periphyton Guideline: Detecting Monitoring and Managing Enrichment of Streams*. Prepared for Ministry of the Environment, NIWA, Christchurch, NZ. June URL: http://www.mfe.govt.nz/publications/water/nz-periphyton-guide-jun00.pdf (accessed 1st Nov 2003).

Billington, N. and Herbert, P.D.N. (eds) 1991. International Symposium on the ecological and genetic implications of fish introductions (FIN). *Canadian Journal of Aquatic Sciences* **48** (Suppl. 1): 181.

Binder, R.C. 1958. *Advanced Fluid Mechanics*. Prentice-Hall, Englewood Cliffs, NJ.

Binnie, Black & Veath Engineering Consultants 1998. Determining the freshwater flow needs of estuaries. *R&D Technical Report W113*. Environment Agency, Bristol. Technical Summary online at URL: http://www.eareports.com/ea/rdreport.nsf/0/f9b3b062221fd7cf802567e80054efd5?OpenDocument (accessed 1st Nov 2003).

Bishop, K. 1996. Review of the 'Expert Panel' (EPAM) process as a mechanism for determining environmental flow releases for freshwater fish. Report to the Centre for Water Policy Research, University of New England, on behalf of the Snowy Mountains Hydro-Electric Authority, Cooma, New South Wales.

Bisson, P.A. and Montgomery, D.R. 1996. Valley segments, stream reaches, and channel units. In Hauer, F.R. and Lamberti, G.A. (eds) *Methods in Stream Ecology*, Academic Press, New York, pp. 23–52.

Bjornn, T.C., Peery, C.A., Hunt, J.P., Tolotti, K.R., Keniry, P.J. and Ringe, R.R. 1999. Evaluation of fishway fences (1991–1998) and spill for adult salmon and steelhead passage (1994) at Snake River dams. Part VI of *Migration of Adult Chinook Salmon and Steelhead Past Dams and Through Reservoirs in the Lower Snake River and into Tributaries*. U.S. Geological Survey, Idaho Cooperative Fish and Wildlife Research Unit, University of Idaho, Moscow. URL: http://www.cnr.uidaho.edu/coop/PDF%20Files/PartVIfwfence.pdf (accessed 1st Nov 2003).

Blackman, D.R. 1969. *SI Units in Engineering*. Macmillan, Melbourne.

Blake, T.A., Blake, M.G. and Kittredge, W. 2000. *Balancing Water: Restoring the Klamath Basin*. University of California Press, Berkeley, CA.

Blanch S.J. 1999. Environmental flows: present and future. Presented at the Australian National Committee on Large Dams 1999 Conference, 15–16 Nov, Jindabyne, N.S.W. Inland Rivers Network. URL: http://www.nccnsw.org.au/member/wetlands/news/media/19991118_EFpaper.html (accessed 1st Nov 2003).

Blanch, S.J., Ganf, G.G. and Walker, K.F. 1999. Tolerance of riverine plants to flooding and exposure indicated by water regime. *Regulated Rivers: Research & Management* **15**(1–3): 43–62.

Blöch, H. 1999. The European Union Water Framework Directive: taking European water policy into the next millennium. *Water Science and Technology* **40**(10): 67–71.

Blom, G. 1958: *Statistical Estimates and Transformed Beta-variables*. John Wiley & Sons, Inc., New York.

Blyth, K. and Rodda, J.C. 1973. A stream length study. *Water Resour. Res.* **9**: 1454–61.

Boedeltje, G., Smolders, A.J.P., Roelofs, J.G.M. and Van Groenendael, J.M. 2001. Constructed shallow zones along navigation canals: vegetation establishment and change in relation to environmental characteristics. *Aquatic Conservation: Marine and Freshwater Ecosystems* **11**(6): 453–471.

Bohn, C. 1986. Biological importance of streambank stability. *Rangelands.* **8**: 55–56.

Bolemon, J. 1989. *Physics: an Introduction*, 2nd edn, Prentice-Hall, Englewood Cliffs, NJ.

Bollweg, A.E. and VanHeerd R. 2001. Measuring and analysing of flood waves in Rhine river; an application of laser altimetry. In vanDijk, A. and Bos, M.G. (eds) *GIS and Remote Sensing Techniques in Land—and Water—Management*. Kluwer Academic Publishers: Dordrecht, The Netherlands. Chapter 5, pp. 55–67.

Bone, Q. and Marshall, N.B. 1982. *Biology of Fishes*. Blackie, Glasgow.

Boon, P.J. 1988. The impact of river regulation on invertebrate communities in the U.K. *Regulated Rivers: Research & Management* **2**: 389–409.

Boon, P.J. 2000. The development of integrated methods for assessing river conservation value. *Hydrobiologia* **422/423**: 413–428.

Boon, P.J., Holmes, N.T.H., Maitland, P.S. and Fozzard, I.R. 2002. Developing a new version of SERCON (System for Evaluating Rivers for Conservation). *Aquatic Conservation: Marine and Freshwater Ecosystems* **12**(4): 439–455.

Boon, P.J., Holmes, N.T.H., Maitland, P.S., Rowell, T.A. and Davies, J. 1997. A system for evaluating rivers for conservation (SERCON): development, structure and function. In Boon, P.J. and Howell, D.L. (eds) *Freshwater Quality: Defining the Indefinable?* The Stationery Office, Edinburgh, pp. 299–326.

Boon, P.J., Wilkinson, J. and Martin, J. 1998. The application of SERCON (System for Evaluating Rivers for Conservation) to a selection of rivers in Britain. *Aquatic Conservation: Marine and Freshwater Ecosystems* **8**: 597–616.

Booth, D.B., Karr, J.R., Schauman, S., Konrad, C.P., Morley, S.A., Larson, M.G., Henshaw, P.C., Nelson, E.J. and Burges, S.R., 2001. Urban Stream Rehabilitation in the Pacific Northwest. Final report of EPA Grant No. R82-5284-010. Center for Urban Water Resources Management, University of Washington, Seattle, March. URL: http://depts.washington.edu/cwws/Research/Reports/final%20rehab%20report.pdf (accessed 1st Nov 2003).

Booth, D.B., Montgomery D.R. and Bethel, J.P. 1996. Large woody debris in urban stream channels of the Pacific Northwest. In Roesner, L.A. (ed) *Effects of Watershed Development and Management of Aquatic Ecosystems. Engineering Foundation Conference Proceedings*, Snowbird, Utah, pp. 4–9.

Borg, D., Rutherfurd, I.D. and Stewardson, M.J. 2003. Rehabilitation of sand bed streams. River and catchment health: presenting current research in the Goulburn Broken Catchment, Melbourne. Cooperative Research Centre for Catchment Hydrology, The University of Melbourne, Parkville, Victoria.

Bostelmann, R., Braukmann, U., Briem, E., Fleischhacker, Humborg, G., Nadolny, I., Scheurlen, K. and Weibel, U. 1998. An approach to classification of natural streams and floodplains in south-west Germany. In DeWaal, L.C., Large, A.R.G. and Wade, P.M. (eds) *Rehabilitation of Rivers: Principles and Implementation*. John Wiley & Sons, Chichester, pp. 31–55.

Boulton, A.J. 1999. An overview of river health assessment methods: philosophies, practice, problems and prognosis. *Freshwater Biology* **41**(2): 469–479.

Bounds, R.L. and Lyons, B.W. 1979. Existing reservoir and stream management recommendations: statewide minimum streamflow recommendations. Statewide Fishery Management Recommendations, Federal Aid Project F-30-R-4. Texas Parks and Wildlife Department, Austin, TX.

Bouyoucos, G.J. 1962. Hydrometer method improved for making particle size analysis of soils. *Agronomy J.* **54**, 464, 465.

Bovee, K.D. 1982. A guide to stream habitat assessment using the Instream Flow Incremental Methodology. *Instream Flow Information Paper No. 12*, FWS/OBS-82/26. Cooperative Instream

Flow Services Group, U.S. Fish and Wildlife Service, Fort Collins, CO.

Bovee, K.D. and Milhous, R. 1978. Hydraulic simulation in instream flow studies: theory and techniques. *Instream Flow Information Paper No. 5*, FWS/OBS-78/33, U.S. Fish and Wildlife Service, Office of Biological Services.

Bovee, K.D. 1974. The determination, assessment, and design of "instream value" studies for the Northern Great Plains region. *No. Great Plains Res. Prog.* (as cited by Stalnaker and Arnette, 1976).

Bovee, K.D. 1986. Development and evaluation of habitat suitability criteria for use in the Instream Flow Incremental Methodology. Instream Flow Information Paper No. 21, *U.S. Fish Wildlife Service Biol. Rep. 86*. U.S. Fish and Wildlife Service, Fort Collins, CO.

Bovee, K.D. 1996. Perspectives on two-dimensional river habitat models: the PHABSIM experience. In Leclerc, M., Capra, H., Valentin, S., Boudreault, A. and Côté, Y. (eds) *Proceedings of the Second IAHR Symposium on Habitat Hydraulics, Ecohydraulics 2000*. Institute National de la Recherche Scientifique-Eau, co-published with FQSA, IAHR/AIRH, Ste-Foy, Québec, Canada, pp. B149–B162.

Bovee, K.D. and Milhous, R. 1978. Hydraulic simulation in instream flow studies: theory and techniques. *Report FWS/OBS-78/33*. U.S. Fish and Wildlife Service, Fort Collins, CO.

Bowen, Z.H., Freeman, M.C. and Watson, D.L. 1996. Index of Biotic Integrity applied to a flow-regulated river system. In *Proceedings of the 50th Annual Conference of the Southeastern Association of Fish and Wildlife Agencies*. U.S. Fish and Wildlife, pp. 26–37.

Bowers, W., Hosford, B., Oakley, A. and Bond, C. 1979. Wildlife habitats in managed rangelands—the Great Basin of southeastern Oregon: Native Trout. *General Technical Report PNW-84*. Forest Service, U.S. Department of Agriculture.

Boyd, A. 1999. Second International Conference on Natural Channel Systems. Stream corridors: adaptive management and design—a conference summary. In *Proceedings of the Second International Conference on Natural Channel Systems*, Conference Summary (CD-ROM), 1–4 Mar, Niagara Falls, Canada.

Boyer, M.C. 1964. Streamflow measurement. In Chow, V.T. (ed) *Handbook of Applied Hydrology* Section 15, McGraw-Hill, New York.

Boyle, T.P., Smilie, G.M., Anderson, J.C. and Beeson, D.R. 1990. A sensitivity analysis of nine diversity and seven similarity indices. *Journal of Water Pollution Control Federation* **62**: 749–762.

Bradshaw, A.D. 1996. Underlying principles of restoration. *Canadian Journal of Fisheries and Aquatic Science* **53**(Suppl. 1): 3–9.

Bradshaw, A.D. 1998. Alternative endpoints for reclamation. In Cairns, J, Jr (ed) *Rehabilitating Damaged Ecosystems*. CRC Press, Boca Raton, FL.

Brakensiek, D.L., Osborn, H.B. and Rawls, W.J. (co-ordinators) 1979. *Field Manual for Research in Agricultural Hydrology*, Agriculture Handbook 224, USDA, Washington, DC.

Branson, F.A., Gifford, G.F., Renard, K.G. and Hadley, F.F. 1981. *Rangeland Hydrology*, 2nd edn, Kendall/Hunt, Iowa.

Brater, E.F. and King, H.W. 1976. *Handbook of Hydraulics*. McGraw-Hill, New York.

Bravard, J.-P., Landon, N., Peiry, J.-L. and Piégay, H. 1999. Principles of engineering geomorphology for managing channel erosion and bedload transport, examples from French rivers. *Geomorphology* **31**: 291–311.

Breen, P., Walsh, C., Nichols, S., Norris, R., Metzeling, L. and Gooderham, J. 2000. Urban AusRivAS: an evaluation of the use of AusRivAS models for urban stream assessment. *Land and Water Resources Research and Development Corporation, Occasional Paper 12/99*. National River Health Program, Urban Sub Program Report No. 5. LWRRDC, Canberra, Australian Capital Territory, p. 33.

Bren, L. J., O'Neill, I. and Gibbs, N.L. 1988. Use of map analysis to elucidate flooding in an Australian riparian river red gum forest. *Water Resour. Res.* **24**: 1152–62.

Brewer, R. 1964. *Fabric and Mineral Analysis of Soils*. John Wiley, New York.

Brierley, G. 1999. River styles: an integrative biophysical template for river management. In Rutherfurd, I.D. and Bartley, R. (eds) *Proceedings of the Second Australian Stream Management Conference*, Adelaide, 8–11 Feb, Cooperative Research Centre for Catchment Hydrology, Monash University, Clayton, pp. 93–100.

Brierley, G.J. and Fitchett, K. 2000. Channel planform adjustments along the Waiau River, 1946–1992; assessment of the impacts of flow regulation. In Bizga, S.O. and Finlayson, B.L. (eds) *River Management: The Australian Experience*. John Wiley & Sons, Chichester, pp. 51–71.

Brierley, G.J., Cohen, T., Fryirs, K. and Brooks, A. 1999. Post-European changes to the fluvial geomorphology of Bega catchment, Australia: implications for river ecology. *Freshwater Biology* **41**: 839–848.

Brierley, G.J., Fryirs, K. and Cohen, T. 1996. Development of a generic geomorphic framework to assess catchment character. Part 1. A geomorphic approach to catchment characterisation. *Working Paper 9603*, Macquarie University, Graduate School of the Environment, Sydney.

Briggs, D. 1977a. *Sources and Methods in Geography: Sediments*. Butterworths, London.

Briggs, D. 1977b. *Sources and Methods in Geography: Soils*. Butterworths, London.

Brinker, R.C. and Minnick, R. (eds) 1995. *The Surveying Handbook*, 2nd edn, Chapman and Hall, New York.

Brinson, M.M. 1993. A hydrogeomorphic classification for wetlands. *Wetlands Research Program Technical Report WRP-DE-4*, U.S. Army Corps of Engineers, Washington, DC, p. 79.

Brinson, M.M., Hauer, F.R., Lee, L.C., Nutter, W.L., Rheinhardt, R.D., Smith, R.D. and Whigham, D. 1995. A guide-book for application of hydrogeomorphic assessments to riverine wetlands. *Technical Report WRP-DE-11*, U.S. Army Engineer Waterways Experiment Station, Vicksburg, MS. URL: http://www.wes.army.mil/el/wetlands/pdfs/wrpde11.pdf (accessed 1st Nov 2003).

Brizga, S.O. 1998. Methods addressing flow requirements for geomorphological purposes. In Arthington, A.H. and Zalucki, J.M. (eds) Comparative Evaluation of Environmental Flow Assessment Techniques: Review of Methods. *Occasional Paper No. 27/98*. Land and Water Resources Research & Development Corporation, Canberra, Australian Capital Territory, pp. 8–46.

Brizga, S.O. and Finlayson, B. (eds) 2000. *River Management, the Australian Experience*. John Wiley & Sons, Chichester.

Brizga, S.O. and Finlayson, B.L. 1990. Channel avulsion and river metamorphosis: the case of the Thompson River, Victoria, Australia. *Earth Surf. Proc. and Landforms* **15**: 391–404.

Brizga, S.O. and Finlayson, B.L. 1994. Interactions between upland catchment and lowland rivers: an applied Australian case study. *Geomorphology* **9**: 189–201.

Brock, M.A. and Casanova, M.T. 1997. Plant life at the edges of wetlands: ecological responses to wetting and drying patterns. In Klomp, N. and Lunt, I. (eds) *Frontiers in Ecology. Building the Links*. Elsevier, Oxford, pp. 181–192.

Brookes A. 1995. The importance of high flows for riverine environments. In Harper, D.M. and Ferguson, A.J.D. (eds) *The Ecological Basis for River Management*. John Wiley & Sons, Chichester, pp. 33–49.

Brookes, A. 1989. Alternative channelization procedures. In Gore, J.A. and Petts, G.E. (eds) *Alternatives in Regulated River Management*. CRC Press, Boca Raton, FL, pp. 139–162.

Brookes, A. 1988. *Channelized Rivers: Perspectives for Environmental Management*. John Wiley & Sons, Chichester.

Brookes, A. 1990. Restoration and enhancement of engineered river channels: some European experiences. *Regulated Rivers: Research & Management* **5**: 45–56.

Brookes, A. and Gregory, K. 1988. Channelization, river engineering and geomorphology. In Hooke, J.M. (ed) *Geomorphology in Environmental Planning*. John Wiley, Chichester, pp. 145–167.

Brookes, A. and Shields, F.D., Jr (eds) 1996. *River Channel Restoration—Guiding Principals for Sustainable Projects*. John Wiley & Sons, Chichester, p. 433.

Brookes, A., Knight, S.S. and Shields, F.D., Jr 1996. Habitat Enhancement. In Brookes, A. and Shields, F.D., Jr (eds) *River Channel Restoration: Guiding Principles for Sustainable Projects*. John Wiley & Sons, Chichester, pp. 103–126.

Brooks, A.P., Abbe, T.B., Jansen, J.D., Taylor, M. and Gippel, C.J. 2001. Putting the wood back into our rivers: an experiment in river rehabilitation. In Rutherfurd, I., Sheldon, F., Brierley, G. and Kenyon, C. (eds) *Proceedings Third Australian Stream Management Conference*, Brisbane, 27–28 Aug, CRC Catchment Hydrology, Monash University, Clayton, Victoria, pp. 73–80.

Brooks, R.P., O'Connell, T.J., Wardrop, D.H. and Jackson, L.E. 1998. Towards a regional index of biological integrity: The example of forested riparian systems. *Environmental Monitoring and Assessment* **51**: 131–143.

Brown, A.L. 1971. *Ecology of Fresh Water*. Heinemann Educational, London.

Brown, C.A. and King, J.M. 2000. A summary of the DRIFT process, environmental flow assessments for rivers. *Southern Waters' Information Report No. 01/00*. Southern Waters, Ecological Research and Consulting, Mowbray, Republic of South Africa, August.

Brown, C.A. and King, J.M. 2003. *Environmental Flows: Concepts and Methods. Water Resources and Environment Technical Note C1,* In Davis, R. and Hirji, R. (eds), The World Bank, Washington, DC. URL: http://lnweb18.worldbank.org/ESSD/ardext.nsf/18ByDocName/EnvironmentalFlowAssessment-NOTEC1EnvironmentalFlowsConceptsandMethods/$FILE/NoteC1EnvironmentalFlowAssessment2003.pdf (accessed 1st Nov 2003).

Brown, C.M., Decker, G.T., Pierce, R.W. and Brandt, T.M. 2003. Applying Natural Channel Design Philosophy to the restoration of inland native fish habitat. *Practical Approaches for Conserving Native Inland Fishes of the West.* Montana Chapter of American Fisheries Society. U.S. Fish and Wildlife Service, Partners for Fish and Wildlife, Denver, Colorado. URL: http://www.r6.fws.gov/pfw/r6pfw2h16.htm (accessed 1st Nov 2003).

Brown, L.E., Hannah, D.M. and Milner, A.M. 2003. Alpine stream habitat classification: an alternative approach incorporating the role of dynamic water source contributions. *Arctic, Antarctic, and Alpine Research* 35(3): 313–322.

Brown, P., Douglas, J., Gooley, G. and Tennant, W. 1999. Biological assessment of aquatic habitat rehabilitation in the Broken River and Ryans Creek, North East Victoria. In Rutherfurd, I.D. and Bartley, R. (eds) *Proceedings of the Second Australian Stream Management Conference.* Adelaide, 8–11 Feb, Cooperative Research Centre for Catchment Hydrology, Monash University, Clayton, pp. 137–142.

Brown, R.M., McClelland, N.I., Deininger, R.A. and Tozer, R.G. 1970. A water quality index—do we dare? *Water and Sewage Works* 117(10): 339–343.

Brown, S.K., Buja, K.R., Jury, S.H. and Monaco, M.E. 2000. Habitat suitability index models for eight fish and invertebrate species in Casco and Sheepscot Bays, Maine. *North American Journal of Fisheries Management* 20: 408–435.

Brown, T.C. and Daniel, T.C. 1991. Landscape aesthetics of riparian environments: relationship of flow quantity to scenic quality along a wild and scenic river. *Water Resources Research* 27(8): 1787–1976.

Brussock, P.P., Brown, A.V. and Dixon, J.C. 1985. Channel form and stream ecosystem models. *Water Resources Bulletin* 21: 859–866.

Brusven, M.A. 1977. Effects of sediments on insects. In Kibbee, L. (ed.) *Transport of Granitic Sediments in Streams and its Effects on Insects and Fish, Wildlife and Range Expt. Sta. Bull. 17D* p. 43, USDA Forest Service, Univ. Idaho, Moscow, Idaho (as cited by Bovee, 1982).

Bryant, B.C. 1999. FERC's dam decommissioning authority under the Federal Power Act. *Washington Law Review* 74: 95–125.

BS. 1973. Methods of measurement of liquid flow in open channels, BS 3680, British Standards Institution, London (publication dates vary as updates are common).

BS. 1975. Methods of testing soils for civil engineering purposes, BS 1377, British Standards Institution, London.

Buchanan, D.V., Hanson, M.L. and Hooton, R.M. 1997. Status of Oregon's bull trout. Oregon Department of Fish and Wildlife, Portland, OR.

Budyko, M.I. 1980. *Global Ecology.* Progress Publishers, Moscow.

Buffangi, A., Kemp, J.L., Erba, S., Belfiorei, C., Hering, D. and Moog, O. 2001. A Europe-wide system for assessing the quality of rivers using macroinvertebrates: the AQEM Project and its importance for southern Europe (with special emphasis on Italy). *Journal of Limnology* 60(Suppl. 1): 39–48.

Buffington, J.M. 1996. An alternative method for determining subsurface grain size distributions of gravel-bedded rivers (abstract). American Geophysical Union 1996 Fall Meeting, Supplement to EOS, AGU Trsnsactions, 77(46): F250.

Bunn, S.E. and Arthington, A.H. 2002. Basic principles and ecological consequences of altered flow regimes for aquatic biodiversity. *Environmental Management,* 30, 492–507.

Bunn, S.E. and Boon, P.I. 1993. What sources of organic carbon drive food webs in billabongs? A study based on stable isotope analysis. *Oecologia* 96: 85–94.

Bunn, S.E., Davies, P.M. and Mosisch, T. 1999. Ecosystem measures of river health and their response to riparian and catchment degradation. *Freshwater Biology* 41: 333–345.

Bunn, S.E., Loneragan, N.R. and Yeates, M. 1998. The influence of river flows on coastal fisheries. In Arthington, A.H. and Zalucki, J.M. (eds). Comparative Evaluation of Environmental Flow Assessment Techniques: Review of Methods. *LWRRDC Occasional Paper No. 27/98.* Land and Water Resources Research & Development Corporation, Canberra, Australian Capital Territory, pp. 106–114.

Bunte, K., and Abt, S.R. 2001. Sampling surface and subsurface particle-size distributions in wadable gravel- and cobble-bed streams for analyses in sediment transport, hydraulics, and streambed monitoring, *Gen. Tech. Rep. RMRS-GTR-74,* USDA Forest Service, Rocky Mountain Research Station, Fort Collins, CO.

Bureau of Reclamation. 1999. Red River Valley Municipal, Rural, and Industrial Water Needs Assessment Phase I, Part B, Instream Flow Needs Assessment. Sheyenne River and Red River of the North, North Dakota and Minnesota, Final Appraisal Report. United States Department of the Interior, Bureau of Reclamation, Dakotas Area Office, Bismarck, ND, August.

Burgman, M.A and Lindenmayer, D.B. 1998. *Conservation Biology for the Australian Environment,* 1st edn., Surry Beatty & Sons, Chipping Norton, Sydney, p. 380.

Burke, D.G., Meyers, E.J., Tiner, R.W. and Groman, H. 1988. What are wetlands, and why are they important? In Protecting Nontidal Wetlands. *Planning Advisory Service Report No. 412/413,* American Planning Association, Chicago, IL, pp. 1–16.

Burley, J.B. 1990. Advancing recreation assessments. *Rivers* 1: 236–239.

Burston, J. 1996. Watercourse revegetation—just a walk in the park. In Rutherfurd, I. and Walker, M. (eds) *First National Conference on Stream Management in Australia,* Merrijig, 19–23 Feb, Cooperative Research Centre for Catchment Hydrology, Monash University, Clayton, pp. 247–251.

Bustard, D.R. 1984. Queen Charlotte Islands stream rehabilitation studies: a review of potential techniques. *Land Management Report No. 28,* B.C. Ministry of Forests, Forest Science Program, September. URL: http://www.for.gov.bc.ca/hfd/pubs/Docs/Mr/Lmr/Lmr028.pdf (accessed 1st Nov 2003).

Cadwallader, P.L. 1986. Flow regulation in the Murray River System and its effect on the native fish fauna. In Campbell,

I.C. (ed) *Stream Protection: The Management of Rivers for Instream Uses*. Water Studies Centre, Chisholm Institute of Technology, East Caulfield, Victoria, pp. 115–134.

Cairns, J., Jr 1988. *Rehabilitating Damaged Ecosystems*, CRC Press, Boca Raton, FL.

Cairns, J., Jr 1990. Lack of theoretical basis for predicting rate and pathways of recovery. *Environmental Management* **14**: 517–526.

Cairns, J., Jr 1995. Chemical versus biological pollution monitoring. In Rana, B.C. (ed) *Pollution and Biomonitoring*. Tata McGraw-Hill, New Delhi, pp. 7–25.

Cairns, J., Jr, Douglas, W.A., Busey, F. and Chaney, M.D. 1968. The sequential comparison index—a simplified method for non-biologists to estimate relative differences in biological diversity in stream pollution studies. *Journal of the Water Pollution Control Federation* **40**: 1607–1613.

Caissie, D. and El-Jabi, N. 2003. Instream flow assessment: from holistic approaches to habitat modelling. *Canadian Water Resources Journal* **28**(2): 173–184.

Caissie, D., El-Jabi, N. and Bourgeois, G. 1998. Évaluation du débit réservé par méthodes hydrologiques et hydrobiologiques. *Rev. Sci. Eau* **11**(3): 347–364.

Calow, P. 1992. Can ecosystems be healthy? critical consideration of concepts. *Journal of Aquatic Ecosystem Health* **1**: 1–5.

Cambray, J.A. 1991. The effects on fish spawning and management implications of impoundment water releases in an intermittent South African River. *Regulated Rivers: Research & Management* **6**: 39–52.

Campbell, I.C., James, K.R. and Edwards, R.T. 1989. Farming and streams—impact, research, and management. *Proceedings, The State of Our Rivers, Issues in Waterway Management II*, 28–29 Sept, Australian National University, Canberra.

Campbell, J.B. 1996. *Introduction to Remote Sensing*, 2nd edn, Guilford Press, New York.

Carling, P.A. 1988. Channel change and sediment transport in regulated U.K. rivers. *Regulated Rivers: Research & Management* **2**: 369–387.

Carpenter, S.C., Walker, B.H., Anderies, M. and Abel, N. 2001. From metaphor to measurement: Resilience of what to what? *Ecosystems* **4**: 765–781.

Carr, G., Lord, R. and Seymour, S. 1999. Rehabilitation of disturbed stream frontages using natural vegetation templates—a case study on the Yarra River, Victoria. In Rutherfurd, I.D. and Bartley, R. (eds) *Proceedings of the Second Australian Stream Management Conference*, Adelaide, 8–11 Feb, Cooperative Research Centre for Catchment Hydrology, Monash University, Clayton, pp. 155–162.

Carroll, J.H., Nolen, S.L. and Peterson, L. 1996. Water quality changes, from 1987 to 1991, in Broken Bow Lake, Oklahoma. *Proceedings of the Oklahoma Academy of Science* **76**: 35–38.

Carson, M.A. and Griffiths, G.A. 1987. Bedload transport in gravel channels. *J. of Hydrol. (NZ)* **26**: 1–151.

Carter, D.J. 1986. *The Remote Sensing Sourcebook: A guide to remote sensing products, services, facilities, publications and other materials*, McCarta Ltd., London.

Carter, R.W. and Davidian, J. 1968. *General Procedure for Gaging Streams*, Techniques of water-resources investigations of the United States Geological Survey. Bk. 3, Applications of hydraulics; ch. A6, Govt. Printing Office, Washington, p. 13.

Carter, V. 1986. An overview of the hydrologic concerns related to wetlands in the United States. *Canadian Journal of Botany* **64**: 364–374.

Castleberry, D.T., Czech, J.J., Erman, D.C., Hankin, D., Healey, M., Kondolf, G.M., Mangel, M., Mohr, M., Moyle, P.B., Nielsen, J., Speed, T.P. and Williams, J.G. 1996. Uncertainty and instream flow standards. *Fisheries* **21**(8): 20–21.

Cattell, R.B. 1965. Factor analysis: an introduction to essentials, 1. The purpose and underlying models. *Biometrics* **21**: 190–215.

Cederholm, C.J., Bilby, R.E., Bisson, P.A., Bumstead, T.W., Fransen, B.R., Scarlett, W.J. and Ward, J.W. 1997. Response of juvenile Coho Salmon and Steelhead to placement of large woody debris in a Coastal Washington stream. *North American Journal of Fisheries Management* **17**(4): 947–963.

CEN (European Committee for Standardization) 2002. A guidance standard for assessing the hydromorphological features of rivers. CEN TC 230/WG 2/TG 5: N32, Fifth revision, Brussels, May. URL: http://www.bygg.ntnu.no/~borsanyi/eamn-web/documents/CEN_TC230-WG2-TG5-N32_05-02.pdf (accessed 29th Feb 2004).

Chandler, J.R. 1970. A biological approach to water quality management. *Water Pollution Control* **69**: 415–422.

Chapman, D., Codrington, S., Blong, R., Dragovich, D., Smith, T.L., Linacre, E., Riley, S., Short, A., Spriggs, J. and Watson, I. 1985. *Understanding Our Earth*, Pitman Publishing, Victoria.

Charbonneau, R. and Resh, V.H. 1992. Strawberry Creek on the University of California, Berkeley Campus: a case history of urban stream restoration. *Aquatic Conservation: Marine and Freshwater Ecosystems* **2**: 293–307.

Chatfield, C. and Collins, A.J. 1980. *Introduction to Multivariate Analysis*. Chapman and Hall, London.

Chen, Y.H. and Cotton, B.A. 1988. Design of roadside channels with flexible linings. Hydraulic Engineering Circular No. 15. Federal Highway Administration, *Publication No. FHWA-IP:87-7*, USDOT/FHWA, McLean, VA, pp. 35–36.

Chessman, B.C. 1985. Estimates of ecosystem metabolism in the La Trobe River, Victoria. *Australian Journal of Marine and Freshwater Research* **36**: 873–880.

Chessman, B.C. 1995. Rapid assessment of rivers using macroinvertebrates: a procedure based on habitat-specific sampling, family-level identification, and a biotic index. *Australian Journal of Ecology* **20**: 122–129.

Chessman, B.C. 2002. Assessing the conservation value and health of New South Wales rivers. The PBH (Pressure-Biota-Habitat) project. Centre for Natural Resources, New South Wales Department of Land and Water Conservation, NSW Government, Parramatta, January. URL: http://www.dlwc.nsw.gov.au/care/water/pdfs/assessing_rivers.pdf (accessed 1st Nov 2003).

Chessman, B.C. 2003a. New sensitivity grades for Australian river macroinvertebrates. *Marine and Freshwater Research* **54**: 95–103.

Chessman, B.C. 2003b. SIGNAL 2—A Scoring System for Macro-invertebrate ('Water Bugs') in Australian Rivers. *Monitoring River Heath Initiative Technical Report No. 31*, Commonwealth of Australia, Canberra. URL: http://www.deh.gov.au/water/rivers/nrhp/signal/index.html#download (accessed 10th Nov 2003).

Chessman, B.C. and Jones, H. 2001. Integrated monitoring of environmental flows: design report. Department of Land and Water Conservation, New South Wales Government, Parramatta, New South Wales.

Chessman, B.C., Growns, J.E. and Kotlash, A.R. 1997. Objective derivation of macroinvertebrate family sensitivity grade numbers for the SIGNAL biotic index: application to the Hunter River system, New South Wales. *Marine and Freshwater Research* **48**: 159–172.

Ching, F.S. 1967. *Basic Sampling Methods*, University of Singapore.

Chow, V.T. 1959. *Open-Channel Hydraulics*. McGraw-Hill, New York.

Chow, V.T. (ed) 1964a. *Handbook of Applied Hydrology*, McGraw-Hill, New York.

Chow, V.T. 1964b. Statistical and probability analysis of hydrologic data. Part I. Frequency analysis. In Chow, V.T. (ed) *Handbook of Applied Hydrology*. McGraw-Hill, New York, pp. 8–1 to 8–42.

Chow, V.T., Maidment, D.R. and Mays, L.W. 1988. *Applied Hydrology*. McGraw-Hill, New York.

Church, M. 1974. *Electrochemical and fluorometric tracer techniques for streamflow measurements, British Geomorphic Research Group Technical Bulletin 12*, Geo. Abstracts Ltd, University of East Anglia, Norwich.

Church, M. 1978. Palaeohydrological reconstructions from a Holocene valley fill. In Miall, A.D. (ed) *Fluvial Sedimentology*. Can. Soc. Petroleum Geologists Memoir 5, Calgary, Alberta, pp. 743–772.

Church, M. 1992. Channel morphology and typology. In Callow, P. and Petts, G.E. (eds) *The Rivers Handbook, Volume 1*. Blackwell Scientific Publications, Oxford, pp. 126–143.

Church, M., McLean, D.G. and Wolcott, J.F. 1987. River bed gravels: sampling and analysis. In Thorne, C.R., Bathurst, J.C. and Hey, R.D. (eds) *Sediment Transport in Gravel-bed Rivers*. John Wiley and Sons, Chichester, pp. 43–88.

Clark, J.W., Viessman, W., Jr. and Hammer, M.J. 1977. *Water Supply and Pollution Control*, IEP, New York.

Clarke, S.J., Bruce-Burgess, L. and Wharton, G. 2003. Linking form and function: towards an eco-hydromorphic approach to sustainable river restoration. *Aquatic Conservation: Marine and Freshwater Ecosystems* **13**(5): 439–450.

Clausen, B. and Biggs, B.J.F. 1997a. Flow indices for describing habitats of benthic biota in streams. *Acta Hydrotechnica* **15/18** – Proceedings of the oral presentations: FRIEND'97, pp. 125–135.

Clausen, B. and Biggs, B.J.F. 1997b. Relationships between benthic biota and hydrological indices in New Zealand streams. *Freshwater Biology* **38**: 327–342.

Clay, C.H. 1951. The engineer's part in fisheries conservation. *The Engineering Journal*, November: 1058–1061.

Clay, C.H. 1995, *Design of Fishways and Other Fish Facilities*. Lewis Publishers, Boca Ration, FL, p. 248.

Clesceri, L.S., Greenberg, A.E. and Eaton, A.D. (eds). 1998. *Standard Methods for the Examination of Water and Wastewater*, 20th edn, American Public Health Association, American Water Works Association, and Water Environment Federation, Washington, DC.

Clifton, C., and Evans, R. 2001. Environmental water requirements of groundwater dependent ecosystems. *Environmental Flows Initiative Technical Report, Report No. 2*. Environment Australia, Commonwealth of Australia, Canberra, Australian Capital Territory. URL: http://www.deh.gov.au/water/rivers/nrhp/groundwater/ (accessed 1st Nov 2003).

Close, A.F. 2002. Options for reducing the risk of closure of the River Murray mouth. *MDBC Technical Report 2002/2, Options, Version 3*. Murray-Darling Basin Commission, Canberra, Australian Capital Territory, April.

Coastal CRC. 2002. Flowing estuaries needed for healthy fisheries. Exploring Coastal Science. Cooperative Research Centre for Coastal Zone, Estuary and Waterway Management, Indooroopilly, Queensland, May. URL: http://www.coastal.crc.org.au/pdf/exploring_coastal_science/flow_estuaries.pdf (accessed 1st Nov 2003).

Coastal Environmental Consultants Pty Ltd., Marine Science and Ecology Pty Ltd., McLean, E.J. and Pollock, T.J. 1989. *Snowy River Estuary Study*. Report to Department of Water Resources, Victoria.

Cochran, W.G. 1977. *Sampling Techniques*, 3rd edn, John Wiley, New York.

Cochran, W.G. and Cox, G.M. 1957. *Experimental Designs*, 2nd edn, John Wiley, New York.

Coe, H.J. 2001. Distribution patterns of hyporheic fauna in a riparian floodplain terrace, Queets River, Washington. Seattle, WA. University of Washington. M.S. thesis, p. 75.

Coleman, N.L. and Alonso, C.V. 1983. Two-dimensional channel flows over rough surfaces. *J. Hyd. Eng.* **109**: 175–188.

Collier, K.J. (ed) 1994. *Restoration of Aquatic Habitats*. Selected papers for the second day of the New Zealand Limnological Society 1993 Annual Conference, 10–12 May, Wellington, N.Z., Department of Conservation, Te Papa Atawhai, Wellington.

Collier, K.J. and McColl, R.H.S. 1992. Assessing the natural values of New Zealand rivers. In Boon, P.J., Calow, P. and Petts, G.E. (eds) *River Conservation and Management*. Wiley, Chichester, pp. 195–211.

Collier, M.P., Webb, R.H. and Andrews, E.D. 1997. Experimental flooding in Grand Canyon. *Scientific American*, pp. 66–73.

Collings, M.R. 1972. A methodology for determining instream flow requirements for fish. *Proceedings of Instream Flow Methodology Workshop*. Washington Dept. of Ecology, Olympia, pp. 72–86. (cited by Wesche and Rechard, 1980).

Collings, M.R. 1974. Generalization of spawning and rearing discharges for several Pacific salmon species in western Washington. *USGS Open File Report*. United States Geological Survey, prepared in cooperation with the Washington Department of Fisheries, Tacoma, Washington, p. 39.

Collins, G.B. and Weber, C.I. 1978. Phycoperiphyton (algae) as indicators of water quality. *Transactions of the American Microscopial Society* **97**: 36–43.

Colwell, R.K. 1974. Predictability, constancy, and contingency of periodic phenomena. *Ecology* 55: 1148–1153.

Colwell, R.N. (ed) 1983. *Manual on Remote Sensing*, 2nd edn, American Society of Photogrammetry, Falls Church, VA.

Commoner, B. 1972. *The Closing Circle*. Alfred Knopf, New York, NY.

Conrick, D. and Ribi, J. 1997. *Urban Stream Rehabilitation*. Brisbane City Council, Department of Recreation & Health, Environment Management Branch, Brisbane.

Cook, P. and Lamontagne, S. 2002. Assessing and protecting water requirements for groundwater dependent ecosystems. *Seminar Proceedings, The Science of Environmental Water Requirements in South Australia*, 24 Sep Adelaide, The Hydrological Society of South Australia Inc., Adelaide, South Australia, pp. 49–54.

Cooling, M.P. and Boulton, A.J. 1993. Aspects of the hyporheic zone below the terminus of a south Australian arid-zone stream. *Australian Journal of Marine and Freshwater Research* 44: 411–426.

Corbett, D.M. 1962. Stream-gaging procedure. *US Geol. Survey Water-Supply Paper 888*, USGS, Washington, DC.

Corbett, R. 1990. A method for determining minimum instream flow for recreational boating. *SAIC Special Report 1-239-91-01*. Science Applications International Corporation. McLean, VA.

Cork, S., Shelton, D., Binning, C. and Parry, R. 2001. A framework for applying the concept of ecosystem services to natural resource management in Australia. In Rutherfurd, I., Sheldon, F., Brierley, G. and Kenyon, C. (eds) *Proceedings of the Third Australian Stream Management Conference*, Brisbane, 27–28 Aug, CRC Catchment Hydrology, Monash University, Clayton, Victoria, pp. 157–162.

Corrick, A.H. 1981. Wetlands of Victoria II. Wetlands and waterbirds of South Gippsland. *Proceedings of the Royal Society of Victoria* 92: 187–200.

Corrick, A.H. 1982. Wetlands of Victoria III. Wetlands and waterbirds between Port Phillip Bay and Mount Emu Creek. *Proceedings of the Royal Society of Victoria* 94: 69–87.

Costa, J.E. 1987. Interpretation of the largest rainfall-runoff floods measured by indirect methods on small drainage basins in the conterminous United States. *Proc. of the US-China Bilateral Symp. on the Analysis of Extraordinary Flood Events*, Oct 1985, Nanjing, China (as cited by Jarrett and Malde, 1987).

Coste, M., Bosca, C. and Dauta, A. 1991. Use of algae for monitoring rivers in France. In Witton, B.A., Rott, E. and Friedrich, G. (eds) *Use of Algae for Monitoring Rivers. Proceedings of an International Symposium*, Düsseldorf, 26–28 May. Institute of Botany, Innsbruck University, Austria, pp. 75–88.

Cottingham, P., Bunn, S., Price, P. and Lovett, S. (eds) 2003. Managing Wood in Streams, *River and Riparian Land Management Technical Guideline Update No. 3*. Land & Water Australia, Canberra, Australian Capital Territory, July. URL: http://www.lwa.gov.au/downloads//PR030531.pdf (accessed 1st Nov 2003).

Cottingham, P., Thoms, M.C. and Quinn, G.P. 2002. Scientific panels and their use in environmental flow assessment in Australia. *Australian Journal of Water Resources* 5(1): 103–111.

Cowan, W.L. 1956. Estimating hydraulic roughness coefficients. *Agricultual Engineering* 37: 473–475.

Cowardin, L.M. 1982. Wetlands and deepwater habitats: a new classification. *Journal of Soil and Water Conservation* 37: 83–85.

Cowardin, L.M., Carter, V., Golet, F.C. and LaRoe, E.T. 1979. Classification of wetlands and deepwater habitats of the United States. U.S. Fish and Wildlife Service, *Pub. FWS/OBS79/31*. Washington, DC, p. 103. URL: http://www.npwrc.usgs.gov/resource/1998/classwet/classwet.htm#table (accessed 1st Nov 2003).

Cowx, I.G. 2001. Factors Influencing Coarse Fish Populations in Rivers: A Literature Review. *R & D Publication 18*, Environment Agency, Bristol, p. 146.

Cowx, I.G. and Welcomme, R.L. 1998. *Rehabilitation of Rivers for Fish*. Fishing News Books, for the Food and Agricultural Organisation of the United Nations (FAO), Oxford, p. 260.

Coysh, J., Nichols, S., Simpson, J., Norris, R., Barmuta, L., Chessman, B. and Blackman, P. 2000. *AusRivAS Predictive Model Manual*. Cooperative Research Centre for Freshwater Ecology, Canberra. URL: http://enterprise.canberra.edu.au/Databases/AusRivAS.nsf/ (accessed 1st Nov 2003).

Cracknell, A. and Hayes, L. (eds) 1988. *Remote Sensing Yearbook 1988/89*. Taylor and Francis, London.

Craig, R.F. 1983. *Soil Mechanics*. Van Nostrand Reinhold, Wokingham.

Cremer, K. 1999. Willow management for Australian rivers. *Natural Resource Management* (journal of The Australian Association of Natural Resource Management, Lyneham, Australian Capital Territory) Special Issue, Dec, pp. 2–22.

Croke, J.C. 1996. Floodplain classification and its relevance to stream management. In Rutherfurd, I. and Walker, M. (eds) *Proceedings of the First National Conference on Stream Management in Australia*, Merrijig, 19–23 Feb, Cooperative Research Centre for Catchment Hydrology, Monash University, Clayton, pp. 41–44.

Cross, R.D. and Williams, D.L. (eds) 1980. Proceedings of the National Symposium on Fresh Water Inflow to Estuaries, 2 Volumes. *Report No. FWS/OBS-81/04*. U.S. Fish and Wildlife Service, Office of Biological Services, Biological Services Program, October.

Crowder, A. and Painter, D.S. 1991. Submerged macrophytes in L. Ontario: current knowledge, importance, threats to stability and needed studies. *Can. J. Fish. Aquat. Sci.* 88: 1530–1545.

Crowder, D.W. and Diplas, P. 2000. Using two-dimensional hydrodynamic models at the scales of ecological importance. *Journal of Hydrology* 230: 172–191.

Crowe, C.T., Elger, D.F. and Roberson, J.A. 2000. *Engineering Fluid Mechanics*, 7th edn, John Wiley, New York.

Csermak, B. and Rakoczi, L. 1987. Erosion and sedimentation. In Starosolszky, O. (ed) *Applied Surface Hydrology*, pp. 760–807, Water Resour. Publ., PO Box 2841, Littleton, CO.

CSU. 1977. *Lecture Notes on River Mechanics*, Colorado State University, Fort Collins, CO.

Cude, C.G. 2001. Oregon Water Quality Index. *Journal of the American Water Resources Association* 37(1): 125–137.

Cummins, K.W. 1986. Riparian influence on stream ecosystems. In Campbell, I.C. (ed) *Stream Protection: The Management of Rivers for Instream Uses*, pp. 45–55, Water Studies Centre, Chisholm Institute of Technology, East Caulfield, Australia.

Cunnane, C. 1978. Unbiased plotting positions—a review. *J. of Hydrol.* **37**: 205–222.

Curran, P.J. 1985. *Principles of Remote Sensing*. Longman Scientific and Technical, Harlow.

Cushman, R.M. 1985. Review of ecological effects of rapidly varying flows downsteam of hydroelectric facilities. *North American Journal of Fisheries Management* **5**: 330–339.

D'Aoust, S. 1998. Ballasting of large woody debris structures—new insights. *Streamline* (British Columbia's Watershed Restoration Technical Bulletin) **3**: 17–18.

Dackombe, R. and Gardiner, V. 1983. *Geomorphological Field Manual*. George Allen & Unwin, London.

Dahm, C.N., Sedell, J.R. and Triska, F.J. 1989. A historical look at streams and rivers in North America. Technical Information Workshop, Stream Rehabilitation and Restoration, 37th Annual North American Benthological Society Meeting, April.

Daily, G.E. 1997. *Nature's Services—Societal Dependence on Natural Ecosystems*. Island Press, Washington.

Daily, J.W. and Harleman, D.R.F. 1966. *Fluid Dynamics*. Addison-Wesley, Reading, MA.

Dalrymple, T. 1960. Flood-frequency analyses. Manual of Hydrology: Part 3. Flood-Flow Techniques, *US Geological Survey Water-Supply Paper 1543-A*, USGS, Washington, DC.

Dalrymple, T. and Benson, M.A. 1967. Measurement of peak discharge by the slope-area method. *US Geol. Survey Techniques of Water-Resources Investigations*, Book 3, Chap. A2, USGS, Washington, DC.

Daugherty, R.L., Franzini, J.B. and Finnemore, E.J. 1985. *Fluid Mechanics with Engineering Applications*, 8th edn, McGraw Hill Book Co., New York.

Davidian, J. 1984. Computation of water-surface profiles in open channels. *US Geol. Survey Techniques of Water-Resources Investigations*, Book 3, Chapter A15, USGS, Washington, DC.

Davies, N.M., Norris, R.H. and Thoms, M.C. 2000. Prediction and assessment of local stream habitat features using large-scale catchment characteristics. *Freshwater Biology* **45**, 343–369.

Davies, P.E. and Kalish, S.R. 1994. Influence of river hydrology on the dynamics and water quality of the Upper Derwent Estuary, Tasmania. *Australian Journal of Marine and Freshwater Research* **45**: 109–130.

Davies, P.E., Sloane, R.D. and Andrew, J. 1988. Effects of hydrological change and the cessation of stocking on a stream population of *Salmo trutta L. Australian Journal of Marine and Freshwater Research* **39**: 337–354.

Davies, P.M. 1997. Assessment of river health by the measurement of community metabolism, *Report No UWA 14*, Final Report to Land and Water Resources Research and Development Corporation, Canberra, Australian Capital Territory.

Davies-Colley, R. 1997. Stream channels are narrower in pasture than in forest. *New Zealand Journal of Marine and Freshwater Research* **31**: 599–608.

Davis, J.A. and Barmuta, L.A. 1989. An ecologically useful classification of mean and near-bed flows in streams and rivers. *Freshwater Biol.* **21**: 271–282.

Davis, J.A. 1986. Boundary layers, flow microenvironments and stream benthos. In De Deckker, P. and Williams, W.D. (eds) *Limnology in Australia*, CSIRO Australia, Melbourne, Victoria, pp. 293–312.

Davis, J.A., Froend, R.H., Hamilton, D.P., Horowitz, P., McComb, A.J., Oldham, C.E. and Thomas, D. 2001. Environmental water requirements to maintain wetlands of National and International Importance. *Environmental Flows Initiative Technical Report, Report No. 1*. Environment Australia, Commonwealth of Australia, Canberra, Australian Capital Territory. URL: http://www.deh.gov.au/water/rivers/nrhp/wetlands/ (accessed 1st Nov 2003).

Davis, W.M. 1899. The geographical cycle. *Geographical J.* **14**: 481–504.

Dawdy, D.A. and Vanoni, V.A. 1986. Modeling alluvial channels. *Water Resour. Res.* **22**, 71S–81S.

Dawson, F.H. and Charlton, F.G. 1988. Bibliography on the hydraulic resistance or roughness of vegetated watercourses. *Occ. Publ. No. 25*, Freshwater Biol. Assoc., Ambleside, Cumbria.

Day, P.R. 1965. Particle fractionation and particle-size analysis. In C.A.B. et al. (eds) *Method of Soil Analysis*. Part I, Agronomy 9, Amer. Soc. of Agronomy, Soil Sci. Soc. of America, Madison, Wisconsin, pp. 545–567.

De Deckker, P. and Williams, W.D. 1986. *Limnology in Australia*, CSIRO Australia, Melbourne, Victoria.

De Pauw, N. and Vanhooren, G. 1983. Method for biological quality assessment of water courses in Belgium. *Hydrobiologia* **100**: 153–168.

de Waal, L.C., Large, A.R.C. and Wade P.M. (eds). *Rehabilitation of Rivers: Principles and Implementation*, Wiley, Chichester.

De Wiest, R.J.M. 1965. *Geohydrology*. New York (as cited by Gregory and Walling, 1973).

DeBano, L.F. and Heede, B.H. 1987. Enhancement of riparian ecosystems with channel structures. *Water Resources Bulletin* **23**: 463–470.

Deeling, D.M. and Paling, E.I. Assessing the ecological health of estuaries in Australia. *LWRRDC Occasional Paper No. 17/99*, Urban Sub-program Report No. 10. Land and Water Resources Research and Development Coorporation, Canberra, Australian Capital Territory.

Demarchi, D.A. 1993. *Ecoregions of British Columbia*. Report prepared for the B.C. Ministry of Environment. Lands and Parks. Wildlife Branch.

Denny, M.W. 1988. *Biology and the Mechanics of the Wave-Swept Environment*. Princeton University Press, Princeton, NJ.

Denny, M.W., Daniel, T.L. and Koehl, M.A.R. 1985. Mechanical limits to size in wave-swept organisms. *Ecological Monographs* **55**: 69–102.

Denslinger, T.L., Gast, W.A., Hauenstein, J.J., Heicher, D.W., Henriksen, J., Jackson, D.R., Lazorchick, G.J., McSparran, J.E., W. Stoe, T.W. and Young, L.M. Instream Flow Studies, Pennsylvania and Maryland. *Susquehanna River Basin Commission Publication 191*, Harrisburg, PA, May. URL: http://www.dep.state.pa.us/dep/deputate/watermgt/Wc/subjects/InstreamFlow/IFSmainreport.PDF (accessed 1st Nov 2003).

Department of Conservation and Environment. 1992. Lower Snowy River Wetlands Proposed Management Plan. Orbost Region and National Parks and Public Land Division, DCE, East Melbourne, Victoria, p. 103.

Department of Ecology. 2003. Instream flow study methods used in Washington State. *The Science Behind Instream Flows,* Ecology, Washington State, Olympia, WA. URL: http://www.ecy.wa.gov/programs/wr/instream-flows/Images/pdfs/if-msum.pdf (accessed 1st Nov 2003).

Department of Land and Water Conservation. 1996. *The NSW Wetlands Management Policy.* New South Wales Government, Clarendon Press, Brookvale. URL: http://www.dlwc.nsw.gov.au/care/wetlands/wetlandmanagement/policy.pdf (accessed 1st Nov 2003).

Department of Land and Water Conservation. 1998. *Stressed Rivers Assessment Report, NSW State Summary.* NSW Department of Land and Water Conservation, Parramatta, NSW.

Department of Land and Water Conservation. 2000. The NSW State groundwater dependent ecosystems policy. A component policy of the NSW state groundwater policy framework document. NSW Department of Land and Water Conservation, Parramatta, NSW. URL: http://www.dlwc.nsw.gov.au/care/water/wr/pdfs/policy_180602.pdf (accessed 1st Nov 2003).

Department of Land and Water Conservation. 2001a. *Integrated Monitoring of Environmental Flows (IMEF): An Overview.* Department of Land and Water Conservation NSW, Parramatta, NSW. URL: http://www.dlwc.nsw.gov.au/care/water/imef/imef_brochure.pdf (accessed 1st Nov 2003).

Department of Land and Water Conservation. 2001b. *Integrated Monitoring of Environmental Flows (IMEF)—Fish Monitoring 1999–2000.* Environmental Notes No. 1, April. Department of Land and Water Conservation NSW, Parramatta, NSW. URL: http://www.dlwc.nsw.gov.au/care/water/imef/fish_brochure.pdf (accessed 1st Nov 2003).

Department of Water Affairs and Forestry. 1999. Resource Directed Measures for Protection of Water Resources. Volume 5: Estuarine Ecosystems Version 1.0, Pretoria, September. URL: http://www.dwaf.gov.za/Documents/Policies/WRPP/Estuarine%20Ecosystems.htm (accessed 1st Nov 2003).

Department of Water, Land and Biodiversity Conservation. 2002. *Lock 5 Flow Enhancement Trial, Murray-Darling Division Report No. 1.* DWLBC, South Australia, September. URL: http://www.dwlbc.sa.gov.au/publications/pdfs/reports/lock_5_flow_enhancement_trial.pdf (accessed 28 Feb 2003).

Deurer, R.H. 1960. Pumping saves water and power in fish-passage facilities. *Civil Engineering* November 38–40.

DEV (Deutsches Institut für Normung e.V.). 1992. Biologisch–ökologische gewässergüteuntersuchung: bestimmung des saprobienindex (M2). *Deutsche Einheitsverfahren zur Wasser–, Abwasser– und Schlammuntersuchung.* VCH Verlagsgesellschaft mbH, Weinheim, pp. 1–13.

Devore, J.L. 1982. *Probability and Statistics for Engineering and the Sciences.* Brooks/Cole, Monterey, CA.

Díaz, S. and Cabido, M. 2001. Vive la différence: plant functional diversity matters to ecosystem processes. *Trends in Ecology & Evolution* 16: 646–655.

Diehl, T.H. 1997. Potential drift accumulation at bridges. *Publication No. FHWA-RD-97-028*, U.S. Department of Transportation Federal Highway Administration, Research and Development, McLean, VA.

Digby, M.J., Saenger, P., Whelan, M.B., McConchie, D., Eyre, B., Holmes, N. and Bucher, D. 1999. A physical classification of estuaries. *LWRRDC Occasonal Paper No. 16/99.* Urban Subprogram Report No. 9. Land and Water Resources Research and Development Corporation, Canberra, Australian Capital Territory.

Dignan, P., Kefford, B., Smith, N., Hopmans, P. and Doeg, T. 1996. The use of buffer strips for the protection of streams and stream dependant biota in forested ecosystems. Centre for Forest Tree Technology, Melbourne. Report to Department of Conservation and Natural Resources, East Melbourne, Victoria.

Dingman, S.L. 1978. Synthesis of flow duration curves for unregulated streams in New Hampshire. *Water Resour. Bull.* **14**: 1481–1502.

Dingman, S.L. 1984. *Fluvial Hydrology.* W.H. Freeman, New York.

Dixon, W.J. (ed.) 1984. *BMDP Statistical Software: 1985 Printing,* University of California Press, Berkeley.

DNRE. 1997. *An Index of Stream Condition: Reference Manual,* Department of Natural Resources and Environment, Government of Victoria, Melbourne.

Dobson, M. and Cariss, H. 1999. Restoration of afforested upland streams—what are we trying to achieve? *Aquatic Conservation: Marine and Freshwater Ecosystems* **9**(1): 133–139.

Dodge, D.P. and Mack, C.C. 1996. Direct control of fauna: role of hatcheries, fish stocking and fishing regulations. In Petts, G. and Calow, P. (eds) *River Restoration.* Blackwell Science, Oxford, pp. 167–181.

Dominick, D.S. and O'Neill, M.P. Effects of flow augmentation on stream channel morphology and riparian vegetation: Upper Arkansas River Basin, CO. *Wetlands* **18**(4): 591–607.

Donald, A., Nathan, R. and Reed, J. 1999. Use of spell analysis as a practical tool to link ecological needs with hydrological characteristics. In Rutherfurd, I.D. and Bartley, R (eds.) *Proceedings of the Second Australian Stream Management Conference,* Adelaide, 8–11 Feb, Cooperative Research Centre for Catchment Hydrology, Monash University, Clayton, pp. 205–210.

Dosskey, M.G. 2001. Toward quantifying water quality pollution abatement in response to installing buffers on crop land. *Environmental Management* **28**(5): 577–598.

Douglas, A.J. and Taylor, J.G. 1999. The economic value of Trinity River water. *International Journal of Water Resources Development* **15**: 309–322.

Douglas, J.F., Gasiorek, J.M. and Swaffield, J.A. 1983. *Fluid Mechanics.* Pitman, London.

Douglas, J.F., Gasiorek, J.M. and Swaffield, J.A. 2001. *Fluid Mechanics,* 4th edn, Prentice-Hall, Upper Saddle River, NJ.

Douglas, M.M. and Pouliot, A.M. 1998. A review of the impacts of the northern Australian grazing industry on wetlands and riparian habitats. In Hook, R. (ed) Catchment Management and Water Quality and Nutrient Flows and the Northern Australian Beef Industry, *NAP Occasional Publication No. 3,* pp. R2.1–R2.1.19.

Downes, B.J., Lake, P.S., Schreiber, E.S.G. and Glaister, A. 1998. Habitat structure and regulation of local species diversity in a stony, upland stream. *Ecological Monographs* **68**: 237–257.

Downes, B.J., Barmuta, L.A., Fairweather, P.G., Faith, D.P., Keough, M.J., Lake, P.S., Mapstone, B.D., Quinn G.P. 2002. *Monitoring Ecological Impacts: Concepts and Practice in Flowing Waters*. Cambridge University Press, Cambridge, p. 434.

Downs, P.W. 1995. River channel classification for channel management purposes. In Gurnell, A. and Petts, G. (eds) *Changing River Channels*. John Wiley and Sons, New York, pp. 347–365.

Doyle, M.W, Miller, D.E. and Harbor, J.M. 1999. Should river restoration be based on classification schemes or process models? Insights from the history of geomorphology. *American Society of Civil Engineers International Conference on Water Resources Engineering* (CD-ROM) Seattle, Washington.

Doyle, M.W., Stanley, E.H. and Harbor, J.M. 2003a. Towards policies and decision making for dam removal. *Environmental Management* **31**: 453–465.

Doyle, M.W., Stanley, E.H., Harbor, J.M. and Grant, G.S. 2003b. Dam removal in the U.S.: Emerging needs for science and policy. *EOS* **84**: 29–33.

Doyle, M.W., Stanley, E.H. and Harbor, J.M. 2003c. Channel adjustments following two dam removals in Wisconsin. *Water Resources Research* **39**(1), 1011, doi10.1029/2002wR001714.

Doyle, P.F. 1992. Performance of alternate methods of bank protection. *Canadian Journal of Civil Engineering* **19**: 1049–1061.

Drummond, I., Tilleard, J. and Dyer, B. 1995. Retards and Groynes Design Guidelines. Guidelines for Stabilising Waterways. Centre for Environmental Applied Hydrology, The University of Melbourne, Melbourne, Victoria.

Drummond, R.R. 1974. When is a stream a stream? *Prof. Geographer* **26**: 34–37.

Drury, S.A. 1990. *A Guide to Remote Sensing: Interpreting Images of the Earth*. Oxford University Press, Oxford.

Duel, H., van der Lee, G.E.M., Penning, W.E. and Baptist, M.J. 2003. Habitat modelling of rivers and lakes in the Netherlands: an ecosystem approach. *Canadian Water Resources Journal* **28**(2): 231–248.

Dunbar, M.J., Gustard, A., Acreman, M.C. and Elliott, C.R.N. 1998. Review of Overseas Approaches to Setting River Flow Objectives. *Environment Agency R8*. URL: http://www.waterandnature.org/pub/FLOW.pdf (accessed 1st Nov 2003).

Dunham, J, Lockwood, J. and Mebane, C., 2001, Salmonid Distributions and Temperature, Issue Paper 2, Prepared as part of EPA Region 10 Temperature Water Quality Criteria Guidance Development Project, US Environment Protection Agency, Region 10, The Pacific Northwest, Seattle, WA, May. URL: http://yosemite.epa.gov/R10/water.nsf/0/5eb9e547ee9e111-f88256a03005bd665/ (accessed 1st Nov 2003).

Dunn, H., 2000, Identifying and Protecting Rivers of High Ecological Value, *Occasional Paper No. 01/00*, Land and Water Resources Research and Development Corporation, Canberra, Australian Capital Territory.

Dunne, T. and Leopold, L.B. 1978, *Water in Environmental Planning*, W.H. Freeman, San Francisco.

Dunteman, G.H., 1984, *Introduction to Multivariate Analysis*. Sage Publications, Beverly Hills, CA.

Dyson, M. Bergkamp. G. and Scanlon, J. (eds) 2003. *Flow: The Essentials of Environmental Flows. Water and Nature Initiative*, International Union for Conservation of Nature and Natural Resources, Gland, Switzerland and Cambridge, UK. p. 118. URL: http://www.waterandnature.org/pub/FLOW.pdf (accessed 1st Nov 2003).

Ebel, W.J., Becker, C.D., Mullan, J.W. and Raymond, H.L. 1989. The Columbia River—toward a holistic understanding. In Dodge, D.P. (ed) Proceedings of the International Large River Symposium. Canadian Special Publication of *Fisheries and Aquatic Sciences* **106**: 205–219.

Ebersole, J.L., Liss, W.J. and Frissell, C.A. 1997. Restoration of stream habitats in the Western United States: restoration as reexpression of habitat capacity. *Environmental Management* **21**: 1–14.

Edgar, G.J., Barrett, N.S. and Graddon, D.J. 1999. A classification of Tasmanian estuaries and assessment of their conservation significance using ecological and physical attributes, population and land use. *Technical Series Report 2*. Tasmanian Aquaculture and Fisheries Institute, p. 205.

Edwards, R.T. 1998. The hyporheic zone. In Naiman, R.J. and Bilby, R.E. (eds) *River Ecology and Management: Lessons from the Pacific Coastal Ecoregion*. Springer-Verlag, New York, pp. 399–429.

Edwards, R.W. and Crisp, D.T. 1982. Ecological implications of river regulation in the United Kingdom. In Hey R.D., Bathurst, J.C. and Thorne, C.R. (eds) *Gravel-bed Rivers: Fluvial Processes, Engineering and Management*. John Wiley & Sons, Chichester, pp. 843–863.

Eiseltová, M. and Biggs, J. (eds) 1995. Restoration of stream ecosystems, an integrated catchment approach. *IWRB Publication 37*, International Waterfowl and Wetland Research Bureau, Slimbridge, Gloucester, UK.

Elliot, J.M. 1977. *Some Methods for the Statistical Analysis of Samples of Benthic Invertebrates*, Sci. Publ. No. 25, Freshwater Biological Assn., Ferry House, UK.

Ellis, J.C. 1989. *Handbook on the Design and Interpretation of Monitoring Programs*. Water Research Centre, Medmenham.

Elmore, W. and Beschta, R.L. 1987. Riparian areas: perceptions in management. *Rangelands* **9**: 260–265.

Emmett, W.W. 1975. The channels and waters of the Upper Salmon River area, Idaho. *US Geol. Survey Prof. Paper 870-A*, USGS, Washington, DC (as cited by Dunne and Leopold, 1978).

Emmett, W.W. 1977. Measurement of bed load in rivers. IAHS Publ. No. 133, *Erosion and Sediment Transport Measurement*, Symposium, Florence (as cited by Csermak and Rakoczi, 1987).

Emmett, W.W., Burrows, R.L. and Chacho, E.F., Jr. 1989. Gravel transport in a gravel-bed river, Alaska. Paper presented at spring meeting of Am. Geophys. Union, May 1989, Baltimore, MD.

Environment Agency. 1997. River Habitat Survey: 1997 Field Survey Guidance Manual, Incorporating SERCON, Centre for Ecology and Hydrology, National Environment Research Council, UK.

Environment and Heritage Service. 2001. A River Water Quality Monitoring Strategy for Northern Ireland. Department of the Environment, May. URL: http://www.ehsni.gov.uk/pubs/publications/River_Monitoring.pdf (accessed 1st Nov 2003).

Environment Protection Authority 1997. Proposed Interim Environmental Objectives for NSW Waters—Inland Rivers. New South Wales EPA, Chatswood, Sydney.

Epple, R.A. 2000. Dam decommissioning: French pilot experiences and the European context. ERN (European Rivers Network), Le Puy, France. URL: http://www.rivernet.org/decom3_e.htm#Damdecom (accessed 1st Nov 2003).

Ernegger, T., Grubinger, H., Vitek, E., Csekits, C., Eitzinger, J., Gaviria, S., Kotek, D., Krisa, H., Nachtnebel, H.P., Pritz, B., Sabbas, T., Schmutz, S., Schreiner, P., Stephan, U., Unfer, G., Wychera, U. and Neudorfer, W. 1998. A natural stream created by human engineering: investigations on the succession of the Marchfeld Canal in Austria. *Regulated Rivers: Research & Management* **14**: 119–139.

Espegren, G.D. 1998. Evaluation of the standards and methods used for quantifying instream flows in Colorado. Colorado Water Conservation Board, Denver, CO, November. URL: http://cwcb.state.co.us/isf/Programs/Docs/evalstan.pdf (accessed 1st Nov 2003)

Estes, C.C. and Orsborn, J.F. 1986. Review and analysis of methods for quantifying instream flow requirements. *Water Resources Bulletin* **22**(3): 389–398.

EUROPA. 2003. The EU Water Framework Directive—integrated river basin management for Europe. The European Union Online. URL: http://europa.eu.int/comm/environment/water/water-framework/index_en.html (accessed 1st Nov 2003).

Everest, F.H., McLemone, C.E. and Ward, J.V. 1980. An improved tri-tube cryogenic gravel sampler. *Res. Note PNW-350*, USDA Forest Service, Pac. NW Forest and Range Expt. Sta., Portland, Oregon (as cited by Hamilton and Bergersen, 1984).

Everitt, B.L. 1968. Use of cottonwood in the investigation of the recent history of a flood plain. *Am. J. Sci.* **206**: 417–439 (as cited by Schumm, 1977).

Everitt, B.L. 1995. Hydrologic factors in regeneration of Fremont Cottonwood along the Fremont River, Utah. In Costa, J.E., Miller, A.J., Potter, K.W. and Wilcock, P.R. (eds) *Natural and Anthropological Influences in Fluvial Geomorphology: The Wolman Volume*. Geophysical Monograph 89, American Geophysical Union, Washington, DC, pp. 197–208.

Extence, C.A., Balbi, D.M. and Chadd, R.P. 1999. River flow indexing using British benthic macro-invertebrates: a framework for setting hydroecological objectives. Environment Agency, Bristol.

Faith, D.P., Minchin, P.R. and Belbin, L. 1987. Compositional dissimilarity as a robust measure of ecological distance. *Vegetation* **69**: 57–68.

Federal Caucus. 2003. SalmonRecovery.gov. Bonneville Power Administration—P, Portland, Oregon. URL: http://www.SalmonRecovery.gov (accessed 1st Nov 2003).

Federal Emergency Management Agency 1999. *National Dam Safety Program*. FEMA, Washington, DC.

Federal Energy Regulatory Commission. 1996. Environmental assessment for new hydropower license: Skagit River Hydro-electric project. *FERC Project No. 553*, Federal Energy Regulatory Commission, Washington, DC, May. URL: http://www.lowimpacthydro.org/Skagit/Skagit_FEA.pdf (accessed 1st Nov 2003).

Fenner, P., Ward, W.B. and Patton, D.R. 1985. Effects of regulated water flows on regeneration of Fremont cottonwood. *Journal of Range Management* **38**: 135–138.

Ferguson, A.J.P. 1996. Estuarine classification schemes. In Digby, M.J. and Ferguson, A.J.P. (eds) *A Physical Classification of Australian Estuaries—Workshop Transcripts*. Centre for Coastal Management, Southern Cross University, Lismore, NSW, pp. 46–66.

Fièvet, E. 2000. Passage facilities for Diadromous freshwater shrimps (*Decapoda: Caridea*) in the Bananier River, Guadeloupe, West Indies. *Regulated Rivers: Research & Management* **16**(2): 101–112.

Findlay, S. 1995. Importance of surface-subsurface exchange in stream ecosystems: the hyporheic zone. *Limnology and Oceanography* **40**(1): 159–164.

Finlayson, B.L. 1979. Electrical conductivity: a useful technique in teaching geomorphology. *Journal of Geography in Higher Education* **3**: 68–87.

Finlayson, B.L. 1981. The analysis of stream suspended loads as a geomorphological teaching exercise. *Journal of Geography in Higher Education* **5**: 23–35.

Finlayson, B.L. and Statham, I. 1980. *Hillslope Analysis*. Butterworths, London.

Finlayson, B.L. and McMahon, T.A. 1988. Australia vs the world: a comparative analysis of streamflow characteristics. In Warner, R.J. (ed) *Fluvial Geomorphology of Australia*. Academic Press, Sydney, pp. 17–40.

Finlayson, B.L., Gippel, C.J. and Brizga, S.O. 1994. Effects of reservoirs on downstream aquatic habitat. *Journal Australian Water Works Association* **21**: 15–20.

FISRWG (Federal Interagency Stream Restoration Working Group). 1998. *Stream Corridor Restoration: Principles, Processes and Practices*. National Technical Information Service, U.S. Department of Commerce, Springfield, VA. URL: http://www.usda.gov/stream_restoration/ (accessed 1st Nov 2003).

Fitzsimmons, A.K. 1996. Sound policy or smoke and mirrors: does ecosystem management make sense? *Water Resources Bulletin* **32**(2): 217–227.

Fjellheim, A. and Raddum, G.G. 1990. Acid precipitation: biological monitoring of streams and lakes. *Sci. Total Environ.* **96**: 57–66.

Folk, R.L. 1980. *Petrology of Sedimentary Rocks*. Hemphill Publishing, Austin, TX.

Fontaine, T.D., III and Bartell, S.M. (eds) 1983. *Dynamics of Lotic Ecosystems*. Ann Arbor Science, MI.

Forest Service, British Columbia, 1997. Total Quality Management and Adaptive Management. *Adaptive Management Newsletter* Summer 1997, pp. 1,3. URL: http://www.for.gov.bc.ca/hfp/amhome/Pubs/AMSummer97.pdf (accessed 1st Nov 2003).

Forman, R.T.T. and Godron, M. 1986. *Landscape Ecology*. John Wiley, New York, p. 619.

Fortner, S.L. and White, D.S. 1988. Interstitial water patterns: a factor influencing the distribution of some lotic aquatic vascular macrophytes. *Aquatic Botany* **31**: 1–12.

Foth, H.D. 1978. *Fundamentals of Soil Science*. John Wiley, New York.

Fox, M.D. and Adamson, D. 1986. The ecology of invasions. In Recher, H.F., Lunney, D. and Dunn, I. (eds) *A Natural Legacy: Ecology in Australia*. Pergamon, Sydney, pp. 131–151.

Franco, J.J. 1967. Research for river regulation dike design. *Journal Waterways and Harbours Division, ASCE* **93**: WW3: 71–88.

Freeman, R. 2002. Harnessing the restoration potential of artifical floods. *Conservation in Practice* **3**(2). Society for Conservation Biology. URL: http://www.conbio.org/InPractice/article32-HAR.html (accessed 1st Nov 2003).

FREMP (Fraser River Estuary Management Program). 1994. *A Living Working River*. An Estuary Management Plan for the Fraser River. FREMP, New Westminster, BC, Canada. URL: http://www.bieapfremp.org/fremp/pdf_files/f_emp_complete.pdf (accessed 1st Nov 2003).

Fresenius, W., Quentin, K.E. and Schneider, W. (eds) 1987. *Water Analysis*. Springer-Verlag, Berlin.

Friedman, J.M., Osterkamp, W.R., Scott, M.L. and Auble, G.T. 1998. Downstream effects of dams on channel geometry and bottomland vegetation: regional patterns in the Great Plains. *Wetlands* **18**(4): 619–633.

Friedmann, J.M, Scott, M.L. and Lewis, W.M., Jr 1995. Restoration of riparian forests using irrigation, disturbance, and natural seedfall. *Environmental Management* **19**: 547–557.

Friedrich, G., Chapman, D. and Beim, A. 1992. The use of biological material. Chapter 5 in *Water Quality Assessments—A Guide to Use of Biota, Sediments and Water in Environmental Monitoring*, 2nd edn, UNESO/WHO/UNEP. URL: http://www.who.int/docstore/water_sanitation_health/wqassess/begin.htm#Contents (accessed 1st Nov 2003).

Fripp, J., Copeland, R. and Jonas, M. 2001. An overview of USACE Stream Restoration Guidelines. *Proceedings of the Seventh Federal Interagency Sedimentation Conference*, March, Reno, NV (CD-ROM). U.S. Subcommittee on Sedimentation.

Frissell, C.A. and Bayles, D. 1996. Ecosystem management and the conservation of aquatic biodiversity and ecological integrity. *Water Resources Bulletin* **32**(2): 229–240.

Frissell, C.A. and Nawa, R.K. 1992. Incidence and causes of physical failure of artificial habitat structures in streams in western Oregon and Washington. *North American Journal of Fisheries Management* **12**: 182–197.

Frissell, C.A. and Ralph, S.C. 1999. Stream and watershed restoration. In Naiman, R.J. and Bilby, R.E. (eds) *River Ecology and Management: Lessons from the Pacific Coastal Ecoregion*. Springer, New York, pp. 599–624. URL: http://www.swf.usace.army.mil/pubdata/regulatory/other/links/stream/streamwatershedrestoration.pdf (accessed 1st Nov 2003).

Frissell, C.A., Liss, W.L., Warren, C.E. and Hurley, M.C. 1986. A hierarchical framework for stream habitat classification, viewing streams in a watershed context. *Environmental Management* **10**: 199–214.

Fritts, H.C. 1976. *Tree-Rings and Climate*. Academic Press, London, reprinted by Blackburn Press, Caldwell, NJ, 2001.

Fruh, E.G. and Lambert, W.P. 1976. Methodology to Evaluate Alternative Coastal Zone Management Policies: Application in the Texas Coastal Zone. Special Report III: A Methodology for Investigating Fresh Water Inflow Requirements of a Texas Estuary, Centre for Research in Water Resources, University of Texas, Austin, TX, p. 348.

Fryirs, K. 1999. The recovery potential of River Styles in the Bega catchment, NSW: a catchment based framework for prioritisation of river rehabilitation strategies. In Rutherfurd, I.D. and Bartley, R. (eds) *Proceedings of the Second Australian Stream Management Conference*, Adelaide, 8–11 Feb, Cooperative Research Centre for Catchment Hydrology, Monash University, Clayton, 279–285.

Fryirs, K. and Brierley, G.J. 1998. The character and age structure of valley fills in Upper Wolumla Creek catchment, South Coast, New South Wales. *Earth Surface Processes and Landforms* **23**: 271–287.

Galay, V.T. 1983. Causes of river bed degradation. *Water Resources Research* **19**: 1057–1090.

Gan, K.C. and McMahon, T.A. 1990a. Comparison of two computer models for assessing environmental flow requirements. Department of Water Resources, Victoria, Australia.

Gan, K.C. and McMahon, T.A. 1990b. Variability of results from the use of PHABSIM in estimating habitat area. *Regulated Rivers: Research & Management* **5**: 233–239.

Gan, K.C., McMahon, T.A. and Finlayson, B.L. 1991. Analysis of periodicity in streamflow and rainfall data by Colwell's indices. *J. Hydrol.* **123**: 105–118

Gan, K.C., McMahon, T.A. and Finlayson, B.L. 1992. Fractal dimensions and lengths of rivers in south east Australia. *The Cartographic Journal*, pp. 31–34.

Gan, K.C., McMahon, T.A. and O'Neill, I.C. 1990. Errors in estimated streamflow parameters and storages for ungauged catchments. *Water Resour. Bull.* **26**: 443–450.

Garcia de Jalon, D. 1995. Management of physical habitat for fish stocks. In Harper, D.M. and Ferguson, A.J.D. (eds) *The Ecological Basis for River Management*. John Wiley & Sons, Chichester, pp. 363–374.

Garde, R.J., Subramanya, K. and Nambudripad, K.D. 1961. Study of scour around spurdykes. *Journal of the Waterways and Harbours Division, ASCE* **87**: 23–27.

Gaudet, C.L., Wong, M.P., Brady, A. and Kent, R. 1997. How are we managing? The transition from environmental quality to ecosystem health. *Ecosystem Health* **3**(1): 3–10.

Gebhardt, K.A., Carolyn, B., Jensen, S. and Platts, W.S. 1989. Use of hydrology in riparian classification. In Gresswell, R.E. Barton, B.A. and Kershner, J.L. (eds) *Proceedings, Practical Approaches to Riparian Resource Management, an Educational Workshop*, 8–11 May, U.S. Department of the Interior, Bureau of Land Management, Billings, MT, pp. 53–59.

Gebhardt, K.A., Leonard, S., Staidl, G. and Prichard, D. 1990. Riparian and wetland classification review: riparian area management. *Technical Reference 1737-5*. U.S. Department of the Interior, Bureau of Land Management, Denver, CO.

Gee, G.W. and Brauder, J.W. 1986. Particle-size analysis. In Klute, A. (ed) *Methods of Soil Analysis, Part 1. Physical and Mineralogical Methods—Agronomy Monograph No. 9*, 2nd edn, pp. 383–411, Amer. Soc. of Agronomy, Soil Sci. Soc. of America, Madison, WI.

Gehrke, P.C., Brown, P., Schiller, C.B., Moffatt, D.B. and Bruce, A.M. 1995. River regulation and fish communities in the Murray-Darling System, Australia. *Regulated Rivers: Research & Management* **11**: 363–375

Gehrke, P.C., Gawne, B. and Cullen, P. 2003. *What is the Status of River Health in the Murray-Darling Basin?* CSIRO Land and Water, Canberra, Australian Capital Territory, October. URL: http://www.clw.csiro.au/priorities/hot_issues/murrayriver_health/murrayriver_health.pdf (accessed 1st Nov 2003).

Gehrke, P.C., Gilligan, D.M. and Barwick, M. 2002. Changes in fish communities of the Shoalhaven River 20 years after construction of Tallowa Dam, Australia. *River Research and Applications* **18**(3): 265–286.

Geller, L.D. 2003. A Guide to Instream Flow Setting in Washington State. *Publication No. 03-11-007.* Water Resources Program, Washington Department of Ecology Habitat Program, Washington Department of Fish and Wildlife, March. URL: http://www.ecy.wa.gov/pubs/0311007.pdf (accessed 1st Nov 2003).

Gerdes, G. 1994. A geomorphological perspective on the importance of a watershed approach to the design of natural channel systems. In Shrubsole, D. (ed) *'Natural' Channel Design: Perspectives and Practice.* Proceedings of the First International Conference on Guidelines for Natural Channel Systems, March, Niagara Falls, Ontario, Canadian Water Resources Association, Cambridge, Ontario, pp. 11–20.

Germanoski, D. and Ritter, D.F. 1988. Tributary response to local base level lowering below a dam. *Regulated Rivers: Research & Management* **2**: 11–24.

Gerrard, A.J. 1981. *Soils and Landforms: An Integration of Geomorphology and Pedology.* George Allen & Unwin. London.

Ghanem, A., Steffler, P., Hicks, F. and Katopodis, C. 1996. Two-dimensional hydraulic simulation of physical habitat conditions in flowing streams. *Regulated Rivers: Research & Management* **12**: 185–200.

Gilbert, G.K. 1877. *Report on the Geology of the Henry Mountains,* U.S. Geographical and Geological Survey of the Rocky Mountain Region, U.S. Government Printing Office, Washington, DC, p. 160.

Gillilan, S. 1996. Utilizing geomorphic analogs for design of natural stream channels, *Proceedings North American Water and Environment Congress* (CD-ROM), (New York: American Society of Civil Engineers) URL: <http://www.interfluve.com/polf_files/1996.2.pdf> (accessed 29th Feb 2000). (New York: American Society of Civil Engineers).

Gilvear, D.J. 1999. Fluvial geomorphology and river engineering: future roles utilizing a fluvial hydrosystems framework. *Geomorphology* **31**: 229–245.

Gippel, C.J. 1989. *The use of turbidity instruments to measure stream water suspended sediment concentration,* Monograph Series No. 4, Dept of Geography and Oceanography, University College, Australian Defence Force Academy, Canberra.

Gippel, C.J. 1995. Environmental hydraulics of large woody debris in streams and rivers. *Journal of Environmental Engineering* **121**: 388–395.

Gippel, C.J. 1999. Developing a focused vision for rehabilitation: The Lower Snowy River, Victoria. In Rutherfurd, I. and Bartley,

R. (eds) *Proceedings Second Australian Stream Management Conference,* Adelaide, Feb 8–11, CRC Catchment Hydrology, Monash University, Clayton, pp. 299–305.

Gippel, C.J. 2000. Managing regulated rivers for environmental values: selected case studies from Southeastern Australia. In Brizga, S. and Finlayson, B. (eds) *River Management, The Australian Experience.* John Wiley & Sons, Chichester, pp. 97–122.

Gippel, C.J. 2001a. Hydrological analyses for environmental flow assessment. In Ghassemi, F. and Whetton, P. (eds) *Proceedings MODSIM 2001.* International Congress on Modelling and Simulation, Modelling & Simulation Society of Australia & New Zealand, The Australian National University, Canberra, Australian Capital Territory, pp. 873–880.

Gippel, C.J. 2001b. Australia's environmental flow initiative: filling some knowledge gaps and exposing others. *Water Science and Technology* **43**(9): 73–88.

Gippel, C.J. 2002a. Geomorphic issues associated with environmental flow assessment in alluvial non-tidal rivers. *Australian Journal of Water Resources* **5**(1): 3–19.

Gippel, C.J. 2002b. The Victorian Snowy River: review of historical environmental change and proposed rehabilitation options. Report by Fluvial Systems Pty Ltd, Stockton in association with EarthTech Pty Ltd, to East Gippsland Catchment Management Authority, Bairnsdale, Victoria.

Gippel, C.J. 2002c. Workshop on environmental water requirements for Australian estuaries: outcomes and future directions. Report by Fluvial Systems Pty Ltd, Stockton, to Environment Australia, Canberra, Australian Capital Territory, June. URL: http://www.deh.gov.au/water/rivers/nrhp/workshop/index.html (accessed 1st Nov 2003).

Gippel, C.J. 2003. Review of achievements and outcomes of environmental flow initiatives undertaken on the extended River Murray System to August 2002. Report by Fluvial Systems Pty Ltd, Stockton, to Murray-Darling Basin Commission, Canberra, Australian Capital Territory, March. URL: http://www.thelivingmurray.mdbc.gov.au/content/item. phtml?itemId=10287&nodeId=file3efacce9b46f8& fn=Review_of_Achievements_OutcomesV10.pdf (accessed 1st Nov 2003).

Gippel, C.J. and Blackham, D. 2002. Review of environmental impacts of flow regulation and other water resorce developments in the River Murray and Lower Darling River System. Report by Fluvial Systems Pty Ltd, Stockton to Murray-Darling Basin Commission, Canberra. Australian Capital Territory. URL: http://www. thelivingmurray.mdbc.gov.au/content/index.phtml/itemld/10239/itemld/4484 (accessed 1st Nov. 2003).

Gippel, C.J. and Collier, K.J. 1998. Degradation and rehabilitation of waterways in Australia and New Zealand. In DeWaal, L.C., Large, A.R.G. and Wade, P.M. (eds) *Rehabilitation of Rivers: Principles and Implementation.* John Wiley & Sons, Chichester, pp. 269–300.

Gippel, C.J. and Finlayson, B.L. 1993. Downstream environmental impacts of regulation of the Goulburn River, Victoria. In *Hydrology and Water Resources Symposium,* 30 June–2 July, Newcastle, Institution of Engineers Australia, Canberra, Australian Capital Territory, pp. 33–38.

Gippel, C.J. and Fukutome, S. 1998. Rehabilitation of Japan's waterways. In de Waal, L.C., Large, A.R.C. and Wade, P.M. (eds) *Rehabilitation of Rivers: Principles and Implementation*. Wiley, Chichester, pp. 301–317.

Gippel, C.J. and Rhodes, T. 2001. Stream velocity distribution: a potentially useful variable for stream health assessment. In Rutherfurd, I., Sheldon, F., Brierley, G. and Kenyon, C. (eds) *Proceedings Third Australian Stream Management Conference*, Brisbane, 27-28 Aug, Cooperative Research Centre for Catchment Hydrology, Monash University, Clayton, pp. 231–237.

Gippel, C.J. and Stewardson, M.J. 1995. Development of an environmental flow management strategy for the Thomson River, Victoria, Australia. *Regulated Rivers: Research & Management* 10: 121–135.

Gippel, C.J. and Stewardson, M.J. 1998. Use of wetted perimeter in defining minimum environmental flows. *Regulated Rivers: Research & Management* 14: 53–67.

Gippel, C.J., Seymour, S., Brizga, S. and Craigie, N. 2001. The application of fluvial geomorphology to stream management in the Melbourne Water area. In Rutherfurd, I., Sheldon, F., Brierley, G. and Kenyon, C. (eds) *Proceedings Third Australian Stream Management Conference*. Brisbane, 27-28 Aug, Cooperative Research Centre for Catchment Hydrology, Monash University, Clayton, pp. 239–244.

Gippel, C.J., Finlayson, B.L. and O'Neill, I.C. 1996a. Distribution and hydraulic significance of large woody debris in a lowland Australian River. *Hydrobiologia* 318(3): 179–194.

Gippel, C.J., O'Neill, I.C., Finlayson, B.L. and Schnatz, I. 1996b. Hydraulic guidelines for the re-introduction and management of large woody debris in lowland rivers. *Regulated Rivers: Research & Management* 12: 223–236.

Gippel, C.J., Jacobs, T. and McLeod, T. 2002. Determining environmental flow needs and scenarios for the River Murray System, Australia. *Australian Journal of Water Resources* 5(1): 61–74.

Gippel, C.J., O'Neill, I.C. and Finlayson, B.L. 1992. *The Hydraulic Basis of Snag Management*. Centre for Environmental Applied Hydrology, Department of Civil and Agricultural Engineering and Department of Geography, The University of Melbourne. Mighty Mouse Publishing Services, Melbourne.

Gleick, J. 1987. *Chaos: Making a New Science*. Viking, New York.

Goethals, P. and De Pauw, N. 2001. Development of a concept for integrated ecological river assessment in Flanders. Belgium. *Journal of Limnology* 60 (Suppl. 1): 7–16.

Goldman, C.R. and Horne, A.J., 1983. *Limnology*. McGraw-Hill, New York.

Goldstein, R.M., Simon, T.P., Bailey, P.A., Ell, M., Pearson, E., Schmidt, K. and Enblom, J.W. 1994. Concepts for an Index of Biotic Integrity for streams of the Red River of the North Basin. In *North Dakota Water Quality Symposium Proceedings*, 30–31 March, Fargo, ND North Dakota State University Extension Service, pp. 169–180.

Goodwin, C.N. 1999. Fluvial classification: Neanderthal necessity or needless normalcy. In Bozeman, Montana, Olson, D.S. and Potyondy, J.P. (eds), *Proceedings of Specialty Conference on Wildland Hydrology*, 30 June–2 July, American Water Resources Association, Herndon, Virginia, pp. 229–236. URL:

http://www.stream.fs.fed.us/publications/PDFs/GOODWIN. PDF (accessed 29th Feb 2004).

Gordon, N. 1995. Summary of technical testimony in the Colorado Water Division 1 Trial. *Gen. Tech. Rep. RM-GTR-270*, U.S. Department of Agriculture, U.S. Forest Service, Rocky Mountain Forest and Range Experiment Station, Fort Collins, CO, September.

Gordon, N.D. 1996. The hydraulic geometry of the Acheron River, Victoria, Australia. Centre for Environmental Applied Hydrology report, University of Melbourne, Australia, February, 1996.

Gore, J.A. and Shields, F.D., Jr 1995. Can large rivers be restored? *BioScience* 45: 142–152.

Gore, J.A. 1978. A technique for predicting in-stream flow requirements of benthic macroinvertebrates. *Freshwater Biol.* 8: 141–151.

Gore, J.A. 1985a. Mechanisms of colonisation and habitat enhancement for benthic macroinvertebrates in restored river channels. In Gore, J.A. (ed) *The Restoration of Rivers and Streams: Theories and Experience*. Butterworth Publishers, Boston, MA, pp. 81–101.

Gore, J.A. (ed) 1985b. *The Restoration of Rivers and Streams: Theories and Experience*. Butterworth Publishers, Boston, MA.

Gore, J.A. and Bryant, F.L. 1988. River and stream restoration. In Cairns, J. (ed) *Rehabilitating Damaged Ecosystems*, Vol. I, CRC Press, Boca Raton, FL, pp. 23–38.

Gore, J.A. and Nestler, J.M. 1988. Instream flow studies in perspective. *Regulated Rivers: Research & Management* 2: 93–101.

Gore, J.A. and Petts, G.E. (eds) 1989. *Alternatives in Regulated River Management*. CRC Press, Boca Raton, FL.

Gore, J.A., Layzer, J.B. and Mead, J. 2001. Macroinvertebrate instream flow studies after 20 years: a role in stream management and restoration. *Regulated Rivers: Research & Management* 17(4,5): 527–542.

Gore, J.A., Nestler, J.M. and Layzer, J.B. 1989. Instream flow predictions and management options for biota affected by peaking-power hydroelectric operations. *Regulated Rivers: Research & Management* 3: 35–48.

Gorman, O.T. and Karr, J.R. 1978. Habitat structure and stream fish communities. *Ecology* 59: 507–515.

Goudie, A. 1981. *Geomorphological Techniques*. George Allen & Unwin, London, edited for the British Geomorphological Research Group.

Gowan, C. and Fausch, K.D. 1996. Long-term demographic responses of trout populations to habitat manipulation in six Colorado streams. *Ecological Applications* 6: 931–946.

Graf, W.H. 1971. *Hydraulics of Sediment Transport*. McGraw-Hill, New York.

Graham, W.J. 1999. A procedure for estimating the loss of life caused by dam failure. *Bureau of Reclamation Report DSO-99-06*, U.S. Department of Interior, Washington DC.

Grant, G.E. 2001. Dam removal: Panacea or Pandora for rivers? *Hydrological Processes* 15: 1531–1532.

Grant, G.E., Schmidt, J.C. and Lewis, S.L. 2003. A geological framework for interpreting downstream effects of dams on rivers. In O'Connor, J.E. and Grant, G.E. (eds) *A Peculiar River: Geology, Geomorphology, and Hydrology of the Deschutes River, Oregon*. Water Science and Application Series,

Vol 7, American Geophysical Union, Washington, DC, pp. 203–219.

Gray, B. 2003. Mapping AusRivAS Scores, Inland Waters, National River Health Program. Environment Australia. Department of the Environment and Heritage. Canberra, Australian Capital Territory. URL: http://www.deh.gov.au/water/rivers/nrhp/pubs/ausrivas-scores.pdf (accessed 1st Nov 2003).

Gray, L.J. 1981. Species composition and life histories of aquatic insects in a lowland sonoran desert stream. *The American Midland Naturalist* **106**: 229–242.

Grayson, R.B., Barling, R.D. and Ogleby, C.L. 1988. A comparison between photogrammetry and a surface profile meter for the determination of surface topography for micro-erosion measurement. Conf. on Ag. Engineering, Hawkesbury, 25–30 Sept 1988, *I.E. Aust. Natl Conf. Publication No. 88/12*, pp. 264–268.

Grayson, R.B., Kenyon, C., Finlayson, B.L. and Gippel, C.J. 1998. Bathymetric and core analysis of the Latrobe River delta to assist in catchment management. *Journal of Environmental Management* **52**: 361–372.

Green, D.L. 1992. Survey of wetlands of the Warrego River. Department of Water Resources, *Technical Services Division Report 92.081*, Environmental Studies Unit, Parramatta.

Green, D.L. 1997. *Wetland Management Technical Manual: Wetland Classification*. Prepared by the Ecological Services Unit for the Water Environments Unit of the Department of Land and Water Conservation, Parramatta.

Green, D.M. and Kauffman, J.B. 1995. Succession and livestock grazing in a northeast Oregon riparian ecosystem. *Journal of Range Management* **48**: 307–313.

Green, R.H. 1979. *Sampling Design and Statistical Methods for Environmental Biologists*. Wiley-Interscience, New York.

Greenwood-Smith, S.L. 2002. The use of rapid environmental assessment techniques to monitor the health of Australian rivers. *Water Science and Technology* **45**: 155–160.

Gregory, K.J. 1976. The determination of river channel capacity. *New England Research Series in Applied Geography*, No. 42, University of New England, Armidale, New South Wales, Australia.

Gregory, K.J. (ed) 1983. *Background to Palaeohydrology: A Perspective*. John Wiley, Chichester.

Gregory, K.J. and Madew, J.R. 1982. Land use change, flood frequency and channel adjustments. In Hey, R.D., J.C. Bathurst and C.R. Thorne (eds) *Gravel-Bed Rivers*. John Wiley, Chichester, pp. 757–781.

Gregory, K.J. and Walling, D.E. 1973. *Drainage Basin Form and Process*. Edward Arnold, London.

Gregory, S. and Staley, K. (eds) 2004. Wood in World Rivers. Proceedings of the International Conference on Wood in World Rivers. American Fisheries Society Special Publication, American Fisheries Society, Bethesda, Maryland, New York.

Gregory, S.V., Swanson, F.J., McKee, W.A. and Cummins, K.W. 1991. An ecosystem perspective of riparian zones. *Bioscience* **41**(8): 540–551.

Griffiths, J.C. 1967. *Scientific Method in Analysis of Sediments*. McGraw-Hill, New York (as cited by Goudie, 1981).

Gringorten, I.I. 1963: A plotting rule foe extreme probability paper. *Jour. Geophy. Res.* **68**(3): 813, 814.

Grossman, G.D. and Freeman, M.C. 1987. Microhabitat use in a stream fish assemblage. *Journal of Zoology (London)* **212**: 151–176.

Growns, I., Gehrke, P.C., Astles, K.L. and Pollard, D.A. 2003. A comparison of fish assemblages associated with different riparian vegetation types in the Hawkesbury-Nepean River system. Fisheries Management and Ecology **10**(4): 209–220.

Growns, J. and Marsh, N. 2000. Characterisation of flow in regulated and unregulated streams in eastern Australia. Cooperative Research Centre for Freshwater Ecology, *Technical Report 3/2000*, University of Canberra, Canberra, Australian Capital Territory.

Grumbine, R.E. 1994. What is ecosystem management? *Conservation Biology* **8**(1): 27–38.

Gumbel, E.J. 1963. Statistical forecast of droughts. *Bull. IASH* **1**: 5–23 (as cited by Shaw, 1988).

Gunderson, L. 1999. Resilience, flexibility and adaptive management—antidotes for spurious certitude? *Conservation Ecology* **3**(1): 7. [online] URL: http://www.consecol.org/vol3/iss1/art7 (accessed 1st Nov 2003).

Gunderson, L.H. and Holling, C.S. (eds) 2002. *Panarchy: Understanding Transformations in Human and Natural Systems*. Island Press, Washington, DC.

Gustard, A. 1992. Analysis of river regimes. In Calow, P. and Petts, G.E. (eds) *The Rivers Handbook. Volume 1*. Blackwell Scientific Publications, Oxford, pp. 29–47.

Guy, H.P. 1969. Laboratory theory and methods for sediment analysis. *Techniques of Water-Resources Investigations of the USGS*. Volume 5, Chapter C1, USGS, Washington, DC.

Haan, C.T. 1977. *Statistical Methods in Hydrology*. The Iowa State University Press, Ames, IA.

Haan, C.T. and Read, H.R. 1970. Prediction of monthly, seasonal and annual runoff volumes for small agricultural watersheds in Kentucky. *Bull. 711*, Kentucky Ag. Expt. Sta., Univ. of Kentucky, Lexington, Kentucky (as cited by Haan, 1977).

Hadley, R.F. and Schumm, S.A. 1961. Sediment sources and drainage basin characteristics in upper Cheyenne River basin. US Geol. Surv. Water-Supply Paper 1531-B, USGS, Washington, DC.

Hagen, V.K. 1989. Flood flow frequency, PC application of Bulletin 17B. Paper presented at IAHS Conf., 10–19 May 1989, Baltimore, MD.

Haines, A.T., Finlayson, B.L. and McMahon, T.A. 1988. A global classification of river regimes. *Applied Geography* **8**: 255–272.

Hall, J.D. and Baker, C.O. 1982. Influence of forest and rangeland management: anadromous fish habitat in western North America: 12. Rehabilitating and enhancing stream habitat: 1. Review and evaluation. Meehan, W.R. (Tech. Ed.), *USDA General Technical Report PNW-138*. U.S. Department of Agriculture Forest Service, Pacific Northwest Forest and Range Experiment Station, Anadromous Fish Habitat Program, Portland, Oregon, June, 29 pp. URL: http://www.krisweb.com/krisweb_kt/biblio/general/usfs/hall82.pdf (accessed 1st Nov 2003).

Hallock, D. 2002. A Water Quality Index for Ecology's Stream Monitoring Program. *Washington State Department of Ecology*

*Publication No. 02-03-052.* Environmental Assessment Program, Olympia WA. URL: http://www.ecy.wa.gov/biblio/0203052.html (accessed 1st Nov 2003).

Haltiner, J.P., Kondolf, G.M. and Williams, P.B. 1996. Restoration approaches in California. In Brookes, A. and Shields, F.D., Jr (eds) *River Channel Restoration—Guiding Principles for Sustainable Projects.* John Wiley & Sons, Chichester, pp. 291–329.

Hamilton, K. and Bergersen, E.P. 1984. *Methods to Estimate Aquatic Habitat Variables*, prepared by Colorado Cooperative Fishery Research Unit, Colorado State University, for the Bureau of Reclamation, Denver, CO.

Hancock, D.A. (ed.), 1993. *Sustainable Fisheries Through Sustaining Fish Habitat.* Australian Society for Fish Biology Workshop, Victor Harbor, SA, 12–13 August 1992. Bureau of Resource Science Proceedings. Australian Government Publishing Service, Canberra, Australian Capital Territory.

Hansen, D.V. and Rattray, M. 1966. New dimensions in estuary classification. *Limnology and Oceanography* **11**: 319–325.

Hardy, T.B. 1998. The future of habitat modeling and instream flow assessment techniques. *Regulated Rivers: Research & Management* **14**: 405–420.

Hardy, T.B. and Addley, R.C. 2003. Instream flow assessment modelling: combining physical and behavioural-based approaches. *Canadian Water Resources Journal* **28**(2): 273–282.

Harmon, M.E., Franklin, J.F., Swanson, F.J., Sollins, P., Gregory, S.V., Lattin, J.D., Anderson, N.H., Cline, S.P., Aumen, N.G., Sedell, J.R., Lienkaemper, G.W., Cromack, J.K. and Cummins, K.W. 1986. Ecology of coarse woody debris in temperate ecosystems. *Advances in Ecological Research* **15**: 133–302.

Harper, D. and Everard, M. 1998. Why should the habitat-level approach underpin holistic river survey management? *Aquatic Conservation: Marine and Freshwater Ecosystems* **8**: 395–413.

Harper, D.M. and Ferguson, A.J.D. (eds) 1995. *The Ecological Basis for River Management.* John Wiley & Sons, Chichester.

Harrelson, C.C., Rawlins, C.L. and Potyondy, J.P. 1994. Stream Channel Reference Sites: An Illustrated Guide to Field Technique, *Gen. Tech. Rep. RM-245*, US Forest Service, Rocky Mountain Forest and Range Experiment Station, Fort Collins, CO.

Harris, G. 2001. A nutrient dynamics model for Australian waterways, land use, catchment biogeochemistry and water quality in Australian rivers, lakes and estuaries. Australia, *State of the Environment Technical Paper Series (Inland Waters) Series 2.* Department of the Environment and Heritage, Canberra, Australian Capital Territory. URL: http://www.deh.gov.au/soe/techpapers/nutrient-dynamics/index.html (accessed 1st Nov 2003).

Harris, J.H. 1995. The use of fish in ecological assessments. *Australian Journal of Ecology* **20**: 65–80.

Harris, N.M., Gurnell, A.M., Hannah, D.M. and Petts, G.E. 2000. Classification of river regimes: A context for hydroecology. *Hydrological Processes* **14**: 16–17, 2831–2848.

Harrison, S.S. and Reid, J.R. 1967. A flood-frequency graph based on tree-scar data. *North Dakota Academy of Science Annual Proc.* pp. 23–33.

Hart, B.T., Bailey, P., Edwards, R., James, K., Swadling, K., Meredith, C., McMahon, A. and Hortle, K. 1990. Effects of saline discharges on aquatic ecosystems. *Water Res.* **24**: 1103–1117.

Hart, B.T., Bailey, P., Edwards, R., James, K., Swadling, K., Meredith, C., McMahon, A. and Hortle, K. 1991. Biological effects of saline discharges to streams and wetlands: a review. *Hydrobiol.* **210**: 105–144.

Hart, D.D. and Finelli, C.M. 1999. Physical-biological coupling in streams: The pervasive effects of flow on benthic organisms. *Annual Review of Ecology and Systematics* **30**: 363–395.

Hartman, K.J. 1999. Models as tools in assessing river biota. In *Appalachian Rivers II Conference*, 28–29 July, 1999, Morgantown, WV. Federal Energy Technology Center and the WMAC Foundation, Morgantown. NETL Publications. URL: http://www.netl.doe.gov/publications/proceedings/99/99apprvr/ar1-2.pdf (accessed 1st Nov 2003).

Harvey, A.M. 1969. Channel capacity and the adjustment of streams to hydrologic regime. *J. of Hydrol.* **8**: 82–98.

Hasfurther, V.R. 1985. The use of meander parameters in restoring hydrologic balance to reclaimed stream beds. In Gore, J.A. (ed) *The Restoration of Rivers and Streams.* Butterworths, Boston, MA, pp. 21–40.

Hatton, T. and Evans, R. 1998. Dependence of ecosystems on groundwater and its significance to Australia. *LWRRDC Occasional Paper No. 12/98.* Land and Water Resources Research & Development Corporation, Canberra, Australian Capital Territory.

Hawkes, H.A. 1975. River zonation and classification. In Whitton, B.A. (ed) *River Ecology.* Blackwell Scientific, Oxford, pp. 312–374.

Hawkes, H.A. 1977. Biological classification of rivers: conceptual basis and ecological validity. In Alabaster, J.S. (ed) *Biological Monitoring of Inland Fisheries.* Applied Science Publishers Ltd, London, pp. 55–67.

Hawkins, C.P., Kershner, J.L., Bisson, P.A., Bryant, M.D., Decker, L.M., Gregory, S.V., McCullough, D.A., Overton, C.K., Reeves, G.H., Steedman, R.J. and Young, M.K. 1993. A hierarchical approach to classifying stream habitat features. *Fisheries* **18**: 3–12.

Hawkins, R.H. 1975. Acoustical energy output from mountain stream channels, *J. Hyd. Div., Proc. ASCE* **101**: 571–575.

Hay, I. 2002. *Communicating in Geography and the Environmental Sciences*, 2nd edn, Oxford University Press, p. 178.

Hayes, D.F. (ed) 1998. *Engineering Approaches to Ecosystem Restoration*, Proceedings of the 1998 Wetlands Engineering and River Restoration Conference, 20–29 March, Denver, CO (CD-ROM), American Society of Civil Engineers, Reston, VA.

Hayes, D.F. (ed) 2001. *Designing Successful Stream and Wetland Restoration Projects*, Proceedings of the 2001 Wetlands Engineering and River Restoration Conference, 27–31 Aug, Reno (CD-ROM), American Society of Civil Engineers, Reston, VA.

Hellawell, J.M. 1977. Biological surveillance and water quality monitoring. In Alabaster, J.S. (ed) *Biological Monitoring of Inland Fisheries.* Applied Science Publishers Ltd, London, pp. 69–88.

Hellawell, J.M. 1986. In Mellanby, K. (ed) Biological indicators of freshwater pollution and environmental management. *Pollution Monitoring Series.* Elsevier Applied Sciences Publishers, London, p. 546.

Heller, D., Roper, B.B., Konnoff, D. and Wieman, K. 2001. Durability of Pacific Northwest instream structures following floods of a five-year or greater return interval. In Rutherfurd, I., Sheldon, F., Brierley, G. and Kenyon, C. (eds) *Proceedings Third Australian Stream Management Conference*, Brisbane, 27-28 Aug, CRC Catchment Hydrology, Monash University, Clayton, Victoria, pp. 309–314.

Helley, E.J. and Smith, W. 1971. Development and calibration of a pressure-difference bedload sampler. *US Geol. Survey Open-File Report*, USGS, Washington, DC.

Henderson, F.M. 1966. *Open Channel Flow*. Macmillan, New York.

Henderson, J.E. and Shields, F.D., Jr 1984. Environmental features for streambank protection measures. *Environmental and Water Quality Operational Studies Technical Report E-84-11*. U.S. Army Engineer Waterways Experiment Station, Environmental Laboratory, Vicksburg, Mississippi. Department of the Army, U.S. Army Corps of Engineers, Washington, DC.

Henrikson, L. and Medin, M. 1986. Biologisk bedömning av försurningspåverkan på Lelångens tillflöden och grundområden, 1986. Aquaekologerna. Report to Älvsborgs County Administrative Board.

Hering, D. and Strackbein, J. 2002. STAR stream types and sampling sites. Framework method for calibrating different biological survey results against ecological quality classifications to be developed for the Water Framework Directive. University of Essen, Germany. URL: http://www.eu-star.at/pdf/FirstDeliverable.pdf (accessed 5th Jan 2004).

Herschy, R.W. (ed) 1978. *Hydrometry, Principles and Practices*, John Wiley, Chichester.

Herschy, R.W. 1985. *Streamflow Measurement*. Elsevier Applied Science, London.

Hey, R.D. 1978. Determinate hydraulic geometry of river channels. *Journal Hydraulics Division ASCE* **104**: 869–885.

Hey, R.D. 1994a. Restoration of gravel-bed rivers: principles and practice. In Shrubsole, D. (ed) *'Natural' Channel Design: Perspectives and Practice*, Proceedings of the First International Conference on Guidelines for 'Natural' Channel Systems, March, Niagara Falls, Ontario, Canadian Water Resources Association, Cambridge, Ontario, pp. 157–173.

Hey, R.D. 1994b. Environmentally sensitive river engineering. In Calow, P. and Petts, G.E. (eds) *The Rivers Handbook. Volume 2*. Blackwell Scientific Publications, Oxford, pp. 337–362.

Hey, R.D. and Thorne, C.R. 1986. Stable channels with mobile gravel beds. *Journal of Hydraulic Engineering* **112**(8): 671–689.

Hey, R.D., Heritage, G.L., Tovey, N.K., Boar, R.R. and Rurner, R.K. 1991. Streambank protection in England and Wales. *Rstream structures*. In Collier, K.J. (ed) *Restoration of Aquatic Habitats*. Selected papers for the second day of the New Zealand Limnological Society 1993 Annual Conference, 10–12 May, Wellington, N.Z., Department of Conservation, Te Papa Atawhai, Wellington, pp. 76–91.

Hicks, B.J. and Reeves, G.H., 1994, Restoration of stream habitat for fish using in-stream structures, in *Restoration of Aquatic Habitats*, Selected papers for the second day of the New Zealand Limnological Society 1993 Annual Conference, 10–12 May,

Collier, K.J. (Ed.) Wellington, N.Z., Department of Conservation, Te Papa Atawhai, Wellington, pp. 76–91.

Higgins, A.L. 1965. *Elementary Surveying*, Longmans, Green and Co., London.

Hilderbrand, R. H., Lemly, A.D., Dolloff, C.A. and Harpster, K.L. 1997. Effects of large woody debris placement on stream channels and benthic macroinvertebrates. *Canadian Journal of Fisheries and Aquatic Science* **54**: 931–939.

Hill, M.T., Platts, W.S. and Beschta, R.L. 1991. Ecological and geomorphological concepts for instream and out-of-channel flow requirements. *Rivers* **2**: 2198–2210.

Hjulstrom, F. 1939. Transportation of detritus by moving water. In Trask, P.D. (ed) *Recent Marine Sediments, a Symposium*. American Assn of Petroleum Geologists, Tulsa, Oklahoma, pp. 5–31.

Hobbs, R.J. and Harris, J.A. 2001. Restoration ecology: repairing the earth's ecosystems in the new millennium. *Restoration Ecology* **9**(2): 239–246.

Hoffmann, M. 2003. River water quality determination using saprobic indices. Independent Laboratory for Environmental Studies in cooperation with the Laboratory of the Main State Ecological Inspectorate, Kiev, Ukraine. URL: http://de.geocities.com/ecology_lab_kiev/Kiev-map-expl.htm (accessed 1st Nov 2003).

Hogan, D.L. and Church, M. 1989. Hydraulic geometry in small, coastal streams: progress toward quantification of salmonid habitat. *Can. J. Fish. Aquat. Sci.* **46**: 844–852.

Hoitsma, T.R. and Payson, E.M. 1998. The use of vegetation in bioengineered streambanks: shear stress resistance of vegetal treatments. *Proceedings of ASCE Wetlands and River Restoration Conference*, Denver. American Society of Civil Engineers, Washington, DC.

Holcik, J. 1991. Fish introductions in Europe with particular reference to its central and eastern part. *Canadian Journal of Fisheries and Aquatic Sciences* **48** (Suppl. 1): 13–23.

Holder, R.L. 1985. *Multiple Regression in Hydrology*, Institute of Hydrology, Wallingford.

Holeman, J.N. 1968. The sediment yield of major rivers of the world. *Water Resour. Res.* **4**: 737–747.

Holling, C.S. (ed), 1978. *Adaptive Environmental Assessment and Management*. John Wiley & Sons, Chichester.

Holmes N.T.H. 1998. A review of river rehabilitation in the UK, 1990–1996. *Environment Agency R&D Technical Report W175*, Environment Agency, Bristol.

Hooke, J.M. 1979. An analysis of the processes of river bank erosion. *Journal of Hydrology* **42**: 39–62.

Hooke, J.M. 1986. The significance of mid-channel bars in an active meandering river. *Sedimentology* **33**: 839–850.

Hooke, J.M. and Kain, R.J.P. 1982. *Historical Change in the Physical Environment: A guide to sources and techniques*. Butterworth Scientific, Guildford.

Hooper, D.U., Solan, M., Symstad, A., Díaz, S., Gessner, M.O., Buchmann, N., Degrange, V., Grime, P., Hulot, F., Mermillod-Blondin, F. Roy, J., Spehn, E. and van Peer, L. 2002. Species diversity, functional diversity and ecosystem functioning. In Loreau, M., Naeem, S. and Inchausti, P. (eds) *Biodiversity and Ecosystem Functioning, Synthesis and Perspectives*. Oxford University Press, Oxford, UK, pp. 195–208.

Hoppe, R.A. 1975. Minimum streamflows for fish. Paper distributed at Soils-Hydrology Workshop, USFS, Montana State University, 26–30 Jan 1976, Bozeman, MT (as cited by Wesche and Reschard, 1980).

Horner, R.R. and Welch, E.B. 1982. *Impacts of Channel Reconstruction in the Pilchuck River.* Washington State Department of Transportation, Olympia, WA.

Horowitz, A.J., Rinella, F.A., Lamothe, P., Miller, T.L., Edwards, T.K., Roche, R.L. and Rickert, D.A. 1989. Cross-sectional variability in suspended sediment and associated trace element concentrations in selected rivers in the US. In Hadley, R.F. and Ongley, E.D. (eds) *Sediment and the Environment*, Proc. of Baltimore Symposium, May 1989, *IAHS Publ. No. 184*, pp. 57–66.

Horton, R.E. 1932. Drainage basin characteristics. *Trans. Am. Geophys. Union* **13**: 350–361.

Horton, R.E. 1945. Erosional development of streams and their drainage basins: hydrophysical approach to quantitative morphology. *Bull. Geol. Soc. Am.* **56**: 275–370.

Horwitz, R.J. 1978. Temporal variability patterns and the distributional patterns of stream fishes. *Ecological Monographs* **48**: 307–321 (as cited by Resh et al., 1988).

Hosking, J.R.M. 1989. The theory of probability weighted moments. *Res. Rep. RC12210 (No. 54860)*, IBM Research Division, T.J. Watson Research Center, Yorktown Heights, NY 10598.

Hosking, J.R.M., Wallis, J.R. and Wood, E.F. 1985. An appraisal of the regional flood frequency procedure in the UK Flood Studies Report. *Hydro. Sci. J.* **30**: 85–109.

Hosmani, S.P. 2002. Biological Indicators of Water Quality. *Lake 2002, Symposium on Conservation, Restoration and Management of Aquatic Ecosystems.* Centre for Ecological Sciences, Indian Institute of Science, Bangalore, Karnataka Environment Research Foundation, Bangalore and Commonwealth of Learning, Canada. URL: http://ces.iisc.ernet.in/energy/Lake2002abs/ses51.html (accessed 1st Nov 2003).

Hötzel, G. and Croome, R. 1998. A phytoplankton methods manual for Australian rivers. *Occasional Paper No. 18/98*. Land and Water Resources Research and Development Corporation, Canberra, Australian Capital Territory.

House, R., Anderson, J., Boehne, P. and Suther, J. (eds) 1988. Stream Rehabilitation Manual Emphasizing Project Design—Construction—Evaluation. Oregon Chapter American Fisheries Society.

Howard-Williams, C. and Pickmere, S. 1994. Long-term vegetation and water quality changes associated with the restoration of a pasture stream. In Collier, K.J. (ed) *Restoration of Aquatic Habitats*. Selected papers for the second day of the New Zealand Limnological Society 1993 Annual Conference, 10–12 May, Wellington, N.Z., Department of Conservation, Te Papa Atawhai, Wellington, pp. 93–107.

Howell, J., Benson, D. and McDougall, L. 1994. Developing a strategy for rehabilitating riparian vegetation of the Hawkesbury-Nepean River, Sydney, Australia. *Pacific Conservation Biology* **1**: 257–271.

Huang, H.Q. and Nanson, G.C. 1997. Vegetation and channel variation; a case study of four small streams in southeastern Australia. *Geomorphology* **18**: 237–249.

Hubbel, P. 1973. A program to identify and correct salmon and steelhead problems in the Trinity River Basin. California, Department of Fish and Game.

Hubbell, D.W. 1964. Apparatus and techniques for measuring bedload. *US Geol. Survey Water Supply Paper No. 1748*, USGS, Washington, DC.

Huet, M. 1949. Apercu dés relations entre la pente et les populations des eaux courantes. *Schweizerische Zeitschrift fur Hydrologie* **11**: 333–351.

Huet, M. 1954. Biologie, profils en long et en travers des eaux courantes. *Bulletin Francais de Pisciculture* **175**: 41–53.

Huet, M. 1959. Profiles and biology of Western European streams as related to fish management. *Transactions of the American Fisheries Society* **88**: 155–163.

Huff, D. 1954. *How to Lie with Statistics*. WW Norton, New York.

Hughes, F.M.R., Adams, W.M., Muller, E., Nilsson, C., Richards, K.S., Barsoum, N., Decamps, H., Foussadier, R., Girel, J., Guilloy, H., Hayes, A., Johansson, M., Lambs, L., Pautou, G., Peiry, J.-L., Perrow, M., Vautier, F. and Winfield, M. 2001. The importance of different scale processes for the restoration of floodplain woodlands. *Regulated Rivers: Research & Management* **17**(4,5): 325–345.

Hughes, J.M. and James, B. 1989. A hydrological regionalization of streams in Victoria, Australia, with implications for stream ecology. *Australian Journal of Marine and Freshwater Research* **40**: 303–326.

Hughes, R.M. and Omernik, J.M. 1983. An alternative for characterizing stream size. In Fontaine, III, T.D. and Bartell, S.M (eds) *Dynamics of Lotic Ecosystems* Ann Arbor Science, MA, pp. 87–101.

Humphries, P. and Lake, P.S. 1996. Environmental flows in lowland rivers: experimental flow manipulation in the Campaspe River, northern Victoria. *Proceedings of the 23rd Hydrology and Water Resources Symposium*. Hobart, Tasmania, 21–24 May, Institution of Engineers, Australia, Australian Capital Territory, pp. 197–202.

Humphries, P. and Lake, P.S. 2000. Fish larvae and the management of regulated rivers. *Regulated Rivers: Research & Management* **16**: 421–432.

Hunsaker, C.T. and Levine, D.A. 1995. Hierarchical approaches to the study of water quality in rivers. *Bioscience* **45**(3): 193–203.

Hunt, R.L. 1988. A compendium of 45 trout stream habitat development evaluations in Wisconsin during 1953–1985. *Technical Bulletin No. 162*. Wisconsin Department of Natural Resources, Madison, WI.

Hunt, W.A. and Graham, R.O. 1975. Evaluation of channel changes for fish habitat. ASCE National Convention meeting reprint (2535), cited in FISRWG (1998).

Hunter, C.J. 1991. *Better Trout Habitat. A Guide to Stream Restoration and Management*. Montana Land Reliance, Island Press, Washington, DC.

Hupp, C.R. and Bryan, B.A. 1989. A dendrogeomorphic approach to flash-flood frequency estimation. Paper presented at spring meeting of Am. Geophys. Union, May 1989, Baltimore, MD.

Hurlbert, S.J. 1984. Pseudoreplication and the design of ecological field experiments. *Ecological Monographs* **54**: 187–211.

Hutchinson, G.E. 1957. Concluding remarks. *Proceedings of Symposium on Quantitative Biology.* Cold Springs Harbour, NY, pp. 415–427.

Hynes, H.B.N. 1970. *The Ecology of Running Waters*, Liverpool University Press, Liverpool.

Hynes, H.B.N. 1975. The stream and its valley. *Internationale Vereinigung für theoretische und angewandte Limnologie, Verhandlungen* **19**: 1–15.

IAC. (1982). Guidelines for determining flood flow frequency. *Bulletin 17B*, Interagency Advisory Committee on Water Data, Hydrology Sub-committee, Office of Water Data Coordination, USGS, Washington, DC.

ICA. 1984. *Basic Cartography for Students and Technicians*, Vol. 1, International Cartographic Association, UK.

ID&A Pty Ltd. 1996. Hydraulic geometry of Brisbane streams: Guidelines for natural channel design. Part I—Guidelines. Final Report, Brisbane City Council, Brisbane, August.

ID&A Pty Ltd. 2001. Scoping Study Waterway Management Plan. Hume to Yarrawonga Reach of the River Murray. Summary Report. Murray-Darling Basin Commission, Canberra, Australian Capital Territory, May. URL: http://www.mdbc.gov.au/river_murray/river_murray_system/yarrawonga/images/Scoping-StudyFinalReport.pdf (accessed 1st Nov 2003).

ID&A Pty Ltd. 2002. Impacts of Grazing on Riparian Zones, Volume 1—Literature Review. Demonstration and Evaluation of Riparian Management. Goulburn Broken Catchment Management Authority in association with Land & Water Australia. Goulburn Broken CMA, Shepparton, Victoria. URL: http://www.rivers.gov.au/acrobat/gbc1/gbc1vol1litreview.pdf (accessed 1st Nov 2003).

IEA. 1998, Australian rainfall and runoff: a guide to flood estimation, Institution of Engineers, Australia, Barton, ACT.

IETC (International Environmental Technology Centre). 2002. Reach and channel unit: habitat quality assessment. *Guidelines for the Integrated Management of the Watershed—Phytotechnology and Ecohydrology. Freshwater Management Series No. 5.* United Nations Environment Programme, Division of Technology, Industry and Economics. Osaka, Japan. URL: http://www.unep.or.jp/ietc/Publications/Freshwater/FMS5/index.asp (accessed 1st Nov 2003).

IH. 1980. *Low Flow Studies: (1) Research Report: (2) Flow Duration Curves Estimation; (3) Flow Frequency Curves Estimation; (4) Catchment Characteristics Estimation Manual*, Institute of Hydrolgy, Wallingford.

Illies, J. 1961. Versuch einer allgemein biozönotischen Gliederung der Fliessgewässer. *Int. Revue ges. Hydrobiol.* **46**: 205–213 (as cited by Hawkes, 1975).

Illies, J. 1962. Die bedeutung der strömung für die biozönose in rithron und potamon. *Schweiz. Z. Hydrol.* **24**: 433–435.

Illies, J. 1978. *Limnofauna Europaea.* Gustav Fischer Verlag, Stuttgart.

Illies, J. and Botosaneanu, L. 1963. Problèmes et methodes de la classification et de la zonation ecologique des eaux courantes, considerées surtout du point de vue faunistique. *Internationale Vereinigung für theoretische und angewandte Limnologie, Mitteilungen* **12**: 1–57.

Industry Canada. (no date). Habitat Improvements: Boulder Placement. Ontario Stream Restoration: Planning and Projects. Prepared by Nolan, J., Division of the Environment, University of Toronto, for Canada's Digital Collections Program, Government of Canada. http://collections.ic.gc.ca/streams/tech/H-boulderplacement.html (accessed 1st Nov 2003).

Institute for Natural Systems Engineering. 1999. Evaluation of Interim Instream Flow Needs in the Klamath River, Phase I Final Report. Prepared to the Department of the Interior by Utah Water Research Laboratory Utah State University, Logan, Utah. URL: http://www.krisweb.com/krisweb_kt/biblio/klamath/misc/hardy99.pdf (accessed 1st Nov 2003).

International Rivers Network. 2001. Reviving the World's Rivers: Dam Removal. Berkeley, CA. URL: http://www.irn.org/revival/decom/brochure/rrpt2.html (accessed 1st Nov 2003).

International Water Management Institute. 2003. Environmental flow assessment for aquatic ecosystems: a database of methodologies. Tharme, R. (Moderator), Ecohydrological Databases, IWMI, Headquaters: Colombo, Sri Lanka. URL: http://www.lk.iwmi.org/ehdb/EFM/efm.asp (accessed 1st Nov 2003).

Irvine, J.R., Jowett, I.G. and Scott. D. 1987. A test of the instream flow incremental methodology for underyearling rainbow trout, *Salmo gairdnerii*, in experimental New Zealand streams. *NZ J. of Mar. and Freshwater Res.* **21**: 35–40.

Irwin, E.R. and Freeman, M.C. 2002. Proposal for adaptive management to conserve biotic integrity in a regulated segment of the Tallapoosa River, Alabama, USA. *The Journal of the Society for Conservation Biology* **16**(5): 1212–1222.

ISO. 1983. *ISO Standards Handbook No. 16*, International Organization for Standardization, Geneva, Switzerland (contains series of standards for flow measurement in open channels).

Iverson, T.M., Kronvang, B., Madsen, B.L., Markmann, P. and Nielsen, M.B. 1993. Re-establishment of Danish streams: restoration and maintenance measures. *Aquatic Conservation: Marine and Freshwater Ecosystems* **3**: 73–92.

Jackson, J. 1997. State of habitat availability and quality in Inland Waters, State of the Environment Technical Paper Series (Inland Waters). Environment Australia, Department of the Environment, Canberra, Australian Capital Territory.

Jamieson, D. 1995. Ecosystem health: some preventative medicine. *Environmental Values* **4**: 333–344.

Jansson, R., Nilsson, C., Dynesius, M. and Andersson, E. 2000. Effects of river regulation on river-margin vegetation: a comparison of eight boreal rivers. *Ecological Applications* **10**(1): 203–224.

Jarrett, R.D. 1984. Hydraulics of high-gradient streams. *J. Hyd. Eng.* **110**: 1519–1539.

Jarrett, R.D. 1985. Determination of roughness coefficients for streams in Colorado. *US Geol. Survey Water-Resources Investigations Report 85-4004*, USGS, Lakewood, CO.

Jarrett, R.D. 1987. Errors in slope-area computations of peak discharges in mountain streams. *J. of Hydrol.* **96**: 53–67.

Jarrett, R.D. 1988. *Hydrologic and hydraulic research in mountain rivers*, prepared for International Workshop on Hydrology of Mountainous Areas, Strbske Pleso, Czechoslovakia, 6–11 June 1988.

Jarrett, R.D. 1990. Paleohydrology and its value in analyzing floods and droughts. *National Water Summary 1988–89— Floods and Droughts: Hydrology*, pp. 105–116, US Geol. Survey Water-Supply Paper 2375, USGS, Denver, CO.

Jarrett, R.D. and Malde, H.E. 1987. Paleodischarge of the late Pleistocene Bonneville flood, Snake River, Idaho, computed from new evidence. *Geol. Soc. of Am. Bull.* **99**: 127–134.

Jarrett, R.D. and Petsch, H.E., Jr 1985. Computer program NCALC user's manual—verification of Manning's roughness coefficient in channels, *US Geol. Survey Water-Resources Investigations Report No. 85–4317*, USGS, Lakewood, CO.

Jarvis, R.S. and Woldenberg, M.J. (eds) 1984. *River Networks*. Hutchinson Ross, Pennsylvania.

Jassby, A.D., Kimmerer, W.J., Monismith, S., Armor, C., Cloern, J.E., Powell, T.M., Schubel, J.R. and Vendlinski, T. 1995. Isohaline position as a habitat indicator for estuarine resources-San Francisco Bay-Delta, California, USA. *Ecological Applications* **5**: 272–289.

Jenkinson, A.F. 1955. The frequency distribution of the annual maximum (or minimum) values of meteorological elements. *Quart. Jour. Roy. Met. Soc.* **81**: 158–171.

Jennings, J.N. 1967. Topographical maps and the geomorphologist. *Cartography* **6**: 73–80.

Jensen, A. 1998. Rehabilitation of the River Murray, Australia: identifying causes of degradation and options for bringing the environment into the management equation. In de Waal, L.C., Large, A.R.C. and Wade, P.M. (eds) *Rehabilitation of Rivers: Principles and Implementation*. Wiley, Chichester, pp. 215–236.

Jensen, A. 2002a. Applying ecohydrology to on-ground management of wetlands and floodplains—'learning by doing'. *Ecohydrology & Hydrobiology* **2**: 67–78.

Jensen, A. 2002b. Repairing wetlands of the Lower Murray: learning from restoration practice. *Ecological Management & Restoration* **3**(1): 5–14.

Jensen, J.G. 2001. Fish passes in Australia—a national success. Can it be copied in the Mekong? *Mekong Fisheries Network Newsletter* **6**(3): 1–4. URL: http://www.mekonginfo.org/mrc_en/doclib.nsf/0/56A4A91F1D2E9AD747256DE20009C208/$FILE/FULLTEXT.pdf (accessed 1st Nov 2003).

Johnson, A.W. and Stypula, J.M. 1993. Guidelines for bank stabilization project in the riverine environments of King County. King County Department of Public Works, Surface Water Management Division, Seattle, WA.

Johnson, C.W., Engelman, R.L., Smith, J.P. and Hanson, C.L. 1977. Helley-Smith bedload samplers. *Proc. ASCE Hyd. Div.* **103**: 1217–1221.

Johnson, C.W., Gordon, N.D. and Hanson, C.L. 1985. Northwest sediment yield analysis by the MUSLE. *Trans. ASAE* **28**: 1885–1895.

Johnson, O.W., Flagg, T.A., Maynard, D.J., Milner, G.B. and Waknitz, F.W. 1991. Status review for Lower Columbia River Coho Salmon. *NOAA Technical Memorandum NOAA F/NWC-202*. U.S. Department of Commerce, National Oceanic and Atmospheric Administration, National Marine Fisheries Service, Northwest Fisheries Science Center, Coastal Zone and Estuarine Studies Division, Seattle, WA, June. URL: http://www.nwfsc.noaa.gov/publications/techmemos/tm202/202.pdf (accessed 1st Nov 2003).

Johnson, T.H., Lincoln, R., Graves, G.R., and Gibbons, R.G. 1997. Status of wild salmon and steelhead stocks in Washington State. In Stouder, D.J., Bisson, P.A. and Niaman, R.J. (eds) 1997. *Pacific Salmon and Their Ecosystems, Status and Future Options*. Chapman and Hall, New York, pp. 127–144.

Johnson, W.C. 1999. Tree recruitment and survival in rivers: influence of hydrological processes. *Hydrological Processes* **14**(16,17): 3051–3074.

Johnston, C.A. and Naiman, R.J. 1990. Aquatic patch creation in relation to beaver population trends. *Ecology* **71**: 1617–1621.

Jones, G. 2003. Towards a national river classification scheme based on human values. *Watershed* Issue **31**, September. Cooperative Research Centre for Freshwater Ecology, University of Canberra, Canberra, Australian Capital Territory. URL: http://enterprise.canberra.edu.au/WWW/www-crcfe.nsf/d87a31d8f4603d1d4a256641000e9021/7e16e5963b71476b4a25664a004a2493?OpenDocument (accessed 1st Nov 2003).

Jones, G., Hillman, T., Kingsford, R., McMahon, T., Walker, K., Arthington, A., Whittington, J. and Cartwright, S. 2002. Independent report of the Expert Reference Panel on environmental flows and water quality requirements for the River Murray System. Cooperative Research Centre for Freshwater Hydrology. Murray-Darling Basin Ministerial Council, Canberra, Australian Capital Territory. URL: http://www.mdbc.gov.au/TLM/pdf/ERPreport.pdf (accessed 1st Nov 2003).

Jones, H.R. and Peters, J.C. 1977. *Physical and Biological Typing of Unpolluted Rivers*. Technical Report TR 41, Medmenham Laboratory, UK.

Joubert, A.R. and Hurley, P.R. 1994. Grouping South African Rivers using flow-derived variables. In Uys, M. (ed) Classification of Rivers, and Environmental Health Indicators. Proceedings of a joint South African Australian Workshop, Cape Town, Feb 7–11, *Water Research Commission Report No. TT 63/94*, Pretoria, Republic South Africa, pp. 27–42.

Jowett, I.G. 1989. RHYHABSIM Computer manual, Freshwater Fisheries Centre, Riccarton, New Zealand.

Jowett, I.G. 1990. Factors related to the distribution and abundance of brown and rainbow trout in New Zealand clear water rivers. *NZ J. of Mar. and Freshwater Res.* **24**: 429–440.

Jowett, I.G. 1997. Instream flow methods: a comparison of approaches. *Regulated Rivers: Research & Management* **13**: 115–127.

Jowett, I.G. and Duncan, M.J. 1990. Flow variability in New Zealand rivers and its relationship to in-stream habitat and biota. *New Zealand Journal of Marine and Freshwater Research* **24**: 305–317.

Jowett, I.G. and Richardson, J. 1989. Effects of a severe flood on instream habitat and trout populations in seven New Zealand rivers. *NZ J. of Mar. and Freshwater Res.* **23**: 11–17.

Junk, W.J., Bayley, P.B. and Sparks R.E. 1989. The flood pulse concept in river-floodplain system. *Canadian Special Publication of Fisheries and Aquatic Sciences* **106**: 110–127.

Juracek, J.E. and Fitzpatrick, F.A. 2003. Limitations and implications of stream classification. *Journal of the American Water Resources Association* **39**(3): 659–670.

Jutila, E. 1992. Restoration of salmonid rivers in Finland. In Boon, P., Calow, P. and Petts, G. (eds) *River Conservation and Management*. John Wiley and Sons, Chichester, UK, pp. 353–363.

Kallis, G. and Butler, D. 2001. The EU Water Framework Directive: measures and directives. *Water Policy* **3**: 125–142.

Kanehl, P.D., Lyons, J. and Nelson, J.E. 1997. Changes in the habitat and fish community of the Milwaukee River, Wisconsin, following removal of the Woolen Mills Dam. *North American Journal of Fisheries Management* **11**: 387–400.

Kapitzke, I.R., Pearson, R.G., Smithers, S.G., Crees, M.R., Sands, L.B., Skull, S.D. and Johnston, A.J. 1998. Stream Stabilisation for Rehabilitation in North-East Queensland. *LWRRDC Occasional Paper 05/98*, Land and Water Resources Research & Development Corporation, Canberra, Australian Capital Territory.

Kapustka, L.A. and Landis W.G. 1998. Ecology: the science versus the myth. *Human and Ecological Risk Assessment* **4**(4): 829–838.

Kareiva, P., Marvier, M. and McClure, M. 2000. Recovery and management options for spring/summer Chinook Salmon in the Columbia River Basin. *Science* **290**: 977–979.

Karim, K., Gubbels, M.E. and Goulter, I.C. 1995. Review of determination of instream flow requirements with special application to Australia. *Water Resources Bulletin* **31**(6): 1063–1077.

Karr, J.R. 1981. Assessment of biotic integrity using fish communities. *Fisheries* **6**: 21–27.

Karr, J.R. 1991. Biological integrity: a long-neglected aspect of water resource management. *Ecological Applications* **1**: 66–84.

Karr, J.R. 1996. Ecological integrity, and ecological health are not the same. In Schulze, P.C. (ed), National Academy of Engineering, *Engineering Within Ecological Constraints*. National Academy Press, Washington, DC, pp. 97–109.

Karr, J.R. 1997. Measuring biological integrity. In Meffe, G.K. and Carroll, C.R. (eds) *Principles of Conservation Biology*. Sinauer, Sunderland, MA, pp. 483–485.

Karr, J.R. 1999. Defining and measuring river health. *Freshwater Biology* **41**(2): 221–234.

Karr, J.R. and Chu, E.W. 1999. *Restoring life in running waters: better biological monitoring*. Island Press, Washington, DC.

Karr, J.R. and Dudley, D.R. 1981. Ecological perspectives on water quality goals. *Environmental Management* **5**: 55–68.

Karr, J.R. and Freemark, K.E. 1985. Disturbance and vertebrates: an integrative perspective. In Pickett, S.T.A. and White, P.S. (eds) *The Ecology of Disturbance and Patch Dynamics*. Academic Press, Orlando, pp. 153–172.

Karr, J.R., Fausch, K.D., Angermeier, P.L., Yant, P.R. and Schlosser, I.J. 1986. Assessing biological integrity in running waters: A method and its rationale. *Special Publication 5*. Illinois Natural History Survey.

Katopodis, C. 1992. Introduction to Fishway Design. Working Document. Freshwater Institute, Central and Arctic Region, Department of Fisheries and Oceans, Winnipeg, Manitoba. URL: http://collection.nlc-bnc.ca/100/200/301/dfo-mpo/introduction_to_fishway/216976.pdf (accessed 1st Nov 2003).

Katopodis, C. 2003. Case studies of instream flow modelling for fish habitat. *Canadian Water Resources Journal* **28**(2): 199–216.

Katsantoni, G. (ed) 1990. Environmental Guidelines for River Management Works. Office of Water Resources, Department of Conservation and Environment, Government of Victoria, East Melbourne, Victoria.

Kauffman, J.B. and Krueger, W.C. 1984. Livestock impacts on riparian ecosystems and streamside management implications—a review. *Journal of Range Management* **37**(5): 430–439.

Kauffman, J.B., Krueger, W.C. and Vavra, M. 1983. Impacts of cattle on stream banks in northeastern Oregon. *Journal of Range Management* **36**: 683–685.

Kaufmann, J.B., Beschta, R.L. and Platts, W.S. 1993. Fish habitat improvement projects in the Fifteen Mile Creek and Trout Creek basins of Central Oregon: field review and management recommendations. U.S. Department of Energy, Bonneville Power Administration, Portland, Oregon.

Kawasaki, H. 1994. River administration and institutional aspects in Japan. *River Management and Planning in Japan. Technical Memorandum of PWRI No. 3265*. Environment Section, River Department, Public Works Research Institute, Ministry of Construction, Asahi, Japan, pp. 20–38.

Keller, E.A. and Melhorn, W.N. 1978. Rhythmic spacing and origin of pools and riffles. *Geological Society of America Bulletin* **89**: 723–730.

Keller, H. 1994. *Grandfather's Dream*. Greenwillow Books, New York.

Keller, R.J. and Peterken, C. (eds) 2002. *Proceedings of Third Australian Technical Workshop on Fishways*, 30 August–1 September 2001, Maroochydore, Queensland. Monash University, Clayton, Victoria.

Kellerhals, R. 1982. Effect of river regulation on channel stability. In Hey, R.D., Bathurst, J.C. and Thorne, C.R. (eds) *Gravel-bed Rivers: Fluvial Processes, Engineering and Management*. John Wiley & Sons, Chichester, pp. 685–705.

Kellerhals, R. and Church, M. 1989. The morphology of large rivers: characterization and management. In Didge, D.P. (ed) *Proc. of the International Large River Symp*. Can. Spec. Publ. Fish. Aquat. Sci. **106**: 31–48.

Kenyon, C. and Rutherfurd, I. 1999. Preliminary evidence for pollen as an indicator of recent floodplain accumulation rates and vegetation changes: the Barmah-Millewa Forest, SE Australia. *Environmental Management* **24**: 359–367.

Kerans, B.L. and Karr, J.R. 1994. A benthic index of biotic integrity (B–IBI) for rivers of the Tennessee Valley. *Ecological Applications* **4**(4): 768–785.

Kern, K. 1992. Restoration of lowland rivers: the German experience. In Carling, P.A. and Petts, G.E. (eds) *Lowland Floodplain Rivers: Geomorphological Perspectives*. John Wiley & Sons, Chichester, pp. 279–297.

Killingtveit, A. and Fossdal, M.L. 1994. The River System Simulator—An integrated model system for water resources planning and operation. Paper presented at HYDROSOFT 94, Greece, 21–23 Sept, Porto Carras.

Killingtveit, A. and Harby, A. 1994. The River System Simulator—Report from testing. Paper presented at Nordic Hydrological Conference, 2–4 Aug, Torshavn, Norway.

Kimmerer, W.J, Burau, J.R. and Bennett, W.A. 1998. Tidally-oriented vertical migration and position maintenance of zooplankton in a temperate estuary. *Limnology and Oceanography* **43**: 1697–1709.

Kimmerer, W.J. 2002a. Physical, biological, and management responses to variable freshwater flow into the San Francisco estuary. *Estuaries* **25**: 1275–1290.

Kimmerer, W.J. 2002b. Effects of freshwater flow on abundance of estuarine organisms: physical effects or trophic linkages? *Marine Ecology Progress Series* **243**: 39–55.

Kimmerer, W.J. and Schubel, J.R. 1994. Managing freshwater flows into San Francisco Bay using a salinity standard: results of a workshop. In Dyer, K.R. and Orth, R.J. (eds) *Changes in Fluxes in Estuaries: Implications from Science to Management*, Olsen and Olsen, Fredensborg, Denmark, pp. 411–416.

Kimmerer, W.J., Cowan, J.H., Jr, Miller, L.W. and Rose, K.A. 2001. Analysis of an estuarine striped bass population: effects of environmental conditions during early life. *Estuaries* **24**: 556–574.

King, C.A.M. 1966. *Techniques in Geomorphology*. Edward Arnold, London.

King J.M. and Tharme, R.E. 1993. Assessment of the instream flow incremental methodology and initial development of alternate instream flow methodologies for South Africa. *Water Research Commission Report No. 295/1/94*. Freshwater Research Unit, University of Cape Town, Cape Town, Republic of South Africa.

King, J.M., Tharme, R.E. and Brown, C.A. 1999. Definition and Implementation of Instream Flows. Thematic Report for the World Commission on Dams. Southern Waters Ecological Research and Consulting, Cape Town, Republic of South Africa. URL: http://www.dams.org/docs/kbase/contrib/env238.pdf (accessed 1st Nov 2003).

King, J.M., Brown, C. and Sabet, H. 2003. A scenario-based holistic approach to environmental flow assessments for rivers. *River Research and Applications* **19**(5-6): 619–639.

King, J.M. and Brown, C.A. 2003. *Environmental Flows: Case Studies. Water Resources and Environment Technical Note C2*, Davis, R. and Hirji, R. (eds), The World Bank, Washington, DC. URL: http://lnweb18.worldbank.org/ESSD/ardext.nsf/18ByDocName/EnvironmentalFlowAssessment-NOTEC2EnvironmentalFlowsCaseStudies/$FILE/NoteC2EnvironmentalFlowAssessment2003.pdf (accessed 1st Nov 2003).

King, J.M., Gorgens, A.H.M. and Holland, J. 1995. In search for ecologically meaningful low flows in Western cape streams. In *Proceedings of the Seventh South Australian National Hydrological Symposium*, Grahamstown, Republic of South Africa.

King, J.M., Tharme, R.E. and De Villiers, M.S. (eds) 2000. Environmental flow assessments for rivers: manual for the Building Block Methodology. *Water Research Commission Technology Transfer Report No. TT 131/00*. Freshwater Research Unit, University of Cape Town. Water Research Commission, Pretoria, Republic of South Africa, July.

King, R.D. and Tyler, P.A. 1982. Downstream effects of the Gordon River power development, south-west Tasmania. *Aus-*

*tralian Journal of Marine and Freshwater Research* **33**: 431–442.

Klassen, H.D. 1991. Operational stream rehabilitation trial at clint creek, Sewell Inlet. *Land Management Report No. 68, B.C. Ministry of Forests, Forest Science Program*. Crown Publications, Victoria, BC.

Klemes, V. 1986. Dilettantism in hydrology: transition or destiny? *Water Resour. Res.* **22**: 177S–188S.

Kleynhans, C.J. 1996. A qualitative procedure for the assessment of the habitat integrity status of the Luvuvhu River (Limpopo System, South Africa). *Journal of Aquatic Ecosystem Health* **5**: 41–54.

Kleynhans, N. and O'Keeffe, J. 2000. Assessment of ecological importance and sensitivity. In King, J.M., Tharme, R.E. and De Villiers, M.S. (eds), Environmental flow assessments for rivers: manual for the building block methodology. *Water Research Commission Technology Transfer Report No. TT131/00*. Water Research Commission, Pretoria, Republic of South Africa, pp. 117–124.

Klimas, C.V. 1988. River regulation effects on floodplain hydrology and ecology. In Hook, D.D., McKee, W.H., Jr, Smith, H.K., Gregory, J., Burrell, V.G., Jr, DeVoe, M.R., Sojka, R.E., Gilbert, S., Banks, R., Stolzy, L.H., Brooks, C., Mathews, T.D. and Shear, T.H. (eds) *The Ecology and Management of Wetlands, Volume 1: Ecology of Wetlands*. Croom Helm, Timber Press, Portland, pp. 40–49.

Klingeman, P.C., Kehe, S.M. and Owusu, Y.A. 1984. Streambank erosion protection and channel scour manipulation using rockfill dykes and gabions. Oregon State University, *Water Resources Research Institute Publication WRRI-98*, Corvallis, OR.

Klute, A. (ed) 1986. *Methods of Soil Analysis, Part 1: Physical and Mineralogical Methods. Agronomy Monograph no. 9*, 2nd edn, Amer. Soc. of Agronomy, Soil Sci. Soc. of America, Madison, WI.

Knapp, B. 1979. *Elements of Geographical Hydrology*, George Allen & Unwin, London.

Knighton, A.D. 1999. Downstream variation in stream power. *Geomorphology* **29**: 293–306.

Knighton, D. 1984. *Fluvial Forms and Processes*. Edward Arnold, London.

Knighton, D. 1998. *Fluvial Forms and Processes: A New Perspective*. Oxford University Press, Oxford.

Knudsen, E.E., MacDonald, D.D. and Steward, C.R. 2000. Setting the stage for a sustainable Pacific Salmon fisheries strategy. In Knudsen, E.E, Stewart, C.R., MacDonald, D.D., Williams, J.E. and Reiser, D.W. (eds) *Sustainable Fisheries Management: Pacific Salmon*. Lewis Publishers, New York. URL: http://www.absc.usgs.gov/research/Fisheries/pdf_files/Chap1.PDF (accessed 1st Nov 2003).

Koehn, J. 1987. Artificial habitat increases abundance of two-spined blackfish (*Gadopsis bispinosis*) in Ovens River, Victoria. *Technical Report Series No. 56*, Arthur Rylah Institute for Environmental Research, Department of Conservation, Forests and Lands, Heidelberg, Victoria.

Koehn, J.D., Brierley, G.J., Cant, B.L. and Lucas, A.M. 2001. River Restoration Framework. *Land & Water Occasional Paper 01/01*. Land and Water Australia, Canberra, Australian Capital

Territory. URL: http://www.lwa.gov.au/downloads/PR010187. pdf (accessed 1st Nov 2003).

Koehn, J.D., Doeg, T.J., Harrington, D.J. and Milledge, G.A. 1995. The effects of Dartmouth Dam on the aquatic fauna of the Mitta Mitta River. Report to Murray-Darling Basin Commission by Arthur Rylah Institute for Environmental Research, Melbourne.

Kolata, G. 1985. Prestidigitator of digits. *Science*, **85**(6): 66–72.

Kolkowitz, R. and Marsson, M. 1908. Okolgie der pflanzlichen Saprobien. *Berichte der Deutschen Botanischen Gesellschaft* **26A**: 505–519.

Komura, S. (ed.) 1990. *Proceedings of the International Symposium on Fishways '90 in Gifu*. Publications Committee of the International Symposium on Fishways '90, Gifu, Japan.

Kondolf, G.M. 1993. Lag in stream channel adjustment to livestock exclosure, White Mountains, California. *Restoration Ecology* **1**: 226–230.

Kondolf, G.M. 1995. Geomorphological stream channel classification in aquatic habitat restoration: uses and limitations. *Aquatic Conservation: Marine and Freshwater Ecosystems* **5**: 1–15.

Kondolf, G.M. 1996. A cross section of stream channel restoration. *Journal of Soil and Water Conservation* **51**(2), 119–125.

Kondolf, G.M. 1998a. Development of flushing flows for channel restoration on Rush Creek, California. *Rivers* **6**(3): 183–193.

Kondolf, G.M. 1998b. Lessons learned from river restoration projects in California. *Aquatic Conservation: Marine and Freshwater Ecosystems* **8**(1): 39–52.

Kondolf, G.M. and Wilcock, P.R. 1993. The flushing flow problem on the Trinity River, CA. In Shen, H.W., Su, S.T. and Wen, F. (eds) *Hydraulic Engineering '93: Proceedings of the Hydraulics Specialty Conference*. Hydraulics Division American Society of Civil Engineers, San Francisco, Volume 1, pp. 1172–1177.

Kondolf, G.M., and Wilcock, P.R. 1996. The flushing flow problem: defining and evaluating objectives. *Water Resources Research* **32**(8): 2589–2599.

Kondolf, G.M., Cata, G.F. and Sale, M.J. 1987. Assessing flushing-flow requirements for brown trout spawning gravels in steep streams. *Water Resources Bulletin* **23**: 927–935.

Konieczki, A.D., Graf, J.B. and Carpenter, M.C. 1997. Streamflow and sediment data collected to determine the effects of a controlled flood in March and April 1996 on the Colorado River between Lees Ferry and Diamond Creek, Arizona. *U.S. Geological Survey Open-File Report 97–224*, p. 55.

Konijn, H.S. 1973. *Statistical Theory of Sample Survey Design and Analysis*, North-Holland, Amsterdam.

Korte, V.L. and Blinn, D.W. 1983. Diatom colonization on artificial substrata in pool and riffle zones studied by light and scanning electron microscopy. *J. Phycol.* **19**, 332–341.

Kotwicki, V. 1986. *Floods of Lake Eyre*. Engineering and Water Supply Dept, Adelaide, South Australia.

Krumbein, W.C. 1941. Measurement and geological significance of shape and roundness of sedimentary particles. *J. of Sedimentary Petrology* **11**: 64–72.

Kurz, J.D. and Rosgen, D.L. 2002. Monitoring restoration effectiveness on the Lower Rio Blanco River, presented at *Monitoring and Modeling from the Peaks to the Prairies, 10th National Nonpoint Source Monitoring Workshop*, Beaver Run Resort,

Breckenridge, CO, 8–12 September. Available from Wildland Hydrology, Pagosa Springs, CO. URL: http://www.wildlandhydrology.com/html/references_.html (accessed 1st March 2004).

Kvanli, A.H. 1988. *Statistics. A Computer Integrated Approach*. West Publishing, St. Paul, MN.

La Violette, N., Richard, Y. and St. Onge, J. 1998. Indicators of biotic integrity in Québec rivers. In Report on the Fourth National Science Meeting, January 21–24, Manoir Richelieu, Charlevoix Biosphere Reserve Pointe-au-Pic/LaMalbaie, Québec. The Ecological Monitoring and Assessment Network (EMAN), Canada Centre for Inland Waters, Burlington, Ontario Canada. URL: http://www.eman-rese.ca/eman/reports/publications/nm98_proceed/part-11.html (accessed 1st Nov 2003).

Laasonen, P., Muotka, T. and Kivijärvi, I. 1998. Recovery of macroinvertebrate communities from stream habitat restoration. *Aquatic Conservation: Marine and Freshwater Ecosystems* **8**: 101–113.

Lackey, R.T. 2001. Values, policy, and ecosystem health. *BioScience* **51**(6): 437–443.

Ladson, A., Gerrish, G., Carr, G. and Thexton, E. 1997. Willows along waterways: towards a willow management strategy. Centre for Environmental Applied Hydrology, University of Mebourne, Published by the Department of Natural Resources and Environment, East Melbourne, Victoria, October.

Ladson, A.R. and Argent, R.M. 2002. Adaptive management of environmental flows: lessons for the Murray-Darling basin from three large North American Rivers. *Australian Journal of Water Resources* **5**: 89–101.

Ladson, A.R. and Finlayson, B.L. 2002. Rhetoric and reality in the allocation of water to the environment: a case study of the Goulburn River, Victoria, Australia. *River Research and Applications* **18**(6): 555–568.

Ladson, A.R. and White, L.J. 1999. *An Index of Stream Condition: Reference Manual*, 2nd edn, Department of Natural Resources and Environment, Government of Victoria, East Melbourne.

Ladson, A.R. and White, L.J. 2000. Measuring stream condition. In: Brizga, S. and Finlayson, B. (eds.) *River Management, The Australasian Experience*. John Wiley and Sons, Chichester, pp. 265–285.

Ladson, A.R., White, L.J., Doolan, J.A., Finlayson, B.L., Hart, B.T., Lake, P.S. and Tilleard, J.W. 1999. Development and testing of an Index of Stream Condition for waterway management in Australia. *Freshwater Biology* **41**: 453–468.

Lake, P.S. 1995. Of floods and droughts: river and stream ecosystems of Australia. In Cushing, C.E., Cummins, K.W. and Minshall, G.W. (eds) *River and Stream Ecosystems*. Elsevier, Oxford, pp. 659–694.

Lake, P.S., Barmuta, L.A., Boulton, A.J., Campbell, I.C. and St. Clair, R.M. 1985. Australian streams and Northern Hemisphere stream ecology: comparisons and problems. *Proc. Ecol. Soc. Aust.* **14**: 61–82.

Lambert, W.P. and Fruh, E.G. 1978. A methodology for investigating freshwater inflow requirements for a Texas estuary. In Wiley, M.L. (ed.) *Estuarine Interactions*. Academic Press, New York, pp. 403–413.

Lammert, M. and Allan, 1999. Assessing biotic integrity of streams: effects of scale in measuring the influence of land use/cover and habitat structure on fish and macroinvertebrates. *Environmental Management* 23(2): 257–270.

Lane, E.W. 1955. The importance of fluvial morphology in hydraulic engineering. *ASCE Proc.* **81**: (745).

Lane, E.W. and Borland, W.M. 1951. Estimating bedload. *Amer. Geophys. Union Trans.* **32**: 121–123 (as cited by Morisawa, 1985).

Lane, E.W. and Lei, K. 1950. Stream flow variability. *Trans. ASCE* **115**: 1084–1134.

Langbein, W.B. and Iseri, K.T. 1960. General introduction and hydrologic definitions. *Manual of Hydrology: Part 1. General Surface-Water Techniques.* US Geol. Survey Water-Supply Paper 1541-A, USGS, Washington, DC.

Langbein, W.B. and Leopold, L.B. 1964. Quasi-equilibrium states in channel morphology. *Amer. Jour. Sci.* **262**: 782–94.

Langbein, W.B. and Schumm, S.A. 1958. Yield of sediment in relation to mean annual precipitation. *Am. Geophys. Union Trans.* **39**: 1076–84.

Langendoen, E.J. 2000. CONCEPTS—Conservation channel evolution and pollutant transport system. *Research Report No. 16.* U.S. Department of Agriculture Agricultural Research Service, National Sedimentation Laboratory, Oxford, MS.

Langendoen, E.J., Simon, A. and Thomas, R.E. 2001. CONCEPTS—A process-based modeling tool to evaluate stream corridor restoration designs. In Hayes, D.F. (ed) *Proceedings 2001 Wetlands and River Restoration Conference*, August 27–31, Reno (CD-ROM). American Society of Civil Engineers, Reston, VA.

LaPerriere, J.D. and Martin, D.C. 1986. Simplified method of measuring stream slope. *Cold Regions Hydrology Symp.*, pp. 143–145, American Water Resources Assn, July 1986.

Large, A.R.G. and Petts, G.E. 1992. *Buffer zones for conservation of rivers and bankside habitats.* International Centre for Landscape Ecology, Loughborough University of Technology, Project Record 340/5Y, National Rivers Authority, Bristol, Avon.

Larinier, M. 1987. Fishways: principles and design criteria. *La Houille Blanche* 1/2: 51–57.

Larsen, D.P., Omernik, J.M., Hughes, R.M., Rohm, C.M., Whittier, T.R., Kinney, A.J., Gallant, A.L. and Dudley, D.R. 1986. Correspondence between spatial patterns in fish assemblages in Ohio streams and aquatic ecoregions. *Environmental Management* **10**: 815–828.

Larson, M. 2000. Effectiveness of large woody debris in stream rehabilitation projects in urban basins. M.Sc. Report. Center for Urban Water Resources Management, University of Washington, Seattle, March. URL: http://depts.washington.edu/cwws/Research/Reports/effeclwd.pdf (accessed 1st Nov 2003).

Leclerc, M.A., Capra, H., Valentin, S., Boudreault, A. and Côté, Y. (eds) 1996. *Proceedings of the 2nd International Symposium on Habitat Hydraulics, Ecohydraulics 2000.* Institute National de la Recherche Scientifique - Eau, co-published with FQSA, IAHR/AIRH, Ste-Foy, Québec.

Leclerc, M.A., Boudreault, J.A., Bechara, J.A., and Corfa, G. 1995. Two-dimensional hydrodynamic modeling: a neglected tool in the instream flow incremental methodolgy. *Transactions of the American Fisheries Society* **124**: 645–662.

Lee, K.N. 1993. *Compass and Gyroscope: Integrating Science and Politics for the Environment*, Island Press, Washington, DC.

Legendre, L., and Legendre, P. 1983. *Numerical Ecology*, Elsevier, Amsterdam.

Lehane, B.M., Giller, P.S., O'Halloran, J., Smith, C. and Murphy, J. 2002. Experimental provision of large woody debris in streams as a trout management technique. *Aquatic Conservation: Marine and Freshwater Ecosystems* **12**(3): 289–311.

Lélé, S. and Norgaard, R.B. 1996. Sustainability and the scientist's burden. *Conservation Biology* **10**(2): 354–365.

Lemly, A.D. and Hilderbrand, R.H. 2000. Influence of woody debris on stream insect communities and benthic detritus. *Hydrobiologia* **421**: 179–185.

Leopold, L.B. 1959. Probability analysis applied to a water-supply problem. *US Geol. Survey Circular 410.* USGS, Washington, DC.

Leopold, L.B. 1960. Ecological systems and the water resource. *Part D, US Geological Survey Circular 414, Conservation and Water Management.* USGS, Washington, DC.

Leopold, L.B. 1790. An improved method for size distribution of stream bed gravel. *Water Resources Research* 6(5): 1357–1366.

Leopold, L.B. and Emmett, W.W. 1976. Bedload measurements, East Fork River, WY. *Nat. Acad. of Sci. Proc.* **73**: 1000–1004.

Leopold, L.B. and Langbein, W.B. 1960. *A Primer on Water.* USGS, Washington, DC.

Leopold, L.B. and Langbein, W.B. 1963. Association and indeterminacy in geomorphology. In Albritton, C.C. (ed.), *The Fabric of Geology*, Addison-Wesley, Reading, MA, pp. 184–192.

Leopold, L.B. and Maddock, T. 1953. The hydraulic geometry of stream channels and some physiographic implications. *U.S. Geological Survey Professional Paper 252.* Washington, DC, pp. 1–57.

Leopold, L.B. and O'Brien Marchand, M. 1968. On the quantitative inventory of the riverscape. *Water Resour. Res.* **4**: 709–717.

Leopold, L.B. and Wolman, M.G. 1957. River channel patters: braided, meandering and straight. *U.S Geological Survey Professional Paper 282-B*, pp. 39–85.

Leopold, L.B., Wolman, M.G. and Miller, J.P. 1964. *Fluvial Processes in Geomorphology*, W.H. Freeman, San Francisco.

Leslie, D.J. 2001. Effect of river management on colonially nesting waterbirds in the Barmah-Millewa Forest, south-eastern Australia. *Regulated Rivers: Research & Management* **17**: 21–36.

Leslie, D.J. and Ward, K.A. 2002. Murray River environmental flows 2000–2001. *Ecological Management & Restoration* 3(3): 221–223.

Levin, S.A. and Paine, R.T. 1974. Disturbance, patch formation and community structure. *Proceedings of the National Academy of Sciences* 71: 2744–2747.

Lewis, B., O'Brien, T. and Perera, S. 1999. Providing for fish passage at small instream structures in Victoria. In Rutherfurd, I.D. and Bartley, R. (eds) *Proceedings of the Second Australian Stream Management Conference*, Adelaide, 8–11 Feb, Cooperative Research Centre for Catchment Hydrology, Monash University, Clayton, pp. 389–393.

Lewis, D.W. 1984. *Practical Sedimentology.* Hutchinson Ross, USA.

Lewis, G. and Williams, G. 1984. *Rivers and Wildlife Handbook—A Guide to Practices Which Further the Conservation of Wildlife on Rivers.* Royal Society for the Protection of Birds and Royal Society for Nature Conservation, UK.

Lichatowich, J.E. and Mobrand, L.E. 1995. Analysis of chinook salmon in the Columbia River from an ecosystem perspective. *U.S. Department of Energy Publication DOE/BP-251-5-2*, Bonneville Power Administration, Portland, OR, May.

Lichthardt, J. 1997. Research Natural Areas on the Clearwater National Forest: a survey of aquatic and riparian plant communities. Cooperative Challenge Cost-share Project, Clearwater National Forest, Idaho Department of Fish and Game, Boise, Idaho, November. URL: http://www2.state.id.us/fishgame/info/cdc/cdc_pdf/cnf97.pdf (accessed 1st Nov 2003).

Likens, G.E., Bormann, F., Pierce, R.S., Eaton, J.S. and Johnson, N.M. 1977. *Biogeochemistry of a Forested Ecosystem.* Springer-Verlag, New York.

Lillesand, T.M. and Kiefer, R.W. 2000. *Remote Sensing and Image Interpretation*, 4th edn, John Wiley, New York.

Limerinos, J.T. 1970. Determination of the Manning coefficient from measured bed roughness in natural channels. *US Geol. Survey Water-Supply Paper 1898-B*. USGS, Washington, DC (as cited by Jarrett, 1985).

Lindloff, S. 2000. *Dam Removal: A Citizen's Guide to Restoring Rivers.* River Alliance of Wisconsin, Madison, WI and Trout Unlimited, Merrifield, VA.

Line, D.E., Harman, W.A., Jennings, G.D., Thompson, E.J. and Osmond, D.L. 2000. Nonpoint source pollutant load reductions associated with livestock exclusion. *Journal of Environmental Quality* 29: 1882–1890.

Linløkken, A., Taugbøl, T., Næss, T., Igland, O.T. and Husebø, Å. 1999. A comparison of genetic variability in artificial and natural populations of brown trout in a regulated river system. *Regulated Rivers: Research & Management* 159(1): 159–168.

Linsley, R. K., Kohler, M.A. and Paulhus, J.L.H. 1975a. *Applied Hydrology*, McGraw-Hill, New Delhi.

Linsley, R.K., Kohler, M.A. and Paulhus, J.L.H. 1975b. *Hydrology for Engineers*, 2nd edn, McGraw-Hill, New York.

Linsley, R.K., Kohler, M.A. and Paulhus, J.L.H. 1982. *Hydrology for Engineers*, 3rd edn, McGraw-Hill, New York.

Lisle, T.E. 1986. Stabilization of a gravel channel by large streamside obstructions and bedrock bends, Jacoby Creek, northwestern California. *Geol. Soc. of Am. Bull.* 97: 999–1011.

Lisle, T.E. 1987. Using "residual depths" to monitor pool depths independently of discharge. *Res. Note PSW-394*. Pac. SW For. and Range Expt. Sta., US Forest Service, Berkeley, CA.

Lisle, T.E. and Hilton, S. 1999. Fine bed material in pools of natural gravel bed channels. *Water Resources Research* 35(4): 1291–1304.

Llansó, R.J., Dauer, D.M., Vølstad, J.H. and Scott, L.C. 2003. Application of the benthic index of biotic integrity to environmental monitoring in Chesapeake Bay. *Environmental Monitoring and Assessment* 81: 163–174.

Lo, C.P. 1986. *Applied Remote Sensing.* Longman Scientific and Technical, Harlow.

Loar, J.M., Sale, M.J. and Cada, G.F. 1986. Instream flow needs to protect fishery resources. In Karamouz, M. Baumi, G.R. and Brick, W.J. (eds) *Proceedings Water Forum '86, World Water Issues in Evolution, Volume 2.* Long Beach, California, 4–6 August 1986, American Society of Civil Engineers, New York, pp. 2098–2105.

Loder, R.T. and Erho, M.W. 1971. Wells hydroelectric project fish facilities. *Journal of the Power Division, Proceedings of the American Society of Civil Engineers* 97(PO 2): 310–316.

Logan, P. 2001. Ecological quality assessment of rivers and integrated catchment management in England and Wales. *Journal of Limnology* 60(1): 25–32.

Loneragan, N.R. and Bunn, S.E. 1999. River flows and estuarine ecosystems: implications for coastal fisheries from a review and a case study of the Logan River, southeast Queensland. *Australian Journal of Ecology* 24(4): 431–440.

Longley, W.L. (ed.) 1994. Freshwater inflows to Texas bays and estuaries: ecological relationships and methods for determination of needs. Water Development Board and Texas Parks and Wildlife Department, Austin, Texas, pp. 386.

Lotspeich, F.B. 1980. Watersheds as the basic ecosystem: this conceptual framework provides a basis for a natural classification system. *Water Resour. Bull.* 16: 581–586.

Lotspeich, F.B. and Platts, W.S. 1982. An integrated land-aquatic classification system. *North American Journal of Fisheries Management* 2: 138–149.

Low, J.W. 1952. *Plane Table Mapping*, Harper and Brothers, New York.

Lowe, R.L. and Pan, Y.D. 1996. Benthic algal communities as biological monitors. In Stevenson, R.J., Bothwell, M.L. and Lowe, R.L. (eds) *Algal Ecology: Freshwater Benthic Ecosystems.* Academic Press, San Diego, pp. 705–739.

Lowe, S. 1996. *Fish Habitat Enhancement Designs: Typical Structures.* Alberta Environmental Protection, Water Resource Management Services, River Engineering Branch, Edmonton, AB.

Lugg, A., Heron, S., Fleming, G. and O'Donnell, T. 1989. Conservation value of the wetlands in the Kerang Lakes area. Report to Kerang Lakes Area Working Group. Report 1, Kerang Lakes Assessment Group, Department of Conservation Forests and Lands, East Melbourne.

Lyon, J.G. 1995. Remote sensing and geographic information systems in hydrology. In Ward A.D. and Elliot W.J. (eds) *Environmental Hydrology*, Chapter 11, Lewis Publishers, Boca Raton, FL, pp. 337–368.

Lyons, J., Navarro-Pérez, S., Cochran, P.A., Santana, E.C. and Guzmán-Arroyo, M. 1995. Index of biotic integrity based on fish assemblages for the conservation of streams and rivers in West-Central Mexico. *Conservation Biology* 9(3): 569–584.

Lyons, J., Trimble, S.W. and Paine, L.K. 2000. Grass versus trees: managing riparian areas to benefit streams of central North America. *Journal of the American Water Resources Association* 36(4): 919–930.

Lyulko, L., Ambalova, T. and Vasiljeva, T. 2001. To integrated water quality assessment in Latvia. In Timmerman, J.G., Cofino, W.P., Enderlein, R.E., Jülich, W., Literathy, P.L., Martin, J.M., Ross, P., Thyssen, N., Turner, R.K. and Ward, R.C. (eds) *MTM (Monitoring Tailor-Made) III, Proceedings of International Workshop on Information for Sustainable Water Management*, 25–28 September 2000, Nunspeet, The Netherlands, pp.

449–452. URL: http://www.mtm-conference.nl/mtm3/docs/Lyulkoa2001.pdf (accessed 1st Nov 2003).

MacArthur, R. 1955. Fluctuations of animal populations and a measure of community stability. *Ecology* **35**: 533–536.

Mack, J.J. 2001. A vegetation Index of Biotic Integrity (VIBI) for wetlands and preliminary wetland aquatic life use designations. Final Report to U.S. EPA Grant No. CD985875-01. *Testing Biological Metrics and Development of Wetland Assessment Techniques using Reference Sites Volume 1.* State of Ohio Environmental Protection Agency, Wetland Ecology Group, Division of Surface Water, Columbus, OH.

MacLean, B.M. 1961. Model and prototype research on fish ladders. *Journal of the Power Division, Proceedings of the American Society of Civil Engineers* **87**(PO 2): 57–68.

Macmillan, L.A. 1986. Criteria for evaluating streams for protection. In Campbell, I.C. (ed) *Stream Protection: The Management of Rivers for Instream Uses.* Water Studies Centre, Chisholm Institute of Technology, East Caulfield, Australia, pp. 199–233.

Macmillan, L.A. 1987. Assessing the nature conservation value of rivers and streams with particular reference to the rivers of East Gippsland, Victoria. Master of Appl. Science thesis, Dept of Chemistry and Biology, Chisholm Institute of Technology, Melbourne, Victoria.

Maddock, I. 1999. The importance of physical habitat assessment for evaluating river health. *Freshwater Biology* **41**: 373–391.

Magilligan, F.J. and McDowell, P.F. 1997. Stream channel adjustments following elimination of cattle grazing. *Journal of the American Water Resources Association* **33**(4): 867–878.

Mahoney, J.M. and Rood, S.B. 1998. Streamflow requirements for Cottonwood seedling requirement—an integrative model. *Wetlands* **18**(4): 634–645.

Maidment, D.R. (ed). 1993. *Handbook of Hydrology*, McGraw-Hill, New York.

Maitland, P.S. 1978. *Biology of Fresh Waters*, Blackie, Glasgow.

Mallen-Cooper, M. 1994. How high can a fish jump? *New Scientist* **142**(1921): 32–36.

Mallen-Cooper, M. and Brand, D. 1992. Assessment of two fishways on the River Murray at Euston (Lock 15) and Murtho (Lock 6). Fisheries Research Institute, NSW Department of Fisheries, Cronulla.

Mallen-Cooper, M., Stuart, I.G., Hides-Pearson, F. and Harris, J.H. 1995. Fish migration in the River Murray and assessment of the Torrumbarry fishway. *Final Report for Natural Resources Management Strategy Project N002.* NSW Fisheries Research Institute and Cooperative Research Centre for Freshwater Ecology. Murray-Darling Basin Commission, Canberra, Australian Capital Territory.

Mallen-Cooper, M., Stuart, I., Hides-Pearson, F. and Harris, J. 1997. Fish migration in the River Murray and assessment of the Torrumbarry fishway. In Banens, R.J. and Lehane, R. (eds) *1995 Riverine Environment Research Forum.* Murray-Darling Basin Commission, Canberra, Australian Capital Territory, pp. 33–37.

Mander, Ü. 1995. Riaparian buffer zones and buffer strips on stream banks: dimensioning and efficiency assessment from catchments in Estonia. In Eiseltová, M. and Biggs, J. (eds) Restoration of Stream Ecosystems, An Integrated Catchment Approach. *IWRB Publication 37*, International Waterfowl and Wetland Research Bureau, Slimbridge, Gloucester, UK, pp. 45–64.

Mann, P.S. 1995. *Introductory Statistics*, 2nd edn, John Wiley, New York.

Mark, D.M. 1983. Relations between field-surveyed channel networks and map-based geomorphometric measures, Inez, Kentucky. *Ann. of the Assn. of Am. Geog.* **73**: 358–372.

Martin, J., Edwards, J. and Edwards, G. 1992. Hatcheries and wild stocks: are they compatible. *Fisheries* **17**: 4.

Martin, W.R. 1964. New lifting device used at Priest Rapids Dam. *Civil Engineering* June: 69–71.

Mathews, R.C., Jr and Bao, Y. 1991. The Texas method of preliminary instream flow assessment. *Rivers* **2**(4): 295–310.

Mathur, D., Bason, W.H., Purdy, E.J., Jr and Silver, C.A. 1985. A critique of the Instream Flow Incremental Methodology. *Canadian Journal of Fisheries and Aquatic Sciences* **42**: 825–31.

Maude, S.H. and Williams, D.D. 1983. Behavior of crayfish in water currents: hydrodynamics of eight species with reference to their distribution patterns in southern Ontario. *Can. J. Fish. Aquat. Sci.* **40**: 68–77.

McBain & Trush 1997. *Trinity River Maintenance Flow Study Final Report.* Hoopa Valley Tribe Fisheries Department, Hoopa, CA, November.

McBain, S. and Trush, B. 1997. Thresholds for managing regulated river ecosystems. In Sommarstrom, S. (ed.) *Proceedings, Sixth Biennial Watershed Management Conference. Water Resources Center Report No. 92.* University of California, Davis, pp. 11–13. URL: http://watershed.org/WMCproceedings96.pdf (accessed 1st November, 2003).

McCarthy, J.H. 2003. Wetted perimeter assessment Shoal Harbour River, Shoal Harbour, Clarenville, Newfoundland. Report by AMEC Earth & Environmental Limited, St. John's, to SGE-Acres, Clarenville, NL, January. URL: http://www.gov.nf.ca/env/Env/EA%202001/pdf%20files/1059%20-%20WettedPerimeterAssessment.pdf (accessed 1st Nov 2003).

McDowell, P.F. and Mowry, A.D. 2002. Measuring and interpreting stream channel response to management changes: a cattle grazing case study. In *Meeting Abstracts, Association of American Geographers Annual Meeting 2002,* Los Angeles (CD-ROM), Washington, DC URL: http://convention.allacademic.com/aag2002/index2.html (accessed 1st Nov 2003).

McKergow, L., Weaver, D., Prosser, I., Grayson, R. and Reed, A. 2001. Before and after riparian management: sediment and nutrient exports from a small agricultural catchment, Western Australia. In Rutherfurd, I., Sheldon, F., Brierley, G. and Kenyon, C. (eds) *Proceedings Third Australian Stream Management Conference*, Brisbane, 27–28 August, CRC Catchment Hydrology, Monash University, Clayton, Victoria, pp. 427–433.

McKnight, T.L. 1990. *Physical Geography*, 3rd edn, Prentice-Hall, Englewood Cliffs, NJ.

McLain, R.J. and Lee, R.G. 1996. Adaptive management: promises and pitfalls. *Environmental Management* **20**(4): 437–448.

McMahon, T.A. 1976. Preliminary estimation of reservoir storage for Australian streams. *Civil Engineering Trans.*, Institution of Engineers, Australia, **CE18**, 55–59.

McMahon, T.A. 1982. *Hydrological Characteristics of Selected Rivers of the World*. UNESCO, Paris.

McMahon, T.A. 1986. Hydrology and management of Australian streams. In Campbell, I.C. (ed) *Stream Protection: The Management of Rivers for Instream Uses* Water Studies Centre, Chisholm Institute of Technology, East Caulfield, Australia, pp. 23–44.

McMahon, T.A. and Diaz Arenas, A. 1982. Methods of computation of low streamflow. *Studies and Reports in Hydrology*. **36**, UNESCO, Paris.

McMahon, T.A. and Finlayson, B.L. 2003. Droughts and anti-droughts: the low flow hydrology of Australian rivers. *Freshwater Biology* **48**: 1147–1160.

McMahon, T.A. and Mein, R.G. 1986. *River and Reservoir Yield*, Water Resources Publications, PO Box 2841, Littleton CO.

McMahon, T.A., Finlayson, B.L., Haines, A. and Srikanthan, R. 1987. Runoff variability: a global perspective. In *The Influence of Climate Change and Climatic Variability on the Hydrologic Regime and Water Resources*. Proc. of the Vancouver Symposium, August 1987, IAHS Publ. No. 168, pp. 3–11.

MDBC (Murray-Darling Basin Commission) 2002. The Living Murray Initiative Proposal for Investment - An Implementation Program of Structural Works and Measures (2003/04 to 2005/06). MDBC (Murray-Darling Basin Commission) 2002, Canberra, Australian Capital Territory, November. URL: http://www.thelivingmurray.mdbc.gov.au/content/item. phtml?itemId=10291&nodeId=file3efacc901a4b8&fn=TLM_ invest_ whole. pdf (accessed 1st Nov 2003).

MDBC (Murray-Darling Basin Commission) 2003. The Living Murray Initiative. Factsheet No. 1. Murray-Darling Basin Commission, Canberra, Australian Capital Territory. URL: http://www.thelivingmurray.mdbc.gov.au (accessed 1st Nov 2003).

MDBMC (Murray-Darling Basin Ministerial Council) 2002. *Draft Native Fish Strategy for the Murray-Darling Basin 2002–2012*. Murray-Darling Basin Ministerial Council, Murray-Darling Basin Commission. Canberrra, Australian Capital Territory. URL: http://www.mdbc.gov.au/naturalresources/native_fish/PDF/nfs_full.pdf (accessed 24 March 2003).

Megahan, W.F. 1982. Channel sediment storage behind obstructions in forested drainage basins draining the granitic bedrock of the Idaho Batholith. In Swanson, F.J., Janda, R.J., Dunne, T. and Swanston, D.N. (eds), Sediment Budgets and Routing in Forested Drainage Basins, *Gen. Tech. Rep. PNW-141*, U.S. Dept of Agriculture, Forest Service, Portland, OR., pp. 114–121.

Melbourne Water 1997. Draft Waterway Strategy: Providing Healthy Waterways. Unpublished report. Melbourne Water, Waterways and Drainage, East Richmond, Victoria.

Melbourne Water 2003. *Melbourne's Water Supply System, Essential Facts*. Melbourne, Victoria, August. URL: http://www.melbournewater.com.au/content/library/publications/fact_sheets/water/water_supply_system/Water%20Supply%20System.pdf (accessed 1st Nov 2003).

Merrill, T. and O'Laughlin, J. 1993. Analysis of methods for determining minimum instream flows. *Idaho Forest, Wildlife and Range Policy Analysis Group Report No. 9*. College of Natural Resources Policy Analysis Group, University of Idaho, Moscow, ID, March. URL: http://www.cnr.uidaho.edu/pag/pag9es.html (accessed 1st Nov 2003).

Metcalfe-Smith, J.L. 1996. Biological water quality assessment of rivers: use of macroinvertebrate communities. In Petts, G. and Calow, P. (eds), *River Restoration*, Blackwell Science, Oxford, pp. 17–43.

Meyer, J.L. 1997. Stream health: incorporating the human dimension to advance stream ecology. *Journal North American Benthological Society* **16**(2): 439–447.

Miall, A.D. 1996. *The Geology of Fluvial Deposits—Sedimentary Facies, Basin Deposits, and Petroleum Geology*. Springer-Verlag, Berlin, p. 582.

Micacchion, M. 2002. Amphibian Index of Biotic Integrity (AmphIBI) for Wetlands. Final Report to U.S. EPA Grant No. CD985875-01 Testing Biological Metrics and Development of Wetland Assessment Techniques Using Reference Sites Volume 3. State of Ohio Environmental Protection Agency, Wetland Ecology Group, Division of Surface Water, Columbus, OH.

Miers, L. 1994. Aquatic habitat classification: literature review towards development of a classification system for BC. discussion document. Prepared by Ministry of Environment, Lands and Parks for the Aquatic Inventory Task Force Resources Inventory Committee. URL: http://srmwww.gov.bc.ca/risc/index.htm (accessed 1st Nov 2003).

Milhous, R.T. 1982. Effect of sediment transport and flow regulation on the ecology of gravel-bed rivers. In Hey, R.D., Bathurst, J.C. and Thorne, C.R. (eds) *Gravel-bed Rivers*, John Wiley, Chichester.

Milhous, R.T. 1986. Development of a habitat time series. *Journal of Water Resources Planning and Management* **112**: 145–148.

Milhous, R.T. 1996. Modeling of instream flow needs: the link between sediment and aquatic habitat. In Leclerc, M. Capra, H., Valentin, S., Boudreault, A. and Côté, Y. (eds) *Proceedings of the Second IAHR Symposium on Habitat Hydraulics, Ecohydraulics 2000*. Institute National de la Recherche Scientifique - Eau, co-published with FQSA, IAHR/AIRH, Ste-Foy, Québec, Canada, pp. A319–A330.

Milhous, R.T. 1998. Modeling of instream flow needs: the link between sediment and aquatic habitat. *Regulated Rivers: Research & Management* **14**: 79–94.

Milhous, R.T., Updike, M.A. and Schneider, D.M. 1989. Physical Habitat Simulation System reference manual—Version II. *Instream Flow Info. Paper No. 26, Biol. Rpt 89(16)*. National Ecology Research Center, U.S. Fish and Wildlife Service, Fort Collins CO.

Millar, R.G. 1997. Ballast requirements for lateral log jams. Technical Memorandum for the Province of British Columbia, Watershed Restoration Program. Province of British Columbia, Watershed Restoration Program, Victoria, BC.

Miller, D.E. 1999. Deformable streambanks: Can we call it a natural channel design without them? *American Water Resources Association, 1999 Annual Conference*, Bozeman, MT, pp. 293–300. URL: http://www.interfluve.com/pdf_files/AWRA-1999.pdf (accessed 1st Nov 2003).

Miller, J.R. and Ritter, J.B. 1996. An examination of the Rosgen classification of natural rivers. *Catena* **27**: 295–299.

Milner, A.M. 1994. System recovery. In Calow, P. and Petts, G.E. (eds) *The Rivers Handbook. Volume 2*. Blackwell Scientific Publications, Oxford, pp. 76–97.

Minchin, P.R. 1987. An evaluation of the relative robustness of techniques for ecological ordination. *Vegetation* **69**: 89–108.

Miner, J.R., Buckhouse, J.C. and Moore, J.A. 1992. Will a water trough reduce the amount of time hay-fed livestock spend in the stream (and therefore improve water quality)? *Rangelands* **14**(1): 35–38.

Ministry of Natural Resources 1994. *Natural Channel Systems: An Approach to Management and Design*. Queen's Printer for Canada, Toronto.

Ministry of the Environment Government of Japan 2003. *Environmental Quality Standards for Water Pollution Establishment of Environmental Quality Standards*, Tokyo. URL: http://www.env.go.jp/en/lar/regulation/wp.html (accessed 1st Nov 2003).

Minshall, G.W. 1984. Aquatic insect-substratum relationships. In Resh, V.M. and Rosenberg D.M. (eds) *Ecology of Aquatic Insects*. Praeger, New York, pp. 358–400.

Mitchell, M.K. and Stapp, W.B. 2000. *Field Manual for Water Quality Monitoring*, 12th Ed, Kendall/Hunt Publishing Co, Dubuque, IA.

Mitchell, P. 1990. *The Environmental Condition of Victorian Streams*, Dept of Water Resources, Victoria, Australia.

Moffett, J.W. and Smith, S.H. 1950. Biological investigations of the fishery resources of the Trinity River, California. *Special Scientific Report No. 12*. U.S. Fish and Wildlife Service, p. 71.

Mohanty, R.P. and Mathew, T. 1987. Some investigations relating to environmental impacts of a water resource project. *Journal of Environmental Management* **24**: 315–336.

Molloy, D.P. and Struble, R.H. 1988. A simple and inexpensive method for determining stream discharge from a streambank. *J. of Freshwater Ecology* **4**: 477–81.

Montana Water Center, 2003. Wildfish Habitat Initiative. Montana Water Center, Partners for Fish & Wildlife, U.S. Fish and Wildlife Service. URL: http://water.montana.edu/wildfish/default.asp (accessed 1st Nov 2003).

Montgomery, D.R. 1997. What's best on the banks? *Nature* **388**: 328–329.

Montgomery, D.R. 2003. Wood in rivers: interactions with channel morphology and processes. *Geomorphology* **51**: 1–5.

Montgomery, D.R. and Buffington, J.M. 1997. Channel-reach morphology in mountain drainage basins. *Geological Society of America Bulletin* **109**: 596–611.

Montgomery, D.R., Bolton, S., Booth, D.B. and Wall, L. 2003. *Restoration of Puget Sound rivers*. Center for Water and Watershed Studies, University of Washington, University of Washington Press, Seattle, WA.

Moog, O., Chovanec, A., Hinteregger, H. and Römer, A. 1999. Richtlinie zur Bestimmung der saprobiologischen Gewässergüte von Fließgewässern-Wasserwirtschaftskataster, Bundesministerium für Land- und Forstwirtschaft, Wien, p. 144.

Moran, R.J. and Rhodes, B.G., 1991. Drought planning and management for Victorian water supplies. *International Hydrology and Water Resources Symposium*, Perth, Western Australia, 2–4 Oct. 1991, Institution of Engineers, Australia, National Conference Publication No. 91/22, Volume 2, pp. 335–341.

Morisawa, M.E. 1958. Measurement of drainage basin outline form. *Jour. Geol.* **66**: 587–591.

Morisawa, M.E. 1968. *Streams, their Dynamics and Morphology*. McGraw-Hill, New York.

Morisawa, M.E. 1985. *Rivers, Form and Process, Geomorphology Texts 7*. Longman, London.

Morris, H.M. 1955. Flow in rough conduits. *Trans. ASCE* **120**: 373–398.

Morris, H.M. 1961. Design methods for flow in rough conduits. *Trans. ASCE* **126**: 454–490.

Mosley, M.P. 1981. Semi-determinate hydraulic geometry of river channels, South Island, New Zealand. *Earth Surf. Proc. and Landforms*, **6**, 127–37.

Mosley, M.P. 1983. Flow requirements for recreation and wildlife in New Zealand rivers—A review. *Journal of Hydrology (NZ)* **22**: 152–174.

Mosley, M.P. 1987. The classification and characterization of rivers. In Richards, K.S. (ed) *River Channels, Environment and Process*. Basil Blackwell, Oxford, pp. 295–320.

Mosley, M.P. and Jowett, I.G. 1985. Fish habitat analysis using river flow simulation. *NZ Journal of Marine and Freshwater Research* **19**: 293–309.

Moss, B. 1988. *Ecology of Fresh Waters, Man and Medium*, 2nd edn, Blackwell Scientific, Oxford.

Moss, B., Carvalho, L. and Plewes, J. 2002. The lake at Llandrindod Wells—a restoration comedy? *Aquatic Conservation: Marine and Freshwater Ecosystems* **12**(2): 229–245.

Mowry, A.D. and McDowell, P.F. 2003. Evaluating geomorphic effects of active stream restoration in the Interior Columbia River Basin, Oregon. *Meeting Abstracts, Association of American Geographers Annual Meeting 2003*, New Orleans, Washington, DC. URL: http://convention.allacademic.com/aag2003/schedule.html (accessed 1st Nov 2003).

Moyle, P.B., Marchetti, M.P., Balrige, J., and Taylor, T.L. 1998. Fish health and diversity: justifying flows for a California stream. *Fisheries* **23**(7): 6–15.

Munné, A., Prat, N., Solà, C., Bonada, N. and Rieradevall, M. 2002. A simple field method for assessing the ecological quality of riparian habitat in rivers and streams: QBR index. *Aquatic Conservation: Marine and Freshwater Ecosystems* **13**(2): 147–163.

Munson, B.R., Young, D.F. and Okiishi, T.H. 2002. *Fundamentals of Fluid Mechanics*, 4th edn, John Wiley, New York.

Murgatroyd, A.L. and Ternan, J.L. 1983. The impact of afforestation on stream bank erosion and channel form. *Earth Surface Processes and Landforms* **8**: 357–369.

Muschenheim, D.K., Grant, J. and Mills, E.L. 1986. Flumes for benthic ecologists: theory, construction and practice. *Mar. Ecol. Prog. Ser.* **28**: 185–96.

Myers, T.J. and Swanson, S. 1992. Variation of stream stability with stream type and livestock bank damage in northern Nevada. *Water Resources Bulletin* **28**: 743–754.

Myers, T.J. and Swanson, S. 1996. Long-term aquatic habitat restoration: Mahogany Creek, Nevada, as a case study. *Water Resources Bulletin* **32**(2): 241–252.

Naiman, R.J. 1998, Biotic Stream Classification. In Naiman, R.J. and Bilby, R.E. (eds) *River Ecology and Management: Lessons from the Pacific Coastal Ecoregion*. Springer-Verlag, New York, pp. 97–119.

Naiman, R.J., Lonzarich, D.G., Beechie, T.J. and Ralph, S.C. 1992. General principles of classification and the assessment of conservation potential in rivers. In Boon, P., Calow, P. and Petts, G. (eds) *River Conservation and Management*. John Wiley and Sons, Chichester, UK, pp. 93–123.

Naiman, R.J., Melillo, J.M. and Hobbie, J.E. 1986. Ecosystem alteration of boreal forest streams by beaver (*Castor Canadensis*). *Ecology* **67**(5): 1254–1269.

Nakamura, S. 1994. Hydraulics of bending-slope fishway. In *Proceedings of the 1st International Symposium on Habitat Hydraulics*, 18–20 Aug, The Norwegian Institute of Technology, Trondheim, Norway. Organized by SINTEF NHL (Norwegian Hydrotechnical Laboratory) in cooperation with The Norwegian Institute of Technology and Norwegian Institute for Nature Research (NINA), IAHR, pp. 143–153.

Nanson, G.C. and Croke, J.C. 1992. A genetic classification of floodplains. *Geomorphology* **4**: 459–486.

Nanson, G.C. and Huang, H.Q. 1999. Anabranching rivers: divided efficiency leading to fluvial diversity. Chapter 19. In Miller, A. and Gupta, A. (eds) *Varieties of Fluvial Form*, Wiley, New York, pp. 477–494.

Nathan, R., Doeg, T. and Voorwinde, L. 2002. Towards defining sustainable limits to winter diversions in Victorian catchments. *Australian Journal of Water Resources* **5**: 49–60.

Nathan, R.J. and McMahon, T.A. 1990a. The estimation of low flow characteristics and yield from small ungauged rural catchments. *AWRAC Research Project 85/105*. Dept of Civil and Agricultural Engineering, University of Melbourne, Victoria.

Nathan, R.J. and McMahon, T.A. 1990b. Evaluation of automated techniques for baseflow and recession analyses. *Water Resour. Res.* **26**(7): 1465–1473.

Nathan, R.J. and McMahon, T.A. 1990c. Practical aspects of low flow frequency analysis. *Water Resour. Res.* **26**(9): 2135–2141.

Nathan, R.J. and McMahon, T.A. 1990d. Identification of homogeneous regions for the purposes of regionalisation. *J. Hydrol.* **121**: 217–238.

National Land and Water Resources Audit 2002. *Australian Catchment, River and Estuary Assessment 2002, Volumes 1 and 2*. Natural Heritage Trust. Australian Natural Resources Atlas V2.0. Land & Water Australia. Department for the Environment and Heritage. Commonwealth of Australia, Canberra, Australian Capital Territory. URL: http://audit.ea.gov.au/ANRA/coasts/docs/estuary_assessment/Est_Ass_Contents.cfm (accessed 1st Nov 2003).

National Park Service 1982. Wild and Scenic Rivers Guidelines. National Wild and Scenic Rivers System; Final Revised Guidelines for Eligibility, Classification and Management of River Areas. *Federal Register 47 (173)*. September 7. National Park Service and Office of the Secretary, Department of the Interior, Forest Service and Office of the Secretary, USDA. URL: http://www.nps.gov/rivers/siteindex.html (accessed 1st Nov 2003).

National Research Council 1991. *Colorado River Ecology and Dam Management. Proceedings of Symposium*, May 24–25, 1990. Santa Fe, NM. National Academy Press, Washington, DC.

National Research Council. 1992. *Restoration of Aquatic Ecosystems: Science, Technology, and Public Policy*. National Academy Press, Washington, DC.

National Wetlands Working Group 1997. *The Canadian Wetland Classification System*. Warner, B.G. and Rubec. 2nd Edition. C.D.A. (eds), Wetlands Research Centre, University of Waterloo. Waterloo, ON. p. 68 URL: http://www.uwwrc.net/web/wetlandsrc/cat/books/books-1053313151236 (accessed 1st Nov 2003).

Nelson, F.A. 1980. Evaluation of selected instream flow methods in Montana. *Proceedings of the Annual Conference of the Western Association of Fish and Wildlife Agencies*, pp. 412–432.

Nelson, M. 1997. Environmental Flows Assessment Methods—A Technical Review. *Report WRA 97/08*. Department of Primary Industry and Fisheries, Tasmania.

NERC 1975. *Flood Studies Report*, 5 volumes, Natural Environment. Research Council, London.

Nestler, J.M., Milhous, R.T. and Layzer, J.B. 1989. Instream habitat modeling techniques. In Gore, J.A. and Petts, G.E. (eds) *Alternatives in Regulated River Management*. CRC Press Inc., Boca Raton, FL, pp. 295–315.

Newbury, R.W. 1984. Hydrologic determinants of aquatic insect habitats. In Resh, V.M. and Rosenberg, D.M. (eds) *Ecology of Aquatic Insects*. Praeger, New York, pp. 323–57.

Newbury, R.W. 1989. Habitats and Hydrology of Streams, workshop conducted by the Centre for Environmental Applied Hydrology, Buxton, Victoria, 15–17 Nov. 1989.

Newbury, R.W. 1995. Rivers and the art of stream restoration. In Costa, J.E., Miller, A.J., Potter, K.W. and Wilcock, P.R. (eds) *Natural and Anthropological Influences in Fluvial Geomorphology: The Wolman Volume*. Geophysical Monograph 89, American Geophysical Union, Washington, DC, pp. 137–149.

Newbury, R.W. and Gaboury, M.N. 1988. The use of natural stream characteristics for stream rehabilitation works below the Manitoba escarpment. *Canadian Water Resources Journal* **13**: 35–51.

Newbury, R.W. and Gaboury, M.N. 1993a. *Stream Analysis and Fish Habitat Design: A Field Manual*. Newbury Hydraulics, Ltd., Gibsons, B.C.

Newbury, R.W. and Gaboury, M.N. 1993b. Exploration and rehabilitation of hydraulic habitats in streams using principles of fluvial behaviour. *Freshwater Biology* **29**: 195–210.

Newson, M. 1994. *Hydrology and the River Environment*. Clarendon Press, Oxford.

Newson, M.D. 2002. Geomorphological concepts and tools for sustainable river ecosystem management. *Aquatic Conservation: Marine and Freshwater Ecosystems* **12**(4): 365–379.

Nezu, I. and Rodi, W. 1986. Open-channel flow measurements with a laser doppler anemometer. *J. Hyd. Eng.* **112**: 335–355.

Nicol, S., Lieschke, J., Lyon, J. and Hughes, V. 2002. River Habitat Rehabilitation through Resnagging. Department of Natural Resources and Environment, Melbourne, Victoria.

Nilsson, C. and Dynesius, M. 1994. Ecological effects of river regulation on mammals and birds—a review. *Regulated Rivers: Research & Management* **9**(1): 45–53.

Noble, R. and Cowx, I. 2002. Development of a river-type classification system (D1), Compilation and harmonization of fish species classification (D2). Development, Evaluation &

Implementation of a Standardised Fish-based Assessment Method for the Ecological Status of European Rivers—A Contribution to the Water Framework Directive (FAME). Final report. University of Hull. URL: http://fame.boku.ac.at/files/WP1a&b_Edited_Report.pdf (accessed 1st Nov 2003).

Nolan, K.M. and Shields, R.R. 2000. Measurement of stream discharge by wading. *US Geol. Survey Water Resources Investigation Report 00–4036*. CD-ROM distributed by USDA Forest Service, Stream Systems Technology Center, Fort Collins, CO.

Nordin, C.F., Jr 1985. The sediment loads of rivers. In Rodda, J.C. (ed) *Facets of Hydrology*, Volume II Chapter 7, John Wiley, Chichester.

Nordin, C.F., Jr and Richardson, E.V. 1971. Instrumentation and measuring techniques. In Shen, H.W. (ed) *River Mechanics* Chapter 14, Fort Collins, CO.

Norris, R.H. and Thoms, M.C. 1999. What is river health? *Freshwater Biology* **41**: 197–209.

Norris, R.H., Liston, P. Davies, N., Dyer, F., Linke, S., Prosser, I. and Young, W. 2001. Snapshot of the Murray-Darling Basin River Condition. Report by Cooperative Research Centre for Freshwater Ecology, University of Canberra and CSIRO Land and Water to Murray-Darling Basin Commission, Canberra, Australian Capital Territory, September.

Northcote, K.H. 1979. *A Factual Key for the Recognition of Australian Soils*. Rellim Technical Publications, Adelaide, South Australia.

Northwest Power and Conservation Council 1994. Adult Salmon migration, Section 6. In *1994/95 Columbia River Basin Fish and Wildlife Program*. Northwest Power and Conservation Council, Portland Oregon. URL: http://www.nwcouncil.org/library/1994/sec6.pdf (accessed 1st Nov 2003).

Nowell, A.R.M. and Jumars, P.A. 1984. Flow environments of aquatic benthos. *Annual Review of Ecology and Systematics* **15**: 303–328.

NSW Fisheries 2001. Snags (large woody debris), DF97, Cronulla, New South Wales, September. URL: http://www.fisheries.nsw.gov.au/hab/snags.htm (accessed 1st Nov 2003).

NSW Government 1997. NSW Weirs Policy. New South Wales Government. URL: http://www.dlwc.nsw.gov.au/care/water/wr/pdfs/weir.pdf (accessed 1st Nov 2003).

NSW Government 2001. Incorporating the results of the weir review into the water sharing plans, No. 13, Advice to Water Management Commitees. New South Wales Government. URL: http://www.dlwc.nsw.gov.au/care/water/sharing/pdf/policy_advice_13-weirreview.pdf (accessed 1st Nov 2003).

Nunnally, N.R. 1978. Stream renovation: an alternative to channelization. *Environmental Management* **2**(5): 403–411.

Nyberg, J.B. 1998. Statistics and the practice of adaptive management. Chapter 1. In Sit, V. and Taylor, B. (eds) *Statistical Methods for Adaptive Management Studies*. Land Management Handbook No. 42. Research Branch. B.C. Ministry of Forests Forestry Division Services Branch, Victoria, BC, pp. 1–7.

Nyberg, J.B. 1999a. Implementing adaptive management of British Columbia's forests—where have we gone right and wrong. In McDonald, G.B., Fraser, J. and Gray, P. (eds) *Adaptive Management Forum: Linking Management and Science to Achieve Ecological Sustainability*. Proceedings of the 1998 Provincial Science Forum, Oct 13–16. Science Development

and Transfer Branch, Ontario Ministry of Natural Resources, Ontario, BC, pp. 25–28.

Nyberg, J.B. 1999b. An introductory guide to adaptive management for project leaders and participants. Forest Services Branch, British Columbia Forest Service, Victoria. URL: http://www.for.gov.bc.ca/hfp/amhome/Pubs/ Introductory-Guide-AM.pdf (accessed 28 Feb 2003).

O'Keeffe, J. 1997. Methods of assessing conservation status for natural fresh waters in the Southern Hemisphere. In Boon, P.J. and Howell, D.L. (eds) *Freshwater Quality: Defining the Indefinable*? Scottish Natural Heritage, Edinburgh, pp. 369–386.

O'Keeffe, J., King, J. and Eekhout, S. 1994. The characteristics and purposes of river classification. In Uys, M. (ed.) Classification of Rivers, and Environmental Health Indicators. Proceedings of a Joint South African Australian Workshop, Cape Town, 7–11 Feb, *Water Research Commission Report No. TT 63/94*, Pretoria, Republic South Africa, pp. 9–17.

O'Keeffe, J.H., Palmer, R.W., Byren, B.A. and Davies, B.R. 1990. The effects of impoundment on the physiochemistry of two contrasting southern African river systems. *Regulated Rivers: Research & Management* **5**: 97–110.

O'Neil, J. and Fitch, L. 1992. Performance audit of in-stream habitat structures created during the period 1982–1990 in southwestern Alberta. In *Proceedings, American Fisheries Society 122nd Annual Meeting*, 14–17 September, Rapid City, South Dakota.

O'Shea, D.T. 1995. Estimating minimum instream flow requirements for Minnesota streams from hydrologic data and watershed characteristics. *North American Journal of Fisheries Management* **15**: 569–578.

Odeh, M. (ed.) 2000. *Advances In Fish Passage Technology—Engineering Design and Biological Evaluation*. American Fisheries Society, Bethesda, MD.

Odgaard, A.J. 1987. Stream bank erosion along two rivers in Iowa. *Water Resources Research* **23**(7): 1225–1236.

Oldani, N.O. and Baigún, C.R.M. 2002. Performance of a fishway system in a major South American dam on the Parana River (Argentina-Paraguay). *River Research and Applications* **18**(2): 171–183.

Olden, J.D. and Poff, N.L. 2003. Redundancy and the choice of hydrologic indices for characterizing streamflow regimes. *River Research and Applications* **19**(2): 101–121.

Olson, R.A. and Hubert, W.A. 1994. Beaver: water resources and riparian habitat manager, Cooperative Extension Service, College of Agriculture, University of Wyoming, Laramie, WY, p. 48.

Olson, T.E. and Knopf, F.L. 1986. Agency subsidization of a rapidly spreading exotic. *Wildlife Society Bulletin* **14**: 492–493.

Omemik, J. 1987. Ecoregions of the conterminous United States. *Annals of the Association of American Geographers* **77**: 118–125.

Ontario Ministry of Natural Resources 1994. *Natural Channel Systems, An Approach to Management and Design*, Ministry of Natural Resources, Natural Resources Information Centre, Toronto, Ontario, June.

Ontario Ministry of Natural Resources and Watershed Science Centre 2001. *Adaptive Management of Stream Corridors in*

*Ontario.* Natural Channel Systems. Watershed Science Center, Trent University, Peterborough, Ontario (CD-ROM).

Ontario Streams 2003. *Ontario's Stream Rehabilitation Manual.* Ontario Streams. URL: http://www.ontariostreams.on.ca/OSRM/toc.htm (accessed 1st Nov 2003).

Oregon Department of Fish and Wildlife 1995. *A Guide to Placement of Large Wood in Streams.* Oregon Department of Fish and Wildlife, Habitat Conservation Division, Portland, OR.

Orth D.J. and Maughan O.E. 1981. Evaluation of the 'Montana Method' for recommending instream flows in Oklahoma streams. *Proc. Okla. Acad. Sci.* **61**: 62–66.

Orth, D.J. and Maughan, O.E. 1982. Evaluation of the Incremental Methodology for recommending instream flows for fishes. *Transactions American Fisheries Society* **111**: 413–445.

Orth, D.J. and Leonard, P.M. 1990. Comparison of discharge methods and habitat optimization for recommending instream flows to protect fish habitat. *Regulated Rivers: Research & Managment* **5**: 129–138.

Osborne, L.L. and Kovacic, D.A. 1993. Riparian vegetated buffer strips in water-quality restoration and stream management. *Freshwater Biology* **29**: 243–258.

Outhet, D., Brooks, A., Hader, W., Corlis, P., Raine, A., Broderick, T., Avery, E. and Babakaiff, S. 2001. Gravel bed riffle restoration in New South Wales. In Rutherfurd, I., Sheldon, F., Brierley, G. and Kenyon, C. (eds) *Proceedings Third Australian Stream Management Conference*, Brisbane, 27–28 Aug, CRC Catchment Hydrology, Monash University, Clayton, Victoria, pp. 483–487.

Outridge, P.M. and Noller, B.N. 1991. Accumulation of toxic trace elements by freshwater vascular plants. *Reviews of Environmental Contamination and Toxicology* **121**: 1–63.

Overmire, T.G. 1986. *The World of Biology.* John Wiley, New York.

Owen. K. 1994. Remarks to the conference on guidelines for 'natural' channel systems. In *'Natural' Channel Design: Perspectives and Practice*, Shrubsole, D. (ed), Proceedings of the First International Conference on Guidelines for 'Natural' Channel Systems, March, Niagara Falls, Ontario. Canadian Water Resources Association, Cambridge, Ontario, pp. 1–4.

Owens, L. B., W. M. Edwards and van Keuren, R.W. 1996. Sediment losses from a pastured watershed before and after fencing. *Journal of Soil and Water Conservation* **51**: 90–94.

Palmer, C.G. 1999. Application of ecological research to the development of a new South African water law. *Journal of the North American Benthological Society* **18**(1): 132–142.

Pantle, R. and Buck, H. 1955. Die Biologische Überwachung der Gewässer und die Darstellung der Ergebnisse. (Biological monitoring of water bodies and the presentation of results). *Gas und Wasserfach* **96**: 604.

Parasiewicz, P. Schmutz, S. and Moog, O. 1996. The effect of managed hydropower peaking on the physical habitat, benthos, and fish fauna in the Bregenzerach, a nival 6th order river in Austria. In Leclerc, M. Capra, H., Valentin, S., Boudreault, A. and Côté, Y. (eds) *Proceedings of the 2nd Internatioanl Symposium on Habitat Hydraulics, Ecohydraulics 2000.* Institute National de la Recherche Scientifique – Eau, co-published with FQSA, IAHR/AIRH, Ste-Foy, Québec, Canada, pp. A685–A697.

Park, C.C. 1977. World wide variations in hydraulic geometry exponents of stream channels: an analysis and some observations. *Journal of Hydrology* **33**: 133–146.

Park, D.L. 1990. Status and Future of Spring Chinook Salmon in the Columbia River Basin—Conservation and Enhancement. *NOAA Technical Memorandum NMFS F/NWC-187.* U.S. Department of Commerce, National Oceanic and Atmospheric Administration, National Marine Fisheries Service, August, p. 130 URL: http://www.nwfsc.noaa.gov/publications/techmemos/tm187/ (accessed 1st Nov 2003).

Parker, G. 1979. Hydraulic geometry of active gravel-bed rivers. *Journal of the Hydraulics Division ASCE* **105**: 1185.

Parker, G., Klingeman, P.C. and McLean, D.G. 1982. Bedload and size distribution in paved gravel-bed streams. *J. Hyd. Div., Proc. ASCE,* **108**(HY4), 544–571.

Parkyn, S.M., Davies-Colley, R.J., Halliday, N.J., Costley, K.J. and Croker, G.F. 2003. Planted riparian buffer zones in New Zealand: do they live up to expectations? *Restoration Ecology* **11**(4): 436–447.

Parliament of Victoria, 1997. *Heritage Rivers Act 1992*, Act No. 36/1992, Version No. 001, Version as at 14 July 1997. URL: http://www.dms.dpc.vic.gov.au/l2d/H/ACT00959/0_1.html (accessed 29th Feb 2004).

Parsons, M., Thoms, M. and Norris, R., 2002a. Australian River Assessment System: Review of Physical River Assessment Methods—A Biological Perspective. *Monitoring River Heath Initiative Technical Report Number 21.* Commonwealth of Australia and University of Canberra. Environment Australia, Canberra, Australian Capital Territory.

Parsons, M., Thoms, M. and Norris, R., 2002b. Australian River Assessment System: AusRivAS Physical Assessment Protocol. *Monitoring River Health Initiative Technical Report Number 22.* Commonwealth of Australia and University of Canberra. Environment Australia, Canberra, Australian Capital Territory.

Patil, G.P., and Taillie, C. 1979. An overview of diversity. In Grassle, J.F., Patil, G.P., Smith, W.K. and Taille (eds) *Ecological Diversity in Theory and Practice.* International Co-operative Publishing House, Fairland, MD, pp. 3–27.

Payne, T.R. 1994. RHABSIM: User-friendly computer model to calculate river hydraulics and aquatic habitat. In *Proceedings of the 1st International Symposium on Habitat Hydraulics*, 18–20 August, The Norwegian Institute of Technology, Trondheim, Norway. Organized by SINTEF NHL (Norwegian Hydrotechnical Laboratory) in cooperation with The Norwegian Institute of Technology and Norwegian Institute for Nature Research (NINA), IAHR, pp. 254–624.

Pegg, M.A. and Pierce, C.L. 2002. Classification of reaches in the Missouri and lower Yellowstone Rivers based on flow characteristics. *River Research and Applications* **18**(1): 31–42.

Peirson, W.L., Bishop, K., Van Senden, D., Horton, P.R. and Adamantidis, C.A. 2002. Environmental Water Requirements to Maintain Estuarine Processes, *Environmental Flows Initiative Technical Report, Report No. 3*, Environment Australia, Commonwealth of Australia, Canberra, Australian Capital Territory. URL: http://www.deh.gov.au/water/rivers/nrhp/estuarine/index.html (accessed 1st Nov 2003).

Pejchar, L. and Warner, K. 2001. A river might run through it again: criteria for consideration of dam removal and interim lessons from California, *Environmental Management* **28**: 561–575.

Pennak, R.W. 1971. Towards a classification of lotic habitats. *Hydrobiologia* **38**: 321–334.

Pepi, D. 1985. *Thoreau's Method: a Handbook for Nature Study*. Prentice-Hall, Englewood Cliffs, NJ.

Petersen, M.S. 1986. *River Engineering*. Prentice-Hall, Englewood Cliffs, NJ.

Petersen, R.C., Jr 1992. The RCE: a Riparian, Channel and Environmental inventory for small streams in the agricultural landscape. *Freshwater Biology* **27**: 295–306.

Peterson, G., Allen, C.R. and Holling. C.S. 1998. Ecological resilience, biodiversity and scale. *Ecosystems* **1**: 6–18.

Petr, T. and Mitrofanov, V.P. 1998. The impact on fish stocks of river regulation in Central Asia and Kazakhstan. *Lakes & Reservoirs: Research and Management* **3**(3–4): 143–164.

Pettit, N.E., Froend, R.H. and Davies, P.M. 2001. Identifying the natural flow regime and the relationship with riparian vegetation for two contrasting western Australian rivers. *Regulated Rivers: Research & Management* **17**(3): 201–215.

Petts, G.E. 1980. Long-term consequences of upstream impoundment, *Environmental Conservation* **7**(4): 325–332.

Petts, G.E. 1984. *Impounded Rivers: Perspectives for Ecological Management*. Wiley-Interscience Publication, John Wiley and Sons, New York.

Petts, G.E. 1989. Perspectives for ecological management of regulated rivers. In Gore, J.A. and Petts, G.E. (eds) *Alternatives in Regulated River Management*. CRC Press Inc., Boca Raton, FL, pp. 3–24.

Petts, G.E. 1996. Water allocation to protect river ecosystems. *Regulated Rivers: Research & Management* **12**: 353–365.

Petts, G.E. and Calow, P. 1996. The nature of rivers. In Petts, G. and Calow, P. (eds) *River Restoration*, Blackwell Science, Oxford, pp. 1–6.

Petts, G.E. and Foster, I. 1985. *Rivers and Landscape*, Edward Arnold, London.

Petts, G.E., Bickerton, M.A., Crawford, C., Lerner, D.N. and Evans, D. 1999. Flow management to sustain groundwater-dominated stream ecosystems. *Hydrological Processes* **13**: 497–513.

Petts, G.E., Large, A.R.G., Greenwood, M.T. and Bickerton, M.A. 1992. Floodplain assessment for restoration and conservation: linking hydrogeomorphology and ecology. In Carling, P.A. and Petts, G.E. (eds) *Lowland Floodplain Rivers: Geomorphological Perspectives*. John Wiley & Sons Ltd, Chichester, pp. 217–234.

Petts, G.E. and Maddock, I. 1996. Flow allocation for in-river needs. In Petts G.E. and Calow, P. (eds). *River Restoration*. Blackwell Science, Oxford, pp. 60–79.

Pettyjohn, W.A. and Henning, Roger, R. 1979. Preliminary estimate of ground-water recharge rates, related streamflow and water quality in Ohio. *Project Completion Rept. No. 552*. Water Resources Center, Ohio State University, Columbus, OH.

Phillips, J.D. 1990. Multiple modes of adjustment in unstable river channel cross-sections. *Journal of Hydrology* **123**: 39–49.

Phillips, J.D. 2003. Toledo Bend reservoir and geomorphic response in the lower Sabine River. *River Research and Applications* **19**(2): 137–159.

Phillips, N., Bennett, J., and Moulton, D. 2001. *Principles and Tools for Protecting Australian Rivers*. Land and Water Australia, Canberra, Australian Capital Territory. URL: http://www.lwa.gov.au/downloads/PR010161.pdf (accessed 1st Nov 2003).

Pielou, E.C. 1977. *Mathematical Ecology*. John Wiley, New York.

Pielou, E.C. 1984. *The Interpretation of Ecological Data*. John Wiley, New York.

Pielou, E.C. 1998. *Fresh Water*. The University of Chicago Press, Chicago.

Pipan, T. 2000. Biological assessment of stream water quality—the example of the Reka River (Slovenia). *Acta Carsologica* **29**(1): 201–222.

Pitlick J. 1994. Coarse sediment transport and the maintenance of fish habitat in the Upper Colorado River. In *ASCE 1994 National Conference on Hydraulic Engineering*, special session on hydraulics of mountain rivers, Buffalo, New York, pp. 855–859.

Pizzuto, J. E. 1986. Flow variability and the bankfull depth of sand-bed streams of the American Midwest. *Earth Surf. Proc. and Landforms* **11**: 441–450.

Plafkin, J.L., Barbour, M.T., Porter, K.D., Gross, S.K. and Hughes, R.M. 1989. *Rapid Bioassessment Protocols for Use in Streams and Rivers: Benthic Macroinvertebrate and Fish*. EPA/444/4-89-001. Washington, DC.

Platts, W.S. 1979. Relationships among stream order, fish populations, and aquatic geomorphology in an Idaho river drainage. *Fisheries* **4**: 5–9.

Platts, W.S. 1983. Vegetation requirement for fisheries habitats. In Monsen, S.B. and Shaw, N. compilers. Managing intermountain rangelands-improvement of range and wildlife habitats. *U.S. Forest Service General Technical Report INT-157*, pp. 184–188.

Platts, W.S. 1991. Livestock grazing. In Meehan, W.R. (ed.) Influences of Forest Rangeland Management on Salmonid Fishes and their Habitats. *American Fisheries Society Special Publication No. 19*, Bethesda, MD, pp. 389–423.

Platts, W.S. and Nelson, R.L. 1985. Stream habitat and fisheries response to livestock grazing and instream improvement structures, Big Creek, Utah. *Journal of Soil and Water Conservation* **40**: 374–379.

Platts, W.S. and Penton, V.E. 1980. A new freezing technique for sampling salmonid redds, *Res. Pap. INF-248*, USDA Forest Service, Intermountain Forest and Range Expt Station, Ogden, Utah.

Platts, W.S. and Rinne, J.N. 1985. Riparian and stream enhancement management and research in the Rocky Mountains. *North American Journal of Fisheries Management* **5**: 115–125.

Platts, W.S., Armour, C., Booth, G.D., Bryant, M., Bufford, J.L., Cuplin, P., Jensen, S., Lienkaemper, G.W., Minshall, G.W., Monsen, S.B., Nelson, R.L., Sedell, J.R. and Tuhy, J.S. 1987. Methods for evaluating riparian habitats with applications to management. *Gen. Tech. Rep. INT-221*. USDA Forest Service, Intermountain Research Station, Ogden, UT.

Platts, W.S., Megahan, W.F. and Minshall, G.W. 1983. Methods for evaluating stream, riparian, and biotic conditions. *Gen. Tech. Rep. INT-138*. USDA Forest Service, Intermountain Forest and Range Expt Station, Ogden, UT.

Poff, N.L., Allan, J.D., Palmer, M.A., Hart, D.D., Richter, B.D., Arthington, A.H., Rogers, K.H., Meyer, J.L., and Stanford, J.A. 2003. River flows and water wars: emerging science for environmental decision making. *Frontiers in Ecology and the Environment* **1**(6): 298–306.

Poff, N.L. 1996. A hydrogeography of unregulated streams in the United States and an examination of scale-dependence in some hydrological descriptors, *Freshwater Biology* **36**: 101–121.

Poff, N.L. 1997. Landscape filters and species traits: towards mechanistic understanding and prediction in stream ecology. *Journal of the North American Benthological Society* **16**: 391–409.

Poff, N.L. and Allan, J.D. 1995. Functional organisation of stream fish assemblages in relation to hydrological variability. *Ecology* **76**: 606–627.

Poff, N.L. and Hart, D.D. 2002. How dams vary and why it matters for the emerging science of dam removal. *BioScience* **52**: 659–668.

Poff, N.L. and Ward, J.V. 1989. Implications of streamflow variability and predictability for lotic community structure: a regional analysis of streamflow patterns. *Canadian Journal of Fisheries and Aquatic Sciences* **46**: 1805–1818.

Poff, N.L., Allan, J.D., Bain, M.B., Karr, J.R., Prestegaard, K.L., Richter, B.D., Sparks, R.E. and Stromberg, J.C. 1977. The natural flow regime, a paradigm for river conservation and restoration. *BioScience* **47**: 769–784.

Pollution Control Department 2002. Classification and Objectives, Surface Water, Water Quality Standards, Bangkok, Thailand. URL: http://www.pcd.go.th/Information/Regulations/Water Quality/SurfaceWater.htm#Class (accessed 1st Nov 2003).

Poole, G.C. 2002. Fluvial landscape ecology: addressing uniqueness within the river discontinuum. *Freshwater Biology* **47**: 641–660.

Porterfield, G. 1972. Computation of fluvial-sediment discharge. *Tech. of Water-Resources Investigations of the USGS*, Chapter C3, USGS, Washington, DC.

Potyondy, J. 1999. Applying channel maintenance concepts and tools to hydropower projects, presented at *Workshop on Integrating Physical and Biological Processes in Evaluating the Effects of Hydropower Projects on Fluvial Systems*, 22–23 September, 1999, Corvallis, Oregon, USDA Forest Service and Region 6 Hydropower Relicensing Team.

Pouilly M., Valentin S., Capra H., Ginot V. and Souchon Y. 1995. Méthode des microhabitats: principles et protocoles d'application. *Bulletin Français de la Pêche et de la Pisciculture* **336**: 41–54.

Powell, G.L. and Matsumoto, J. 1994. Texas estuarine mathematical programming model: a tool for freshwater inflow management. In Dyer, K.R. and Orth, R.J. (eds) *Changes in Fluxes in Estuaries: Implications from Science to Management*. Olsen and Olsen, Fredensborg, Denmark, pp. 401–404.

Power, M.E., Dietrich, W.E. and Finlay, J.C. 1996. Dams and downstream aquatic biodiversity: potential food web consequences of hydrologic and geomorphic change. *Environmental Management* **20**(6): 887–895.

Prat, N., Munné, A., Bonada, N., Rieradevall, M., Solà, C., Vila-Escalé, M., Casanovas-Berenguer, R., PuntÚ, T., Plans, M., and Múrria, C. 2004. Ecostrimed Index (Ecological Status River Mediterranean). Diputació de Barcelona, Barcelona. URL: http://www.diba.es/mediambient/ecostrimeduk.asp (accessed 5th Jan 2004).

Prestegaard, K.L. 1989. Selective unravelling and the downstream fining of streambed material. Paper presented at spring meeting of Am. Geophys. Union, May 1989, Baltimore, MD.

Pretty, J.L., Harrison, S.S.C., Shepherd, D.J., Smith, C., Hildrew, A.G. and Hey, R.D. 2003. River rehabilitation and fish populations: assessing the benefit of instream structures. *Journal of Applied Ecology* **40**(2): 251–265.

Prewitt, C.G. and Carlson, C.A. 1980. An evaluation of four instream flow methodologies. *Biological Sciences Series Number 2*. US Bureau of Land Management, Denver, CO.

Pritchard, D.W. 1952a. Estuaries hydrography. *Advances in Geophysics* **1**: 243–280.

Pritchard, D.W. 1952b. Salinity distribution and circulation in the Chesapeake Bay Estuaries System. *Journal, Marine Research* **11**: 106–123.

Prosser, I., Bunn, S., Mosisch, T., Ogden, R. and Karssies, L. 1999. The delivery of sediment and nutrients to streams. In Lovett, S. and Price, P. (eds) *Riparian Land Management Technical Guidelines, Volume One: Principles of Sound Management*. Land and Water Resources Research and Development Corporation, Canberra, Australian Capital Territory, pp. 37–61.

Puckridge, J.T., Sheldon, F., Walker, K.F. and Boulton, A.J. 1998. Flow variability and the ecology of large rivers. *Marine and Freshwater Research* **49**: 55–72.

Purcell, E.M. 1977. Life at low Reynolds number. *Am. J. of Physics*. **45**(1): 3–11.

Pusey, B.J. 1998. Methods addressing the flow requirements of fish. In Arthington A.H. and Zalucki, J.M. (eds) Comparative evaluation of environmental flow assessment techniques: review of methods. *LWRRDC Occasional Paper No 27/98*. Land and Water Resources Research & Development Corporation, Canberra, Australian Capital Territory, pp. 66–105.

Pusey, B.J. and Arthington, A.H. 1990. Limitations to the valid application of the instream flow incremental methodology (IFIM) for determining instream flow requirements in highly variable Australian lotic environments. *Verh. Int. Verein. Limnol.* **24**, Proc. of SIL Congress, August 1989, Munich.

Quigley, T.M., Sanderson, H.R., Tiedemann, A.R. and McInnes, M.L. 1990. Livestock control with electrical and audio stimulator. *Rangelands* **12**(3): 152–155.

Quinn, F. 1991. As long as the rivers run: the impacts of corporate water development on native communities in Canada. *The Canadian Journal of Native Studies* **11**(1): 137–154.

Quinn, J.M., Cooper, A.B., Stroud, M.J. and Burrell, G.P. 1997. Shade effects on stream periphyton and invertebrates: an experiment in streamside channels. *New Zealand Journal of Marine and Freshwater Research* **31**: 665–683.

Quinn, J.M., Williamson, R.B., Smith, K.R. and Vickers, M.L. 1992. Effects on riparian grazing and channelisation on streams in Southland, New Zealand. 2. Benthic invertebrates. *New Zealand Journal of Marine and Freshwater Research* **26**: 259–273.

Raborn, S.W. and Schramm, H.L., Jr 2003. Fish assemblage response to recent mitigation of a channelized warmwater stream. *River Research and Applications* **19**(4): 289–301.

Raddum, G.G., Fjellheim, A. and Hesthagen, T. 1988 Monitoring of acidification through the use of aquatic organisms. *Verh. Int. Verein. Limnol.* **23**: 2291–2297.

Railsback, S. 1999. Reducing uncertainties in instream flow studies. *Fisheries* **24**(4): 24–26.

Raine, A.W. and Gardiner, J.N. 1995. RIVERCARE. Guidelines for ecologically sustainable management of rivers and riparian vegetation. *LWRRDC Occasional Paper No. 03/95*. Land and Water Resources Research & Development Corporation, Canberra, Australian Capital Territory.

Rajaratnam, N. and Katopodis, C. 1984. Hydraulics of Denil fishways. *Journal of the Hydraulcs Division ASCE* **110**: 1219–1233.

Rajaratnam, N., Katopodis, C. and Solanki, S. 1989. New designs for vertical slot fishways. *Tech. Rep. WRE 89–1*. Department of Civil Eng, University of Alberta, Edmonton. p. 50.

Raleigh, R.F., Hickman, T., Soloman, R.C. and Nelson, P.C. 1984. Habitat suitability information: rainbow trout. *Biological Report 82 (10.60)*, *FWS/OBS-82/10.60*, U.S. Fish and Wildlife Service, Washington, DC.

Raleigh, R.F., Miller, W.J. and Nelson, P.C. 1986. Habitat suitability index models and instream flow suitability curves: chinook salmon. *Biological Report 82(10.122)*. U.S. Fish and Wildlife Service, Washington, DC.

Ramsar Convention Bureau 1996. *The Criteria for Identifying Wetlands of International Importance*. Ramsar, Gland.

Rao, A.M. and Hamed, K.H. 2000. *Flood Frequency Analysis*. CRC Press, Washington, D.C.

Rapport, D.J. 1999. On the transformation from healthy to degraded aquatic ecosystems. *Aquatic Ecosystem Health and Management* **2**: 97–103.

Rasid, H. 1979. The effects of regime regulation by the Gardner Dam on downstream geomorphic processes in the Saskatchewan River. *Canadian Geographer* **23**: 140–156.

Raudkivi, A.J. 1967. *Loose Boundary Hydraulics*, Pergamon Press, Oxford.

Raudkivi, A.J. 1979. *Hydrology: An Advanced Introduction to Hydrological Processes and Modelling*. Pergamon Press Ltd., Oxford.

Raven, P.J., Fox, P., Everard, M., Holmes, N.T.H. and Dawson, F.H. 1997. River habitat survey: a new system for classifying rivers according to their habitat quality. In Boon, P.J. and Howell, D.L. (eds.) *Freshwater Quality: Defining the Indefinable?* The Stationery Office, Edinburgh, pp. 215–234.

Raven, P.J., Holmes, N.T.H., Dawson, F.H., Fox, P.J.A., Everard, M., Fozzard, I.R. and Rowen, K.J. 1998. *River Habitat Quality: The Physical Character of Rivers and Streams in the UK and Isle of Man*. Environment Agency, Bristol, UK.

Ravera, O. 1998. Utility and limits of biological and chemical monitoring of the aquatic environment. *Annali di Chimica* **88**: 909–913.

Ravera, O. 2001. Ecological monitoring for water body management. In Timmerman, J.G., Cofino, W.P., Enderlein, R.E., Jülich, W., Literathy, P.L., Martin, J.M., Ross, P., Thyssen, N., Turner, R.K. and Ward, R.C. (eds) *MTM (Monitoring Tailor-Made) III, Proceedings of International Workshop on Information for Sustainable Water Management*, 25–28 Sept 2000, Nunspeet, The Netherlands, pp. 157–167. URL: http://

www.mtm-conference.nl/mtm3/docs/Ravera2001.pdf (accessed 1st Nov 2003).

Read, M. and Barmuta, L.A. 1999. Comparisons of benthic communities adjacent to rparian native eucalypt and introduced willow vegetation. *Freshwater Biology* **42**: 359–374.

Reckhow, K.H., and Chapra, S.C. 1983. *Engineering Approaches for Lake Management, Volume 1: Data Analysis and Empirical Modeling*. Butterworth Publishers, Boston, MA.

Rees, W.G. 2001. *Physical Principles of Remote Sensing*, 2nd edn, Cambridge University Press, Cambridge.

Reeve, R.C. 1986. Water potential: piezometry. In Klute, A. (ed) *Methods of Soil Analysis, Part 1: Physical and Mineralogical Methods, Agronomy Monograph no. 9*, 2nd edn, Amer. Soc. of Agronomy, Soil Sci. Soc. of America, Madison, WI, pp. 545–561.

Reeves, G.H., Roelofs, T.D. 1982. Influence of forest and rangeland management on anadromous fish habitat in western North America: 13. Rehabilitating and enhancing stream habitat: 2. Field applications. Meehan, W.R. (Tech. Ed.) *USDA General Technical Report PNW-140*. U.S. Department of Agriculture Forest Service, Pacific Northwest Forest and Range Experiment Station, Anadromous Fish Habitat Program, Portland, OR, pp. 38.

Regier, H.A. 1993. The notion of natural and cultural integrity. In *Ecological Integrity and the Management of Ecosystems*. Woodley, S.J., Kay, J.J. and Francis, G. (eds), St. Lucie Press, Delray Beach, FL, pp. 3–18.

Reid, M.A. and Brooks, J.J. 2000. Detecting effects of environmental water allocations in wetlands of the Murray-Darling basin, Australia. *Regulated Rivers: Research & Management* **16**: 479–496.

Reinfelds, I., Haeusler, T., Brooks, A.J. and Williams, S. 2004. Refinement of the wetted perimeter breakpoint method for setting cease-to-pump limits or minimum environmental flows. *River Research and Applications* (in press).

Reiser, D.W., Ramey, M.P. and Lambert, T.R. 1985. Review of flushing flow requirements in regulated streams. Pacific Gas and Electric Co. Department of Engineering Research, San Ramon, CA, February.

Reiser, D.W., Ramey, M.P. and Wesche, T.A. 1989a. Flushing flows. In Gore, J.A. and Petts, G.E. (eds) *Alternatives in Regulated River Management*. CRC Press, FL, pp. 91–138.

Reiser, D.W., Wesche, T.A. and Estes, C. 1989b. Status of instream flow legislation and practise in North America. *Fisheries* **14**(2): 22–29.

Resh, V.H., Brown, A.V., Covich, A.P., Gurtz, M.E., Li, H.W., Minshall, G.W., Reice, S.R., Sheldon, A.L., Wallace, J.B. and Wissmar, R.C. 1988. The role of disturbance in stream ecology. *Journal of the North American Benthological Society* **7**: 433–455.

Resh, V.M., and Rosenberg, D.M. (eds) 1984. *Ecology of Aquatic Insects*, Praeger, New York.

Reynolds, O. 1883. An experimental investigation of the circumstances which determine whether the motion of water shall be direct or sinuous, and the law of resistance in parallel channels. *Trans. Roy. Soc. Lond.* **174**: 935–982.

Reynoldson, T.B., Day, K.E. and Pascoe, T. 2000. The development of the BEAST: a predictive approach for assessing sediment quality in the North American Great Lakes. In Wright, J.F.,

Sutcliffe, D.W. and Furse, M.T. (eds) *Assessing the Biological Quality of Freshwaters: RIVPACS and other techniques*. Freshwater Biological Association, Ambleside, pp. 165–180.

Reynoldson, T.B., Norris, R.H., Resh, V.H., Day, K.E. and Rosenberg, D.M. 1997. The reference condition: a comparison of multimetric and multivariate approaches to assess water-quality impairment using benthic macroinvertebrates. *Journal of the North American Benthological Society* **16**: 833–852.

Richards, K.S. 1976. The morphology of riffle-pool sequences. *Earth Surf. Proc.* **1**: 71–88.

Richards, K.S. 1982. *Rivers, Form and Process in Alluvial Channels*. Methuen, London.

Richards, K.S. (ed) 1987. *River Channels: Environment and Process*. Basil Blackwell, Oxford.

Richardson, B.A. 1986. Evaluation of instream flow methodologies for freshwater fish in New South Wales. In Campbell, I.C. (ed) *Stream Protection: The Management of Rivers for Instream Uses*. Chisholm Institute of Technology, Victoria, Australia, pp. 143–67.

Richardson, J.S. and Hinch, S.G. 1998. Ecological objectives for stream and watershed restoration along the Pacific Coast of North America. In *Proceedings International Workshop on Environmental Hydrodynamics and Ecological River Restoration in Cold Regions*. 21–23 Sept 1998, Trondheim, Norway, SINTEF Civil and Environmental Engineering, Norway, pp. 47–56.

Richter, B.D. Baumgartner, J.V., Powell, J. and Braun, D.P. 1996. A method for assessing hydrological alteration within ecosystems. *Conservation Biology* **10**(4): 1163–1174.

Richter, B.D., Baumgartner, J.V., Wigington, R. and Braun, D.P. 1997. How much water does a river need? *Freshwater Biology* **37**: 231–249.

Ridley, J.E. and Steel, J.A. 1975. Ecological aspects of river impoundments. In Whitton, B.A. (ed.) *River Ecology*, Blackwell Scientific, Oxford, pp. 565–587.

Ridley, S.J. 1972. A comparison of morphometric measures of bankfull. *J. Hydrol* **17**: 23–31.

Riggs, H.C. 1968. Some statistical tools in hydrology: U.S. Geological Survey Techniques of Water-Resources Investigations, Book 4, Chap. A1, p. 39.

Riggs, H.C. 1976. A simplified slope area method for estimating flood discharges in natural channels. *Jour. Research US Geol. Survey* **4**: 285–291.

Riley, A.L. 1998. *Restoring Streams in Cities: A Guide for Planners, Policymakers, and Citizens*. Island Press, Covelo, CA.

Risser, R.J. and Harris, R.R. 1989. Mitigation for impacts to riparian vegetation on western montaine streams. In Gore, J.A. and Petts, G.E. (eds) *Alternatives in Regulated River Management*, CRC Press, Inc., Boca Raton, FL, pp. 235–250.

River Health Programme 2002. State of Rivers Report: The uMngeni River and Neighbouring Rivers and Streams. *WRC Report No. TT 200/02*, Water Research Commission, Pretoria. URL: http://www.csir.co.za/rhp/state_of_rivers/state_of_umngeni_02/umngeni.html (accessed 1st Nov 2003).

Rivinoja, P., McKinnell, S. and Lundqvist, H. 2001. Hindrances to upstream migration of atlantic salmon (*Salmo salar*) in a northern Swedish river caused by a hydroelectric power-station. *Regulated Rivers: Research & Management* **17**(2): 101–115.

Roberson, J.A. and Crowe, C.T. 1990. *Engineering Fluid Mechanics*, 2nd edn, Houghton Mifflin, Boston, MA.

Roberts, J., Young, W.J. and Marston, F. 2000. Estimating the water requirements for plants of floodplain wetlands: a guide. *LWRRDC Occasional Paper No. 04/00*. Land and Water Resources Research and Development Corporation, Canberra, Australian Capital Territory.

Rolauffs, P., Hering, D., Sommerhäuser, M., Rödiger, S. and Jähnig, S. 2003. Entwicklung eines leitbildorientierten Saprobienindexes für die biologische Fließgewässerbewertung. Umweltforschungsplan des Bundesministeriums für Umwelt, Naturschutz und Reaktorsicherheit. Forschungsbericht 200 24 227 UBA-FB 000366. URL: http://www.umweltbundesamt.org/fpdf-k/2253.pdf (accessed 1st Nov 2003).

Romesburg, H.C. 1984. *Cluster Analysis for Researchers*. Lifetime Learning Publications, Belmont, CA.

Rood, S.B. and Mahoney, J.M. 2000. Revised instream flow regulation enables cottonwood recruitment along the St. Mary River, Alberta, Canada. *Rivers* **7**(2): 109–125.

Rood, S.B. and Tymensen, W. 2001. Recreational flows for paddling along rivers in southern Alberta. Chinook Environmental Resources and Dept. Biological Sciences, University of Lethbridge, Letbridge, AB, Canada. Alberta Environment, Lethbridge, AB, Feb. URL: http://www3.gov.ab.ca/env/water/regions/ssrb/pdf_phase2/OldmanR%20Rec%20Flows%20report%20FINAL1.pdf (accessed 1st Nov 2003).

Rood, S.B., Tymensen, W. and Middleton, R. 2003. A comparison of methods for evaluating instream flow needs for recreation along rivers in southern Alberta, Canada. *River Research and Applications* **19**(2): 123–135.

Roper, B.R., Konnof, D. Heller, D. and Wieman, K. 1998. Durability of Pacific Northwest instream structures following floods. *North American Journal of Fisheries Management* **18**: 686–693.

Rose, T. and Bevitt, R. 2003. Snowy River Benchmarking and Environmental Flow Response Monitoring Project: Summary Progress Report on Available Data From 1999–2001. *DIPNR Publication No. Y03/1115*, Report to Environment Australia. The Department of Infrastructure Planning and Natural Resources, Cooma, NSW, June. URL: http://www.deh.gov.au/water/rivers/nrhp/snowy/pubs/snowy.pdf (accessed 1st Nov 2003).

Rosenberg, D.M. and Resh, V.H. (eds) 1993. *Freshwater Biomonitoring and Benthic Macroinvertebrates*. Chapman and Hall, London.

Rosenberg, D.M., Reynoldson, T.B. and Resh, V.H. 2000. Establishing reference conditions in the Fraser River catchment, British Columbia, Canada, using the BEAST (BEnthic Assessment of SedimenT) predictive model. In Wright, J.F., Sutcliffe, D.W. and Furse, M.T. (eds) *Assessing the Biological Quality of Freshwaters: RIVPACS and other techniques*. Freshwater Biological Association, Ambleside. pp. 181–194.

Rosgen, D.L., 1989. Conversion of a braided river pattern to meandering–a landmark restoration project. In Protection, Management and Restoration for the 1990's, Proceedings of the California Riparian Systems Conference, 22–24 September, 1988, Abell, D. (Tech. Coord.), Davis, CA, *UDSA Forest*

*Service Gen. Tech. Rep. PSW-110*, Berkeley, CA, U.S. Department of Agriculture, Forest Service, Pacific Southwest Forest and Range Experiment Station.

Rosgen, D.L. 1994. A classification of natural rivers. *Catena* 22: 169–199.

Rosgen, D.L. 1996a. *Applied River Morphology*. Wildland Hydrology, Pagosa Springs, CO.

Rosgen, D.L. 1996b. A classification of natural rivers: reply to the comments by J.R Miller and J.B. Ritter. *Catena* 27: 301–307.

Rosgen, D.L., 1998. The reference reach—a blueprint for natural channel design. In Hayes, D.F. (ed) *Proceedings Wetlands and River Restoration Conference*, March, Denver, CO, American Society of Civil Engineers, Reston, VA. Can be downloaded from URL: <http://www.wildlandhydrology.com/html/references_.html> (accessed 1st March 2004).

Rosgen, D.L., Silvey, H.L. and Potyondy, J.P. 1986. The use of channel maintenance flow concepts in the Forest Service. *Hydrological Science and Technology: Short Papers*, Volume 2. American Institute of Hydrology, pp. 19–26.

Roth, N.E, Southerland, M.T., Chaillou, J.C., Kazyak, P.F. and Stranko, S.A. 2000. *Refinement and Validation of a Fish index of Biotic Integrity for Maryland Streams*. Versar, Inc. Columbia and Maryland Department of Natural Resources, Annapolis, MD, October. URL: http://www.esm.versar.com/pprp/bibliography/MD_Biological_Streams_Survey_Reports/CBWP- MANTA-EA-00-2.pdf (accessed 1st Nov 2003).

Roux, D., de Moor, F., Cambray, J. and Barber-James, H. 2002. Use of landscape-level river signatures in conservation planning: a South African case study. *Conservation Ecology* 6(2): 6. [online]. URL: http://www.consecol.org/vol6/iss2/art6 (accessed 1st Nov 2003).

Rowntree, K.M. and Wadeson, R.A. 1998. A geomorphological framework for the assessment of instream flow requirements, *Aquatic Ecosystem Health and Management* 1: 125–141.

Rowntree K.M. and Wadeson R.A. 1999. A hierarchical framework for categorising the geomorphology of selected river systems. *Water Research Commission Report No 497/1/99*. Pretoria.

Rowntree, K.M. and Wadeson, R.A. 2000. An index of stream geomorphology for the assessment of River Health. Field manual for channel classification and condition assessment. *NAEBP Report Series No 13*, South African River Health Programme. Institute for Water Quality Studies, Department of Water Affairs and Forestry, Pretoria, South Africa. URL: http://www.csir.co.za/rhp/reports/reportseries13.html (accessed 1st Nov 2003).

Roy, A.G. and Abrahams, A.D. 1980. Discussion of rhythmic spacing and origin of pool sand riffles. *Geological Society of America Bulletin* 91: 248–250.

Rushton, C.D. 2000. *Instream flows in Washington State, past, present and future (White paper)*. Water Resources Program, Department of Ecology, Olympia, July. URL: http://www.olympus.net/community/dungenesswc/InstreamFlowversion12.PDF (accessed 1st Nov 2003).

Rutherford, J.C., Blackett, S., Blackett, C., Saito, L. and Davies-Colley, R.J. 1997. Predicting the effects of shade on water temperature in small streams. *New Zealand Journal of Marine and Freshwater Research* 31: 707–721.

Rutherfurd, I.D. 2000. Some human impacts on Australian stream channel morphology. In Brizga, S.O. and Finlayson, B.L. (eds) *River Management: the Australasian Experience*. John Wiley & Sons, Chichester, pp. 11–49.

Rutherfurd, I.D. and Bartley, R. (eds) 1999. *Proceedings of the Second Australian Stream Management Conference*, Adelaide, 8–11 Feb, Cooperative Research Centre for Catchment Hydrology, Monash University, Clayton.

Rutherfurd, I.D. and Gippel, C.J. 2001. Australia versus the world: do we face special opportunities and challenges in restoring Australian streams? *Water Science and Technology* 43(9): 165–174.

Rutherfurd, I.D., Jerie, K., Walker, M. and Marsh, N. 1999. Don't raise the Titanic: how to set priorities for stream rehabilitation. In Rutherford, I. and Bartley, R. (eds) *Proceedings Second Australian Stream Management Conference*, Adelaide, 8–11 Feb, CRC Catchment Hydrology, Monash University, Clayton, pp. 527–532.

Rutherfurd. I.D., Jerie, K. and Marsh, N. 2000. *A Rehabilitation Manual for Australian Streams, Volume 1: Concepts and Planning, Volume 2: Rehabilitation Tools*. Cooperative Research Centre for Catchment Hydrology and Land and Water Resources Research and Development Corporation, Canberra, Australian Capital Territory. URL: http://www.rivers.gov.au/publicat/rehabmanual.htm (accessed 1st Nov 2003).

Rutherfurd, I.D., Sheldon, F., Brierley, G. and Kenyon, C. (eds), 2001, *Proceedings Third Australian Stream Management Conference*, 27–28 Aug, Brisbane, CRC Catchment Hydrology, Monash University, Clayton, Victoria.

Ryan, B.F., Joiner, B.L. and Ryan, T.A., Jr 1985. *Minitab Handbook*, 2nd edn, PWS Publishers, Boston, MA.

Ryder, R.A. 1990. Ecosystem health, a human perception: definition, detection, and the dichotomous key. *Journal of Great Lakes Research* 16(4): 619–624.

S. Brizga & Associates Pty Ltd 1998. Fluvial Geomorphology of the Bunyip River: Princes Highway to Cora Lynn, with N.M Craigie. Melbourne Water Corporation, Waterways and Drainage Group, Richmond, Victoria.

Sack, D. 1992. New wine in old wine bottles: the historiography of a paradigm change. *Geomorphology* 5: 251–263.

Sarver, R. 2000. *Stream Habitat Assessment: Project Procedure*. Missouri Department of Natural Resources, Division of Environmental Quality, Environmental Services Program, Jefferson City, MO. November. URL: http://www.cpcb.ku.edu/archives/methods/MO/Streamhab.pdf (accessed 1st Nov 2003).

SAS 1987. *SAS User's Guide: Statistics, Version 6 Edition*, SAS Institute, Cary, NC.

Savage, N.L. and Rabe, F.W. 1979. Stream types in Idaho: an approach to classification of streams in natural areas. *Biological Conservation* 15: 301–315.

Scarnecchia, D.L. 1988. The importance of streamlining in influencing fish community structure in channelized and unchannelized reaches of a prairie stream. *Regulated Rivers: Research & Application* 2: 155–166.

Scatena, F.N. 2004. A survey of methods for setting minimum instream flow standards in the Caribbean basin. *River Research & Applications* 20(2), 127–135.

Schaff, S.D., Pezeshki, S.R. and Shields, F.D. Jr 2002. Effects of pre-planting soaking on growth and survival of Black Willow cuttings. *Restoration Ecology* **10**(2): 267–274.

Schauman, S. and Salisbury, S. 1998. Restoring nature in the city: Puget Sound experiences. *Landscape and Urban Planning* **42**: 287–295.

Scheaffer, R.L., Mendenhall, W. and Ott, L. 1979. *Elementary Survey Sampling*, 2nd edn, PWS Publishers, MA.

Scheidegger, A.E. 1965. The algebra of stream-order numbers. *US Geol. Survey Prof. Paper 525B*, USGS, Washington, DC, pp. 187–189.

Schlichting, H. 1961. Boundary layer theory. In Streeta, V.L. (ed) *Handbook of Fluid Dynamics*, Chapter 9, McGraw-Hill, New York.

Schmedtje, U., Sommerhäuser, M., Braukmann, U., Briem, E., Haase, P. and Hering, D. 2001. 'Top down–bottom up'— Konzept einer biozönotisch begründeten Fließgewässerty pologie Deutschlands. Deutsche Gesellschaft für Limnologie (DGL) – Tagungsbericht 2000 (Magdeburg), Tutzing, pp. 147–151.

Schmetterling, D.A. and Pierce, R.W. 1999. Success of instream habitat structures after a 5-year flood in Gold Creek, Montana. *Restoration Ecology* **7**(4): 369–375.

Schmidt, J.C. 1999. Summary and synthesis of geomorphic studies conducted during the 1996 controlled flood in Grand Canyon. In Webb, R.H., Schmidt, J.C., Marzolf, G.R. and Valdez, R.A. (eds) *The Controlled Flood in the Grand Canyon. Geophysical Monograph 110*. American Geophysical Union, Washington, pp. 329–341.

Schmidt, J.C. and Rubin, D.M. 1995. Regulated streamflow, fine-grained deposits, and effective discharge in canyons with abundant debris fans. In Costa, J.E., Miller, A.J., Potter, K.W. and Wilcock, P.R. (eds) *Natural and Anthropological Influences in Fluvial Geomorphology: The Wolman Volume. Geophysical Monograph 89*. American Geophysical Union, Washington, D.C, pp. 177–195.

Schoor, M.M. and Sorber, A.M. 1999. *Morphology, Naturally*. Ministry of Transport, Public Works and Water Management, Directorate-General for Public Works and Water Management, RIZA Institute for Inland Water Management and Waste Water Treatment, Arnhem, The Netherlands, July.

Schueler, T. 1995. The importance of imperviousness. *Watershed Protection Techniques* **1**(3): 100–111.

Schulz, T.T. and Leininger, T.S. 1990. Differences in riparian vegetation structure between grazed areas and exclosures. *Journal of Range Management* **43**: 295–299.

Schumm, S.A. 1954. The relation of drainage basin relief to sediment loss. *Internat. Assoc. Sci. Hyd. Pub.* **36**: 216–219 (as cited by Gregory and Walling, 1973).

Schumm, S.A. 1956. Evolution of drainage systems and slopes in badlands at Perth Amboy, New Jersey. *Geol. Soc. Amer. Bull.* **67**: 597–646.

Schumm, S.A. 1977. *The Fluvial System*. John Wiley, New York.

Schumm, S.A. 1985. Patterns of alluvial rivers. *Annual Review of Earth and Planetary Sciences* **13**: 5–27.

Schumm, S.A., Harvey, M.D. and Watson, C.C. 1984. *Incised Channels: Morphology, Dynamics and Control*. Water Resources Publications, Littleton, CO.

Scientific Reference Panel 2003. Ecological Assessment of Environmental Flow Reference Points for the River Murray System. Interim report for the Murray-Darling Basin Commission, The Living Murray. Canberra, Australian Capital Territory, October. URL: http://www.thelivingmurray.mdbc.gov.au/content/index. phtml/itemId/15978/fromItemId/4484 (accessed 1st Nov 2003).

Scott, D. and Shirvell, C.S. 1987. A critique of the instream flow incremental methodology and observations on flow determination in New Zealand. In Kemper, J.B.and Craig, J. (eds) *Regulated Streams—Advances in Ecology*. Plenum Press, New York, pp. 27–44.

Scottish Environment Protection Agency 2000. River Classification Scheme. SEPA Corporate Office, Stirling. URL: http://www.sepa.org.uk/data/classification/classification_scheme_rivers_2000.htm (accessed 1st Nov 2003).

Scrimgeour, G.J. and Wicklum, D. 1996. Aquatic ecosystem health and integrity: problems and potential solutions. *Journal of North American Benthological Society* **15**(2): 254–261.

Scruton, D.A. and LeDrew, L.J. 1996. A retrospective assessment of a regulated flow regimen for a Newfoundland (Canada) river. In Leclerc, M. Capra, H., Valentin, S., Boudreault, A. and Côté, Y. (eds) *Proceedings of the 2nd Internatioanl Symposium on Habitat Hydraulics, Ecohydraulics 2000*. Institute National de la Recherche Scientifique – Eau, co-published with FQSA, IAHR/AIRH, Ste-Foy, Québec, Canada, pp. A533-A546.

Sear, D.A. 1996. The sediment system and channel stability. In Brookes, A. and Shields, F.D. Jr (eds) *River Channel Restoration—Guiding Principals for Sustainable Projects*. John Wiley & Sons, Chichester, pp. 149–177.

Searcy, J.K. and Hardison, C.H. 1960. Double-mass curves. *US Geological Survey Water-Supply Paper 1541-B*. USGS, Washington, DC.

Searcy, J.K. 1959. Flow-duration curves. *Manual of Hydrology: Part 2. Low-Flow Techniques, US Geol. Survey Water-Supply Paper 1542-A*. USGS, Washington, DC.

Searcy, J.K. 1960. Graphical correlation of gaging-station records. *US Geol. Survey Water-Supply Paper 1541-C*. USGS, Washington, DC.

Selby, M.J. 1985. *Earth's Changing Surface, An Introduction to Geomorphology*. Oxford University Press, Oxford.

Shafroth, P.B., Friedman, J.M., Auble, G.T., Scott, M.L. and Braatne, J.H. 2002. Potential responses of riparian vegetation to dam removal. *BioScience* **52**: 703–712.

Sharma, S. and Moog, O. 1996. The applicability of Biotic indices and scores in water quality assessment of Nepalese rivers. In *Proceedings of the Ecohydrology Conference on High Mountain Areas*, March 23–26, Kathmandu, Nepal, pp. 641–657.

Shaw, E. M. 1988. *Hydrology in Practice*, 2nd edn, Van Nostrand Reinhold International, Wokingham.

Shear, T.H., Lent, T.J. and Fraver, S. 1996. Comparison of restored and mature bottomland hardwood forests of Southwestern Kentucky. *Restoration Ecology* **4**(2): 111–123.

Shelton, D., Cork, S., Binning, C., Parry, R., Hairsine, P., Vertessy, R. and Stauffacher, M. 2001. Application of an ecosystem services inventory approach to the Goulburn Broken catchment. In Rutherfurd, I., Sheldon, F., Brierley, G. and Kenyon, C. (eds) *Proceedings Third Australian Stream Management Conference*, Brisbane, 27–28 Aug, CRC Catchment Hydrology, Monash University, Clayton, Victoria, pp. 571–576.

Sherman, B. 2000. Scoping options for mitigating cold water discharges from dams. *Consultancy Report 00/21*. Report to Agriculture, Fisheries and Forestry Australia, NSW Fisheries, CRC for Freshwater Ecology, and Department of Land and Water Conservation NSW. CSIRO Land and Water, Canberra, Australian Capital Territory, May. URL: http://www.clw.csiro.au/publications/consultancy/2000/cr00-21.pdf (accessed 1st Nov 2003).

Shields, F.D., Jr 1982. Environmental features for flood control channels. *Water Resources Bulletin* 18: 779–784.

Shields, F.D., Jr and Gippel, C J. 1995. Prediction of effects of woody debris removal on flow resistance. *Journal of Hydraulic Engineering* 121: 341–354.

Shields, F.D., Jr, Cooper, C.M. and Knight, S.S. 1992. Rehabilitation of aquatic habitats in unstable streams. *Proceedings Fifth International Symposium on River Sedimentation*, Karlsruhe, pp. 1093–1102.

Shields, F.D., Jr, Cooper, C.M. and Knight, S.S. 1995. Experiment in stream restoration. *Journal of Hydraulic Engineering* 121: 494–502.

Shields, F.D., Jr, Knight, S.S. and Cooper, C.M. 1998. Rehabilitation of aquatic habitats in warmwater streams damaged by channel incision in Mississippi. *Hydrobiologia* 382: 63–86.

Shields, F.D., Jr, Knight, S.S. and Cooper, C. 2000a. Cyclic perturbation of lowland river channels and ecological response. *Regulated Rivers: Research & Management* 16: 307–325.

Shields, F.D., Jr, Simon, A. and Steffen, L.J. 2000b. Reservoir effects on downstream river channel migration. *Environmental Conservation* 27(1): 54–66.

Shields, F.D., Jr, Knight, S.S. and Cooper, C.M. 2000c. Warmwater stream bank protection and fish habitat: a comparative study. *Environmental Management* 26(3): 317–328.

Shields, F.D., Jr, Knight, S.S., Cooper, C. and Testa, S. 2000d. Large woody debris structures for incised channel rehabilitation. In *Proceedings of ASCE 2000: Joint Conference on Water Resources Engineering and Water Resources Planning and Management* (CD-ROM), American Society of Civil Engineers, Reston, VA.

Shields, F.D., Jr, Knight, S.S. and Cooper, C.M. 2000e. Warmwater stream bank protection and fish habitat: a comparative study. *Environmental Management* 26(3): 317–328.

Shields F.D., Jr, Morin, N. and Cooper, C.M. 2001. Design of large woody debris structures for channel rehabilitation. In *Proceedings Fifth Federal Interagency Sedimentation Conference*, Reno, Nevada (CD-ROM), U.S. Federal Interagency Subcommittee on Sedimentation.

Shields, F.D., Jr, Copeland, R.R., Klingeman, P.C., Doyle, M.W. and Simon, A. 2003a. Design for stream restoration. *Journal of Hydraulic Engineering* 129(8): 575–584.

Shields, F.D., Jr, Knight, S.S., Morin, N. and Blank, J. 2003b. Response of fishes and aquatic habitats to sand-bed stream restoration using large woody debris. *Hydrobiologia* 494: 251–257.

Shields, N.D. 1936. Anwendung der ahnlickeit Mechanik und der Turbulenzforschung auf die Geschiebelerwegung. *Mitt. Preoss Versuchanstalt fur Wasserbau und Schiffbau*, 26.

Shreve, R.L. 1967. Infinite topologically random channel networks. *J. of Geology*. 75: 178–86.

Shrubsole, D. (ed.) 1994. '*Natural*' *Channel Design: Perspectives and Practice*. Proceedings of the First International Conference on Guidelines for 'Natural' Channel Systems, March, Niagara Falls, Ontario. Canadian Water Resources Association, Cambridge, Ontario.

Shuman, J.R. 1995. Environmental considerations for assessing dam removal alternatives for river restoration. *Regulated Rivers: Research & Management* 11: 249–261.

Siebentritt, M.A., Ganf, G.G. and Walker, K.F. 2004. Effects of an enhanced flood on riparian plants of the River Murray, South Australia. *River Research and Applications* (in press).

Silvester, N.R. and Sleigh, M.A. 1985. The forces on microorganisms at surfaces in flowing water. *Freshwater Biol.* 15: 433–448.

Simmons, H.B. 1955. Some effects of upland discharge on estuarine hydraulics. *Proceedings of the American Society of Civil Engineers* 81: 792.

Simon, A. 1989. A model of channel response in disturbed alluvial channels. *Earth Surface Processes and Landforms* 14(1): 11–26.

Simon, A. and Downs, P.W. 1995. An interdisciplinary approach to evaluation of potential instability in alluvial channels. *Geomorphology* 12: 215–232.

Simon, A. and Hupp, C.R. 1986. Channel evolution in modified Tennessee channels. In *Proceedings of the Fourth Federal Interagency Sedimentation Conference*, 24–27 March, Las Vegas. Subcommittee on Sedimentation of the Interagency Advisory Committee on Water Data, U.S. Government Printing Office, Washington DC, 2(5): 71–82.

Simon, A. and Thomas, R.E. 2002. Processes and forms of an unstable alluvial system with resistant, cohesive streambeds. *Earth Surface Processes and Landforms* 27: 699–718.

Simon, T.P. and J. Lyons. 1995. Application of the index of biotic integrity to evaluate water resource integrity in freshwater ecosystems. In Davis, W.S. and Simon, T.P (eds) *Biological Assessment and Criteria: Tools for Water Resource Planning and Decision Making*. Lewis Publishers, Boca Raton, FL, pp. 245–262.

Simons, D.B. and Richardson, E.V. 1961. Forms of bed roughness in alluvial channels. *J. Hyd. Div., ASCE* 87: 87–105.

Simons, D.B. and Richardson, E.V. 1966. Resistance to flow in alluvial channels. *United States Geological Survey Professional Paper* 422J.

Simons, J.H.E.J., Bakker, C., Schropp, M.H.I., Jans, L.H., Kok, F.R. and Grift, R.E. 2001. Man-made secondary channels along the River Rhine (The Netherlands); results of post-project monitoring. *Regulated Rivers: Research & Management* 17(4–5): 473–491.

Simpson, E.H. 1949. Measurement of diversity. *Nature* 163: 688.

Simpson, J.C. and Norris, R.H. 2000. Biological assessment of river quality: development of AusRivAS models and outputs. In Wright, J.F., Sutcliffe, D.W. and Furse, M.T. (eds) *Assessing the Biological Quality of Freshwaters. RIVPACS and Other Techniques*. Freshwater Biological Association, Ambelside, pp. 125–142.

Singhall, H.S.S. *et al.* 1977. Sediment sampling in rivers and canals. *Erosion and Sediment Transport Measurement, Sympo-*

*sium, IAHS Publ. No. 133*, Florence (as cited by Starosolszky, 1987).

SINTEF NHL 1994. *Proceedings of the 1st International Symposium on Habitat Hydraulics*. 18–20 Aug, The Norwegian Institute of Technology, Trondheim, Norway. Organized by SINTEF NHL (Norwegian Hydrotechnical Laboratory) in cooperation with The Norwegian Institute of Technology and Norwegian Institute for Nature Research (NINA), IAHR.

Sit, V. and Taylor, B. (eds) 1998. *Statistical Methods for Adaptive Management Studies*. Land Management Handbook No. 42. Research Branch. British Columbia Ministry of Forests Forestry Division Services Branch, Victoria, BC.

Skaggs, R.W., Amatya, D., Evans, R.O., and Parsons, J.E. 1994. Characterization and evaluation of proposed hydrologic criteria for wetlands. *Journal of Soil and Water Conservation* **49**: 501–510.

Skidmore, P. B. and D.E. Miller. 1998. Application of deformable stream bank concepts to Natural Channel Design. In *Proceedings of the River Management Society 1998 Symposium*, Anchorage, AK. URL: http://www.interfluve.com/pdf_files/RMS%201998%20Paper.pdf (accessed 1st Nov 2003).

Skidmore, P.B., Shields, F.D. Jr, Doyle, M.W. and Miller, D.E. 2001. A categorization of approaches to Natural Channel Design. In Hayes, D.F. (ed.) *Proceedings 2001 Wetlands and River Restoration Conference*, Aug 27–31, Reno (CD-ROM). American Society of Civil Engineers, Reston, VA. URL: http://www.interfluve.com/pdf_files/skidmore_2001_ncd.pdf (accessed 1st Nov 2003).

Skinner, J. 2000. A study of methods used in measurement and analysis of sediment loads in reservoirs. *Report NN*. Federal Interagency Sedimentation Project, Vicksburg, Mississippi.

Sklar, F.H. and Browder, J.A. 1998. Coastal environmental impacts brought about by alterations to freshwater flow in the Gulf of Mexico. *Environmental Management* **22**(4): 547–562.

Skriver, J., Friberg, N. and Kirkegaard, J. 2000. Biological assessment of watercourse quality in Denmark: introduction of the Danish Stream Fauna Index (DSFI) as the official biomonitoring method. *Verhandlungen der Internationalen Vereinigung für Theoretische und Angewandte Limnologie* **27**: 1822–1830.

Slaney, P.A. and Zaldokas, D. (eds) 1997. Fish Habitat Rehabilitation Procedures. *Watershed Restoration Technical Circular No. 9*. Province of British Columbia, Ministry of Environment, Lands and Parks, Vancouver, BC.

Slaney, P.A., Finnigan, R.J. and Millar, R.G. 1997. Accelerating the recovery of log-jam habitats: large woody debris-boulder complexes. In Slaney, P.A. and Zaldokas, D. (eds) *Fish Habitat Rehabilitation Procedures, Watershed Restoration Technical Circular No. 9*. Ministry of Environment, Lands and Parks, and Ministry of Forests, Victoria, BC, pp. 9-1–9-24.

Smakhtin, V.U. 2001. Low flow hydrology: a review. *Journal of Hydrology* **240**: 147–186.

Smart, P.L. 1984. A review of the toxicity of twelve fluorescent dyes used for water tracing. *NSS Bulletin* **46**: 21–33.

Smith, D.G. 1976. Effect of vegetation on lateral migration of anastomosed channels of a glacier meltwater river. *Bull. of the Geol. Soc. of Am.* **87**: 857–856 (as cited by Knighton, 1984).

Smith, D.I. and Stopp, P. 1978. *The River Basin*. Cambridge University Press, Cambridge.

Smith, E.E. and Medin, D.L. 1981. *Categories and Concepts*. Harvard University Press, Cambridge, MA, p. 203.

Smith, I. and Lyle, A. 1978. *Distribution of Freshwaters in Great Britain*. Institute of Terrestrial Ecology, Edinburgh.

Smith, I.R. 1975. *Turbulence in Lakes and Rivers*. Scientific Publication No. 29, Freshwater Biological Association, UK.

Smith, J.D. 1999. Flow and sediment transport in the Colorado River near National Canyon. In Webb, R.H., Schmidt, J.C., Marzolf, G.R. and Valdez, R.A. (eds) The controlled flood in the Grand Canyon. *Geophysical Monograph 110*, American Geophysical Union, Washington, pp. 99–115.

Smock, L.A., Gladden, J.E., Riekenberg, J.L., Smith, L.C. and Black, C.R. 1992. Lotic macroinvertebrate production in three dimensions: channel surface, hyporheic, and floodplain environments. *Ecology* **73**(3): 867–886.

Sneath, P.H.A. and Sokal, R.R. 1973. *Numerical Taxonomy*. W.H. Freeman, San Francisco.

Snedecor, G.W. and Cochran, W.G. 1989. *Statistical Methods*. 8th edn, Iowa State University, Ames.

Snelder, T.H. and Biggs, B.J.F. 2002. Multi-scale river environment classification for water resources management. *Journal of the American Water Resources Association* **38**(5): 1225–1240.

Snelder, T.H., Biggs, B.J.F., Weatherhead, M. and Niven, K. 2003. A brief overview of New Zealand's River Environment Classification. National Institute of Water and Atmospheric Research, Riccarton, New Zealand. URL: http://www.niwa.cri.nz/ncwr/tools/rec/nzrec.pdf (accessed 1st Nov 2003).

Snowy Water Inquiry 1998. *Final Report*. Snowy Water Inquiry, Sydney.

Soar, P.J., Copeland, R.R. and Thorne C.R. 2001. Channel restoration design for meandering rivers. In *Proceedings of the Seventh Federal Interagency Sedimentation Conference*, March, Reno, NV (CD-ROM). U.S. Subcommittee on Sedimentation.

Sokal, R.R. 1966. Numerical taxonomy. *Scientific American* **165**: 106–116.

Sokal, R.R. and Rohlf, F.J. 1969. *Biometry*, W.H. Freeman, San Francisco.

Sokal, R.R. and Sneath, P.H.A. 1963. *Principles of Numerical Taxonomy*. W.H. Freeman, San Francisco.

Spellerberg, I.F. and Fedor, P.J. 2003. A tribute to Claude Shannon (1916-2001) and a plea for more rigorous use of species richness, species diversity and the 'Shannon-Wiener' Index. *Global Ecology & Biogeography* **12**(3): 177–179.

SPSS 1986. *spss$^x$ User's Guide, Edition 2*, SPSS Inc., McGraw-Hill, New York.

Srikanthan, R. and McMahon, T.A. 1981. Log Pearson III distribution—an empirically-derived plotting position. *J. of Hydrol.* **52**, 161–163.

Stalnaker, C.B. 1979. The use of habitat structure preferenda for establishing flow regimes necessary for maintenance of fish habitat. In Ward, J.V. and Stanford, J.A. (eds) *The Ecology of Regulated Streams*. Plenum Press, New York, pp. 321–337.

Stalnaker, C.B. and Arnette, S.C. 1976. *Methodologies for the Determination of Stream Resource Flow Requirements: An Assessment*. Utah State University, Logan for the U.S. Department of the Interior, Fish and Wildlife Service, Office of Biological Services, Western Water Association.

Stanford, J.A. and Ward, J.V. 1988. The hyporheic habitat of river ecosystems. *Nature* **335**(1): 64–66.

Stanford, J.A. and Ward, J.V. 1993. An ecosystem perspective of alluvial rivers: connectivity and the hyporheic corridor. *Journal of the North American Benthological Society* **12**: 48–60.

Stanford, J.A., Ward, J.V. and Ellis, B.K. 1994. Ecology of the alluvial aquifers of the Flathead River, Montana. In Gibert, J., Danielopol, D.L. and Stanford, J.A. (eds) *Groundwater Ecology*. Academic Press, New York, pp. 367–390.

Stanford, J.A., Ward, J.V., Liss, W.J., Frissell, C.A., Williams, R.N., Lichatowich, J.A. and Coutant, C.C. 1996. A General protocol for restoration of regulated rivers. *Regulated Rivers: Research & Management* **12**: 391–413.

Stankey, G.H. 2003. Adaptive management at the regional scale: breakthrough innovation or mission impossible? a report on an American experience. In Wilson, B.P. and Curtis, A. (eds) *Agriculture for the Australian Environment. Proceedings of the 2002 Fenner Conference on the Environment*, 30 July – 1 August 2002, Canberra, Johnstone Centre, Charles Sturt University, Albury. URL: http://www.csu.edu.au/special/fenner/papers/ref/13%20Stankey%20George.pdf (accessed 1st Nov 2003).

Stanley, E.H. and Doyle, M.W. 2002. A geomorphic perspective on nutrient retention following dam removal. *BioScience* **52**: 693–702.

Stanley, E.H. and Doyle, M.W. 2003. Trading off: the ecological effects of dam removal. *Frontiers in Ecology and the Environment* **1**: 15–22.

Stanley, E.H., Luebke, M.A., Doyle, M.W. and Marshall, D.W. 2002. Short-term changes in channel form and macroinvertebrate communities following low-head dam removal. *Journal of the North American Benthological Society* **21**: 172–187.

Statzner, B. 1981. The relation between 'hydraulic stress' and microdistibution of benthic macroinvertebrates in a lowland running water system, the Schierenseebrooks (North Germany). *Arch. Hydrolbiol.* **91**: 192–218.

Statzner, B. 1988. Growth and Reynolds number of lotic macroinvertebrates: a problem for adaptation of shape to drag. *Oikos* **51**: 84–87.

Statzner, B. and Higler, B. 1986. Stream hydraulics as a major determinant of benthic invertebrate zonation patterns. *Freshwater Biol.* **16**: 127–139.

Statzner, B. and Holm, T.F. 1982. Morphological adaptations of benthic invertebrates to stream flow—An old question studied by means of a new technique (Laser doppler anemometry). *Oecologia* **53**: 290–292.

Statzner, B. and Holm, T.F. 1989. Morphological adaptation of shape to flow: microcurrents around lotic macroinvertebrates with known Reynolds numbers at quasi-natural flow conditions. *Oecologia* **78**: 145–157.

Statzner, B. and Müller, R. 1989. Standard hemispheres as indicators of flow characteristics in lotic benthos research. *Freshwater Biol.*, **21**: 445–460.

Statzner, B., Bis, B., Doledec, S. and Usseglio-Polatera, P. 2001. Perspectives for biomonitoring at large spatial scales: a unified measure for the functional diversity of invertebrate communities in European running waters. *Basic and Applied Ecology* **2**: 73–85.

Statzner, B., Gore, J.A. and Resh, V.H. 1988. Hydraulic stream ecology: observed patterns and potential applications. *J. N. Am. Benthol. Soc.* **7**: 307–60.

Statzner, B., Resh, V.H. and Roux, A.L. 1994. The synthesis of long-term ecological research in the context of concurrently developed ecological theory: design of a research strategy for the Upper Rhone River and its floodplain. *Freshwater Biology* **31**: 253–263.

Steel, R.G.D. and Torrie, J.H. 1960. *Principles and Procedures of Statistics with Special Reference to the Biological Sciences*, McGraw-Hill, New York.

Stelczer, K. 1987. Physical, chemical and biological properties of water. In Starosolszky, O. (ed) *Applied Surface Hydrology*, pp. 150–74, Water Resources Publications, Littleton, CO.

Stephens, P.S. 1974. *Patterns in Nature*, Little-Brown, Boston, MA.

Stewardson, M., Rutherfurd, I. and Tennant, W. 1999. Physical assessment of rehabilitation works in Broken River and Ryans Creek, North East Victoria. In Rutherfurd, I.D. and Bartley, R. (eds) *Proceedings of the Second Australian Stream Management Conference*, Adelaide, 8–11 Feb, Cooperative Research Centre for Catchment Hydrology, Monash University, Clayton, pp. 595–600.

Stewardson, M.J. and Cottingham, P. 2002. A demonstration of the flow events method: environmental flow requirements of the Broken River. *Australian Journal of Water Resources* **5**: 33–48.

Stewardson, M.J. and Gippel, C.J. 1997. *In-stream Environmental Flow Design: A Review*. Report to the Hydro-Electric Corporation, Tasmania. Department of Civil and Environmental Engineering, The University of Melbourne, p. 99.

Stewardson, M.J. and Gippel, C.J. 2003. Incorporating flow variability into environmental flow regimes using the Flow Events Method. *River Research and Applications* **19**(5-6): 459–472.

Stillwater Sciences, 2003. Environmental Water Program: Restoring ecosystem processes through geomorphic high flow prescriptions, Final Draft. Prepared for CALFED Bay-Delta Program and Jones and Stokes, Berkeley, CA, May. URL: http://www.calfedewp.org/pdf/high_flow/text.pdf (accessed 1st Nov 2003).

Stouder, D.J., Bisson, P.A. and Niaman, R.J. (eds) 1997. *Pacific Salmon and Their Ecosystems, Status and Future Options*. Chapman and Hall, New York.

Strahler, A.N. 1952. Hypsometric (area-altitude) analysis of erosional topography. *Bull. Geol. Soc. Am.* **63**: 1117–1142.

Strahler, A.N. 1964. Geology. Part II. Quantitative geomorphology of drainage basins and channel networks, in V.T. Chow (ed) *Handbook of Applied Hydrology*, pp. 4–39 to 4–76, McGraw-Hill, New York.

Strahler, A.N. 1952. Dynamic basis of geomorphology. *Bulletin of the Geological Society of America* **63**: 923–938.

Straub, L.G. 1934. Types of fishway construction used in Europe. *Civil Engineering* **4**(2): 96.

Stream Systems Technology Centre, 2001. Forest Service stream classification: adoption of a first approximation. *Stream Notes* April 2001. Stream Systems Technology Center, USDA Forest Service, Rocky Mountain Research Station, U.S. Department of Agriculture, Fort Collins, Colorado, pp. 1–2

STREAMS 2003. Proceedings of the STREAMS Channel Protection and Restoration Conference. 6–7 October 2003, Columbus OH, The Ohio State University, Columbus, Ohio. Ohio Department of Natural Resources, Ohio State University and North Carolina State University. Stream Restoration, Ecology & Aquatic Management Solutions.

Street, R.L., Watters, G.Z. and Vennard, J.K. 1995. *Elementary Fluid Mechanics*, 7th edn, John Wiley, New York.

Streeter, V.L. (ed) 1961. *Handbook of Fluid Dynamics*, McGraw-Hill, New York.

Streeter, V.L. and Wylie, E.B. 1979. *Fluid Mechanics*, 7th edn, McGraw-Hill International, Sydney.

Strickler, A. 1923. Beitrage zur Frage der Geschwindigheitsformel und der Rauhigkeiszahlen für Strome, Kanale und Geschlossene Leitungen', *Mitteilungen des Eidgenössischer Amtes fur Wasserwirtschaft*, no. 16, Bern (as cited by Dackombe and Gardiner, 1983).

Strom, H.G. 1962. *River Improvement and Drainage in Australia and New Zealand.* State Rivers and Water Supply Commission, Victoria, Australia.

Stromberg, J.C. and Richter, B.D. 1991. Flood flows and dynamics of Sonoran riparian forests. *Rivers* 2(3): 221–235.

Stuart, I.G. and Mallen-Cooper, M. 1999. An assessment of the effectiveness of a vertical-slot fishway for non-salmonid fish at a tidal barrier on a large tropical/subtropical river. *Regulated Rivers: Research & Management* 15(6): 575–590.

Stuebner, S. 1988. Riparian renewal on range. *Idaho Statesman*: 8 May.

Sullivan, K., Lisle, T.E., Dolloff, C.A., Grant, G.E. and Reid, L.M. 1987. Stream channels: the link between forests and fishes. In Salo, E.O. and Cundy, T.W. (eds) *Streamside Management: Forestry and Fishery Interactions*. Institute of Forest Resources, University of Washington, Seattle, pp. 39–97.

Suter, G.W. 1993. A critique of ecosystem health concepts and indexes. *Environmental Toxicology and Chemistry* 12:1533–1539.

Sutherland, L., Ryder, D. and Watts, R. 2002. Ecological assessment of variable flow releases in the Mitta Mitta River, Victoria. Johnstone Centre for Research in Natural Resources & Society, Charles Sturt University, *Environmental Consulting Report No. 27*. Report to Murray-Darling Basin Commission, Canberra, Australian Capital Territory.

Swales, S. and Harris, J. 1995. The expert panel assessment method (EPAM): a new tool for determining environmental flows in regulated rivers. In Harper, D.M. and Ferguson, A.J.D. (eds) *The Ecological Basis for River Management*. Wiley, Chichester, pp. 125–134.

Swales, S. and O'Hara, K. 1980. Instream habitat improvement devices and their use in freshwater fisheries management. *Journal of Environmental Management* 10: 167–179.

Swedish Environmental Protection Agency 2002. Environmental Quality Criteria for Lakes and Watercourses. Swedish EPA (Naturvårdsverket), Stockholm, Sweden. URL: http://www.internat.naturvardsverket.se/index.php3?main=/documents/legal/assess/assess.htm (accessed 1st Nov 2003).

Swift, C.H., III 1976. Estimation of Stream Discharges Preferred by Steelhead Trout for Spawning and Rearing in Western Washington. *USGS Open-File Report 75–155*. United Stated

Geological Survey, Tacoma, Washington. URL: http://www.ecy.wa.gov/programs/wr/instream-flows/Images/pdfs/if-twstl.pdf (accessed 1st Nov 2003).

Swift, C.H., III 1979. Preferred Stream Discharges for Salmon Spawning and Rearing in Washington. *USGS Open-File Report 77–422*. United Stated Geological Survey, Tacoma, Washington. URL: http://www.ecy.wa.gov/programs/wr/instream-flows/Images/pdfs/if-twsal.pdf (accessed 1st Nov 2003).

Tabacchi, E., Correll, D.L., Hauer, R., Pinay, G., Planty-Tabacchi, A. and Wissmar, R.C. 1998. Development, maintenance and role of riparian vegetation in the river landscape. *Freshwater Biology* 40: 497–516.

Tabachnick, B.G. and Fidell, L.S. 1989. *Using Multivariate Statistics*, 2nd edn, Harper and Row, New York.

Takahashi, G. 2000. On evaluation of effectiveness of fishway. *Ecology and Civil Engineering* 3(2): 199–208.

Telfer, D. and Miller, D. 2001. Zen and the art of river management: a socio-geomorphological experiment exposed. In Rutherfurd, I., Sheldon, F., Brierley, G. and Kenyon, C. (eds) *Proceedings Third Australian Stream Management Conference*, Brisbane, 27–28 August, CRC Catchment Hydrology, Monash University, Clayton, Victoria, pp. 589–595.

Tennant, D.L. 1976. Instream flow regimens for fish, wildlife, recreation and related environmental resources. *Fisheries* 1(4): 6–10.

Teti, P. 1998. The effects of forest practices on stream temperature: a review of the literature. B.C. Ministry of Forests, Cariboo Forest Region, Williams Lake, B.C. URL: http://www.for.gov.bc.ca/cariboo/research/hydro/StreamTemperatureReview.pdf (accessed 1st Nov 2003).

Teti, P. 2000. Riparian management and stream temperature. In Hollstedt, C., Sutherland, K. and Innes, T. (eds) *Proceedings From Science to Management and Back: A Science Forum for Southern Interior Ecosystems of British Columbia*. Southern Interior Forest Extension and Research Partnership, Kamloops, B.C., pp. 27–28.

Texas Parks and Wildlife Department 2003. Environmental Flow Targets: Planning Criteria of the Consensus State Water Plan. Austin, TX. URL: http://www.tpwd.state.tx.us/texaswater/sb1/enviro/envwaterneeds/envwaterneeds.html (accessed 1st Nov 2003).

Texas Parks and Wildlife Department, Texas Commission on Environmental Quality and Texas Water Development Board 2003. Texas Instream Flow Studies: Technical Overview. Draft, Texas Water Development Board, Austin, TX, August. URL: http://www.twdb.state.tx.us/assistance/InstreamFlows/InstreamFlows-Draft-TechnicalOverviewForNAS.pdf (accessed 1st Nov 2003).

Tharme, R.E. 2000. An overview of environmental flow methodologies, with particular reference to South Africa. In King, J.M., Tharme, R.E. and De Villiers, M.S. (eds) *Environmental Flow Assessments for Rivers: Manual for the Building Block Methodology*. Water Research Commission Technology Transfer Report No. TT131/00. Water Research Commission, Pretoria, Republic of South Africa, pp. 15–40.

Tharme, R.E. 2003. A global perspective on environmental flow assessment: emerging trends in the development and application of environmental flow methodologies for rivers. *River Research and Applications* **19**(5-6): 397–441.

Tharme, R.E. and King, J.M. 1998. Development of the Building Block Methodology for instream flow assessments and supporting research on the effects of different magnitude flows on riverine ecosystems, Freshwater Research Unit. *WRC Report no. 576/1/98*. University of Capetown, Capetown, Republic of South Africa.

The Environment Agency 2003. The General Quality Assessment of Rivers – Biology Method. GQA methodologies for the classification of river and estuary quality, Environmental Monitoring, Science and Research. United Kingdom. URL: http://www.environment-agency.gov.uk/science/219121/monitoring/184353/?version=1<_e (accessed 1st Nov 2003).

The National Wetlands Research Center 2003. Habitat Suitability Index Models. Digital library collection. USGS Biological Resources Division. Lafayette, LA. URL: http://www. nwrc.usgs.gov/wdb/pub/hsi/hsiintro.htm (accessed 1st Nov 2003).

The Pacific Coast Federation of Fishermen's Associations 2003. Dams and Salmon, Why Some Dams Must Go! San Francisco, CA. URL: http://www.pcffa.org/dams.htm (accessed 1st Nov 2003).

The River Restoration Centre 1999. *Manual of River Restoration Techniques. Restoring the River Cole and River Skerne—U.K.* 1st ed. RRC, Silsoe, Bedfordshire, UK URL: http://www.therrc.co.uk/manual.php (accessed 1st Nov 2003).

The River Restoration Centre 2002. *Manual of River Restoration Techniques 2002 Update*. RRC, Silsoe, Bedfordshire, UK.

The Surface Water Regulations. 1994. *Surface Waters (River Ecosystem) (Classification) Regulations*. SI, 1994 (1057), UK (as cited by Maddock, 1999).

Thexton, E. 1999. Rehabilitation of the Genoa River, far east Gippsland, Victoria, with assisted regeneration. In Rutherfurd, I.D. and Bartley, R. (eds) *Proceedings of the Second Australian Stream Management Conference*, Adelaide, 8–11 Feb, Cooperative Research Centre for Catchment Hydrology, Monash University, Clayton, pp. 623–628.

Thomas, I.L., Benning, V.M. and Ching, N.P. 1987. *Classification of Remotely Sensed Images*, Adam Hilger, Bristol.

Thomas, W.O. 1985. A uniform technique for flood frequency analysis. *Jour. Water Resources Planning & Management* **111**(3): 321–337.

Thoms, M.C. and Sheldon, F., 2002, an ecosystem approach for determining environmental water allocations in Australian dryland river systems: the role of geomorphology, *Geomorphology* **47**: 153–168.

Thoms, M.C., Sheldon, F., Roberts, J., Harris, J. and Hillman, T.J. 1996. Scientific Panel Assessment of Environmental Flows for the Barwon-Darling River. Report to the Technical Services Division of the New South Wales Department of Land and Water Conservation, Parramatta.

Thoms, M.C., Suter, P., Roberts, J., Koehn, J., Jones, G., Hillman, T. and Close, A. 2000. Report of the River Murray Scientific Panel on Environmental Flows: River Murray - Dartmouth to Wellington and the Lower Darling River, River Murray Scientific Panel on Environmental Flows, Murray- Darling Basin Commission, Canberra, Australian Capital Territory.

Thomson, J.R., Taylor, M.P. and Brierley, G.J. 2004. Are River Styles ecologically meaningful? a test of the ecological significance of a geomorphic river characterization scheme. *Aquatic Conservation: Marine and Freshwater Ecosystems* **14**(1): 25–48.

Thomson, J.R., Taylor, M.P., Fryirs, K.A. and Brierley, G.J. 2001. A geomorphological framework for river characterization and habitat assessment. *Aquatic Conservation: Marine and Freshwater Ecosystems* **11**(5): 373–389.

Thorn, W.C. and Anderson, C.S. 2001. Comparison of two methods of habitat rehabilitation for brown trout in a southeast Minnesota stream. Minnesota Department of Natural Resources, *Investigational Report 488*. Section of Fisheries, St. Paul, Minnesota. URL: http://files.dnr.state.mn.us/publications/fisheries/investigational_reports/488.pdf (accessed 1st Nov 2003).

Thorncraft, G. and Harris, J.H. 2000. Fish Passage and Fishways in New South Wales: A Status Report. *Technical Report No. 1/2000*. Office of Conservation, NSW Fisheries, Sydney. Cooperative Research Centre for Freshwater Ecology, Canberra, Australian Capital Territory, May.

Thorne, C.R. 1998. *Stream Reconnaissance Handbook*, John Wiley, Chichester.

Thorne, C.R. 1997. Channel types and morphological classification. In Thorne, C.R., Hey, R.D. and Newson, M.D. (eds) *Applied Fluvial Geomorphology for River Engineering and Management*. John Wiley and Sons, Chichester, pp. 175–222.

Thorne, C.R., Bathurst J.C. and Hey R.D. (eds) 1987. *Sediment Transport in Gravel-Bed Rivers*, John Wiley, Chichester.

Thorne, C.R., Hey, R.D and Newson, M D (eds) 1997. *Applied Fluvial Geomorphology for River Engineering and Management*, John Wiley & Sons, Chichester, p. 384.

Thorp, J.H. and Delong, M.D. 1994. The riverine productivity model: a heuristic view of carbon sources and organic processing in large river systems. *Oikos* **70**: 305–308.

Thurow, R.F., Lee, D.C. and Rieman, B.E. 2000. Status and distribution of chinook salmon and steelhead in the interior Columbia River Basin and portions of the Klamath River Basin. In Knudsen, E.E, Stewart, C.R., MacDonald, D.D., Williams, J.E. and Reiser, D.W. (eds) *Sustainable Fisheries Management: Pacific Salmon*. Lewis Publishers, New York. URL: http://home.teleport.com/~salmo/docs/Status/status.htm (accessed 1st Nov 2003).

Tickner, D.P., Angold, P.G., Gurnell, A.M. and Mountford, J.O. 2001. Riparian plant invasions: hydrogeomorphological control and ecological impacts. *Progress in Physical Geography* **25**: 22–52.

Tikkanen, P., Laasonen, P., Muotka, T., Huhta, A. and Kuusela, K. 1994. Short-term recovery of benthos following disturbance from stream habitat rehabilitation. *Hydrobiologia* **273**: 121–130.

Tinkler, K.J. and Wohl, E. (eds) 1998. *Rivers Over Rock: Fluvial Processes in Bedrock Channels*, American Geophysical Union, Geophysical Monograph Series, Vol. 107, p. 340.

Tockner, K., Malard, F. and Ward, J.V. 2000. An extension of the flood pulse concept. *Hydrological Processes* **14**: 2861–2883.

Todd, D.K. 1959. *Ground Water Hydrology*, John Wiley, New York.

Townsend C.R. 1991. Community organization in marine and freshwater environments. In Barnes, R.S.K. and Mann, K.H. (eds) *Fundamentals of Aquatic Ecology*. Blackwell Scientific, London, pp. 125–144.

Townsend, C.R. 1980. *The Ecology of Streams and Rivers*, Institute of Biology, Studies in Biology 122, London.

Townsend, C.R. and Hildrew, A.G. 1994. Species traits in relation to a habitat templet for river systems. *Freshwater Biology* **31**: 265–275.

Travnichek, V.H., Bain, M.B. and Maceina, M.J. 1995. Recovery of a warmwater fish assemblage after the initiation of a minimum-flow release downstream from a hydroelectric dam. *Transactions of the American Fisheries Society* **124**: 836–844.

Treadwell, S.A., Campbell, I.C. and Edwards, R.T. 1997. Organic matter dynamics in Keppel Creek, southeastern Australia. *Journal of the North American Benthological Society* **16**: 58–61.

Triest, L., Kaur, P., Heylen, S. and De Pauw, N. 2001. Comparative monitoring of diatoms, macroinvertebrates and macrophytes in the Woluwe River (Brussels, Belgium). *Aquatic Ecology* **35**(2): 183–194.

Trieste, D.J. and Jarrett, R.D. 1987. Roughness coefficients of large floods, in James, L.G. and English, M.J. (eds) *Irrigation and Drainage Division Specialty Conference Proceedings* ASCE, New York, pp. 32–40.

Trimble, S.W. 1997. Stream channel erosion and change resulting from riparian forests. *Geology* **25**: 467–469.

Trimble, S.W. and Mendel, A.C. 1995. The cow as a geomorphic agent—a critical review. *Geomorphology* **13**: 233–253.

Triska, F.J., Kennedy, V.C., Avanzino, R.J., Zellweger, G.W. and Bencala, K.E. 1989a. Retention and transport of nutrients in a third-order stream: channel processes. *Ecology* **70**: 1877–1892.

Triska, F.J., Kennedy, V.C., Avanzino, R.L., Zellweger, G.W. and Bencala, K.E. 1989b. Retention and transport of nutrients in a third-order stream in northwestern California: hyporheic processes. *Ecology* **70**(6): 1893–1905.

Trotsky, H.M. and Gregory, R.W. 1974. The effects of water flow manipulation below a hydroelectric power dam on the bottom fauna of the upper Kennebec River, Maine. *Transactions of the American Fisheries Society* **103**: 318–324.

Trotter, E. 1990. Woody debris, forest-stream succession, and catchment geomorphology. *Journal of the North American Benthological Society* **9**: 141–56.

Tunbridge, B.R. and Glenane, T.G. 1982. Fisheries value and classification of fresh and estuarine waters in Victoria. Fisheries and Wildlife Division, Ministry for Conservation, Melbourne, Australia, p. 88.

Tunbridge, B.R. and Glenane, T.J. 1988. A study of environmental flows necessary to maintain fish populations in the Gellibrand River and estuary. *Technical Report Series No. 25*. Arthur Rylah Institute for Environmental. Research, Department of Conservation, Forests and Lands, Heidelberg, Victoria.

U.S. Army Corps of Engineers 1991. Hydraulic Design of Flood Control Channels. USACE Headquarters, EM1110-2-1601, Washington, DC.

U.S. Army Corps of Engineers 2002a. Columbia River Basin: Dams and Salmon. Northwestern Division, Pacific Salmon Coordination Office, Portland Oregon. URL: http://www.nwd.usace.army.mil/ps/colrvbsn.htm (accessed 1st Nov 2003).

U.S. Army Corps of Engineers 2002b. Abstracts, 2002 AFEP Annual Review. Anadromous Fish Evaluation Program (AFEP) Annual Program Review, 18–21 November, Portland, Oregon. USACE, Northwestern Division, Portland, Oregon. URL: https://www.nwp.usace.army.mil/pm/e/AFEP/AbstractsPage.html (accessed 1st Nov 2003).

U.S. Army Corps of Engineers 2003. Anadromous Fish Evaluation Program Reports. USACE, Northwestern Division, Portland, Oregon. URL: https://www.nwp.usace.army.mil/pm/e/afep_reports.htm (accessed 1st Nov 2003).

U.S. Army Corps of Engineers, Bonneville Power Administration and U.S. Bureau of Reclamation 2000. Biological Opinion. Effects to Listed Species from Operations of the Federal Columbia River Power System. U.S. Fish and Wildlife Service (Regions 1 and 6), December. URL: http://www.r1.fws.gov/finalbiop/biop.html (accessed 1st Nov 2003).

U.S. Army Corps of Engineers, U.S. Bureau of Reclamation and Bonneville Power Administration 1999. Biological Assessment for Effects of FCRPS Operations on Columbia Basin Bull Trout and Kootenai River White Sturgeon, Federal Columbia River Power System, Project Operations, June. URL: http://www.efw.bpa.gov/REPORTS/BAS/BullTroutSturgeon/19990617.bulltrout.BA.pdf (accessed 1st Nov 2003).

U.S. Department of Agriculture 1996. Streambank and shoreline protection. In *Engineering Field Handbook*. USDA Natural Resources Conservation Service, Washington, DC, December. URL: http://water.montana.edu/wildfish/manuals/EFH-Ch16.pdf (accessed 1st Nov 2003).

U.S. Environmental Protection Agency 2002. *National Water Quality Inventory: 2000 Report. Report No.* EPA-841-R-02-001. U.S. EPA Office of Water, Washington, DC. URL: http://www.epa.gov/305b/2000report/ (accessed 1st Nov 2003).

U.S. Fish and Wildlife Service 1980a. *Habitat Evaluation Procedure (HEP) Manual*. 102 ESM. U.S. Fish and Wildlife Service, Washington, DC.

U.S. Fish and Wildlife Service 1980b. *Environmental Impact Statement on the Management of River Flows to Mitigate the Loss of the Anadromous Fishery of the Trinity River, California.* Volumes I and II. U.S. Fish and Wildlife Service, Division of Ecological Services, Sacramento, CA.

U.S. Fish and Wildlife Service 1981. *Standards for the development of habitat suitability index models for use in the Habitat Evaluation Procedures*. 103 ESM. U.S. Fish and Wildlife Service, Washington, DC.

U.S. Fish and Wildlife Service 2000. Trinity River Mainstem Fishery Restoration Environmental Impact Statement/Report. U.S. Fish and Wildlife Service, Sacramento, CA, October. URL: http://www.ccfwo.r1.fws.gov/treis/default.htm (accessed 1st Nov 2003).

U.S. Fish and Wildlife Service 2001. Policy on streambank stabilization projects. Region 6, U.S. Fish and Wildlife Service, Denver, Colorado. URL: http://www.r6.fws.gov/pfw/PDFiles/bankersos10threv.pdf (accessed 1st Nov 2003).

U.S. Fish and Wildlife Service and Hoopa Valley Tribe 1999. *Trinity River Flow Evaluation, Final Report*. Report to U.S. Department of Interior by USFWS, Arcata Fish and Wildlife Office, Arcata, CA, and Hoopa Valley Tribe, Hoopa, CA, June. URL: http://arcata.fws.gov/fisheries/trfefinal.html (accessed 1st Nov 2003).

Uhlmann, D. 1979. *Hydrobiology*, John Wiley, Chichester.

UNESCO 1970. *International Legend for Hydrogeological Maps*, Int. Assoc. of Sci. Hydrology and Int. Assoc. of Hydrogeologists, Cook, Hammond and Kell Ltd, England.

UNESCO/WHO/UNEP 1992. *Water Quality Assessments—A Guide to Use of Biota, Sediments and Water in Environmental Monitoring*. 2nd edn URL: http://www.who.int/docstore/water_sanitation_health/wqassess/begin.htm#Contents (accessed 1st Nov 2003).

Uren, J., and Price, W.F. 1994. *Surveying for Engineers*, 3rd edn, Palgrave Macmillan, Hampshire.

USBR 1977. *Design of Small Dams*, US Bureau of Reclamation, Denver, CO.

USEPA 1987. *A Guide to the Sampling and Analysis of Water and Wastewater*, US Environmental Protection Agency.

USFS 1995. *A Guide to Field Identification of Bankfull Stage in the Western United States* (31-minute video), USDA Forest Service, Stream Systems Technology Center, Fort Collins, CO.

USGS 1977. *National Handbook of Recommended Methods for Water Data Acquisition*, US Government Printing Office, Washington, DC.

USGS 2003. National field manual for the collection of water-quality data. *US Geol. Survey Techniques of Water-Resources Investigations*, Book 9, Chap. A1–A9, USGS, Washington, DC. URL: http://water.usgs.gov/owq/FieldManual/ (accessed 13th March 2004).

USGS 2001. Water resources applications software: DOTABLES. U.S. Geological Survey, Hydrologic Analysis Software Support Program, Reston, VA. URL: http://water.usgs.gov/software/dotables.html (accessed 13th March 2004).

Uys, M. (ed) 1994. Classification of Rivers, and Environmental Health Indicators. *Proceedings of a joint South African Australian Workshop*, Cape Town, Feb 7–11, *Water Research Commission Report No. TT 63/94*, Pretoria, Republic South Africa.

V.A. Poulin & Associates 1991. Stream rehabilitation using LOD placements and off-channel pool development. *Land Management Report No. 61*, B.C. Ministry of Forests, Crown Publications, Victoria, B.C.

Valdez, R.A., Shannon, J.P. and Blinn, D.W. 1999. Biological implications of the 1996 controlled flood, in Webb, R.H., Schmidt, J.C., Marzolf, G.R. and Valdez, R.A. *The Controlled Flood in the Grand Canyon. Geophysical Monograph 110*, American Geophysical Union, Washington, pp. 343–350.

Van de Kraats, J.A. (ed) 1994. Rehabilitation of the River Rhine. *Water Science & Technology* **29**: 1–390.

Van Haveren, B.P. 1986. *Water Resource Measurements: A Handbook for Hydrologists and Engineers*, American Water Works Assn, Denver, CO.

Van Niekerk, A.W., Heritage, G.L. and Moon, B.P. 1995. River classification for management: the geomorphology of the Sabie River in the Eastern Transvaal. *South African Geographical Journal* **77**(2): 68–76.

Van Rijen, J.P.M. 1998. Practical approaches for nature development: let nature do its own thing. In de Waal, L. C., Large, A.R.C. and Wade, P.M. (eds) *Rehabilitation of Rivers: Principles and Implementation*. Wiley, Chichester, pp. 113–130.

Van Velson, R. 1979. Effects of livestock grazing upon rainbow trout in Otter Creek. In Cope, O.B. (ed), *Forum—Grazing and Riparian/Stream Ecosystems*, Trout Unlimited, Vienna, VA, pp. 53–55.

Van Winkle, W., Coutant, C.C., Jager, H.I., Mattice, J.S., Orth, D.J., Otto, R.G., Railsback, S.F. and Sale, M.J. 1997. Uncertainty and instream flow standards: perspectives based on hydropower research and assessment. *Fisheries* **22**(7): 21–22.

Vannote, R.L., Minshall, G.W., Cummins, K.W., Sedell, J.R. and Cushing, C.E. 1980. The river continuum concept, *Can. J. Fish. Aquat. Sci.* **37**: 130–37.

Vanoni, V.A. 1984. Fifty years of sedimentation. *Journal of Hydraulic Engineering* **10**(8): 1022–1057.

Vennard, J.K. and Street, R.L. 1982. *Elementary Fluid Mechanics*, 6th edn, John Wiley, New York.

Verdonschot, P.F.M. 2000. Integrated ecological assessment methods as a basis for sustainable catchment management. *Hydrobiologia* **422/423**: 389–412.

Versar Inc. 1998. Resolution of Winter Minimum Flow Issues at Conowingo Dam: Summary of Study Findings and Recommendation. Maryland Department of Natural Resources, Power Plant Research Program, Columbia, MD. URL: http://www.esm.versar.com/pprp/bibliography/Cono-1/Cono-1_text.pdf(accessed 1st Nov 2003).

Vogel, S. 1981. *Life in Moving Fluids, The Physical Biology of Flow*, Princeton University Press, New Jersey.

Vogel, S. 1988. *Life's Devices: The Physical World of Animals and Plants*, Princeton University Press, Princeton, NJ.

Vollenweider, R.A. 1968. Scientific fundamentals of the eutrophication of lakes and flowing waters, with particular reference to nitrogen and phosphorus as factors in eutrophication. Organisation for Economic Cooperation and Development, Paris, *Technical Report DAS/SCI/68.27*, p. 182.

Von Gunten, G.H., Smith, H.A. and Maclean, B. 1956. Fish passage facilities at McNary Dam. *Journal, Power Division, Proceedings of the American Society of Civil Engineers* **82**(PO1): 1–27.

Waddle, T.J. (ed) 2001. PHABSIM for Windows: User's Manual and Exercises: U.S. Geological Survey, Fort Collins, CO, 288 p. URL: http://www.fort.usgs.gov/products/Publications/15000/preface.html (accessed 1st Nov 2003).

Waddle, T.J., Steffler, P., Ghanem, A., Katopodis, C. and Locke, A. 1996. Comparison of one and two-dimentional hydrodynamic models for a small habitat stream. In Leclerc, M.A. Capra, H., Valentin, S., Boudreault, A. and Côté, Y. (eds) *Proceedings of the 2nd International Symposium on Habitat Hydraulics, Ecohydraulics 2000*. Institute National de la Recherche Scientifique – Eau, co-published with FQSA, IAHR/AIRH, Ste-Foy, Québec. Addendum, p. 12.

Wadell, H. 1932. Volume, shape, and roundness of rock particles, *Jour. Geol.* **40**: 443–451 (as cited by Krumbein,1941).

Wadell, H. 1933. Sphericity and roundness of rock particles, *Jour. Geol.* **41**: 310–31 (as cited by Krumbein, 1941).

Wadeson, R.A. and Rowntree, K.M. 1994. A hierarchical geomorphological model for the classification of South African river systems. In Uys, M. (ed) *Classification of Rivers, and Environmental Health Indicators. Proceedings of a Joint South African Australian Workshop*, Cape Town, Feb 7–11, *Water Research Commission Report No. TT 63/94*, Pretoria, Republic South Africa, pp. 49–67.

Walker, D. 2002. What is possible: hydrology and morphology. In *The Murray Mouth: Exploring the Implications of Closure or Restricted Flow*. Murray-Darling Basin Commission and Department of Water, Land and Biodiversity Conservation, Adelaide, South Australia. URL: http://www.dwlbc.sa.gov.au/publications/pdfs/reports/murray_mouth_exploring_implications.pdf (accessed 1st Nov 2003).

Walker, K.F. 1985. A review of the ecological effects of river regulation in Australia, *Hydrobiologia* **125**: 111–129.

Walker, K.F., Hillman, T.J. and Williams, W.D. 1978. Effects of impoundments on rivers: an Australian case study. *Verhandlungen der Internationalen Vereinigung fur Theoretische und Angwandte Limnologie* **2**: 1695–1701.

Walker, K.F., Sheldon, F. and Puckridge, J.T. 1995. A perspective on dryland river ecosystems, *Regulated Rivers: Research & Management* **11**: 85–104.

Walling, D.E. and Kleo, A.H.A. 1979. Sediment yields of rivers in areas of low precipitation: a global view, in *The Hydrology of Areas of Low Precipitation, Proc. of IAHS Symposium*, Canberra, December 1979, IAHS-AISH Publ. 128, 479–93 (as cited by Knighton, 1984).

Wallis, J.R. 1965. Multivariate statistical methods in hydrology—a comparison using data of known functional relationship, *Water Resour. Res.* **1**: 447–461.

Wallis, J.R. and O'Connell, E. 1972. Small sample estimates of $\rho1$, *Water Resour. Res.* **8**: 707–712.

Walsh, C.J. and Breen, P.F. 2001. A biological approach to assessing the potential success of habitat restoration in urban streams. *Verhandlungen Internationale Vereinigung fur Theoretische und Angewandte Limnologie* **27**(6): 3654–3658.

Walters C. 1997. Challenges in adaptive management of riparian and coastal ecosystems. *Conservation Ecology* [online] **1**(2):1. URL: http://www.consecol.org/vol1/iss2/art3 (accessed 1st Nov 2003).

Walters, C. and Holling, C.S. 1990. Large scale management experiments and learning by doing. *Ecology* **71**(6): 2060–2068.

Walton, M. 1986. *The Deming Management Method*. Perigree Books, New York.

Wang, Q.J. 1997: LH moments for statistical analysis of extreme events. *Water Resources Research* **33**(12): 2841–2848.

Ward, J.R. and Harr, C.A. (eds). 1990. Methods for collection and processing of surface water and bed-material samples for physical and chemical analyses, *US Geol. Survey Open-File Report 90–140*, USGS, Washington, DC.

Ward, J.V. 1982. Ecological aspects of stream regulation: responses in downstream lotic reaches, *Water Pollution and Management Reviews (New Delhi)* **2**: 1–26.

Ward, J.V. 1984. Ecological perspectives in the management of aquatic insect habitat, in Resh V.H. and Rosenberg D.M:, (eds), *The Ecology of Aquatic Insects*, Praeger, New York. pp. 558–577.

Ward, J.V. 1989. Riverine-wetland interactions. In Sharitz, R.R. and Gibbons, J.W. (eds) *Freshwater Wetlands and Wildlife, 1989 Conf. 8603101, DOE Symposium Series No. 61*, U.S. DOE Office of Scientific and Technical Information, Oak Ridge, TN, pp. 385–400.

Ward, J.V. and Stanford, J.A. (eds) 1979. *The Ecology of Regulated Streams*, Plenum Press, New York.

Ward, J.V. and Stanford, J.A. 1983a. The intermediate-disturbance hypothesis: an explanation for biotic diversity patterns in lotic ecosystems. In Fontaine, T.D. III and Bartell, S.M. (eds) *Dynamics of Lotic Ecosystems*. Ann Arbor Press, Ann Arbor, Michigan, pp. 347–356.

Ward, J.V. and Stanford, J.A. 1983b. Serial discontinuity concept of lotic ecosystems. In Fontaine T.D. III and Bartell, S.M. (eds) *Dynamics of Lotic Systems*. Ann Arbor Science, Ann Arbor, pp. 29–42.

Ward, J.V., Tockner, K., Uehlinger, U. and Malard, F. 2001. Understanding natural patterns and processes in river corridors as the basis for effective river restoration. *Regulated Rivers: Research & Management* **17**(4-5): 311–323.

Ward, R.C., Loftis, J.C. and McBride, G.B. 1990. *Design of Water Quality Monitoring Systems*. VanNostrand Reinhold, New York.

Wardlaw, A.C. 1985. *Practical Statistics for Experimental Biologists*. John Wiley, Chichester.

Washington Department of Fish and Wildlife 1995. *Restoring the Watershed. A Citizen's Guide to Riparian Restoration in Western Washington*. Washington Department of Fish and Wildlife, Olympia, WA.

Washington, H.G. 1984. Diversity, biotic and similarity indices: a review with special relevance to aquatic ecosystems. *Water Research* **18**: 653–694.

Water and Environmental Consultants Inc. 1980. Flushing flow discharge evaluation for 18 streams in the Medicine Row National Forest. Completion Report for Environmental Research and Technology, Fort Collins, CO.

Water and Rivers Commission 2000. The management and replacement of large woody debris in waterways. *Water Notes No. 13*, Advisory notes for land managers on river and wetland restoration, WRC, East Perth, Western Australia. URL: http://www.wrc.wa.gov.au/public/waternotes/pdf/13.pdf (accessed 1st Nov 2003).

Water and Rivers Commission 2002. *River Restoration: A Guide to the Nature, Protection, Rehabilitation and Long Term Management of Waterways in Western Australia*. River Restoration Series. WRC, East Perth, Western Australia, September. URL: http://www.wrc.wa.gov.au/public/RiverRestoration/index.htm (accessed 1st Nov 2003).

Water Science and Technology Board 2002. *The Missouri River Ecosystem: Exploring the Prospects for Recovery*. The National Academy of Sciences, National Academy Press, Washington, DC. URL: http://books.nap.edu/books/0309083141/html/index.html (accessed 1st Nov 2003).

Watershed Science Center, Trent University 1999. *Proceedings of the Second International Conference on Natural Channel*

*Systems*, March 1–4, 1999, Niagara Falls, Canada. URL: http://www.trentu.ca/wsc/PDFfiles/natproceeds.pdf (accessed 1st Nov 2003).

Watson, C.C., Biedenharn, D.S. and Bledsoe, B.P. 2002. Use of incised channel evolution models in understanding rehabilitation alternatives. *Journal of American Water Resources Association* **38**(1): 151–160.

Webb, A.A. and Erskine, W.D. 2003. A practical scientific approach to riparian vegetation rehabilitation in Australia. *Journal of Environmental Management* **68**: 329–341.

Webb, P.W. 1984. Form and function in fish swimming, *Scientific American* **251**: 58–68.

Webb, R.H., Schmidt, J.C., Marzolf, G.R. and Valdez, R.A. 1999. *The Controlled Flood in the Grand Canyon. Geophysical Monograph 110*. American Geophysical Union, Washington.

Webster, R.G. and McLellan, G. 1965. Snowy River Diversion, Effect on Lower River in Victoria, 3rd Progress Report, Conditions as Recorded Between 1959 and 1965. State Rivers and Water Supply Commission, Major Works Branch, Armadale, Victoria.

Weisberg, S.B., Janicki, A.J., Gerritson, J. and Wilson, H.T. 1990. Enhancement of benthic macroinvertebrates by minimum flow from a hydroelectric dam. *Regulated Rivers: Research & Management* **5**: 265–277.

Welch, P.S. 1935. *Limnology*, McGraw-Hill, New York.

Welcomme, R.L. 1976. Some general and theoretical considerations on fish yields of African rivers. *Journal of Fisheries Biology* **8**: 351–364.

Welcomme, R.L. 1989. Floodplain fisheries management. In Gore, J.A. and Petts, G.E. (eds) *Alternatives in Regulated River Management*. CRC Press Inc., Boca Raton, FL, pp. 209–223.

Wells, F. and Newall, P. 1997. *An Examination of an Aquatic Ecoregion Protocol for Australia*. Australian and New Zealand Environment and Conservation Council (ANZECC). Inland Waters, National River Health Program, Department of the Environment and Heritage, Australian Government, Canberra, Australian Capital Territory. URL: http://www.deh.gov.au/water/rivers/nrhp/ecoregion/ (accessed 1st Nov 2003).

Welsch, D.A., Smart, D.L., Boyer, J.N., Minkin, P. Smith, H.C. and McCandless, T.l. 1995. Forested Wetlands, Functions, Benefits and the Use of Best Management Practices. *Report NA-PR-01-95*. United States Department of the Interior Fish and Wildlife Service. Radnor, PA. URL: http://www.na.fs.fed.us/spfo/pubs/n_resource/wetlands/index.htm#Table%20of%20Contents (accessed 1st Nov 2003).

Werren, G. and Arthington, A. 2002. The assessment of riparian vegetation as an indicator of stream condition, with particular emphasis on the rapid assessment of flow related impacts. In Franks, A., Playford. J. and Shapcott, A. (eds) *Landscape Health of Queensland*. Royal Society of Queensland, St Lucia, pp. 194–222.

Wesche, T.A. 1985. Stream channel modifications and reclamation structures to enhance fish habitat. In Gore, J.A. (ed.) *The Restoration of Rivers and Streams*. Butterworths, Boston, MA, pp. 103–159.

Wesche, T.A. and Rechard, P.A. 1980. A Summary of Instream Flow Methods for Fisheries and Related Needs. *Eisenhower Consortium Bulletin No. 9*. Water Resources Research Institute, University of Wyoming for the Eisenhower Consortium for Western Environmental Forestry Research.

Wesche, T.A., Goertler, C.M. and Hubert, W.A. 1987. Modified habitat suitability index model for brown trout in southeastern Wyoming. *North American Journal of Fisheries Management* **7**: 232–237.

West, C.J. 1994. Wild Willows in New Zealand. *Proceedings of a Willow Control Workshop*. 24–26 November 1993, Hamilton, Department of Conservation, Te Papa Atawhai, Wellington, New Zealand.

Wetmore, S.H., Mackay, R.J. and Newbury, R.W. 1990. Characterization of the hydraulic habitat of Brachycentrus occidentalis, a filter feeding caddisfly, *J. No. Am. Benth. Soc.* **9**: 157–169.

Wharton, G. 1995. Information from channel geometry-discharge relations. In Gurnell, A.M. and Petts G.E. (eds) *Changing River Channels*. John Wiley and Sons, Chichester, pp. 325–346.

White, D.S., Elzinga, C.H. and Hendricks, S.P. 1987. Temperature patterns within the hyporheic zones of a northern Michigan river, *J. No. Am. Benthol. Soc.* **6**: 85–91.

White, F.M. 1986. *Fluid Mechanics*, 2nd edn, McGraw-Hill, New York.

White, F.M. 1999. *Fluid Mechanics*, 4th edn, McGraw-Hill College Div., New York.

White, L.J. and O'Brien, T.A. 1999. Fishways—an element of stream rehabilitation: some issues, recent successes, and research needs. In Rutherfurd, I.D. and Bartley, R. (eds) *Proceedings of the Second Australian Stream Management Conference*, Adelaide, 8–11 Feb, Cooperative Research Centre for Catchment Hydrology, Monash University, Clayton, pp. 691–695.

Whiting, P.J. and Bradley, J.B. 1993. A process based classification system for headwater streams. *Earth Surface Processes and Landforms* **18**: 603–612.

Whittaker, D., Shelby, B., Jackson, W. and Beschta, R. 1993. *Instream Flows for Recreation: A Handbook on Concepts and Research Methods*. U.S. Department of the Interior, National Parks Service, Alaska Region, Anchorage, AK, p. 104.

Whitton, B.A. and Rott, E. 1996. *Use of Algae for Monitoring rivers II*. E. Rott, Institut für Botanik, Universität Innsbruck, Innsbruck, Austria.

Whitton, B.A., Rott, E. and Friedrich, G. 1991. *Use of Algae for Monitoring Rivers*. E. Rott, Institut für Botanik, Universität Innsbruck, Innsbruck, Austria.

Wicklum, D. and Davies, R.W. 1995. Ecosystem health and integrity? *Canadian Journal of Botany* **73**: 997–1000.

Wiederholm, T. 1980. Use of benthos in lake monitoring. *Journal of the Water Pollution Control Federation* **52**: 537–547.

Wiens J.A. 2002. Riverine landscapes: taking landscape ecology into the water. *Freshwater Biology* **47**: 501–515.

Wilcock, P.R., Kondolf, G.M., Matthews, W.V. and Barta, A.F. 1996. Specification of sediment maintenance flows for a large gravel-bed river. *Water Resources Research* **32**(9): 2911–2921.

Wildman, L., Parasiewicz, P., Katopodis, C., and Dumont, U. 2002. An Illustrative Handbook on Nature-Like Fishways – Summarized Version. American Rivers, Glastonbury, CT. URL:

http://www.amrivers.org/docs/AFS_Paper.pdf (accessed 1st Nov 2003).

Wiley, M.J. and Seelbach, P.W. 1997. An introduction to rivers—the conceptual basis for the Michigan River Inventory (MRI) Project. *Michigan Department of Natural Resources Fisheries Special Report No. 20*. Ann Arbor, Michigan, December. URL: http://rivers.snre.umich.edu/mri/20sr.pdf (accessed 1st Nov 2003).

Wilkinson, L. 1986. *SYSTAT: The System for Statistics*, SYSTAT, Inc., Evanston, IL.

Williams, B.K. 1983. Some observations on the use of discriminant analysis in ecology. *Ecology* **64**: 1283–1291.

Williams, G.P. 1978, Bank-full discharge of rivers, *Water Resources Research* **14**: 1141–1154.

Williams, G.P. 1986. River meanders and channel size. *Journal of Hydrology* **88**: 147–164.

Williams, D.D. 1987. The Ecology of Temporary Waters. Timber Press, Oregon, p. 205.

Williams, G.P. and Wolman, M.G. 1984. Downstream effects of dams on alluvial rivers. *U.S. Geological Survey Professional Paper 1286*. U.S. Government Printing Office, Washington.

Williams, J.E, Wodd, C.A and Domeback, M.P (eds) 1997. *Watershed Restoration: Principles and Practices*. American Fisheries Society, Bethesda, MD, p. 561.

Williamson, R.B., Smith, C.M. and Cooper, A.B. 1996. Watershed riparian management and its benefits to a eutrophic lake. *Journal of Water Resource Planning and Management* **122**(1): 24–33.

Williamson, R.B., Smith, R K. and Quinn, J.M. 1992. Effects of riparian grazing and channelisation on streams in Southland, New Zealand. 1. *Channel form and stability*. New Zealand Journal of Marine and Freshwater Research **26**: 241–258.

Wilson, E.M. 1969. *Engineering Hydrology*, Macmillan London.

Wilson, J.F., Jr, Cobb, E.D. and Kilpatrick, F.A. 1984. Flurometric procedures for dye tracing, *Open File Report 84–234*, USGS, Washington, DC.

Winegar, H.H. 1977. Camp Creek channel fencing plant, wildlife, soil and water response. *Rangeman's Journal* **4**: 10–12.

Wissmar, R.C., Smith, J.E., McIntosh, B.A., Li, H.W., Reeves, G.H. and Sedell, J.R. 1994. A history of use and disturbance in riverine basins of eastern Oregon and Washington (Early 1800s – 1900s). *Northwest Science* **68**:1–35.

WMO 1980. *Manual on Stream Gauging*, Volume 1: *Fieldwork*, and Volume II: *Computation of Discharge*, Operational Hydrology Report No. 13, WMO No. 519, Secretariat of the World Meteorological Organization, Geneva, Switzerland.

WMO 1981. *Guide to Hydrological Practices*, Volume I: *Data Acquisition and Processing*, WMO No. 168, Secretariat of the World Meteorological Organization, Geneva, Switzerland.

WMO 1983. *Guide to Hydrological Practices*, Volume II: *Analysis, Forecasting and Other Applications*, WMO No. 168, Secretariat of the World Meteorological Organization, Geneva, Switzerland.

Wohl, E. 2000. *Mountain Rivers. Water Resources Monograph 14*, Am. Geophys. Union, Washington, DC.

Wolman, M.G. 1955. The natural channel of Brandywine Creek, Pennsylvania, *US Geol. Survey Prof. Paper 271*, USGS, Washington, DC (as cited by Harvey, 1969).

Wolman, M.G. and Leopold, L.B. 1957. River flood plains: some observations on their formation. *U.S. Geological Survey Professional Paper 282C*. U.S. Government Printing Office, Washington, pp. 145–181.

Woo, S. 1999. Habitat modeling not enough to save fish. . .or rivers. *Stream Notes*, April. Stream Systems Technology Center, Rocky Mountains Research Station, USDA Forest Service, pp. 5–7.

Woodfull, J., Finlayson, B., and McMahon, T. (eds) 1993. *The Role of Buffer strips in the Management of Waterway Pollution from Diffuse Urban and Rural Sources. LWRRDC Occasional paper No 01/93*. Centre of Applied Hydrology, University of Melbourne and Land and Water Resources Research and Development Corporation, Canberra, Australian Capital Territory.

Woodiwiss, F.S. 1964. The biological system of stream classification used by the Trent River Board. *Chemistry and Industry* **11**: 443–447.

Woods, A.J., Omernik, J.M. and Brown, D.D. 1996. *Level III and IV Ecoregions of Pennsylvania and the Blue Ridge Mountains, the Ridge and Valley, and the Central Appalachians of Virginia, West Virginia, and Maryland*, (Addenda May, 1999). U.S. Environmental Protection Agency, National Health and Environmental Effects Research Laboratory, Corvallis, Oregon.

Woodyer, K.D. 1968. Bankfull frequency in rivers. *J. of Hydrol.* **6**: 114–142.

Woolhiser, D.A. and Saxton, K.E. 1965. Computer program for the reduction and preliminary analyses of runoff data, *ARS 41–109*, USDA Agricultural Research Service.

Wright, J.F., Furse, M.T. and Armitage, P.D. 1993. RIVPACS—a technique for evaluating the biological quality of rivers in the UK. *European Water Pollution Control* **3**(4): 15–25.

Wright, J.F., Moss, D., Armitage, P.D. and Furse, M.T. 1984. A preliminary classification of running-water sites in Great Britain based on macroinvertebrate species and the prediction of community type using environmental data. *Freshwater Biology* **14**: 221–256.

Wu, J. and Loucks, O.L. 1995. From balance of nature to hierarchical patch dynamics: a paradigm shift in ecology. *Quarterly Review of Biology* **70**: 439–466.

Yalin, M.S. 1972. *Mechanics of Sediment Transport*, Pergamon Press, Oxford.

Yalin, M.S. and Karahan, E. 1979. Inception of sediment transport *J. Hyd. Div.*, ASCE **105**: 1433–1443.

Yang, C.T. 1973. Incipient motion and sediment transport. *J. Hyd. Div., Proc. ASCE* **99**(HY10), 1679–1704.

Yevjevich, V.M. 1972. *Probability and Statistics in Hydrology*, Water Res. Publ., Fort Collins, CO.

Yoder, C.O. and Kulik, B.H. 2003. The development and application of multimetric indices for the assessment of impacts to fish assemblages in large rivers: a review of current science and applications. *Canadian Water Resources Journal* **28**(2): 301–328.

Young, W.J. (ed) 2001. *Rivers as Ecological Systems: The Murray Darling Basin*. Murray-Darling Basin Commission, Canberra, Australian Capital Territory.

Young, W.J., Dyer, F. and Thoms, M.C. 2001. An index of river hydrology change for use in assessing river condition. In

Rutherfurd, I., Sheldon, F., Brierley, G. and Kenyon, C. (eds) *Proceedings Third Australian Stream Management Conference*, Brisbane, 27–28 August, CRC Catchment Hydrology, Monash University, Clayton, Victoria, pp. 695–700.

Young, W.J., Ogden, R.W., Hughes, A.O. and Prosser, I.P. 2002. Predicting channel type from catchment and hydrological variables. In Dyer, F.J., Thoms, C. and Olley, J.M. (eds) The Structure, Function and Management Implications of Fluvial Sedimentary Systems. Proceedings the IAHS Symposium, Alice Springs, Australia, Sept. *IAHS Publication no. 276*, Wallingford, pp. 53–60.

Yrjänä, T. 1998. Efforts for in-stream fish habitat restoration within the River Iijoki, Finland—goals, methods and test results. In DeWaal, L.C., Large, A.R.G. and Wade, P.M. (eds) *Rehabilitation of Rivers: Principles and Implementation*. John Wiley & Sons, Chichester, pp. 240–250.

Zar, J.H. 1974. *Biostatistical Analysis*, Prentice-Hall, Englewood Cliffs, NJ.

Zelinka, M. and Marvan, P. 1961 Zur Präzisierung der biologischen Klassifikation der Reinheit fliessender Gewässer. (Making a precise biological classification of the quality of running waters). *Arch. Hydrobiol.* **57**: 389–407.

Zellweger, G.W., Avanzino, R.J. and Bencala, K.E. 1989. Comparison of tracer-dilution and current-meter discharge measurements in a small gravel-bed stream, Little Lost Man Creek, California, *Water-Res. Invest. Rep. 89–4150*, US Geol. Survey, Menlo Park, California.

Zingg, T. 1935. Beitrag zur Schotteranalyse, *Schweiz. Min. u. Pet. Mitt.* **15**: 39–140 (as cited by Krumbein, 1941).

# Index

---

Stream Hydrology: An Introduction for Ecologists, Second Edition.
Nancy D. Gordon, Thomas A. McMahon, Brian L. Finlayson, Christopher J. Gippel, Rory J. Nathan
© 2004 John Wiley & Sons, Ltd ISBNs: 0-470-84357-8 (HB); 0-470-84358-6 (PB)